Recovery Methodologies: Regularization and Sampling

Mathematical
Surveys
and
Monographs

Volume 274

Recovery Methodologies: Regularization and Sampling

Willi Freeden

M. Zuhair Nashed

American Mathematical Society

Providence, Rhode Island

2020 *Mathematics Subject Classification.* Primary 65J20, 62D05; Secondary 65R32, 78A50, 94A20, 86A22.

For additional information and updates on this book, visit
www.ams.org/bookpages/surv-274

Library of Congress Cataloging-in-Publication Data

Names: Freeden, W. (Willi), author. | Nashed, M. Zuhair, 1936- author.
Title: Recovery methodologies : regularization and sampling / Willi Freeden, M. Zuhair Nashed.
Description: Providence, Rhode Island : American Mathematical Society, [2023] | Series: Mathematical surveys and monographs, 0076-5376 ; volume 274 | Includes bibliographical references and index.
Identifiers: LCCN 2023014083 | ISBN 9781470473457 (paperback) | ISBN 9781470474515 (ebook)
Subjects: LCSH: Sampling (Statistics) | Inverse problems (Differential equations)–Numerical solutions. | AMS: Numerical analysis – Numerical analysis in abstract spaces – Improperly posed problems; regularization. | Statistics – Sampling theory, sample surveys – Sampling theory, sample surveys. | Numerical analysis – Integral equations, integral transforms – Inverse problems. | Optics, electromagnetic theory – General – Antennas, wave-guides. | Information and communication, circuits – Communication, information – Sampling theory. | Geophysics – Geophysics – Inverse problems.
Classification: LCC QA276.6 .F75 2023 | DDC 519.5/2–dc23/eng20230708
LC record available at https://lccn.loc.gov/2023014083

∞ The paper used in this book is acid-free and falls within the guidelines
established to ensure permanence and durability.
Visit the AMS home page at https://www.ams.org/

10 9 8 7 6 5 4 3 2 1   28 27 26 25 24 23

It is by logic that we prove, but by intuition that we discover. To know how to criticize is good, to know how to create is better."

Henri Poincaré, Science and Method, 1908 (translated by F. Maitland, Dover Publications, 2003).

Contents

Preface

Some decades ago the connections among the areas of inverse problems and sampling were rather tenuous. Researchers in one of the areas were often unfamiliar with the techniques and the capabilities of the other area. Today, the situation, however, has changed, not least because of the intensive study of regularization and sampling methodologies. It turned out that the common thread among these problems led to key constituents in both inverse and sampling problem theory, so that it is time now to pass over to a unifying description of the underlying mathematical concepts. The canonical result is a superordinate framework concerned with the recovery of an object (function, signal, picture) from partial or indirect information about the object.

Recovery problems are characterized by the fact that the transformation from data to object constituents (or vice versa) is a result of the interaction of a certain system with the object that we wish to infer properties about. As a consequence, a good understanding of the system using recovery methods enables the formulation of a mathematical problem by reduction in theoretic sense such that a new, more concrete situation can be efficiently attacked within a well-structured context, remarkably often in reproducing kernel Hilbert and/or Banach spaces. Coherently, by solution processes of recovery problems, the capacity is provided to recognize the causality between the virtuality of the mathematical apparatus and the impact to the scientific reality.

The book allows the reader to concentrate on the essential topics of the *methodologies of recovery problems*. The approach is selective, although a broad palette of recent work including heterogeneous applications is specified in the contents. In order to keep the book manageable, the proofs (available in the literature) are omitted. The special accentuation of the work thus lies in the highlighting of content relationships both intrinsically for inverse and sampling problems as well as in the synopsis of interdependencies. Despite the avoidance of extreme technicalities and elaborate proof techniques, the book requires some basic knowledge in functional analysis, Fourier theory, geometric number theory, constructive approximation, special function theory, etc. For the convenience of the reader, essential cornerstones of these areas are collected as background material (usually without proofs, too). All chapters (including those concerned with the background tools) supply the reader with an adequate overview of relevant references.

The work is a result of the research on recovery problems due to the authors during the last decades. The purpose is to give a consistent overview of scattered results and developments in the field of recovery problems in the form of a monograph. Although the book contains original work, we have tried to make it accessible to students and scientists not only from applied mathematics, but also from all branches of engineering and science. Without being complete a major objective is to provide an addition to the library of any individual interest in recovery theory and to show how advances in this theory lead to new discoveries in other mathematical as well as scientific branches.

It is our great pleasure to thank our colleagues, especially P.L. Butzer, E. Hlawka, C. Müller, G.G. Walter for their specific influence on our scientific thinking. Without their great preliminary achievements, it would not have been possible to complete this book in its present shape.

The authors owe a great debt of gratitude to the unknown reviewers for their valuable comments and suggested additions. Valuable remarks and comments by Dr. Helga Nutz are gratefully acknowledged. The AMS-book consultants Dr. Ina Mette and Dr. Eriko Hironaka deserve special thanks for their excellent support of the book project. The authors thank Emily Lannon, AMS Production Editor, for making final corrections.

March 2023,

W. F., M. Z. N.

Part 1

Introductory remarks

CHAPTER 1

Constituents of the univariate antenna problem

There is no doubt that the study of the 1D bandlimited function of the form

$$t \mapsto \int_{\mathcal{F}_\mathbb{Z}} F(x)\, e^{-2\pi i t\cdot x}\, dx = F^\wedge_{\mathcal{F}_\mathbb{Z}}(t), \quad t \in \mathbb{R}, \quad F \in L^2(\mathcal{F}_\mathbb{Z}), \tag{1.1}$$

where $\mathcal{F}_\mathbb{Z} = [-\frac{1}{2}, \frac{1}{2})$ is the one-dimensional fundamental cell of the 1D unit lattice $\Lambda = \mathbb{Z}$, has significantly influenced the mathematical development during the last decades. In fact, the specific properties of the 1D Fourier integral (1.1) has led to a deepened mathematical discussion in two essential research areas initiated by engineering, namely

- regularization methods in inverse problems,
- sampling methods in signal analysis/processing.

As pointed out in the Preface, thirty years ago, the connections among the areas of regularization and sampling were rather tenuous. Researchers in one of the areas were often unfamiliar with the techniques and relevance of the other area. Today the situation, however, has changed, not least because of the 1D study of the bandlimited functions (1.1) as a bridge between two specific problem fields, namely

- antenna problem oriented approach to inverse problem theory,
- Shannon sampling oriented approach to signal analysis.

It turned out that the common thread among these problems led to key constituents in both inverse as well as sampling problem theory, so that it is time now to pass over to a description of the underlying mathematical ideas and concepts resulting canonically in a superordinate unifying framework:

Recovering an object (function, signal, picture) from partial or indirect information about the object.

As a matter of fact, a substantial amount of the machinery from functional analysis, Fourier analysis, theory of special functions, potential theory, lattice point theory, approximation theory, and numerical analysis has been brought to bear on the resolution and understanding of a recovery problem, and the interdisciplinary character of many recovery problems has emerged very clearly. The concept is to provide a new outlook within which scientific results can be better motivated and understood. Within this context, criteria can be given relative to which the scope and limitations of the various methods can be assessed. These aspects are important both in theory and practice.

In what follows, for purposes of motivation, our first interest is in showing that the univariate treatment of (1.1) within the antenna problem has provided deep and profound results in the theory of inverse problems. Simultaneously, the 1D Shannon sampling theory of the last century experienced far-reaching generalizations and innovative developments. Today, in a synopsis of both research fields, the concept of recovery becomes obvious and further developable, so that a common view has started to take a new prominent role within the traditional branches of mathematics, thereby leading to appropriate multivariate extensions and discoveries in other areas of mathematics and engineering.

1.1. Problem induced regularization

The following approach to a trend setting philosophy to treat ill-posed problems of mathematics is essentially influenced by M. Z. Nashed's understanding of the 1D antenna problem published in "IEEE Transactions on Antennas and Propagation, Vol. AP-29, No. 2, 220-231, 1981" under the title "Operator-Theoretic and Computational Approaches in Ill-Posed Problems with Applications to Antenna Theory" (cf. [**625**]).

In 1D antenna theory, the essential problem is the one in which the *far-field radiation pattern* $F^{\wedge}_{\mathcal{F}_{\mathbb{Z}}}$ is known, either exactly or approximately, by means of physical measurements. The problem is to find the *aperture distribution* F producing this given pattern $F^{\wedge}_{\mathcal{F}_{\mathbb{Z}}}$. More explicitly, in a mathematical formalism, determine the aperture distribution $F \in L^2(\mathcal{F}_{\mathbb{Z}})$ from the far field $F^{\wedge}_{\mathcal{F}_{\mathbb{Z}}}$ via the Fredholm integral equation of the first kind

$$AF(t) = \int_{\mathcal{F}_{\mathbb{Z}}} F(x)\, e^{-2\pi i t\cdot x}\, dx = F^{\wedge}_{\mathcal{F}_{\mathbb{Z}}}(t), \quad t \in \mathbb{R}. \tag{1.2}$$

In other words, the antenna problem (1.2) of determining the aperture distribution from far-field input requires particular solvability procedures which come from the theory of inverse problems (IPs).

Schematically, a *direct (forward) problem* can be formulated as follows:

object (e.g., aperture distribution) $\longrightarrow$ data information of the object (e.g., far-field).

The *inverse problem* is considered the "inverse" to the forward problem which relates the object (sub)information to the object:

data information of the object (e.g., far-field information) $\longrightarrow$ object (e.g., aperture distribution).

An object may be understood to be the systematic relationship of all data (sub)information, object parameters, and other auxiliary information.

In a mathematical abstraction of equations such as (1.2) involving an operator A we are usually confronted with the following situation: Given spaces X, Y equipped with the settings of "distance" (i.e., norm) and "angle" (i.e., inner product), i.e., X, Y are (ideally) assumed to be Hilbert spaces (especially, for the 1D antenna problem, we are led to the Hilbert spaces $X = L^2(\mathcal{F}_{\mathbb{Z}})$ and $Y = L^2(\mathbb{R})$). Consider a general mapping A from X to Y,

$$A : X \to Y. \tag{1.3}$$

The *direct problem* is as follows: Given $x \in X$, find $y = Ax \in Y$. The *inverse problem* is as follows: Given an observed output y, find an input x that produces it, i.e., $Ax = y \in Y$, or given a desired output z, find an input x that produces an output $y = Ax \in Y$ that is as "close" to z as possible, e.g., in the least squares sense:

$$\min_{x \in X} \|Ax - z\|^2. \tag{1.4}$$

A *well-posed (properly-posed) problem in the sense of Hadamard* is as follows: For $y \in Y$, the operator equation $X \ni x \mapsto Ax = y \in Y$ has one and only one solution $x \in X$, and the solution depends continuously on y. So, a mathematical problem is well-posed in the sense of Hadamard, if it satisfies the following properties:

(H1) *Existence: For given data, there exists a solution of the problem (in an appropriate sense).*

(H2) *Uniqueness: For given data, the solution is unique.*

(H3) *Stability: The solution depends continuously on the data.*

In accordance with this definition, *a problem is ill-posed in the sense of Hadamard*, if at least one of these three conditions is violated.

In this respect it should be noted that J. Hadamard (1865-1963) dismissed ill-posed problems as irrelevant to real world applications, but he was proven wrong four decades after his declaration. In fact, it turned out that Hadamard's classification (cf. [**390, 391**]) had a tremendous influence on the development of practice-oriented mathematics.

Initially, the theory of inverse problems was influenced by two essential cornerstones, the *method of least squares* and the *concept of the pseudoinverse*: The actual reasoning for an adjustment theory with its core, the method of least squares, was already done by Gauss and Legendre around 1800. Gauss probably hit on the fundamental idea in the autumn 1794 reading a publication of Lambert concerning applications of mathematics (see A. Galle [**342**]), but Gauss published his method not before 1809 (cf. "Theoria motus corporum coelestium in sectionibus conicis solem ambientium"[**345**]). He closed this publication by giving an explanation of his method ("Determinatio orbitae observationibus quotcumque quam proximae satisfacientis"). Further works followed in the year 1810 ("Disquisitio de elementis ellipticis Palladis" [**346**]), in 1816 ("Bestimmung der Genauigkeit der Beobachtungen" 1880) and finally, in 1821 and 1823 in a systematic presentation ("Theoria combinationis observationum erroribus minimis obnoxiae, pars prior et pars posterior" [**350**]). Three years later these works were completed by the note "Supplementum theoriae combinationis observationum erroribus minimis obnoxiae" [**353**]. From then on, the concept of a pseudoinverse (or generalized inverse) $A^\dagger$ has rated considerable attention in mathematics as well as geosciences (a bibliography, for example, listing over 1700 references on the subject is due to M.Z. Nashed [**618**]).

Early interest in the first half of the last century in the subject of pseudoinverses was associated to a paper on matrices by R. Penrose [**684**]. However, this concept had been considered somewhat earlier. For example, E. H. Moore [**588**] presented a development of the notion (see also R. Baer [**57**], A. Bjerhammar [**89**], K. Friedrich

[**336**], E.H. Moore [**589**], R. Rado [**706**], and C.L. Siegel [**774**]). Moreover, in the setting of integral and differential operators the concept was considered even earlier by I. Fredholm [**247**] and W.A. Hurwitz [**452**], and by D. Hilbert [**433**] (see W.T. Reid [**716**] for a discussion of generalized inverses in classical analysis, and see also A. Ben-Israel and T.N.E. Greville [**77**], T.L. Boullion, P.L. Odell [**108**], M.Z. Nashed [**618**], W. Freeden, T. Sonar, B. Witte [**328**], and W. Freeden, B. Witte [**332**] for brief historical sketches on the subject). The relation between Gauss's ideas and the concept of the pseudoinverse was discussed by D.W. Robinson [**724**]. His paper attempts to show that although Gauss did not formalize the notion of a pseudoinverse, he provided the essential ingredients to produce one.

Today, the approach to the generalized inverse is well-described within the notational framework of Linear Algebra, more explicitly, within the reference structure of the linear space of $n \times m$ matrices $\mathbb{K}^{n\times m}$ (with $\mathbb{K} = \mathbb{R}$ for the set of real and $\mathbb{K} = \mathbb{C}$ for the set of complex numbers). Within this reference framework, the foundation is laid for the operator nomenclature of range, nullspace (kernel), adjoint, conventionally denoted by $\mathcal{R}(A)$, $\mathcal{N}(A)$, and A^*, respectively.

The *pseudoinverse (or generalized inverse)* $A^\dagger$ is the linear operator which assigns to each element of $\mathcal{D}(A^\dagger) = \mathcal{R}(A) + \mathcal{R}(A)^\perp$ the unique solution in $\mathcal{N}(A)^\perp$ of the *normal equation*

$$A^*Ax = A^*y \tag{1.5}$$

(the equation obtained by setting the first variation of $\|Ax - y\|^2$ equal to zero). It also follows that $A^\dagger$ can be characterized as the linear operator with the properties:

$$\mathcal{D}(A^\dagger) = \mathcal{R}(A) + \mathcal{R}(A)^\perp, \quad \mathcal{N}(A^\dagger) = \mathcal{R}(A)^\perp = \mathcal{N}(A^*) \tag{1.6}$$

and

$$\mathcal{R}(A^\dagger) = \mathcal{N}(A)^\perp. \tag{1.7}$$

Moreover,

$$AA^\dagger A = A, \quad A^\dagger AA^\dagger = A, \quad (AA^\dagger)^* = AA^\dagger, \quad (A^\dagger A)^* = A^\dagger A. \tag{1.8}$$

The equivalence of these characterizations of $A^\dagger$ is established and generalized by M. Z. Nashed (see, e.g., [**627**]) to unbounded operators such as the integral operator A, given in (1.2).

As an evolution of Hadamard's classification, M.Z. Nashed's work [**618**], [**625**], [**627**] also shows the advantage of adopting a notion of well-posedness that is focused on infinite-dimensional problems (e.g., an inconsistent finite system of linear algebraic equations will not be ill-posed in Nashed's sense, while it is ill-posed in the sense of Hadamard). Indeed, it follows immediately from the open mapping theorem in functional analysis (see, e.g., [**39**]) that the following statements are equivalent:

(N1) *the inverse problem is well-posed,*

(N2) *the range $\mathcal{R}(A)$ of the operator A is closed,*

(N3) *the pseudoinverse $A^\dagger$ of the operator A is bounded.*

Thus, we are led to call an IP *well-posed in the sense of Nashed*, if $\mathcal{R}(A)$ is closed. If $\mathcal{R}(A)$ is not closed (such as (1.2) in the L^2-nomenclature), IP is called *ill-posed in the sense of Nashed.*

Regularization strategies. The rationale in most methods for the resolution of an inverse problem IP such as (1.2) is to construct a "solution" that is acceptable physically as a meaningful approximation and is sufficiently stable from the computational standpoint. The main dilemma of modeling inverse problems is that most of them are ill-posed. The characteristic of such problems is that the closer the mathematical model describes the ill-posed problem (IPP), the worse is the "condition" of the associated computational problem (i.e., the more sensitive to errors). Therefore, the indispensable problem is to bring additional information about the desired solution, compromises, or new outlooks as aids to the resolution of ill-posed problems (IPPs). It is conventional to use the phrase *"regularization of an ill-posed problem"* to refer to various approaches to circumvent the lack of continuous dependence as well as to bring about existence and uniqueness if necessary. Roughly speaking, this entails an analysis of an IPP via an analysis of an associated well-posed problem, i.e., a family (usually a sequence or a net) of well-posed problems (WPPs), yielding meaningful answers to the IPP.

Nowadays, from the standpoint of mathematical and numerical analysis, one can roughly group "regularization methods" into several (canonically overlapping) categories:

- *Regularization methods in function spaces* is one category. This includes Tikhonov-type regularization (see, e.g., [**618**], [**821**]), the method of quasi-reversibility, the use for certain function spaces such as scale spaces in multiresolutions, the method of generalized inverses (pseudoinverses), e.g., in reproducing kernel Hilbert spaces (cf. [**640**], [**641**]), and multiscale wavelet regularization (see, e.g., [**266**], [**289, 292**], [**313**]).
- *Resolution of ill-posed problems by "control of dimensionality"* is another category (see, e.g., [**629**] and the literature therein). This includes projection methods and moment-discretization schemes. The success of these methods hinges on the possibility of obtaining an approximate solution while keeping the dimensionality of the finite dimensional problem within the "range of numerical stability". It also depends on deriving error estimates for the approximate solutions that is crucial to the control of the dimensionality.
- A third category is formed by *iterative methods* (see, e.g., [**487**], [**488**]) which can be applied either to the problem in function spaces or to a discrete version of it. The crucial ingredient in iterative methods is to stop the iteration before instability creeps into the process. Thus iterative methods have to be modified or accelerated so as to provide a desirable accuracy by the time a stopping rule is applied.
- A fourth category constitutes *filter methods.* Filter methods refer to procedures where, for example, values producing highly oscillatory solutions are eliminated. Various "lowpass" filters can, of course, be used. They are also crucial for the determination of a stopping rule. *Mollifiers* (see, e.g., [**218**], [**215**], [**296**], [**291**], [**270**]) are known as smooth filter functions with special properties to create sequences of smooth functions approximating a nonsmooth or singular function. Thus, we compromise by changing the problem into a more well-posed one, namely that of trying to determine a mollified version of the solution. Once more, the heuristic motivation

often is that the trouble comes from high frequency components of the data and of the solution, which are damped out by mollification.

- The root of the *Backus-Gilbert method (BG method)* was geophysical (cf. [**54**], [**55**], [**56**]). The characterization involved in the model is known as moment problem in the mathemetical literature. The BG method can be thought of as resulting from discretizing an integral equation of the first kind. Where other regularization methods, such as the frequently used Tikhonov regularization method, seek to impose smoothness constraints on the solution, the BG method instead realizes stability constraints (cf. [**867**]). As a consequence, the solution is varying as little as possible if the input data were resampled multiple times. The common feature between the mollifier and the BG method is that an approximate inverse is determined independently from the right side of the equation.
- The usual determination of parameters by inverse analysis assumes that the parameters are deterministic, that is that they are unknown but constants. The statistical approach to estimating parameters assumes that the parameters are random and that some knowledge of their statistical properties is known via a prior probability density distribution. Originally, *statistical regularization* is the name given to a form of estimation which resembles the usual Tikhonov-Phillips regularization method in which the regularization parameter is related to the variance of the prior. In modern nomenclature, the designation *stochastic regularization* is used to reduce the sensitivity to the degree of uncertainty in the parameters implying an estimation of the data, for example, a mean square estimation of a parameter which has a Gaussian distribution when the input data is also Gaussian distributed.
- Linear models for nonlinear relationships usually describe the reality in a very limited way. Thus it is required to work with nonlinear concepts. Because of the diversity of nonlinear phenomena, there is no closed inverse theory. As a consequence, nonlinear inverse problems are mainly based on the specific properties of the individual case.

Each method for resolution involves a critical "parameter" whose "optimal value" is crucial to the amenability and numerical implementation of the method. In *Tikhonov-type regularization* it is the regularization parameter, or more generally the choice of the regularization operator. In *multiscale methods* it is the scale parameter to determine the scale space in which multiresolution is realizable relative to the data width and noise. In *projection* and other discrete methods, it is the optimal dimension of the approximating subspaces. *Mollifier regularization* of integral operators (cf. [**270**], [**100**]) uses a kernel function that convolved with a particular function yields a filtered function, which is "close" to the function but "smoother". Usually, a mollifier is not only one kernel function but a sequence, or usually a one-parameter (i.e., scale dependent) family of kernels (also known as an approximate identity), so that the kernels of a mollifier have to tend to the Dirac kernel. As a consequence, mollifier kernels become more and more space-localized, and this make them extremely attractive for numerical purposes. In addition, filtering (convolution) with mollifier wavelets constituting the difference of two subsequent mollifier kernels provides band-pass information which, with increasing scales, deliver better and better details up to a certain scale. In discretization methods it is the choice of the mesh size beyond which a further refinement will lead to instability. In *iterative methods* it is the moment at which one should terminate the iteration (i.e., it is the number of iterates). In *filter-truncation methods* it is the number of terms to be included, etc. Our work intends to delineate suitable principles that quantify the choice of the parameter, the type of estimates, and a priori information that are needed to arrive at an "optimal" value for this parameter.

Expansion theorems play a fundamental role for an ill-posed *inverse problem*. Two problems are particularly significant: The determination of an (object) function from the inner product with a given set of functions (i.e., the *moment problem* (MP)), and the recovery of a function from its values approaches on a subset of its domain (i.e., the *reconstruction from samples* from sampling expansions). Moreover, new directions in expansion approaches are of major interest, such as in reproducing kernel Hilbert spaces, translation/rotation-invariant spaces, Sobolev spaces, etc. As a consequence, our textbook is concerned with a variety of function spaces in which functions admit representative recovery expansions.

Dilemmas and methodologies of the (re)solution of ill-posed problems and their numerical implementations are examined in this framework with particular reference to the problem of finding minimum weighted-norm least-squares solutions of first kind integral equations (and more generally of linear operator equations with nonclosed range). A common problem in all these methods is delineated: Each method reduces the problem of resolution to a "nonstandard" minimization problem involving an unknown critical "parameter" whose "optimal" value is crucial to the numerical realization and amenability of the method. The "nonstandardness" results from the fact that one does not have explicitly, or a priori, the function to be minimized: It has to built up using additional information, convergence rate estimates, and robustness conditions, etc. Several results are discussed that represent and complement recent advances in numerical analysis and regularization of inverse and ill-posed problems. An emphasis is placed on the role of constraints, function space methods, the role of generalized inverses, and reproducing kernels in the regularization and stable computational resolution of these problems.

The thrust of the work is devoted to the interdisciplinary character of operator-theoretic and numerical methods for ill-posed problems. In fact, our purpose is to provide a new outlook within which technical results can be better motivated and understood. Within this framework, criteria can be given relative to which the scope and limitations of the various methods can be assessed. This is important both in theory and practice since there is no cure-all method for ill-posed problems. Therefore it is imperative to be able to clarify why a certain method works in some context as well as when not to use that method. The work discusses at length the intuitive principles that underlie the various methods and establishes some results within this framework. The development in the first part of the work mostly based on a functional analytic background is fairly general in scope and theory, and the applicability covers a wide range.

Univariate antenna problem in L^2-nomenclature. The particular genealogical role of Nashed's 1D antenna note [**625**] in the recovery context is as follows: The operator A as given by (1.2) is a linear mapping from the space $L^2(\mathcal{F}_{\mathbb{Z}})$ to the range

$$\mathcal{R}(A) = \left\{ \mathbb{R} \ni x \mapsto F^{\wedge}_{\mathcal{F}_{\mathbb{Z}}}(x) = \int_{\mathcal{F}_{\mathbb{Z}}} F(t)\, e^{-2\pi i t \cdot x}\, dt : F \in L^2(\mathcal{F}_{\mathbb{Z}}) \right\}. \tag{1.9}$$

Central for the understanding of the regularization as well as sampling constituents in our recovery context below is the fact that the space $\mathcal{R}(A)$ is nonclosed in $L^2(\mathbb{R})$. To show this we give a specific Fourier theoretic and a general functional analytic based verification:

- *Fourier theoretic verification.* For every sufficiently small $\varepsilon > 0$, the characteristic function $\chi_{[-\varepsilon,\varepsilon]}$ given by

$$\chi_{[-\varepsilon,\varepsilon]}(x) = \begin{cases} 0 & , \quad x > \varepsilon, \\ 1 & , \quad x \leq \varepsilon, \end{cases} \tag{1.10}$$

  is in the class $L^2(\mathcal{F}_{\mathbb{Z}})$, and $\frac{1}{2\varepsilon}(\chi_{[-\varepsilon,\varepsilon]})^{\wedge}_{\mathcal{F}_{\mathbb{Z}}}$ explicitly expressed by

$$\frac{1}{2\varepsilon}(\chi_{[-\varepsilon,\varepsilon]})^{\wedge}_{\mathcal{F}_{\mathbb{Z}}}(x) = \frac{1}{2\varepsilon}\int_{\mathcal{F}_{\mathbb{Z}}} \chi_{[-\varepsilon,\varepsilon]}\, e^{-2\pi i t\cdot x}\, dt = \frac{\sin(2\pi\varepsilon x)}{2\pi\varepsilon x} \tag{1.11}$$

  represents a function in $\mathcal{R}(A)$. The limit function of (1.11) with $\varepsilon \to 0$ is equal to 1 and, therefore, not in $L^2(\mathbb{R})$. In other words, the range of the linear operator $A : L^2(\mathcal{F}_{\mathbb{Z}}) \to L^2(\mathbb{R})$, as defined by (1.2) mapping the space $L^2(\mathcal{F}_{\mathbb{Z}})$ to the range $\mathcal{R}(A)$ is nonclosed in $L^2(\mathbb{R})$.
- *Functional analytic verification.* It is a well-known result in functional analysis (see, e.g., [**435**]) that $A : L^2(\mathcal{F}_{\mathbb{Z}}) \to L^2(\mathbb{R})$, as defined by the integral equation (1.2), is a compact linear operator mapping the space $L^2(\mathcal{F}_{\mathbb{Z}})$ to the infinite-dimensional range $\mathcal{R}(A)$, so that A is nonclosed understood in the framework of $L^2(\mathbb{R})$. The nonclosedness indeed is a consequence of the open mapping theorem, and it is valid not only for the operator in (1.2), but also for all compact operators with infinite-dimensional range.

In order to overcome the nonclosedness of the range of the linear operator $A : L^2(\mathcal{F}_{\mathbb{Z}}) \to L^2(\mathbb{R})$, as defined by (1.2), a way out (see [**625**]) is the observation that the range $\mathcal{R}(A)$ can be endowed with a new topology that would make this space a reproducing kernel Hilbert space (RKHS) with useful properties, that means, we are able to shrink the image space $L^2(\mathbb{R})$ to the range $\mathcal{R}(A)$ and understand the latter space as the so-called Paley-Wiener space $B_{\mathcal{F}_{\mathbb{Z}}}$ with an associated topology to guarantee that the range is closed. As a consequence, in analyzing the 1D antenna problem, we are led to a remarkable transition, namely from an ill-posed problem in the sense of Hadamard (in the nomenclature of $L^2(\mathbb{R})$) to a well-posed problem in the Paley-Wiener reproducing kernel Hilbert space $B_{\mathcal{F}_{\mathbb{Z}}}$ in a canonically defined topology (see (1.14)):

Paley-Wiener reproducing kernel Hilbert space. The *Paley-Wiener space* $B_{\mathcal{F}_{\mathbb{Z}}}$ defined by (1.12)

$$B_{\mathcal{F}_{\mathbb{Z}}} = \left\{ \mathbb{R} \ni y \mapsto F^{\wedge}_{\mathcal{F}_{\mathbb{Z}}}(y) = \int_{\mathcal{F}_{\mathbb{Z}}} F(t)\, e^{-2\pi i t\cdot y}\, dt : F \in L^2(\mathcal{F}_{\mathbb{Z}}) \right\} \subset C^{(0)}(\mathbb{R}) \tag{1.12}$$

is a reproducing kernel Hilbert space (RKHS) with the uniquely determined reproducing kernel (RK)

$$K_{\mathcal{F}_{\mathbb{Z}}}(x-y) = \int_{\mathcal{F}_{\mathbb{Z}}} e^{2\pi i t\cdot(x-y)}\, dt = \frac{\sin(\pi(x-y))}{\pi(x-y)} = \operatorname{sinc}(\pi(x-y)), \tag{1.13}$$

so that

$$F^{\wedge}_{\mathcal{F}_{\mathbb{Z}}}(y) = \int_{\mathbb{R}} F^{\wedge}_{\mathcal{F}_{\mathbb{Z}}}(x) \left(\int_{\mathcal{F}_{\mathbb{Z}}} e^{2\pi i a\cdot(x-y)}\, da \right) dx = \int_{\mathbb{R}} F^{\wedge}_{\mathcal{F}_{\mathbb{Z}}}(x)\, K_{\mathcal{F}_{\mathbb{Z}}}(x-y)\, dx \tag{1.14}$$

for all $y \in \mathbb{R}$.

Univariate antenna problem in $B_{\mathcal{F}_\Lambda}$-nomenclature. Under the assumption that $\mathcal{R}(A)$ is the subspace $B_{\mathcal{F}_{\mathbb{Z}}}$ of $L^2(\mathbb{R})$, the operator A defined by the integral equation (1.2) becomes *onto and*, hence, by the open mapping theorem of functional analysis, the inverse operator $A^{-1} : B_{\mathcal{F}_{\mathbb{Z}}} \to L^2(\mathcal{F}_{\mathbb{Z}})$ is *bounded.* Therefore, methods of constructive approximation in reproducing kernel Hilbert spaces (for example, associated spline methods as proposed in [**298**]) become efficiently applicable.

1.2. Problem induced sampling

The essential objective of sampling is to deal with the reconstruction of a bandlimited function, usually given in the form (1.1), by a sum in terms of the function itself at a discrete set of points. Sampling is a keystone of signal processing. The most common form of sampling today is lattice point sampling of a bandlimited signal $F^{\wedge}_{\mathcal{F}_{\mathbb{Z}}}$ over the fundamental cell $\mathcal{F}_{\mathbb{Z}} = [-\frac{1}{2}, \frac{1}{2})$ of the unit lattice $\mathbb{Z}$ for a square-integrable function F on $\mathcal{F}_{\mathbb{Z}}$.

Classical Fourier analytic methodology. Fourier analysis offers different ways to the *1D Shannon sampling theorem* (based, e.g., on tools already known to L.J. Mordell [**590**], [**591**], [**592**], and H.G. Hardy [**412**], [**413**], [**414**]). Continuing the nomenclature used in the discussion of (1.1) we are able to explain the sampling theorem as follows:

Let $F^{\wedge}_{\mathcal{F}_\Lambda}$ be the Fourier transform for some $F \in L^2([-\frac{1}{2}, \frac{1}{2}))$ extended $\mathbb{Z}$-periodically to the real line $\mathbb{R}$. We expand the extension in a series of the form

$$F(x) = \sum_{h\in\mathbb{Z}} c_h\, \Phi_h(x), \tag{1.15}$$

where the system $\{\Phi_h\}_{h\in\mathbb{Z}}$ of one-dimensional "periodic polynomials" is given by

$$\Phi_h(x) = e(h\cdot x) = e^{2\pi i h\cdot x}, \quad h \in \mathbb{Z}, \tag{1.16}$$

and $\sum_{h\in\mathbb{Z}} \ldots$ means that the sum is extended over all h of the lattice $\mathbb{Z}$. It is not hard to see that

(1.17)

$$\int_{\mathcal{F}_{\mathbb{Z}}} \Phi_h(x)\, \overline{\Phi_{h'}(x)}\, dx = \int_{\mathcal{F}_{\mathbb{Z}}} e^{2\pi i h\cdot x}\, e^{-2\pi i h'\cdot x}\, dx = \delta_{h,h'} = \begin{cases} 1, & h = h', \\ 0, & h \neq h'. \end{cases}$$

Hence, the system $\{\Phi_h\}_{h\in\mathbb{Z}}$ is orthonormal with respect to the L^2-inner product ($\delta_{h,h'}$ is the Kronecker symbol). As a consequence, the coefficients c_h are given in the form

$$c_h = \int_{\mathcal{F}_{\mathbb{Z}}} F(x)\, \overline{\Phi_h(x)}\, dx = \int_{\mathcal{F}_{\mathbb{Z}}} F(x)\, e^{-2\pi i h\cdot x}\, dx, \tag{1.18}$$

i.e., $c_h = F^{\wedge}_{\mathcal{F}_{\mathbb{Z}}}(h)$. Substituting (1.15) into (1.18) we obtain the identity

$$F^{\wedge}_{\mathcal{F}_{\mathbb{Z}}}(t) = \int_{\mathcal{F}_{\mathbb{Z}}} \sum_{h\in\mathbb{Z}} c_h \, e^{2\pi i h\cdot x} \, e^{-2\pi i x\cdot t} \, dx, \quad t \in \mathbb{R}, \tag{1.19}$$

such that by observing (1.2) and interchanging the order of integration and summation

$$F^{\wedge}_{\mathcal{F}_{\mathbb{Z}}}(t) = \sum_{h\in\mathbb{Z}} F^{\wedge}_{\mathcal{F}_{\mathbb{Z}}}(h) K_{\mathcal{F}_{\mathbb{Z}}}(h-t) = \sum_{h\in\mathbb{Z}} F^{\wedge}_{\mathcal{F}_{\mathbb{Z}}}(h) \int_{\mathcal{F}_{\mathbb{Z}}} e^{2\pi i x\cdot(h-t)} \, dx. \tag{1.20}$$

This identity yields the celebrated classical expansion of a bandlimited signal for the "base band" $\mathcal{F}_{\mathbb{Z}} = [-\frac{1}{2}, \frac{1}{2})$:

$$F^{\wedge}_{\mathcal{F}_{\mathbb{Z}}}(t) = \sum_{h\in\mathbb{Z}} F^{\wedge}_{\mathcal{F}_{\mathbb{Z}}}(h) \, \frac{\sin(\pi(h-t))}{\pi(h-t)}. \tag{1.21}$$

Hence, documented in the lattice point nomenclature as used in this work, Shannon's sampling result (1.21) can be rephrased as follows:

Any function $F^{\wedge}_{\mathcal{F}_{\mathbb{Z}}}$ bandlimited to the fundamental cell $\mathcal{F}_{\mathbb{Z}} = \left[-\frac{1}{2}, \frac{1}{2}\right)$ of the unit lattice $\mathbb{Z}$, i.e., the Fourier transform (1.1) *may be reconstructed from the sequence of samples*

$$\{F^{\wedge}_{\mathcal{F}_{\mathbb{Z}}}(h)\}_{h\in\mathbb{Z}} \tag{1.22}$$

via the so-called cardinal series

$$\sum_{h\in\mathbb{Z}} F^{\wedge}_{\mathcal{F}_{\mathbb{Z}}}(h) \int_{\mathcal{F}_{\mathbb{Z}}} e^{2\pi i x\cdot(h-t)} \, dx, \quad t \in \mathbb{R}, \tag{1.23}$$

where the series is absolutely and uniformly convergent on any compact set of the real line $\mathbb{R}$.

Obviously, the Fourier transform $F^{\wedge}_{\mathcal{F}_{\mathbb{Z}}}$ (considered as a one-dimensional signal) is an infinitely repeated replication of the samples at points h of the lattice $\mathbb{Z}$.

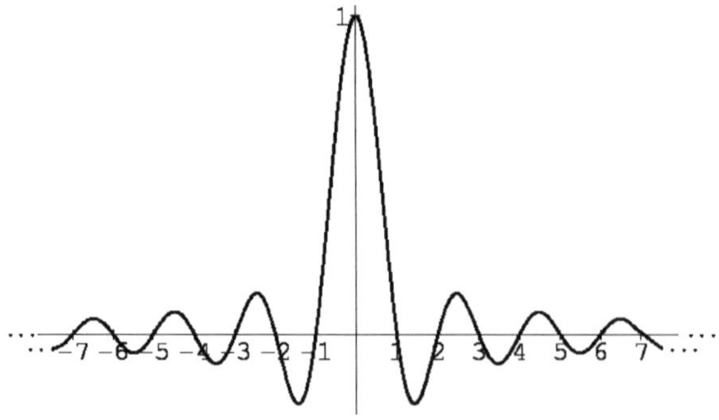

FIGURE 1. Illustration of the function $t \mapsto \mathrm{sinc}(\pi t) = \frac{\sin(\pi t)}{\pi t}$, $t \in \mathbb{R}$.

The *interpolatory background of bandlimited sampling* can be seen briefly as follows: A mathematical way to interpolate samples $F^{\wedge}_{\mathcal{F}_{\mathbb{Z}}}(h')$, $h' \in \mathbb{Z}$, may be realized by use of sinc-functions (cf. Figure 1), since the sequence

$$\{\mathrm{sinc}(\pi\,(h-\cdot))\}_{h\in\mathbb{Z}} \tag{1.24}$$

shows the discrete orthogonality property:

$$\text{sinc}(\pi(h-h')) = \delta_{h,h'} = \begin{cases} 1, & h = h', \\ 0, & h \neq h'. \end{cases} \tag{1.25}$$

As a consequence, $F^{\wedge}_{\mathcal{F}_{\mathbb{Z}}}$ given by the so-called *cardinal series*

$$F^{\wedge}_{\mathcal{F}_{\mathbb{Z}}}(t) = \sum_{h\in\mathbb{Z}} F^{\wedge}_{\mathcal{F}_{\mathbb{Z}}}(h)\ \text{sinc}(\pi\,(h-t)), \quad t \in \mathbb{R}, \tag{1.26}$$

satisfies the interpolatory property

$$F^{\wedge}_{\mathcal{F}_{\mathbb{Z}}}(t)|_{t=h'} = \sum_{h\in\mathbb{Z}} F^{\wedge}_{\mathcal{F}_{\mathbb{Z}}}(h)\ \text{sinc}(\pi\,(h-t))|_{t=h'} = F^{\wedge}_{\mathcal{F}_{\mathbb{Z}}}(h'), \quad h' \in \mathbb{Z}. \tag{1.27}$$

It should be noted that the classical proof of the Shannon sampling theorem is indeed rigorous. The interchange of integration and summation from (1.19) to (1.20) can be easily justified. However, the proof is very revealing (as already pointed out by M.Z. Nashed [**629**]): We perform this interchange and a theorem pops up, but what is its essential mathematical core?

Sampling strategies. In fact, many questions arise when thinking about generalizations and modifications of the 1D Shannon sampling context (see also some aspects in P.L. Butzer [**123, 128**], M. Unser [**827**], M. Unser, A. Aldroubi [**828**], W. Freeden, M.Z. Nashed [**298**]):

- *Engineering reflected aspects.* Engineering uses the 1D Shannon sampling theorem to express a continuous (time-dependent) signal in a discrete context, but what are the essential ingredients how to do so, seen from a mathematical point of view? Is there a deeper interrelationship between Shannon sampling and a periodically reflected Dirac distributional expression (Dirac impulse) leading to the specification of lattice reflected Green functions (lattice functions) and resulting Euler and Poisson summation formulas?
- *Functional analytic manifestations.* Are there functional analytic strategies to transfer reproducing kernel Hilbert/Banach space structures to sampling and/or to induce structures in reproducing kernel Hilbert/Banach spaces from sampling concepts? In this respect, it is of tremendous importance that both continuous and discrete variants of reproducing kernel Hilbert spaces arise from given positive definite kernels, i.e., a corresponding pre-Hilbert form and subsequently a Hilbert space completion. What are the specific characteristics of sampling procedures for continuous and discrete variants? While the better known and classical sampling algorithms (Shannon and others) are based on interpolation, are there theories to go beyond this? For example, is it possible to make precise the notion of "nonuniform point sampling" in a discrete reproducing kernel Hilbert/Banach space context?
- *Geometric number theoretic manifestations.* Are there straightforward multivariate counterparts of the univariate Shannon sampling theorem, not only by Parzen's iteration of 1D framework, but also on general qD "base bands" independent of the choice of the general qD lattice? In other words, is a specifically multivariate Shannon sampling variant valid for a

class of bandlimited sample functions related not only to the base band determined by the fundamental cell of the underlying lattice, but also to "finite bandwidth bands" such as, e.g., certain "potato-like" (i.e., regular) regions $\mathcal{G}$ in Euclidean space $\mathbb{R}^q$? How can we describe geometrically resulting procedures of over- and undersampling in sampling related to "base band" regions $\mathcal{G}$ by explicitly determining the aliasing error? How can we come to the explicit knowledge about the fortunate situation that there is no aliasing error between the cardinal series and the (Fourier transformed) functional value in bandlimited sampling?

- *Fourier theoretic manifestations.* Are there certain conditions at infinity involving Fourier transforms to enable multivariate nonbandlimited Shannon-type sampling, i.e., how can we concretize the asymptotic properties of functions at infinity reflecting infinite bandwidth sample methodologies and explicit representations of the aliasing error?
- *Potential theoretic manifestations.* Harmonic kernel structures are known to be involved in thin plate and cardinal spline concepts. Are there metaharmonic reproducing kernel structures to allow Shannon-type sampling procedures and Paley-Wiener related framework to deduce, for example, Paley-Wiener and Slepian splines for inverse sampling application in multivariate antenna theory?
- *Special functions summability based manifestations.* Are there bandlimited/nonbandlimited methodologies to decorrelate a certain signal (signature) in an adequate Shannon-type wavelet context provided by special function originated features such as multiscale Gauss-Weierstrass (averaging) sampling?

Most of these questions will be studied in this book by choosing standard as well as nonstandard approaches to sampling.

1.3. Recovery as common thread

Since the middle of the last century, the rudiments of sampling are covered in almost every engineering textbook on signal analysis, but sampling theory could be found only in rare exceptions in the mathematical literature. The connections among the areas (inverse problems, sampling in signal processing, and image analysis) were rather tenuous. Not least the results for the antenna problem have shown, that our understanding today of inverse problems and sampling in signal analysis is very clear (cf. [**629**]):

- *Inverse problems* deal with determining for a given input-output system an input that produces an observed output, or of determining an input that produces a desired output (or comes as close to it as possible), often in the presence of noise. Most inverse problems are ill-posed and demand regularization. In particular, *Moment problems* deal with the recovery of a function or signal from its moments, and the construction of efficient stable algorithms for determining or approximating the function.
- *Signal analysis/processing* deals with digital representations of signals and their analog reconstructions from digital representations. Sampling

methodologies such as expansions, filters, constituents of reproducing kernel spaces, various function space specifications, and techniques of functional and Fourier analysis, computational and (meta)harmonic analysis play pivotal roles in this area. In particular, *Image analysis/processing* deals with image refinement and recovery, and include all facets of imaging.

The common thread among all these disciplines canonically is the *problem of recovering an object (function, signal, picture) from partial or indirect information about the object.*

1.4. Organization of the work

Reflecting the 1D framework described in the Introduction for the antenna problem we are led to organize the parts of the book as follows:

- Part I is devoted to the study of the Fourier transform over the one-dimensional fundamental cell of the 1D unit lattice. It is documented that the Fourier integral bandlimited with respect to a fundamental cell of the 1D unit lattice has led to a deepened mathematical discussion in two essential research areas both initiated by engineering, namely regularization methods in inverse problems and sampling methods in signal analysis/processing.
- Part II presents some essential tools, that serve as auxiliary material for the first part of the book concerned with inverse problems and their regularization. It contains selected key items of classical functional and Fourier analysis.
- Part III is concerned with inverse problems and regularization strategies. Resolution methodologies involving generalized inverses and singular value decompositions are discussed for matrix equations. A broad collection of reconstruction and regularization methods for compact operator equations are outlined mostly with the aid of functional analytic means.
- Part IV illustrates some regularization methodologies in three-dimensional Euclidean space, applied to three significant inverse potential problems. The presented examples are of great interest in today's geoengineering, namely (geodetic) satellite gravitation gradiometry as well as inverse (exploration) gravimetry and magnetometry.
- Part V presents some essential tools, that serve as preparatory material for the second part of the book concerned with sampling. It contains selected key items of geometric number theory, metaharmonic special function theory such as Bessel and Kelvin functions, and Gauss-Weierstrass as well as Abel-Poisson variants in the Fourier inversion.
- Part VI deals with sampling problems and solution strategies. As prototype for various developments, the 1D Shannon sampling theorem is deduced in a twofold way, on the one hand by classical Fourier analytic concepts, on the other hand by lattice point theoretically based Euler/Poisson summation formulas. A palette of multivariate sampling methodologies using harmonic and metaharmonic tools and structures is given resulting in both finite and infinite bandwidth sampling.

- Part VII explains two important examples of sampling methodology: The inversion of the gravimetry problem in geoexploration is documented by certain expressions, which reproduce exponentials. The solvability of the multivariate antenna problem is analyzed within the framework of Paley-Wiener spline theory in multi-dimensional Euclidean spaces.
- Part VIII explains the idea of recovery as an interconnecting whole of inverse and sampling theory in essential publication organs and its manifestations in today's literature.

Part 2

Regularization tools

CHAPTER 2

Functional and Fourier analytic auxiliaries

In this preparatory chapter we collect essential material to be needed for the notational background as well as the readability and the understanding of regularization induced recovery.

2.1. Functional analytic tools

First we recapitulate some basic results of functional analysis. Our collection is partially oriented on the synopsis given by M. Augustin et al. [**50**] in the "Handbook of Mathematical Geodesy" initiated by its editor W. Freeden [**295**] . More details can be found in any textbook about functional analysis, e.g., R.A. Adams [**5**], M. Dobrowolski, [**190**], H. Heuser [**429**], L.W. Kantorowitsch, G.P. Akilow [**489**], L.A. Lieusternik, V.J. Sobolev [**537**], F. Riesz, B. Sz.-Nagy [**720**], L.I. Rudin [**731**], A.E. Taylor, D. Lay [**816**], D. Werner [**861**], K. Yoshida [**875**].

First we are interested in the notions of metric and norm. For that purpose we consider $\mathbb{K}$-linear spaces X, Y, and Z over the field $\mathbb{K} \in \{\mathbb{R}, \mathbb{C}\}$.

Metric space. A function $\rho : \ X \times X \to [0, \infty)$, $x, y \mapsto \rho(x, y)$ is called a *metric* if it has the following properties:

(a) for all $x, y \in V$, $\rho(x, y) \geq 0$, and $\rho(x, y) = 0$ if and only if $x = y$,
(b) for all $x, y \in V$, $\rho(x, y) = \rho(y, x)$,
(c) for all $x, y, z \in V$ we have the triangle inequality $\rho(x, y) \leq \rho(x, z) + \rho(z, y)$.

A sequence $\{x_n\}_{n\in\mathbb{N}}$ in X is called *convergent with limit* $x_0 \in X$ if

$$\lim_{n\to\infty} \rho(x_n, x_0) = 0.$$

A sequence $\{x_n\}_{n\in\mathbb{N}}$ in X is called *Cauchy sequence* if, for any $\varepsilon > 0$, there exists a positive number $N (= N(\varepsilon))$ such that for all $n, m \in \mathbb{N}$, $n \geq m \geq N$, we have $\rho(x_n, x_m) \leq \varepsilon$.

A metric space X is called *complete* if every Cauchy sequence is convergent.

Let Y, Z be two sets in a metric space X such that $Y \subset Z \subset X$. Y is said to be *dense in* Z if, for every $z \in Z$ and every $\varepsilon > 0$, there exists a $y \in Y$ such that $\rho(y, z) \leq \varepsilon$.

A metric space X is called *separable* if it contains a countable, dense subset.

Normed space. A function $\|\cdot\| : X \to [0,\infty)$, $y \mapsto \|y\|$ is called a *norm* if, for all $x, y \in X$ and $\alpha \in \mathbb{K}$, it has the following properties:

(a) $\|y\| \geq 0$, and $\|y\| = 0$ if and only if $y = 0$,

(b) $\|\alpha y\| = |\alpha| \|y\|$,

(c) $\|x + y\| \leq \|x\| + \|y\|$.

From every norm $\|\cdot\|$ on X, a metric can be constructed by

$$\rho(x,y) = \|x - y\| = \|y - x\|, \quad x, y \in X. \tag{2.1}$$

Additionally to the defining properties of a metric, a metric defined from a norm also satisfies for all $x, y \in X$, $\alpha \in \mathbb{K}$:

$$\rho(x+z, y+z) = \rho(x,y), \ z \in X, \tag{2.2}$$

$$\rho(\alpha x, \alpha y) = |\alpha| \rho(x,y). \tag{2.3}$$

Vice versa, a metric that also has the properties (2.2) and (2.3) can be used to define a norm via $\|y\| = \rho(y, 0)$, $y \in X$.

If there is no risk of confusion, we often do not state the norm explicitly and also omit to index the norm by the corresponding space.

Let $(X, \|\cdot\|_X)$ be a normed space and I a suitable set to index a system of elements $\{x_i : i \in I\} \subset X$. The system $\{x_i : i \in I\}$ is called *closed* if span$\{x_i : i \in I\}$, i.e., the set of all finite linear combinations of the elements $\{x_i : i \in I\}$, is dense in X.

Any finite dimensional normed space is closed. Any finite dimensional subspace of a normed space X is complete.

Banach space. A normed space $(X, \|\cdot\|_X)$ is called Banach space if X is complete with respect to $\|\cdot\|_X$.

The space $\ell^p(\mathbb{N})$, $1 \leq p < \infty$, is the space of all $\mathbb{K}$-sequences $x = \{x_n\}_{n\in\mathbb{N}}$ for which $\sum_{n=1}^{\infty} |x_n|^p$ is finite. Equipped with the norm

$$\|x\|_{\ell^p(\mathbb{N})} = \left(\sum_{n=1}^{\infty} |x_n|^p \right)^{1/p} \tag{2.4}$$

the space $\ell^p(\mathbb{N})$ is a Banach space. The space $\ell^\infty(\mathbb{N})$, which is the set of bounded real or complex sequences equipped with the norm

$$\|x\|_{\ell^\infty(\mathbb{N})} = \sup_{n\in\mathbb{N}} |x_n|, \tag{2.5}$$

also constitutes a Banach space.

Let $(X, \|\cdot\|_X)$ be a *Banach space.* A sequence $\{x_n\}_{n\in\mathbb{N}}$ in X is called a *basis*, if every element $x \in X$ can be written in the form $x = \sum_{n=1}^{\infty} a_n x_n = \lim_{N\to\infty} \sum_{n=1}^{N} a_n x_n$ with coefficients $a_n \in \mathbb{K}$ and the elements of every finite subset of the sequence $\{x_n\}_{n\in\mathbb{N}}$ are linearly independent.

Whereas every Banach space with a countable basis is separable, not every separable Banach space has a countable basis.

Operators and functionals. Let X, Y be two spaces and U an arbitrary subset of X. A function $A: X \to Y, U \ni x \mapsto Ax \in Y$ that maps every element $x \in U$ on a unique element $y = Ax \in Y$ is called an *operator*. The set U is called the *domain* of the operator A and denoted by $\mathcal{D}(A)$. The set $\mathcal{R}(A) = \{y \in Y : y = Ax, x \in \mathcal{D}(L)\}$ is called the *range* of A. The set $\mathcal{N}(A) = \{x \in D(A) : Ax = 0\}$ is called the *kernel* or *nullspace* of A. If $\mathcal{R}(A)$ is a subset of $\mathbb{K}$, the operator is called a *functional*.

Let (X, ρ_X) and (Y, ρ_Y) be two metric spaces. An operator $A: X \supset \mathcal{D}(A) \to \mathcal{R}(A) \subset Y$ is called *continuous in* $x \in \mathcal{D}(A)$ if, for every sequence $\{x_n\}_{n\in\mathbb{N}} \subset \mathcal{D}(A)$ with $\lim_{n\to\infty} \rho_X(x_n, x) = 0$, we have $\lim_{n\to\infty} \rho_Y(Ax, Ax_n) = 0$.

A is said to be *continuous* on $\mathcal{D}(A)$ if A is continuous for all $x \in \mathcal{D}(A)$.

Let $\mathcal{D}(A) \subset X$ be $\mathbb{K}$-linear spaces and $A: \mathcal{D}(A) \mapsto \mathcal{R}(A) \subset Y$. A is called a *linear operator* if, for arbitrary $\alpha \in \mathbb{K}$ and $x_1, x_2 \in \mathcal{D}(A)$, we have

$$A(\alpha x_1 + x_2) = \alpha A x_1 + A x_2. \tag{2.6}$$

It follows that $\mathcal{N}(A)$ is a linear subspace of $\mathcal{D}(A)$ and $\mathcal{R}(A)$ is a linear subspace of Y.

Let $(X, \|\cdot\|_X)$, $(Y, \|\cdot\|_Y)$ be two normed spaces and $A: X \supset \mathcal{D}(A) \mapsto \mathcal{R}(A) \subset Y$ a linear operator. A is called *isometric* if, for every $x \in \mathcal{D}(A)$, we have

$$\|Ax\|_Y = \|x\|_X . \tag{2.7}$$

Inverse operator. In the following, we assume every operator to be linear. Let $A: X \supset \mathcal{D}(A) \mapsto \mathcal{R}(A) \subset Y$ such that, for every $y \in \mathcal{R}(A)$, there exists a unique $x \in \mathcal{D}(A)$ with $Ax = y$. The operator that assigns to every $y \in \mathcal{R}(A)$ its inverse image $x \in \mathcal{D}(A)$ is called the *inverse operator to* A, denoted by A^{-1}, and is a linear operator from $\mathcal{R}(A)$ to $\mathcal{D}(A)$. We have

$$A^{-1}(Ax) = x \quad \text{for all } x \in \mathcal{D}(A), \tag{2.8}$$

$$A(A^{-1}y) = y \quad \text{for all } y \in \mathcal{R}(A). \tag{2.9}$$

As the existence of the inverse operator requires the uniqueness of the inverse image, a linear operator A has an inverse operator A^{-1} if and only if $\mathcal{N}(A) = \{0\}$.

Let $(X, \|\cdot\|_X)$, $(Y, \|\cdot\|_Y)$ be two normed spaces and $A: X \supset \mathcal{D}(A) \mapsto \mathcal{R}(A) \subset Y$ a linear operator. A is *bounded* if there exists a $C > 0$, such that

$$\|Ax\|_Y \leq C \|x\|_X \quad \text{for all } x \in \mathcal{D}(A). \tag{2.10}$$

Let $(X, \|\cdot\|_X)$, $(Y, \|\cdot\|_Y)$ be normed spaces and $A: X \to Y$ be a linear operator. Then the following statements are equivalent:

(i) A is continuous on X.
(ii) A is continuous in $0 \in X$.
(iii) A is bounded.

The *space of all linear bounded operators with domain* $\mathcal{D}(A)$ *and range* $\mathcal{R}(A)$ is denoted by $\mathfrak{L}(\mathcal{D}(A), \mathcal{R}(A))$. This space can be equipped with the *operator norm*, given by

$$\|A\|_{\mathfrak{L}(\mathcal{D}(A),\mathcal{R}(A))} = \sup_{\substack{v\in\mathcal{D}(A)\\ x\neq 0}} \frac{\|Ax\|_Y}{\|x\|_X} = \sup_{\substack{x\in\mathcal{D}(A)\\ \|x\|_X=1}} \|Ax\|_Y = \sup_{\substack{x\in\mathcal{D}(A)\\ \|x\|_X\leq 1}} \|Ax\|_Y . \tag{2.11}$$

We have

$$\|Ax\|_Y \leq \|A\|_{\mathfrak{L}(D(A),R(A))} \|x\|_X . \tag{2.12}$$

The space $X' = \mathfrak{L}(X, \mathbb{K})$ is called the *(topological) dual space to* X. X' consists of all continuous linear functionals on X. A norm on X' is defined in the same way as the general operator norm. The dual space of X' is denoted by X'' and called the *bidual space of* X. A space is called *reflexive*, if X'' and X are isomorph.

Dual space adjoint operator. Let $(X, \|\cdot\|_X)$, $(Y, \|\cdot\|_Y)$ be two normed spaces and $A \in \mathfrak{L}(X, Y)$. The *dual space adjoint operator* $A' : Y' \to X'$ is defined by

$$(A'y')(x) = y'(Ax), \quad x \in X, y \in Y'. \tag{2.13}$$

Because of

$$|(A'y')x| \leq \|y'\|_{Y'} \|A\|_{\mathfrak{L}(X,Y)} \|x\|_X \tag{2.14}$$

we have $A' \in \mathfrak{L}(Y', X')$ and $\|A'\|_{\mathfrak{L}(Y',X')} \leq \|A\|_{\mathfrak{L}(X,Y)}$.

Sesquilinear forms, inner products, and Hilbert spaces. Inner product spaces (also called pre-Hilbert spaces) provide the concepts of norm (distance) and angle. Inner products are introduced by positive definite, hermitian sesquilinear forms (as will be explained next):

$a(\cdot,\cdot) : X \times X \to \mathbb{K}$ is called a *sesquilinear form* if for every $x, x_1, x_2, y, y_1, y_2 \in X$ and every $\alpha \in \mathbb{K}$ we have

$$a(\alpha\, x, y) = \alpha\, a(x, y), \tag{2.15}$$

$$a(x, \alpha\, y) = \overline{\alpha}\, a(x, y), \tag{2.16}$$

$$a(x_1 + x_2, y) = a(x_1, y) + a(x_2, y), \tag{2.17}$$

$$a(u, y_1 + y_2) = a(x, y_1) + a(x, y_2). \tag{2.18}$$

If, instead of (2.16), we have $a(x, \alpha\, y) = \alpha\, a(x, y)$, then $a(\cdot,\cdot)$ is called a *bilinear form.*

A sesquilinear form is called *hermitian* if $a(x, y) = \overline{a(y, x)}$. A bilinear form is called *symmetric* if $a(x, y) = a(y, x)$. A hermitian sesquilinear form is called *positive definite* if

$$a(x, x) \geq 0 \text{ and } a(x, x) = 0 \text{ if and only if } x = 0. \tag{2.19}$$

A function $\langle\cdot,\cdot\rangle_X : X \times X \to \mathbb{K}$ is called an *inner product* if it is a positive definite, hermitian sesquilinear form. The space X equipped with $\langle\cdot,\cdot\rangle_X$ is called an *inner product space.*

Cauchy-Schwarz inequality. Let $(X, (\cdot,\cdot)_X)$ *be an inner product space. For every* $x, y \in V$, *we have*

$$|\langle x, y\rangle_X|^2 \leq \langle x, x\rangle_X \langle y, y\rangle_X . \tag{2.20}$$

Pre-Hilbert space and Hilbert space. An inner product space $(X, \langle\cdot,\cdot\rangle_X)$ is called a *pre-Hilbert space* if it is equipped with the norm defined by $\|x\|_X = \sqrt{\langle x, x\rangle_X}$. If X is complete with respect to the norm $\|\cdot\|_X$ defined by its inner product $\langle\cdot,\cdot\rangle_X$, it is called a *Hilbert space.*

Let $(X, \langle\cdot,\cdot\rangle_X)$ be a pre-Hilbert space. A sesquilinear form $a(\cdot,\cdot) : X \times X \to \mathbb{K}$ is called

(a) *bounded or continuous,* if there is a $C>0$ such that $|a(x,y)| \le C\,\|x\|_X\;\|y\|_X$ for all $x, y \in X$,

(b) *coercive,* if there exist a $C > 0$ such that $a(x,x) \ge C\;\|x\|_X^2$ for all $x \in X$.

The norm of a bounded sesquilinear form is defined as

$$\|a\|_{\mathcal{L}(X\times X,\mathbb{K})} = \sup_{\|x\|_Y \le 1, \|y\|_X \le 1} |a(x,y)|\,. \tag{2.21}$$

For any two nonzero elements x, y in a real pre-Hilbert space $(X, \langle\cdot,\cdot\rangle_X)$ we have

$$-1 \le \frac{\langle x, y\rangle_X}{\|x\|_X\;\|y\|_X} \le 1. \tag{2.22}$$

There is a uniquely determined value (called *angle*) $\phi \in [-\frac{\pi}{2}, \frac{\pi}{2}]$, that satisfies

$$\cos(\phi) = \frac{\langle x, y\rangle_X}{\|x\|_X\;\|y\|_X}. \tag{2.23}$$

Let $(X, \langle\cdot,\cdot\rangle_X)$ be a pre-Hilbert space. Then the inner product is continuous in both components and the norm satisfies the parallelogram identity

$$\|x+y\|_X^2 + \|x-y\|_X^2 = 2\left(\|x\|_X^2 + \|y\|_X^2\right) \tag{2.24}$$

for all $x, y \in X$.

Let $(X, \langle\cdot,\cdot\rangle_X)$ be a pre Hilbert space. Two elements x and y of X are orthogonal, if $\langle x, y\rangle_X = 0$. A finite or countable set of elements, $\{x_n : n \in N\} \subset X$, $N \subset \mathbb{N}$, is called an *orthonormal system* if $\langle x_i, x_k\rangle_X = \delta_{i,k}$ for all $i, k \in \mathbb{N}$.

Two subsets $U, V \subset X$ of $(X, \langle\cdot,\cdot\rangle_X)$ are called *orthogonal* (symbolized by $U \perp V$), if $\langle x, y\rangle_X = 0$ for all $x \in U$ and all $y \in V$. The set $U^\perp = \{x \in X : \{x\} \perp U\}$ is called *orthogonal complement* of U in X.

Let $(X, \langle\cdot,\cdot\rangle_X)$ be a pre-Hilbert space.

(1) *We have $\{0\}^\perp = X$ and $X^\perp = \{0\}$.*

(2) *For every subset $U \subset X$ the orthogonal complement $U^\perp$ is a closed subspace of X.*

(3) *If $U \subset V \subset X$, then $X^\perp \subset V^\perp \subset U^\perp$.*

Let U be a closed subspace of the pre-Hilbert space $(X, \langle\cdot,\cdot\rangle_X)$. If $\{u_i\}_{i\in I}$ is an orthonormal basis in U, then, for every $x \in X$, its *orthogonal projection* $P_U x$ is defined by

$$P_U x = \sum_{i\in I} \langle x, u_i\rangle_X\; u_i\,. \tag{2.25}$$

Let U be a closed subspace of the pre-Hilbert space $(X, \langle\cdot,\cdot\rangle_X)$. The orthogonal projection P_U has the following properties:

(1) $\langle P_U x, y\rangle_X = \langle x, P_U y\rangle_X$ for all $x, y \in X$,
(2) $P_U^2 = P_U$,
(3) $\|P_U\| = 1$,
(4) $\|x - P_U x\|_X = \min_{u\in U} \|x-u\|_U$, i.e., $P_U x$ is the *best approximation to* x in U,
(5) $I - P_U$ is the orthogonal projection to $U^\perp$,
(6) $\mathcal{R}(P_U) = \mathcal{N}(P_U)^\perp$.

Let $(X, \langle\cdot,\cdot\rangle_X)$ be a pre-Hilbert space. Suppose that $\{u_k\}_{k=1,\ldots,n} \subset X$, $n \in \mathbb{N}$, is an orthonormal system. For every $x \in X$, the element $a = \sum_{k=1}^n \langle x, u_k\rangle_X\, u_k$ satisfies the inequality $\|x-a\|_X \le \|x-y\|_X$ for all $y \in \operatorname{span}_{k=1,\ldots,n}\{u_k\}$. The coefficients $\langle x, u_k\rangle_X$ are called *Fourier coefficients* of x with respect to $\{u_k\}_{k=1,\ldots,n}$. Moreover, we obtain *Bessel's inequality*

$$\sum_{k=1}^{n} |\langle x, u_k\rangle_X|^2 \le \|x\|_X^2 . \tag{2.26}$$

Every separable Hilbert space contains an orthonormal basis.

If $\{x_k\}_{k\in\mathbb{O}} \subset X$, $\mathbb{O} \subset \mathbb{N}$, is an orthonormal system in a separable Hilbert space X which is not complete, we can find a system $\{y_k\}_{k\in\mathbb{N}\setminus\mathbb{O}} \subset X$ such that both systems together form an orthonormal basis.

Let $(X, \langle\cdot,\cdot\rangle_X)$ be a separable Hilbert space. If $\{u_n\}_{n\in\mathbb{N}}$ is a countable orthonormal system in X, then the following relations are equivalent:

(1) The system $\{u_n\}_{n\in\mathbb{N}}$ is *closed* in X, i.e. for every $x \in X$ and for every $\varepsilon > 0$ there exists an $N \in \mathbb{N}$ and coefficients $a_1, \ldots, a_N \in \mathbb{K}$ such that

$$\left\| x - \sum_{n=1}^{N} a_n u_n \right\|_X < \varepsilon .$$

(2) The orthogonal expansion of every element $x \in X$ converges in norm to x, i.e.,

$$\lim_{N\to\infty} \left\| x - \sum_{n=1}^{N} \langle x, u_n\rangle_X\, u_n \right\|_X = 0 .$$

(3) Parseval's identity is valid. That is, for every $x \in X$,

$$\|x\|_X^2 = \langle x, x\rangle_X = \sum_{n=1}^{\infty} |\langle x, u_n\rangle_X|^2 .$$

(4) The extended Parseval's identity holds true, i.e., for every $x, y \in X$,

$$\langle x, y\rangle_X = \sum_{n=1}^{\infty} \langle x, u_n\rangle_X\, \langle u_n, y\rangle_X .$$

(5) There is no strictly larger orthonormal system in X containing the orthonormal system $\{u_n\}_{n\in\mathbb{N}}$.

(6) The system $\{u_n\}_{n\in\mathbb{N}}$ is *complete* in X, i.e. $x \in X$ with $\langle x, u_n\rangle_X = 0$ for all $n \in \mathbb{N}$ implies $x = 0$.

(7) Every element $x \in X$ is uniquely determined by its orthogonal coefficients, i.e. if, for $x, y \in X$, $\langle x, u_n\rangle_X = \langle y, u_n\rangle_X$ for all $n \in \mathbb{N}$, then $x = y$.

Riesz Representation Theorem. Let $(X, \langle\cdot,\cdot\rangle_X)$ be a Hilbert space. For every bounded linear functional $\mathfrak{L} \in X'$ exists a unique element $y \in X$ such that $\mathfrak{L}(x) = \langle x, y\rangle_X$ for every $x \in X$. The function $J_X : X \to X'$, $z \mapsto J_X(z) = \langle\,\cdot, z\rangle_X$ is an isometric, conjugate linear isomorphism. Consequently, $\|\mathfrak{L}\|_{X'} = \|z\|_X$.

Let $(X, \langle\cdot,\cdot\rangle_X)$, $(Y, \langle\cdot,\cdot\rangle_Y)$ be Hilbert spaces, J_X and J_Y the corresponding Riesz isomorphisms according to the Riesz Theorem. Suppose that A is of class $\mathfrak{L}(X, Y)$. The *Hilbert space adjoint operator* A^* of A is defined by $A^* = J_X^{-1}\, A'\, J_Y$, where A' is the *dual space adjoint operator to* A. A is called *self-adjoint*, if $A^* = A$.

Let $(X, \langle\cdot,\cdot\rangle_X)$, $(Y, \langle\cdot,\cdot\rangle_Y)$ be Hilbert spaces. Suppose that A is of class $\mathfrak{L}(X, Y)$. Then the Hilbert space adjoint operator A^* is of class $\mathfrak{L}(Y, X)$, and $\langle x, A^*y\rangle_X = \langle Ax, y\rangle_Y$. Moreover, we have $A^{**} = A$, $\|A^*\|_{\mathfrak{L}(Y,X)} = \|A\|_{\mathfrak{L}(X,Y)}$, as well as $\|A^*A\|_{\mathfrak{L}(X,X)} = \|A\|^2_{\mathfrak{L}(X,Y)}$.

Let X, Y and Z be Hilbert spaces: (1) For $A, B \in \mathfrak{L}(X, Y)$ and $\alpha \in \mathbb{K}$, we have $(A + B)^* = A^* + B^*$ and $(\alpha A)^* = \overline{\alpha} A^*$. (2) For $A \in \mathfrak{L}(X, Y)$ and $B \in \mathfrak{L}(Y, Z)$, we have $(BA)^* = A^*B^*$. Moreover, the following relations are true:

$$\mathcal{R}(A)^\perp = \mathcal{N}(A^*), \quad \overline{\mathcal{R}(A)} = \mathcal{N}(A^*)^\perp, \tag{2.27}$$

$$\mathcal{R}(A^*)^\perp = \mathcal{N}(A), \quad \overline{\mathcal{R}(A^*)} = \mathcal{N}(A)^\perp. \tag{2.28}$$

The operator A is bijective if and only its adjoint A^* is bijective.

Lax-Milgram Theorem. Let $(X, \langle\cdot,\cdot\rangle_X)$ be a Hilbert space and $a(\cdot,\cdot) : X \times X \to \mathbb{K}$ a sesquilinear form. If $a(\cdot,\cdot)$ is continuous, i.e., bounded with the constant $B > 0$, then there exists a unique bounded linear operator $A \in \mathfrak{L}(X)$ such that

$$a(x, y) = \langle x, Ay\rangle_X \quad \text{for all } x, y \in X. \tag{2.29}$$

We have $\|A\|_{\mathfrak{L}(X)} \le B$ or, more precisely, $\|A\|_{\mathfrak{L}(X)} = \|a(\cdot,\cdot)\|_{\mathfrak{L}(X\times X,\mathbb{K})}$.

If $a(\cdot,\cdot)$ is also coercive with coercivity constant $C > 0$, then A is bijective, i.e., there exists a bounded linear operator $A^{-1} \in \mathfrak{L}(X)$ which is the inverse operator to A and $\|A^{-1}\|_{\mathfrak{L}(X)} \le \frac{1}{C}$.

Let $(X, \langle\cdot,\cdot\rangle_X)$ be a Hilbert space and $a(\cdot,\cdot) : X \times X \to \mathbb{K}$ a continuous, coercive sesquilinear form, A the corresponding operator according to the Lax-Milgram Theorem and $J : X \to X'$ the Riesz isomorphism according to the Riesz Theorem.

For $F \in X'$, let $x = A^{-1}J^{-1}F$. The so-defined x is the unique solution to the problem

$$a(x, y) = F(y) \quad \text{for all } y \in X \tag{2.30}$$

and satisfies

$$\|x\|_X \le \frac{1}{C}\,\|F\|_{X'} \tag{2.31}$$

with $C > 0$ being the coercivity constant to the sesquilinear form. If, additionally, $a(\cdot,\cdot)$ is hermitian, i.e., $a(\cdot,\cdot)$ is an inner product, then x is also the unique minimizer of the functional

$$X \ni y \mapsto F(y) = \frac{1}{2}a(y,y) - \Re(F(y)) \in \mathbb{R}. \tag{2.32}$$

Weak and weak⋆ convergence. We start from normed spaces X, Y.

A bilinear form $\langle\cdot,\cdot\rangle : X \times Y \to \mathbb{K}$ is called *duality*, if the following statements hold true:

(i) for all $0 \neq x \in X$, there exists a $y \in Y$ such that $\langle x, y\rangle \neq 0$
(ii) for all $0 \neq y \in Y$, there exists a $x \in X$ such that $\langle x, y\rangle \neq 0$.

The most common duality is the duality between a linear space X and its dual space X'. For $x' \in X'$ and a $x \in X$ this duality is defined by $\langle x', x\rangle = x'(x)$.

Weak and weak⋆ convergence and compactness. Let X be a Banach space.

(a) A sequence $\{x_n\}_{n\in\mathbb{N}}$ in X *converges weakly* to $x \in X$, if and only if

$$\lim_{n\to\infty} \langle x_n, x'\rangle = \langle x, x'\rangle \quad \text{for all } x' \in X'. \tag{2.33}$$

(b) A sequence $\{x'_n\}_{n\in\mathbb{N}}$ in X' *converges weakly⋆* to $x' \in X$, if and only if

$$\lim_{n\to\infty} \langle x, x'_n\rangle = \langle x, x'\rangle \quad \text{for all } x \in X. \tag{2.34}$$

(c) Weak and weak⋆ Cauchy sequences are defined correspondingly.
(d) A subset $U \subset X$ is called *weak sequentially compact* if and only if each sequence in U possesses a weak convergent subsequence whose weak limit is also in U.
(e) Weak⋆ compactness for subsets of X' is defined analogously.
(f) Convergence with respect to the norm is subsequently called *strong convergence.*

Let X be a Banach space.

(a) *Weak and weak⋆ limits are unique and weakly or weakly⋆ convergent sequences are bounded in the norms of the corresponding spaces.*
(b) *There exists an isometric mapping $\widetilde{J} \in \mathfrak{L}(X, X'')$ so that $\left\langle x', \widetilde{J}x\right\rangle = \langle x, x'\rangle$ for any $x \in X$ and every $x' \in X'$. X is called reflexive if and only if $\widetilde{J}$ is surjective.*
(c) *A sequence $\{x_n\}_{n\in\mathbb{N}}$ in X converges weakly to $x \in X$ if and only if $\left\{\widetilde{J}x_n\right\}_{n\in\mathbb{N}}$ converges weakly⋆ to $\widetilde{J}x \in X''$.*
(d) *$\lim_{n\to\infty}\langle x_n, x'\rangle = \langle x, x'\rangle$ for all $x' \in X'$ implies $\|x\|_X \leq \liminf_{n\to\infty} \|x_n\|_X$ and $\lim_{n\to\infty}\langle x, x'_n\rangle = \langle x, x'\rangle$ for all $x \in X$ implies $\|x'\|_{X'} \leq \liminf_{n\to\infty} \|x'_n\|_{X'}$.*

The following results should be mentioned:

(i) If X is a Hilbert space, then the Riesz Theorem tells us that $\lim_{n\to\infty}\langle x_n, x'\rangle = \langle x, x'\rangle$ for all $x' \in X'$ means $\lim\limits_{n\to\infty} (x_n, u) = (x, u)$ for all $u \in X$.
(ii) Strong convergence implies weak or weak⋆ convergence, respectively.

(iii) In the dual space X' of a Banach space X, we have three different concepts of convergence:
(a) strong convergence: $\lim\limits_{n\to\infty} \|x'_n - x'\|_{X'} = 0$,
(b) weak* convergence: $\lim_{n\to\infty}\langle x, x'_n\rangle = \langle x, x'\rangle$ for all $x \in X$,
(c) weak convergence: $\lim_{n\to\infty}\langle x'_n, x''\rangle = \langle x', x''\rangle$ for all $x'' \in X''$.
However, as X is always isomorph to at least a subset of X'', weak convergence in X' always implies weak* convergence in X'. If X is reflexive, both concepts are identical on X'.

(iv) In a finite dimensional normed space, strong and weak convergence coincide.

Moreover we have the following results:

Let X be a Banach space, $Y \subset X$ dense, $Z' \subset X'$ dense, $x \in X$, $x' \in X'$, $\{x_n\}_{n\in\mathbb{N}}$ in X and $\{x'_n\}_{n\in\mathbb{N}}$ in X'.

(a)
$$\lim_{n\to\infty} \langle x_n, x'\rangle = \langle x, x'\rangle$$
for all $x' \in X'$ holds if and only if there is a $C > 0$, such that $\|x_n\|_X \le C$ for all $n \in \mathbb{N}$ and $\lim_{n\to\infty}\langle x_n, x'\rangle = \langle x, x'\rangle$ for all $x' \in Z'$.

(b)
$$\lim_{n\to\infty} \langle x, x'_n\rangle = \langle x, x'\rangle$$
for all $x \in X$ holds if and only if there is a $C' > 0$, such that $\|x'_n\|_{X'} \le C'$ for all $n \in \mathbb{N}$ and $\lim_{n\to\infty}\langle x, x'_n\rangle = \langle x, x'\rangle$ for all $x \in Y$.

Distributions. If $\mathcal{G}$ is a region i.e., an open and connected pointset in $\mathbb{R}^q$, $\partial\mathcal{G}$ denotes its *boundary*. The set $\overline{\mathcal{G}} = \mathcal{G} \cup \partial\mathcal{G}$ is called the *closure* of $\mathcal{G}$. A region is called bounded if there is a value $R > 0$ such that $\mathcal{G}$ is a subset of the ball around the origin with radius R.

A function U has a *compact support in* $\mathcal{G}$, if there is a compact set $\mathcal{K} \subset \mathcal{G}$ such that

$$\text{supp}(U) = \overline{\{x \in \mathcal{G} : U(x) \neq 0\}} \subset \mathcal{K}\,. \tag{2.35}$$

Let $\mathcal{G} \subset \mathbb{R}^q$, $q \in \mathbb{N}$, be a bounded region. The subspace of functions in $C^{(k)}(\mathcal{G})$ that have a compact support in $\mathcal{G}$ is denoted by $C_c^{(k)}(\mathcal{G})$.

The spaces $C^{(k)}(\overline{\mathcal{G}})$ contain all functions $U \in C^{(k)}(\mathcal{G})$ for which $D^\gamma U$ is bounded and uniformly continuous for all $\gamma \in \mathbb{N}_0^k$ with $0 \le |\gamma| \le k$, i.e., it possesses a unique, bounded, continuous extension to the closure $\overline{\mathcal{G}}$. These spaces are Banach spaces when equipped with the norm

$$\|U\|_{C^{(k)}(\overline{\mathcal{G}})} = \max_{0\le|\gamma|\le k} \sup_{x\in\mathcal{G}} |(D^\gamma U)(x)|\,, \tag{2.36}$$

where the partial derivative $D^\gamma u$ is defined by

$$(D^\gamma u) = \left(\frac{\partial^{|\gamma|} u}{\partial x_1^{\gamma_1} \dots \partial x_q^{\gamma_q}}\right), \quad \gamma \in \mathbb{N}_0^q\,, \tag{2.37}$$

with $|\gamma| = \sum_{i=1}^q \gamma_i$ being the order of the derivative.

Convergence in $C_c^{(\infty)}(\mathcal{G})$ (see, e.g., [**5**]). Let $\mathcal{G} \subset \mathbb{R}^q$, $q \in \mathbb{N}$, be a bounded region, $\{\phi_l\}_{l\in\mathbb{N}}$ in $C_c^{(\infty)}(\mathcal{G})$, and $\phi_0 \in C_c^{(\infty)}(\mathcal{G})$. The sequence $\{\phi_l\}_{l\in\mathbb{N}}$ is said to converge towards ϕ_0 in $C_c^{(\infty)}(\mathcal{G})$ for $l \to \infty$ if there is a compact subset $\mathcal{K} \subset \mathcal{G}$ such that

$$\text{supp}(\phi_l) \subset \mathcal{K}, \quad l \in \mathbb{N}, \ \text{supp}(\phi_0) \subset \mathcal{K}\,, \tag{2.38}$$

as well as all partial derivatives of ϕ_l of arbitrary order converge uniformly to those of ϕ_0, i.e.,

$$\sup_{x\in\mathcal{G}} |(D^\gamma \phi_l)(x) - (D^\gamma \phi_0)(x)| \xrightarrow{l\to\infty} 0 \quad \text{for all } \gamma \in \mathbb{N}_0^q\,. \tag{2.39}$$

(i) $C_c^{(\infty)}(\mathcal{G})$ is often denoted by $\mathfrak{D}(\mathcal{G})$ and called the *space of test functions*, although the latter identification is not unique. It is a topological vector space, but not normable (see, e.g., [**5**], [**730**]).

(ii) A function ϕ defined on $\mathbb{R}^q$ is called *finite* if it vanishes outside a bounded set. The space $C_c^{(\infty)}(\mathbb{R}^q)$ consist of all finite, arbitrary often continuously differentiable functions on $\mathbb{R}^q$.

The above definition allows us to define distributions:

Let $\mathcal{G} \subset \mathbb{R}^q$, $n \in \mathbb{N}$, be a bounded region. A *distribution (or generalized function)* is a linear functional $\mathfrak{F} : C_c^{(\infty)}(\mathcal{G}) \to \mathbb{R}$ which is continuous in the following sense: If a sequence $\{\phi_l\}_{l\in\mathbb{N}_0}$ in $C_c^{(\infty)}(\mathcal{G})$ converges for $l \to \infty$ towards $\phi \in C_c^{(\infty)}(\mathcal{G})$, then $\mathfrak{F}(\phi_l) = \langle \mathfrak{F}, \phi_l \rangle$ converges for $l \to \infty$ towards $\mathfrak{F}(\phi)$. The set of all distributions is denoted by $\left(C_c^{(\infty)}(\mathcal{G})\right)'$ (note that the spaces of vector/tensor-valued distributions may be defined accordingly). The space of distributions is often denoted by $\mathfrak{D}'(\mathcal{G})$. If we consider $C_c^{(\infty)}(\mathcal{G})$ as a topological vector space, $\left(C_c^{(\infty)}(\mathcal{G})\right)'$ is its topological dual (cf. [**5**], [**730**]).

Let $\mathcal{G} \subset \mathbb{R}^q$, $n \in \mathbb{N}$, be a region. A function U is called locally integrable on $\mathcal{G}$ if, for every compact subset $\mathcal{D} \subset \mathcal{G}$ we have $\int_\mathcal{D} |U(x)|\, dx < \infty$. For every locally integrable function U, we can define a corresponding distribution $\mathfrak{T}_U \in \left(C_c^{(\infty)}(\mathcal{G})\right)'$ simply by letting

$$\mathfrak{T}_U(\phi) = \int_\mathcal{G} U(x)\, \phi(x)\, dx, \quad \phi \in \mathrm{C}_c^{(\infty)}(\mathcal{G})\,. \tag{2.40}$$

Usually, the notation is a little bit abused by also using U instead of $\mathfrak{T}_U$ to denote the corresponding distribution. Distributions that correspond in that way to a locally integrable function are called *regular*. There are many distributions for which no corresponding locally integrable function can be found. The most prominent example is the evaluation of a function ϕ at a certain point x, known as *Dirac's delta distribution*. If $0 \in \mathcal{G}$, the evaluation of a function $\phi \in C_c^{(\infty)}(\mathcal{G})$ is given by

$$\delta(\phi) = \phi(0). \tag{2.41}$$

It is easy to prove that there is no locally integrable function for which

$$\int_\mathcal{G} \delta(x)\, \phi(x)\, dx = \phi(0), \quad \phi \in C_c^{(\infty)}(\mathcal{G})\,. \tag{2.42}$$

It is obvious how addition of two distributions and multiplication with a constant should be defined on distributions. Distributions may even be multiplied by smooth functions (cf. [**5**]). For $\mathfrak{T} \in \left(C_c^{(\infty)}(\mathcal{G})\right)'$ and $U \in C^{(\infty)}(\mathcal{G})$, the product $U\mathfrak{T} \in \left(C_c^{(\infty)}(\mathcal{G})\right)'$ is defined by $(U\mathfrak{T})(\phi) = \mathfrak{T}(U\phi)$, $\phi \in C_c^{(\infty)}(\mathcal{G})$. The support of a distribution is defined as follows (see, e.g., [**190**]):

Suppose $\mathfrak{T} \in \left(C_c^{(\infty)}(\mathcal{G})\right)'$ for an open bounded region $\mathcal{G} \subset \mathbb{R}^q$, $n \in \mathbb{N}$. The support of $\mathfrak{T}$ is defined as

$$\operatorname{supp}(\mathfrak{T}) = \left\{x \in \overline{\mathcal{G}} : \mathfrak{T}|\mathcal{G} \cap \mathbb{B}_\tau^q(x) \neq 0, \tau > 0\right\}. \tag{2.43}$$

The support of a distribution is closed (see, e.g., [**190**]), and the concept of compact support is directly transferable to distributions.

Convolution. An operation which we need not only on distributions is the convolution [**730**]: Let U be a function defined on $\mathbb{R}^q$, $q \in \mathbb{N}$, and $x, y \in \mathbb{R}^q$. We define

$$(\mathsf{T}_x U)(y) = U(y - x)\,, \tag{2.44}$$

$$\breve{U}(y) = U(-y)\,, \tag{2.45}$$

$$(\mathsf{T}_x \breve{U})(y) = U(x - y)\,. \tag{2.46}$$

Let V be another function on $\mathbb{R}^q$. The convolution $U * V$ is defined as

$$(U * V)(x) = \int_{\mathbb{R}^q} U(y)\, V(x - y)\, dy = \int_{\mathbb{R}^q} U(y)(\mathsf{T}_x \breve{V})(y)\, dy \tag{2.47}$$

if the integral exists for V-almost all $x \in \mathbb{R}^q$, i.e., the set of all points x for which the integral does not exist has vanishing measure.

For a distribution $U \in \left(\mathrm{C}_c^{(\infty)}(\mathbb{R}^n)\right)'$ and $\phi \in C_c^{(\infty)}(\mathbb{R}^q)$, the function $U * \phi$ is defined by

$$(U * \phi)(x) = U(\mathsf{T}_x \breve{\phi}). \tag{2.48}$$

Let $U \in \left(C_c^{(\infty)}(\mathbb{R}^q)\right)'$, $\phi, \psi \in C_c^{(\infty)}(\mathbb{R}^q)$. Then the following holds:

(i) $\mathsf{T}_x(U * \phi) = (\mathsf{T}_x U) * \phi = U * (\mathsf{T}_x \phi)$ for all $x \in \mathbb{R}^q$,

(ii) $u * \phi \in C^{(\infty)}(\mathbb{R}^q)$ and $U * (\phi * \psi) = (U * \phi) * \psi$,

(iii) the operator L, defined by

$$L\phi = U * \phi, \quad \phi \in C_c^{(\infty)}(\mathbb{R}^q)\,, \tag{2.49}$$

is a linear mapping of $C_c^{(\infty)}(\mathbb{R}^q)$ into $C^{(\infty)}(\mathbb{R}^q)$ which satisfies $\mathsf{T}_x L = L \mathsf{T}_x$, $x \in \mathbb{R}^q$.

Until now, convolutions for distributions are only declared if a distribution is convolved with an element of $C_c^{(\infty)}(\mathbb{R}^q)$. Next we extend the convolution to elements of $C^{(\infty)}(\mathbb{R}^q)$ (cf. [**730**]).

Let $U \in \left(C_c^{(\infty)}(\mathbb{R}^q)\right)'$ *have compact support,* $\phi \in C^{(\infty)}(\mathbb{R}^q)$, $\psi \in C_c^{(\infty)}(\mathbb{R}^q)$. *The convolution* $U * \phi \in C^{(\infty)}(\mathbb{R}^q)$ *is well-defined. Moreover,*

(i) $\mathsf{T}_x(U * \phi) = (\mathsf{T}_x u) * \phi = U * (\mathsf{T}_x \phi)$ *for all* $x \in \mathbb{R}^q$,

(ii) $U * \psi \in C_c^{(\infty)}(\mathbb{R}^q)$,

(iii) $U * (\phi * \psi) = (U * \phi) * \psi = (U * \psi) * \phi$.

Convolutions may also be defined between distributions:

Let $U, V, W \in \left(C_c^{(\infty)}(\mathbb{R}^q)\right)'$, $q \in \mathbb{N}$.

(i) If at least one of U, V has compact support, the convolution $U*V$ is defined by $(U * V) * \phi = U * (V * \phi)$ for all $\phi \in C_c^{(\infty)}(\mathbb{R}^q)$ and $U * V = V * U$.

(ii) If at least one of the supports $\operatorname{supp}(U)$, $\operatorname{supp}(V)$ is compact, we have $\operatorname{supp}(U * V) \subset \operatorname{supp}(U) + \operatorname{supp}(V)$.

(iii) If at least two of the supports $\operatorname{supp}(U)$, $\operatorname{supp}(V)$, $\operatorname{supp}(W)$ are compact, we have $(U * V) * W = U * (V * W)$.

We are now able to define weak derivatives by defining derivatives of distributions (see, e.g., [**5**], [**730**]).

Let $\mathcal{G}$ be a bounded region in $\mathbb{R}^q$, $q \in \mathbb{N}$, and $U \in \left(C_c^{(\infty)}(\mathcal{G})\right)'$. The weak derivative of u with respect to x_i, $i \in \{1, \ldots, n\}$, is defined by

$$\left\langle \frac{\partial}{\partial x_i} U, \phi \right\rangle = -\left\langle U, \frac{\partial}{\partial x_i} \phi \right\rangle, \quad \phi \in C_c^{\infty}(\mathcal{G}) \,. \tag{2.50}$$

For a multi-index $\gamma \in \mathbb{N}_0^q$, we have the generalization

$$\langle D^\gamma U, \phi \rangle = (-1)^{|\gamma|} \langle U, D^\gamma \phi \rangle, \quad \phi \in C_c^{(\infty)}(\mathcal{G}) \,. \tag{2.51}$$

It should be noted that the same notation is used for weak derivatives and classical (strong) partial derivatives (based on the limit of difference quotients). If a continuous strong derivative exists, it coincides with the weak derivative as can be seen by integration by parts (cf. [**730**]).

(i) Suppose $U \in \left(C_c^{(\infty)}(\mathbb{R}^q)\right)'$ and $\phi \in C_c^{(\infty)}(\mathbb{R}^q)$, $q \in \mathbb{N}$, or $U \in \left(C_c^{(\infty)}(\mathbb{R}^q)\right)'$ with compact support and $\phi \in C^{(\infty)}(\mathbb{R}^q)$, then $D^\gamma(U * \phi) = (D^\gamma U) * \phi = U * (D^\gamma \phi)$ for all $\gamma \in \mathbb{N}_0^q$.

(ii) Suppose $U \in \left(C_c^{(\infty)}(\mathbb{R}^q)\right)'$ and δ is the delta distribution, then $D^\gamma U = (D^\gamma \delta) * U$ for all $\gamma \in \mathbb{N}_0^q$. In particular, $U = \delta * U$.

(iii) Suppose $U, V \in \left(C_c^{(\infty)}(\mathbb{R}^q)\right)'$ and at least one of them has compact support, then $D^\gamma(U * V) = (D^\gamma U) * V = U * (D^\gamma V)$ for all $\gamma \in \mathbb{N}_0^q$.

Now we are allowed to give an informal integral expression for *Dirac's delta distribution* and its derivatives, with a slight abuse of notation, by

$$\int_{\mathbb{R}^q} D^\gamma \delta(x - y) \phi(y) \, dy = (-1)^{|\gamma|} (D^\gamma \phi)(x) \,. \tag{2.52}$$

It is easily seen that the weak derivative of a distribution is also a distribution. Thus, for every distribution, there exist weak derivatives of arbitrary order. Nevertheless, classes of distributions and their derivatives can be distinguished if we introduce a new concept of regularity based on integrability.

Lebesgue spaces. Let $\mathcal{G}$ be a bounded region in $\mathbb{R}^q$, $q \in \mathbb{N}$, and $p > 0$. The Lebesgue space $L^p(\mathcal{G})$ consists of all equivalence classes with respect to the Lebesgue measure almost everywhere identical functions on $\mathcal{G}$, whose representatives U satisfy

$$\int_{\mathcal{G}} |U(x)|^p \, dx < \infty \,. \tag{2.53}$$

Moreover, the space $L^\infty(\mathcal{G})$ contains all such equivalence classes whose representatives are measurable, essentially bounded functions, i.e., $U : \mathcal{G} \to \mathbb{R}$, i.e., $\operatorname{ess\,sup}_{x\in\mathcal{G}} |U(x)| < \infty$. It is convenient to identify a function with its respective equivalence class. So, it is generally not possible to evaluate a function from $L^p(\mathcal{G})$ at a point $x \in \mathcal{G}$.

We summarize a few properties of the Lebesgue spaces (see, e.g., [**5**] for more details).

Let $\mathcal{G}$ be a bounded measurable region in $\mathbb{R}^q$, $q \in \mathbb{N}$, and $1 \le p < \infty$.

(i) *The Lebesgue space $L^p(\mathcal{G})$ is a Banach space with respect to the norm*

$$\|U\|_{L^p(\mathcal{G})} = \left(\int_{\mathcal{G}} |U(x)|^p \, dx \right)^{\frac{1}{p}} \,. \tag{2.54}$$

$L^\infty(\mathcal{G})$ is a Banach space with respect to the norm

$$\|U\|_{L^\infty(\mathcal{G})} = \operatorname*{ess\,sup}_{x\in\mathcal{G}} |U(x)| \,. \tag{2.55}$$

(ii) *The space $L^2(\mathcal{G})$ if equipped with the scalar product*

$$\langle U, V \rangle_{L^2(\mathcal{G})} = \int_{\mathcal{G}} U(x) \, \overline{V(x)} \, dx \tag{2.56}$$

is a Hilbert space.

(iii) *For arbitrary $1 \le p_1, p_2 < \infty$ with $p_1 \ge p_2$, we have $L^{p_1}(\mathcal{G}) \subset L^{p_2}(\mathcal{G})$ and $L^{p_1}(\mathcal{G}) \subset L^1_{\mathrm{loc}}(\mathcal{G})$, with $L^1_{\mathrm{loc}}(\mathcal{G})$ being the space of locally integrable functions.*

(iv) *Let p_1 such that $1 < p_1 < \infty$ and p_2 such that $\frac{1}{p_1} + \frac{1}{p_2} = 1$. For $U \in L^{p_1}(\mathcal{G})$, $V \in L^{p_2}(\mathcal{G})$, we have $UV \in L^1(\mathcal{G})$ and*

$$\|UV\|_{L^1(\mathcal{G})} \le \|U\|_{L^{p_1}(\mathcal{G})} \, \|V\|_{L^{p_2}(\mathcal{G})} \,. \tag{2.57}$$

This is known as Hölder's inequality. Hölder's inequality also holds for $U \in L^1(\mathcal{G})$ and $V \in \mathrm{L}^\infty(\mathcal{G})$. Then we have $UV \in L^1(\mathcal{G})$.

(v) *Let p_1 such that $1 < p_1 < \infty$ and p_2 such that $\frac{1}{p_1} + \frac{1}{p_2} = 1$. Then the dual space $(L^{p_1}(\mathcal{G}))'$ of $(\mathrm{L}^{p_1}(\mathcal{G}))$ is isometrically isomorph to $L^{p_2}(\mathcal{G})$. Moreover, $\left(L^1(\mathcal{G})\right)'$ is isometrically isomorph to $L^\infty(\mathcal{G})$, but $(L^\infty(\mathcal{G}))'$ is not isometrically isomorph to $L^1(\mathcal{G})$.*

(vi) *$C_c^{(0)}(\mathcal{G})$and $C_c^{(\infty)}(\mathcal{G})$ are dense subspaces of $L^p(\mathcal{G})$ for all $1 \le p \le \infty$.*

The space $C^{(0)}(\mathcal{G})$ with $\mathcal{G} \subset \mathbb{R}^q$ together with the inner product

$$\langle F, G\rangle_{L^2(\mathcal{G})} = \int_{\mathcal{G}} F(a)\, \overline{G(a)}\; da \tag{2.58}$$

is an incomplete inner product space. Its completion with respect to the $\|\cdot\|_{L^2(\mathcal{G})}$-norm is the space $L^2(\mathcal{G})$, i.e.,

$$\overline{C^{(0)}(\mathcal{G})}^{\|\cdot\|_{L^2(\mathcal{G})}} = L^2(\mathcal{G}). \tag{2.59}$$

Forming the dual space of the Hilbert space $L^2(\mathcal{G})$ is easy, because it is the Hilbert space itself.

The definition of Lebesgue spaces allows us to evaluate the regularity of a distribution by asking if it is also an element of some Lebesgue space. It comes naturally to extend this to a derivative of a distribution. This leads to the definition of Sobolev spaces.

Sobolev spaces. Let $\mathcal{G}$ be a bounded region in $\mathbb{R}^q$, $q \in \mathbb{N}$, and $1 \leq p \leq \infty$. The *Sobolev space* $W^{k,p}(\mathcal{G})$, $k \in \mathbb{N}_0$, is defined as the subspace of $L^p(\mathcal{G})$ with

$$W^{k,p}(\mathcal{G}) = \{U \in L^p(\mathcal{G}) :\; D^\gamma U \in L^p(\mathcal{G}) \quad \text{for all } \gamma \in \mathbb{N}_0^q, |\gamma| \leq k\}\ . \tag{2.60}$$

$W^{k,p}(\mathcal{G})$ is a separable Banach space with respect to the norm

$$\|U\|_{W^{k,p}(\mathcal{G})} = \left(\sum_{|\gamma|\leq k} \int_{\mathcal{G}} |D^\gamma U(x)|^p \; dx \right)^{\frac{1}{p}} . \tag{2.61}$$

For $p = 2$, we set $H^k(\mathcal{G}) = \mathrm{W}^{k,2}(\mathcal{G})$. These spaces are separable Hilbert spaces with inner product

$$\langle U, V\rangle_{H^k(\mathcal{G})} = \sum_{|\gamma|\leq k} \langle D^\gamma U, D^\gamma V\rangle_{L^2(\mathcal{G})}\ . \tag{2.62}$$

Compact operators. Next we discuss a certain class of operators, viz. compact operators, that turn out to characterize specific prototypes of a large class of ill-posed problems.

Let X, Y be normed spaces. An operator $A : X \to Y$ is called *compact*, if one of the following equivalent conditions is fulfilled:

(i) For every bounded sequence $\{x_n\}_n \subset X$ the sequence $\{Ax_n\}_n$ possesses a subsequence that converges in Y.

(ii) Every bounded subset $U \subset X$ possesses an image in Y, which is *relatively compact*, i.e., $\overline{A(U)}$ is a compact set.

We introduce the abbreviations $\mathfrak{K}(X,Y) = \{A : X \to Y : A \text{ is linear and compact}\}$ and $\mathfrak{K}(X) = \mathfrak{K}(X,X)$.

Let X, Y, Z be normed spaces. Then the following statements hold true:

(i) $\mathfrak{K}(X,Y) \subset \mathfrak{L}(X,Y)$.

(ii) If $A \in \mathfrak{L}(X,Y)$ with $\dim \mathcal{R}(A) < \infty$, then A is compact.

(iii) If $A : X \to Y$ is compact and $B : Y \to Z$ is continuous or $A : X \to Y$ is continuous and $B : Y \to Z$ is compact, then $AB : X \to Z$ is compact.

(iv) The identity operator $I : X \to X$ is compact if and only if X is finite dimensional.

(v) Let Y be a Banach space. Then $\mathfrak{K}(X,Y)$ is closed, i.e., a sequence of compact operators $\{A_n\}_n \subset \mathfrak{K}(X,Y)$ with $\lim_{n\to\infty} \|A_n - A\| = 0$ has a compact limit, i.e., the limit operator A is compact.

(vi) If A is compact and invertible and X is not finite dimensional, then A^{-1} is *not* continuous.

The *Theorem of Arzela-Ascoli* of Mathematical Physics (see, e.g., [**585**]) provides two equivalent properties to the relative compactness of a subset $U \subset C^{(0)}(\overline{\mathcal{G}})$:

Let $\mathcal{G} \neq \emptyset$ be bounded. A subset $U \subset C^{(0)}(\overline{\mathcal{G}})$ is relatively compact if and only if the following two statements are valid:

(1) *U is equicontinuous, i.e., for every $\varepsilon > 0$ there exists $\delta(\varepsilon) > 0$, such that, for all $F \in U$, $|F(x) - F(y)| < \varepsilon$ for all $x, y \in G$ with $\|x - y\| < \delta(\varepsilon)$.*

(2) *U is bounded, i.e., there exists an $M > 0$ with $\|F\|_{C^{(0)}(\overline{\mathcal{G}})} < M$ for all $F \in U$.*

Suppose that K is of class $C^{(0)}(\overline{\mathcal{G}} \times \overline{\mathcal{G}})$, where $\mathcal{G}$ is a measurable bounded region in $\mathbb{R}^q$. We introduce the *integral operator* $A : C^{(0)}(\overline{\mathcal{G}}) \to C^{(0)}(\overline{\mathcal{G}})$ by letting

$$(AF)(x) = \int_{\mathcal{G}} K(x,y) F(y)\, dy, \quad F \in C^{(0)}(\overline{\mathcal{G}}). \tag{2.63}$$

In accordance with the Theorem of Arzela-Ascoli we are able to apply the conditions (1) and (2) to the context of the integral operator introduced in (2.63):

(1) K is uniformly continuous in $\overline{\mathcal{G}} \times \overline{\mathcal{G}}$. Therefore, AU is equicontinuous.

(2) Suppose that $U \subset C^{(0)}(\overline{\mathcal{G}})$ is *bounded*. Assume that F is of class U. Then we have

$$|AF(x)| = \left| \int_{\mathcal{G}} K(x,y) F(y)\, dy \right| \leq M \sup_{x,y \in \overline{\mathcal{G}}} |K(x,y)| \, \|\mathcal{G}\| < \infty. \tag{2.64}$$

In other words, AU is bounded. As a consequence, by virtue of the Theorem of Arzela-Ascoli, we are able to conclude that AU is relatively compact, so that the operator A defined by (2.63) is compact.

Example. Let $\mathcal{G}$ be a measurable bounded region in $\mathbb{R}^q$, and K be of class $L^2\left(\overline{\mathcal{G}}\times\overline{\mathcal{G}}\right)$, then $A : L^2\left(\overline{\mathcal{G}}\right) \to L^2\left(\overline{\mathcal{G}}\right)$ given by

$$AF = \int_{\mathcal{G}} K(\cdot, y)\, F(y)\, dy, \quad F \in L^2\left(\overline{\mathcal{G}}\right) \tag{2.65}$$

is compact (the proof can be found in, e.g., [**429**], [**513**]). Moreover, we have $\langle AF, G\rangle_{L^2(\overline{\mathcal{G}})} = \langle F, A^*G\rangle_{L^2(\overline{\mathcal{G}})}$ with $A^* : L^2\left(\overline{\mathcal{G}}\right) \to L^2\left(\overline{\mathcal{G}}\right)$ given by

$$A^*G = \int_{\mathcal{G}} \overline{K(y,\cdot)}\, G(y)\, dy. \tag{2.66}$$

Example. Let $\mathcal{G}$ be a measurable bounded region in $\mathbb{R}^q$. We introduce the operator $A : L^2\left(\overline{\mathcal{G}}\right) \to L^2\left(\overline{\mathcal{G}}\right)$ given by

$$AF(x) = \int_{\mathcal{G}} K(x, y)\, F(y)\ dy, \quad F \in L^2(\overline{\mathcal{G}}), \tag{2.67}$$

where K is continuous for $x \neq y$ and *weakly singular*, i.e., there exist a value $\alpha \in (0, q)$ and a constant $C > 0$ such that

$$|K(x,y)| \leq C\, \frac{1}{|x-y|^{q-\alpha}}. \tag{2.68}$$

Then A is compact (for the proof see, e.g., [**489**]).

We mention some important examples occurring in modern geosciences, in particular in Physical Geodesy and geoexploration:

- The *Newton volume integral*

$$V(x) = AF(x) = \frac{1}{4\pi} \int_{\mathcal{G}} \frac{1}{|x-y|}\ F(y)\, dy, \quad F \in L^2(\overline{\mathcal{G}}), \tag{2.69}$$

  occurring in the so-called *inverse gravimetry problem* of determining the scalar density distribution F inside $\mathcal{G}$ from the gravitational potential V forms a compact operator A.
- Let $\mathcal{G}$ be a bounded region in $\mathbb{R}^3$. The *inverse magnetometry problem*

$$B(x) = Am\ (x) = \frac{1}{4\pi} \int_{\mathcal{G}} \nabla_y \frac{1}{|x-y|} \cdot m(y)\ dy \tag{2.70}$$

  of determining the vectorial (square-integrable) magnetization distribution m inside $\mathcal{G}$ from the magnetic dipole potential B defines a compact operator A .
- The *tensorial satellite gravitational gradiometry problem*

$$\mathbf{v}(x) = AF\ (x) = \frac{1}{4\pi} \int_{\partial\mathcal{G}} (\nabla_x \otimes \nabla_x)\, \frac{1}{|x-y|}\ F(y)\ dy, \quad F \in L^2(\partial\mathcal{G}), \tag{2.71}$$

  of determining the gravitational potential F on the surface $\partial\mathcal{G}$ from tensorial measurements $\mathbf{v}(x) = (\nabla_x \otimes \nabla_x)\, V(x)$ on satellite orbits defines a compact operator A.

The gravitational constant as well as the vacuum permeability are set to one in the above formulas for simplicity. In numerical calculations, however, they have to be considered accordingly.

Reproducing kernel Hilbert spaces. Reproducing kernel Hilbert space framework plays an important role in both inverse ill-posed problems and in sampling problems. This is the reason why we collect some essential properties of the theory of reproducing kernel Hilbert spaces (RKHSs). Occasionally, we use the abbreviations CRKHS and DRKHS, respectively, for an RKRH defined over a continuously connected pointset and a discrete pointset.

Reproducing kernel Hilbert space over continuously connected pointsets (in brief, CRKHS). A Hilbert space H of $\mathbb{K}$-valued functions on a continuously connected pointset, say a bounded region $\mathcal{G} \subset \mathbb{R}^q, q \geq 1$, is called an RKHS, if the evaluation functionals $H \ni F \mapsto F(t) \in \mathbb{K}$ are continuous (bounded) for each fixed $t \in \mathcal{G}$, i.e., there exists a positive constant C_t for each $t \in \mathcal{G}$ such that $|F(t)| \leq C_t \, \|F\|_H$ for all $F \in H$. By the Riesz representation theorem, for each $t \in \mathcal{G}$, there exists a unique element K_t such that $F(t) = \langle F, K_t \rangle_H$ for all $F \in H$. The *reproducing kernel* (RK) $K(\cdot,\cdot) : \mathcal{G} \times \mathcal{G} \mapsto \mathbb{K}$ of an RKHS H is defined by

$$K(s,t) = \langle K_s, K_t \rangle_H, \qquad s,t \in \mathcal{G}. \tag{2.72}$$

We list some basic properties of RKHSs that are particularly relevant (for further details see, e.g., N. Aronszajn [**47**], E. Hille [**434**], H.S. Shapiro [**766**]):

- $K(s,t) = \overline{K(t,s)}$ for all $t, s \in \mathcal{G}$.
- $K(t,t) \geq 0$ for all $t \in \mathcal{G}$. Furthermore, if $K(t_0, t_0) = 0$ for some $t_0 \in \mathcal{G}$, then $F(t_0) =< F, K_{t_0} >_H = 0$ for all $F \in H$.
- By virtue of the Cauchy-Schwarz inequality we have

$$|K(s,t)| \leq \sqrt{K(s,s)} \, \sqrt{K(t,t)}, \quad s,t \in \mathcal{G}.$$

- The reproducing kernel $K(s,t)$ on $\mathcal{G} \times \mathcal{G}$ is a positive definite symmetric kernel. Conversely, by the Aronszajn-Moore Theorem (cf. N. Aronszajn [**47**]), every positive definite symmetric function $K(\cdot,\cdot)$ on $\mathcal{G} \times \mathcal{G}$ determines a unique Hilbert space $H (= H_K)$ for which $K(\cdot,\cdot)$ is a reproducing kernel (note that a $\mathbb{K}$-valued kernel K on $\mathcal{G} \times \mathcal{G}$ is positive definite if, for any n points $t_1, \ldots, t_N \in \mathcal{G}$, the matrix $K = (K(t_i, t_j))_{1 \leq i,j \leq N}$ is nonnegative definite, i.e., $u^H K u = \sum_{i,j=1}^{N} \overline{u_i} \; K(t_i, t_j) \; u_j \geq 0$ for all $u = (u_1, \ldots, u_n) \in \mathbb{K}^N$).
- A closed subspace $\tilde{H}$ of a RKHS H constitudes also an RKHS. Moreover, the orthogonal projector P of H onto $\tilde{H}$ and the reproducing kernel $\tilde{K}(s,t)$ of the RKHS $\tilde{H}$ are related by $PF(s) = \langle F, \tilde{K}_s \rangle, s \in \mathcal{G}$ for all $F \in H$, where $\tilde{K}_s = PK_s$.
- If an RKHS space H with kernel $K(\cdot,\cdot)$ has a direct orthogonal decomposition $H = H_1 \oplus H_2$ for some complementary orthogonal closed subspaces H_1 and H_2, then $K = K_1 + K_2$, where K_1, K_2 are reproducing kernels of the reproducing kernel Hilbert space H_1 and H_2, respectively.
- For $s = (s_1, s_2), t = (t_1, t_2)$ with $s_1, t_1 \in \mathcal{G}_1$ and $s_2, t_2 \in \mathcal{G}_2$,

$$K(s,t) = K_1(s_1, t_1) K_2(s_2, t_2)$$

  constitutes a RKHS on $\mathcal{G}_1 \times \mathcal{G}_2$, where $K_1(\cdot,\cdot)$ and $K_2(\cdot,\cdot)$ define RHKSs on $\mathcal{G}_1$ and $\mathcal{G}_2$, respectively.

- In an RKHS, the element representing a given bounded linear functional $\mathcal{L}$ can be expressed by means of the reproducing kernel $\mathcal{L}(F) = \langle F, Q\rangle_H$, where $Q = \mathcal{L}(K)$. For a bounded linear operator $\mathcal{L}$ on H to H, we have
$$\mathcal{L}F = \langle \mathcal{L}F, Q\rangle_H = \langle F, \mathcal{L}^* Q\rangle_H.$$
- Every finite-dimensional function space is a RKHS H with reproducing kernel
$$K(s,t) = \sum_{i=1}^{m} U_i(s)\, \overline{U_i(t)},$$
where $\{U_i\}_{i=1,\ldots,m}$ is an orthonormal basis for H (notice that the sum in the above definition of the kernel K is invariant under the choice of an orthonormal basis).
- If
$$\int_{\mathcal{G}\times\mathcal{G}} |K(s,t)|^2 \, ds\, dt < \infty,$$
then $K(\cdot,\cdot)$ has a countable sequence of eigenvalues and eigenfunctions (*Mercer Theorem*).
- $L^2(\mathcal{G})$, $\mathcal{G}$ measurable bounded region, the space of all square-integrable functions on $\mathcal{G}$ is <u>not</u> an RKHS.
- Let H be a separable RKHS, then its reproducing kernel $K(\cdot,\cdot)$ has the expansion
$$K(s,t) = \sum_{n\in\mathbb{N}} \Phi_n(t)\, \overline{\Phi_n(s)},$$
where $\{\Phi_n\}_{n\in\mathbb{N}}$ is an orthonormal basis for H (we remark that for a general separable Hilbert space H, the series $\sum_{n\in\mathbb{N}} \Phi_n(t)\, \overline{\Phi_n(s)}$ does not constitute a reproducing kernel and also that the Φ_n's do not generally correspond to sampling expansions). If they do, i.e., if $\Phi_n(t) = K(t_n, t)$ for some sequence $\{t_n\}$, then
$$F(t) = \sum_{n\in\mathbb{N}} F(t_n)\, \Phi_n(t),$$
this indeed is a *sampling theorem.*
- Let $\{\Phi_n\}_{n\in\mathbb{N}}$ be a sequence of $\mathbb{K}$-valued functions defined on a set $\mathcal{G}$ such that, for every $s \in \mathcal{G}$, $\sum_{n\in\mathbb{N}} |\Phi_n(s)|^2 < \infty$. Then, for every sequence $\{c_n\}_{n\in\mathbb{N}}$ with $\sum_{n\in\mathbb{N}} |c_n|^2 < \infty$, the series $\Sigma_{n\in\mathbb{N}}$, $c_n\Phi_n(s)$ is convergent in $\mathbb{K}$ for every $x \in \mathcal{G}$. The functions which are the sums of such series form a linear subspace H, on which we are able to define the structure of a separable Hilbert space by taking as scalar product, for
$$F = \sum_{n\in\mathbb{N}} c_n \Phi_n, \quad G = \sum_{n\in\mathbb{N}} d_n \Phi_n,$$
the number
$$\langle F, G\rangle_H = \sum_{n\in\mathbb{N}} c_n\, \overline{d_n}.$$
This space has a reproducing kernel, namely
$$K(s,t) = \sum_{n\in\mathbb{N}} \Phi_n(s)\, \overline{\Phi_n(t)}, \quad s,t \in \mathcal{G}\times\mathcal{G}.$$

- If the reproducing kernel $K(s,t)$ of an RKHS H is continuous on $\mathcal{G} \times \mathcal{G}$, then H is a space of continuous functions (uniformly continuous on a bounded $\mathcal{G}$). This follows from

$$|F(t) - F(s)| = |\langle F, K_t - K_s \rangle_H| \leq \|F\|_H \; \|K_t - K_s\|_H$$

and

$$\|K_t - K_s\|^2 = K(t,t) - 2K(t,s) + K(s,s) \tag{2.73}$$

for all $s, t \in \mathcal{G}$.

- Strong convergence in an RKHS H implies pointwise convergence and uniform convergence on compact sets, because of the fact

$$|F(t) - F_n(t)| = |\langle F - F_n, K_t \rangle_H| \leq \|F - F_n\|_H \; \sqrt{K(t,t)}.$$

- For every positive definite kernel $K(\cdot,\cdot)$ on $\mathcal{G} \times \mathcal{G}$, there exist a zero mean Gaussian process with $K(\cdot,\cdot)$ as its covariance, giving rise to the relation between Bayes estimates, Gaussian processes, and optimization processes in an RHKS (see, e.g., K.R. Koch [**501**], G. Wahba [**848**]).

Interest in reproducing kernel Hilbert spaces have increased in recent years, as the computer capacity has made solutions of ever larger and more complex problems practicable. Indeed, new reproducing kernel representations and applications (e.g., in geosciences and medicine) are being contributed at a rapid rate. For example, a certain RKHS in terms of outer harmonics allows the adequate determination of the Earth's gravitational and/or geomagnetic potential in consistency with gravitational and/or observables of heterogeneous type that are interpretable as (bounded) linear functionals on the RKHS under consideration (see, e.g., W. Freeden [**256, 257, 270**], L. Shure et al. [**773**], C. Blick et al. [**100**]).

Gramians, frames, and RKHSs. Let H be a separable Hilbert space. Suppose that $\{\Phi_n\}_{n\in\mathbb{N}}$ is a system of elements in H. Then we say that the matrix

$$G = \left[\langle \Phi_k, \Phi_i \rangle_H\right]_{k,i=1,\ldots,\infty} \tag{2.74}$$

is a *Gramian (Gram matrix)*. In order to understand frames, it is canonical to assume that $\{\Phi_n\}_{n\in\mathbb{N}}$ spans a dense subspace in H. Furthermore, it is necessary to interpret G as an operator in $\ell^2(\mathbb{N})$, i.e., we assume that

$$\sum_{k\in\mathbb{N}} |\langle \Phi_k, \Phi_i \rangle_H|^2 \; < \infty, \quad i \in \mathbb{N}. \tag{2.75}$$

In that case, there is an operator T_G given by

$$(T_G \; a)_i = \sum_{k\in\mathbb{N}} \langle \Phi_i, \Phi_k \rangle_H \; a_k, \qquad a = (a_k)_{k\in\mathbb{N}} \in \ell^2(\mathbb{N}). \tag{2.76}$$

By virtue of (2.75), $T_G : \ell^2(\mathbb{N}) \to \ell^2(\mathbb{N})$ defines a hermitian semi-bounded operator with dense domain (finite sequences) in $\ell^2(\mathbb{N})$. However, in general, T_G is unbounded, only its domain is dense in $\ell^2(\mathbb{N})$.

Let H be a separable Hilbert space. Suppose that $\{\Phi_n\}_{n\in\mathbb{N}}$ spans a dense subspace in H. Let C', C'' be given positive constants such that $C' \leq C''$. A closed subspace $\tilde{H} \subset H$ is called a (C', C'')-*frame subspace* if

$$C' \; \|F\|_H^2 \leq \sum_{k\in\mathbb{N}} |\langle F, \Phi_k \rangle_H|^2 \leq C'' \; \|F\|_H^2 \tag{2.77}$$

holds for all $F \in \tilde{H}$.

The following results due to [**477**] provide the connection between Gramians and reproducing kernel Hilbert spaces:

(i) Let H be a separable Hilbert space. Suppose that $\{\Phi_n\}_{n\in\mathbb{N}}$ spans a dense subspace in H. Assume further that $\{\Phi_n\}_{n\in\mathbb{N}}$ satisfies the conditions:

- *For each $i \in \mathbb{N}$,*

$$\sum_{k\in\mathbb{N}} |\langle \Phi_k, \Phi_i \rangle|^2 < \infty. \tag{2.78}$$

- *For $d = (d_k)_{k\in\mathbb{N}} \in \ell^2(\mathbb{N})$ with $\sum_{k\in\mathbb{N}} (\langle \Phi_i, \Phi_k \rangle_H + \delta_{i,k})\ d_k = 0$ holds*

$$d = 0 \tag{2.79}$$

in $\in \ell^2(\mathbb{N})$.

Then, for every C', C'' with $C' \leq C''$, there is a unique maximal subspace H (dependent on (C',C")), which satisfies the frame condition (2.77).

(ii) Let H be a separable Hilbert space. Suppose that $\{\Phi_n\}_{n\in\mathbb{N}}$ is a frame with frame bounds C', C'' such that $C' \leq C''$, i.e., $\{\Phi_n\}_{n\in\mathbb{N}}$ spans a (C', C'')-frame subspace in H. Suppose further that there is a set $\mathcal{G}$ such that each Φ_n is a function on $\mathcal{G}$, and that $\{\Phi_n(x)\}_{n\in\mathbb{N}} \in \ell^2(\mathbb{N})$ for all $x \in \mathcal{G}$. Then, H is a reproducing kernel Hilbert space (RKHS) of functions on $\mathcal{G}$.

For further properties of RKHSs the reader is referred to [**47**], [**434**] (see also [**859**] for an edited collection of reprints on theory and applications of RKHSs).

Reproducing kernel Hilbert space over discrete pointsets (in brief, DRKHS). There is a related reproducing kernel notion on prescribed countably infinite discrete sets $\mathcal{S}$. A choice of points in, e.g., relevant models based on kernel theory suggests the need for a better understanding of point masses as they are accounted for a CRKHS at hand. Following P. Jorgenson, F. Tian [**478**] almost literally, this is a leading theme of our next explanations, and so far we feel (as the authors of [**478**]) that the dichotomy CRKHS versus DRKHS has not yet received sufficient attention.

The point of departure for the discrete RKHS-approach is the following definition:

Let $\mathcal{S}$ be a countable and infinite discrete set, and $\mathcal{F}(\mathcal{S})$ the set of all finite subsets $\mathcal{A}$ of $\mathcal{S}$. A kernel $K : \mathcal{S} \times \mathcal{S} \to \mathbb{C}$ is said to be *positive definite*, if

$$\sum_{(x,y)\in\mathcal{A}\times\mathcal{A}} K(x,y)\ \overline{C_x}\ C_y \geq 0 \tag{2.80}$$

holds for all coefficients $\{C_x\}_{x\in\mathcal{A}} \subset \mathbb{C}$ and all sets $\mathcal{A} \in \mathcal{F}(\mathcal{S})$.

The definitoric settings (S1), ..., (S5) are concerned with the discrete structure of an RKHS H.

(S1) For $x \in \mathcal{S}$, by convention, we again use the abbreviation $K_x = K(x, \cdot)$. H is understood as the completion of the span of $\{K_x\}_{x\in\mathcal{S}}$

$$H = \overline{\text{span}\{K_x : x \in \mathcal{S}\}} \tag{2.81}$$

with respect to the inner product

$$\left\langle \sum_{x\in\mathcal{A}} C_x\, K_x, \sum_{y\in\mathcal{A}} D_y\, K_y \right\rangle_H = \sum_{(x,y)\in\mathcal{A}\times\mathcal{A}} K(x,y)\, \overline{C_x}\, D_y \tag{2.82}$$

modulo the subspace of functions of zero H-norm (note that the summations in (2.82) are all finite).

(S2) For $\mathcal{A} \in \mathcal{L}(\mathcal{S})$ we understand by $H_{\mathcal{A}}$ the closed span of $\{K_x\}_{x\in\mathcal{A}} \subset H$, i.e.,

$$H_{\mathcal{A}} = \overline{\text{span}\{K_x : x \in \mathcal{A}\}}, \tag{2.83}$$

note that closed is automatic if $\mathcal{A}$ is finite.

(S4) $P_{\mathcal{A}}$ denotes the orthogonal projection of H onto $H_{\mathcal{A}}$.

(S5) For $\mathcal{A} \in \mathcal{F}(\mathcal{B})$, we denote by $K_{\mathcal{A}}$ the $\sharp\mathcal{A} \times \sharp\mathcal{A}$ matrix

$$K_{\mathcal{A}} = [K(x,y)]_{(x,y)\in\mathcal{A}\times\mathcal{A}}. \tag{2.84}$$

In conclusion, starting with finitely supported summations in (2.82), the discrete RKHS H is obtained by an Hilbert space completion. From (S1) it immediately follows that H is a reproducing kernel Hilbert space satisfying the reproducing property:

$$\langle F, K_x\rangle_H = F(x), \quad x \in \mathcal{S}, \quad F \in H. \tag{2.85}$$

Central for our considerations is the positive definiteness (cf. [**478**]):

Let $K : \mathcal{S}\times\mathcal{S} \to \mathbb{C}$ be positive definite, and let H be the corresponding discrete RKHS. Let F be an arbitrary function on $\mathcal{S}$. Then F is a member of H, if and only if there is a constant $C(= C_F) < \infty$ such that, for every finitely supported function $\Xi : \mathcal{S} \to \mathbb{C}$, we have

$$\left| \sum_{x\in\mathcal{S}} \overline{\Xi(x)}\, F(x) \right|^2 \leq C \sum_{x,y\in\mathcal{S}} K(x,y)\, \overline{\Xi(x)}\, \Xi(y) \tag{2.86}$$

with a constant $C = C_F$ independent of Ξ.

Finally it should be mentioned that both the continuous and discrete variant of reproducing kernel Hilbert spaces arise from a given positive definite kernel K.

2.2. Fourier asymptotics and limit relations

The following considerations can be regarded as classical realizations of the *Riemann-Lebesgue theorem* of 1D Fourier theory. All results are well-known, for example, from the monograph I.P. Natanson [**645**]. Nonetheless, they play an essential role in our bivariate lattice and sampling theory (cf. Section 13.2):

Let F be continuous in the interval $(-1,1)$ such that

$$\int_{-1}^{1} |F(t)|\, dt < \infty. \tag{2.87}$$

Then

$$\lim_{r\to\infty} \int_{-1}^{1} e^{irt} F(t)\, dt = 0. \tag{2.88}$$

Of particular interest in the metaharmonic theory (see Section 8.3) is the following limit relation:

Let F be continuous in $[-1,1]$ and continuously differentiable in $(-1,1)$ such that

$$\int_{-1}^{1} |F'(t)|\, dt < \infty. \tag{2.89}$$

Then we have for $r \to \infty$

$$\sqrt{\frac{2r}{\pi}} \int_{-1}^{1} e^{irt} F(t)(1-t^2)^{-\frac{1}{2}}\, dt = i^{-\frac{1}{2}} e^{ir} F(1) + i^{\frac{1}{2}} e^{-ir} F(-1) + o(1). \tag{2.90}$$

One-dimensional Fourier theory allows a large number of extensions of the aforementioned asymptotic results.

Boundary point asymptotics. The following asymptotic relation holds true:

Let G be of class $C^{(m)}([a,b])$, $a,b \in \mathbb{R}$ with $a < b$. Suppose that F is given in the form

$$F(t) = G(t)(t-a)^{\lambda-1}(b-t)^{\mu-1}, \quad t \in [a,b], \tag{2.91}$$

$0 < \lambda,\ \mu \leq 1$. *Then, for $r \to \infty$,*

$$\int_a^b F(t) e^{irt}\, dt = -A_m(r) + B_m(r) + O\left(\frac{1}{r^m}\right), \tag{2.92}$$

where

$$A_m(r) = \sum_{n=0}^{m-1} \frac{\Gamma(n+\lambda)}{n!\ r^{n+\lambda}} e^{ira + i\frac{\pi}{2}(n+\lambda-2)} \left(\frac{\partial}{\partial u}\right)^n \left((b-u)^{\mu-1} G(u)\right)\big|_{u=a} \tag{2.93}$$

and

$$B_m(r) = \sum_{n=0}^{m-1} \frac{\Gamma(n+\mu)}{n!\ r^{n+\mu}} e^{irb + i\frac{\pi}{2}(n-\mu)} \left(\frac{\partial}{\partial v}\right)^n \left((v-a)^{\lambda-1} G(v)\right)\big|_{v=b}. \tag{2.94}$$

(note that, $\lambda = \mu = 1$ the term $O(r^{-m})$ can be replaced by $o(r^{-m})$).

An easy consequence is the following asymptotic expansion:

Suppose that $F \in C^{(1)}([a,b])$, $H \in C^{(2)}([a,b])$ such that $H'(t) > 0$ for $t \in [a,b]$. Then, for $r \to \infty$,

$$\int_a^b F(t) e^{irH(t)}\, dt = \frac{1}{ir}\left(\frac{F(b)}{H'(b)} e^{irH(b)} - \frac{F(a)}{H'(a)} e^{irH(a)}\right) + o\left(\frac{1}{r}\right). \tag{2.95}$$

The last relation is based on the assumption that the function H does not possess a stationary point, i.e., $\tau \in [a,b]$ such that $H'(\tau) = 0$; hence, the asymptotic relation (2.95) depends only on the values of the endpoints of the interval $[a,b]$.

Stationary point asymptotics. The following asymptotic relation (2.98) includes influences from stationary points.

Let G be of class $C^{(m)}([a,b])$. Suppose that $\tilde{H}$ is of class $C^{(m)}([a,b])$ with $\tilde{H}(t) > 0$ for all $t \in [a,b]$. Furthermore, we let

$$F(t) = G(t)(t-a)^{\lambda-1}(b-t)^{\mu-1}, \quad t \in [a,b], \tag{2.96}$$

$0 \le \lambda$, $\mu \le 1$, and $H \in C^{(1)}([a,b])$ with

$$H'(t) = \tilde{H}(t)(t-a)^{\rho-1}(b-t)^{\sigma-1}, \quad t \in [a,b], \tag{2.97}$$

$\rho, \sigma \ge 1$. Then, for $r \to \infty$,

$$\int_a^b F(t)e^{ikH(t)}\,dt = -A_m(r) + B_m(r) + O\left(\frac{1}{r^{\frac{m}{\rho}}}\right) + O\left(\frac{1}{r^{\frac{m}{\sigma}}}\right), \tag{2.98}$$

where

$$A_m(r) = -e^{irH(a)}\sum_{n=0}^{m-1}\frac{\psi^{(n)}(0)}{\rho}\frac{\Gamma(\frac{n+\lambda}{\rho})}{n!}e^{i\frac{\pi}{2}\frac{n+\lambda}{\rho}}\frac{1}{r^{\frac{n+\lambda}{\rho}}} \tag{2.99}$$

and

$$B_m(r) = -e^{irH(b)}\sum_{n=0}^{m-1}\frac{\chi^{(n)}(0)}{\sigma}\frac{\Gamma(\frac{n+\mu}{\sigma})}{n!}e^{-i\frac{\pi}{2}\frac{n+\mu}{\sigma}}\frac{1}{r^{\frac{n+\mu}{\sigma}}} \tag{2.100}$$

with

$$\psi(u) = F(t)u^{1-\lambda}\frac{dt}{du}, \quad u^\rho(t) = H(t) - H(a), \tag{2.101}$$

and

$$\chi(v) = F(t)v^{1-\mu}\frac{dt}{dv}, \quad v^\sigma(t) = H(b) - H(t). \tag{2.102}$$

If $\lambda = \mu = 1$, then the "O" symbol can be replaced by the "o" symbol.

A point $\tau \in [a,b]$ is called a *stationary point of order n in $[a,b]$*, if the following conditions are satisfied: (i) $H \in C^{(n+1)}([a,b])$, (ii) $H(\tau) = \ldots = H^{(n)}(\tau) = 0$, and (iii) $H^{(n+1)}(\tau) \neq 0$. Note that a stationary point of order 1 is simply called a *stationary point.*

The definition of a stationary point leads to the following corollaries:

Suppose that $F \in C^{(1)}([a,b])$ and $H \in C^{(3)}([a,b])$. If there is a stationary point $\tau \in (a,b)$ such that $H'(\tau) = 0$ and $H''(\tau) > 0$, then for $r \to \infty$

$$\int_a^b F(t)e^{irH(t)}\,dt = F(\tau)\sqrt{\frac{2\pi}{H''(\tau)r}}\;e^{irH(\tau)+i\frac{\pi}{4}} + o\left(\frac{1}{r^{\frac{1}{2}}}\right). \tag{2.103}$$

If $\tau = b$ and/or $\tau = a$ we obtain the same result, however, equipped with a factor $\frac{1}{2}$.

Suppose that $F \in C^{(1)}([a,b])$ and $H \in C^{(3)}([a,b])$. If there is a stationary point $\tau \in (a,b)$ such that $H'(\tau) = 0$ and $H''(\tau) < 0$, then for $r \to \infty$

$$\int_a^b F(t)e^{irH(t)}\,dt = F(\tau)\sqrt{\frac{-2\pi}{H''(\tau)r}}e^{irH(\tau)-i\frac{\pi}{4}} + o\left(\frac{1}{r^{\frac{1}{2}}}\right). \tag{2.104}$$

Gauß-Weierstraß and Abel-Poisson limits. Next, some limit relations are collected that are needed as auxiliary tools in the theory of Gauss-Weierstrass and Abel-Poisson transforms. The considerations are well known (see, e.g., [**607**]); they are listed here for a better understanding of the multi-dimensional integral transforms of Section 13.4).

Suppose that $\varphi : r \mapsto \varphi(r)$, $r \geq 0$ *is continuous with* $\lim_{r\to\infty} \varphi(r) = 0$. *Then the integral* $\int_0^\infty e^{-\pi\tau r^2} \varphi(r) r^{l-1}\, dr$ *exists for all* $\tau > 0$, $l \in \mathbb{N}$, *and we have*

$$\lim_{\substack{\tau\to 0\\ \tau>0}} \int_0^\infty e^{-\pi\tau r^2} \varphi(r)\, r^{l-1}\, dr = \int_0^\infty \varphi(r)\, r^{l-1}\, dr \tag{2.105}$$

if the integral on the right side exists in the sense $\int_0^\infty \cdots = \lim_{T\to\infty} \int_0^T \cdots$.

The integral (2.105), which provides a limit for $\tau \to 0$, should be discussed for the limit $\tau \to \infty$, too.

Suppose that $\psi : r \mapsto \psi(r)$, $r \geq 0$, *is continuous such that, for fixed* $n \geq 1$,

$$\psi(r) = a_0 + \ldots + a_n r^n + O\left(r^{n+1}\right) \tag{2.106}$$

for all $r \in [0, 1)$ *and* $\psi(r) = O\left(r^n\right)$ *for* $r \to \infty$. *Then, for* $l \geq 2$ *and* $\tau \to \infty$, *we have*

$$\int_0^\infty e^{-\tau r}\, r^{\frac{l-3}{2}}\, \psi(r)\, dr = \left(\frac{1}{\tau}\right)^{\frac{l-1}{2}} \left(b_0 + \ldots + \frac{b_n}{\tau^n}\right) + O\left(\frac{1}{\tau^{n+1}}\right), \tag{2.107}$$

where the coefficients a_k *and* b_k *satisfy* $b_k = \Gamma\left(k + \frac{l-1}{2}\right)\, a_k$, $k = 1, \ldots, n$.

The next limit relation and its consequences are of significance in the Fourier analysis of the Gauss-Weierstrass integral transform.

Suppose that $\varphi : r \mapsto \varphi(r)$, $r \geq 0$, *is integrable on* $[0, \infty)$ *and continuous at the origin. Moreover, assume that* $\int_0^\infty |\varphi(r)|\, r^{l-1}\, dr$, $l \in \mathbb{N}$, *exists. Then*

$$\lim_{\substack{\tau\to 0\\ t>0}} \left(\frac{1}{\tau}\right)^{\frac{l}{2}} \int_0^\infty e^{-\frac{\pi}{\tau} r^2} \varphi(r)\, r^{l-1}\, dr = \frac{\varphi(0)}{\|\mathbb{S}^{l-1}\|}. \tag{2.108}$$

Finally we come to an auxiliary result, which is of importance in the context of the Abel-Poisson integral transform.

Suppose that $\varphi : r \mapsto \varphi(r)$, $r \geq 0$, *is integrable on* $[0, \infty)$ *and continuous at the origin. Moreover, assume that* $\int_0^\infty |\varphi(r)|\, r^{l-1}\, dr$, $l \in \mathbb{N}$, *exists. Then*

$$\lim_{\substack{\tau\to 0\\ \tau>0}} \frac{2}{\|\mathbb{S}^l\|} \int_0^\infty \frac{\tau\, \varphi(r)\, r^{l-1}}{\left(\tau^2 + r^2\right)^{\frac{l+1}{2}}}\, dr = \frac{\varphi(0)}{\|\mathbb{S}^{l-1}\|}. \tag{2.109}$$

Fourier transform and Fourier inversion in Lebesgue spaces. We present some specific material on Fourier theory in Euclidean spaces (influenced by P.L. Butzer, R. Nessel [**124**] and E.M. Stein, G. Weiss [**790**]): For reasons of notational consistency we start with classical results on the Fourier transform for Lebesgue and Sobolev spaces. Of particular interest, however, is the relation between functions being integrable over $\mathbb{R}^q$ as well as periodical with respect to the unit lattice $\mathbb{Z}^q \subset \mathbb{R}^q$. As a consequence, we are immediately led to the process of "periodization" as a "bridging tool", i.e., in the conventional nomenclature to the Poisson summation formula in $\mathbb{R}^q$. The essential calamity in periodization is the convergence behavior of the Fourier series expansion of the "lattice Green's function" showing the same singular behavior as the fundamental solution of the Laplacian in all points of the lattice $\mathbb{Z}^q$, $q \geq 2$. The pointwise convergence shows a striking difference to the one-dimensional theory, where the singular behavior is characterized just by a finite jump in the first derivative, and the pointwise convergence of the lattice Green functions is guaranteed. In contrast to the one-dimensional case, another serious difficulty is the Fourier inversion formula. As a powerful multivariate remedy, the inversion formula of Fourier integrals can be understood in pointwise sense within the terminology of certain means such as Gauss-Weierstrass transform, which will be used later in our multivariate sampling approach.

The *Fourier transform* of $F : \mathbb{R}^q \to \mathbb{C}$ at $x \in \mathbb{R}^q$ is defined by

$$\mathcal{F}_{\mathbb{R}^q}(F)(x) = F^{\wedge}_{\mathbb{R}^q}(x) = \int_{\mathbb{R}^q} F(y)\, e^{-2\pi i x\cdot y}\, dy. \tag{2.110}$$

If F is of class in $L^1(\mathbb{R}^q)$, then it is known (for more details see, e.g., [**790**]) that the function $\mathcal{F}_{\mathbb{R}^q}(F)$ is uniformly continuous. Moreover, the Riemann-Lebesgue theorem of Fourier analysis tells us that $F^{\wedge}_{\mathbb{R}^q}(x) \to 0$ as $|x| \to \infty$.

As already known from Section 2.1, if F, G belong to $L^1(\mathbb{R}^q)$, their convolution $H = F * G$ is the function, whose functional value at a point $x \in \mathbb{R}^q$ is given by $H(x) = \int_{\mathbb{R}^q} F(x-y)\, G(y)\, dy$. Clearly, $H \in L^1(\mathbb{R}^q)$ and $\|H\|_{L^1(\mathbb{R}^q)} \leq \|F\|_{L^1(\mathbb{R}^q)}\|G\|_{L^1(\mathbb{R}^q)}$. The operation is commutative and associative. Even more, if $F \in L^p(\mathbb{R}^q)$, $1 \leq p < \infty$, and $G \in L^1(\mathbb{R}^q)$, then $H = F * G$ belongs to $L^p(\mathbb{R}^q)$, and we have $\|H\|_{L^p(\mathbb{R}^q)} \leq \|F\|_{L^p(\mathbb{R}^q)}\, \|G\|_{L^1(\mathbb{R}^q)}$.

An essential feature is that the Fourier transform of the convolution of two functions is the product of their Fourier transforms. More concretely, if F and G belong to $L^1(\mathbb{R}^q)$, then

$$\mathcal{F}_{\mathbb{R}^q}(F * G) = (F * G)^{\wedge}_{\mathbb{R}^q} = F^{\wedge}_{\mathbb{R}^q}\, G^{\wedge}_{\mathbb{R}^q} = \mathcal{F}_{\mathbb{R}^q}(F)\mathcal{F}_{\mathbb{R}^q}(G). \tag{2.111}$$

Many other important operations have simple relations with the Fourier transform: If T_d denotes the *translation* by $d \in \mathbb{R}^q$ (by this we mean the operator mapping the function $F \in L^1(\mathbb{R}^q)$ into the function $(\mathrm{T}_d F) = F(\cdot - d)$), then

$$(\mathrm{T}_d F)^{\wedge}_{\mathbb{R}^q}(x) = e^{-2\pi i d\cdot x} F^{\wedge}_{\mathbb{R}^q}(x), \quad x \in \mathbb{R}^q, \tag{2.112}$$

and

$$(e^{2\pi i d\cdot} F)^{\wedge}_{\mathbb{R}^q}(x) = (\mathrm{T}_d F^{\wedge}_{\mathbb{R}^q})(x), \quad x \in \mathbb{R}^q. \tag{2.113}$$

If $d_\tau, \tau > 0$, denotes the *dilation* (by this we mean the operator mapping $F \in L^1(\mathbb{R}^q)$ into the function $d_\tau F = F(\tau\cdot)$), then

$$\tau^q (d_\tau F)^{\wedge}_{\mathbb{R}^q}(x) = F^{\wedge}_{\mathbb{R}^q}(\tau^{-1}x), \quad x \in \mathbb{R}^q. \tag{2.114}$$

If $x \mapsto P(x)$, $x \in \mathbb{R}^q$, is a polynomial in q variables $x = (x_1, \ldots, x_q)^T$ and $P(\nabla)$ is the associated differential operator (i.e., replace formally $x_i = x \cdot \varepsilon^i$ by $\frac{\partial}{\partial x_i} = \nabla \cdot \varepsilon^i$, $i \in \{1, \ldots, q\}$), then

$$P(\nabla) F^{\wedge}_{\mathbb{R}^q}(x) = (P(-2\pi i \cdot) F)^{\wedge}_{\mathbb{R}^q}(x), \quad x \in \mathbb{R}^q, \tag{2.115}$$

and

$$(P(\nabla) F)^{\wedge}_{\mathbb{R}^q}(x) = P(-2\pi i x) F^{\wedge}_{\mathbb{R}^q}(x), \quad x \in \mathbb{R}^q. \tag{2.116}$$

Applying Fubini's theorem we immediately obtain the identity

$$\int_{\mathbb{R}^q} F^{\wedge}_{\mathbb{R}^q}(x)\, G(x)\, dx = \int_{\mathbb{R}^q} F(x)\, G^{\wedge}_{\mathbb{R}^q}(x)\, dx \tag{2.117}$$

provided that F, G belong to $L^1(\mathbb{R}^q)$. Moreover, if $F \in L^1(\mathbb{R}^q)$, $G \in L^p(\mathbb{R}^q)$, $1 \le p \le 2$, then $H = F * G \in L^p(\mathbb{R}^q)$ satisfies $H^{\wedge}_{\mathbb{R}^q}(x) = F^{\wedge}_{\mathbb{R}^q}(x)\, G^{\wedge}_{\mathbb{R}^q}(x)$ (for almost every x).

The *Fourier inversion formula*

$$F(x) = \int_{\mathbb{R}^q} F^{\wedge}_{\mathbb{R}^q}(y)\, e^{2\pi i x \cdot y}\, dy \tag{2.118}$$

(for almost every x) may be formulated in its standard framework under the assumption that $F, F^{\wedge}_{\mathbb{R}^q}$ belong to $L^1(\mathbb{R}^q)$ (see, e.g., [**790**]). In operator notation the inversion formula is briefly given by

$$\mathcal{F}_{\mathbb{R}^q}\left(\mathcal{F}_{\mathbb{R}^q}(F)\right)(-x) = F(x). \tag{2.119}$$

It should be noted that the assumption that $F, F^{\wedge}_{\mathbb{R}^q}$ belong to $L^1(\mathbb{R}^q)$ is not senseful in lattice point and sampling theory (see [**268**]). This is the reason why we actually are not interested in a Lebesgue-variant of the Fourier inversion formula, but we need a pointwise manifestation, that is applicable, for example, for alternating functions (such as sinc-type functions) that do not define a function of the class $L^1(\mathbb{R}^2)$.

Schwartz space of rapidly decreasing functions and Fourier transform. It is not possible to define the Fourier transform for all distributions. We need the concept of rapidly decreasing functions (see, e.g., [**190**]).

For $\phi \in C^{(\infty)}(\mathbb{R}^q)$, $k, l \in \mathbb{N}_0$, $q \in \mathbb{N}$, $\alpha \in \mathbb{N}_0^q$, we introduce the expression

$$p_{k,l}(\phi) = \sup_{x \in \mathbb{R}^q} \left(|x|^k + 1\right) \sum_{|\alpha| \le l} |D^\alpha \phi(x)|\,. \tag{2.120}$$

ϕ is called *rapidly decreasing* if $p_{k,l}(\phi) < \infty$ for all $k, l \in \mathbb{N}_0$. The space $\mathfrak{S}(\mathbb{R}^q)$ of all rapidly decreasing functions is called the *Schwartz space.* The convergence of a sequence $\{\phi_j\}_{j \in \mathbb{N}}$ in $\mathfrak{S}(\mathbb{R}^q)$ is defined by

$$\phi_j \overset{\mathfrak{S}}{\longrightarrow} \phi \quad \text{if and only if} \quad \lim_{j \to \infty} p_{k,l}(\phi_j - \phi) = 0 \text{ for all } k, l \in \mathbb{N}_0\,. \tag{2.121}$$

It is clear that $C_c^{(\infty)}(\mathbb{R}^q) \subset \mathfrak{S}(\mathbb{R}^q)$.

The Fourier transformation for rapidly decreasing functions (see, e.g., [**190**]) is defined as follows:

Let $\phi \in \mathfrak{S}(\mathbb{R}^q)$. The Fourier transform $\mathcal{F}_{\mathbb{R}^q}\phi$ of ϕ is defined by

$$\mathcal{F}_{\mathbb{R}^q}(\phi)(\xi) = \phi^{\wedge}_{\mathbb{R}^q}(\xi) = \int_{\mathbb{R}^n} \phi(x)\, e^{-2\pi i x\cdot\xi}\, dx, \quad \xi \in \mathbb{R}^q . \tag{2.122}$$

Properties of Fourier Transformation. Let $\phi \in \mathfrak{S}(\mathbb{R}^q)$, $\alpha \in \mathbb{N}_0^q$:

(i) $x^\alpha\phi, D^\alpha\phi, \mathcal{F}_{\mathbb{R}^q}\phi, D^\alpha\mathcal{F}_{\mathbb{R}^q}\phi, \mathcal{F}_{\mathbb{R}^q}D^\alpha\phi \in \mathfrak{S}(\mathbb{R}^q)$,

(ii) $D^\alpha\mathcal{F}_{\mathbb{R}^q}\phi = (-2\pi i)^{|\alpha|}\mathcal{F}_{\mathbb{R}^q}(x^\alpha\phi)$, $x^\alpha\mathcal{F}_{\mathbb{R}^q}\phi = (-2\pi i)^{|\alpha|}\mathcal{F}_{\mathbb{R}^q}(\nabla^\alpha\phi)$,

(iii) $\mathcal{F}_{\mathbb{R}^q} : \mathfrak{S}(\mathbb{R}^q) \to \mathfrak{S}(\mathbb{R}^q)$ is bijective and bicontinuous with inverse

$$\mathcal{F}^{-1}_{\mathbb{R}^q}(\phi)(x) = \phi^{\wedge}_{\mathbb{R}^q}(-x) = \int_{\mathbb{R}^q} \phi(a)\, e^{2\pi i x\cdot a}\, da. \tag{2.123}$$

We consider the dual of the Schwartz space, $\mathfrak{S}'(\mathbb{R}^q)$. Since $C_c^{(\infty)}(\mathbb{R}^q) \subset \mathfrak{S}(\mathbb{R}^q)$, we have

$$\mathfrak{S}'(\mathbb{R}^q) \subset \left(C_c^{(\infty)}(\mathbb{R}^q)\right)' . \tag{2.124}$$

The elements of $\mathfrak{S}'(\mathbb{R}^q)$ are called *tempered distributions*. Any distribution with compact support can be extended to a tempered distribution.

Regular distributions are tempered distributions if their corresponding locally integrable function is of class $L^p(\mathbb{R}^q)$ for some $1 \le p \le \infty$. A counter-example would be the distribution corresponding to the Gauss function (see 2.125)

Fourier transformation of tempered distributions. We can extend the Fourier transformation to $\mathfrak{S}'(\mathbb{R}^q)$ (see, e.g., [**190**]): Let $\mathfrak{T}$ be of class $\mathfrak{S}'(\mathbb{R}^q)$. The Fourier transform $\mathcal{F}_{\mathbb{R}^q}\mathfrak{T}$ of $\mathfrak{T}$ is defined by

$$\mathcal{F}_{\mathbb{R}^q}\mathfrak{T}(\phi) = \mathfrak{T}\left(\mathcal{F}_{\mathbb{R}^q}\phi\right), \quad \phi \in \mathfrak{S}(\mathbb{R}^q). \tag{2.125}$$

We summarize some properties of the Fourier transformation for tempered distributions (cf. [**190**]):

(i) $\mathfrak{T} \in \mathfrak{S}'(\mathbb{R}^n) \Rightarrow \mathcal{F}_{\mathbb{R}^q}\mathfrak{T} \in \mathfrak{S}'(\mathbb{R}^q)$,

(ii) for regular distributions $\mathfrak{T}_U \in \mathfrak{S}'(\mathbb{R}^q)$, we have $\mathcal{F}_{\mathbb{R}^q}\mathfrak{T}_U = \mathfrak{T}_{\mathcal{F}_{\mathbb{R}^q}}U$,

(iii) $\mathcal{F}_{\mathbb{R}^q}, \mathcal{F}^{-1}_{\mathbb{R}^q} : \mathfrak{S}'(\mathbb{R}^n) \to \mathfrak{S}'(\mathbb{R}^q)$ are bijective, and bicontinuous with

$$\mathcal{F}_{\mathbb{R}^q}\mathcal{F}^{-1}_{\mathbb{R}^q}\mathfrak{T} = \mathcal{F}^{-1}_{\mathbb{R}^q}, \mathcal{F}_{\mathbb{R}^q}\mathfrak{T} = \mathfrak{T} . \tag{2.126}$$

(iv) $\mathfrak{T} \in \mathfrak{S}'(\mathbb{R}^q), \phi \in \mathfrak{S}(\mathbb{R}^q) \Rightarrow \mathcal{F}^{-1}_{\mathbb{R}^q}\mathfrak{T}(\phi(\xi)) = \mathcal{F}_{\mathbb{R}^q}\mathfrak{T}(\phi(-\xi))$.

The Fourier transform can also be used to characterize Sobolev spaces (see, e.g., [**190**]).

Let $\mathcal{G} \subset \mathbb{R}^q$, $q \in \mathbb{N}$, be a bounded region:

(i) For $u \in H^k(\mathbb{R}^q)$, $k \ge 0$, there exist $C', C'' > 0$ such that

$$C' \|U\|^2_{H^k(\mathbb{R}^q)} \le \int_{\mathbb{R}^q} (1+|a|)^{2k} |\mathcal{F}_{\mathbb{R}^q}U(a)|^2\, da \le C'' \|U\|^2_{H^k(\mathbb{R}^q)} .$$

(ii) If there exists a continuous operator $Q : H^k(\mathcal{G}) \to H^k(\mathbb{R}^q)$ with $QU|\mathcal{G} = U$ for all $U \in H^k(\mathcal{G})$, the space $H^k(\mathcal{G})$ coincides with the restriction of functions in $H^k(\mathbb{R}^q)$ on $\mathcal{G}$ and the norm

$$\|U\|_{H^k(\mathcal{G})} = \inf_{\substack{\widetilde{U} \in H^k(\mathbb{R}^q) \\ \widetilde{U}|\mathcal{G}}} \left\| (1 + |\cdot|)^k \, \mathcal{F}_{\mathbb{R}^q} \widetilde{U} \right\|_{L^2(\mathbb{R}^q)} \tag{2.127}$$

is equivalent to the norm in $H^k(\mathcal{G})$.

For a region $\mathcal{G}$ to satisfy an extension condition as above, it is sufficient that $\mathcal{G}$ is bounded and has the uniform cone property. This result is due to Calderon and Zygmund (see [**866**]).

Part 3

Regularization methodologies

CHAPTER 3

Matricial methodologies of resolution

Since any numerical approximation procedure usually leads to finite-dimensional problems, we begin with the recapitulation of finite systems of linear equations (see, e.g., A. Ben-Israel, T.N.E. Greville [**77**], G.H. Golub, C.F. van Loan [**367**], M.Z. Nashed [**618**], H. Engl et al. [**215**] and the list of references therein for more details). After having treated the finite-dimensional situation, we turn to a corresponding theory in an infinite-dimensional operator framework using functional analytic means.

3.1. Least squares method and pseudoinverse

We start with a linear matrix equation of the form

$$Ax = y, \tag{3.1}$$

with $A \in \mathbb{K}^{n\times n}$ ($\mathbb{K} = \mathbb{R}$ or $\mathbb{K} = \mathbb{C}$) being a hermitian positive definite matrix with n rows and n columns. From the spectral theory it is well-known that there exist eigenvalues $0 < \lambda_1 \leq \ldots \leq \lambda_n$ and a corresponding unitary matrix $U = (u_1, \ldots, u_n)$ of eigenvectors $u_i \in \mathbb{C}^n\backslash\{0\}$ (i.e., $u_i^H u_j = u_i \cdot u_j = \delta_{i,j}$) such that A has a representation of the form

$$U^H A U = \underbrace{\begin{pmatrix} \lambda_1 & & & 0 \\ & \ddots & & \\ & & \ddots & \\ 0 & & & \lambda_n \end{pmatrix}}_{n\times n} = \operatorname{diag}(\lambda_1, \ldots, \lambda_n). \tag{3.2}$$

The *condition number* of A is given by the quotient of the largest and smallest eigenvalue, i.e., $\kappa = \frac{\lambda_n}{\lambda_1}$ (note that $\lambda_1 > 0$). For the sake of simplicity and coherence with the analysis for the infinite-dimensional case, we may assume here that the scaling is such that $\lambda_n = 1$, so that $\kappa = \lambda_1^{-1}$. The condition number is a measure for the stable solvability of the problem (3.1).

Assume that we have *noisy data* y^ε instead of y, which satisfy the condition

$$\|y^\varepsilon - y\| \leq \varepsilon \tag{3.3}$$

in the Euclidean norm on $\mathbb{K}^n$. Let x^ε denote the solution with right-hand side y^ε. Then it follows from the spectral representation that

$$x^\varepsilon - x = U \operatorname{diag}\left(\frac{1}{\lambda_1}, \ldots, \frac{1}{\lambda_n}\right) U^H (y^\varepsilon - y). \tag{3.4}$$

Hence, observing the orthogonality of eigenvectors we are led to the estimate

$$\|x^\varepsilon - x\|^2 = \sum_{i=1}^{n} \lambda_i^{-2} |u_i^H (y^\varepsilon - y)|^2 \le \lambda_1^{-2} \|y^\varepsilon - y\|^2. \tag{3.5}$$

In other words, we have

$$\|x^\varepsilon - x\| \le \kappa \|y^\varepsilon - y\| \le \kappa\, \varepsilon. \tag{3.6}$$

The sharpness of this estimate can be immediately seen for $y^\varepsilon - y = \varepsilon u_1$. It is clear that with increasing condition number of the matrix A, the noise amplification increases in the worst-case. For large κ one speaks of an *"ill-conditioned problem"* (ICP). However, it should be remarked that a finite-dimensional linear problem is never ill-posed (in the sense that the third condition in Hadamard's classification is violated), but for κ large one certainly comes close to this case.

We also observe that errors in low frequencies (i.e., corresponding to eigenvectors with large eigenvalues) are amplified less. Following our nomenclature we see that an error in the lowest frequency, i.e.,

$$y^\varepsilon - y = \varepsilon\, u_n \tag{3.7}$$

is not amplified at all. In fact, we just obtain $\|x^\varepsilon - x\| = \varepsilon$ from the spectral representation. This is a typical effect for inverse problems. It means that not all possible versions of noise of the same size are equally bad, high-frequency noise corresponding to low eigenvalues is always worse than low-frequency noise. However, in practice, we are able to make any assumption on the noise only in rare exceptions, so that a regularization method has to deal with arbitrary noise.

Until now, we have assumed that the matrix A is hermitian positive definite, i.e., the minimal eigenvalue is positive. If this is not the case, the matrix has a nontrivial nullspace. If λ_r denotes the minimal nonzero eigenvalue, then the solution formula becomes $x = \sum_{i=r}^{n} \lambda_i^{-1} u_i u_i^H y$, and the problem is solvable if and only if $u_i^H y = 0$ for $i < r$. If the data set is noisy, i.e, instead of y we have y^ε, we are led to use the projection Py^ε onto the range of A. In doing so we obtain for the corresponding solution x^ε with data Py^ε that $x^\varepsilon - x = \sum_{i=r}^{n} \lambda_i^{-1} u_i u_i^H (Py^\varepsilon - y)$. Since $u_i^H P y^\varepsilon = u_i^H y^\varepsilon$ for $i \ge r$ we thus can estimate similarly as described above $\|x^\varepsilon - x\| \le \lambda_r\, \varepsilon$. Consequently, there is no error propagation in the nullspace components and the noise amplification is actually determined by the minimal nonzero eigenvalue.

Let $A \in \mathbb{K}^{n\times m}$ ($\mathbb{K} = \mathbb{R}$ or $\mathbb{K} = \mathbb{C}$) be a matrix with n rows and m columns, $y \in \mathbb{K}^m$. Note that A is not required to be square, no rank assumptions are made at this stage. Then the linear system

$$Ax = y, \quad x \in \mathbb{K}^n,\ y \in \mathbb{K}^m \tag{3.8}$$

need not have a (unique) solution. If the system (3.8) is unsolvable, a reasonable generalized notion of a solution is a " least squares solution", which minimizes the residual $Ax - y$ in the Euclidean norm (note that $\|\cdot\|$ stands for the Euclidean norm in this section): A vector $x \in \mathbb{K}^n$ is a

(1) *least squares solution of* (3.8) if and only if

$$\|Ax - y\| = \inf\{\|Az - y\| : z \in \mathbb{K}^n\}, \tag{3.9}$$

(2) *best-approximate solution (or minimal norm solution) of* (3.8) if and only if x is a least squares solution and, in addition, satisfies

$$\|x\| = \inf\{\|z\| : z \text{ is a least squares solution}\}. \tag{3.10}$$

The following results are well-known from classical Linear Algebra: A vector x is a solution of (3.9) if and only if the *normal equations*

$$A^H A x = A^H y \tag{3.11}$$

are satisfied, where A^H is the hermitian transpose matrix of A (in the case of $\mathbb{K} = \mathbb{R}$ also denoted by A^T). The problem (3.9) possesses a unique solution if and only if A has full rank. If $A^\dagger \in \mathbb{K}^{m \times n}$ may be understood as the matrix which assigns to each $y \in \mathbb{K}^m$ the best-approximate solution of (3.8), then it is called the *pseudoinversee (also designated, Moore-Penrose inverse or generalized inverse) of A*.

3.2. Singular value decomposition

In order to construct $A^\dagger$ and, hence, best-approximate solutions via the so-called *singular value decomposition (SVD) of A* we recall the definition of *singular values of a matrix A*:

Let $\sigma_1, \ldots, \sigma_r > 0$ be such that $\sigma_1^2 \geq \sigma_2^2 \geq \ldots \geq \sigma_r^2 > 0$ are the positive eigenvalues of the matrix $A^H A$ (each one written down as often as its multiplicity is). Then $\sigma_1, \ldots, \sigma_r$ are denoted the (nonzero) singular values of A.

This setting makes sense, since the matrix $A^H A$ is positive semidefinite. Obviously, $r \leq \min\{n, m\}$, where r is the rank of A. We know that a hermitian matrix can be diagonalized, where the diagonal elements are its eigenvalues. The following theorem generalizes this result to the nonhermitian case.

Let $\sigma_1 \geq \sigma_2 \geq \ldots \geq \sigma_r > 0$ be the singular values of A. Then there exist unitary matrices $U \in \mathbb{K}^{m \times m}$ and $V \in \mathbb{K}^{n \times n}$ such that

$$V^H A U = \underbrace{\begin{pmatrix} \sigma_1 & & & & & 0 \\ & \ddots & & & & \\ & & \sigma_r & & & \\ & & & 0 & & \\ & & & & \ddots & \\ 0 & & & & & 0 \end{pmatrix}}_{m \times n} \tag{3.12}$$

holds true. The columns of U and V are eigenvectors of $A^H A$ and AA^H, respectively. The expression (3.12) *is the so-called singular value decomposition (SVD) of A.*

The singular value decomposition (SVD) is not unique, since the unitary matrices U and V are not. Obviously, from (3.12), we obtain

$$A = V \begin{pmatrix} \sigma_1 & & & & & 0 \\ & \ddots & & & & \\ & & \sigma_r & & & \\ & & & 0 & & \\ & & & & \ddots & \\ 0 & & & & & 0 \end{pmatrix} U^H, \tag{3.13}$$

since V and U are invertible, the rank of A is r, the number of nonzero singular values (counted with multiplicity). Note that with $U = (u_1, \dots, u_m)$, $V = (v_1, \dots v_n)$, we have, for $i \in \{1, \dots, r\}$,

$$Au_i = \sigma_i v_i \tag{3.14}$$

and

$$A^H v_i = \sigma_i u_i, \tag{3.15}$$

which follows from the singular value decomposition (3.12) via multiplication by V and U^H, respectively. The system $\{\sigma_i; u_i, v_i : i \in \{1, \dots, r\}\}$ is the so-called *singular system* for A. The system $\{v_1, \dots, v_r\}$ is an orthonormal basis for $\mathcal{R}(A) = \{y : Ax = y\}$ for any $x \in \mathbb{K}^n$,

$$Ax = \sum_{i=1}^{r} \langle Ax, v_i \rangle v_i = \sum_{i=1}^{r} \langle x, A^H v_i \rangle v_i, \tag{3.16}$$

which implies, in connection with (3.15), that

$$Ax = \sum_{i=1}^{r} \sigma_i \langle x, u_i \rangle \; v_i \tag{3.17}$$

holds true. Analogously, for all $y \in \mathbb{K}^m$, we have

$$A^H y = \sum_{i=1}^{r} \sigma_i \langle y, v_i \rangle \; u_i. \tag{3.18}$$

Note that if A has real entries, so U and V have (we remark that the notion of a singular system and the expansions (3.17) and (3.18) generalize to compact operators on infinite-dimensional spaces, e.g., integral operators, as we will see later on).

Let A have the SVD (3.12). *Then*

$$A^\dagger = U \underbrace{\begin{pmatrix} \frac{1}{\sigma_1} & & & & & 0 \\ & \ddots & & & & \\ & & \frac{1}{\sigma_r} & & & \\ & & & 0 & & \\ & & & & \ddots & \\ 0 & & & & & 0 \end{pmatrix}}_{n \times m} V^H. \tag{3.19}$$

This also implies the existence and uniqueness of a best-approximate solution. Since $A^\dagger y$ is the least squares solution of minimal norm, we obtain that $A^\dagger y$ is a solution of the normal equations $A^H Ax = A^H y$ with minimal norm, i.e., $A^\dagger y = (A^H A)^\dagger A^H y$. This means that in order to approximate $A^\dagger y$ we may as well compute an approximation to the minimal-norm solution in the normal equations, a fact we will heavily use in the construction of regularization methods, later on.

3.3. Regularization methods

The decomposition (3.19), more concretely, the identity

$$A^\dagger y = \sum_{i=1}^{r} \frac{\langle y, v_i\rangle}{\sigma_i} u_i \tag{3.20}$$

also shows how errors in y affect the result $A^\dagger y$: Errors in components of y corresponding to small singular values are amplified by the large factor of the singular value, so that such data errors are dangerous. This explains the numerical instability of (3.20), if A has small singular values. Although the problem of computing the best-approximate solution is well-posed, it shows numerical unstability.

Truncated singular value regularization. The first idea to reduce the instability of computing the best-approximate solution is to replace (3.20) by

$$x_\alpha = \sum_{\substack{i=1\\ \sigma_i^2\geq a}}^{r} \frac{\langle y, v_i\rangle}{\sigma_i} u_i \tag{3.21}$$

with an appropriately chosen value $\alpha > 0$; this truncation is the first example of a regularization, where the original problem is replaced by a neighboring one, which is more stable. However, the choice of the "regularization parameter" α is quite crucial.

If we use (3.21) with perturbed data y^ε (such that $\|y - y^\varepsilon\| \leq \varepsilon$), we obtain as the "regularized solution"

$$x_\alpha^\varepsilon = \sum_{\substack{i=1\\ \sigma_i^2\geq\alpha}}^{r} \frac{\langle y^\varepsilon, v_i\rangle}{\sigma_i} u_i. \tag{3.22}$$

We estimate the total error between x_α^ε and the sought-for quantity $A^\dagger y$, thereby observing that the elements u_i are orthonormal:

$$\begin{aligned}\|x_\alpha^\varepsilon - A^\dagger y\|^2 &= \left\| \sum_{\substack{i=1\\ \sigma_i^2\geq\alpha}}^{r} \frac{\langle y^\varepsilon, v_i\rangle}{\sigma_i} u_i - \sum_{i=1}^{r} \frac{\langle y, v_i\rangle}{\sigma_i} u_i \right\|^2 \\ &= \left\| \sum_{i=1}^{r} \frac{\langle y, v_i\rangle}{\sigma_i} u_i - \sum_{\substack{i=1\\ \sigma_i^2\geq\alpha}}^{r} \frac{\langle y, v_i\rangle}{\sigma_i} u_i \right\|^2 + \left\| \sum_{\substack{i=1\\ \sigma_i^2\geq\alpha}}^{r} \frac{\langle y, v_i\rangle - \langle y^\varepsilon, v_i\rangle}{\sigma_i} u_i \right\|^2 .\end{aligned}$$

Now, we have
(3.23)

$$\left\| \sum_{i=1}^{r} \frac{\langle y, v_i \rangle}{\sigma_i} u_i - \sum_{\substack{i=1 \\ \sigma_i^2 \geq \alpha}}^{r} \frac{\langle y, v_i \rangle}{\sigma_i} u_i \right\|^2 = \left\| \sum_{\substack{i=1 \\ \sigma_i^2 < \alpha}}^{r} \frac{\langle y, v_i \rangle}{\sigma_i} u_i \right\|^2 = \sum_{\substack{i=1 \\ \sigma_i^2 < \alpha}}^{r} \frac{|\langle y, v_i \rangle|^2}{\sigma_i^2} \to 0$$

for $\alpha \to 0$. Hence, for sufficiently small α, the last sum is empty such that
(3.24)

$$\left\| \sum_{\substack{i=1 \\ \sigma_i^2 \geq \alpha}}^{r} \frac{\langle y, v_i \rangle - \langle y^\varepsilon, v_i \rangle}{\sigma_i} u_i \right\|^2 = \sum_{\substack{i=1 \\ \sigma_i^2 \geq \alpha}}^{r} \frac{|\langle y - y^\varepsilon, v_i \rangle|^2}{\sigma_i^2} \leq \frac{1}{\alpha} \sum_{i=1}^{r} |\langle y - y^\varepsilon, v_i \rangle|^2 \leq \frac{\varepsilon^2}{\alpha}.$$

The second error term does not blow up as $\alpha \to 0$, since the sum has always at most r terms, hence, it can be estimated by $\varepsilon^2 \left(\min\{\sigma_i : i \in \{1, \ldots, r\}\}\right)^{-1}$. The sum (3.22) is called *truncated singular value expansion.* It can be interpreted as applying a lowpass filter to the input vector.

Tikhonov regularization. Another way of making (3.20) more stable would be to replace it by the sum

$$x_\alpha^\varepsilon = \sum_{i=1}^{r} \frac{\sigma_i}{\sigma_i^2 + \alpha} \langle y^\varepsilon, v_i \rangle u_i, \quad \alpha > 0. \tag{3.25}$$

This is a classical variant of the famous *Tikhonov regularization method* for matrix equations, which we shall consider in more detail in infinite dimensions. It is helpful to characterize it in a different way: Let x_α^ε be defined by (3.25). Then, by the orthonormality of the u_i, we have, for all $j \in \{1, \ldots, r\}$,

$$\langle x_\alpha^\varepsilon, u_j \rangle = \frac{\sigma_j}{\sigma_j^2 + \alpha} \langle y^\varepsilon, v_j \rangle \tag{3.26}$$

and

$$\sigma_j^2 \langle x_\alpha^\varepsilon, u_j \rangle + \alpha \langle x_\alpha^\varepsilon, u_j \rangle = \sigma_j \langle y^\varepsilon, v_j \rangle. \tag{3.27}$$

Now, because of (3.14) and (3.18), it follows that

$$\sum_{j=1}^{r} \sigma_j \langle y^\varepsilon, v_j \rangle u_j = \sum_{j=1}^{r} \langle y^\varepsilon, A u_j \rangle u_j = \sum_{j=1}^{r} \langle A^H y^\varepsilon, u_j \rangle u_j = A^H y^\varepsilon \tag{3.28}$$

and

$$\begin{aligned} \sum_{j=1}^{r} \left(\sigma_j^2 \langle x_\alpha^\varepsilon, u_j \rangle + \alpha \langle x_\alpha^\varepsilon, u_j \rangle\right) u_j &= \sum_{j=1}^{r} \left(\langle x_\alpha^\varepsilon, A^H A u_j \rangle + \alpha \langle x_\alpha^\varepsilon, u_j \rangle\right) u_j \\ &= A^H A x_\alpha^\varepsilon + \alpha x_\alpha^\varepsilon, \end{aligned} \tag{3.29}$$

note that x_α^ε is in the linear span of $\{u_1, \ldots, u_r\}$ which follows from (3.25). Now, the identity (3.27) implies in vector nomenclature that

$$\left(A^H A + \alpha I\right) x_\alpha^\varepsilon = A^H y^\varepsilon, \tag{3.30}$$

which is an alternative characterization of the *Tikhonov regularization.* From this "*regularized normal equation*", we can obtain still another characterization of x_α^ε, namely as the unique minimizer of the so-called *Tikhonov functional*

$$x \mapsto \|Ax - y^\varepsilon\|^2 + \alpha\|x\|^2, \tag{3.31}$$

which can be seen by putting the first derivative of the functional in (3.31) to 0, resulting exactly in the linear equation (3.30). The minimization of (3.31) can be seen as a combination of the two minimizations that appear in the definition of a best-approximate solution. It has also interpretations as a penalty method, e.g., via Lagrange multipliers.

The computation explained above can be formally performed for $\alpha = 0$, too. In this case it shows that x is the solution of minimal norm of the normal equation $A^H Ax = A^H y$ which was already attacked by C.F. Gauss [**345**]. If $A^H A$ is invertible (and hence positive definite), the normal equation $A^H Ax = A^H y$ can be solved by standard Cholesky decomposition, which leads to an alternative method for computing best-approximation solutions, for which no SVD is needed. However, forming $A^H A$ may seriously decrease the numerical stability. Hence, it should be avoided in ill-conditioned cases.

CHAPTER 4

Compact operator methodologies of resolution

Next we go over to operator equations: Let X and Y be Hilbert spaces and let $A : X \to Y$ be a bounded linear operator whose range $\mathcal{R}(A)$ is not necessarily closed. Then we have the orthogonal decompositions

$$X = \mathcal{N}(A) \oplus \mathcal{N}(A)^{\perp}, \tag{4.1}$$

$$Y = \overline{\mathcal{R}(A)} \oplus \mathcal{R}(A)^{\perp}, \tag{4.2}$$

and

$$\mathcal{N}(A^*) = \mathcal{R}(A)^{\perp}, \tag{4.3}$$

where $\mathcal{N}(A)^{\perp}$ is the orthogonal complement of $\mathcal{N}(A)$, $\overline{\mathcal{R}(A)}$ is the closure of the range of A, and A^* is the adjoint operator of A.

4.1. Least squares method and pseudoinverse

We consider the operator equation

$$Ax = y, \quad x \in X,\ y \in Y, \tag{4.4}$$

in more detail: Four (mutually exclusive) situations arise (cf. [**625**]):

(1) $\mathcal{R}(A)$ is dense in Y, hence, $\mathcal{N}(A^*) = \{0\}$, and $y \in \mathcal{R}(A)$;

(2) $\mathcal{R}(A)$ is dense in Y, and $y \notin \mathcal{R}(A)$;

(3) $\overline{\mathcal{R}(A)}$ is a proper subspace of Y, and $y \in \mathcal{R}(A) + \mathcal{R}(A)^{\perp}$;

(4) $\overline{\mathcal{R}(A)} \neq Y$, and $y \notin \mathcal{R}(A) + \mathcal{R}(A)^{\perp}$.

In case (1) one has, of course, a solution in the classical sense; in case (2) and case (4) a classical solution does not exist, while in case (3) a solution need not exist.

We say x is a *"least-squares solution" of* (4.4) if

$$\inf\{\|Au - y\| : u \in X\} = \|Ax - y\|. \tag{4.5}$$

Since

$$\|Au - y\|^2 = \|Au - Qy\|^2 + \|y - Qy\|^2, \tag{4.6}$$

where Q is the orthogonal projector of Y onto $\overline{\mathcal{R}(A)}$, it is clear that a least-squares solution exists if and only if

$$y \in \mathcal{R}(A) + \mathcal{R}(A)^{\perp}, \tag{4.7}$$

where $\mathcal{R}(A) + \mathcal{R}(A)^{\perp}$ is a dense set in Y. For such y the set of all least-squares solutions of (4.4), denoted by $\mathcal{L}(y)$, is a nonempty closed convex set (indeed $\mathcal{L}(y)$ is the translate of $\mathcal{N}(A)$ by a fixed element of $\mathcal{N}(y)$), hence, it has a unique element of minimal norm, denoted by $A^{\dagger}y$. The *pseudoinverse (or generalized inverse)* $A^{\dagger}$ is the linear operator which assigns to each $y \in \mathcal{D}(A^\dagger) = \mathcal{R}(A) + \mathcal{R}(A)^\perp$, the unique element in $\mathcal{L}(y) \cap \mathcal{N}(A)^\perp$, so that

$$\mathcal{L}(y) = A^\dagger y + \mathcal{N}(A). \tag{4.8}$$

It is easy to show that $A^\dagger y$ is the minimal norm solution (equivalently the unique solution in $\mathcal{N}(A)^\perp$) of the *normal equation*

$$A^* A x = A^* y \tag{4.9}$$

(the equation obtained by setting the first variation of $\|Ax - y\|^2$ equal to zero). It also follows that $A^\dagger = (A|\mathcal{N}(A)^\perp)^{-1} Q$ so that $A^\dagger$ ischaracterized as the linear operator with the properties:

$$\mathcal{D}(A^\dagger) = \mathcal{R}(A) + \mathcal{R}(A)^\perp, \quad \mathcal{N}(A^\dagger) = \mathcal{R}(A)^\perp = \mathcal{N}(A^*) \tag{4.10}$$

and

$$\mathcal{R}(A^\dagger) = \mathcal{N}(A)^\perp. \tag{4.11}$$

The equivalence of these characterizations of $A^\dagger$ is established, e.g., in M.Z. Nashed [**613**] (see also [**379**], [**627**] for a lucid exposition and [**639**], [**627**] for the generalization to unbounded operators).

In case (1) above, $A^\dagger$ gives the minimal-norm solution of (3). In case (3), Equation (4.4) has a least-squares solution (which is unique if and only if $\mathcal{N}(A) = \{0\}$). In both cases the infimum in (4.5) is attained and is equal to zero and $\|y - Qy\|$, respectively. Case (2) and case (4) are pathological and usually are not under discussion in inverse theory, since in both cases $y \notin \mathcal{D}(A^\dagger)$, and the infimum in (4.5) is not attained.

Weighted least-squares problems. Very often we are interested in weighted minimal-norm least-squares solutions. Let $\mathcal{L}_A(y)$ be the set of all least-squares solutions of $Ax = y$, where A is a member of the *class* $\mathfrak{L}(X, Y)$ *of all bounded linear operators from* X *into* Y. Let Z be a Hilbert space and $L : \mathcal{D}_L \subset X \to Z$ be a closed linear operator with dense domain and closed range. For $y \in \mathcal{D}(A^\dagger)$, we consider the following problem: Find $w \in \mathcal{L}_A(y)$ such that

$$\|Lw\| \leq \|Lu\| \quad \text{for all } u \in \mathcal{L}_A(y). \tag{4.12}$$

If $L(\mathcal{N}(A))$ is closed and $\mathcal{N}(A) \cap \mathcal{N}(L) = \{0\}$, then the relation (4.12) has a unique solution $w(y)$. We denote by $A_L^\dagger$ the linear map induced by $y \to w(y)$ and call it the *weighted generalized inverse* of A. We define a new inner product and norm on $\mathcal{D}(L)$ by

$$[u, v]_L = (Au, Av) + (Lu, Lv) \tag{4.13}$$

with

$$\|u\|_L = \sqrt{[u, u]_L}. \tag{4.14}$$

We denote the space $\mathcal{D}(L)$ with this new inner product by X_L. It is easy to show that $A_L^\dagger y$ is the least-squares solution of X_L-minimal norm of $Ax = y$. Finally, let

$$\mathcal{M} = \{x \in X : L^* L x \in \mathcal{N}(A)^\perp\}. \tag{4.15}$$

Then $\mathcal{M}$ is the orthogonal complement of $\mathcal{N}(A)$ with respect to $[\cdot, \cdot]_L$, and $\mathcal{R}(A_L^\dagger) = \mathcal{M}$.

4.2. Singular value decomposition

Let $(X, \|\cdot\|_X)$ be a normed space. Suppose that A is of class $\mathfrak{L}(X)$. We are concerned with the introduction of eigenvalues and eigenfunctions corresponding to an operator $A \in \mathfrak{L}(X)$.

(1) $\lambda \in \mathbb{C}$ is called a *regular value* of A if and only if $\lambda I - A$ is continuously invertible. $\rho(A) = \{\lambda \in \mathbb{C} : \lambda I - A \text{ continuously invertible}\}$ is called the *resolvent set.*
(2) $\sigma(A) = \mathbb{C} \setminus \rho(A)$ is called the *spectrum* of A.
(3) $\lambda \in \sigma(A)$ is called an *eigenvalue* of A if $\mathcal{N}(\lambda I - A) \neq \{0\}$. The elements of $\mathcal{N}(\lambda I - A) \setminus \{0\}$ are called *eigenvectors* of A corresponding to the eigenvalue λ.

The following results are standard for a Banach space X (see, e.g., [**429**]):
(a) If $\lambda \in \sigma(A)$, then $|\lambda| \leq \|A\|$, i.e., the spectrum is bounded.
(b) $\sigma(A) \subset \mathbb{C}$ is compact.

Let X be a normed space. Suppose that A is a compact operator on X (i.e., $A \in \mathfrak{K}(X)$).
(1) If $\lambda \in \sigma(A) \setminus \{0\}$, then λ is an eigenvalue of A.
(2) If λ is an eigenvalue of A, then $\dim \mathcal{N}(\lambda I - A) < \infty$.
(3) $\sigma(A)$ is at most countable. Furthermore, $0 \in \sigma(A)$.
(4) 0 is the only accumulation point of $\sigma(A)$.

Singular value decomposition. Central in our considerations about compact operator is the following *spectral theorem for compact self-adjoint operators* that can be seen in parallel to the finite-dimensional case of matrix operators:

Let X be a Hilbert space. Assume that A is of class $\mathfrak{K}(X)$ and that A is self-adjoint, i.e., $A^* = A$. Then there exists an orthonormal system $\{x_i\}_{i\in\mathbb{N}} \subset X$ and a sequence $\{\mu_i\}_{i\in\mathbb{N}} \subset \mathbb{R}$ (finite or countable) with $|\mu_1| \geq |\mu_2| \geq \ldots > 0$, such that

$$Ax = \sum_{i=1}^{\infty} \mu_i \langle x, x_i \rangle x_i \tag{4.16}$$

holds true for all $x \in X$ (once again, note that we usually write $\langle x, x_i\rangle$ instead of $\langle x, x_i\rangle_X$ if no confusion is likely to arise).

Suppose that X and Y are Hilbert spaces. Furthermore, let A be of class $\mathfrak{K}(X,Y)$. Then A^*A is also compact and obviously self-adjoint. Due to the spectral theorem there exist a sequence $\{\lambda_i\}_{i\in\mathbb{N}} \subset \mathbb{R}$ and an orthonormal system $\{x_i\}_{i\in\mathbb{N}} \subset X$ such that

$$A^*Ax = \sum_{i=1}^{\infty} \lambda_i \langle x, x_i \rangle x_i, \quad x \in X. \tag{4.17}$$

Suppose that $\lambda_i \in \sigma(A^*A) \setminus \{0\}$ and denote, as usual, by x_i its corresponding eigenvector. It follows that

$$\lambda_i \|x_i\|^2 = \lambda_i \langle x_i, x_i \rangle = \langle \lambda_i x_i, x_i \rangle_X = \langle A^*Ax_i, x_i \rangle_X = \langle Ax_i, Ax_i \rangle_Y = \|Ax_i\|_Y^2 . \tag{4.18}$$

Therefore we are able to conclude that $\lambda_i > 0$.

Singular values. In the sequel, we assume that the eigenvalues are listed in the chronological order as follows:

$$\lambda_1 \geq \lambda_2 \geq \cdots \geq \lambda_i \geq \lambda_{i+1} \geq \cdots \geq 0. \tag{4.19}$$

Set $\sigma_j = \sqrt{\lambda_j}$. Moreover, let $y_i = \frac{1}{\sigma_i} A x_i$ i.e., $Ax_i = \sigma_i y_i$, $i \in \mathbb{N}$, and

$$A^* y_i = A^* \left(\frac{1}{\sigma_i} A x_i \right) = \frac{1}{\sigma_i} A^* A x_i = \frac{1}{\sigma_i} \lambda_i x_i = \sigma_i x_i. \tag{4.20}$$

It is not hard to see that

$$\begin{aligned} \langle y_i, y_k \rangle_Y &= \frac{1}{\sigma_i \sigma_k} \langle A x_i, A x_k \rangle_Y = \frac{1}{\sigma_i \sigma_k} \langle A^* A x_i, x_k \rangle_X = \frac{1}{\sigma_i \sigma_k} \langle \lambda_i x_i, x_k \rangle_X \\ &= \frac{\lambda_i}{\sigma_i \sigma_k} \langle x_i, x_k \rangle_X = \frac{\sigma_i}{\sigma_k} \delta_{i,k} = \delta_{i,k}. \end{aligned}$$

Thus, $\{y_i\}_{i \in \mathbb{N}} \subset Y$ forms a complete orthonormal system (ONS) in $\overline{\mathcal{R}(A)}$, so that the system $\{x_i\}_{i \in \mathbb{N}}$ is a complete ONS in $\mathcal{N}(A)^{\perp}$. Now, assume that x is a member of $\mathcal{N}(A)^{\perp}$. Then it follows that

$$x = \sum_{i=1}^{\infty} \langle x, x_i \rangle_X \, x_i \tag{4.21}$$

and

$$Ax = \sum_{i=1}^{\infty} \langle x, x_i \rangle_X \, A x_i = \sum_{i=1}^{\infty} \sigma_i \langle x, x_i \rangle_X \, y_i \tag{4.22}$$

for all $x \in \mathcal{N}(A)^{\perp}$.

Let X, Y be Hilbert spaces. The set $\{\sigma_i; x_i, y_i\}_{i \in \mathbb{N}} \subset (0, \infty) \times X \times Y$ is called the singular system of an operator $A \in \mathfrak{K}(X, Y)$. The values σ_i are called the singular values of A. The elements x_i, y_i are called the singular vectors. Furthermore, the series

$$Ax = \sum_{i=1}^{\infty} \sigma_i \langle x, x_i \rangle_X \, y_i, \quad x \in X \tag{4.23}$$

is called the singular value decomposition (SVD) of A.

Picard-condition. The following Picard condition plays an essential role in the solvability of inverse problems:

If $A: X \to Y$ is compact with singular value decomposition (SVD) $\{\sigma_i; x_i, y_i\}_{i \in \mathbb{N}}$, then $y \in \overline{\mathcal{R}(A)}$ is an element of $\mathcal{R}(A)$ if and only if

$$\sum_{i=1}^{\infty} \frac{|\langle y, y_i \rangle|^2}{\sigma_i^2} \tag{4.24}$$

is convergent.

Obviously, from the Picard condition, it follows that $\sigma_i^{-2} |\langle y, y_i \rangle|^2 \to 0$ if $i \to \infty$, so that information about the decay of the Fourier coefficients of an element y becomes available.

Let A be of class $\mathfrak{K}(X,Y))$ with SVD $\{\sigma_i; x_i, y_i\}_{i\in\mathbb{N}}$. Then the generalized inverse (or pseudoinverse) of a compact operator is representable in the form

$$A^\dagger y = \sum_{i=1}^{\infty} \frac{1}{\sigma_i} \langle y, y_i\rangle x_i \tag{4.25}$$

for all $y \in \mathcal{D}(A^\dagger)$ (if $\mathcal{R}(A)$ is finite dimensional, i.e., there exist only finitely many element y_i, then $\mathcal{R}(A^\dagger)$ is finite. Therefore, $A^\dagger$ is compact. In particular, $A^\dagger$ is continuous).

The representation of the generalized inverse in terms of the singular value decomposition (4.25) opens the perspective to classify ill-posed problems. Indeed, the summands $\sigma_i^{-1}\langle y, y_i\rangle x_i$ occurring in the series (4.25) depend closely on the singular values. If the values σ_i are small, then the contribution by the series (4.25) becomes large. The existence of the SVD can be guaranteed for all compact operators. Nevertheless, the concrete knowledge of the SVD is critical. Only in rare exceptions, SVD is explicitly known in practice. All in all, singular systems are theoretically nice, but their calculation might be rather tricky.

4.3. Regularization methods

Obviously, the first two criteria (H1) and (H2) determining a well-posed problem in the sense of Hadamard, can always be enforced by considering the generalized inverse $A^\dagger$. A violating the third criterion, means that instability arises if the spectrum of the operator A is not bounded away from zero. Thus, it seems to be natural to construct regularizing approximations via modifying the smallest singular values. In accordance with the singular value decomposition of the generalized inverse, it follows that such a modification of small values and, hence, a construction of regularization operators can be obtained in the form

$$x_\alpha = R_\alpha y = \sum_{i=1}^{\infty} \sigma_i F_\alpha(\sigma_i^2)\langle y, y_i\rangle x_i \quad y \in Y, \tag{4.26}$$

with some function $F_\alpha : \mathbb{R}_+ \to \mathbb{R}_+$ such that

$$F_\alpha(\lambda) \to \frac{1}{\lambda}, \quad \lambda > 0,\ \alpha \to 0. \tag{4.27}$$

Such an operator R_α as defined by (4.26) and (4.27) may be understood as a *regularization operator* if

$$\lambda |F_\alpha(\lambda)| \le C_{F_\alpha} < \infty, \quad \lambda > 0. \tag{4.28}$$

If (4.28) is satisfied, then we are able to see that

$$\|R_\alpha y\|^2 = \sum_{i=1}^{\infty} \sigma_i^2 (F_\alpha(\sigma_i))^2 |\langle y, y_i\rangle|^2 \le C_{F_\alpha}^2 \sum_{i=1}^{\infty} |\langle y, y_i\rangle|^2 \le C_{F_\alpha}^2 \|y\|^2, \tag{4.29}$$

where C_{F_α} is a bound for the norm of R_α. Note that the pointwise convergence of F_α immediately implies the pointwise convergence of R_α to $A^\dagger$.

Truncated singular value decomposition. Let A be of class $\mathfrak{K}(X,Y)$ with SVD $\{\sigma_i; x_i, y_i\}_{i\in\mathbb{N}}$. The main idea of a truncated singular value decomposition is to ignore all singular values below a certain threshold value, which we can identify with the regularization parameter α, hence, the representation of the regularized solution is given by

$$F_\alpha(\lambda) = \begin{cases} \frac{1}{\lambda} & , \quad \lambda \geq \alpha \\ 0 & , \quad \lambda < \alpha \end{cases} \tag{4.30}$$

i.e.,

$$x_\alpha = R_\alpha y = \sum_{\sigma_i \geq \alpha} \frac{1}{\sigma_i} \langle y, y_i \rangle\, x_i, \quad y \in Y, \tag{4.31}$$

which explains the name truncated singular value decomposition, since all terms in the sum corresponding to small singular values are truncated. Since 0 is the only accumulation point of the singular values of a compact operator, the sum in (4.31) is always finite for $\alpha > 0$. In particular, only a finite number of singular values and singular vectors has to be computed in order to realize this method. On the other hand it should be mentioned that, for α being sufficiently small, the number of singular values that need to be computed can increase strongly. Obviously, $C_{F_\alpha} = \alpha^{-1}$.

Lavrentiev regularization. The main idea of this regularization method is to shift all singular values by α, i.e.,

$$F_\alpha(\lambda) = \frac{1}{\sqrt{\lambda}(\sqrt{\lambda}+\alpha)}, \quad \lambda > 0,\ \alpha > 0, \tag{4.32}$$

so that

$$x_\alpha = R_\alpha y = \sum_{i=1}^{\infty} \frac{1}{\sigma_i + \alpha} \langle y, y_i \rangle\, x_i, \quad y \in Y. \tag{4.33}$$

In this case, the sum is really infinite and the full singular system is needed in order to compute the solution. However, if A is a positive definite operator (and, thus, $x_i = y_i$), we obtain

$$(A + \alpha I)\, x_\alpha = \sum_{i=1}^{\infty} \langle y, y_i \rangle\, x_i = y. \tag{4.34}$$

Hence, the regularized solution can be obtained in this case without any knowledge of the singular system as the solution of the linear equation $(A + \alpha I)\, x_\alpha = y$. Clearly, $(\sigma + \alpha)^{-1} \leq \alpha^{-1}$, hence, $C_{F_\alpha} = \alpha^{-1}$.

Tikhonov regularization. The regularized solution is given by

$$F_\alpha(\lambda) = \frac{1}{\lambda + \alpha}, \quad \lambda > 0,\ \alpha > 0, \tag{4.35}$$

so that

$$x_\alpha = R_\alpha y = \sum_{i=1}^{\infty} \frac{\sigma_i}{\sigma_i^2 + \alpha} \langle y, y_i \rangle\, x_i, \quad y \in Y. \tag{4.36}$$

As in the case of Lavrentiev regularization, we can compute x_α defined by (4.36) without any knowledge of the singular system. In fact, it is easy to see that $(A^*A + \alpha I)\, x_\alpha = A^* y$ and, hence, we can solve a well-posed linear system to obtain x_α. From this representation it also follows that Tikhonov regularization is just Lavrentiev regularization applied to the normal equation. It is not hard to see that $\lambda^2 + \alpha \geq 2\lambda\sqrt{\alpha}$, hence, C_{F_α} can be chosen as $2\alpha^{-1/2}$.

Asymptotic regularization. Asymptotic regularization is usually constructed from the solution x of the initial value problem

$$x'(t) = -A^*(Ax(t) - y), \quad t > 0, \tag{4.37}$$

$$x(0) = 0, \tag{4.38}$$

as $x_\alpha = x(\frac{1}{\alpha}\ \cdot)$. By representing x in terms of the singular vectors x_i in the form

$$x(t) = \sum_{i=1}^{\infty} \alpha_i(t)\ x_i \tag{4.39}$$

with $\alpha_i(0) = 0$, we obtain from the singular value decomposition

$$\alpha_i'(t) = -\sigma_i^2 \alpha_i(t) + \sigma_i\ \langle y_i, y\rangle. \tag{4.40}$$

This ordinary differential equation can be solved analytically by

$$\alpha_i(t) = \left(1 - \exp(-\sigma_i^2 t)\right) \frac{1}{\sigma_i} \langle\ y_i, y\rangle. \tag{4.41}$$

Hence, the regularized solution is given by

$$F_\alpha(\lambda) = \left(1 - \exp\left(-\frac{\lambda}{\alpha}\right)\right) \frac{1}{\lambda} \tag{4.42}$$

i.e.,

$$x_\alpha = \sum_{i=1}^{\infty} \left(1 - \exp\left(-\frac{\sigma_i^2}{\alpha}\right)\right) \frac{1}{\sigma_i} \langle y, y_i\rangle\ x_i, \qquad y \in Y. \tag{4.43}$$

The specific amount of the growth of the singular values finally leads us to classify ill posedness

Classification of ill-posed problems for compact operators. Let A be of class $\mathfrak{K}(X,Y))$ with SVD $\{\sigma_i; x_i, y_i\}_{i\in\mathbb{N}}$.

(1) If there exists $\alpha > 0$ such that
$$\sigma_i = O(i^{-\alpha}),$$
then the operator A is called *ill-posed of order* α (geomathematical examples are the operators for gravimetry and magnetometry).

(2) If there exists $\rho > 0$ such that
$$|\ln \sigma_i| \geq c\ i^{\rho},$$
then the operator A is called *exponentially ill-posed* (a geomathematical example is satellite gravitational gradiometry).

Note that the aforementioned classification is senseful only for linear problems reflecting the representation of $A^\dagger y$ by its superposition in terms (4.25).

4.4. Approximation and data error

For the error between $x^\dagger$ and x_α^ε in the case of noisy data y^ε, we are able to write (with $x_\alpha^\varepsilon = R_\alpha y^\varepsilon$)

$$x^\dagger - x_\alpha^\varepsilon = (x^\dagger - x_\alpha) + (x_\alpha - x_\alpha^\varepsilon). \tag{4.44}$$

The first term $x^\dagger - x_\alpha$ is the *approximation error of the regularization method*, which is independent of the noise. The second term $x_\alpha - x_\alpha^\varepsilon$ corresponding to the *propagation of data noise* in the regularized case. By aid of the triangle inequality it follows that

$$\|x^\dagger - x_\alpha^\varepsilon\| \leq \|x^\dagger - x_\alpha\| + \|x_\alpha - x_\alpha^\varepsilon\|. \tag{4.45}$$

Thus, the estimate of the error between the regularized solution and the exact solution, can be handled by two error terms separately. It is clear that such an estimation provides a guideline for the parameter choice, namely by choosing α such that the terms on the right-hand side are balanced.

Next we deal with an *estimate of the approximation* error, which is independent of the noise level ε:

Let $F_\alpha : \mathbb{R}^+ \to \mathbb{R}^+$ be a piecewise continuous function satisfying the assumptions

$$F_\alpha(\lambda) \to \frac{1}{\lambda}, \quad \lambda > 0, \quad \alpha \to 0, \tag{4.46}$$

$$|F_\alpha(\lambda)| \leq M_\alpha < \infty, \quad \lambda > 0, \tag{4.47}$$

and

$$\sup_{\alpha,\lambda}(\lambda F_\alpha(\lambda)) \leq C_F < \infty \tag{4.48}$$

for some constant $C_F > 0$. Moreover, let the regularization operator be defined by (4.26). Then, for all $y \in \mathcal{D}(A^\dagger)$, we have

$$R_\alpha y \to A^\dagger y, \quad \alpha \to 0. \tag{4.49}$$

The function $t \mapsto F_\alpha(t)$, $t \in \mathbb{R}^+$, converges in pointwise sense to the function

$$F(t) = \begin{cases} 0 & , \quad t > 0 \\ 1 & , \quad t = 0. \end{cases} \tag{4.50}$$

Due to the discontinuity at zero, the convergence of $tF_\alpha(t) - 1$ to zero is becoming slower and slower as t decreases to zero. Since it is allowed to specify an arbitrarily small singular value σ_i and the minimal norm solution $x^\dagger = x_i$, the convergence of regularized solutions is arbitrarily slow. On the other hand, we observe that there is a possibly faster convergence if the components $\langle x^\dagger, x_i\rangle$ decay sufficiently fast compared to the eigenvalues. For example, if we have $|\langle x^\dagger, x_i\rangle| \leq c\sigma_i^\mu$ for some constant $c > 0$ and $\mu > 0$, then it follows

$$\begin{aligned}\limsup_{\alpha\to 0} \|R_\alpha y - A^\dagger y\|^2 &\le \limsup_{\alpha\to 0} c^2 \sum_{n=1}^{\infty} (\sigma_i F_\alpha(\sigma_i) - 1)^2 \sigma_i^{2\mu} \\ &\le c^2 \sum_{n=1}^{\infty} \lim_{\alpha} (\sigma_i^{1+\mu} F_\alpha(\sigma_i) - \sigma_i^\mu)^2.\end{aligned} \tag{4.51}$$

In other words, one has to consider the limit of the function $t \mapsto |t^{1+\mu}F_\alpha(t) - t^\mu|$ as $t \to \infty$ instead, which is usually much faster. For example, in case of the truncated singular value decomposition, we obtain

$$|t^{1+\mu}F_\alpha(t) - t^\mu| = \begin{cases} 0 & , \quad t \ge \alpha \\ t^\mu & , \quad t < \alpha. \end{cases} \tag{4.52}$$

If the singular values of the operator decay sufficiently fast (which is the typical case for ill-posed problems), e.g., $\sum_{n=1}^{\infty} \sigma_i^\mu < \infty$, we are confronted with the situation

$$\|R_\alpha y - A^\dagger y\|^2 \le c^2 \sum_{\sigma_i < \alpha} \sigma_i^{2\mu} \le c^2 \alpha^\mu \sum_{i=1}^{\infty} \sigma_i^\mu \tag{4.53}$$

so that the term $\|R_\alpha y - A_y^\dagger\|$ is of order $\alpha^{\mu/2}$. Consequently, we somehow need smoothness of the solution (in terms of the smoothing properties of the operator) in order to obtain a convergence rate in terms of α. We shall pursue this idea by introducing spaces of smoothness involving the absolute value of a compact operator.

Next we are concerned with the *propagation of the data error* through the regularization. Let F_α and C_F be as given above, and let $x_\alpha = R_\alpha y$, $x_\alpha^\varepsilon = R_\alpha y^\varepsilon$. From the singular value decomposition it follows directly

$$\begin{aligned}\|Ax_\alpha - Ax_\alpha^\varepsilon\|^2 &\le \sum_{i=1}^{\infty} \left(\sigma_i^2 F_\alpha(\sigma_i^2)\right)^2 |\langle y - y^\varepsilon, y_i\rangle|^2 \\ &\le C_F^2 \sum_{n=1}^{\infty} |\langle y - y^\varepsilon, y_i\rangle|^2 = C_F^2 \|y - y^\varepsilon\|^2 \le (C_F\varepsilon)^2,\end{aligned} \tag{4.54}$$

so that

$$\|Ax_\alpha - Ax_\alpha^\varepsilon\| \le C_F\varepsilon. \tag{4.55}$$

In the same way we obtain

$$\begin{aligned}\|x_\alpha - x_\alpha^\varepsilon\|^2 &\le \sum_{i=1}^{\infty} (F_\alpha(\sigma_i))^2 |\langle y - y^\varepsilon, y_i\rangle|^2 \\ &\le M_\alpha^2 \sum_{i=1}^{\infty} |\langle y - y^\varepsilon, y_i\rangle|^2 = M_\alpha^2 \|y - y^\varepsilon\|^2 \le (M_\alpha\varepsilon)^2,\end{aligned} \tag{4.56}$$

so that

$$\|x_\alpha - x_\alpha^\alpha\| \le M_\alpha\varepsilon \tag{4.57}$$

is implied (note that (4.57) estimates the norm of R_α by C_F).

CHAPTER 5

Example realizations light: univariate differentiation

For a continuous function F on the interval $[0,1] \subset \mathbb{R}$ we consider an integral operator A defined by

$$AF(x) = \int_0^x F(t)\,dt = G(x), \quad x \in [0,1]. \tag{5.1}$$

5.1. Regularization by change of space and/or topology

Unperturbed right side. We are interested in solving the integral equation $AF = G$ under the condition $G(0) = 0$. We introduce the class

$$C_*^{(1)}([0,1]) = \left\{ G \in C^{(1)}([0,1]) : G(0) = 0 \right\} \tag{5.2}$$

equipped with the norm

$$\|G\|_{C_*^{(1)}([0,1])} = \sup_{x\in[0,1]} |G(x)| + \sup_{x\in[0,1]} |G'(x)|. \tag{5.3}$$

It is not hard to see that $C_*^{(1)}([0,1])$ equipped with this norm is a Banach space. For $G \in C_*^{(1)}([0,1])$, set $F = G'$. Then it follows that

$$\begin{aligned} \|AF\|_{C_*^{(1)}([0,1])} &= \sup_{x\in[0,1]} |AF(x)| + \sup_{x\in[0,1]} \left|(AF)'(x)\right| \\ &= \sup_{x\in[0,1]} |AF(x)| + \sup_{x\in[0,1]} |F(x)| \\ &\geq \sup_{x\in[0,1]} |F(x)| \\ &= \|F\|_{C^{(0)}([0,1])}. \end{aligned} \tag{5.4}$$

Hence, we have

$$\|AF\|_{C_*^{(1)}([0,1])} \geq \|F\|_{C^{(0)}([0,1])}. \tag{5.5}$$

Thus it follows that

$$\left\|A^{-1}\right\|_{C_*^{(1)}([0,1]) \to C^{(0)}([0,1])} \leq 1, \tag{5.6}$$

i.e., A^{-1} is continuous. All in all, we are led to the following conclusion:

The problem $\left(A; C^{(0)}([0,1]), C_*^{(1)}([0,1])\right)$ is well-posed in the sense of Hadamard.

Perturbed right side. In reality, we measure G^ε instead of G such that

$$G^\varepsilon = G + n^\varepsilon, \quad x \in [0,1] \tag{5.7}$$

where n^ε denotes the noise of the measurements and ε is the noise level. Note that we generally have $n^\varepsilon(0) \neq 0$ and $n^\varepsilon(1) \neq 0$, in practice. Usually, n^ε cannot be classified to be of class $C^{(1)}([0,1])$. Thus, we are confronted with the situation that $G^\varepsilon \notin C^{(1)}_*([0,1])$. The operator A^{-1} is not applicable to G^ε. In order to overcome this situation we have to modify our mathematical framework, i.e., we have to enlarge the space $C^{(0)}([0,1])$.

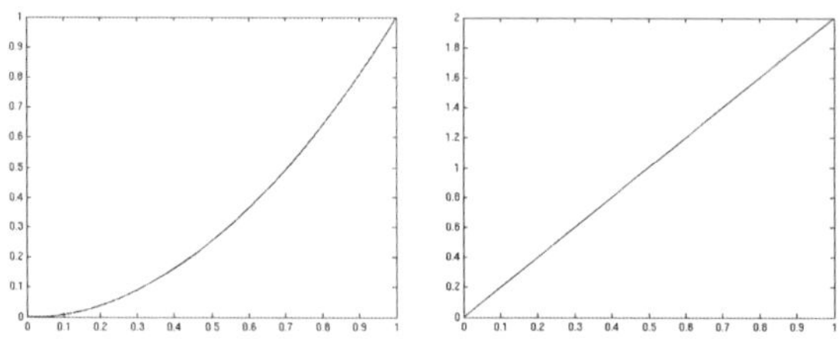

FIGURE 1. $G(x) = x^2$ (left) and $F(x) = 2x$ (right).

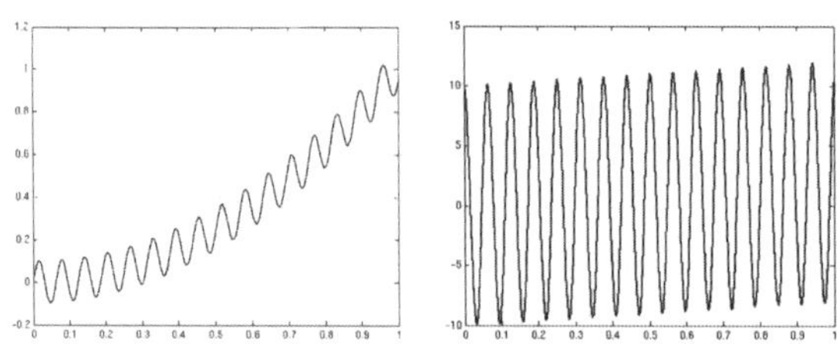

FIGURE 2. $G(x) = x^2 + n^\varepsilon(x)$ (left) and $F(x) = 2x + \frac{dn^\varepsilon}{dx}(x)$ (right) for $\varepsilon = \frac{1}{100}, k = 100$.

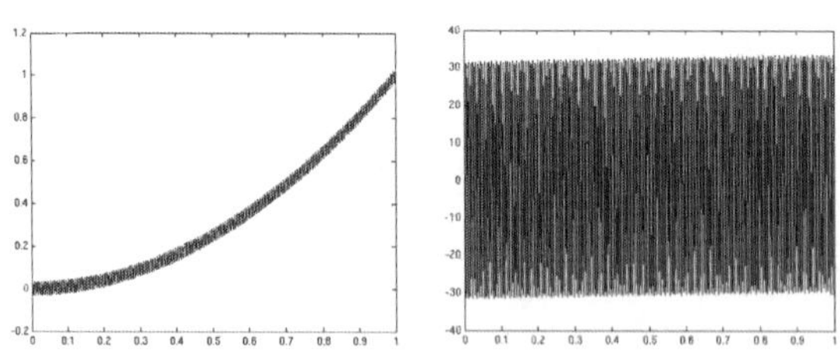

FIGURE 3. $G(x) = x^2 + n^\varepsilon(x)$ (left) and $F(x) = 2x + \frac{dn^\varepsilon}{dx}(x)$ (right) for $\varepsilon = \frac{1}{1000}, k = 1000$.

$L^\infty([0,1])$*-context:* Our aim is to show that the problem $(A, L^\infty([0,1]), L^\infty([0,1]))$ with A given by (5.1) does not satisfy Hadamard's condition of continuous dependence. To be more concrete, taking as special choice for the noise

$$G^\varepsilon(x) - G(x) = n^\varepsilon(x) = \sqrt{2}\sin(2\pi k x)\,\varepsilon, \quad k \in \mathbb{N} \tag{5.8}$$

with $\varepsilon > 0$ and $G \in C_*^{(1)}([0,1])$, we are led to the following inequalities for the constituents $F^\varepsilon, G^\varepsilon$ of the operator equation $AF^\varepsilon = G^\varepsilon$

$$\|G^\varepsilon - G\|_{L^\infty([0,1])} = \left\|\sqrt{2}\sin(2\pi k\cdot)\,\varepsilon\right\|_{L^\infty([0,1])} = \sqrt{2}\,|\varepsilon| = \sqrt{2}\;\varepsilon \tag{5.9}$$

and

$$\begin{aligned} \|F^\varepsilon - F\|_{L^\infty([0,1])} &= \left\|(G^\varepsilon)' - G'\right\|_{L^\infty([0,1])} \\ &= \left\|\sqrt{2}\;2\pi k\cos(2\pi k\cdot)\,\varepsilon\right\|_{L^\infty([0,1])} \;=\; 2\sqrt{2}\pi k\;\varepsilon. \end{aligned} \tag{5.10}$$

If we choose $\varepsilon_k = \frac{1}{\sqrt{2k}}$, $k \in \mathbb{N}$, we are able to conclude that

$$\|G^\varepsilon - G\|_{L^\infty([0,1])} = \varepsilon_k = \frac{1}{\sqrt{k}} \xrightarrow{k\to\infty} 0, \tag{5.11}$$

whereas

$$\|F^\varepsilon - F\|_{L^\infty([0,1])} = 2\sqrt{2}\pi k\varepsilon_n = 2\pi\sqrt{k} \xrightarrow{k\to\infty} \infty. \tag{5.12}$$

So, $A : L^\infty([0,1]) \to L^\infty([0,1])$ is not continuously invertible, i.e., the problem of one-dimensional differentiation $(A; X, Y)$ is ill-posed in the framework of the spaces $X = L^\infty([0,1])$ and $Y = L^\infty([0,1])$ (the Figures 1, 2, and 3 graphically illustrate the instability for increasing integers k).

$L^2([0,1])$*-context.* The bottleneck in handling noisy data in H is that, for k arbitrarily large, the value εk can be arbitrarily large. This observation also causes difficulties when the $L^2([0,1])$-error

(5.13)

$$\|F^\varepsilon - F\|_{L^2([0,1])} = \|(G^\varepsilon)' - G'\|_{L^2([0,1])} = \left(\int_0^1 \left(\frac{dG^\varepsilon}{dx}(x) - \frac{dG}{dx}(x)\right)^2 dx\right)^{1/2} = 2\pi\varepsilon k$$

is discussed: Without regularization and without further information, the error between the exact and noisy solution can be arbitrarily large, even if the noise is small. The question is to find additional information that helps to bound the error. Of course, a viable way would be that the noise is bounded in a stronger norm, e.g.,

$$\|(n^\varepsilon)'\|^2_{L^2([0,1])} = \int_0^1 \left(\frac{dn^\varepsilon}{dx}(x)\right)^2 dx \le \varepsilon^2. \tag{5.14}$$

In doing so we arrive in a trivial way at the error estimate

$$\|(G^\varepsilon)' - G'\|_{L^2([0,1])} = \left(\int_0^1 \left(\frac{dG^\varepsilon}{dx}(x) - \frac{dG}{dx}(x)\right)^2 dx\right)^{1/2} \le \varepsilon. \tag{5.15}$$

However, we already saw that this result is not realistic for practical applications, where an estimate for $\frac{dn^\varepsilon}{dx}$ can be hardly provided. A way out is to require stronger smoothness of G, e.g., $G \in C^{(2)}([0,1])$. The error then remains large, but regularisation can come into play. A simply structured example is the problem of determining a *Tikhonov-type regularization* G_α from a differential equation of the type

$$-\alpha \frac{d^2 G_\alpha}{dx^2}(x) + G_\alpha(x) = G^\varepsilon(x), \quad G_\alpha(0) = G_\alpha(1) = 0, \quad \alpha > 0. \tag{5.16}$$

In fact, this approach turns out to be *Tikhonov regularization.* In more detail, we are interested in the variational problems of performing a least-squares fit of G_α

$$\Phi_\alpha(G) = \int_0^1 (G_\alpha(x) - G^\varepsilon(x))^2 \, dx + \alpha \int_0^1 \left(\frac{dG_\alpha}{dx}(x)\right)^2 \, dx. \tag{5.17}$$

involving functions G_α of class $C^2([0,1])$ by imposing a penalty term enforcing that $\frac{dG_\alpha}{dx}$ be bounded. Under the assumption $G \in C^{(2)}([0,1])$ it easily follows from (5.16) that

$$-\alpha \frac{d^2}{dx^2}(G_\alpha(x) - G(x)) + (G_\alpha(x) - G(x)) = (G^\varepsilon(x) - G(x)) + \alpha \frac{d^2 G}{dx^2}(x). \tag{5.18}$$

Consequently, multiplying by $G_\alpha(x) - G(x)$ and integrating by parts we obtain

$$\begin{aligned} \int_0^1 \left[\alpha \left(\frac{dG_\alpha}{dx}(x) - \frac{dG}{dx}(x)\right)^2 + (G_\alpha(x) - G(x))^2\right] \, dx \\ = \int_0^1 \left(G^\varepsilon(x) - G(x) + \alpha \frac{d^2 G}{dx^2}\right)(G_\alpha(x) - G(x)) \, dx. \end{aligned} \tag{5.19}$$

Now, the Cauchy-Schwarz inequality applied to the right-hand side of (5.19) yields

$$\begin{aligned} \int_0^1 \left[\alpha \left(\frac{dG_\alpha}{dx}(x) - \frac{dG}{dx}(x)\right)^2 + (G_\alpha(x) - G(x))^2\right] \, dx \\ \le \int_0^1 (G^\varepsilon(x) - G(x))^2 \, dx + \alpha^2 C^2 \le \varepsilon^2 + \alpha^2 C^2, \end{aligned} \tag{5.20}$$

where $C = \|G\|_{C^2([0,1])}$. Thus, from (5.20), it can be readily deduced that

$$\int_0^1 \left(\frac{dG_\alpha}{dx}(x) - \frac{dG}{dx}(x)\right)^2 \, dx \le \frac{\varepsilon^2}{\alpha} + \alpha \, C^2, \tag{5.21}$$

i.e., a bound for the error is obtained in terms of α and ε.

Clearly, the next calamity is the choice of the appropriate regularization parameter: How to choose α such that the error in the solution becomes as small as possible? In general, it will not be possible to really minimize the error, but in accordance with the estimate (5.20) we are at least able to minimize the right-hand side which happens for the choice $\alpha = \frac{\varepsilon}{C}$, so that the error estimate takes the form

$$\int_0^1 \left(\frac{dG_\alpha}{dx}(x) - \frac{dG}{dx}(x)\right)^2 \, dx \le \; 2C \, \varepsilon. \tag{5.22}$$

Taking the square root in (5.22) to establish the norm on the left-hand side, the error amounts to $\sqrt{2C\varepsilon}$ hence, it turns out to be of the order $O(\sqrt{\varepsilon})$. In fact, it is much larger than the data error ε. This is another typical feature occurring in ill-posed problems: *Even with regularization, we will not be able to achieve an error in the reconstruction which is as slow as the error in the data* (note that the error bound $\sqrt{2\varepsilon}$ was only achieved under the "additional smoothness" $G \in C^{(2)}([0,1])$.

If we only assume that $G \in C^{(1)}([0,1])$, which actually seems to be much more natural. In this case, we are led to estimate alternatively

$$\int_0^1 \alpha \frac{d^2G}{dx^2}(x)\,(G_\alpha(x))\,dx = -\alpha \int_0^1 \frac{dG}{dx}(x)\left(\frac{dG_\alpha}{dx}(x) - \frac{dG}{dx}(x)\right)\,dx$$
$$\leq \frac{\alpha}{2}\int_0^1 \left(\frac{dG_\alpha}{dx}(x) - \frac{dG}{dx}(x)\right)^2 dx + \frac{\alpha}{2}\int_0^1 \left(\frac{dG}{dx}(x)\right)^2 dx.$$

We see that the second integral can be estimated by $\frac{\alpha}{2}D^2$ with $D = \|G\|_{C^{(1)}([0,1])}$ so that the final estimate is of the form

$$\int_0^1 \left(\frac{dG_\alpha}{dx}(x) - \frac{dG}{dx}(x)\right)^2 dx \leq 2\,\frac{\varepsilon^2}{\alpha} + D^2. \tag{5.23}$$

We notice that the right-hand side is larger than D no matter how we choose the regularization parameter α. Later on, within the context of this work, we will be able to show by different arguments that the integral on the left hand side of (5.23), in fact, tends to 0, but the convergence is arbitrarily slow, so that we are confronted with another general statement for ill-posed problems: Without additional and specific smoothness assumptions on the exact solution, the convergence of regularized solutions turns out to be arbitrarily slow.

5.2. Singular value decomposition

Until now, inverse problems are motivated as the inversion of some kind of the direct problem. For numerical differentiation, we have started with the inverse problem immediately. However, the direct problem can easily be obtained by integration. Indeed, if $G(0) = 0$, then the direct problem in L^2-nomenclature is given by a compact integral equation of the first kind, namely

$$G(x) = \int_0^x G'(y)\,dy = \int_0^x \frac{dG}{dx}(y)\,dy. \tag{5.24}$$

The resulting integral operator $A : L^2([0,1]) \to L^2([0,1])$ is given in the form

$$AF(x) = \int_0^1 K(x,t)\,F(t)\,dt \tag{5.25}$$

with

$$K(x,t) = \begin{cases} 1, & 0 \leq t \leq x \leq 1 \\ 0, & 0 \leq x \leq t \leq 1, \end{cases} \tag{5.26}$$

so that an analogue to (5.1) is considered in the context of $L^2([0,1])$. It is commonly known from functional analysis that the inversion of a compact linear operator leads to an ill-posed problem with characteristic features involving the solvability in terms of a pseudoinverse as a certain series expansion. Continuing with the Hilbert space famework of $L^2([0,1])$ we are canonically led to introduce the adjoint operator of A given by (5.25). For $F_1, F_2 \in L^2([0,1])$ it readily follows that

$$\langle AF_1, F_2\rangle = \int_0^1 \int_0^x F_1(t)\,dt\,F_2(x)\,dx = \int_0^1 \int_t^1 F_1(t)\,F_2(x)\,dx\,dt \tag{5.27}$$
$$= \int_0^1 F_1(t)\left(\int_t^1 F_2(x)\,dx\right)dt = \int_0^1 F_1(t)\,A^*F_2(t)\,dt.$$

This shows that the *adjoint operator* A^* *of* A in $L^2([0,1])$ is given by

$$A^*G\,(x) = \int_x^1 G\,(t)\,dt, \quad G \in L^2([0,1]). \tag{5.28}$$

Moreover, we have

$$A^*AF\,(x) = \int_x^1 \int_0^t F\,(y)\,dy\;dt, \quad F \in L^2([0,1]). \tag{5.29}$$

In the nomenclature of differential equations, the operator equation $A^*AF\,(x) = A^*G\,(x)$ amounts to

$$\frac{d^2F}{dx^2}(x) = \frac{dG}{dx}(x), \quad x \in (0,1). \tag{5.30}$$

In order to handle the operator equation $A^*AF = A^*G$ we are first interested in solving the eigenvalue equation

$$A^*AF(x) = \lambda F(x), \quad \lambda > 0, \quad x \in [0,1]. \tag{5.31}$$

for determining the eigenvalues of A^*A. It immediately follows from (5.28) that $F\,(1) = 0$. Differentiation of (5.29) yields

$$-\int_0^t F\,(y)\,dy = \lambda \frac{dF}{dt}\,(t)\,, \quad t \in (0,1),$$

so that $F'\,(0) = 0$ (note that $\lambda > 0$). Another differentiation yields $-F = \lambda F''$ so that we end up with the 1D-boundary-value problem

$$\frac{d^2F}{dx^2}(x) + \frac{1}{\lambda}F\,(x) = 0, \quad x \in (0,1), \tag{5.32}$$

$$F\,(1) = 0, \quad F'\,(0) = 0. \tag{5.33}$$

The values $\lambda_i, i \in \mathbb{N}$, given by

$$\lambda_i = \left(\left(i - \frac{1}{2}\right)^2 \pi^2\right)^{-1}, \ i \in \mathbb{N}, \tag{5.34}$$

are the eigenvalues of the operator A^*A, and the eigenvectors are the functions F_i given by

$$F_i\,(x) = \sqrt{2}\cos\left(\frac{1}{\sqrt{\lambda_i}}x\right), \quad i \in \mathbb{N} \tag{5.35}$$

(note that the factor $\sqrt{2}$ implies that $\|F_i\|_{L^2([0,1])} = 1$ for all $i \in \mathbb{N}$), i.e., we set

$$F_i\,(x) = \sqrt{2}\cos\left(\left(i - \frac{1}{2}\right)\pi x\right). \tag{5.36}$$

In IPP, the *singular values of the operator* A are understood to be the square-roots of the eigenvalues of A^*A, i.e.,

$$\sigma_i = \sqrt{\lambda_i} = \frac{1}{\left(i - \frac{1}{2}\right)\pi}, \ i \in \mathbb{N}. \tag{5.37}$$

What we still need is the third part of the so-called singular system, which can be computed as follows:

$$G_i(x) = \frac{1}{\sigma_i} A F_i(x) = \left(i - \frac{1}{2}\right)\pi \int_0^x \sqrt{2}\cos\left(\left(i-\frac{1}{2}\right)\pi t\right) dt = \sqrt{2}\sin\left(\left(i-\frac{1}{2}\right)\pi x\right). \tag{5.38}$$

All in all, the operator A has the triple $\{\sigma_i; F_i, G_i\}_{i\in\mathbb{N}}$ as *singular system* with

$$\sigma_i = \frac{1}{(i-\frac{1}{2})\pi}, \quad i \in \mathbb{N}, \tag{5.39}$$

$$F_i = \sqrt{2}\cos\left(\left(i-\frac{1}{2}\right)\pi t\right), \quad i \in \mathbb{N}, \tag{5.40}$$

$$G_i = \sqrt{2}\sin\left(\left(i-\frac{1}{2}\right)\pi t\right), \quad i \in \mathbb{N}. \tag{5.41}$$

Now we are able to understand the concept of the *"generalized inverse"* $A^\dagger$ *of* A by solving the noise-free operator equation $AF = G$ as follows:

(5.42)

$$\begin{aligned} A^\dagger G(x) &= \sum_{i=1}^{\infty} \frac{1}{\sigma_i}\langle G, G_i\rangle_{L^2([0,1])}\, F_i \\ &= \sum_{i=1}^{\infty} (i-\frac{1}{2})\pi\, \sqrt{2}\int_0^1 G(t)\sin\left(\left(i-\frac{1}{2}\right)\pi t\right) dt\, \sqrt{2}\cos\left(\left(i-\frac{1}{2}\right)\pi x\right). \end{aligned}$$

It should be noted that the sequence $\{F_i\}_{i\in\mathbb{N}}$ forms an orthonormal basis in $L^2([0,1])$ so that $\overline{\mathcal{R}(A)} = L^2([0,1])$. Moreover, $G \in L^2([0,1])$ is in the range $\mathcal{R}(A)$ of the operator A if and only if

$$\sum_{i=1}^{\infty}\left(i-\frac{1}{2}\right)^2\left(\sqrt{2}\int_0^1 G(t)\sin\left(\left(i-\frac{1}{2}\right)\pi t\right) dt\right)^2 < \infty. \tag{5.43}$$

The property (5.43) is the *Picard-condition for* G in terms of the singular system $\{\sigma_i; F_i, G_i\}_{i\in\mathbb{N}}$ given by (5.39), (5.40), and (5.41).

5.3. Tikhonov regularization

The identity (5.42) also shows how errors in G^ε affect the result $A^\dagger G^\varepsilon$: Errors corresponding to small singular values are amplified by a large factor of the singular value, so that such data errors are dangerous. This explains numerical instability of (5.42). A way of making (5.42) more stable is to replace it by the series

$$F_\alpha = \sum_{i=1}^{\infty}\frac{\sigma_i}{\sigma_i^2+\alpha}\langle G^\varepsilon, G_i\rangle_{L^2([0,1])}\, F_i \tag{5.44}$$

for some real value $\alpha > 0$. This is the famous *Tikhonov regularization method* in terms of the singular system. A comparison of (5.42) with (5.44) shows the stabilization effect: Errors in $\langle G, G_i\rangle_{L^2([0,1])}$ are not propagated with the factor $\frac{1}{\sigma_i}$, but only with the factor $\frac{\sigma_i}{\sigma_i^2+\alpha}$, these factors stay bounded if $i \to \infty$. It is helpful to characterize (5.44) in a different way: By the orthonormality of the functions $F_j, j \in \mathbb{N}$, we obtain

$$\langle F_\alpha, F_j \rangle_{L^2([0,1])} = \frac{\sigma_j}{\sigma_j^2 + \alpha} \langle G^\varepsilon, G_j \rangle_{L^2([0,1])}, \tag{5.45}$$

so that

$$\sigma_j^2 \langle F_\alpha, F_j \rangle_{L^2([0,1])} + \alpha \langle F_\alpha, F_j \rangle_{L^2([0,1])} = \sigma_j \langle G^\varepsilon, G_j \rangle_{L^2([0,1])}. \tag{5.46}$$

Now, it follows that

$$\begin{aligned} \sum_{j=1}^{\infty} \sigma_j \langle G^\varepsilon, G_j \rangle_{L^2([0,1])} G_j &= \sum_{j=1}^{\infty} \langle G^\varepsilon, AF_j \rangle_{L^2([0,1])} F_j \\ &= \sum_{j=1}^{\infty} \langle A^* G^\varepsilon, F_j \rangle_{L^2([0,1])} F_j = A^* G^\varepsilon \end{aligned} \tag{5.47}$$

and

$$\begin{aligned} &\sum_{j=1}^{\infty} \left(\sigma_j^2 \langle F_\alpha, F_j \rangle_{L^2([0,1])} + \alpha \langle F_\alpha, F_j \rangle_{L^2([0,1])} \right) F_j \\ &\qquad = \sum_{j=1}^{\infty} \left(\langle F_\alpha, A^* A F_j \rangle_{L^2([0,1])} + \langle F_\alpha, F_j \rangle_{L^2([0,1])} \right) F_j \\ &\qquad = A^* A F_\alpha + \alpha F_\alpha, \end{aligned} \tag{5.48}$$

i.e., we arrive at the *Tikhonov regularized normal equation* that formally leads back to the normal equation (cf. (5.30)) by choosing $\alpha = 0$ (note that the normal equations do not need the explicit knowledge of the singular value decomposition $\{\sigma_i; F_i, G_i\}_{i \in \mathbb{N}}$). Finally it should be mentioned that F_α^ε as given by $F_\alpha^\varepsilon = (A^*A + \alpha I)^{-1} A^* G^\varepsilon$ is the *unique minimizer of the Tikhonov functional* (5.17)

$$F \mapsto \|AF - G^\varepsilon\|^2_{L^2([0,1])} + \alpha \|F\|^2_{L^2([0,1])}. \tag{5.49}$$

5.4. Multiscale regularization

Starting from the singular system

$$\left\{ \frac{1}{(n-\frac{1}{2})\pi}; \sqrt{2}\cos\left(\left(n - \frac{1}{2}\right)\pi x\right), \sqrt{2}\sin\left(\left(n - \frac{1}{2}\right)\pi x\right) \right\}_{n \in \mathbb{N}} \tag{5.50}$$

we form the J-scale dependent kernel functions

$${}^d\Phi_J(x,y) = 2\sum_{n=1}^{\infty} ((\Phi_J)^\wedge(n)) \cos\left(\left(n - \frac{1}{2}\right)\pi x\right) \sin\left(\left(n - \frac{1}{2}\right)\pi y\right), \tag{5.51}$$

$${}^r\Phi_J(x,y) = 2\sum_{n=1}^{\infty} ((\Phi_J)^\wedge(n)) \cos\left(\left(n - \frac{1}{2}\right)\pi x\right) \cos\left(\left(n - \frac{1}{2}\right)\pi y\right) \tag{5.52}$$

with $\{(\Phi_J)^\wedge(n)\}_{n \in \mathbb{N}, J \in \mathbb{Z}}$ given, e.g., by

$$(\Phi_J)^\wedge(n) = \left(\frac{\sigma_n}{\sigma_n^2 + 2^{-J}} \right)^{\frac{1}{2}}. \tag{5.53}$$

As we shall see later in our work, these settings suffice to realize a multiscale regularization method in the following way: Let G be a member of the range $\mathcal{R}(A)$ of A. Suppose that $G^\varepsilon \in L^2([0,1])$ denotes the right side of the problem

$$AF(x) = \int_0^x F(t)\, dt = G^\varepsilon(x), \quad x \in [0,1], \tag{5.54}$$

with noise level satisfying

$$\|G - G^\varepsilon\|_{L^2([0,1])} \leq \varepsilon. \tag{5.55}$$

Moreover, assume that the Tikhonov parameter $J(= J(\varepsilon))$ satisfies

(i) $\lim_{\varepsilon\to 0} J(\varepsilon) = \infty$,

(ii) $\lim_{\varepsilon\to 0} \varepsilon((\Phi_J)^\wedge(n))^2 = 0$.

Then we obtain

$$\lim_{\varepsilon\to 0} \left\| \left(\int_0^1 \int_0^1 {}^r\Phi_J(x,z)\, {}^d\Phi_J(z,y)\, dz\, G^\varepsilon(y)\, dy \right) - A^\dagger G \right\|_{L^2([0,1])} = 0. \tag{5.56}$$

Equivalently, in the convolution " * "-nomenclature, we have

$$\lim_{\varepsilon\to 0} \left\| ({}^r\Phi_0 * {}^d\Phi_0) * G^\varepsilon + \sum_{j=0}^{J} ({}^r\tilde{\Psi}_j * {}^d\Psi_j) * G^\varepsilon - A^\dagger G \right\|_{L^2([0,1])} = 0, \tag{5.57}$$

where

$$({}^r\Phi_j * {}^d\Phi_j) * G^\varepsilon = \int_0^1 \int_0^1 {}^d\Phi_j(\cdot, z)^r\Phi_j(z,y)\, dz\, G^\varepsilon(y)\, dy \tag{5.58}$$

and

$$({}^r\Psi_j * {}^d\Psi_j) * G^\varepsilon = \int_0^1 \int_0^1 {}^d\Psi_j(\cdot, z)^r\Phi_j(z,y)\, dz\, G^\varepsilon(y)\, dy. \tag{5.59}$$

5.5. Finite difference schemes

Next we come to numerical differentiation of sampled data from a function G, e.g., by finite difference schemes. In this case, the difference scheme has the effect of a regularization method and the grid size h plays the role of a (control) regularization parameter. More concretely, we approximate the derivative G' in terms of the central difference quotient

$$D_h G = \frac{G(x+h) - G(x-h)}{2h} \tag{5.60}$$

with step size $h > 0$. We additionally assume that G is of class $C^{(i)}[0,1]$, $i \geq 3$. We apply a Taylor expansion in the usual way:

$$G(x+h) = G(x) + h\, G'(x) + \frac{h^2}{2} G''(x) + \frac{h^3}{6} G'''(\xi), \quad \xi \in (x, x+h), \tag{5.61}$$

$$G(x-h) = G(x) - h\, G'(x) + \frac{h^2}{2} G''(x) - \frac{h^3}{6} G'''(\eta), \quad \eta \in (x-h, x). \tag{5.62}$$

Let F be the solution of $AF = G$, i.e., $F = G'$. Then we find that

$$D_h G = \frac{G(x+h) - G(x-h)}{2h} = G'(x) + \frac{h^2}{12}\left(G'''(\xi) + G'''(\eta)\right) \tag{5.63}$$

and

$$\begin{aligned}\|F - D_h G\|_{L^\infty([0,1])} &= \|G' - D_h G\|_{L^\infty([0,1])} \\ &\leq \frac{h^2}{6} \|G'''\|_{L^\infty([0,1])} = \frac{h^2}{6} \|F''\|_{L^\infty([0,1])} \,.\end{aligned}$$

If G is noisy such that $G^\varepsilon = G + n^\varepsilon$ with $\|\varepsilon\|_{L^\infty([0,1])} \leq \tilde{\varepsilon}$, then it follows that

$$\begin{aligned}\|D_h G - D_h G^\varepsilon\|_{L^\infty([0,1])} &= \|D_h (G - G^\varepsilon)\|_{L^\infty([0,1])} \\ &= \left\| \frac{n^\varepsilon(\cdot + h) - n^\varepsilon(\cdot - h)}{2h} \right\|_{L^\infty([0,1]))} \\ &\leq \frac{2\,\|n^\varepsilon\|_{L^\infty([0,1])}}{2h} \leq \frac{\tilde{\varepsilon}}{h}.\end{aligned} \tag{5.64}$$

Therefore we are able to deduce that

$$\begin{aligned}\|F - D_h G^\varepsilon\|_{L^\infty([0,1])} &\leq \|F - D_h G\|_{L^\infty([0,1])} \quad + \quad \|D_h G - D_h G^\varepsilon\|_{L^\infty((0,1))} \\ &\leq \underbrace{\frac{h^2}{6} \|F''\|_{L^\infty([0,1])}}_{\text{approximation error}} \quad + \quad \underbrace{\frac{\tilde{\varepsilon}}{h}}_{\text{data error}} \quad .\end{aligned}$$

In other words, the total error consists of two ingredients, namely the approximation error and the data error. While the approximation error for small values h decreases, the data error increases for small h. As a consequence, it is impossible that the total error decreases for small h. In conclusion, the data error $\|F - D_h G^\varepsilon\|_{L^\infty([0,1])}$ cannot become arbitrarily small. However, there exists a value h_{opt} (cf. Figure 4),

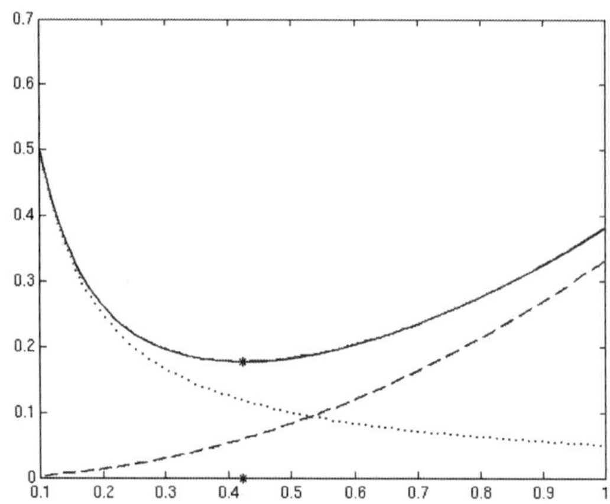

FIGURE 4. Error behavior characterizing one-dimensional numerical differentiation, the dashed line illustrates the data error, the pointed line shows the approximation error, and the marked point denotes h_{opt} of the total error.

so that $\left\|F - D_{h_{\text{opt}}} G^\varepsilon\right\|_{L^\infty([0,1])}$ can be estimated minimally as follows

$$h_{\text{opt}} = \left(\frac{3\overline{\varepsilon}}{\|F''\|_{L^\infty([0,1])}} \right)^{\frac{1}{3}} \tag{5.65}$$

with

$$\|F - D_{h_{\text{opt}}} G\|_{L^\infty([0,1])} \leq \frac{3^{\frac{2}{3}}}{2} \varepsilon^{\frac{2}{3}} \|F''\|^{\frac{1}{3}}_{L^\infty([0,1])} . \tag{5.66}$$

The example of numerical differentiation offers a large number of inherent effects occurring in ill-posed problems which can be recovered within the context of the general theory, viz. amplification of high frequency errors, restoration of stability by use of a priori information and assumptions, two error terms of different nature, i.e., the approximation and data error, the desire of an optimal discretization parameter, and the loss of information even under optimized circumstances.

5.6. Fredholm integral equation with strong noise

We take a more concrete look at the issue of well- and ill-posedness of some specific Fredholm integral equations of the first and second kind (cf. [**204**], [**205**]). For simplicity, in this preparatory part, we consider a completely understood one-dimensional example set in the Hilbert space $L^2([0,1])$ of square integrable functions on the interval $[0,1]$ (cf. (10.117)). Let the kernel $K(\cdot,\cdot)$ be defined on $[0,1] \times [0,1]$ by

$$K(s,t) = \min(s,t) - st \ , \quad s,t \in [0,1]. \tag{5.67}$$

Moreover, let us introduce the operator $K : L^2([0,1]) \to L^2([0,1])$ by

$$KF(s) = \int_0^1 K(s,t)\, F(t)\, dt \ , \quad s \in [0,1] \ . \tag{5.68}$$

In fact, K as introduced by (5.68) is the Green function operator for the two-point boundary value problem

$$\left\{ \begin{array}{rcl} -U''(x) & = & x, \ \ x \in (0,1), \\ U(0) & = & U(1) = 0. \end{array} \right. \tag{5.69}$$

As a consequence, the solution of (5.69) is given by $U = KF$. From functional analysis it is well-known that the operator K is compact.

$L^2([0,1])$*-(strong) noise model.* We easily show that the Fredholm equation of the first kind $KF = G$ is an ill-posed problem: Consider the data $G \in L^2([0,1])$ following the model $G = KF_0 + \eta$, for some unknown $F_0 \in L^2([0,1])$ and noise η with $\|\eta\|_{L^2([0,1])} = \varepsilon$, , with ε "small". Taking, for example, $G = 0$ and $\eta(t) = (\pi k)^{-2} \sin(\pi k t)$,$0 \leq t \leq 1$, where $k \in \mathbb{N}$ is "large", one easily has $\|\eta\|_{L^2([0,1])} = \sqrt{2}/(\pi k)^2$, and the solution is $F = (\pi k)^2\, \eta$. This example shows that any inequality of the form

$$\|F - F_0\|_{L^2([0,1])} \leq C \ \|\eta\|_{L^2([0,1])},$$

with C, not depending on η, must fail. Consequently, F does not depend continuously on the data, and the problem is ill-posed (for more aspects on the ill-posed context of Fredholm integral equations of the first kind as well equations of the second kind, see, e.g., [**513**]).

Now, consider integral equations of the second kind. Consider the data function $G \in L^2([0,1])$ according to the model

$$G = F_0 + KF_0 + \eta, \tag{5.70}$$

with the operator K as specified by (5.6), as before, with some unknown $F_0 \in L^2([0,1])$, and with η the noise. If we assume that η is small as supposed in (5.6), then we are able to show that the problem

$$\text{find} \quad F \quad \text{such that} \quad F + KF = G \tag{5.71}$$

is well-posed: The solution F exists and is unique, and it satisfies

$$\|F - F_0\|_{L^2([0,1])} \leq \|\eta\|_{L^2([0,1])} \,. \tag{5.72}$$

This result may be seen as follows: First, it readily follows that K is positive definite. Since $U = KF$ solves (5.69), then we obtain by integration by parts

$$\langle F, KF\rangle_{L^2([0,1])} = \langle -U'', U\rangle_{L^2([0,1])} = \|U'\|^2_{L^2([0,1])} > 0, \tag{5.73}$$

unless U is a constant function, which must then be identically zero (note that it must vanish at the endpoints $t = 0$ and $t = 1$). Thus, K is positive definite. Therefore we have on the one hand

$$\langle F - F_0, F - F_0 + K(F - F_0)\rangle_{L^2([0,1])} \geq \|F - F_0\|^2_{L^2([0,1])}, \tag{5.74}$$

since K is positive definite, and on the other hand we are led to

$$\begin{aligned} &\langle F - F_0, F - F_0 + K(F - F_0)\rangle_{L^2([0,1])} \\ &\quad = \langle F - F_0, \eta\rangle_{L^2([0,1])} \leq \|F - F_0\|_{L^2([0,1])}\|\eta\|_{L^2([0,1])}, \end{aligned} \tag{5.75}$$

so that

$$\|F - F_0\|^2_{L^2([0,1])} \leq \|F - F_0\|_{L^2([0,1])} \, \|\eta\|_{L^2([0,1])},$$

and our assertion follows. In other words, the problem (5.74) is well-posed. Thus, we come to the conclusion that *Fredholm integral equations of the first kind are ill-posed, and Fredholm integral equations of the second kind are well-posed*, where the aforementioned *noise model is the classical or strong noise model* for noise in $L^2([0,1])$, viz. $\|\eta\|_{L^2([0,1])} = \varepsilon$, with ε supposed to be "small".

Our aim now is discuss an alternative noise assumption. We investigate how changing this assumption affects the well/ill-posedness of the problems.

Super-strong noise model. Consider again the Fredholm integral equation of the first kind, $KF = G$, with K the Green function operator for the boundary value problem (5.69). As already said, $K : L^2([0,1]) \to L^2([0,1])$ is compact, but much more is true. Let $H_0^2([0,1])$ be the Sobolev space of square integrable functions on the interval $(0,1)$ which vanish at the endpoints, and with square integrable (weak) first and second derivatives. The norm on $H_0^2([0,1])$ is given by

$$\|U\|_{H^2([0,1])} = \|U''\|_{L^2([0,1])}. \tag{5.76}$$

Then, since $\|KF\|_{H^2([0,1])} = \|F\|_{L^2([0,1])}$, the mapping $K : L^2([0,1]) \to H_0^2([0,1)]$ is a homeomorphism, and the inverse mapping $K^{-1} : H_0^2([0,1]) \to L^2([0,1])$ exists and is a bounded linear transformation.

Consider the data function G according to the model

$$G = KF_0 + \eta \tag{5.77}$$

for some unknown $F_0 \in L^2([0,1])$ and noise η, but now with

$$\|\eta\|_{H^2([0,1])} = \varepsilon, \tag{5.78}$$

with ε "small". We refer to this as a *super-strong noise model.* Then the solution of the problem

$$\text{(5.79)} \qquad \textit{find} \quad F \textit{ such that} \quad KF = G$$

satisfies $\|F - F_0\|_{L^2([0,1])} = \|K^{-1}\eta\|_{L^2([0,1])} = \|\eta\|_{H^2([0,1])}$, thus showing that the problem (5.79) is *well-posed.*

5.7. Fredholm integral equation with weak noise

Next we have a look at the integral equations of the second kind

$$\text{(5.80)} \qquad F + KF = G\ ,$$

with the data function G following the model

$$\text{(5.81)} \qquad G = F_0 + KF_0 + \eta\ ,$$

with the operator K as before, with some unknown $F_0 \in L^2([0,1])$, and with η the noise. Note that so far, the noise is assumed to be small. As such, it does not represent noise as one usually understands it: high-frequency or random (without necessarily implying a firm probabilistic foundation). In this interpretation, the size of the noise is then better measured by the size of averages of the noise over small intervals (but not too small) or by the size of smoothed versions of the noise.

Weak noise model. The size of the noise must be measured by something like $\varepsilon = \|S\,\eta\|_{L^2([0,1])}$, for a suitable compact smoothing operator $S : L^2([0,1]) \to L^2([0,1])$. We refer to such a model as *weak noise model* (note that if η converges weakly to 0, then $S\,\eta$ converges strongly to 0 since S is compact).

We now come to the crux of the matter regarding the problem (5.80). Since the solution is given by $F = (I + K)^{-1}G$, then $F - F_0 = (I+K)^{-1}\eta$ but any inequality of the form

$$\text{(5.82)} \qquad \|(I+K)^{-1}\eta\|_{L^2([0,1])} \le C\ \|S\,\eta\|_{L^2([0,1])}$$

with the constant C not depending on η, must necessarily fail. Consequently, there is no constant C such that

$$\text{(5.83)} \qquad \|F - F_0\|_{L^2([0,1])} \le C\ \|S\eta\|_{L^2([0,1])}.$$

It follows that F does not depend continuously on the data. In other words, in the *weak noise setting the integral equation of the second kind* (5.80) is *ill-posed.*

Altogether, the Fredholm integral equation of the first kind turns out to be well-posed, and the Fredholm integral of the second kind is ill-posed (of course, this actually shows that the designations of Fredholm integral equations as being of the first or second kind are not the most informative with regard to their well- or ill-posedness).

CHAPTER 6

Reconstruction and regularization methods

The concept of a *regularizer* plays an important role in obtaining an approximate solution of an IPP in the presence of contamination in the data. We shall show that it is, indeed, a generic concept that can be used to unify some of the principles occurring in various regularization-approximation schemes (Tikhonov's regularization, truncated or filtered singular-value expansions, iterative methods, projection methods, multiscale techniques, mollifier techniques, iterative methods, etc). More explicitly, let X, Y be normed spaces and let $A : X \to Y$ be a one-to-one mapping (not necessarily linear). Note that the assumption that A is one-to-one is imposed for the convenience of the linear case. In fact, it can be dropped if one uses generalized inverses, as is done in several papers of M.Z. Nashed [**616, 618**]. This assumption will also be dropped in this contribution after we will have motivated and explained the auxiliary procedure for the case when A is assumed to be one-to-one.

6.1. Concept of regularization revisited

A *regularizer to the operator equation* $Ax = y$ is a one-parameter family of operators $\{R_t : t \in \Gamma\}$, where Γ is an index set of real numbers with $0 \in \overline{\Gamma}$ (the closure of Γ) satisfying the following conditions:

(C1) For each $t \in \Gamma, R_t$ is a continuous operator on all of Y into X.

(C2) For each $x \in X$,

$$\lim_{t \to 0} \|R_t Ax - x\| = 0. \tag{6.1}$$

Strong noise. Ill-posed problems of mathematical practice are usually regularizable in following sense: For $y \in \mathcal{R}(A)$, let y^ε be known with "noise level", $\|y^\varepsilon - y\| \leq \varepsilon$ (*"strong noise condition")*. Consider the operator equation between normed spaces X, Y

$$Ax = y^\varepsilon. \tag{6.2}$$

The existence of a regularizer enables us to calculate an "approximate solution", i.e., $\|x^\varepsilon - A^{-1}y\| \to 0$ and $\|Ax^\varepsilon - y\| \to 0$. We explain this fact in more detail: Let x^ε be given by $x^\varepsilon = R_t y^\varepsilon$. Then

$$\|x^\varepsilon - A^{-1}y\| \leq \|R_t y - A^{-1}y\| + \|R_t y^\varepsilon - R_t y\|, \tag{6.3}$$

where we assume that both norms on the right side of (6.3) are known. Note that $\|R_t y - A^{-1}y\|$ gives a *rate of convergence* of $R_t Ax$ to x and $\|R_t y^\varepsilon - R_t y\|$ is the *modulus of continuity of the operator* R_t at y. Observe that $\|R_t y^\varepsilon - R_t y\| \to 0$ as $\varepsilon \to 0$ for any fixed t. For given $\varepsilon > 0$ we choose $t = t(\varepsilon)$ to minimize the right side of (6.3). Then $x^\varepsilon = R_{t(\varepsilon)}y^\varepsilon$ has the claimed property since $\|R_{t(\varepsilon)}y^\varepsilon - A^{-1}y\| \to 0$ as $\varepsilon \to 0$.

At this stage we shall be primarily interested in the case when A will be assumed to be linear. For simplicity, R_t will be required to be linear, too. We then have

$$\|x^\varepsilon - A^{-1}y\| \le \|R_t y - A^{-1}y\| + M(t)\,\varepsilon \tag{6.4}$$

where $\|R_t\| \le M(t)$ (note that the operators R_t are not uniformly bounded since A^{-1} is unbounded, so $M(t) \to \infty$ as $t \to 0$).

The notion of a regularizer can be easily extended to weighted least-squares problems: We say that the problem (6.2), or equivalently $A_L^\dagger$, is regularizable if there exists a one-parameter family of linear operators $\{R_t : t \in \Gamma\}$ with $\mathcal{R}(R_t) \subset \mathcal{M}$ such that

$$\lim_{t\to 0} \|R_t A x - x\| = 0 \tag{6.5}$$

for $x \in \mathcal{M}$ and for each $t > 0$, R_t is bounded. Again, the family $\{R_t\}_{t\in\Gamma}$ is not uniformly bounded in t since $\mathcal{R}(A)$ is nonclosed. As before, the existence of a regularizer provides us with a family of approximate solutions determined by a well-posed problem. In the presence of contamination in y, say $\|y^\varepsilon - y\| \le \varepsilon$, the error $\|R_t y^\varepsilon - A_L^\dagger y\|_X \to 0$ as $t \to 0$ (in fact, it blows up). The criterion then is to choose t to minimize the error:

$$\begin{aligned}\|R_t y^\varepsilon - A_L^\dagger y\| &\le \|R_t y - A_L^\dagger y\| + \|R_t(y^\varepsilon - y)\|\\ &\le \|R_t g - A_L^\dagger y\| + \|R_t\|\,\varepsilon.\end{aligned} \tag{6.6}$$

The first term (*regularization error*) tends to zero as $t \to 0$, while the second term (*magnification of contamination error* due to ill-posedness) tends to ∞:

(6.7)

	↗	$\left\|A_L^\dagger y - R_t y\right\|$	$\longrightarrow 0$
$t \to 0$			
	↘	$\|R_t\,(y - y^\varepsilon)\| \le \|R_t\|\,\varepsilon$	$\longrightarrow \infty$

(6.8)

	↗	$\left\|A_L^\dagger y - R_t y\right\|$	$\longrightarrow \infty$
$t \to \infty$			
	↘	$\|R_t\,(y - y^\varepsilon)\| \le \|R_t\|\,\varepsilon$	$\longrightarrow 0$

If we know an *error estimate* for the first term and a growth estimate for $M(t)$, a suitable t can be determined. Such estimates can be obtained for particular regularizers R_t using additional information on the solution $A_L^\dagger y$, e.g., smoothness, and some robustness condition on $M(t)$, e.g., one might estimate that $M(t) \le d(t)$ is a known function which tends to ∞ as $t \to 0$, and $\|R_t y - A_L^\dagger y\| \le b(t)$, where $b(t) \to 0$ as $t \to 0$. Then, an optimal $t(\varepsilon)$ can be easily calculated, and for this $t(\varepsilon)$, $R_{t(\varepsilon)}y^\varepsilon \to A_L^\dagger y$, as $\varepsilon \to 0$. Concrete realizations of regularizers abound in regularization methods, projection and iterative methods, etc., as we shall see in the remaining work. In the case of an iterative scheme, $t = 1/n$, $x_n = R_t y$ represents the n-th iterate of a process which converges to $A_L^\dagger$ (or A^{-1}) in the absence of contamination in y. In a projection method $t = 1/n$, where n is the dimension of the approximating subspace. In finite differences, t represents the mesh size h. The preceding results show how to obtain stable approximate solutions in the presence of error in y if the needed estimates in (6.6) are available.

Weak noise. Let $K : X \to Y$ be a linear compact operator between the Hilbert spaces X and Y. The inner products and norms of X and Y are denoted by $\langle \cdot, \cdot \rangle_X$, $\langle \cdot, \cdot \rangle_Y$ and $\| \cdot \|_X, \| \cdot \|_Y$ (once more, usually we do not use subscripts if they are clear from the context, here and elsewhere). Consider the data $y \in Y$ according to the equation

$$y = Kx_0 + \eta \ , \tag{6.9}$$

where $\eta \in Y$ is the unknown noise and $x_0 \in X$ is an unknown element one wishes to recover from the data y. The following model is imposed on the noise. Let $A : Y \to Y$ be linear, compact, hermitian, and positive definite, and let

$$\varepsilon^2 = \langle \eta, A\eta \rangle. \tag{6.10}$$

We assume that ε is "small" and investigate what happens when $\varepsilon \to 0$. The operator A introduced above is not meant to be arbitrary. In fact, it must be connected with K in the sense that, for some $m \geq 1$ (not necessarily integer), the range of K is continuously embedded into the range of A^m, so that

$$A^{-m} K : X \to Y \quad \text{is continuous.} \tag{6.11}$$

If η satisfies (6.10), (6.11), it is referred to as *weakly bounded noise.*

Some comments should be made: In a deterministic setting, a reasonable model for the noise is that it is "high-frequency", and we would like to investigate what happens when the frequency tends to ∞, but without the noise tending to 0 strongly, that is without assuming that $\|\eta\|_Y \to 0$. Thus, $\eta \to 0$ weakly begins capturing the essence of "noise". Then, for any linear compact operator $S : Y \to Y$, we would have $\|S\eta\|_Y \to 0$. So, in this sense, there is nothing unusual about (6.10) and (6.11). Moreover, we would like (6.10) to capture the whole truth, i.e., the relations $\langle \eta, A^p \eta \rangle = o(\varepsilon^2)$ and $\langle \eta, A^q \eta \rangle = O(\varepsilon^2)$ fail for $p > 1$ and $q < 1$ as $\varepsilon \to 0$. This may be a tall order, although examples of operators A and noises η satisfying (6.10)-(6.1) are easily constructed (for more details see [**204**]). At the same time A is supposed to capture the smoothing effect of K in the sense of (6.11). Ideally, one would like $A^{-m}K$ to be continuous with a continuous inverse. The natural choice $A = (KK^*)^{1/2m}$ would achieve this, but would have to be reconciled with (6.10) and possibly (6.1). The condition (6.11) is not unreasonable.

Eggermont et al. [**207**] show how the weak noise model leads to simple bounds on expressions like $\langle \eta, y \rangle_Y$ for $y \in A^m(Y)$, the range of A^m. For $\beta > 0$, they introduce an inner product on $A^m(Y)$, by letting

$$\langle y, z \rangle_{m,\beta} = \langle y, z \rangle_Y + \beta^{2m} \langle A^{-m} y, A^{-m} z \rangle, \quad z \in A^m(Y), \tag{6.12}$$

and denote the associated norm by $\|\cdot\|_{m,\beta}$. The following result is of interest in itself, but it also later on plays a crucial role in the context of Tikhonov regularization with weakly bounded noise we have:

Suppose that $m \geq 1$. Under the assumptions (6.10), (6.11) on the weakly bounded noise we have for all $y \in A^m(Y)$ and all $\beta > 0$

$$|\langle \eta, y \rangle| \leq \beta^{-1/2} \varepsilon \, \|y\|_{m,\beta} \tag{6.13}$$

(note that the factor $\beta^{-1/2}$ stays the same, regardless of m).

6.2. Use of a priori bounds

The use of *a priori bounds* (more generally, a priori information) about the solution of an ill-posed problem has long been recognized to play a significant role in bringing about continuous dependence (i.e., of providing a regularizing effect). Early in the study of ill-posed problems, a fundamental observation was made by Tikhonov (cf. [**817**] and [**821**] for original references) that the restriction to a compact set insures well-posedness. More precisely, suppose X and Y are metric spaces and $F : X \to Y$ is a continuous injection, and let $C \subset X$ be compact. Then $F^{-1} : F(C) \to C$ is continuous: To show this, let $W \subset C$ be open in the relative topology, then the complement of W relative to C, denoted by W^c, is closed and, hence, compact since C is compact. Continuity of F implies that $F(W^c)$ is compact and, therefore, closed. From this it is follows that $F(W)$ is open, for F is injective (and hence $F(W^c) \cap F(W)$ is empty).

It should be remarked that the *use of differential operators as smoothing conditions* often leads to a setting in which the restriction of the domain of the operator to a compact set automatically prevails. For example, consider the simple situation treated by A.N. Tikhonov [**822**], where

$$J_\alpha(F) = \|AF - G\|^2 + \alpha\, \Omega(F), \quad \alpha > 0, \tag{6.14}$$

with

$$\Omega(F) = \int_0^1 \{p(x)(F'(x))^2 + q(x)(F(x))^2\}\, dx, \tag{6.15}$$

when p and q are positive, p is continuous, and q has a continuous derivative. Then, the set

$$C_r = \{F \in L^2([0,1]) : \Omega(F) \leq r\} \tag{6.16}$$

is compact in X for each $r > 0$. The smoothing function $\Omega(F)$ can be considered to be induced by a differential operator L, i.e., $\Omega(F) = \|LF\|^2$, where $L^*LF = -(pF')' + qF$ on $[0,1]$ with the boundary conditions $F'(0) = F'(1) = 0$. In particular, taking $p = q = 1$, it follows that C_r is a ball in the Sobolev space $W^{1,2}([0,1])$. Thus by restricting solutions to lie in a ball in $W^{1,2}[0,1]$ the problem is no longer ill-posed. Similar results can be obtained using more general differential operators and related spaces that are compactly embedded in X. The success of the aforementioned approach hinges on the fact that the unit ball of $W^{1,2}([0,1])$ is a compact set in the topology of $L^2([0,1])$ (note that it is, of course, not compact in the topology of $W^{1,2}([0,1])$, for more details the reader is referred, e.g., to [**625**], [**627**]).

6.3. Tikhonov regularization

Let $L : \mathcal{D}(L) \subset X \to Z$, where Z is a Hilbert space, be a closed linear operator with dense domain and closed range. We first assume (see also [**616**], [**819**], [**821**]), [**822**]) that $\mathcal{N}(L)$ is finite dimensional and that $\mathcal{N}(L) \cap \mathcal{N}(A) = \{0\}$. We endow $\mathcal{D}(L)$ with the topology induced by the graph norm $|u| = (\|u\|^2 + \|Lu\|^2)^{1/2}$. Then L becomes a bounded operator on $\mathcal{D}(L)$. We define a new inner product on $\mathcal{D}(L)$ by letting

$$[u, v]_L = \langle Au, Av\rangle_Y + \langle Lu, Lv\rangle_Z \tag{6.17}$$

Then the induced norm $\|u\|_L = \sqrt{[u, u]}$ is equivalent to the graph norm of u. Thus, both A and L are bounded operators on $\mathcal{D}(L)$ equipped with the inner product (6.17); we denote this Hilbert space by X_L. The discussion shows that, under the hypotheses listed above, without loss of generality we may restrict ourselves to the case in which $A : X \to Y$ and $L : X \to Z$ are both bounded. For each $y \in D(A^\dagger)$, there is a unique element $x \in \mathcal{L}(y)$ which minimizes $\|Lu\|$. Let $A_L^\dagger y = x$, and define

$$\mathcal{M} = \{x \in X : L^* Lu \in \mathcal{N}(A)^\perp\}. \tag{6.18}$$

Then $\mathcal{M}$ is the orthogonal complement of $\mathcal{N}(A)$ with respect to the inner product (6.17), so that $A_L^\dagger$ is the generalized inverse relative to the decompositions:

$$X : \mathcal{N}(A) \oplus \mathcal{M}, \quad Y = \overline{\mathcal{R}(A)} \oplus \mathcal{R}(A)^\perp. \tag{6.19}$$

Under the above assumptions for each $\alpha > 0$, there exists a unique x_α which minimizes

$$J_\alpha(x) := \|Ax - y\|_Y^2 + \alpha\|Lx\|_Z^2. \tag{6.20}$$

Furthermore,

$$x_\alpha = (A^* A + \alpha L^* L)^{-1} A^* y \tag{6.21}$$

and x_α converges to $A_L^\dagger y$ for $y \in \mathcal{D}(A^\dagger)$ as $\alpha \to 0$, and diverges otherwise. Thus it follows that in the presence of contamination, say y is replaced by y_ε, where $\|y_\varepsilon - y\| \le \varepsilon$ for some $\varepsilon > 0$, the norm of the corresponding x_α^ε may well diverge. It should be noted that $(A^*A + \alpha L^*L)^{-1}A^*$ does not converge in the uniform operator topology as $\alpha \to 0$; in fact, $\|(A^*A + \alpha L^*L)^{-1}A^*\| \to \infty$. Furthermore, the system

$$(A^*A + \alpha L^*L)\, x = A^* y \tag{6.22}$$

is poorly conditioned for small α, and hence numerically unstable. Thus, both contamination and numerical approximation dictate that a choice for α has to be made which would be a suitable compromise between accuracy and stability. Several procedures for the choice of an "optimal" value α are available, some of which take into consideration various a priori information. A simple choice that works would be $\alpha = \varepsilon^2$.

Strongly bounded noise in Tikhonov's regularization. We come back to the situation where $K : X \to Y$ is a linear compact operator between the Hilbert spaces X and Y. Consider the data $y \in Y$ according to the equation $y = Kx_0 + \eta$, where $\eta \in Y$ is the unknown noise and $x_0 \in X$ is an unknown element one wishes to recover from the data y. We study Tikhonov's regularization as a scheme to recover x_0 from the data y in the *strong noise model*

$$y = Kx_0 + \eta \quad \text{with} \quad \|\eta\|_Y \le \varepsilon\ . \tag{6.23}$$

The interest is in what happens when $\varepsilon \to 0$. It should be noted that, in the Tikhonov regularization scheme, the unknown x_0 is estimated by specifying the solution $x = x^{\alpha,\varepsilon}$ of the problem

$$\text{minimize} \quad \|Kx - y\|_Y^2 + \alpha\|x\|_X^2 \quad \text{over} \quad x \in X, \tag{6.24}$$

for some regularization parameter α, $\alpha > 0$, yet to be specified. This procedure dates back to [**693**], [**821**]. Its minimizer exists and is unique. Moreover, it is well-known (see, e.g., [**379**]) that convergence rates on the error $\|x^{\alpha,\delta} - x_0\|_X$ can be obtained from a source condition. For simplicity, it is assumed there that there exists a $z_0 \in X$ such that "source condition"

$$x_0 = (K^*K)^{\frac{\nu}{2}} z_0 \quad \text{for some} \quad 0 < \nu \le 2. \tag{6.25}$$

holds true. Precise necessary and sufficient conditions are given in [**652**]. In the study of convergence rates under the source condition (6.25), it is assumed here that ν is known and that α is chosen accordingly. Clearly, one wants to obtain bounds on the error $\|x^{\alpha,\varepsilon} - x_0\|_X$. As usual, this is broken up into two parts

$$\|x^{\alpha,\varepsilon} - x_0\|_X \le \|x^{\alpha,\varepsilon} - x^{\alpha,0}\|_X + \|x^{\alpha,0} - x_0\|_X, \tag{6.26}$$

where $x^{\alpha,\varepsilon}$ is the "noiseless" estimator, i.e., the minimizer of $\|Kx - y\|_Y^2 + \alpha\|x\|_X^2$. Thus, $x^{\alpha,\varepsilon} - x^{\alpha,0}$ is the noise part of the error and $x^{\alpha,0} - x_0$ is the error introduced by the regularization.

The following results (see, e.g., [**204**], [**379**]) are well-known:

(1) *There exists a constant* c *such that for all* α, $0 < \alpha \le 1$,

$$\|x^{\alpha,\varepsilon} - x^{\alpha,0}\|_X \le c\,\alpha^{-\frac{1}{2}}\|\eta\|_Y. \tag{6.27}$$

(2) *Under the source condition* (6.25), *there exists a constant such that for all values* α, $0 < \alpha \le 1$,

$$\|x^{\alpha,\varepsilon} - x_0\|_X \le c\,\alpha^{\frac{\nu}{2}}. \tag{6.28}$$

As a consequence, the two results (6.27), (6.28) above then provide the following convergence rates:

Assuming the source condition (6.25) *and the condition* (6.23) *on the noise for* $\alpha \to 0$ *we have*

$$\|x^{\alpha,\varepsilon} - x_0\|_X = O\left(\alpha^{-\frac{1}{2}}\,\varepsilon + \alpha^{\frac{\nu}{2}}\right). \tag{6.29}$$

Moreover, if $\alpha \asymp \varepsilon^{\frac{2}{(\nu+1)}}$, *then*

$$\|x^{\alpha,\varepsilon} - x_0\|_X = O\left(\varepsilon^{\frac{\nu}{(\nu+1)}}\right). \tag{6.30}$$

Weak noise in Tikhonov's regularization. Tikhonov's regularization may also be considered (cf. [**203**], [**207**], [**206**]) as the scheme to recover x_0 from the data y in the *weak noise model*

$$y = Kx_0 + \eta\,. \tag{6.31}$$

Thus, we assume that there is a smoothing operator A such that the noise η and A satisfy (6.10) and (6.11). In particular, $\langle \eta, A\eta\rangle_Y = \varepsilon^2$, and the discussion (cf. [**207**]) focusses on what happens when $\varepsilon \to 0$. Formally, Tikhonov regularization does not depend on the noise being strongly or weakly bounded. Thus x_0 is estimated by the solution $x = x^{\alpha,\varepsilon}$ of the problem (6.24) for some positive regularization parameter α yet to be specified. Again we want to obtain bounds on the error $\|x^{\alpha,\varepsilon} - x_0\|_X$, and it is broken up as $\|x^{\alpha,\varepsilon} - x_0\|_X \le \|x^{\alpha,\varepsilon} - x^{\alpha,0}\|_X + \|x^{\alpha,0} - x_0\|_X$, where $x^{\alpha,0}$ is the "noiseless" estimator, i.e., the minimizer of $\|Kx - y\|_Y^2 + \alpha\|x\|_X^2$. Thus, $x^{\alpha,\varepsilon} - x^{\alpha,0}$ is the noise part of the error and $x^{\alpha,0} - x_0$ is the error caused by the regularization. It is useful to introduce a new norm on X by way of

$$\|x\|_{\alpha,X}^2 = \|Kx\|_Y^2 + \alpha\,\|x\|_X^2\ . \tag{6.32}$$

Assuming again the source condition (6.25) we see that the noiseless part $x^{\alpha,0} - x_0$ can be covered as before, but the treatment of the noise part is markedly different from the case of strong noise (see P.N. Eggermont et al. [**207**]):

(1) *Under the conditions* (6.310), (6.11) *on the noise* η, *there exists a constant* C *depending on* A *only such that for* $\alpha \to 0$ *we have the estimate* $\|x^{\alpha,\varepsilon} - x^{\alpha,0}\|_{\alpha,X}^2 \le C\ \alpha^{-\frac{1}{4m}}\varepsilon$. *This leads to the convergence rates* (6.3) *and* (6.3) *specified below, where it is shown in P.N. Eggermont, V.N. LaRiccia* [**204**] *that they are optimal, following arguments from F. Natterer* [**648**], *but assuming, in addition, that* $A^{-m}K$ *has a continuous inverse.*

(2) *Assuming the conditions* (6.25) *and* (6.310) *and* (6.11) *on the noise for* $\alpha \to 0$, *we have* $\|x^{\alpha,\varepsilon} - x_0\|_X = O\left(\alpha^{-\frac{1}{2}-\frac{1}{4m}}\varepsilon + \alpha^{\frac{\nu}{2}}\right)$. *Moreover, if* $\alpha \asymp \varepsilon^{4m/(2m\nu+2m+1)}$, *then* $\|x^{\alpha,\varepsilon} - x_0\|_X = O\left(\varepsilon^{2m\nu/(2m\nu+2m+1)}\right)$.

6.4. Characterization of regularizers

Let A be of class $\mathfrak{L}(X, Y)$ with nonclosed range. In what follows we introduce concepts of regularizing families for the ill-posed problem $Ax = y$ based on bounded outer inverses of the operator A:

A linear operator $B : Y \to X$ is called an *inner inverse of* A if $ABA = A$. A (nonzero) linear operator $B : Y \to X$ is called an *outer inverse of* A if $BAB = B$.

In what follows, B is always taken to be a nonzero operator. In the case of Hilbert spaces, the regularizers will approximate least-squares solutions of $Ax = y$. In the case of Banach spaces, we assume that A is injective, and $\mathcal{R}(A)$ is dense in Y (otherwise the regularizers would apply to the equation $Ax = Qy$, where Q is a continuous projector of Y onto $\mathcal{R}(A)$, whose existence has to be assumed).

We will classify ill-posed in Banach spaces according to the type of regularizing families that they admit (cf. M.Z. Nashed [**627**] [**628**]). In the case of Hilbert spaces this classification will particularly distinguish the set of all compact operators with infinite dimensional range within the set of all bounded operators with nonclosed range.

Many of the operator-theoretic aspects of ill-posed linear equation (including regularization and stabilization methods) are really problems in operator ranges and operator factorizations. In particular, ranges of outer inverses play several roles:

(1) Very often the constructed approximation solution is in the range of some outer inverse (or an approximate outer inverse).

(2) The problem of finding an "optimal" outer inverse with a prescribed rank can be solved for several classes of operators.

(3) Outer inverses with a prescribed finite dimensional range can be easily constructed; this cannot always be done in case the prescribed range is infinite-dimensional and the outer inverse is required to be bounded.

(4) Outer inverses have desirable "stability/continuous dependence" properties which inner inverses or the generalized inverse lack.

Our purpose is to introduce notions of regularizers in form of bounded outer inverses with infinite-dimensional range (within this class, convergent regularizers can be selected to provide "optimal" resolution), approximate outer inverses and approximate right inverses in scales of norms.

The set of all operators in $\mathfrak{L}(X, Y)$ that have bounded outer inverses with infinite dimensional range and the set of full-rank $m \times n$ matrices share several common properties: Each of them is both *open* and *dense*, and all elements of each of the sets have outer inverses with the maximal possible rank (namely, the same as the rank of A). These properties and other results to be analyzed next indicate that, in Hilbert space, an equation involving a bounded noncompact operator with nonclosed range is "less" ill-posed than an equation with a compact operator with infinite dimensional range. In comparison with least-squares or generalized inverse problems for $m \times n$ matrices, one may say that for operators with nonclosed range, the case of a noncompact operator corresponds to the full-rank case for matrices, while the case of a (nondegenerate) compact operator is the infinite dimensional analog of the rank-deficient case for matrices.

Our results are closely based on the note M.Z. Nashed [**627**]. It addresses only operator-theoretic and classification characterization aspects of the notions of regularizers introduced. Convergence analysis for specific families or regularizers and the choice of an "optimal" or "suitable" regularizer within a specific family in the presence of contamination (i.e., what corresponds to a "regularizing parameter" and its choice in the present setting) will not be addressed here.

Outer inverses in "solvability" and "regularization" of ill-posed problems. We illustrate the role and properties of outer inverses for approximate solutions of ill-posed problems in standard and familiar settings. It should be pointed pout that*regularizers are never inner inverses.* Inner inverses are connected with solvability of the equation $AF = G$. More precisely, for a given transformation $A : X \to Y$, a transformation $B : Y \to X$ has the property that $x = By$ is a solution of $Ax = y$ for every $y \in \mathcal{R}(A)$ if and only if B is an inner inverse of A. Outer inverses of A are related to "approximate solvability" and "stability".

More explicitly, if X is of dimension m and Y is of dimension n, it follows from the property $ABA = A$ that the rank of any inner inverse of A cannot be less than the rank of A. Similarly it follows from the property $BAB = B$ that the rank of any outer inverse of A cannot exceed the rank of A. Moreover, if $r = \operatorname{rank} A$, then one can construct outer inverses of rank s for any $s \leq r$ and inner inverses of rank t for any $r \leq t \leq \min(m, n)$ (see, e.g., [**618**]). Even in the finite dimensional case the possibility of approximating the generalized inverse (or the least-squares solution of minimal norm in Hilbert space settings) by an outer inverse of rank $s < r$ is an attractive feature of outer inverses. A similar approximation by inner inverses is of course not possible.

The situation is more drastic in the infinite-dimensional case as can be seen from the following known result: *If the range of $A \in \mathfrak{L}(X,Y)$ is nonclosed, then A has no bounded inner inverse.* In fact, if B is any inner inverse of A, then AB and BA are linear idempotents, with $\mathcal{N}(BA) = \mathcal{N}(A)$ and $\mathcal{R}(A) = \mathcal{R}(AB)$. Thus the following algebraic decompositions hold:

$$X = \mathcal{N}(A)\dot{+}\mathcal{R}(BA), \tag{6.33}$$

$$Y = \mathcal{R}(A)\dot{+}\mathcal{N}(AB), \tag{6.34}$$

where $\dot{+}$ denotes the algebraic direct sum. Now, suppose B is bounded, then the projectors BA and AB are continuous (equivalently, the decomposition in (6.33) and (6.34) are topological) and so $\mathcal{R}(A)$ is closed, which contradicts the assumption. Thus, we are led to the statement: No regularizer can be an inner inverse.

The nonboundedness of the inner inverse of A in case of a nonclosed range of $A \in \mathfrak{L}(X,Y)$ is actually a part of the following known statement (see [**639**]):

Let $A \in \mathfrak{L}(X,Y)$, where X and Y are Banach spaces. Then A has a bounded inner inverse B if and only if the decompositions (6.33) *and* (6.34) *are topological. Equivalently, A has a bounded inner inverse if and only if $\mathcal{N}(A)$ and $\mathcal{R}(A)$ have topological complements in X and Y, respectively, (i.e., the projectors on $\mathcal{N}(A)$ and $\mathcal{R}(A)$ are continuous).*

If X and Y are Hilbert spaces, then $A \in \mathfrak{L}(X,Y)$ has a bounded inner inverse if and only if $\mathcal{R}(A)$ is closed. It should be noted in all these cases that it does not mean that all inner inverses are bounded, unless A is invertible.

Truncated singular value expansions as outer inverses. Let H_1 and H_2 be Hilbert spaces and let $K : H_1 \to H_2$ be a (nonzero) compact linear operator. Let K^* denote the adjoint of K. Since K^*K is a nonnegative symmetric compact linear operator on H_1 we have in terms of the singular system $\{\sigma_k; x_k, y_k\}$ for K the following spectral representation

$$K^*Kx = \sum_{k=1}^{\infty} \sigma_k^2 \langle x, x_k\rangle x_k, \tag{6.35}$$

where $\{x_k\}$ is an orthonormal set of eigenvectors of K^*K with $K^*Kx_k = \sigma_k^2 x_k$ with $\sigma_1 \geq \sigma_2 \geq \ldots > 0$. Set $y_k := \sigma_k^{-1} K x_k$. Then the y_k's form an orthonormal set in H_2, and it is easy to show that

$$Kx = \sum_{k=1}^{\infty} \sigma_k \langle x, x_k\rangle y_k. \tag{6.36}$$

Obviously, the series (6.36) is the singular value expansion (SVD) of K (note that, if $\operatorname{rank}(K) = r$, then the number of nonzero singular values is r and the summation in (6.36) extends from $k = 1$ to r).

From here on, we assume, unless stated otherwise, that the range K is infinite dimensional. Then we have an infinite number of nonzero singular values with $\sigma_n \to 0$ as $n \to \infty$.

Let m be a fixed positive integer and define the operator $B_m : H_2 \to H_1$ by

$$B_m y = \sum_{k=1}^{m} \sigma_k^{-1} \langle y,\ y_k \rangle x_k. \tag{6.37}$$

It then follows that

$$K B_m\ y = \sum_{k=1}^{m} \sigma_k^{-1} \langle y, y_k \rangle K x_k = \sum_{k=1}^{m} \langle y, y_k \rangle y_k \tag{6.38}$$

and

$$B_m\ K\ B_m\ y = \sum_{k=1}^{m} \sigma_k^{-1} \langle \sum_{i=1}^{m} \langle y, y_i \rangle y_i, y_k \rangle\ x_k = B_m\ y. \tag{6.39}$$

Thus, for each $m \in \mathbb{N}$, B_m is an outer inverse of rank m. For each $y \in \mathcal{D}(K^\dagger) := \mathcal{R}(K) + \mathcal{R}(K)^\perp$,

$$\|B_m\ y - K^\dagger y\| \to 0 \tag{6.40}$$

as $m \to \infty$, where $K^\dagger y = \sum_{i=1}^{\infty} \sigma_i^{-1} \langle y, y_i \rangle x_i$, but the convergence is not uniform and the operators B_m are not uniformly bounded.

We mention the following example: The SVD is particularly useful because it permits a quantification of the notion of near rank deficiency. It is well known from linear algebra that for any $m \times n$ real or complex matrix A of rank r and any $k < r$,

$$\inf \{\|A - B\|_{Frob} : \text{rank}(B) = k\} = \|A - A_k\|_{\text{Frob}}, \tag{6.41}$$

where A_k is the truncated singular value decomposition (with k terms) of A and $\|\cdot\|_{\text{Frob}}$ is the Frobenius norm (see, e.g., [**367**]). In view of the above observation that the TSVD is an outer inverse and by use of the well-known relation between the singular values and singular vectors of A and those of the Moore-Penrose of A, it follows that

$$\inf \left\{\|A^\dagger - B\|_{\text{Frob}} : BAB = B, \text{rank}(B) = k\right\} = \|A^\dagger - B_k\|_{\text{Frob}}, \tag{6.42}$$

where

$$B_k\ y = \sum_{i=1}^{k} \sigma_i^{-1} \langle y, y_i \rangle x_i. \tag{6.43}$$

The same analysis and properties of outer inverses can be easily extended to linear inverse problems with discrete data (such as those that arise from moment discretization of ill-posed linear integral and operator equations in Hilbert space (see, e.g., [**78**], [**618**]) or when projection methods on finite dimensional subspaces are used (see, e.g., [**621**], [**378**]).

Outer inverses are not "equation solvers" (cf. [**627**]), i.e., if B is an outer inverse for A, which is not also an inner inverse, then for $y \in \mathcal{R}(A)$, $x = By$ is not a solution to $Ax = y$, and for $y \notin \mathcal{R}(A)$, $x = By$ is not a least-squares solution (in the case of a Hilbert space). One finds in some books statements like: "Since almost every application of various generalized inverses involves subsets of 1-inverses (inner inverses), we will mainly consider inner inverses that satisfy additional conditions …". For ill-posed problems (see [**627**]), the situation is precisely the opposite. We are not interested in a generalized or inner inverse that would be an "equations solver", since such an "inverse" will be unbounded. Rather we seek a bounded operator that has some "inverse-like" properties, and that can serve simultaneously as an "approximate inverse" and stabilizer to the inverse problem. Indeed, outer inverses possess these qualities:

(a) If B is an outer inverse of A, then B is also an inner inverse of $\tilde{A} = A|\mathcal{R}(B)$.

(b) For all $y \in \mathcal{R}(AB)$, $x = By$ is the unique solution in $\mathcal{M} = \mathcal{R}(B)$ of the equation $Ax = y$.

From these properties (see [**627**] for more details), it follows that the unique solution of $Ax = y$ in $\mathcal{R}(B)$ is a "regularized" solution and can be constructed in a stable way. If $\mathcal{R}(BA)$ or $\mathcal{R}(B)$ is infinite-dimensional, then we have in a sense the possibility of "infinite resolution", and the equation with bounded outer inverses of infinite rank are not as ill-posed as those for which an outer inverse with infinite rank does not exist.

Approximate outer and approximate right inverses. The concept of a regularizer plays an important role in obtaining an approximate solution of an ill-posed problem. Let $A : X \to Y$ be a one-to-one mapping (not necessarily linear). In the already known definition by Tikhonov (see e.g. [**821**], [**598**]) a regularizer is a bounded "approximate" inverse or "approximate" generalized inverse. Any regularizer must also satisfy the defining equations of an inner as well as outer inverse approximately, but it can not satisfy the defining equation of an inner inverse exactly. This motivates considering regularizers (cf. [**627**]) that satisfy the defining equation of an outer inverse and, in addition, have "maximum" rank.

A family $\mathcal{F}$ of regularizers by outer inverses for the ill-posed problem $Ax = y$ is said to be of type I if each $B \in \mathcal{F}$ satisfies the following conditions:

(i) $B \in \mathfrak{L}(Y, X)$

(ii) $BAB = B$,

(iii) *the range of B is infinite-dimensional.*

An ill-posed problem that does not admit a family or regularizers of type I is said to be of type II.

Any $B \in \mathcal{F}$ is called a regularizer. Of course, $u = By$ is not necessarily a good approximation to the "solution" x of $Ax = y$ for each $B \in \mathcal{F}$, just like, say, $(A^*A + \alpha I)^{-1}A^*y$ is not necessarily a good approximation for each $\alpha > 0$. As already mentioned, every "regularization method" involves a critical "parameter", whose optimal value – or at least a suitable choice of it – is crucial to the approximation of the solution. Thus criteria or strategies have to be developed for selecting a suitable or "optimal" outer inverse from our class of regularizers. We will not address this problem here. Instead we focus only on the operator-theoretic aspects of outer inverses as a class of regularizers.

As an example, we note that *Tikhonov's regularization provides an approximate outer inverse, but not an outer inverse.*

Let

$$B_\alpha = (K^*K + \alpha I)^{-1}K^*, \qquad \alpha > 0. \tag{6.44}$$

Then

$$B_\alpha - B_\alpha K B_\alpha = \alpha(K^*K + \alpha I)^{-1}B_\alpha, \quad \alpha > 0. \tag{6.45}$$

Note that

$$B_\alpha - B_\alpha K B_\alpha \to 0, \quad \alpha \to 0. \tag{6.46}$$

Regularization operators obtained via spectral families (such as those in [**67**], [**214**], [**376**], [**620**]) are usually not outer inverses. The same is true for the "regularized" truncated SVD.

Tikhonov's regularization operator and other regularizers obtained by spectral families are approximate outer inverses in the following sense: $A \in \mathfrak{L}(X, Y)$ is *approximately outer-invertible* if, for each $\mu \in (0, 1)$, there exists a $B_\mu \in \mathfrak{L}(Y, X)$ with the following properties:

$$\|(B_\mu A B_\mu - B_\mu)y\| \le C(\mu)\|B_\mu y\| \tag{6.47}$$

and

$$\|B_\mu y\| \le \Gamma_\mu \|y\| \quad \text{for all} \quad y \in Y. \tag{6.48}$$

Each such B_μ is called an approximate outer inverse of A. Properties of $C(\mu)$ and Γ_μ are to be prescribed for convergence analysis.

For a compact linear operator $K : H_1 \to H_2$, *the filtered truncated SVD*

$$B_r y = \sum_{i=1}^{r} \frac{\sigma_i}{\sigma_i^2 + \alpha} \langle y, y_i \rangle \, x_i, \quad \alpha > 0 \tag{6.49}$$

is an approximate outer inverse, but not an outer inverse (choose $\mu = r^{-1}$).

It is also useful to introduce a notion of "approximate right-invertibility", which abstracts some characteristics of regularization methods:

$A \in \mathfrak{L}(X, Y)$ is called *approximately right-invertible* if, for each $\mu \in (0, 1)$, there exists a norm $\|\cdot\|_\mu$ on X and a $B_\mu : Y \to X$ such that for all $y \in Y$ and all $x \in X$,

$$\|AB_\mu y - y\| \le \mu \, \|y\| \tag{6.50}$$

$$\|B_\mu y\|_\mu \le \Gamma(\mu) \, \|y\|, \tag{6.51}$$

and $\|x\|_\mu \to \|x\|$ as $\mu \to 0^+$. Each such B_μ is called an *approximate right inverse* of A. Properties of $C(\mu)$ and Γ_μ and B_μ are to be prescribed for convergence analysis.

Each such B_μ is called an *approximate right inverse* of A with a bound $\Gamma(\mu)$ (note that B_μ need not be linear). For regularization one requires $\Gamma(\mu) = O(\mu^{-\gamma})$ or a similar behaviour. Again $\mu \, \|y\|$ in (6.50) may be replaced by $C(\mu) \, \|y\|$.

For regularizers of type I we have bounded outer inverses with infinite-dimensional range. For ill-posed problems for which such regularizers do not exist, we may use approximate outer inverses as regularizers. Again, these can only be approximate inner inverses or approximate right inverses.

Characterizations of ill-posed problems. Next we deal with characterizations of ill-posed problems of so-called type I and II: *Let* $\mathrm{Out}(L)$ *denote that set of all (nonzero) outer inverses to a (nonzero) linear transformation* $L : V \to W$, where V and W are vector spaces over the same field. The following proposition is immediate:

Let $L : V \to W$ be a (nonzero) linear transformation. Then the following statements are equivalent for any (nonzero) linear transformation $M : W \to V$ (where $\dotplus$ again denotes algebraic direct sum):

(a) $M \in \mathrm{Out}(L)$.

(b) ML is idempotent and $V = \mathcal{R}(M) \dotplus \mathcal{N}(ML)$.

(c) LM is idempotent and $W = N(M) \dotplus \mathcal{R}(LM)$.

(d) LM is idempotent and $\mathcal{R}(M) \cap \mathcal{N}(L) = \{0\}$.

The next proposition establishes the existence of (algebraic) outer inverses with a prescribed range: Let V_1 be a subspace of V such that $V_1 \cap \mathcal{N}(L) = \{0\}$ and W_2 be an algebraic complement of LV_1. Then there exists an outer inverse M to L such that $\mathcal{R}(M) = V_1$ and $\mathcal{N}(M) = W_2$. Under these conditions $M|LV_1 = (L|V_1^{-1})$.

We now consider bounded outer inverses of $A \in L(X, Y)$, $A \neq 0$, and their connections with certain topological complements. A topological direct sum will be denoted by $\oplus$. We are again interested in conditions under which there exists a bounded outer inverse B with a range and a nullspace prescribed.

Let $A : X \to Y$ *be a bounded linear operator with nonclosed range. Then the ill-posed problem* $Ax = y$ *is of type I if and only if the following two conditions hold:*

(a) $\mathcal{R}(A)$ *contains a closed infinite-dimensional subspace, say* M.

(b) $\mathcal{N}(A)$ *is (topologically) complemented in the subspace* $A^{-1}(M)$, *the inverse image of* M *under* A.

In the case of Hilbert spaces, these characterizing conditions take an explicit and simple form.

Let A *be a bounded linear operator on a Hilbert space* H_1 *into a Hilbert space* H_2, *and let the range of* A *be nonclosed.* Then the following statements are equivalent:

(i) *The ill-posed problem* $Ax = y$ *is of type I,*

(ii) $\mathcal{R}(A)$ *contains a closed infinite dimensional subspace,*

(iii) A *is not compact.*

An operator $A \in \mathfrak{L}(X, Y)$ is *strictly singular* if the subspaces $Z \subset X$ for with the restriction $A|Z$ has a bounded inverse on AZ, the image of Z under A, are necessarily finite dimensional (see, e.g., [**491**]). This notion of a strictly singular operator is clearly the right generalization of a compact operator in Hilbert space. It is natural to ask if it is possible to characterize ill-posed problems of type I in Banach spaces by the condition that "the operator is not strictly singular". However, this is not possible (cf. [**627**], [**628**]).

Finally some comments should be made (following M. Z. Nashed [**627**]):

(i) For various aspects of operator factorizations and operator ranges, M.Z. Nashed [**625**] has initiated the study of bounded (or densely defined closed) linear operators which have bounded outer inverses of infinite rank within the framework of operator factorization and operator ranges.

(ii) The classification of ill-posed linear problems as proposed here provides also a classification of ill-posed nonlinear problems based on properties of outer inverses of the Fréchet or Hadamard derivative of the nonlinear operator (cf. [**614**]). Approximate outer inverses have been used by [**138**] in the context of inverse function theorems when the derivative does not have a bounded inverse of a bounded generalized inverse.

(iii) Examples of operator equations with a noncompact bounded linear operator with nonclosed range arise from various integral and convolution operators on the half-line and for certain generalized Wiener-Hopf operators. They also arise if zero belongs to the continuous spectrum of a linear, bounded selfadjoint and nonnegative operator on a Hilbert space.

(iv) For various aspects of construction and computation of outer inverses of a prescribed rank $s \leq r$ for a matrix of rank r, see [**77**] and several papers and the annotated bibliography in [**617**]. For constructions of outer inverses of bounded linear operators in Banach spaces the reader is referred to [**588**], [**589**], [**620**]. For stability properties for outer inverses, see [**625**], [**627**], [**628**]. For perturbation analysis of outer inverses, see [**589**], [**620**], [**627**], and [**628**]. Convergence analysis of regularization operators hinges upon perturbation and stability properties of outer inverses (see [**627**], [**628**]).

6.5. Reconstruction optimality

Assume that A is of class $\mathfrak{L}(X,Y)$, X,Y Hilbert spaces. Let $\{R_t\}_{t>0}$ be a family of continuous operators (regularizers) from Y to X with $R_t 0 = 0$. If there exists a mapping $\alpha : (0,\infty) \times Y \to (0,\infty)$, such that, for all elements $y \in \mathcal{R}(A)$ and regularization parameter $t = \alpha(\varepsilon, y^\varepsilon)$,

$$\sup\left\{\left\|A^\dagger y - R_{\alpha(\varepsilon,y^\varepsilon)} y^\varepsilon\right\| : y^\varepsilon \in Y \text{ with } \|y - y^\varepsilon\| \leq \varepsilon\right\} \longrightarrow 0, \tag{6.52}$$

then the pair $\left(\{R_t\}_{t>0}, \alpha\right)$ is called a *regularization method* (or simply *a regularization*) of $A^\dagger$.

If $\{R_t\}_{t>0}$ is a subset of $\mathfrak{L}(Y,X)$, then the regularization $\left(\{R_t\}_{t>0}, \alpha\right)$ is called linear. The mapping α is known as the *parameter choice* provided that

$$\sup\left\{\alpha(\varepsilon, y^\varepsilon) : y^\varepsilon \in Y \text{ with } \|y - y^\varepsilon\| \leq \varepsilon\right\} \xrightarrow{\varepsilon \to 0} 0. \tag{6.53}$$

If α is only dependent on ε, i.e., $\alpha(\varepsilon, y^\varepsilon) = \alpha(\varepsilon)$, we speak of an *a priori parameter choice.* If α is dependent on ε and y^ε, i.e., $\alpha = \alpha(\varepsilon, y^\varepsilon)$, it is called an *a posteriori parameter choice.*

The limit relation (6.52) can be equivalently written in the form

$$\sup\left\{\left\|x - R_{\alpha(\varepsilon,y^\varepsilon)} y^\varepsilon\right\| : y^\varepsilon \in Y \text{ with } \|Ax - y^\varepsilon\| \leq \varepsilon\right\} \xrightarrow{\varepsilon \to 0} 0 \tag{6.54}$$

for all $x \in \mathcal{N}(A)^{\perp}$. A direct consequence of a regularization method is that the limit relation

$$\lim_{\varepsilon \to 0} \left\| A^{\dagger} y - R_{\alpha(\varepsilon, y)} y \right\| = 0 \tag{6.55}$$

holds true for all $y \in \mathcal{R}(A)$ (note that, in Equation (6.55), the regularization is applied to y instead of y^{ε}). It is usual to collect all regularization parameters that are relevant in the limit relation (6.55) in a set $\Gamma = \{\alpha(\varepsilon, y) : \varepsilon > 0, y \in \mathcal{R}(A)\}$. Because of the fact that $\lim_{\varepsilon \to 0} \alpha(\varepsilon, y) = 0$ the set Γ possesses an accumulation point at 0. Moreover, $\lim_{\Gamma \ni \lambda \to 0} \left\| A^{\dagger} y - R_{\lambda} y \right\| = 0$ for all $y \in \mathcal{R}(A)$. All in all, we are led to the following result:

Assume that A is of class $\mathfrak{L}(X, Y)$. Let $\left(\{R_t\}_{t>0}, \alpha\right)$ be a regularization method of $A^{\dagger}$. Then, the subfamily $\{R_{\lambda}\}_{\lambda \in \Gamma}$ converges pointwise in $\mathcal{R}(A)$ to $A^{\dagger}$ for $\lambda \to 0$.

As a direct consequence we obtain the statement (see, e.g., Louis [**547**], Rieder [**723**]):

Assume that A is of class $\mathfrak{L}(X, Y)$. Let $\left(\{R_t\}_{t>0}, \alpha\right)$ be a regularization of $A^{\dagger}$. If $\mathcal{R}(A) \neq \overline{\mathcal{R}(A)}$, then $\{\|R_t\|\}_{t>0}$ is unbounded.

As already known, for a linear regularization, the *reconstruction error*, i.e., $\|A^{\dagger} y - R_t y^{\varepsilon}\|$ can be split by use of a family $\{R_t\}_{t>0}$ in the following way:

$$\left\| A^{\dagger} y - R_t y^{\varepsilon} \right\| \leq \underbrace{\left\| A^{\dagger} y - R_t y \right\|}_{\text{approximation error}} + \underbrace{\left\| R_t y - R_t y^{\varepsilon} \right\|}_{\substack{\text{data error} \\ \leq \|R_t\| \, \|y - y^{\varepsilon}\| \leq \|R_t\| \cdot \varepsilon}} \tag{6.56}$$

Once again, it should be pointed out, that both parts of the reconstruction error, i.e., the approximation error and the data error, exhibit an opposed behavior in limit considerations for $t \to 0$ and $t \to \infty$. Hence, an essential task is to search for a value t_{opt} that balances both errors. In fact, the regularization parameter should be chosen in such a way that $\alpha(\varepsilon, y^{\varepsilon}) \approx t_{\mathrm{opt}}$.

Classification of regularization methods. The question (cf. A.K. Louis [**547**]) arises how to classify regularization methods. Since all methods need to converge as $\varepsilon \to 0$, an obvious criterion of specifying their characteristics is the speed of the convergence, i.e., we make the attempt to introduce a classification with respect to the speed of convergence of the total error:

$$\sup\left\{ \left\| A^{\dagger} y - R_{\alpha(\varepsilon, y^{\varepsilon})} y^{\varepsilon} \right\| : y \in \mathcal{R}(A), y^{\varepsilon} \in Y \text{ with } \|y - y^{\varepsilon}\| \leq \varepsilon \right\} \xrightarrow{\varepsilon \to 0} 0. \tag{6.57}$$

Unfortunately, it turns out that the speed of the convergence is arbitrarily slow for <u>all</u> regularization methods.

Let A be of class $\mathfrak{L}(X, Y)$. Assume that $\mathcal{R}(A) \neq \overline{\mathcal{R}(A)}$. Let $(\{R_t\}_{t>0}, \alpha)$ be a regularization of $A^{\dagger}$. Then there exists <u>no</u> function $h : [0, \infty) \to [0, \infty)$ with $\lim_{\varepsilon \to 0} h(\varepsilon) = 0$ such that

$$\sup\left\{ \left\| A^{\dagger} y - R_{\alpha(\varepsilon, y^{\varepsilon})} y^{\varepsilon} \right\| : y \in \mathcal{R}(A), \|y\| \leq 1, y^{\varepsilon} \in Y \text{ with } \|y - y^{\varepsilon}\| \leq \varepsilon \right\} \leq h(\varepsilon). \tag{6.58}$$

Next our interest is to show that the concept of smoothness can be used for the classification of regularization methods. An auxiliary tool is the *absolute value* $|A|$ *of the operator* A. In order to motivate the setting $|A|$ we start with the explanation of a *functional calculus for compact operators* which also helps us to introduce filters for purposes of regularization in the next section. Even better, the functional calculus for compact operators enables us to reduce spectral features to a study in terms of real functions.

Let A be of class $\mathfrak{K}(X,Y)$ with the singular system $\{\sigma_j; x_j, y_j\}_{j\in\mathbb{N}}$. Let $\Phi : [0,\infty) \to \mathbb{R}$ be a piecewise continuous function defined on the interval $\left[0, \|A\|^2\right]$. Then we understand the operator $\Phi(A^*A)$ to be given as

$$\Phi(A^*A)\,x = \sum_{j=1}^{\infty} \Phi\left(\sigma_j^2\right) \langle x, x_j\rangle\, x_j + \Phi(0)\, P_{\mathcal{N}(A)}x. \tag{6.59}$$

Note that the series on the right side of (6.59) is convergent on the interval $\left[0, \|A\|^2\right]$.

Let A be of class $\mathfrak{K}(X,Y)$ with the singular system $\{\sigma_j; x_j, y_j\}_{j\in\mathbb{N}}$. Suppose that $\Phi : [0,\infty) \to \mathbb{R}$ is piecewise continuous. Then the following properties hold true:

(a) $\|A\| = \sigma_1$, *i.e., the norm coincides with the largest singular value of* A,

(b) $\Phi(A^*A)A^* = A^*\Phi(AA^*)$,

(c) $\|\Phi(A^*A)\| = \sup_{j\in\mathbb{N}} |\Phi(\sigma_j^2)| \le \sup_{0\le\lambda\le\|A\|^2} |\Phi(\lambda)|$,

(d) $\|\Phi(A^*A)A^*\| = \|A^*\Phi(AA^*)\| = \sup_{j\in\mathbb{N}} \left(\sigma_j|\Phi(\sigma_j^2)|\right) \le \sup_{0\le\lambda\le\|A\|^2} \left(\sqrt{\lambda}\,|\Phi(\lambda)|\right)$.

Example. If $\Phi = 1$, then it is clear that

$$\Phi(A^*A)\,x = \sum_{j=1}^{\infty} \langle x, x_j\rangle\, x_j + P_{\mathcal{N}(A)}x = P_{\overline{\mathcal{R}(A^*)}}x + P_{\mathcal{N}(A)}x = P_X x = x. \tag{6.60}$$

This explains the occurrence of the term $\Phi(0)\,P_{\mathcal{N}(A)}x$ in (6.59).

Example. If $\Phi(t) = t^{1/2}, t \ge 0$, then

$$(A^*A)^{1/2}\, x = \sum_{j=1}^{\infty} \sigma_j \langle x, x_j\rangle\, x_j + \underbrace{\Phi(0)}_{=0}\left(P_{\mathcal{N}(A)}x\right) = \sum_{j=1}^{\infty} \sigma_j \langle x, x_j\rangle\, x_j, \tag{6.61}$$

holds true for all $x \in X$.

The operator $|A| = (A^*A)^{1/2}$ is called *the absolute value of* A. Analogously, $|A^*|$ is given by

$$|A^*|\, y = (AA^*)^{1/2}\, y = \sum_{j=1}^{\infty} \sigma_j \langle y, y_j\rangle\, y_j, \quad y \in Y. \tag{6.62}$$

It is easy to see that

$$\|\,|A|\,x\|^2 = \langle |A|x, |A|x\rangle = \langle A^*Ax, x\rangle = \langle Ax, Ax\rangle = \|Ax\|^2. \tag{6.63}$$

We remark that our notation (6.61) can be used to introduce fractional powers $(A^*A)^\mu, \mu \geq 0$, of A^*A:

$$|A|^{2\mu}x = (A^*A)^\mu x = \sum_{j=1}^{\infty} \sigma_j^{2\mu} \langle x, x_j \rangle x_j. \tag{6.64}$$

Let X, Y be Hilbert spaces. Suppose that A is of class $\mathfrak{K}(X, Y)$. Then

(1) $\mathcal{R}(A^*) = \mathcal{R}(|A|) = \mathcal{R}\left((A^*A)^{\frac{1}{2}}\right)$,

(2) $\mathcal{R}(A) = \mathcal{R}(|A^*|) = \mathcal{R}\left((AA^*)^{\frac{1}{2}}\right)$.

Next we introduce subspaces of the Hilbert space X involving on the concept of powers of the absolute value $|A|$ of the operator A: For $\nu \geq 0$, let X_ν be defined by

$$X_\nu = \mathcal{R}(|A|^\nu) = \left\{ |A|^\nu x : x \in \mathcal{N}(A)^\perp \right\}. \tag{6.65}$$

Then the following properties can be verified by straightforward arguments:

(1) $X_\nu \subset \mathcal{N}(A)^\perp \subset X$ for all $\nu \geq 0$,

(2) $X_\nu \subset X_\mu$ for $\nu > \mu \geq 0$,

(3) $X_0 = \mathcal{N}(A)^\perp$.

In connection with $x = |A|^\nu z = \sum\limits_{k=1}^{\infty} \sigma_k^\nu \langle z, x_k \rangle x_k$ we are able to impose the following norm on X_ν :

$$\|x\|_\nu^2 = \|z\|^2 = \sum_{k=1}^{\infty} \frac{|\langle z, x_k \rangle|^2}{\sigma_k^{2\nu}} \sigma_k^{2\nu} = \sum_{k=1}^{\infty} \frac{|\langle x, x_k \rangle|^2}{\sigma_k^{2\nu}}. \tag{6.66}$$

Alternatively, the spaces X_ν can be characterized by the norms (6.66), i.e.,

$$X_\nu = \left\{ x \in \mathcal{N}(A)^\perp : \|x\|_\nu < \infty \right\}. \tag{6.67}$$

The spaces X_ν impose conditions on the smoothness of the elements $x \in X$.

After these preliminaries about powers of absolute values we come back to the discussion of the speed of the convergence of a regularization method involving the concept of smoothness, i.e., the solution of an operator equation will be assumed to be a member of a subspace X_ν, $\nu > 0$, of X.

Indeed, in a large number of ill-posed problems $(A; X, Y)$ the operator A shows the property that the image Ax is smoother than x. In concrete situations this leads us to functions x_k in the singular system with strongly growing oscillations for increasing k. Looking at the higher frequency parts of an element $x \in X$, i.e., the inner products $\langle x, x_k \rangle$ for large k, we notice that there is a damping effect on Ax by the factor σ_k. Hence, the norm $\|x\|_\nu$ can be interpreted in the sense that $\langle x, x_k \rangle / \sigma_k^\nu \to 0$ for $k \to \infty$ is demanded. In addition, the larger the value of ν is chosen, the faster $\langle x, x_k \rangle$ has to converge to 0. Therefore it can be concluded that the element x does not contain relevant high frequency components. In other words, x can be regarded as "smooth".

Optimality of reconstruction methods. A stable *reconstruction method* for the solution of the operator equation involving $A \in \mathfrak{L}(X,Y)$ is understood to be a continuous (not-necessarily linear) mapping $T : Y \to X$ with $T\,0 = 0$. The question is which reconstruction error does occur in the *best worst case*, if the data are noisy.

The *worst case error of a reconstruction method* T for A corresponding to the noise level ε and the additional information $\|A^\dagger y\|_\nu \le \rho$ is defined as

(6.68)

$$E_\nu(\varepsilon,\rho,T) = \sup\left\{\|Ty^\varepsilon - A^\dagger y\| : \ y \in \mathcal{R}(A), y^\varepsilon \in Y, \|y - y^\varepsilon\| \le \varepsilon, \|A^\dagger y\|_\nu \le \rho\right\}.$$

Another expression (avoiding the occurrence of the operator $A^\dagger$) can be formulated by

$$E_\nu(\varepsilon,\rho,T) = \sup\left\{\|Ty^\varepsilon - x\| : \ x \in X_\nu, y^\varepsilon \in Y, \|Ax - y^\varepsilon\| \le \varepsilon, \|x\|_\nu \le \rho\right\}. \tag{6.69}$$

Clearly, the smaller the worst case error, the better the reconstruction method.

The *best worst case error* for A corresponding to the noise level ε and the additional restriction $\|A^\dagger y\|_\nu \le \rho$ is understood to be

$$E_\nu(\varepsilon,\rho) = \inf\left\{E_\nu(\varepsilon,\rho,T) : \ T : Y \to X \text{ continuous}, T\,0 = 0\right\}. \tag{6.70}$$

Note that the best worst case error, i.e., $E_\nu(\varepsilon,\rho)$, depends on the problem (i.e., on the operator A), but not on the reconstruction method.

The next result tells us about the quantity of the best worst case error.

Let A be of class $\mathfrak{L}(X,Y)$. Then we have

$$E_\nu(\varepsilon,\rho) = e_\nu(\varepsilon,\rho), \tag{6.71}$$

where

$$e_\nu(\varepsilon,\rho) = \sup\left\{\|x\| : \ x \in X_\nu, \|Ax\| \le \varepsilon, \|x\|_\nu \le \rho\right\}. \tag{6.72}$$

Clearly, our results characterize the best worst case error independently of the knowledge of a specific reconstruction method.

Let A be of class $\mathfrak{L}(X,Y)$. Then, for $\nu > 0$,

$$e_\nu(\varepsilon,\rho) \le \rho^{\frac{1}{\nu+1}} \varepsilon^{\frac{\nu}{\nu+1}}. \tag{6.73}$$

Furthermore, there exists a sequence $\{\varepsilon_k\}_{k\in\mathbb{N}}$ with $\varepsilon_k \to 0$ for $k \to \infty$ such that

$$e_\nu(\varepsilon,\rho) = \rho^{\frac{1}{\nu+1}} \varepsilon^{\frac{\nu}{\nu+1}}. \tag{6.74}$$

In other words, the estimate (6.73) is sharp, i.e., it cannot be improved.

6.6. Order optimality of regularizing filters

If A is an injective operator of class $\mathfrak{K}(X,Y)$, then $A^\dagger$ can be expressed in the form $(A^*A)^{-1}A^*$. The noncontinuity is caused by the term $(A^*A)^{-1}$, that has to be stabilized. In connection with the functional calculus for compact operators we are therefore led to filters as appropriate tools for regularization.

Let $\{F_t\}_{t>0}$, $F_t : \left[0, \|A\|^2\right] \to \mathbb{R}$ be a family of piecewise continuous functions satisfying the conditions

(F1) $\lim_{t\to 0} F_t(\lambda) = \frac{1}{\lambda}$ for all $\lambda \in \left(0, \|A\|^2\right]$,

(F2) $\lambda |F_t(\lambda)| \leq C_F$ for all $\lambda \in \left[0, \|A\|^2\right]$ and $t > 0$.

Then the family $\{F_t\}_{t>0}$ is called a *filter relative to* A.

By virtue of Condition (F1), $F_t(A^*A)$ becomes a continuous operator, which converges in pointwise sense to $(A^*A)^{-1}$ as $t \to 0$. This is the reason why we let

$$R_t y = F_t(A^*A) A^* y, \quad y \in Y. \tag{6.75}$$

As a consequence, $\{F_t\}_{t>0}$ filters the influence of small singular values of A on the operator R_t in (6.75).

In terms of the singular system $\{\sigma_k; x_k, y_k\}_{k\in\mathbb{N}}$ we are able to write

$$\begin{aligned} F_t(A^*A) A^* y &= \sum_{k=1}^{\infty} F_t\left(\sigma_k^2\right) \sigma_k \langle y, y_k\rangle x_k + F_t(0) P_{\mathcal{N}(A)} A^* y \\ &= \sum_{k=1}^{\infty} F_t\left(\sigma_k^2\right) \sigma_k \langle y, y_k\rangle x_k \end{aligned} \tag{6.76}$$

due to fact that $P_{\mathcal{N}(A)} A^* y = 0$ (note that $A^* y \in \mathcal{R}(A^*) = \mathcal{N}(A)^{\perp}$). Considering the approximation error we obtain, for $y \in \mathcal{R}(A)$, that

$$\begin{aligned} A^\dagger y - R_t y &= A^\dagger y - F_t(A^*A) A^* y \\ &= A^\dagger y - F_t(A^*A) A^* A A^\dagger y \\ &= (I - F_t(A^*A) A^* A) A^\dagger y \\ &= p_t(A^*A) A^\dagger y, \end{aligned} \tag{6.77}$$

where the function $p_t : \lambda \mapsto p_t(\lambda), \lambda \in \left[0, \|A\|^2\right]$ is given by

$$p_t(\lambda) = 1 - \lambda F_t(\lambda), \quad \lambda \in \left[0, \|A\|^2\right]. \tag{6.78}$$

The identity (6.77) leads us to the formulation of the following result:

Assume that $A \in \mathfrak{K}(X,Y)$. *Let* $\{F_t\}_{t>0}$ *be a filter. Then*

$$\lim_{t\to 0} R_t y = \begin{cases} A^\dagger y & y \in \mathcal{D}(A^\dagger), \\ \infty & y \notin \mathcal{D}(A^\dagger), \end{cases} \tag{6.79}$$

where

$$R_t y = F_t(A^*A) A^* y$$

for $y \in Y$.

The next result concretizes the stability of $R_t y^\varepsilon$ under the noise level ε in more detail:

Let $\{F_t\}_{t>0}$ *be a filter so that* $(F1), (F2)$ *hold true. For* $y, y^\varepsilon \in Y$ *with* $\|y - y^\varepsilon\| \leq \varepsilon$, *set* $x_t = R_t y$ *and* $x_t^\varepsilon = R_t y^\varepsilon$. *Then, for the residual term, we have*

$$\|Ax_t - Ax_t^\varepsilon\| \leq C_F \varepsilon, \tag{6.80}$$

while, for the error term, it follows that

$$\|x_t - x_t^\varepsilon\| \le \varepsilon\sqrt{C_F M(t)}$$

where we have used the abbreviation

$$M(t) = \sup_{0\le\lambda\le\|A\|^2} |F_t(\lambda)|. \tag{6.81}$$

Let us continue with the *estimate of the total error*, thereby using $x^\dagger = A^\dagger y$, $y \in D(A^\dagger)$. We base our considerations on the usual splitting into the approximation error and the data error in the form

$$\begin{aligned} \|A^\dagger y - R_t y^\varepsilon\| &\le \|A^\dagger y - R_t y\| + \|R_t y - R_t y^\varepsilon\| \\ &= \|x^\dagger - x_t\| + \|x_t - x_t^\varepsilon\| \\ &\overset{t\to 0}{\le} \underbrace{\|x^\dagger - x_t\|}_{\to 0} + \varepsilon\sqrt{C_F M(t)}. \end{aligned} \tag{6.82}$$

Obviously, the approximation error $\|x^\dagger - R_t y\|$ tends to 0. The discussion of the data error is much more problematic than the approximation error: From the limit relation $\lim_{t\to 0} F_t(\lambda) = \frac{1}{\lambda}$ imposed on a filter within the interval $[0, \|A\|^2]$ it follows that $M(t) \overset{t\to 0}{\longrightarrow} \infty$. Therefore, for the total error, we are confronted with a divergent behavior as $t \to 0$. Nevertheless, convergence properties can be forced by a suitable coupling of t and ε. In fact, it can be deduced that, under the a priori parameter choice indicated above, the approximation error as well as the data error converge to 0, if the noise level ε tends to zero:

Let $\{F_t\}_{t>0}$ be a filter. If we choose $\alpha : (0,\infty) \to (0,\infty), \varepsilon \mapsto \alpha(\varepsilon)$, such that $\alpha(\varepsilon) \overset{\varepsilon\to 0}{\longrightarrow} 0$ as well as $\varepsilon\sqrt{M(\alpha(\varepsilon))} \overset{\varepsilon\to 0}{\longrightarrow} 0$, then $(\{R_t\}_{t>0}, \alpha)$ with $R_t = F_t(A^*A)A^*$ is a regularization of $A^\dagger$ (by convention, $\{F_t\}_{t>0}$ is called a regularizing filter).

Next we mention which additional requirement imposed on filters $R_t = F_t(A^*A)A^*$ will be necessary to guarantee the order optimality.

Let $\{F_t\}_{t>0}$ be a regularizing filter for $A \in \mathfrak{L}(X,Y)$. Assume there are $t_0 > 0$, $\mu > 0$, and a function $\omega_\mu : (0,t_0] \to \mathbb{R}$ such that

$$\sup_{0\le\lambda\le\|A\|^2} \lambda^{\mu/2}|p_t(\lambda)| \le \omega_\mu(t) \tag{6.83}$$

for all $t \in (0,t_0]$ (with $p_t(\lambda) = 1 - \lambda F_t(\lambda)$). Let $y \in \mathcal{R}(A)$ and let $x^\dagger = A^\dagger y$ in X_μ with $\|x^\dagger\|_\mu \le \varrho$. Then the following estimates

(a) $\|x^\dagger - x_t\|_X \le \omega_\mu(t)\varrho$,

(b) $\|Ax^\dagger - Ax_t\|_Y \le \omega_{\mu+1}(t)\varrho$

hold true for $x_t = R_t y = F_t(A^*A)A^*y$ and $0 < t < t_0$.

Moreover, let

(1) $\omega_\mu(t) \le C_p t^{\frac{\mu}{2}}$ *for* $t \to 0$,

(2) $M(t) = \sup_{0\le\lambda\le\|A\|^2} |F_t(\lambda)| \le C_M t^{-1}$ *for* $t \to 0$,

where $\mu > 0$, $C_p, C_M > 0$ are constants. Let the a priori parameter choice $\alpha : (0,\infty) \to (0,\infty)$ fulfill

$$C_1 \left(\frac{\varepsilon}{\rho}\right)^{\frac{2}{\mu+1}} \leq \alpha(\varepsilon) \leq C_2 \left(\frac{\varepsilon}{\rho}\right)^{\frac{2}{\mu+1}}, \quad \varepsilon \to 0, \tag{6.84}$$

where C_1, C_2 are positive constants. Then, $(\{R_t\}_{t>0}, \alpha)$, $R_t = F_t(A^*A)A^*$, is an order optimal regularization for $A^\dagger$ with respect to X_μ.

Of course, we need to know the values ρ and μ to guarantee the order optimality. Without the availability of ρ but based on the knowledge of μ, we are led to $\alpha(\varepsilon) = C\varepsilon^{\frac{2}{\mu+1}}$ with C being a positive constant to find an order optimal regularization. Without any information of both parameters ρ as well as μ we have to deal with a posteriori parameter choices.

An asymptotic behavior of ω_μ determines the speed of convergence for the reconstruction error. An important feature is the so-called *qualification.*

Let $\{F_t\}_{t\geq 0}$ be a regularizing filter for $A^\dagger$, where $A \in \mathfrak{L}(X,Y)$ satisfying the asymptotic relation

$$M(t) \leq \frac{C_M}{t}, \quad t \to 0. \tag{6.85}$$

where C_M is a positive constant. The maximal value μ_0, such that there exists, for all values $\mu \in (0, \mu_0]$, a constant $C_p > 0$ satisfying

$$\sup \lambda^{\frac{\mu}{2}} |p_t(\lambda)| \leq C_p\, t^{\frac{\mu}{2}}, \quad t \to 0, \tag{6.86}$$

is called the *qualification of the filter* (remember $p_t(\lambda) = 1 - \lambda F_t(\lambda)$).

In other words, the qualification can be regarded as the maximal rate of decay. If the qualification is finite, there exists a parameter choice $\alpha(\varepsilon)$ such that

$$\left\| A^\dagger y - R_{\alpha(\varepsilon)} y^\varepsilon \right\| = O\left(\varepsilon^{\frac{\mu_0}{\mu_0+1}}\right), \quad \varepsilon \to 0, \tag{6.87}$$

holds true for $A^\dagger y \in X_{\mu_0}$. If the qualification is infinite, there exists a parameter choice $\alpha(\varepsilon)$ for which the error decay comes arbitrarily close to $O(\varepsilon)$. As a consequence, filters with infinite qualification are more advantageous than others.

In the following we recover important examples of regularization methods constituted by filters, i.e., the truncated singular value decomposition SVD and the Tikhonov-Phillips regularization.

Truncated singular value decomposition revisited. The SVD of $A^\dagger$ is

$$A^\dagger y = \sum_{k=1}^{\infty} \frac{1}{\sigma_k} \langle y, y_k\rangle\, x_k. \tag{6.88}$$

We choose the filter

$$F_t(\lambda) = \begin{cases} \frac{1}{\lambda}, & \lambda \geq t, \\ 0, & \lambda < t. \end{cases} \tag{6.89}$$

Then it follows that

$$R_t y = F_t(A^*A)\, A^* y = \sum_{k=1}^{\infty} F_t\left(\sigma_k^2\right) \sigma_k \langle y, y_k\rangle\, x_k = \sum_{\sigma_k \geq \sqrt{t}} \frac{1}{\sigma_k} \langle y, y_k\rangle\, x_k \tag{6.90}$$

is the truncated SVD of $A^\dagger$ (with finitely many summands). Clearly we have

(1) $\lim_{t\to 0} F_t(\lambda) = \frac{1}{\lambda}$ for all $\lambda > 0$,

(2) $C_F = \sup_{0\le\lambda\le\|A\|^2} \lambda |F_t(\lambda)| = 1$ for all $t \le \|A\|^2$,

(3) $M(t) = \sup_{0\le\lambda\le\|A\|^2} |F_t(\lambda)| = \frac{1}{t}$ for all $t > 0$.

The total error can be described as follows:

$$\begin{aligned}\left\|A^\dagger y - R_t y^\varepsilon\right\| &\le \left\|A^\dagger y - R_t y\right\| + \|R_t y - R_t y^\varepsilon\| \\ &= \left\|\sum_{\sigma_k<\sqrt{t}} \frac{1}{\sigma_k} \langle y, y_k\rangle x_k\right\| + \varepsilon\sqrt{C_F M(t)} \\ &\overset{\text{Parseval}}{=} \left(\sum_{\sigma_k<\sqrt{t}} \frac{1}{\sigma_k^2} |\langle y, y_k\rangle|^2\right)^{\frac{1}{2}} + \frac{\varepsilon}{\sqrt{t}}.\end{aligned} \tag{6.91}$$

Moreover, for all $\mu > 0$ and $0 \le t \le \|A\|^2$, we have

$$\sup_{0\le\lambda\le\|A\|^2} \lambda^{\frac{\mu}{2}} |p_t(\lambda)| = \sup_{0\le\lambda\le\|A\|^2} \lambda^{\frac{\mu}{2}} |1 - \lambda F_t(\lambda)| = \sup_{0\le\lambda\le t} \lambda^{\frac{\mu}{2}} = t^{\frac{\mu}{2}}.$$

Together with $M(t) = \frac{1}{t}$ we find that this filter possesses an infinite qualification, and it is order optimal for all $\mu > 0$. However, one can show that the TSVD is *not* optimal for *any* $\mu > 0$.

Asymptotic regularization revisited. We choose the following filter

$$F_t(\lambda) = \begin{cases} \frac{1-\exp\left(-\frac{\lambda}{t}\right)}{\lambda}, & \lambda > 0 \\ \frac{1}{t}, & \lambda = 0. \end{cases} \tag{6.92}$$

Then we obtain

$$R_t y = \sum_{k=1}^\infty F_t\left(\sigma_k^2\right) \sigma_k \langle y, y_k\rangle x_k = \sum_{k=1}^\infty \frac{1 - \exp\left(-\frac{\sigma_k^2}{t}\right)}{\sigma_k} \langle y, y_k\rangle\ x_k.$$

It is not difficult to prove the following properties:

(1) $\lim_{t\to 0} F_t(\lambda) = \frac{1}{\lambda}$ for $\lambda > 0$

(2) $\lambda |F_t(\lambda)| = \begin{cases} 1 - \exp\left(-\frac{\lambda}{t}\right), & \lambda > 0 \\ \frac{\lambda}{t}, & \lambda = 0 \end{cases} \le 1 \quad , \quad t > 0,$

(3) $\sup_{0\le\lambda\le\|A\|^2} \lambda |F_t(\lambda)| = 1 = C_F$,

(4) $M(t) = \sup_{0\le\lambda\le\|A\|^2} |F_t(\lambda)| = \frac{1}{t}$, $t > 0$ (note that $F_t(\lambda)$ is monotonically decreasing in λ and $\lim_{\lambda\to 0} F_t(\lambda) = \frac{1}{t}$).

The qualification of the filter again is infinite.

Tikhonov regularization revisited. Using the filter

$$F_t(\lambda) = \frac{1}{\lambda + t}, \quad t > 0, \tag{6.93}$$

we find that

$$R_t y = F_t(A^* A) A^* y = \sum_{k=1}^\infty \frac{\sigma_k}{\sigma_k^2 + t} \langle y, y_k\rangle x_k \tag{6.94}$$

and

$$(A^*A+tI)\,R_t y = \sum_{k=1}^{\infty} \frac{\sigma_k}{\sigma_k^2+t}\,\langle y, y_k\rangle\left(\sigma_k^2+t\right)x_k = \sum_{k=1}^{\infty}\sigma_k\,\langle y, y_k\rangle\,x_k$$
$$= \sum_{k=1}^{\infty}\langle y, \sigma_k y_k\rangle\,x_k = \sum_{k=1}^{\infty}\langle y, Ax_k\rangle\,x_k = \sum_{k=1}^{\infty}\langle A^*y, x_k\rangle\,x_k = A^*y,$$

i.e., $x_t = R_t y$ is the unique solution of the equations $(A^*A+tI)\,x_t = A^*y$. Such equations are called *regularized normal equations.* It can be easily seen that

(1) $\lim_{t\to 0} F_t(\lambda) = \frac{1}{\lambda}$, $\lambda > 0$,
(2) $\lambda\,|F_t(\lambda)| = \frac{\lambda}{\lambda+t} \le 1 = C_F$ for all $t > 0$,
(3) $M(t) = \sup_{0\le\lambda\le\|A\|^2} |F_t(\lambda)| = \sup_{0\le\lambda\le\|A\|^2} \frac{1}{\lambda+t} = \frac{1}{t}$, $t > 0$.

In order to compute the qualification for Tikhonov's regularization we take a look at the term

$$\sup_{0\le\lambda\le\|A\|^2} \lambda^{\frac{\mu}{2}}\,|p_t(\lambda)| = \sup_{0\le\lambda\le\|A\|^2} \underbrace{t^{\frac{\mu}{2}}\,\frac{\left(\frac{\lambda}{t}\right)^{\frac{\mu}{2}}}{\frac{\lambda}{t}+1}}_{=h_\mu(\lambda,t)}. \tag{6.95}$$

In fact, we are led to distinguish two cases:

Case 1: For $\mu > 2$: $h_\mu(\lambda,t)$ is strictly monotonically increasing in λ.

Case 2: For $\mu \le 2$: we find the estimate

$$\sup_{0\le\lambda\le\|A\|^2} h_\mu(\lambda,t) \le t^{\frac{\mu}{2}} \underbrace{\sup_{0\le z<\infty} \frac{z^{\frac{\mu}{2}}}{z+1}}_{=C_p<\infty}. \tag{6.96}$$

Summarizing our considerations we obtain

$$\sup_{0\le\lambda\le\|A\|^2} \lambda^{\frac{\mu}{2}}\,|p_t(\lambda)| \le \begin{cases} C_p t^{\frac{\mu}{2}}, & 0<\mu\le 2,\\ \|A\|^{\mu-2}t, & \mu > 2. \end{cases} \tag{6.97}$$

In other words, the qualification of the Tikhonov filter is $\mu_0 = 2$, i.e., we arrive at the order optimality and even at the optimality for $0 < \mu \le 2$, if we use the parameter choice $\alpha(\varepsilon) = \frac{1}{\mu}\left(\frac{\varepsilon}{\rho}\right)^{\frac{2}{\mu+1}}$.

Morozov's discrepancy principle. Let $y \in \mathcal{D}\left(A^\dagger\right)$ and $y^\varepsilon \in Y$ with $\|y - y^\varepsilon\| < \varepsilon$ and $x_t^\varepsilon = F_t(A^*A)\,A^*y^\varepsilon$. The *idea of the discrepancy principle* can be explained as follows: Choose the parameter $\alpha = \alpha(\varepsilon, y^\varepsilon)$ such that

$$\|Ax_t^\varepsilon - y^\varepsilon\| \approx \varepsilon. \tag{6.98}$$

In other words, the *residual* or *discrepancy* of y^ε is assumed to have the order of the data error.

In order to realize the assumption (6.98) we consider the defect function d : $t \mapsto d(t) = \|Ax_t^\varepsilon - y^\varepsilon\|$. It is not hard to see that

$$
\begin{aligned}
(6.99) \qquad d(t) &= \|AF_t(A^*A)A^*y^\varepsilon - y^\varepsilon\| \\
&= \|p_t(AA^*)y^\varepsilon\| \\
&= \left(\sum_{k=1}^{\infty} p_t^2(\sigma_k^2)|\langle y^\varepsilon, y_k\rangle|^2 + \underbrace{p_t(0)}_{=1}\left\|P_{\mathcal{N}(A^*)}y^\varepsilon\right\|^2\right)^{\frac{1}{2}},
\end{aligned}
$$

so that

$$
(6.100) \qquad \lim_{t\to 0, t>0} d(t) = \|P_{\mathcal{N}(A^*)}y^\varepsilon\| = \left\|P_{\mathcal{R}(A)^\perp}y^\varepsilon\right\|.
$$

If $y \notin \mathcal{R}(A)$, then $\left\|P_{\mathcal{R}(A)^\perp}y^\varepsilon\right\|_Y$ can be arbitrarily large. However, if $y \in \mathcal{R}(A)$, then we have

$$
(6.101) \qquad \left\|P_{\mathcal{R}(A)^\perp}y^\varepsilon\right\| = \left\|P_{\mathcal{R}(A)^\perp}(y - y^\varepsilon)\right\| \le \|y - y^\varepsilon\| \le \varepsilon
$$

and, therefore, $\lim_{t\to 0, t>0} d(t) \le \varepsilon$ In other words, for all $\tau > 1$ exists a t_0 with $d(t) < \tau\varepsilon$ for all $t \le t_0$.

Let $\tau > 1$ be chosen (fixed) and $\{t_k\}$ be a strictly monotonically decreasing sequence with limit zero. Determine k^*, such that

$$
(6.102) \qquad d(t_{k^*}) \le \tau\varepsilon \le d(t_i), \quad i = 1, \dots, k^* - 1
$$

Set $\alpha(\varepsilon, y^\varepsilon) = t_{k^*}$.

Let A be of class $\mathfrak{L}(X,Y)$. Suppose that $\{F_t\}_{t>0}$ is a regularizing filter with qualification $\mu_0 > 1$. Moreover, assume that $M(t) \le \frac{C_M}{t}$ for $t \to 0$. Let the parameter choice $\alpha : (0,\infty) \times Y \to (0,\infty)$ be taken in accordance with the discrepancy principle, such that the sequence $\{t_k\}_k$ satisfies $t_k = \theta_k t_{k-1}$, where $0 < \vartheta \le \theta_k < 1$ for all k. Furthermore, suppose that $\tau > \sup\{|p_t(\lambda)| : t > 0, 0 \le \lambda \le \|A\|^2\} \ge p_t(0) = 1$. Then $\left(\{R_t\}_{t>0}, \gamma\right)$ with $R_t = F_t(A^*A)A^*$ is an order optimal regularization of $A^\dagger$ with respect to X_μ for all $(0, \mu_0 - 1]$.

Note that, for the values $\mu \in (\mu_0 - 1, \mu_0]$ we obtain nothing, whereas, the Tikhonov case leads to the order of convergence $O(\varepsilon^{\frac{1}{2}})$ instead of $O(\varepsilon^{\frac{2}{3}})$. This is the reason why the discrepancy principle may produce suboptimal rates of convergence in case of a finite qualification. However, in connection with an infinite qualification, no such problems arise.

Generalized discrepancy principle. Next we are interested in a generalization of the discrepancy principle. To this end we make a reformulation: Let $\{t_k\}$ be a strictly monotonically decreasing sequence with limit zero. Then, t_{k^*} is chosen as follows:

$$
\begin{aligned}
(6.103) \qquad t_{k^*} &= \sup\left\{t_k : \left\|Ax_{t_k}^\varepsilon - y^\varepsilon\right\|^2 \le \tau\varepsilon^2\right\} \\
&= \sup\left\{t_k : \|p_{t_k}(AA^*)y^\varepsilon\|^2 \le \tau\varepsilon^2\right\} \\
&= \sup\left\{t_k : \langle y^\varepsilon, p_{t_k}^2(AA^*)y^\varepsilon\rangle \le \tau\varepsilon^2\right\}.
\end{aligned}
$$

In doing so we have used the function $s_t = p_t^2$.

For a generalization we allow arbitrary functions s_t in the following sense: Let $\tau > 1$ be chosen (fixed) and $\{t_k\}$ as before. Determine k^* such that

$$t_{k^*} = \sup\left\{t_k : \langle y^\varepsilon, s_{t_k}(AA^*) y^\varepsilon\rangle \leq \tau\varepsilon^2\right\}. \tag{6.104}$$

Set $\alpha(\varepsilon, y^\varepsilon) = t_{k^*}$. We have to look for functions s_t which yield order optimal methods for the whole parameter domain $(0, \mu_0]$. An answer is given by the example:

$$s_t(\lambda) = p_t^{2+\frac{2}{\mu_0}}(\lambda) = p_t^3(\lambda) = \left(\frac{t}{t+\lambda}\right)^3. \tag{6.105}$$

The generalized discrepancy principle with this family of functions s_t together with the Tikhonov regularization is an order optimal method with respect to X_μ for $\mu \in (0, 2]$.

In the previous considerations we have studied a number of *a posteriori parameter choice rules* which all depend in one way or the other on the computed approximation - and on the given data error level ε. A perfect example to illustrate this general reasoning is the discrepancy principle where reconstructions are discarded unless their data fit has the order to the noise level ε. In practical examples such noise level information is not always available (or reliable). For instance, a given discrete data vector may consist of a finite number of measurements, for each of which we may or may not know the standard deviation and/or a worst-case error bound. Typically, the worst-case bound will be a severe overestimation, while the standard deviation might underestimate the true error. both estimates may therefore lead to a significant loss of accuracy when used in these parameter choice rules. Another uncertainty problem arises if we are going to embed the discrete data into a continuous model by some interpolation or approximation process. Then we have to estimate the L^2-norm of the difference between the constructed function and the true data function from the discrete noise information, and from a-priori assumed smoothness properties of the data.

Often it is necessary to consider alternative (a-posteriori) parameter choice rules that avoid knowledge of the noise level, and to determine some realistic regularization parameter on the basis of the actual performance of the regularization method under consideration. Such heuristic parameter choice rules will be called *error free*. A good reference to these strategies is [**215**]. It must be emphasized, however, that error free parameter choice rules cannot provide a convergent regularization method in the strict sense. Still, there are examples where an error free rule leads to better reconstructions than some sophisticated order-optimal rule, (see, e.g., [**400**] for some numerical comparisons). Another heuristic parameter choice rule which can be interpreted via some kind of error estimation is the method of generalized cross-validation introduced by G. Wahba (see [**844**] for the history of this method and a more detailed exposition). It applies to problems where A is an operator into a finite-dimensional data space, e.g., a generalized moment problem. Another very popular error-free parameter choice rule has been advocated by P. C. Hansen [**406**]. This method is based on an inspection of the residual norms of the computed approximations, this time by relating them to the norms of the approximations themselves (cf. [**400**]). In spite of its use in several applications, for example, in satellite to satellite tracking, satellite gravitational gradiometry (see, e.g., [**732**], [**733**], [**266**], [**302**], [**323**], [**307**] and the references therein) there still lacks a sound mathematical foundation of the L-curve method.

Tikhonov-Phillips regularization. The Tikhonov-Phillips filter is given by

$$F_t(\lambda) = \frac{1}{\lambda + t}, \quad t > 0, \quad \lambda \in \left[0, \|A\|^2\right]. \tag{6.106}$$

Let A be of class $\mathfrak{L}(X, Y)$. Without loss of generality, suppose that A is injective (otherwise we have to replace X by $\mathcal{N}(A)^{\perp}$). Let Z be a Hilbert space and $B \in \mathfrak{L}(X, Z)$ be continuously invertible, i.e., there exists a value $\beta > 0$ such that

$$\beta \|x\|_X \leq \|Bx\|_Z \quad \text{for all } x \in X. \tag{6.107}$$

Before we deal with the generalization of the Tikhonov-Phillips regularization we mention some preparatory results:

(1) *Lax-Milgram lemma.* Suppose that $L \in \mathfrak{L}(X)$. Assume there exists a value $\lambda > 0$ such that $\langle Lx, x \rangle \geq \lambda \|x\|^2$ holds for all $x \in X$. Then L is continuously invertible and $\|L^{-1}\| \leq \frac{1}{\lambda}$.

(2) The stabilized normal equation

$$(A^*A + tB^*B)\, x = A^*y \tag{6.108}$$

with $y \in Y$ possesses a unique solution for all $t > 0$ which continuously depends on y. Our aim is to show that the solution of (6.108) can be equivalently obtained by minimizing the argument of the *Tikhonov-Phillips functional* given by

$$J_{t,y}(x) = \|Ax - y\|^2 + t\|Bx\|^2. \tag{6.109}$$

Note that the second term on the right hand side of (6.109) is called the *penalty term of the Tikhonov-Phillips functional*:

Let A and B be given as indicated above. Moreover, assume that $y \in Y$ and $t > 0$. Then the following statements are equivalent:

(1) $(A^*A + tB^*B)\, x_t = A^*y$,

(2) *x_t minimizes the functional*

$$J_{t,y}(x) = \|Ax - y\|^2 + t\|Bx\|^2, \tag{6.110}$$

i.e., $x_t = \arg\min \{J_{t,y}(x) : x \in X\}$.

Our purpose is to comment on this result in more detail: For $y \in Y$ and the family of generalized Tikhonov-Phillips regularizations R_t given by

$$\begin{aligned} x_t = R_t y &= (A^*A + tB^*B)^{-1} A^*y \\ &= \arg\min \{J_{t,y}(x) : x \in X\}, \end{aligned} \tag{6.111}$$

the penalty term satisfies the estimate

$$\|Bx_t\| \leq \frac{1}{\sqrt{t}} \|y\|. \tag{6.112}$$

In fact, the property (6.112) explains the role of the penalty term. If t is large, then $\|Bx_t\|$ is small compared to $\|Ax_t - y\|_Y$. If $t \ll 1$, then $\|Bx_t\|$ becomes large in comparison with the residual term $\|Ax_t - y\|$. All in all, the choice of the operator influences the character of x_t. Some features can be strengthened, where others can be weakened.

Let A and B be given as before. Furthermore, suppose that $y \in \mathcal{D}\left(A^\dagger\right)$ and $r > 0$. Set

$$\delta_r = \inf\left\{\left\|\frac{1}{\beta^2}B^*BA^\dagger y - A^*y'\right\| : y' \in Y, \|y'\| \le r\right\}. \tag{6.113}$$

Then the following statements hold true for $x_t = (A^*A + tB^*B)^{-1}A^*y$:

(1) $\left\|x_t - A^\dagger y\right\|^2 \le \delta_r^2 + t\beta^2 r^2, \quad r, t > 0,$
(2) $\lim_{t\to 0} x_t = A^\dagger y.$

Some additional effort is needed to formulate regularizations under a priori parameter choice.

Let A, B given as before. If we choose $\gamma : (0,\infty) \to (0,\infty)$ such that

$$\lim_{\varepsilon\to 0}\gamma(\varepsilon) = 0 \quad \textit{and} \quad \lim_{\varepsilon\to 0}\frac{\varepsilon}{\sqrt{\gamma(\varepsilon)}} = 0, \tag{6.114}$$

then $\left(\{R_t\}_{t>0}, \gamma\right)$ with $R_t = (A^*A + tB^*B)^{-1}A^*$ is a regularization of $A^\dagger$.

In order to derive further convergence results we impose further assumtions on B: If $\{\sigma_n; x_n, y_n\}_{n\in\mathbb{N}}$ is the singular system of A and if, for $B \in \mathfrak{L}(X, Z)$, we set

$$B^*Bx = \sum_{k=1}^{\infty}\beta_k^2\langle x, x_k\rangle x_k, \quad \beta_k \in \left[\beta, \|B\|^2\right] \tag{6.115}$$

(this is a particular specialization), then we obtain

$$R_t y = (A^*A + tB^*B)^{-1}A^*y = \sum_{k=1}^{\infty}\frac{\sigma_k}{\sigma_k^2 + t\beta_k^2}\langle y, y_k\rangle x_k. \tag{6.116}$$

Note that the classical Tikhonov-Phillips regularization uses $B = I$, i.e., $\beta_k = 1$ for all k. By β_k we control which singular value gets damped and how strongly it is regularized.

In general, R_t cannot be written as a filter, i.e., in the form $F_t(A^*A)A^*$. However, the following helpful estimates can be easily verified (see [**723**]):

$$\left\|F_{\|B\|^2 t}(A^*A)A^*y\right\| \le \|R_t y\| \le \left\|F_{\beta^2 t}(A^*A)A^*y\right\| \qquad \text{for all } y \in Y, \tag{6.117}$$

and

$$\left\|p_{\beta^2 t}(A^*A)x\right\| \le \|(I - R_tA)x\| \le \left\|p_{\|B\|^2 t}(A^*A)x\right\| \qquad \text{for all } x \in X, \tag{6.118}$$

where $p_t(\lambda) = 1 - \lambda F_t(\lambda) = \frac{t}{\lambda+t}$ and $F_t(\lambda) = \frac{1}{\lambda+t}$ (as for the classical Tikhonov filter).

The family $\{R_t\}_{t>0}$ possesses the same asymptotic behavior for $t \to 0$ as the classical Tikhonov filter, the behavior for $\varepsilon \to 0$ is independent of B.

Suppose that $A \in \mathfrak{K}(X, Y)$ and $B \in \mathfrak{L}(X, Z)$. Assume that the representation (6.115) *holds true. Let R_t be given in the form $R_t = (A^*A + tB^*B)^{-1}A^*$, $t > 0$.*

(a) *If the a priori parameter γ is chosen such that*

$$C_\gamma \left(\frac{\varepsilon}{\rho}\right)^{\frac{2}{\mu+1}} \le \gamma(\varepsilon) \le C_\Gamma \left(\frac{\varepsilon}{\rho}\right)^{\frac{2}{\mu+1}}, \quad \varepsilon \to 0, \tag{6.119}$$

(as in (6.84)), where C_γ, C_Γ are positive constants, then the method $(\{R_t\}_{t>0}, \gamma)$ is an order optimal regularization of $A^\dagger$ with respect to X_μ, $\mu \in (0, 2]$.

(b) *If we choose γ according to the discrepancy principle, then $(\{R_t\}_{t>0}, \gamma)$ is an order optimal regularization of $A^\dagger$ with respect to X_μ, $\mu \in (0, 1]$. The order of decay of the error $O(\varepsilon^{\frac{1}{2}})$ is maximal.*

(c) *If we choose γ in accordance with the generalized discrepancy principle with $t_k^* = \sup\{t_k : \eta(t_k) \le \tau\varepsilon^2\}$, where*

$$\eta(t) = \|Ax_t^\varepsilon - y^\varepsilon\|^2 - \langle A^*(Ax_t^\varepsilon - y^\varepsilon), (A^*A + tB^*B)^{-1} A^*(Ax_t^\varepsilon - y^\varepsilon)\rangle, \tag{6.120}$$

with $y^\varepsilon \in Y$ and $x_t^\varepsilon = R_t y^\varepsilon$, then $(\{R_t\}_{t>0}, \gamma)$ is an order optimal regularization of $A^\dagger$ with respect to X_μ, $\mu \in (0, 2]$.

Once again, it is possible to prove that $O(\varepsilon^{\frac{2}{3}})$ is the maximal order of decay for the error. In other words, the qualification does not increase based on this generalization. However, it should be remarked that the reconstructions using $B \ne I$ may lead to strongly different results. In particular, it is possible to adapt β_k to the spectrum of the noise, if this information is known or it can be estimated by other methods.

6.7. Pseudoinverses in reproducing kernel Hilbert spaces

Interest in reproducing kernel Hilbert spaces have increased in recent years, as the computer capacity has made solutions of ever larger and more complex problems practicable. Indeed, new reproducing kernel representations and new applications (for example, in physical geodesy and geophysics) are being contributed at a rapid rate. For example, a certain RHKS in terms of outer harmonics allows the adequate determination of the Earth's gravitational potential (see, e.g., [**256**–**258**], [**773**] for early spline approaches generated by reproducing kernel framework) in consistency with gravitational observables of heterogeneous type (that are interpretable as (bounded) linear functionals on the RKHS under consideration).

Hilbert-Schmidt operator theory. Let $\mathcal{G}$ be a measurable bounded region. A reproducing kernel Hilbert space H_Q with reproducing kernel Q determines a selfadjoint *Hilbert-Schmidt operator* (also denoted by Q) on $L^2(\mathcal{G})$ to $L^2(\mathcal{G})$ by letting

$$(Qx)(s) = \int_{\mathcal{G}} Q(s, s')\, x(s')\, ds', \quad x \in L^2(\mathcal{G}). \tag{6.121}$$

Since Q is assumed to be continuous, then by the Theorem of Mercer (see, e.g., [**513**]), the operator Q has an $L^2(\mathcal{G})$-complete orthonormal system of eigenfunctions $\{\Phi_i\}_{i=1}^\infty$ and corresponding eigenvalues $\{\lambda_i\}_{i=1}^\infty$ with $\lambda_i \ge 0$ and $\sum_{i=1}^\infty \lambda_i < \infty$. Thus Q is a *trace-class* operator (see, e.g., [**197**]) so that $Q(\cdot,\cdot)$ has the uniformly convergent Fourier expansions

$$Q(s, s') = \sum_{i=1}^\infty \lambda_i\, \Phi_i(s)\, \Phi_i(s') \tag{6.122}$$

and

$$Qx = \sum_{i=1}^{\infty} \lambda_i \langle x, \Phi_i \rangle_{L^2(\mathcal{G})} \, \Phi_i, \tag{6.123}$$

where, as always, $\langle \cdot, \cdot \rangle_{L^2(\mathcal{G})}$ is the inner product in $L^2(\mathcal{G})$. It is well known (see, for example, [**821**]) that the space H_Q consists of all $x \in L^2(\mathcal{G})$ satisfying the condition

$$\sum_{i=1}^{\infty} \lambda_i^{-1} \left(\langle x, \Phi_i \rangle_{L^2(\mathcal{G})} \right)^2 < \infty \tag{6.124}$$

(note that the notational convention $0/0 = 0$ is being adopted here) with inner product $\langle \cdot, \cdot \rangle_{H_Q}$ given by

$$\langle x_1, x_2 \rangle_{H_Q} = \sum_{i=1}^{\infty} \lambda_i^{-1} \langle x_1, \Phi_i \rangle_{L^2(\mathcal{G})} \langle x_2, \Phi_i \rangle_{L^2(\mathcal{G})}. \tag{6.125}$$

The operator Q has a well-defined symmetric square root $Q^{1/2}$ which is a Hilbert-Schmidt operator (see, e.g., [**640–642**]):

$$Q^{1/2} x = \sum_{i=1}^{\infty} \sqrt{\lambda_i} \langle x, \Phi_i \rangle_{L^2(\mathcal{G})} \, \Phi_i. \tag{6.126}$$

Thus, since $\mathcal{N}(Q) = \mathcal{N}(Q^{1/2})$, we have

$$H_Q = Q^{1/2}(L^2(\mathcal{G})) = Q^{1/2} \left(L^2(\mathcal{G}) \ominus \mathcal{N}(Q) \right). \tag{6.127}$$

$(Q^{1/2})^\dagger$ has the representation

$$(Q^{1/2})^\dagger x = \sum_{i=1}^{\infty} (\sqrt{\lambda_i})^\dagger \langle x, \Phi_i \rangle_{L^2(\mathcal{G})} \, \Phi_i \tag{6.128}$$

on $H_Q \oplus H_Q^\perp$ ($\perp$ in $L^2(\mathcal{G})$), where, for Θ a real number, $\Theta^\dagger = \Theta^{-1}$ for $\Theta \neq 0$; $\Theta^\dagger = 0$ for $\Theta = 0$. Similarly, $Q^\dagger$ has the representation

$$Q^\dagger x = \sum_{i=1}^{\infty} \lambda_i^\dagger \langle x, \phi_i \rangle_{L^2(\mathcal{G})} \, \phi_i. \tag{6.129}$$

For any operator Q on $L^2(\mathcal{G})$, induced by an RK $Q(s, s')$ as defined in (6.121), we shall adopt the notational conventions (cf. [**640–642**])

$$Q^{-1} = Q^\dagger, \tag{6.130}$$

$$Q^{-1/2} = (Q^{1/2})^\dagger. \tag{6.131}$$

This leads to the relations

$$\|x\|_{H_Q} = \inf \left\{ \|p\|_{L^2(\mathcal{G})} = p \in L^2(\mathcal{G}) : x = Q^{1/2} p \right\}, \quad x \in H_Q. \tag{6.132}$$

and

$$\langle x_1, x_2 \rangle_{H_Q} = \langle Q^{1/2} x_1, Q^{1/2} x_2 \rangle_{L^2(\mathcal{G})}, \quad x_1, x_2 \in H_Q, \tag{6.133}$$

and, if $x_1 \in H_Q$ and $x_2 \in H_Q$ with $x_2 = Q\rho$ for some $\rho \in L^2(\mathcal{G})$, then

$$\langle x_1, x_2 \rangle_{H_Q} = \langle x_1, \rho \rangle_{L^2(\mathcal{G})} \, . \tag{6.134}$$

Generalized inverses and L^2-topology. We are now ready to explore properties of the generalized inverse of a linear operator between two RK spaces (cf. [**640–642**]). To this end we let $X = L^2(\mathcal{G})$ and $Y = L^2(\mathcal{H})$ denote the Hilbert spaces of square-integrable real-valued functions on bounded regions $\mathcal{G}$ and $\mathcal{H}$, respectively. Let A be a linear operator from X into Y. Let $\subset$ denote point set inclusion only, and suppose that A has the following properties:

$$H_Q \subset \mathcal{D}(A) \subset X, \tag{6.135}$$

where H_Q is an RKHS with continuous RK on $\overline{\mathcal{G}} \times \overline{\mathcal{G}}$;

$$A(H_Q) = H_{\overline{R}} \subset H_R \subset Y, \tag{6.136}$$

where $H_{\overline{R}}$ and H_R are RKHS's with continuous RK's on $\overline{H} \times \overline{H}$, so that

$$\mathcal{N}(A) \text{ in } H_Q \text{ is closed in } H_Q. \tag{6.137}$$

We emphasize, in particular, that the space $H_{\overline{R}}$ is not necessarily closed in the topology of H_R.

Let $A^\dagger_{(X,Y)}$ denote the generalized inverse of A, when A is considered as a mapping from X into Y, and let $A^\dagger_{(Q,R)}$ denote the generalized inverse of A when A is considered as a mapping from H_Q into H_R. Now, the topologies in (X, Y) are not the same as the topologies in (H_Q, H_R). Thus, the generalized inverses $A^\dagger_{(X,Y)}$ and $A^\dagger_{(Q,R)}$ show distinct continuity properties, in general. We shall develop the relation between $A^\dagger_{(Q,R)}$ and certain (X, Y) and (Y, Y) generalized inverses. In the sequel, the operators $R : Y \to Y$ and $R^{1/2} : Y \to Y$ are defined from the RK of H_R analogous to Q and $Q^{1/2}$ (see (6.123) and (6.126)). We continue the notational convention of (6.130), i.e., $R^{-1} = R^\dagger = R^\dagger_{(Y,Y)}$ and $R^{-1/2} = (R^{1/2})^\dagger_{(Y,Y)}$.

Let $A^\dagger_{(X,Y)}$ denote the generalized inverse of A, when A is considered as a mapping from X into Y, and let $A^\dagger_{(Q,R)}$ denote the generalized inverse of A when A is considered as a mapping from H_Q into H_R. Now, the topologies in (X, Y) are not the same as the topologies in (H_Q, H_R). Thus, the generalized inverses $A^\dagger_{(X,Y)}$ and $A^\dagger_{(Q,R)}$ show distinct continuity properties, in general. We shall develop the relation between $A^\dagger_{(Q,R)}$ and certain (X, Y) and (Y, Y) generalized inverses. In the sequel, the operators $R : Y \to Y$ and $R^{1/2} : Y \to Y$ are defined from the RK of H_R analogous to Q and $Q^{1/2}$ (see (6.123) and (6.126)). We continue the notational convention of (6.130), i.e., $R^{-1} = R^\dagger = R^\dagger_{(Y,Y)}$ and $R^{-1/2} = (R^{1/2})^\dagger_{(Y,Y)}$.

Moreover, it follows that $A^\dagger_{(Q,R)}$ is bounded, provided that $A(H_Q) = H_R$.

It should be noted that an operator A may satisfy the assumption $A(H_Q) = H_R$ while failing to have a closed range in the space Y. This is, for example, the case if A is a Hilbert-Schmidt linear integral operator (with nondegenerate kernel) on X. It is this observation which makes RKHS useful in the context of regularization and approximation of ill-posed linear operator equations.

Explicit representation of minimum-norm solutions. We assume that H_Q is chosen so that the linear functionals $\{\mathcal{E}_t : t \in \mathcal{H}\}$ defined by

$$\mathcal{E}_t x = (Ax)(t) \tag{6.138}$$

are continuous in H_Q. Then, by the Riesz representation theorem, there exists a family

$$\{\eta_t\}_{t\in\mathcal{H}} \subset H_Q\mathcal{H} \tag{6.139}$$

such that

$$(Ax)(t) = \langle \eta_t, x\rangle_{H_Q}, \quad t \in \overline{\mathcal{H}}, \quad x \in H_Q, \tag{6.140}$$

where η_t is explicitly given by

$$\eta_t(s) = \langle \eta_t, Q_s\rangle_Q = (AQ_s)(t) \tag{6.141}$$

($\eta_t(s)$ is readily obtained in a more explicit form from (6.141) if A is a differential or integral operator).

Let $R(t,t')$ be the nonnegative definite kernel on $\overline{\mathcal{H}} \times \overline{\mathcal{H}}$ given by

$$R(t,t') = \langle \eta_t, \eta_i\rangle_{H_Q}, \quad t,t' \in \mathcal{H}. \tag{6.142}$$

Assume that H_R is the RKHS with RK R given by (6.142). Let R_t be the element of H_R defined by $R_t(t') = R(t,t')$, and let $\langle\cdot,\cdot\rangle_R$ be the inner product in H_R. Suppose that V is the closure of the span of $\{\eta_t, t \in \overline{\mathcal{H}}\}$ in H_Q. Now, $\{R_t, t \in \overline{\mathcal{H}}\}$ spans H_R, and by the properties of RKHS, we have

$$\langle \eta_t, \eta_{t'}\rangle_Q = R(t,t') = \langle R_t, R_{t'}\rangle_{H_R}. \tag{6.143}$$

Thus there is an isometric isomorphism between the subspace V and H_R, generated by the correspondence

$$\eta_t \in V \sim R_t \in H_R. \tag{6.144}$$

Then, $x \in V \sim y \in H_R$ if and only if $\langle \eta_t, x\rangle_Q = y(t) = \langle R_t, g\rangle_{H_R}$, $t \in \overline{\mathcal{H}}$, i.e., if and only if $y(t) = (Ax)(t), t \in \overline{\mathcal{H}}$. Thus, $A(H_Q) = A(V) = H_R$. The nullspace of A in H_Q is $\{x : x \in H_Q, \|Ax\|_R = 0\}$. Since

$$\langle \eta_t, x\rangle_Q = 0, \quad t \in \overline{\mathcal{H}} \quad \text{and} \quad x \in H_Q \Rightarrow x \in V^\perp, \tag{6.145}$$

and $x \in V$ implies $\|x\|_Q = \|Ax\|_V$, it follows that the nullspace of A in H_Q is $V^\perp$ ($\perp$ in H_Q). Hence, (6.138) entails that the nullspace of $A : H_Q \to H_R$ in H_Q is always closed, irrespective of the topological properties of $A : X \to Y$.

We list the following table of corresponding sets and elements, under the correspondence "$\sim$" of (6.144), where the entries on the left are in H_Q:

$$V \sim H_R, \tag{6.146}$$

$$x \sim y, \tag{6.147}$$

$$\eta_t \sim R_t, \tag{6.148}$$

$$P_V Q_s \sim \eta_s^*. \tag{6.149}$$

Here, P_V is the projector from H_Q onto the (closed) subspace V, $y(t) = \langle \eta_t, x\rangle_Q$, $t \in \overline{\mathcal{H}}$, and $\eta_s^* = AQ_s = A(P_V Q_s)$, i.e.,

$$\eta_s^*(t) = \langle \eta_i, P_V Q_s\rangle_{H_Q} = \eta_t(s). \tag{6.150}$$

This leads to the following result (the proof is given in [**640, 641**]):

Let A and H_Q satisfy (6.138), and let R be given by (6.143), where η_i is defined by (6.140). Let $\eta_s^* = AQ_s$. Then, for $y \in H_R$,

$$(A^\dagger_{(Q,R)}y)(x) = \langle \eta_s^*, y\rangle_{H_R}, \quad s \in \overline{\mathcal{G}}. \tag{6.151}$$

We also obtain another operator representation of $A^\dagger_{(Q,R)}$:

Under the assumptions

(i) *$\mathcal{D}(A^*)$ is dense in Y, where A^* is the adjoint of A considered as an operator from X to Y,*

(ii) *A and H_Q satisfy (6.140),*

(iii) *H_Q and $H_R = A(H_Q)$ possess continuous reproducing kernels,*

we have $y \in H_R$, with

$$(A^\dagger_{(Q,R)}y)(s) = (QA^*(AQA^*)^\dagger_{(Y,Y)}y)(s), \quad s \in \overline{\mathcal{G}}. \tag{6.152}$$

Poorly conditioned operator equations. For $A : X \to Y$, the *pseudocondition number* of A (relative to the norms of X and Y) is given by

$$\gamma(A; X, Y) = \sup_{\substack{x\neq 0\\ x\in\mathcal{D}(A)}} \frac{\|Ax\|_Y}{\|x\|_X} \sup_{\substack{y\neq 0\\ y\in\mathcal{D}(A^\dagger)}} \frac{\|A^\dagger y\|_X}{\|y\|_Y}. \tag{6.153}$$

The equation $Ax = y$ is said to be *poorly conditioned in the spaces X, Y* if the number $\gamma(A; X, Y)$ is much greater than 1 (note that $1 \le \gamma(A; X, Y)$; for ill-posed problems, γ is not finite).

Suppose H_Q is an RKHS with $H_Q \subset \mathcal{D}(A)$, and A and H_Q satisfy (6.138) with $A(H_Q) = H_R$, R given by (6.143). Then $\gamma(A; H_Q, H_R) = 1$. To see this, write $x \in H_Q$ in the form $x = x_1 + x_2$, where $x_2 \in V^\perp$. Then $Ax = Ax_1 = y_1$ and $\|y_1\|_R = \|x_1\|_Q$. Thus (by using $\|\cdot\|_{H_R}$ instead of $\|\cdot\|_R$)

$$\gamma(A; H_Q, H_R) = \sup_{x\neq 0} \frac{\|y_1\|_{H_R}}{\|x\|_{H_Q}} \sup_{y_1\neq 0} \frac{\|x_1\|_{H_Q}}{\|y_1\|_{H_R}} = 1. \tag{6.154}$$

On the other hand, the number $\gamma(A; X, Y)$ may be large. Thus, the casting of the operator equation $Ax = y$ in the reproducing kernel spaces H_Q, H_R always leads to a well-conditioned (indeed, optimally-conditioned) problem.

Regularization of pseudosolutions in reproducing kernel spaces. We study properties of *regularized pseudosolutions* (in RKHS) x_α of the operator equation $Ax = y$, where y is not necessarily in the range of the operator A. By a *regularized pseudosolution* we mean a solution to the variational problem: Find x_α in H_Q to minimize

$$J_y(x) = \|y - Ax\|^2_{H_P} + \alpha\|x\|^2_{H_Q}, \qquad \alpha > 0, \tag{6.155}$$

where H_Q is an RKHS in the domain of A, $\|\cdot\|_{H_P}$ denotes the norm in an RKHS H_P with reproducing kernel P, $H_P \subset Y$, $J_y(x)$ is assigned the value $+\infty$ if $y - Ax \notin H_P$, and $\alpha > 0$.

We suppose A and H_Q satisfy (6.138), hence, $A(H_Q) = H_R$ possesses a reproducing kernel. As before, A may be unbounded, invertible, or compact considered as an operator from $X(= L^2(\mathcal{G}))$ to $Y(= L^2(\mathcal{H}))$. It is assumed that y possesses a (not necessarily unique) representation $y = y_0 + \xi$, for some $y_0 \in A(H_Q)$ and $\xi \in H_P$, where ξ may be thought of as a "disturbance".

For $\alpha > 0$, let $H_{\alpha P}$ be the RKHS with RK $\alpha P(t,t')$, where $P(t,t')$ is the RK on $\overline{\mathcal{H}} \times \overline{\mathcal{H}}$ associated with H_P. We have $H_P = H_{\alpha P}$ and

$$\|\cdot\|^2_{H_P} = \alpha\|\cdot\|^2_{H_{\alpha P}}. \tag{6.156}$$

Let $R(\alpha) = R + \alpha P$, and let $H_{R(\alpha)}$ be the RKHS with RK $R(\alpha) = R(\alpha, t, t')$. According to [**47**], $H_{R(\alpha)}$ is the Hilbert space of functions of the form $y = y_0 + \xi$, where $y_0 \in H_R$ and $\xi \in H_P$. Following [**47**], we note that this decomposition is not unique unless H_R and H_P have no element in common except the zero element. The norm in $H_{R(\alpha)}$ is given by

$$\|y\|^2_{R(\alpha)} = \min\{\|y_0\|^2_{H_P} + \|\xi\|^2_{H_{\alpha R}} : y_0 \in H_R, \xi \in H_P, y_0 + \xi = y\}, \tag{6.157}$$

where, however, the y_0 and ξ attaining the minimum in (6.157) are easily shown to be unique by the strict convexity of the norm.

Consider the problem of finding $x_\alpha \in H_Q$ to minimize $J_y(x)$ in (6.155) for $y \in H_{R(\alpha)}$. Then $y - Ax_\alpha$ must be in H_P and it is obvious that $x_\alpha \in V$, the orthogonal complement of the nullspace of A in H_Q. For any $x \in V$, $\|x\|_{H_Q} = \|Ax\|_{H_R}$ by the isometric isomorphism between V and H_R, and (6.155) may be written in the equivalent form: Find $x_\alpha \in V$ to minimize

$$\alpha\|Ax\|^2_{H_R} + \|y - Ax\|^2_{H_P}. \tag{6.158}$$

Comparing (6.157) and (6.158) with the aid of (6.156), we see that y_0 and ξ attaining the minimum on the right-hand side of (6.157) are related to the solution x_α of the minimization problem (6.158), by

$$y_0 = Ax_\alpha \quad \text{and} \quad \xi = y - Ax_\alpha. \tag{6.159}$$

We call the (linear) mapping which assigns to each $y \in H_{R(\alpha)}$ the unique minimizing element x_α the *regularization operator* of the equation $Ax = y$. A representation of the solution x_α is given (see M.Z. Nashed, G. Wahba [**640–642**]) as follows:

Suppose $\mathcal{D}(A^*)$ is dense in $Y, H_Q \subset \mathcal{D}(A)$ and A and H_Q satisfy (6.138). Suppose $H_Q, H_R (= A(H_Q))$ and $H_P \subset Y$ all have continuous RK's. Then, for $y \in H_{R(\alpha)}$, the unique minimizing element $x_\alpha \in H_Q$ of the functional $J_y(x)$ is given by

$$\langle \eta_s^*, y\rangle_{R(\alpha)} = x_\alpha(s) = (QA^*(AQA^* + \alpha P)^\dagger_{(Y,Y)} y)(s) \quad s \in \overline{\mathcal{G}}, \tag{6.160}$$

where $\eta_s^* = AQ_s$. We call the (linear) mapping which assigns to each $y \in H_{R(\alpha)}$ the unique minimizing element x_α the *regularization operator* of the equation $Ax = y$.

The most useful situations occur, of course, when H_R is strictly contained in $H_{R(\alpha)}$. For example, H_R may be a dense subset of Y in the Y-topology and $H_{R(\alpha)}$ a bigger dense subset. We deal with this situation later. If $H_R^\perp$ (in Y) is not empty, then P may be chosen so that the closure of H_P in the Y-topology equals $H_R^\perp$ in Y. Then $H_P \cap H_R = \{0\}$, $H_{\alpha P}$ and H_R are orthogonal subspaces of $H_{R(\alpha)}$ (see, e.g., [**47**]), and the decomposition (6.7) is unique. In this case we have the following theorem which shows that the regularization operator is indeed a generalized inverse in an appropriate RKHS: *If $H_P \cap H_R = \{0\}$, then the minimizing element x_α of* (6.155) *is the solution to the problem: Find $x \in \mathcal{L}(y)$ to minimize*

$$\|x\|_{H_Q}, \tag{6.161}$$

where

$$\mathcal{L}(y) = \{x : x \in H_Q, \|y - Ax\|_{H_{R(\alpha)}} = \inf_{z \in H_Q} \|y - Az\|_{H_{R(\alpha)}}\}. \tag{6.162}$$

It should be remarked that, in our approach, we have $A(H_Q) = H_R \subset H_{R(\alpha)} \subset Y$. Replacing $H_{\overline{R}}$ and H_R in (6.136) by H_R and $H_{R(\alpha)}$, respectively, we obtain

$$A^\dagger_{(Q,R(\alpha))} y = Q^{1/2}[(R + \alpha P)^{-1/2} A Q^{1/2}]^\dagger_{(X,Y)} (R + \alpha P)^{-1/2} y \tag{6.163}$$

for $y \in \mathcal{D}(A^\dagger_{(Q,R(\alpha))})$.

It is helpful to remember that the topology on H_R is not, in general, the restriction of the topology of $H_{R(\alpha)}$, with the notable exception of the case $H_R \cap H_P = \{0\}$. In [**459**] a concrete example is provided arising in the approximate solution of boundary value problems, where H_R is a closed subspace of $H_{R(\lambda)}$ and we have

$$A^\dagger_{(Q,R(\alpha))} = Q A^* (R + \alpha P)^{-1}. \tag{6.164}$$

Note that in this case, the generalized inverse and the regularization operator coincide. If $H_R = A(H_Q)$ is not closed in $H_{R(\alpha)}$, then the regularization operator and the generalized inverse are different. Also, the right-hand side of (6.163) and (6.164) are not the same: (6.164) has maximal domain $H_{R(\alpha)}$, while (6.163) has maximal domain $\mathcal{H}_R \oplus H_R^\perp$ ($\perp$ in $H_{R(\alpha)}$).

Rates of convergence to the generalized inverse. We note some properties of x_α as $\alpha \to 0$ when $H_R \subset H_P$. If $y \in H_R = A(H_Q)$, then we have $x_\alpha \to A^\dagger_{(Q,R)} y$ as $\alpha \to 0$; here we may say something about the rate of convergence if certain additional conditions are satisfied (compare also with [**459**]). However, y may not be in the domain of $A^\dagger_{(Q,R)}$. This situation can occur if, for example, H_R is dense in $H_{R(1)}$. In this case, $\lim_{\alpha \to 0} \|x_\alpha\|_Q = \infty$: *Assume that $y = Ax_0 + \xi_0$, where $x_0 \in V$, $\xi_0 \in H_P$ and suppose that $H_R \subset H_P$.* Then the following properties hold true:

(i) *$B = P^{-1/2} R^{1/2}$ is a bounded operator on $Y = L^2(\mathcal{H})$;*

(ii) *if $\xi_0 = 0$ and $\|(B^*B)^{-1} R^{-1/2}(Ax_0)\|_{L^2(\mathcal{H})} < \infty$, then*

$$\|A^\dagger_{(Q,R)} y - x_\alpha\|^2_{H_Q} = O(\alpha^2); \tag{6.165}$$

(iii) *if $\xi_0 = 0$ and $\|(B^*B)^{-1/2} R^{-1/2}(Ax_0)\|_{L^2(\mathcal{H})} < \infty$, then*

$$\|A^\dagger_{(Q,R)} y - x_\alpha\|^2_{H_Q} = O(\alpha); \tag{6.166}$$

(iv) *if $\xi_0 \notin H_R$, then $\lim_{\alpha \to 0} \|x_\alpha\|_{H_Q} = \infty$.*

Here, inverses indicated by "– " are the generalized inverses in the topology of L^2-spaces.

6.8. Iterative methods as regularization schemes

From classical Linear Algebra we borrow the idea to use iterative solvers of the equation $Ax = y, y \in Y$, in the following way: Suppose that the matrix $A \in \mathbb{K}^{n\times n}$ is written in the form $A = B - C$ with an invertible B. Then we are able to write $Ax = y$, in equivalent form

$$Ax = y \Leftrightarrow (B - C)\,x = y \Leftrightarrow Bx = Cx + y \Leftrightarrow x = B^{-1}Cx + B^{-1}y. \tag{6.167}$$

Let us base the determination of a (uniquely determined) fixed point on the following iteration:

$$\begin{aligned} &x^{(0)} \in \mathbb{R}^n, \quad \text{arbitrary}, \\ &Bx^{(n+1)} = Cx^{(n)} + y, \quad n \in \mathbb{N}_0. \end{aligned} \tag{6.168}$$

Clearly, if $\{x^{(n)}\}$ converges to $x^* \in \mathbb{R}^n$, then it follows that $Ax^* = y$.

In numerical Linear Algebra the following procedures are convenient which should be recapitulated here: Let $A \in \mathbb{R}^{n\times n}$ be given. Let us decompose the matrix A in the form $A = L + D + U$ such that

$$L = \begin{pmatrix} 0 & & 0 \\ & \ddots & \\ * & & 0 \end{pmatrix} \quad D = \begin{pmatrix} a_{11} & & 0 \\ & \ddots & \\ 0 & & a_{nn} \end{pmatrix} \quad U = \begin{pmatrix} 0 & & * \\ & \ddots & \\ 0 & & 0 \end{pmatrix}. \tag{6.169}$$

We distinguish the following concepts:

(1) *Jakobi method.* We let $B = D,\ C = -L - U$. Then we are led to

$$x^{(n+1)} = D^{-1}(L + U)\,x^{(n)} + D^{-1}y, \quad n \in \mathbb{N}. \tag{6.170}$$

(2) *Gauss-Seidel method.* We choose $B = D + L,\ C = -U$. Then we are led to

$$x^{(n+1)} = (D + L)^{-1}\,Ux^{(n)} + (D + L)^{-1}\,y, \quad n \in \mathbb{N}. \tag{6.171}$$

(3) *Richardson method.* For $\omega > 0$, let $B = \frac{1}{\omega}I,\ C = \frac{1}{\omega}I - A$. Then we are led to the recursion

$$\begin{aligned} x^{(n+1)} &= \left(\frac{1}{\omega}I\right)^{-1}\left(\frac{1}{\omega}I - A\right)x^{(n)} + \left(\frac{1}{\omega}I\right)^{-1} y \\ &= \omega\left(\frac{1}{\omega}I - A\right)x^{(n)} + \omega y \\ &= (I - \omega A)\,x^{(n)} + \omega y, \quad n \in \mathbb{N}. \end{aligned} \tag{6.172}$$

Since iteration methods known from Linear Algebra that use the upper or lower triangular part of a matrix cannot be adequately transferred to operator equations, we are not able to use the first two of the aforementioned iteration procedures. However, the third method (i.e., Richardson method or successive relaxation method) can be applied leading to a variant called Landweber iteration.

Landweber iteration. In order to solve the normal equation for operators

$$A^*Ax = A^*y, \quad y \in Y,\ x \in X, \tag{6.173}$$

the *Landweber iteration* starting from the initial value $x^{(0)} \in X$ is defined by

$$\begin{aligned} x^{(m+1)} &= (I - \omega A^*A)\, x^{(m)} + \omega A^* y \\ &= x^{(m)} + \omega A^* \left(y - Ax^{(m)}\right), \quad m = 0, 1, \ldots \end{aligned} \tag{6.174}$$

For simplicity, we introduce the family of operators $\{R_m\}_{m\in\mathbb{N}} \subset \mathfrak{L}(Y, X)$ by $R_m y = x^{(m)}$:

Let $0 < \omega < \frac{2}{\|A\|^2}$ and $A \in \mathfrak{L}(X, Y)$. Then, for $x^{(0)} \in X$, we have

$$\lim_{m\to\infty} R_m y = \begin{cases} A^\dagger y + P_{\mathcal{N}(A)} x^{(0)}, & y \in D\left(A^\dagger\right), \\ \infty, & y \notin D\left(A^\dagger\right). \end{cases} \tag{6.175}$$

The Landweber iteration is characterized by

$$x^{(m)} = R_m y = F_m\left(A^*A\right) A^* y + \left(I - \omega A^*A\right)^m x^{(0)}, \tag{6.176}$$

where

$$F_m(\lambda) = \omega \sum_{j=0}^{m-1} (1 - \omega\lambda)^j = \frac{1 - (1 - \omega\lambda)^m}{\lambda}. \tag{6.177}$$

Thus, Landweber iteration represents a regularization with a filter that cannot be used as such if we do not know the singular values of our problem.

The polynomial $\lambda \mapsto p_m(\lambda) = 1 - \lambda F_m(\lambda) = (1 - \omega\lambda)^m$ is called the *residual polynomial.*

- For $x^{(0)} = 0$ it follows that $A^\dagger y$ is the solution of the normal equation

$$A^*Ax = A^*y \tag{6.178}$$

  showing minimal norm (as usual). This result is also obtained for $x^{(0)} \in \mathcal{N}(A)^\perp$, since then $P_{\mathcal{N}(A)} x^{(0)} = 0$.
- For $x^{(0)} \neq 0$ we see that $A^\dagger y + P_{\mathcal{N}(A)} x^{(0)}$ is the solution of the normal equation

$$A^*Ax = A^*y \tag{6.179}$$

  with minimal distance to $x^{(0)}$, i.e., the $x^{(0)}$-minimum norm solution.

$$\left\|A^\dagger y + P_{\mathcal{N}(A)} x^{(0)} - x^{(0)}\right\| = \min\left\{\left\|x - x^{(0)}\right\| : A^*Ax = A^*y\right\} \tag{6.180}$$

Obviously, the choice of the initial value of the Landweber iteration for inverse problems is of great importance for the solution. We investigate the same choice for a noiseless and noisy right hand side.

Let $A \in \mathfrak{L}(X, Y)$, $y, y^\varepsilon \in Y$ with $\|y - y^\varepsilon\|_Y < \varepsilon$. Let $\left\{x^{(m)}\right\}$ and $\left\{x^{(m),\varepsilon}\right\}$ be the Landweber iteration sequences with respect to y and y^ε corresponding to the same initial value $x^{(0)} = x^{(0),\varepsilon} \in X$ and the same parameter $\omega \in \left(0, \frac{2}{\|A\|^2}\right)$. Then

$$\left\|x^{(m)} - x^{(m),\varepsilon}\right\| \le \varepsilon \begin{cases} \sqrt{2\omega}, & m = 1, \\ \sqrt{m\omega}, & m \ge 2. \end{cases} \tag{6.181}$$

The estimate (6.181) suggests that the parameter ω should be chosen as small as possible. We have seen, however, that the convergence of the approximation error depends on the estimate $|1-\omega\lambda| < 1$. As a consequence, for small ω the approximation error converges very slowly.

In more detail, we separate the total error in the usual way

$$\left\|A^\dagger y - R_m y^\varepsilon\right\| = \underbrace{\left\|A^\dagger y - R_m y\right\|}_{\substack{\text{approximation}\\ \text{error}}} + \underbrace{\left\|R_m y - R_m y^\varepsilon\right\|}_{\substack{\text{data}\\ \text{error}}} \tag{6.182}$$

with $\left\|A^\dagger y - R_m y\right\| \xrightarrow{m\to\infty} 0$ if $x^{(0)}=0$ and $\|R_m y - R_m y^\varepsilon\| \le \sqrt{m\omega}\,\varepsilon$, i.e., the Landweber iteration is a regularization with a suitably chosen m. Our aim now is to find a stopping rule for m. In fact, we obtain for the residual term

$$\begin{aligned} Ax^{(m),\varepsilon} - y^\varepsilon &= A\left(\left(I-\omega A^*A\right)x^{(m-1),\varepsilon} + \omega A^* y^\varepsilon\right) - y^\varepsilon \\ &= Ax^{(m-1),\varepsilon} - \omega AA^*Ax^{(m-1),\varepsilon} + \omega AA^*y^\varepsilon - y^\varepsilon \\ &= \left(I-\omega AA^*\right)Ax^{(m-1),\varepsilon} - \left(I-\omega AA^*\right)y^\varepsilon \\ &= \left(I-\omega AA^*\right)\left(Ax^{(m-1),\varepsilon} - y^\varepsilon\right). \end{aligned} \tag{6.183}$$

Under the choice $0 < \omega < \frac{2}{\|A\|^2}$ we have

$$\begin{aligned} \left\|Ax^{(m),\varepsilon} - y^\varepsilon\right\| &\le \left\|I - \omega AA^*\right\| \cdot \left\|Ax^{(m-1),\varepsilon} - y^\varepsilon\right\| \\ &\le \left\|Ax^{(m-1),\varepsilon} - y^\varepsilon\right\|, \end{aligned} \tag{6.184}$$

i.e., the norms of the residuals are monotonically decreasing in m. Furthermore, if $y^\varepsilon \notin \mathcal{D}\left(A^\dagger\right)$, it follows that $Ax^{(m-1),\varepsilon} - y^\varepsilon \notin \mathcal{N}\left(A^*\right)$ (due to the fact that $y^\varepsilon \in \mathcal{D}(A^\dagger)$).

If $0 < \omega < \frac{2}{\|A\|^2}$ *and* $y \notin \mathcal{N}\left(A^*\right)$, *then*

$$\left\|\left(I - \omega AA^*\right)y\right\| < \|y\|. \tag{6.185}$$

If $y^\varepsilon \notin D\left(A^\dagger\right)$, then $\left\|Ax^{(m),\varepsilon} - y^\varepsilon\right\| < \left\|Ax^{(m-1),\varepsilon} - y^\varepsilon\right\|$, i.e., the residual is strictly monotonically decreasing. Hence, we are confronted with the typical dilemma of ill-posed problems that a small residual term does not imply a small error. The monotonicity of the residual term suggests to use a discrepancy principle as a kind of "stopping rule". This observation goes back to M. Defrisc, C. de Mol [**177**]. In more detail, let $\tau > 1$ be fixed. We are interested in determining $m^* \in \mathbb{N}_0$, such that

$$\left\|Ax^{(m^*),\varepsilon} - y^\varepsilon\right\| \le \tau\varepsilon < \left\|Ax^{(m),\varepsilon} - y^\varepsilon\right\| \qquad m = 0,1,\dots,m^*-1. \tag{6.186}$$

The discrepancy principle (6.186) seems to be particularly suitable for Landweber iteration, since a residual term larger than 2ε implies the monotonicity of the error:

Suppose that $A \in \mathfrak{L}(X,Y)$, $y \in \mathcal{R}(A)$, *and* $y^\varepsilon \in Y$ *with* $\|y - y^\varepsilon\| < \varepsilon$. *If* $\left\|Ax^{(m),\varepsilon} - y^\varepsilon\right\| > 2\varepsilon$ *and* $0 < \omega < \frac{1}{\|A\|^2}$, *then*

$$\left\|A^\dagger y - x^{(m+1),\varepsilon}\right\| < \left\|A^\dagger y - x^{(m),\varepsilon}\right\|, \quad m \in \mathbb{N}_0, \tag{6.187}$$

i.e., the error is also strictly monotonically decreasing.

In other words, as already announced, the error decreases monotonically just like the residual as long as the residual stays larger than 2ε.

An upper bound for the number of iterations when using the discrepancy principle is as follows (cf. [**723**]):

Assume that $A \in \mathfrak{L}(X,Y)$, $y \in \mathcal{R}(A)$, *and* $y^\varepsilon \in Y$ *with* $\|y - y^\varepsilon\| < \varepsilon$. *Suppose that* $0 < \omega < \frac{1}{\|A\|^2}$. *The discrepancy principle* (6.186) *with* $\tau > 1$ *yields the stopping index* $m^* = m^*(\varepsilon, y^\varepsilon) \le C_L \varepsilon^{-2}$ *with the constant* $C_L > 0$ *for the Landweber iteration.*

Note that this result does not require any assumptions on the smoothness of the solution. With such information the discrepancy principle enables us to stop the iteration much earlier.

Suppose that $A \in \mathfrak{L}(X,Y)$, $0 < \omega < 2/\|A\|^2$, $y \in \mathcal{R}(A)$, $y \in Y$ *with* $\|y - y^\varepsilon\| < \varepsilon$ *and* $x^{(0)} = 0$. *Then the Landweber iteration together with the discrepancy principle* (6.186) *is an order optimal regularization of* $A^\dagger$, *i.e., the Landweber iteration possesses infinite qualification. The stopping index can be estimated as follows*

$$m^* = m^*(\varepsilon, y^\varepsilon) \le C_\mu \varepsilon^{\frac{-2}{\mu+1}} \tag{6.188}$$

with $C_\mu > 0$.

If we choose as a initial value $0 \ne x^{(0)}$ with $x^{(0)} \in \mathcal{N}(A)^\perp$, it suffices to study the convergence of the sequence $\{\tilde{x}^{(m),\varepsilon}\}$ with $\tilde{x}^{(0),\varepsilon} = 0$ which results from the Landweber method applied to the equation $Ax = y^\varepsilon - Ax^{(0)}$. The minimum norm solution of this equation is $A^\dagger y - x^{(0)}$ and if $A^\dagger y \in X_\mu$, we also need to have that $x^{(0)} \in X_\mu$ to obtain the optimal order of decay for the error, i.e., $O(\varepsilon^{\frac{\mu}{\mu+1}})$.

Unfortunately, for unknown μ we have to chose 0 as starting value.

The disadvantage of the Landweber iteration is that its convergence is rather slow, i.e., the stopping index m^* is often large. This is the reason why semi-iterative methods (see, e.g., A. Rieder [**723**] and the references therein) come into play to accelerate the convergence.

Semi-iterative methods. The characteristics of Landweber iteration are as follows;

$$x^{(0)} = 0 \tag{6.189}$$

and

$$x^{(m)} = F_m(A^*A)\,A^*y \tag{6.190}$$

with $F_m(\lambda)$ given by (6.177), i.e.,

$$F_m(\lambda) = \omega \sum_{j=0}^{m-1} (1-\omega\lambda)^j = \frac{1-(1-\omega\lambda)^m}{\lambda}. \tag{6.191}$$

For $\lambda \in [0, \|A\|^2]$ we have

$$F_m(\lambda) \to \frac{1}{\lambda}, \qquad m \to \infty. \tag{6.192}$$

Moreover, we are able to show that

$$\sup_{\lambda\in[0,\|A\|^2]} \lambda|F_m(\lambda)| = \sup_{\lambda\in[0,\|A\|^2]} |1-(1-\omega\lambda)^m| \leq 2, \tag{6.193}$$

so that $\{F_m\}_{m\in\mathbb{N}}$ is a regularizing filter. The *residual polynomials* corresponding to the *filter polynomials* are of degree m:

$$p_m(\lambda) = 1 - \lambda F_m(\lambda) = (1-\omega\lambda)^m. \tag{6.194}$$

All in all, the Landweber iteration procedure is as follows:

$$y - Ax^{(m)} = p_m(AA^*)\, y. \tag{6.195}$$

In order to accelerate the Landweber iteration we are led to the idea to replace the polynomial filter by another filter family, that shows a faster convergence to $1/\lambda$.

To this end we consider an alternative polynomial F_m of degree $m-1$, so that its residual polynomial $\lambda \mapsto p_m(\lambda) = 1 - \lambda F_m(\lambda)$ is of degree m. Letting $x^{(m)} = F_m(A^*A)\, A^*y$ we obtain as residuum $y - Ax^{(m)} = p_m(AA^*)\, y$. Now, if $\{F_m\}_{m\in\mathbb{N}}$ is a regularizing filter, the corresponding family $\{p_m\}_{m\in\mathbb{N}}$ has the following properties:

If $\{F_m\}$ is a regularizing filter, the residual polynomials $\{p_m\}$ satisfy the following properties:

(1) $\lim_{m\to\infty} p_m(\lambda) = 0$ *for* $\lambda \in [0, \|A\|^2]$

(2) $\{p_m\}$ *is uniformly bounded on the interval* $\left[0, \|A\|^2\right]$ *(by* $1 + C_F$*).*

(3) $p_m(0) = 1$ *for all* m.

Conversely, if $\{p_m\}$ is a family of polynomials satisfying the properties (1), (2) (3), *then*

$$F_m(\lambda) = \frac{1 - p_m(\lambda)}{\lambda}, \quad m \in \mathbb{N}_0 \tag{6.196}$$

constitutes a regularizing filter.

From the theory on special functions of Mathematical Physics (see, e.g., [**812**]) we borrow the following result:

Let the family $\{p_m\}_{m\in\mathbb{N}_0}$ fulfill the following conditions:

(1) p_m is a polynomial of degree m on the interval $\left[0, \|A\|^2\right]$.

(2) $\int_0^{\|A\|^2} p_m(\lambda)\, p_n(\lambda)\, w(\lambda)\, d\lambda = 0$ if $n \neq m$, where $w(\lambda) > 0$ for $\lambda \in \left[0, \|A\|^2\right]$ and w is piecewise continuous.

(3) $p_m(0) = 1$, $m \in \mathbb{N}_0$.

Then there exist $A_m, B_m \in \mathbb{R}$ satisfying

$$p_m(\lambda) = p_{m-1}(\lambda) + A_m(p_{m-1}(\lambda) - p_{m-2}(\lambda)) - B_m \lambda p_{m-1}(\lambda). \tag{6.197}$$

The polynomials $\{p_m\}$ are *orthogonal polynomials* on the interval $\left[0, \|A\|^2\right]$ with respect to the weight function $w(\lambda)$ note that ($w(\lambda)$ induces a measure on $\left[0, \|A\|^2\right]$). In transition to iteration we are led to

$$x^{(0)} = 0, \tag{6.198}$$

$$x^{(1)} = F_1(A^*A)A^*y \quad \text{with } F_1(\lambda) = \frac{1 - p_1(\lambda)}{\lambda}, \tag{6.199}$$

(6.200)
$$x^{(m)} = x^{(m-1)} + A_m\left(x^{(m-1)} - x^{(m-2)}\right) + B_m A^*\left(y - Ax^{(m-1)}\right), \; m \geq 2.$$

This is the reason why methods of type (6.198), (6.199), (6.200) are called semi-iterative (note that, for each iteration step, two previous iteration values are required).

For simplicity, assume now that $\|A\| \leq 1$ which can be achieved by scaling the operator. As a consequence, we only need to consider polynomials on the interval $[0, 1]$. Then the following results are known for semi-iterative methods (see, e.g., [**723**]):

(1) *Let $\{p_m\}$ be a sequence of residual polynomials, so that it is uniformly bounded on $[0, 1]$, normalized by $p_m(0) = 1$, and it converges pointwise to 0 on $(0, 1]$. If $\{F_m\}$ is the corresponding sequence of filters, then for $A \in \mathfrak{L}(X, Y)$ and $x^{(0)} \in X$ it follows that*

$$\begin{aligned} \lim_{m\to\infty} x^{(m)} &= \lim_{m\to\infty}\left(x^{(0)} + F_m(A^*A)A^*(y - Ax^{(0)})\right) \\ &= \begin{cases} A^\dagger y + P_{\mathcal{N}(A)}x^{(0)}, & y \in \mathcal{D}(A^\dagger), \\ \infty, & y \notin \mathcal{D}(A^\dagger). \end{cases} \end{aligned}$$

(2) *Suppose that A is of class $\mathfrak{L}(X, Y)$. Let $\{x^{(m)}\}$, $\{x^{(m),\varepsilon}\}$ be the iterates of a semi-iterative method with respect to $y \in \mathcal{R}(A)$ and $y^\varepsilon \in Y$ using the same initial value. Let the residual polynomials of the semi-iterative method be uniformly bounded by $C_p > 0$. Then, for the data error, we have*

$$\left\|x^{(m)} - x^{(m),\varepsilon}\right\| \leq 2\, C_p m\, \varepsilon. \tag{6.201}$$

Together with a stopping rule that fulfills

$$m^*(\varepsilon) \to \infty\,, \qquad \varepsilon m^*(\varepsilon) \to 0 \;\; \text{for } \varepsilon \to 0 \tag{6.202}$$

the semi-iterative method is a regularization of $A^\dagger$.

In order to to determine the speed of convergence we investigate

$$\omega_\mu(m) = \sup_{0\leq\lambda\leq 1} \lambda^{\mu/2}|p_m(\lambda)| \tag{6.203}$$

(3) *Each sequence of polynomials* $\{p_m\}_{m\in\mathbb{N}_0}$, $p_m(0) = 1$, *which satisfies the best asymptotic behavior*

$$\omega_\mu(m) = O(m^{-\mu}) \quad \textit{for } m \to \infty \tag{6.204}$$

for some $\mu > 0$, *is uniformly bounded on* $[0,1]$ *and converges pointwise to* 0 *on* $(0,1]$. *In other words* $\{p_m\}_{m\in\mathbb{N}_0}$ *given in such a way is a sequence of residual polynomials, for which, in addition,* $\omega_\alpha(m) = O(m^{-\alpha})$ *for* $0 < \alpha \le \mu$.

The discrepancy principle (cf. (6.186)) can also be used as stopping rule.

(4) *Let* $A \in \mathfrak{L}(X,Y)$ *and* $y \in \mathcal{R}(A)$. *Let the normalized polynomials* $\{p_m\}_{m\in\mathbb{N}_0}$, $p_m(0) = 1$, *satisfy* (6.204) *for some* $\mu > 1$. *Then the corresponding semi-iterative method with starting value* $x^{(0)} = 0$ *is an order optimal regularization of* $A^\dagger$ *with respect to* X_α *for* $0 < \alpha \le \mu - 1$ *if it is combined with the discrepancy principle* (6.186) *as stopping rule with* $\tau > \sup\{\|p_m\|_{C[0,1]} : m \in \mathbb{N}_0\} \ge 1$. *The stopping index satisfies* $m^* = m^*(\varepsilon, y^\varepsilon) = O\left(\varepsilon^{-1/(\alpha+1)}\right)$ *for* $\varepsilon \to 0$.

Normalized polynomials that fulfill (6.204) automatically lead to semi-iterative order optimal regularization methods. The reason for this is that (6.204) implies the uniform boundedness of the polynomials on $[0,1]$ as well as

$$\sup\{|F_m(\lambda)|, \lambda \in [0,1]\} \le 2\tau m^2$$

for the corresponding filters.

In the general case that $p_t(\lambda) = 1 - \lambda F_t(\lambda)$ we are confronted with the situation (see, e.g., [**723**]) that $\omega_\mu(t) \le C_p t^{\frac{\mu}{2}}$ for $t \to 0$ which is the analogue of (6.204). However, this neither implies the uniform boundedness of $\{p_t\}_{t>0}$ nor an estimate like $|F_t(\lambda)| \le C_F t^{-\alpha}$.

Gradient method. The method of successive approximation gradient, and related iterative methods can be used for finding approximate solutions of ill-posed problems (see, e.g., C.W. Groetsch, O. Scherzer [**380**], W.J. Kammerer, M.Z. Nashed [**487, 488**], M.Z. Nashed [**612, 620**] and the references therein for more details).

We let X and Y be two Hilbert spaces, (both over $\mathbb{K} \in \{\mathbb{R}, \mathbb{C}\}$), and let A be a bounded linear operator on X into Y. As already known, the linear equation

$$Ax = y, \quad y \in Y \tag{6.205}$$

may or may not have a solution depending on whether or not y is in $\mathcal{R}(A)$, the range of A, and even if $y \in \mathcal{R}(A)$ the solution of (6.205) need not be unique. For any bounded linear operator $A : X \to Y, \mathcal{R}(A)$ and $\mathcal{R}(A^*)$ are closed subspaces of X and Y, respectively, hence, $X = \mathcal{R}(A) \oplus \mathcal{R}(A)^\perp$ and $Y = \mathcal{R}(A^*) \oplus \mathcal{R}(A^*)^\perp$. The relations $\overline{\mathcal{R}(A)} = \mathcal{R}(A^*)^\perp$, $\overline{\mathcal{R}(A^*)} = \mathcal{R}(A)^\perp$, $\mathcal{R}(A)^\perp = \mathcal{R}(A^*), \overline{\mathcal{R}(A)} = \overline{\mathcal{R}(AA^*)}$ are also valid (see, for instance, [**875**]). Let P denote the orthogonal projection of X onto $\overline{\mathcal{R}(A^*)}$ and let Q denote the orthogonal projection of A on $\overline{\mathcal{R}(A)}$. Then $Ax = APx$ for all $x \in X, A^*y = A^*Qy$ for each $y \in Y$, and the restriction of A to $\mathcal{R}(A)^\perp$ has an inverse, which is not necessarily continuous.

In either case, i.e., $y \in \mathcal{R}(A)$ unique or nonunique solution of (6.205), one can seek a best approximate solution, i.e., a solution which minimizes the quadratic function $J(x) = \|Ax - y\|^2$. Such a solution always exist for all $y \in Y$ if $\mathcal{R}(A)$ is closed. If $\mathcal{R}(A)$ is arbitrary, a best approximation does not exist for all $y \in Y$, however, it does exist for all $y \in \mathcal{R}(A)^\perp$.

We consider the *conjugate gradient* (CG) method (cf. W.J. Kammerer, M.Z. Nashed [**487, 488**]) that minimizes $J(x)$ at each step. That is, choose an initial vector $x_0 \in X$, then compute $r_0 = p_0 = A^*(Ax_0 - y)$, where as always A^* is the adjoint of A. If $p_0 \neq 0$, compute $x_1 = x_0 - \alpha_0 p_0$, where $\alpha_0 = \|r_0\|^2/\|Ap_0\|^2$. For $i = 1, 2, \ldots$, compute

$$r_i = A^*(Ax_i - y) = r_{i-1} - \alpha_{i-1}A^*Ap_{i-1}, \tag{6.206}$$

where

$$\alpha_{i-1} = \frac{\langle r_{i-1}, p_{i-1}\rangle}{\|Ap_{i-1}\|^2}, \tag{6.207}$$

and if $r_i \neq 0$, then compute

$$p_i = r_i + \beta_{i-1}p_{i-1}, \tag{6.208}$$

where

$$\beta_{i-1} = -\frac{\langle r_i, A^*Ap_{i-1}\rangle}{\|Ap_{i-1}\|^2} \tag{6.209}$$

and set

$$x_{i+1} = x_i - \alpha_i p_i. \tag{6.210}$$

We examine some properties of the CG algorithm, for the case of an arbitrary bounded linear operator. In this case, the domain of the generalized inverse of A is $\mathcal{D}(A^\dagger) = \mathcal{R}(A) + \mathcal{R}(A)^\dagger$. If $y \in \mathcal{D}(A^\dagger)$, then $Qy = \bar{y}$ is in the range of A and $v = A^\dagger y = A^\dagger\bar{y}$ and $\bar{y} = Qy = Av = AA^\dagger y$. Since Q is an orthogonal projection, the functional J can be written as $J(x) = \|Ax - y\|^2 = \|Ax - \bar{y}\|^2 + \|\bar{y} - y\|^2$. Thus, minimizing J is equivalent to minimizing the functional $\|Ax - \bar{y}\|^2$ which we shall denote by $g(x)$. Setting

$$u = v + (I - P)x_0 = A^\dagger y + (I - P)x_0 \tag{6.211}$$

one can define the error vector $e = x - u$ and the vector $r = A^*(Ax - y) = A^*(Ax - \bar{y})$. Then

$$(A^*A)e = r \tag{6.212}$$

and

$$\langle r, e\rangle = \|Ax - \bar{y}\|^2 = g(x). \tag{6.213}$$

The sequence of iterates $\{x_i\}$ generated by the CG method (6.206)–(6.210) is contained in the flat $x_0 + \mathcal{R}(A^*)$ with both r_i and p_i, for $i = 0, 1, 2, \ldots$, in $\mathcal{R}(A^*)$. Moreover, $p_0, p_1, \ldots, p_{i-1}$ form a A^*A-orthogonal set of vectors and their span in an i-dimensional subspace of $\mathcal{R}(A^*)$. If at the ith step, $r_i = 0$, then both $Ax_i - y$ and $Ax_i - \bar{y}$ are vectors in $\mathcal{R}(A^*) = \overline{\mathcal{R}(A)^\perp}$. However, Ax_i and $\bar{y}$ are also in $\overline{\mathcal{R}(A)}$, and therefore $Ax_i = \bar{y}$, implying that $g(x_i) = 0$. In this case, the iteration terminates at the ith step and

$$x_i = x_0 - \sum_{k=0}^{i-1} \alpha_k p_k = u, \tag{6.214}$$

and

$$A^\dagger y = px_0 - \sum_{k=0}^{i-1} \alpha_i p_k. \tag{6.215}$$

Therefore, unless explicitly mentioned otherwise, we shall assume that the conjugate gradient method does not terminate in a finite number of steps, that is, $r_i \neq 0$ for $i = 0, 1, \ldots$.

We list some known identities for the conjugate gradient method: For indices satisfying $k = 0, 1, 2, \ldots$, and $i = 0, 1, 2, \ldots$, we have

$$\langle r_i, r_i \rangle = \langle p_i, r_k \rangle, \tag{6.216}$$

$$\|Ap_i\| \leqq \|Ar_i\|, \tag{6.217}$$

$$\beta_i = \frac{\|r_{i+1}\|^2}{\|r_i\|^2}, \tag{6.218}$$

$$\langle p_i, p_k \rangle = \frac{\|r_i\|^2 \|p_k\|^2}{\|r_k\|^2}, \tag{6.219}$$

$$p_i = \|r_i\|^2 \sum_{j=0}^{i} \frac{r_j}{\|r_j\|^2} \tag{6.220}$$

and

$$\|r_i\|^2 \leqq \|p_i\|. \tag{6.221}$$

Setting $g(x_i) = \langle r_i, e_i \rangle = \|Ax_i - \bar{y}\|^2$, where $e_i = x_i - u$, we find that

$$g(x_i) - g(x_{i+1}) = \alpha_i \|r_i\|^2. \tag{6.222}$$

Finally, x_i minimizes the functionals $J(x)$ and $g(x)$ on the i–dimensional flat

$$x_0 + \operatorname{span}\{p_0, p_1, \ldots, p_{i-1}\}. \tag{6.223}$$

It is worth mentioning the following three properties (see [**488**]):

(i) For $k = 0, 1, 2, \ldots$,

$$g(x_i) = \langle e_i, r_k \rangle. \tag{6.224}$$

For $i = 0, 1, 2, \ldots$,

$$\langle p_i, e_i \rangle \|r_i\|^2 = g(x_i) \|p_i\|^2. \tag{6.225}$$

(ii) The inequality

$$\|e_{i+1}\|^2 \leqq \|e_i\|^2 - \alpha_i g(x_i) \tag{6.226}$$

holds for $i = 0, 1, 2, \ldots$.

(iii) For any nonnegative integers i and j, both $\langle p_i, e_i \rangle$ and $\langle e_i, e_j \rangle$ are nonnegative numbers.

The main result on the CG method for bounded linear operators with closed range depends heavily upon the following observation:

Let X and Y be two Hilbert spaces over the same field $\mathbb{K}$ and let A be a bounded linear transformation mapping X into Y. If $\mathcal{R}(A)$ is closed, then $S = A|\mathcal{R}(A^*)$, the restriction of A to $\mathcal{R}(A^*)$, has a bounded inverse and $U = A^*A|\mathcal{R}(A^*)$ is a positive definite operator onto $\mathcal{R}(A^*)$.

Let $\mu(x) = \langle Ux, x\rangle / \langle x, x\rangle, x \neq 0$, be the *Rayleigh quotient of* U. Since U is a bounded symmetric positive definite linear operator on the Hilbert space $\mathcal{R}(A^*)$, the spectral bounds

$$m = \inf\{\mu(x) : x \in \mathcal{R}(A^*)\} \tag{6.227}$$

and

$$M = \sup\{\mu(x) : x \in \mathcal{R}(A^*)\} \tag{6.228}$$

are positive and finite.

Let X and Y be two Hilbert spaces over the real field and let A be a bounded linear transformation mapping X into Y. If the range of A is closed then the conjugate gradient method (6.206)-(6.210) *converges monotonically to the least-squares solution*

$$u = A^\dagger y + (I - P)x_0 \tag{6.229}$$

of $Ax = y$. Moreover, m and M are the spectral bounds of $U = \{A^*A|\mathcal{R}(A^*)\}$, then

$$\|x_k - u\|^2 \leqq \frac{g(x_0)}{m}\left(\frac{M-m}{M+m}\right)^{2k}, \quad k = 0, 1, 2, \ldots. \tag{6.230}$$

Altogether, let X and Y be two Hilbert space over the same field. If A is a linear transformation mapping X into Y of rank r, then the conjugate gradient method associated with the system $Tx = y$ converges in at most r steps to the least squares solution $u = A^\dagger y + (I - P)x_0$. When $\mathcal{R}(A^*)$ is not closed, then the operator $U = A|\mathcal{R}(A^*)$ need not be positive definite, and therefore the generalization of an inequality can not longer be utilized. However, it is still possible to establish convergence of the CG method under mild restrictions.

Let X and Y be two Hilbert spaces over the real field, and let A be a bounded linear operator mapping X to Y. If $Qy \in \mathcal{R}(AA^*A)$, then the conjugate gradient method (6.206)-(6.210), *with initial value $x_0 \in \mathcal{R}(A^*A)$, converges monotonically to the least squares solution of minimal norm $u = A^\dagger y$. In fact,*

$$\|x_k - u\|^2 \leqq \frac{\|A\|^2\|x_0 - A^\dagger y\|^2\|A^{*\dagger}x_0 - (AA^*)^\dagger\|^2}{\|A\|^2\|A^{*\dagger}x_0 - (AA^*)^\dagger y\|^2 + k\|x_0 - A^\dagger y\|^2}, \quad k = 1, 2, \ldots. \tag{6.231}$$

6.9. Projection methods

First, we recall the definition of a projection operator known from functional analysis: Let X be a normed space over the field $\mathbb{K}$ ($\mathbb{K}$ is equal to $\mathbb{R}$ or $\mathbb{C}$). Let $U \subset X$ be a closed subspace. A linear bounded operator $P : X \to X$ is called a *projection operator* on U if $Px \in U$ for all $x \in X$ and $Px = x$ for all $x \in U$. Every nontrivial projection operator satisfies $P^2 = P$ and $\|P\| \geq 1$.

The following two examples provide some important projection operators:

(a) *(Orthogonal projection.)* Let X be a pre-Hilbert space. Suppose that $U \subset X$ is a complete subspace. Let $Px \in U$ be the *best approximation* to x in U, i.e., Px satisfies the relation

$$\|Px - x\| \leq \|u - x\| \quad \text{for all} \quad u \in U. \tag{6.232}$$

$P : X \to U$ is linear and $Px \in U$ is characterized by $\langle x - Px, u\rangle_X = 0$ for all $u \in U$, i.e., $x - Px \in U^\perp$. Therefore,

$$\begin{aligned}\|x\|^2 &= \|Px + (x - Px)\|^2 \\ &= \|Px\|^2 + \|x - Px\|^2 + 2\,\underbrace{\Re\langle x - Px, Px\rangle}_{=0} \geq \|Px\|^2,\end{aligned}$$

i.e., $\|P\| = 1$.

(b) (*Interpolation operator.*) Let $X = C^{(0)}([a, b])$ be the space of real valued continuous functions on an interval $[a, b]$ supplied with the supremum norm $\|\cdot\|_{C^{(0)}[a,b]}$. Then X is a normed space over $\mathbb{R}$. Let $U = \operatorname{span}\{u_1, \ldots, u_n\}$ be an n–dimensional subspace and $t_1, \ldots, t_n \in [a, b]$ such that the interpolation problem in U is uniquely solvable, i.e., $\det(u_j(t_k)) \neq 0$. We define $Px \in U$ by the interpolant of $x \in C^{(0)}([a, b])$ in U, i.e., $u = Px \in U$ satisfies $u(t_i) = x(t_i)$ for all $i = 1, \ldots, n$. Then $P : X \to U$ is a projection operator.

Examples for U in (b) are spaces of algebraic or trigonometric polynomials. As a drawback of these choices, we note that from the results of Faber (see, for example, [**645**]) the interpolating polynomials of continuous functions x do not, in general, converge to x as the degree of the polynomials tends to infinity. Nonetheless, trigonometric interpolation at equidistant points converges with optimal order of convergence.

Next we are concerned with a certain class of projection methods in Banach spaces, essentially following the monograph of A. Kirsch [**497**]: Let X and Y be Banach spaces and $A : X \to Y$ be bounded and one-to-one. Furthermore, let $X_n \subset X$ and $Y_n \subset Y$ be finite-dimensional subspaces of dimension n and $Q_n : Y \to Y_n$ be a projection operator. For $y \in Y$, the projection method for solving the equations $Ax = y$ is to solve the equations

$$Q_n A x_n = Q_n y \quad \text{for} \quad x_n \in X_n. \tag{6.233}$$

Assume that $\{\hat{x}_1, \ldots, \hat{x}_n\}$ and $\{\hat{y}_1, \ldots, \hat{y}_n\}$ are bases of X_n and Y_n, respectively. Then $Q_n y$ and every $Q_n A\hat{x}_j$, $j = 1, \ldots, n$, is representable in the forms

$$Q_n y = \sum_{i=1}^n \beta_i \hat{y}_i \quad \text{and} \quad Q_n A\hat{x}_j = \sum_{i=1}^n B_{ij} \hat{y}_i, \quad j = 1, \ldots, n, \tag{6.234}$$

with $\beta_i, B_{ij} \in \mathbb{K}$. The linear combination $x_n = \sum_{j=1}^n \alpha_j \hat{x}_j$ solves (6.233) if and only if $\alpha = (\alpha_1, \ldots, \alpha_n)^T \in \mathbb{K}^n$ solves the finite system of linear equations

$$\sum_{i=1}^n B_{ij}\alpha_j = \beta_i, \quad i = 1, \ldots, n. \tag{6.235}$$

We are led to the following important classes of projection methods for $A : X \to Y$ being a bounded and one-to-one operator: Let X and Y be pre-Hilbert spaces and $X_n \subset X$ and $Y_n \subset Y$ be finite-dimensional subspaces with $\dim X_n = \dim Y_n = n$. Let $Q_n : Y \to Y_n$ be the *orthogonal projection*. Then the projected equation $Q_n A x_n = Q_n y$ is equivalent to

$$\langle Ax_n, z_n\rangle = \langle y, z_n\rangle \quad \text{for all} \quad z_n \in Y_n \tag{6.236}$$

We let $X_n = \operatorname{span}\{\hat{x}_1, \dots, \hat{x}_n\}$ and $Y_n = \operatorname{span}\{\hat{y}_1, \dots, \hat{y}_n\}$. Looking for a solution of (6.236) in the form of a linear combination $x_n = \sum_{j=1}^n \alpha_j \hat{x}_j$ we arrive at

$$\sum_{j=1}^{n} \alpha_j \langle A\hat{x}_j, \hat{y}_i\rangle = \langle y, \hat{y}_i\rangle \quad \text{for} \quad i = 1, \dots, n, \tag{6.237}$$

or in matrix-vector nomenclature $A\alpha = \beta$, where $A_{ij} = \langle A\hat{x}_j,\ \hat{y}_i\rangle$ and $\beta_i = \langle \hat{y},\ \hat{y}_i\rangle$, $i = 1, \dots, n$.

A priori assumption. For the remaining part of this section about projection methods, it is helpful to make the following *a priori assumption (APA)*:

(i) *Let $A : X \to Y$ be a linear, bounded, and injective operator between Banach spaces, $X_n \subset X$ and $Y_n \subset Y$ be finite-dimensional subspaces of dimension n and $Q_n : Y \to Y_n$ be a projection operator. We assume that $\bigcup_{n\in\mathbb{N}} X_n$ is dense in X and the $Q_n A|X_n : X_n \to Y_n$ is one-to-one and, thus, invertible.*

(ii) *Let $x \in X$ be the solution of*

$$Ax = y. \tag{6.238}$$

By $x_n \in X_n$, we understand the unique solutions of the equations

$$Q_n A x_n = Q_n y, \ n \in \mathbb{N}. \tag{6.239}$$

As a consequence of (APA) we are allowed to represent the solutions x_n of (6.239) in the form $x_n = R_n y$, where $R_n : Y \to X_n \subset X$ is defined by

$$R_n = (Q_n A|X_n)^{-1} Q_n : Y \to X_n \subset X. \tag{6.240}$$

Suppose that (APA) is valid. The *projection method is called convergent* if the approximate solutions $x_n \in X_n$ of (6.239) converge to the exact solution $x \in X$ of (6.238) for every $y \in A(X)$, i.e., if the limit relation

$$R_n A x = (Q_n A|X_n)^{-1} Q_n A x \to x, \quad n \to \infty, \tag{6.241}$$

holds true for every $x \in X$. Obviously, this definition of convergence coincides with the definition of a regularization strategy for the equation $Ax = y$. Therefore, the projection method converges if and only if R_n is a regularization strategy for the equation $Ax = y$.

Convergence can only be expected if we require that $\bigcup_{n\in\mathbb{N}} X_n$ is dense in X and $Q_n y \to y$ for all $y \in A(X)$. For a compact operator A, however, this property is not sufficient for the convergence. In fact we have to assume an additional boundedness condition:

Let the assumption (APA) be satisfied. The solution $x_n = R_n y \in X_n$ of (6.239) converges to x for every $y = Ax$ if and only if there exists $c > 0$ such that

$$\|R_n A\| \leq C \quad \textit{for all} \quad n \in \mathbb{N}. \tag{6.242}$$

If (6.242) is satisfied, the following error estimate can be shown to be valid

$$\|x_n - x\| \leq (1 + C) \min_{z_n \in X_n} \|z_n - x\| \tag{6.243}$$

with the same constant C as in (6.242).

The estimates (6.242) and (6.243) can be verified in straightforward way.

Suppose that $\|R_n A\|$ is bounded. The operator $R_n A$ is a projection operator onto X_n since for $z_n \in X_n$ we have $R_n A z_n = (Q_n A|X_n)^{-1} Q_n A z_n = z_n$. Thus we are able to deduce that

$$x_n - x = (R_n A - I)x = (R_n A - I)(x - z_n) \quad \textit{for all} \quad z_n \in X_n, \tag{6.244}$$

so that

$$\|x_n - x\| \leq (c+1)\|x - z_n\| \quad \textit{for all} \quad z_n \in X_n. \tag{6.245}$$

Thus, (6.243) is valid. The convergence $x_n \to x$ follows from the fact that $\bigcup_{n \in \mathbb{N}} X_n$ is dense in X.

Thus far, we were concerned with the case, where the right-hand side y is exactly known. Next, we consider the case where the right-hand side is known only approximately, i.e., we start from an element $y^\varepsilon \in Y$ with $\|y^\varepsilon - y\| \leq \varepsilon$. To this end we understand the operator R_n from (6.240) as a regularization operator in the usual sense so that we are led to distinguish two kinds of errors for the right-hand side. A straightforward application of the triangle inequality yields with $x_n^\varepsilon = R_n y^\varepsilon$ the inequality

$$\begin{aligned} \|x_n^\varepsilon - x\| &\leq \|x_n^\varepsilon - R_n y\| + \|R_n y - x\| \\ &\leq \|R_n\| \, \|y^\varepsilon - y\| + \|R_n A x - x\|. \end{aligned} \tag{6.246}$$

As usual, we are confronted with the dilemma of IP: The error ε of the right-hand side is multplied by the norm of R_n. The second term describes the discretization error against exact data. In practice, one solves the discrete system (6.235), where the vector β is replaced by a perturbed vector $\beta^\varepsilon \in \mathbb{K}^n$ with

$$|\beta^\varepsilon - \beta|^2 = \sum_{j=1}^{n} |\beta_j^\varepsilon - \beta_j|^2 \leq \varepsilon^2. \tag{6.247}$$

We will call this fact the *discrete perturbation* of the right hand side. Instead of (6.235) one solves $B\alpha^\varepsilon = \beta$ and defines $x_n^\varepsilon \in X_n$ by $x_n^\varepsilon = \sum_{j=1}^n \alpha_j^\varepsilon \hat{x}_j$. Note that the choice of the basis functions $\hat{x}_j \in X_n$ and $\hat{y}_j \in Y_n$ are essential rather than the norm of Y. Unfortunately, it turns out that the condition number of B reflects the ill-conditioning of the equation $Ax = y$. In this respect it should be mentioned that it suffices to study the question of convergence for the "principal part" of the operator A under discussion. More concretely, if the projection method converges for an operator A, then convergence and error estimates also hold true for $A + C$, where C is compact relative to A (i.e., $A^{-1}C$ is compact).

The proof of the following result can be found, e.g., in the monograph [**497**]:

Let (APA) be satisfied. Let $C : X \to Y$ be a linear operator with $C(X) \subset A(X)$ such that $A + C$ is one-to-one and $A^{-1}C$ is compact in X. Assume, furthermore, that the projection method converges for A, i.,e., that $R_n A x \to x, n \to \infty$, for every $x \in X$, where

$$R_n = (Q_n A|X_n)^{-1} Q_n.$$

Then it also converges for $A + C$ i.e.,

$$(Q_n(A+C)|X_n)^{-1} Q_n (A+C) x \to x, \quad n \to \infty,$$

for all $x \in X$. Let $x \in X$ be the solution of $(A+C)x = y$ and $x^\varepsilon \in X_n$ be the solution of the corresponding projected equation $Q_n(A+C)x_n^\varepsilon = y_n^\varepsilon$ for some $y_n^\varepsilon \in Y_n$. Then there exists a constant $c > 0$ such that

$$\|x - x_n^\varepsilon\| \le c\left[\|A^{-1}Cx - R_nCx\| + \|A^{-1}y - R_ny_n^\varepsilon\|\right] \tag{6.248}$$

for all sufficiently large n and $\varepsilon > 0$.

It should be mentioned that the first term on the right hand side of (6.248) is just the error of the projection methods for the equations $Ax = Cx$ without perturbation of the right hand side. This allows to assure the estimate as follows:

$$\|A^{-1}Cx - R_nCx\| \le (1+c) \min_{z_n \in X_N} \|A^{-1}Cx - z_n\|.$$

The second term on the right hand side of (6.248) is the error for the equation $Ax = y$. Hence, our results include both the continuous and the discrete perturbations of the right hand side. For the continuous case we set $y_n^\varepsilon := Q_ny^\varepsilon$, while in the discrete case, we set $y_n^\varepsilon = \sum_{i=1}^n \beta_i^\varepsilon y_i$.

All in all, a framework for reduction of inverse and identification problems to finite-dimensional problems exists within the concept of projection methods (see also [**625**]). Projection methods (e.g., spline or finite-element functions) can be either applied directly to the IPP or to the regularized problem, i.e., to the problem of minimizing a Tikhonov type functional. The moment-discretization (or semidiscretization) method for integral equations of the first kind and for IPP in reproducing kernel Hilbert spaces (see M.Z. Nashed, G. Wahba [**640–642**], M.Z. Nashed [**618**]), viewed as a projection method, is both quasi-optimal and robust. It is also particularly suited when values of the data function $y(s)$ are known only at a finite number of points. These properties, together with the convergence and commutativity properties established in [**618**] may account for the favorable behavior of computer implementations of this method (cf. M.L. Baart [**53**]).

Galerkin methods. We deal with the situation that X and Y are (real or complex) Hilbert spaces. Moreover, $A : X \to Y$ is assumed to be linear, bounded, and one-to-one; $X_n \subset X$ and $Y_n \subset Y$ are assumed to be finite-dimensional subspaces with $\dim X_n = \dim Y_n = n$; and $Q_n : Y \to Y_n$ is the orthogonal projection operator onto Y_n. Then, $Q_nAx_n = Q_ny$ reduces to the so-called *Galerkin equations*

$$\langle Ax_n, z_n\rangle = \langle y, z_n\rangle \quad \text{for all} \quad z_n \in Y_n. \tag{6.249}$$

Choosing bases $\{\hat{x}_1, \dots, \hat{x}_n\}$ and $\{\hat{y}_1, \dots, \hat{y}_n\}$ of X_n and Y_n, respectively, we are led to a finite system in the coefficients of $x_n = \sum_{j=1}^n \alpha_j \hat{x}_j$ (see (6.237))

$$\sum_{i=1}^n B_{ij}\alpha_j = \beta_i, \quad i = 1, \dots, n, \tag{6.250}$$

where we have used the abbreviations $B_{ij} = \langle A\hat{x}_j, \hat{y}_i\rangle_Y$ and $\beta_i = \langle y, \hat{y}_i\rangle_Y$. We observe that B_{ij} and β_i coincide with the settings in (6.234) only if the set $\{\hat{y}_j : j = 1, \dots, n\}$ forms an orthonormal basis of Y_n.

It should be remarked that the Galerkin method is also known as the *Petrov-Galerkin method* (see G.I. Petrov [**692**]), since Petrov was the first to consider the general situation of (6.249). The special case $X = Y$ and $X_n = Y_n$ was studied by Bubnov in 1914 and later by Galerkin in 1915 (see [**341**]). For this reason, this special case is also known as the *Bubnov-Galerkin method*. In the case when the operator A is self-adjoint and positive definite, we will see that the Bubnov-Galerkin method coincides with the *Rayleigh-Ritz method* (see W. Ritz [**722**] and L. Rayleigh [**712**]).

The following *error estimates for the Galerkin method* of the form (6.246) (see, e.g., [**497**]) only in the first term, which corresponds to the perturbation of the right-hand side. The second term bounds the error for the exact right-hand side and tends to zero, provided the boundedness assumption (6.242) is satisfied.

Assume that the Galerkin equations (6.249) *are uniquely solvable for every right-hand side of the equation $Ax = y$.*

(a) Let $y^\varepsilon \in Y$ with $\|y - y^\varepsilon\|_Y \le \varepsilon$ be given and $x_n^\varepsilon \in X_n$ be the solution of

$$\langle Ax_n^\varepsilon, z_n\rangle = \langle y^\varepsilon, z_n\rangle \quad \text{for all} \quad z_n \in Y_n. \tag{6.251}$$

Then the following error estimate holds true:

$$\|x_n^\varepsilon - x\| \le \varepsilon\|R_n\| + \|R_nAx - x\|. \tag{6.252}$$

(b) Let B and β be given by (6.250) and $\beta^\varepsilon \in \mathbb{K}^n$ with $|\beta - \beta^\varepsilon| \le \varepsilon$, where $|\cdot|$ denotes the Euclidean norm in $\mathbb{K}^n$. Let $\alpha^\varepsilon \in \mathbb{K}^n$ be the solution of $B\alpha^\varepsilon = \beta^\varepsilon$. Set $x_n^\varepsilon = \sum_{j=1}^n \alpha_j^\varepsilon \hat{x}_j \in X_n$. Then the following error estimate holds true:

$$\|x_n^\varepsilon - x\| \le \frac{a_n}{\lambda_n}\,\varepsilon + \|R_nAx - x\|, \tag{6.253}$$

$$\|x_n^\varepsilon - x\| \le b_n\|R_n\|\varepsilon + \|R_nAx - x\|, \tag{6.254}$$

where

$$a_n = \max\left\{\left\|\sum_{j=1}^n \rho_j\hat{x}_j\right\|_X : \sum_{j=1}^n |\rho_j|^2 = 1\right\}, \tag{6.255}$$

$$b_n = \max\left\{\sqrt{\sum_{j=1}^n |\rho_j|^2} : \left\|\sum_{j=1}^n \rho_j\hat{y}_j\right\| - 1\right\}, \tag{6.256}$$

and $\lambda_n > 0$ denotes the smallest singular value of the matrix B.

Next we are interested in deriving error estimates for three particularly interesting choices for the finite-dimensional subspaces X_n and Y_n . The cases, where X_n and Y_n are coupled by $Y_n = A(X_n)$ or $X_n = A^*(Y_n)$ will lead to the least squares method or the dual least squares method, respectively. Clearly, $A^* : Y \to X$ is the adjoint of A. In addition we will study the Bubnov-Galerkin method for the case where, A additionally satisfies Gårding's inequality. In all cases, we formulate the Galerkin equations for the perturbed cases first without using particular bases and then with respect to given bases in X_n and Y_n.

Least-squares method. For a finite-dimensional subspace $X_n \subset X$, determine $x_n \in X_n$ such that

$$\|Ax_n - y\| \le \|Az_n - y\| \quad \text{for all} \quad z_n \in X_n. \tag{6.257}$$

Clearly, existence and uniqueness of $x_n \in X_n$ can be guaranteed easily since X_n is finite-dimensional and A is assumed to be one-to-one. The solution $x_n \in X_n$ of the least-squares problem is characterized by

$$(6.258) \qquad \langle Ax_n, Az_n \rangle = \langle y, Az_n \rangle \quad \text{for all} \quad z_n \in X_n.$$

We notice that this method is a special case of the Galerkin method, where we have $Y_n = A(X_n)$. Choosing a basis $\{\hat{x}_j, j = 1, \ldots, n\}$ of X_n leads to the finite linear system

$$(6.259) \qquad \sum_{j=1}^{n} \alpha_j \langle A\hat{x}_j, Ax_i \rangle = \beta_i = \langle y, A\hat{x}_i \rangle \quad \text{for all} \quad i = 1, \ldots, n,$$

i.e., in matrix-vector nomenclature $B\alpha = \beta$. The corresponding matrix $B \in \mathbb{K}^{n \times n}$ with $B_{ij} = \langle A\hat{x}_j, A\hat{x}_i \rangle$ is Hermitean and positive definite, since A is assumed to be one-to-one.

Of practical interest is the case where the right hand side is perturbed by an error. Let $x_n^\varepsilon \in X_N$ solve the equations

$$(6.260) \qquad \langle Ax_n^\varepsilon, Az_n \rangle = \langle y^\varepsilon, Az_n \rangle \quad \text{for all} \quad z_n \in X_n,$$

where $y^\varepsilon \in Y$ is the perturbed right hand side satisfying $\|y^\varepsilon - y\|_Y \leq \varepsilon$. For the discrete perturbation, we assume that $\beta \in \mathbb{K}^n$ is replaced by $\beta^\varepsilon \in \mathbb{K}^n$ with $|\beta^\varepsilon - \beta| \leq \varepsilon$, where $|\cdot|$ denotes the Euclidean norm in $\mathbb{K}^n$. This leads to the following finite system of equations in the unknown coefficients of $x_n^\varepsilon = \sum_{j=1}^n a_j^\varepsilon \hat{x}_j$:

$$(6.261) \qquad \sum_{j=1}^{n} \alpha_j^\varepsilon \langle A\hat{x}_j, A\hat{x}_i \rangle = \beta_i^\varepsilon \quad \text{for all} \quad i = 1, \ldots, n.$$

The system (6.261) is uniquely solvable, since the matrix B is positive definite.

Obviously, for least-squares methods, the boundedness condition (6.242) is not satisfied without imposing additional assumptions (for more details we refer, e.g., to R. Kress [**513**], T.I. Seidmann [**760**]):

Let $A : X \to Y$ be a linear, bounded, and injective operator between Hilbert spaces. Suppose that $X_n \subset X$ form finite-dimensional subspaces such that $\bigcup_{n\in\mathbb{N}} X_n$ is dense in X. Let $x \in X$ be the solution of $Ax = y$ and $x_n^\varepsilon \in X_n$ be the least squares solution from (6.260) *or* (6.261). *Denote by σ_n the quantity*

$$(6.262) \qquad \sigma_n = \max\{\|z_n\| : z_n \in X_n, \|Az_n\| = 1\}.$$

Suppose that there exists $c > 0$ independent of n, such that

$$(6.263) \qquad \min_{z_n \in X_n} \{\|x - z_n\| + \sigma_n \|A(x - z_n)\|\} \leq c\|x\| \quad \textit{for all} \quad x \in X.$$

Then, the least-squares method is convergent, and we have $\|R_n\| \leq \sigma_n$. Moreover, we have the error estimate

$$(6.264) \qquad \|x - x_n^\varepsilon\| \leq r_n \sigma_n\, \varepsilon + \bar{c}\, \min\{\|x - z_n\| : z_n \in X_n\}$$

for some $\bar{c} > 0$. Here, $r_n = 1$, if $x_n^\varepsilon \in X_n$ solves (6.260), *i.e., ε measures the continuous perturbation $\|y^\varepsilon - y\|_Y$. If ε measures the discrete error $|\beta^\varepsilon - \beta|$ in the*

Euclidean norm and $x_n^\varepsilon = \sum_{j=1}^n \alpha_j^\varepsilon \hat{x}_j \in X_n$, *where the vector* $\alpha^\varepsilon = (\alpha_1^\varepsilon, \ldots, \alpha_n^\varepsilon)^T$ *solves* (6.261), *then* r_n *is given by*

$$r_n = \max \left\{ \sqrt{\sum_{j=1}^n |\rho_j|^2} : \left\| A \left(\sum_{j=1}^n \rho_j \hat{x}_j \right) \right\| = 1 \right\}. \tag{6.265}$$

For further numerical aspects of least-squares method, we refer, e.g., to [**210**, **211**, **487**, **488**, **559**, **613**, **618**].

Dual least-squares method. As another variant of the Galerkin method, we come to the dual least squares method. In this case the boundedness condition (6.242) is indeed always satisfied: Given some finite-dimensional subspaces $Y_n \subset Y$, determine $u_n \in Y_n$ such that

$$\langle A^* u_n, z_n \rangle = \langle y, z_n \rangle \quad \text{for all} \quad z_n \in Y_n, \tag{6.266}$$

where $A^* : Y \to X$ represents the adjoint of A. Then $x_n = A^* z_n$ is called the *dual least squares solution.* It is a special case of the Galerkin method, where $X_n = A^*(Y_n)$. Writing (6.266) for $y = Ax$ in the form

$$\langle A^* u_n, A^* z_n \rangle = \langle x, A^* z_n \rangle \quad \text{for all} \quad z_n \in Y_n, \tag{6.267}$$

we see that the dual least-squares method is just the least squares method for the equation $A^* u = x$. This explains the standard terminology in the literature.

Suppose now that the right hand side is perturbed. Let $y^\varepsilon \in Y$ be given such that $\|y^\varepsilon - y\|_Y \leq \varepsilon$. Instead of the linear equation (6.266), we determine $x_n^\varepsilon = A^* u_n^\varepsilon \in X_n$ via

$$\langle A^* u_n^\varepsilon, A^* z_n \rangle = \langle y^\varepsilon, z_n \rangle \quad \text{for all} \quad z_n \in Y_n. \tag{6.268}$$

For discrete perturbations, we specify a basis $\{\hat{y}_j\}_{j=1,\ldots,n}$ of Y_n and assume that the right-hand side $\beta_i = \langle y, \hat{y}_i \rangle_Y, i = 1, \ldots, n$, of the resulting linear equations are perturbed by a vector $\beta^\varepsilon \in \mathbb{K}^n$ with $|\beta^\varepsilon - \beta| \leq \varepsilon$, where $|\cdot|$ denotes the Euclidean norm in $\mathbb{K}^n$. Instead of (6.266) we are led to

$$x_n^\varepsilon = A^* u_n^\varepsilon = \sum_{j=1}^n \alpha_j^\varepsilon A^* \hat{y}_j, \tag{6.269}$$

where $\alpha^\varepsilon \in \mathbb{K}^n$ solves the linear equation

$$\sum_{j=1}^n \alpha_j^\varepsilon \langle A^* \hat{y}_j, A^* \hat{y}_i \rangle = \beta_i^\varepsilon, \quad i = 1, \ldots, n. \tag{6.270}$$

Results on convergence and error estimates are listed, e.g., in the textbook A. Kirsch [**497**].

Let X *and* Y *be Hilbert spaces. Suppose that* $A : X \to Y$ *is linear, bounded, an one-to-one such that the range* $A(X)$ *is dense in* Y. *Let* $Y_n \subset Y$ *form finite-dimensional subspaces such that* $\bigcup_{n \in \mathbb{N}} Y_n$ *is dense in* Y. *Assume that* $x \in X$ *is the solution of* $Ax = y$. *Then the linear equations* (6.268) *and* (6.270) *are uniquely solvable for every right hand side and every* $n \in \mathbb{N}$. *Furthermore, the dual least-squares method is convergent, and we have*

$$\|R_n\| \leq \sigma_n = \max\{\|z_n\| : z_n \in Y_n, \|A^* z_n\| = 1\}. \tag{6.271}$$

Moreover, we have the error estimate

$$\|x - x_n^\varepsilon\| \leq r_n \sigma_n\ \varepsilon + c\ \min\{\|x - z_n\| : z_n \in A^*(Y_n)\} \tag{6.272}$$

for some $c > 0$. Here, $r_n = 1$ if $r_n^\varepsilon \in X_n$ solves (6.268), i.e., ε measures the norm $\|y^\varepsilon - y\|_Y$ in Y. If ε measures the discrete error $|\beta^\varepsilon - \beta|$ and $x_n^\delta = \sum_{j=1}^n \alpha_j^\varepsilon A^* \hat{y}_j \in X_n$, where α^ε solves (6.270), then r_n is given by

$$r_n = \max\left\{\sqrt{\sum_{j=1}^n |\rho_j|^2} : \left\|\sum_{j=1}^n \rho_j \hat{y}_j\right\| = 1\right\} \tag{6.273}$$

(note that $r_n = 1$ if $\{\hat{y}_j\}_{j=1,\ldots,n}$ constitutes an orthonormal system in Y).

Bubnov-Galerkin method. We assume that $A : X \to X$ is a linear and bounded operator and $X_n, n \in \mathbb{N}$, are finite-dimensional subspaces. The Galerkin method amounts to the problem of determining $x_n \in X_n$ such that

$$\langle Ax_n, z_n\rangle = \langle y, z_n\rangle \quad \text{for all} \quad z_n \in X_n. \tag{6.274}$$

This special case is called the *Bubnov-Galerkin method.* Again, we consider the perturbation of the right hand side. If $y^\varepsilon \in Y$ is chosen such that $\|y^\varepsilon - y\| \leq \varepsilon$ represents a perturbed right hand side, then instead of (6.274) we study the equation system

$$\langle Ax_n^\varepsilon, z_n\rangle = \langle y^\varepsilon, z_n\rangle \quad \text{for all} \quad z_n \in X_n. \tag{6.275}$$

An alternative is to choose a basis $\{\hat{x}_j\}_{j=1,\ldots,n}$ of X_n. We assume that the right hand side $\beta_i =< y, \hat{x}_i >$, $i = 1, \ldots, n$ of the Galerkin equations are perturbed by a vector $\beta^\varepsilon \in \mathbb{K}^n$ with $|\beta^\varepsilon - \beta| \leq \varepsilon$, where $|\cdot|$ denotes again the Euclidean norm in $\mathbb{K}^n$. In this case, instead of (6.274), we have to solve

$$\sum_{j=1}^n a_j^\varepsilon \langle A\hat{x}_j, \hat{x}_i\rangle = \beta_i^\varepsilon \quad \text{for} \quad i = 1, \ldots, n. \tag{6.276}$$

For $\alpha^\varepsilon \in \mathbb{K}^n$ we set $x_n^\varepsilon = \sum_{j=1}^n \alpha_j^\varepsilon \hat{x}_j$.

Next we show that the *Rayleigh-Ritz method,* in fact, it is a special case of the Bubnov-Galerkin method.

Rayleigh-Ritz method. Let $A : X \to X$ be also self-adjoint and positive definite, so that $\langle Ax, y\rangle_X = \langle x, Ay\rangle_X$ and $\langle Ax, x\rangle_X > 0$ for all $x, y \in X$ with $x \neq 0$. We introduce the functional

$$J(z) = \langle Az, z\rangle - 2\Re\langle y, z\rangle \quad \text{for} \quad z \in X. \tag{6.277}$$

The identity

$$J(z) - J(x) = 2\Re\langle Ax - y, z - x\rangle + \langle A(z - x), z - x\rangle \tag{6.278}$$

and the positivity of A tells us that $x \in X$ is the unique minimum of J if and only if x solves $Ax = y$. The Rayleigh-Ritz method is to minimize J over the finite-dimensional subspace X_n. From (6.278), we see that if $x_n \in X_n$ minimizes J on X_n, then, for $z_n = x_n \pm \varepsilon u_n$ with $u_n \in X_n$ and $\varepsilon > 0$, it follows that

$$0 \leq J(z_n) - J(x_n) = \pm\varepsilon \quad 2\Re\langle Ax_n - y, u_n\rangle + \varepsilon^2\langle Au_n, u_n\rangle$$

for all $u_n \in X_n$. By dividing by $\varepsilon > 0$ and afterwards letting $\varepsilon \to 0$ we find that $x_n \in X_n$ satisfies the system (6.274). If, on the other hand, $x_n \in X_n$ solves (6.274), then we obtain from (6.278),

$$J(z_n) - J(x_n) = \langle A(z_n - x_n), z_n - x_n \rangle \geq 0$$

for all $z_n \in X_n$. Therefore, the *Rayleigh-Ritz method coincides with the Bubnov-Galerkin method.*

Finally we are interested in the Bubnov-Galerkin method for the important class of coercive operators. As preparatory material we briefly recapitulate some settings (see, e.g., [**497**]):

(i) A *Gelfand triple* (V, X, V^*) consists of a reflexive Banach space V, an Hilbert space X, and the dual space V^* of V such that
   (a) V is a dense subspace of X,
   (b) the imbedding $J : V \to X$ is bounded.

   It is conventional to write (see, e.g., [**497**]) $V \subset X \subset V^*$ since we can identify X with a dense subspace of V^*. This identification is given by the dual operator $J^* : X \to V^*$ of J, where we identify the dual of the Hilbert space X by itself. From $(x, y) = \langle J^* x, y \rangle$ for all $x \in X$ and $y \in V$ we see that with this identification the dual pairing $\langle \cdot, \cdot \rangle$ in (V^*, V) is an extension of the inner product $\langle \cdot, \cdot \rangle$ in X, i.e., we are led to write

$$\langle x, y \rangle_V = \langle x, y \rangle_X \quad \text{for all} \quad x \in Y \quad \text{and} \quad y \in V.$$

   Furthermore, we have the estimates

$$|\langle x, y \rangle| \leq \|x\|_{V^*} \|y\|_V \quad \text{for all} \quad x \in V^*, y \in V,$$

   thus,

$$|\langle x, y \rangle| \leq \|x\|_{V^*} \|y\|_V \quad \text{for all} \quad x \in X, y \in V.$$

   It is well-known that J^* is one-to-one and has a dense range.

(ii) Let V be a reflexive Banach space with dual space V^*. We denote the norms in V and V^* by $\|\cdot\|_V$ and $\|\cdot\|_{V^*}$, respectively. A linear bounded operator $A : V^* \to V$ is *coercive* if there exists $\gamma > 0$ with

$$\Re\langle x, Ax \rangle \geq \gamma \, \|x\|_{V^*}^2 \quad \text{for all} \quad x \in V^*, \tag{6.279}$$

   where $\langle \cdot, \cdot \rangle$ is the dual pairing in (V^*, V).

(iii) The operator A satisfies *Gårding's inequality*, if there exists a linear compact operator $C : V^* \to V$ such that $A + C$ is coercive, i.e.,

$$\Re\langle x, Ax \rangle \geq \gamma \, \|x\|_{V^*}^2 - \Re\langle x, Cx \rangle \quad \text{for all} \quad x \in V^*. \tag{6.280}$$

   Note that, by the same argument as in the Lax-Milgram theorem of functional analysis, it can be shown that every coercive operator is an isomorphism from V^* onto V. Coercive operators play an important role in the study of partial differential equations and integral equations by variational methods. In the conventional definition, the roles of V and V^* are interchanged. For integral operators that are "smoothing", our definition seems more appropriate. However, both definitions are equivalent in the sense that the inverse operator $A^{-1} : V \to V^*$ is coercive in the usual sense with γ replaced by $\gamma/\|A\|^2$ in (6.279)

Convergence of the Bubnov-Galerkin method. After these preparations we are in the position to formulate the convergence of the Bubnov-Galerkin method for coercive operators (see [**497**]).

Let (V, X, V^*) be a Gelfand triple, and $X_n \subset V$ be finite-dimensional subspaces such that $\bigcup_{n\in\mathbb{N}} X_n$ is dense in X. Let $K : V^* \to V$ be coercive with constant $\gamma > 0$. Let $x \in X$ be the solution of $Ax = y$. Then we have the following results:

(a) *There exist unique solutions of the Galerkin equations* (6.274)-(6.276), *and the Bubnov-Galerkin method converges in V^* such that*

$$\|x - x_n\|_{V^*} \leq c \, \min\{\|x - z_n\|_{V^*} : z_n \in X_n\} \tag{6.281}$$

for some $c > 0$.

(b) *Define the quantity $\rho_n > 0$ by*

$$\rho_n = \max\{\|u\| : u \in X_n, \; \|u\|_{V^*} = 1\} \tag{6.282}$$

and the orthogonal projection operator P_n from X onto X_n. The Bubnov-Galerkin method converges in X, if there esists $c > 0$ with

$$\|u - P_n u\|_{V^*} \leq \frac{c}{\rho_n}\|u\| \quad \text{for all} \quad u \in X. \tag{6.283}$$

In this case, we have the estimates

$$\|R_n\| \leq \frac{1}{\gamma}\rho_n^2 \tag{6.284}$$

and

$$\|x - x_n^\varepsilon\| \leq c(r_n \rho_n^2 + \min\{\|x - z_n\| : z_n \in X_n\}) \tag{6.285}$$

for some $c > 0$. Here $r_n = 1$ if $x_n^\varepsilon \in X_n$ solves (6.275), *i.e., ε measures the norm $\|y^\varepsilon - y\|$ in X. If ε measures the discrete error $|\beta^\varepsilon - \beta|$ in the Euclidean norm and $x_n^\varepsilon = \sum_{j=1}^n \alpha_j^\varepsilon \hat{x}_j \in X_n$, where α^ε solves* (6.276), *then r_n is given by*

$$r_n = \max\left\{\sqrt{\sum_{j=1}^n |\rho_j|^2} : \left\|\sum_{j=1}^n \rho_j \hat{x}_j\right\| = 1\right\}. \tag{6.286}$$

Again, we note that $r_n = 1$ if $\{\hat{x}_j, j = 1, \ldots, n\}$ forms an orthonormal system in X. For further details, we refer to [**647**] and the monographs [**70, 497, 513, 547**].

6.10. Multiscale spectral regularization

Next a compact operator equation is dealt within spectral regularization methods, based on filtering techniques by means of wavelets. In a general setup a singular integral approach to regularization is established, decomposition/reconstruction regularization wavelets are introduced in the frequency space which allow the regularization in form of a multiresolution analysis. Two different types of regularization wavelets are discussed in more detail, namely nonbandlimited Tikhonov-Phillips regularization wavelets and bandlimited truncated singular value decomposition wavelets. Our considerations closely follow W. Freeden, F. Schneider [**313**], W. Freeden [**266**] about regularization and multiresolution.

Let $(X, \langle\cdot,\cdot\rangle_X)$ and $(Y, \langle\cdot,\cdot\rangle_Y)$ be separable real functional Hilbert spaces over domains $\mathcal{G}_X$ and $\mathcal{G}_Y$, respectively, i.e., X, respectively, Y consists of functions

$x : \mathcal{G}_X \to \mathbb{R}$, respectively, $y : \mathcal{G}_Y \to \mathbb{R}$. We consider a linear, compact operator $A : X \to Y$ satisfying $\mathcal{N}(A) = \{0\}$, $\mathcal{R}(A) \subsetneq Y$, $\overline{\mathcal{R}(A)} = Y$, so that $A^*A : X \to X$ is self-adjoint.

In the sequel, we denote by $\{x_n\}_{n=0,1,\ldots}$ a complete orthonormal system in $(X, \langle\cdot,\cdot\rangle_X)$ and by $\{y_n\}_{n=0,1,\ldots}$ a complete orthonormal system in $(Y, \langle\cdot,\cdot\rangle_Y)$ such that the singular values $\{\sigma_n\}_{n=0,1,\ldots}$ of A satisfy $Ax_n = \sigma_n y_n$, $A^*y_n = \sigma_n x_n$, $n \in \mathbb{N}_0$. Since A is supposed to be injective it follows that $\sigma_n > 0$ for all $n \in \mathbb{N}_0$. Any $p \in P$, $P \in \{X, Y\}$, can be represented, respectively, in terms of an orthonormal (Fourier) expansion with respect to $\{p_n\}_{n=0,1,\ldots}$

$$p = \sum_{n=0}^{\infty} p_P^{\wedge}(n) p_n \tag{6.287}$$

with

$$p_P^{\wedge}(n) = \langle p, p_n \rangle_P, \tag{6.288}$$

where the equality in (6.287) is understood in the $\|\cdot\|_P$-sense. In conclusion, any element of $\mathcal{R}(A)$ admits an expansion of the form

$$Ax = \sum_{n=0}^{\infty} \sigma_n p_X^{\wedge}(n)\, y_n. \tag{6.289}$$

As is well-known, the Picard-condition tells us that the problem

$$Ax = y, \quad x \in X, \quad y \in Y, \tag{6.290}$$

has a solution if and only if $y \in Y$ satisfies

$$\sum_{n=0}^{\infty} (\sigma_n^{-1}\, p_Y^{\wedge}(n))^2 < \infty. \tag{6.291}$$

In this case it is known that the solution of (6.290) is representable in the form

$$x = \sum_{n=0}^{\infty} \sigma_n^{-1}\, p_Y^{\wedge}(n)\, x_n \;=\; A^{\dagger} y. \tag{6.292}$$

Since the right hand side y in (6.290) is error affected (due to the inaccuracy of the measuring instrument) in any practical application, the series (6.292) will not converge in general. For that purpose we are interested in regularized solutions.

The idea (cf. [**313**]) we follow is to represent the J-level regularization of the problem

$$Ax \;=\; y, \qquad x \in X, \ y \in Y, \tag{6.293}$$

by means of a wavelet analysis. The overall advantage of such a method is that we obtain a $J+1$-level regularization by starting with the J-level regularization and adding so-called detail information. It becomes clear that any classical regularization method based on a filtered singular value decomposition can be reformulated in terms of our wavelet method. Thus, any known parameter choice strategy depending on the special method is also applicable and, moreover, any of the corresponding error estimates holds true, too. For that reason we omit these discussions here.

Now, a family $\{R_J\}_{J\in\mathbb{Z}}$ of linear operators $R_J : Y \to X$, $J \in \mathbb{Z}$, is a regularization of $A^{\dagger}$ if it satisfies the following properties:

(i) R_J is bounded on Y for all $J \in \mathbb{Z}$,

(ii) for any member $y \in \mathcal{R}(A)$, the limit relation $\lim_{J\to\infty} R_J y = A^\dagger y$ holds in the $\|\cdot\|_X$-sense.

$x_J = R_J y$ is called the J-level regularization of the problem (6.290).

A function $\Gamma^{P,Q} : \mathcal{G}_P \times \mathcal{G}_Q \to \mathbb{R}$, $P, Q \in \{X, Y\}$, of the form

(6.294)

$$\Gamma^{P,Q}(x,y) = \sum_{n=0}^{\infty} \Gamma^\wedge(n)\; p_n(x)\; q_n(y), \quad x \in \mathcal{G}_P, y \in \mathcal{G}_Q,\; \Gamma^\wedge(n) \in \mathbb{R}, n \in \mathbb{N}_0,$$

is called a (P,Q)-*(product) kernel.* Note that the indices P and Q in $\Gamma^{P,Q}$ are associated to the variables x and y, respectively, such that $\Gamma^{P,Q}(x,y) = \Gamma^{Q,P}(y,x)$. The sequence $\{\Gamma^\wedge(n)\}_{n=0,1,\ldots}$ is called the *symbol of the* (P,Q)*-kernel.* For brevity, a (P,P)-kernel is simply said to be a P*-kernel.*

An important question for our investigation is as follows: Fix one variable of a product kernel, what are the conditions for the product kernel (as a function of the remaining variable) to be of class $P \in \{X, Y\}$? The answer is provided by the concept of symbol admissibility.

A symbol $\{\Gamma^\wedge(n)\}_{n=0,1,\ldots}$ is called P*-admissible,* $P \in \{X,Y\}$, if it satisfies the following conditions:

(Ai)

$$\sum_{n=0}^{\infty} (\Gamma^\wedge(n))^2 < \infty, \tag{6.295}$$

(Aii)

$$\sup_{x\in\mathcal{G}_P} \left(\sum_{n=0}^{\infty} (\Gamma^\wedge(n) p_n(x))^2 \right) < \infty. \tag{6.296}$$

A symbol $\{\Gamma^\wedge(n)\}_{n=0,1,\ldots}$ is called ((P,Q)-admissible, $P,Q \in \{X,Y\}$, or simply) *admissible*, if it is P-admissible as well as Q-admissible.

From the definition of admissibility we immediately obtain the following result:

Let $\{\Gamma^\wedge(n)\}_{n=0,1,\ldots}$ be the symbol of a (P,Q)-kernel, $P, Q \in \{X,Y\}$.

(α) If $\{\Gamma^\wedge(n)\}_{n=0,1,\ldots}$ is P-admissible, then $\Gamma^{P,Q}(x,\cdot) \in Q$ for every (fixed) $x \in \mathcal{G}_{\mathrm{P}}$,

(β) If $\{\Gamma^\wedge(n)\}_{n=0,1,\ldots}$ is Q-admissible, then $\Gamma^{P,Q}(\cdot,y) \in P$ for every (fixed) $y \in \mathcal{G}_{\mathrm{Q}}$.

A fundamental tool for our wavelet theory is the concept of a *convolution* we introduce below: Let $\Gamma^{X,X}$ be an X-kernel with X-admissible symbol. Suppose that x is of class X. Then we understand the *convolution of* $\Gamma^{X,X}$ *and* x to be the function given by

$$(\Gamma^{X,X} * x)(t) = \left\langle \Gamma^{X,X}(t,\cdot), x \right\rangle_X = \sum_{n=0}^{\infty} \Gamma^\wedge(n) p_X^\wedge(n) x_n(t), \quad t \in \mathcal{G}_X. \tag{6.297}$$

We immediately see that $(\Gamma^{X,X} * x)^{\wedge}_X(n) = \Gamma^{\wedge}(n)p^{\wedge}_X(n)$ and $\Gamma^{X,X} * x \in X$. In analogous way we define the *convolution* of an (X, Y)-kernel $\Gamma^{X,Y}$ having an X-admissible symbol *with a function* $y \in Y$ to be the expression

$$(\Gamma^{X,Y} * y)(t) = \langle \Gamma^{X,Y}(t, \cdot), y \rangle_Y = \sum_{n=0}^{\infty} \Gamma^{\wedge}(n)p^{\wedge}_Y(n)x_n(t), \quad t \in \mathcal{G}_X, \tag{6.298}$$

and it follows that $\Gamma^{X,Y} * y \in X$. Lastly, the *convolution* of an (X, Y)-kernel $\Gamma^{X,Y}$ having an Y-admissible symbol *with a function* $x \in X$ is given by

$$(\Gamma^{X,Y} * x)(s) = \langle \Gamma^{X,Y}(\cdot, s), x \rangle_X = \sum_{n=0}^{\infty} \Gamma^{\wedge}(n)p^{\wedge}_X(n)\ y_n(s), \quad s \in \mathcal{G}_Y, \tag{6.299}$$

and we have $\Gamma^{X,Y} * x \in Y$. Next we proceed with the convolution of two product kernels leading to the following result: Let $\Gamma^{X,X}$ be an X-kernel with X-admissible symbol and let $\Upsilon^{X,Y}$ be an (X, Y)-kernel with Y-admissible symbol. Then

$$\begin{aligned}(\Gamma * \Upsilon)^{X,Y}(t, s) &= (\Gamma^{X,X} * \Upsilon^{X,Y}(\cdot, s))(t) \\ &= \langle \Gamma^{X,X}(t, \cdot), \Upsilon^{X,Y}(\cdot, s) \rangle_X \\ &= \sum_{n=0}^{\infty} \Gamma^{\wedge}(n)\Upsilon^{\wedge}(n)\ x_n(t)y_n(s), \quad t \in \mathcal{G}_X, s \in \mathcal{G}_Y\end{aligned} \tag{6.300}$$

represents an (X, Y)-kernel with admissible symbol

$$((\Gamma * \Upsilon)^{X,Y})(n) = \Gamma^{\wedge}(n)\Upsilon^{\wedge}(n). \tag{6.301}$$

In order to prepare the fundamentals of the forthcoming wavelet theory we are now interested in countable families $\{\Gamma_J^{P,Q}\}$, $J \in \mathbb{Z}$, of product kernels $\Gamma_J^{P,Q}$, $P, Q \in \{X, Y\}$. Observing our notations we are able to define a dilation operator acting on these families in the following way: let $\Gamma_J^{P,Q}$ be a member of the family of product kernels.

Then the *dilation operator* D_K, $K \in \mathbb{Z}$ is defined by $D_K\Gamma_J^{P,Q} = \Gamma_{J+K}^{P,Q}$. Especially, we obtain $\Gamma_J^{P,Q} = D_J\Gamma_0^{P,Q}$, $J \in \mathbb{Z}$. Thus we refer $\Gamma_0^{P,Q}$ to as a "mother kernel".

Moreover, we define a *shifting operator* S_t^P, $t \in \mathcal{G}_P$, $P \in \{X, Y\}$, by $S_t^P\Gamma_J^{P,Q} = \Gamma_J^{P,Q}(t, \cdot)$, $t \in \mathcal{G}_P, J \in \mathbb{Z}$, resp. $S_s^Q\Gamma_J^{P,Q} = \Gamma_J^{P,Q}(\cdot, s)$, $s \in \mathcal{G}_Q, J \in \mathbb{Z}$. In doing so we consequently get $\Gamma_J^{P,Q}(t, \cdot) = S_t^P D_J\Gamma_0^{P,Q}$, $t \in \mathcal{G}_P, J \in \mathbb{Z}$, resp. $\Gamma_J^{P,Q}(\cdot, s) = S_s^Q D_J\Gamma_0^{P,Q}$, $s \in \mathcal{G}_Q, J \in \mathbb{Z}$.

Next we are concerned with a wavelet based regularization technique of problem (6.293): Let $\{(\Phi_J)^{\wedge}(n)\}_{n=0,1,\ldots}$, $J \in \mathbb{Z}$, be an (X, Y)-admissible symbol of a family of product kernels which additionally satisfies the following conditions:

(i) $\lim_{J\to\infty} \sigma_n((\Phi_J)^{\wedge}(n))^2 = 1$, $n \in \mathbb{N}$,
(ii) $((\Phi_{J+1})^{\wedge}(n))^2 \geq ((\Phi_J)^{\wedge}(n))^2$, $J \in \mathbb{Z}$, $n \in \mathbb{N}$,
(iii) $\lim_{J\to-\infty}((\Phi_J)^{\wedge}(n))^2 = 0$, $n \in \mathbb{N}$,
(iv) $\sigma_0((\Phi_J)^{\wedge}(0))^2 = 1$, $J \in \mathbb{Z}$.

Then $\{(\Phi_J)^\wedge(n)\}_{n=0,1,\ldots}$ is said to be the *generating symbol of a regularization scaling function* (with respect to (6.290)). The (X,Y)-kernel

$$ {}^d\Phi_J^{X,Y} = \sum_{n=0}^{\infty} (\Phi_J)^\wedge(n) x_n y_n \tag{6.302} $$

is called a *decomposition regularization scaling function*, the (X,X)-kernel

$$ {}^r\Phi_J^{X,X} = \sum_{n=0}^{\infty} (\Phi_J)^\wedge(n) x_n x_n \tag{6.303} $$

is called a *reconstruction regularization scaling function.*

From the results of the previous section it is clear that ${}^d\Phi_J^{X,Y}(t,\cdot) \in Y, t \in \mathcal{G}_X, J \in \mathbb{Z}$, ${}^r\Phi_J^{X,X}(t,\cdot) \in X$, $t \in \mathcal{G}_X, J \in \mathbb{Z}$, and $({}^r\Phi_J * {}^d\Phi_J)^{X,Y}$ is an (X,Y)-kernel with (X,Y)-admissible symbol $\{((\Phi_J)^\wedge(n))^2\}_{n=0,1,\ldots}$. Observing these properties we are able to verify the following result, which is central for our considerations:

Let $\{(\Phi_J)^\wedge(n)\}_{n=0,1,\ldots}$, $J \in \mathbb{Z}$, be the generating symbol of a regularization scaling function. Then, for any $y \in Y$,

$$ x_J = {}^r\Phi_J^{X,X} * ({}^d\Phi_J^{X,Y} * y) \tag{6.304} $$

represents the J-level regularization of problem (6.293). *If, in addition, $y \in \mathcal{R}(A)$, then*

$$ \lim_{J\to\infty} \|x_J - A^\dagger y\| = 0. \tag{6.305} $$

For $R_J : Y \to X$ defined by

$$ R_J y = \left({}^r\Phi_J * {}^d\Phi_J\right)^{X,Y} * y, \ \ y \in Y \tag{6.306} $$

we have (see **[313]**)

$$ R_J y = \sum_{n=0}^{\infty} ((\Phi_J)^\wedge(n))^2 \; y_Y^\wedge(n) \; x_n $$

and

$$ \|R_J\|^2 \le \sum_{n=0}^{\infty} ((\Phi_J)^\wedge(n))^4, \quad J \in \mathbb{Z}. \tag{6.307} $$

As an immediate consequence we obtain the following result: Let y be a member of $\mathcal{R}(A)$. Suppose that $y^\varepsilon \in Y$ denotes the right hand side of problem (6.293) with noise level $\|y - y^\varepsilon\|_Y \le \varepsilon$. Moreover, assume that $\{(\Phi_J)^\wedge(n)\}_{n=0,1,\ldots}$, $J \in \mathbb{Z}$, is the generating symbol of a regularization scaling function, where the parameter $J = J(\varepsilon)$ is assumed to satisfy

(1) $\lim_{\varepsilon\to 0} J(\varepsilon) = \infty$,
(2) $\lim_{\varepsilon\to 0} \varepsilon((\Phi_J)^\wedge(n))^2 = 0$.

Then we have

$$ \lim_{\varepsilon\to 0} \|({}^r\Phi_J * {}^d\Phi_J)^{X,Y} * y^\varepsilon - A^\dagger y\| = 0. \tag{6.308} $$

Condition (iii) seems to be unnecessary for the proof of (6.308) and, in fact, it is. Nevertheless, in what follows we need this assumption for our multiresolution analysis and the (spectral) introduction of wavelets.

For any $y \in \mathcal{R}(A)$ each $({}^r\Phi_J * {}^d\Phi_J)^{X,Y} * y$ provides a regularization of the solution $A^\dagger y$ at scale J by "smoothing" the Fourier coefficients of $A^\dagger y$ with the symbol $\{((\Phi_J)^\wedge(n))^2\}_{n=0,1,\ldots}$. In terms of filtering, $({}^r\Phi_J * {}^d\Phi_J)^{X,Y}$ may be interpreted as a low-pass filter. Accordingly we understand the scale spaces $\mathcal{V}_J$ to be the image of $\mathcal{R}(A)$ under the operator R_J:

$$\mathcal{V}_J = R_J(\mathcal{R}(A)) = \{({}^r\Phi_J * {}^d\Phi_J)^{X,Y} * y : y \in \mathcal{R}(A)\}. \tag{6.309}$$

This leads to the following statement: The scale spaces satisfy the properties

(i) $\bigcap_{J=-\infty}^{\infty} \mathcal{V}_J = \{X_0\}$,

(ii) $\{X_0\} \subset \mathcal{V}_J \subset \mathcal{V}_{J'} \subset X$, $J \leq J'$, i.e., for any right hand side $y \in \mathcal{R}(A)$ of problem (6.293), all J-level regularizations with fixed parameter J are sampled in a scale space $\mathcal{V}_J$ with the above property,

(iii) $\overline{\bigcup_{J=-\infty}^{\infty} \mathcal{V}_J}^{\|\cdot\|_X} = X$,

(iv) if $x_J \in \mathcal{V}_J$, then $D_{-1}x_J \in \mathcal{V}_{J-1}$, $J \in \mathbb{Z}$.

If a collection of subspaces of X satisfies the above conditions we call them a *regularization multiresolution analysis (RMRA)*.

The definition of the regularization scaling function allows us to introduce regularization wavelets. An essential point is the definition of a decomposition and a reconstruction regularization wavelet associated to regularization mother wavelets. This definition, of course, has to be formulated in close relation to a prescribed regularization scaling function.

Let $\{(\Phi_j)^\wedge(n)\}_{n=0,1,\ldots}$, $j \in \mathbb{Z}$, be the generating symbol of a regularization scaling function. Then the (X,Y)-admissible generating symbol $\{(\Psi_j)^\wedge(n)\}_{n=0,1,\ldots}$, $j \in \mathbb{Z}$, and the (X,X)-admissible generating symbol $\{(\tilde{\Psi}_j)^\wedge(n)\}_{n=0,1,\ldots}$, $j \in \mathbb{Z}$, respectively, are defined by the "scaling equation"

$$(\tilde{\Psi}_j)^\wedge(n)(\Psi_j)^\wedge(n) = ((\Phi_{j+1})^\wedge(n))^2 - ((\Phi_j)^\wedge(n))^2. \tag{6.310}$$

Correspondingly, the (X,Y)-kernel

$${}^d\Psi_J^{X,Y} = \sum_{n=0}^{\infty} (\Psi_J)^\wedge(n)\; x_n\; y_n \tag{6.311}$$

is called the *decomposition regularization wavelet*, while the (X,X)-kernel

$${}^r\tilde{\Psi}_J^{X,X} = \sum_{n=0}^{\infty} (\tilde{\Psi}_J)^\wedge(n)\; x_n\; x_n \tag{6.312}$$

is called the *reconstruction regularization wavelet*. The corresponding regularization mother wavelets are denoted by ${}^d\Psi_0^{X,Y}$ and ${}^r\tilde{\Psi}_0^{X,X}$, respectively.

Using this notation, any decomposition regularization wavelet, respectively, any reconstruction regularization wavelet can be interpreted as a dilated and shifted copy of the corresponding mother wavelet.

$$
{}^d\Psi_J^{X,Y}(t,\cdot) = S_t^X D_J {}^d\Psi_0^{X,Y}, \tag{6.313}
$$

$$
{}^r\tilde{\Psi}_J^{X,X}(t,\cdot) = S_t^X D_J {}^r\Psi_0^{X,X}. \tag{6.314}
$$

At this stage it becomes obvious why we required Condition (iv). A conclusion of (6.310) is a vanishing 0-th moment of the regularization wavelets. Moreover, from (6.310) it can be readily deduced that

(6.315)

$$
((\Phi_0)^\wedge(n+1))^2 = \sum_{j=-\infty}^{J} (\tilde{\Psi}_j)^\wedge(n)(\Psi_j)^\wedge(n) = ((\Phi_0)^\wedge(n))^2 + \sum_{j=0}^{J}(\tilde{\Psi}_j)^\wedge(n)(\Psi_j)^\wedge(n).
$$

Thus, we easily see in connection with (6.310) that

(6.316)

$$
\left({}^r\tilde{\Phi}_{J+1} * {}^d\Phi_{J+1}\right)^{X,Y} = \sum_{j=-\infty}^{J} ({}^r\tilde{\Psi}_j * {}^d\tilde{\Psi}_j)^{X,Y} = ({}^r\Phi_0 * {}^d\Phi_0)^{X,Y} + \sum_{j=0}^{J}({}^r\tilde{\Psi}_j * {}^d\tilde{\Psi}_j)^{X,Y}.
$$

In analogy to the definition of the operator R_J we consider now convolution operators $S_J : Y \to X$, $J \in \mathbb{Z}$, defined by

$$
S_J\, y = ({}^r\tilde{\Psi}_J * {}^d\Psi_J)^{X,Y} * y. \tag{6.317}
$$

It describes the "detail information" of the right hand side y at scale J. From Equation (6.316) it follows that the operator R_{J+1} can be decomposed in the following way:

$$
R_{J+1} = R_0 + \sum_{j=0}^{J} S_j. \tag{6.318}
$$

But this gives rise to introduce the *detail spaces* as follows:

$$
\mathcal{W}_J = S_J(\mathcal{R}(A)) = \{({}^r\tilde{\Psi}_J * {}^d\Psi_J)^{X,Y} * y : \ y \in \mathcal{R}(A)\}. \tag{6.319}
$$

The space $\mathcal{W}_J$ contains the "detail information" needed to go from a regularization at level J to a regularization at level $J+1$. Note that

$$
\mathcal{V}_J = \mathcal{V}_{J-1} + \mathcal{W}_{J-1}, \tag{6.320}
$$

$$
\sum_{j=-\infty}^{J} \mathcal{W}_j = \mathcal{V}_0 + \sum_{j=0}^{J} \mathcal{W}_j = \mathcal{V}_{J+1}. \tag{6.321}
$$

It is worth mentioning that, in general, the sum decomposition is neither direct nor orthogonal. Two examples leading to orthogonal and nonorthogonal multiresolution are introduced below.

Any right hand side $y \in Y$ can now be decomposed as follows. Starting with $R_0 y$ we find

$$
R_{J+1}y = R_0 y + \sum_{j=0}^{J} S_j y \tag{6.322}
$$

for any $J \in \mathbb{Z}$. In other words, the partial reconstruction $R_J y$ is nothing else than the "difference of two smoothings" at two consecutive scales,

$$
S_J y = R_{J+1}y - R_J y. \tag{6.323}
$$

For what follows we define the *regularization wavelet transform at scale* $J \in \mathbb{Z}$ *and position* $t \in \mathcal{G}_X$ by letting

$$RWT(y)(J;t) = \langle {}^d\Psi_J^{X,Y}(t,\cdot), y\rangle_Y, \quad y \in Y. \tag{6.324}$$

From (6.316) it is not hard to verify the main result in this context which is now formulated in more detail:

Let $\{(\Phi_j)^\wedge(n)\}_{n=0,1,\ldots}$, $j \in \mathbb{Z}$, be the generating symbol of a regularization scaling function. Suppose that $\{(\Psi_j)^\wedge(n)\}_{n=0,1,\ldots}$, $j \in \mathbb{Z}$, $j \in \mathbb{Z}$, $\{(\tilde{\Psi}_j)^\wedge(n)\}_{n=0,1,\ldots}$, $j \in \mathbb{Z}$ are the generating symbols of the corresponding regularization wavelets. Furthermore, let y be of class Y. Then,

$$x_J = ({}^r\Phi_0 * {}^d\Phi_0)^{X,Y} * y + \sum_{j=0}^{J} {}^r\tilde{\Psi}_j * RWT(y)(j;\cdot) \tag{6.325}$$

denotes the J-level regularization of the problem (6.293) satisfying

$$\lim_{J\to\infty} \|x_J - A^\dagger y\| = 0 \tag{6.326}$$

provided that $y \in \mathcal{R}(A)$.

Equation (6.325) shows the essential characteristic of regularization wavelets. By adding the so-called detail information of level J as the difference of two smoothings of two consecutive scales $J+1$ and J, we change the regularized solution from x_J to x_{J+1} thereby satisfying $\lim_{J\to\infty} x_J = A^\dagger y$ in the case that $y \in \mathcal{R}(A)$. Of course, this can be understood as a kind of a permanence principle.

The multiscale method as presented here can be illustrated by the following scheme:

$$\begin{array}{ccccccccc}
R_0(y) & & R_1(y) & & R_2(y) & \ldots & \underset{j\to\infty}{\longrightarrow} & A^\dagger y \\
\cap & & \cap & & \cap & & & \cap \\
\mathcal{V}_0 & \subset & \mathcal{V}_1 & \subset & \mathcal{V}_2 & \ldots & = & A^\dagger X \\
\hline
\mathcal{V}_0 + & \mathcal{W}_0 & + & \mathcal{W}_1 & + \ \mathcal{W}_2 & \ldots & = & A^\dagger X \\
\cup & \cup & & \cup & \cup & & & \cup \\
R_0(y) + & S_0(y) & + & S_1(y) & + S_2(y) & \ldots & = & A^\dagger y
\end{array}$$

FIGURE 1. Multiresolution regularization scheme.

The singular values $\{\sigma_n\}_{n=0,1,\ldots}$ of A satisfy $Ax_n = \sigma_n y_n$, $\quad A^* y_n = \sigma_n x_n$, $n \in \mathbb{N}_0$. Keeping these facts in mind we are led to introduce the following examples of generating symbols of a regularization scaling function:

(i) *Truncated singular value decomposition (bandlimited regularization):*
  (a) orthogonal:

$$(\Phi_J)^\wedge(n) = \begin{cases} \sigma_n^{-1/2}, & n = 0,\ldots,N_J, \\ 0, & n \geq N_J + 1, \end{cases} \tag{6.327}$$

$$N_J = \begin{cases} 0, & J \in \mathbb{Z}, J < 0, \\ 2^J - 1, & J \in \mathbb{Z}, J \geq 0, \end{cases} \tag{6.328}$$

(b) nonorthogonal:

(6.329)
$$(\Phi_J)^\wedge(n) = \begin{cases} \sigma_n^{-1/2}, & n = 0, \ldots, M_J, \\ \sigma_n^{-1/2}(\tau_J(n))^{1/2}, & n = M_J + 1, \ldots, N_J, \\ 0, & n \geq N_J + 1, \end{cases}$$

(6.330)
$$N_J = \begin{cases} 0, & J \in \mathbb{Z}, J < 0, \\ 2^{J+1} - 1, & J \in \mathbb{Z}, J \geq 0, \end{cases}$$

(6.331)
$$M_J, = \begin{cases} 0, & J \in \mathbb{Z}, J < 0, \\ 2^J - 1, & J \in \mathbb{Z}, J \geq 0, \end{cases}$$

and

(6.332)
$$\tau_J(n) = 2 - 2^{-J}(n+1), \quad n \in [2^J - 1, 2^{J+1} - 1], \ J \in \mathbb{N}_0.$$

It is easy to see that case (a) leads to an orthogonal RMRA, i.e., the detail and the scale spaces satisfy the orthogonality conditions

(6.333)
$$\mathcal{V}_{J+1} = \mathcal{V}_J \oplus \mathcal{W}_J, \quad \mathcal{W}_J \perp \mathcal{W}_K, \ K \neq J, \ K, J \geq 0.$$

In case (b) the scale and detail spaces are still finite dimensional, but the detail spaces are no longer orthogonal.

(ii) *Tikhonov's regularization (nonbandlimited regularization):*

(a) *classical*

(6.334)
$$(\Phi_J)^\wedge(n) = \left(\frac{\sigma_n}{\sigma_n^2 + \gamma_J^2} \right)^{\frac{1}{2}}, \quad n \in \mathbb{N}, J \in \mathbb{Z},$$

(b) *Tikhonov-Phillips*

(6.335)
$$(\Phi_J)^\wedge(n) = \left(\frac{\sigma_n}{\sigma_n^2 + \gamma_J^2 (n + \frac{1}{4})^4} \right)^{\frac{1}{2}}, \quad n \in \mathbb{N}, J \in \mathbb{Z}$$

with $\{\gamma_J\}$, $J \in \mathbb{Z}$, being a sequence of real numbers satisfying $\lim_{J\to\infty} \gamma_J = 0$ and $\lim_{J\to-\infty} \gamma_J = \infty$. Case (a) leads to the minimization of the Tikhonov functional, where the penalty term is given by the norm of the regularized solution of problem (6.293). Case (b) leads to the minimization of the Tikhonov-Phillips functional, where the penalty term is given by the linearized bending energy of the regularized solution of problem (6.293). Both cases lead to infinite-dimensional scale and detail spaces and, furthermore, the RMRA is neither direct nor orthogonal.

6.11. Multiscale space mollifier schemes

The original idea of the mollifier method can be characterized as follows: We are interested in the solution $x^\dagger$ of $Ax = y$, but we realize that the problem is "too ill-posed" for being able to determine $x^\dagger$ accurately. Thus, we compromises by changing the problem into a more well-posed one, namely that of trying to determine a mollified version $E_\tau x^\dagger$ of the solution, where E_τ is a suitable "mollifier operator" depending on a parameter τ. The heuristic motivation is that the trouble usually comes from high frequency components of the data and of the solution, which are damped out by mollification (which, in fact, defines mollification).

In an abstract nomenclature, early mollifier methods for ill-posed problems were studied in A.K. Louis [**547**]. Our approach involving the determination of a mollifier pseudoinverse is based on H.W. Engl et al. [**215**]: We start from the operator equation

$$Ax = y, \quad x \in X, \ y \in Y \tag{6.336}$$

with X, Y Hilbert spaces and $\mathcal{R}(A)$ nonclosed. Our aim is to introduce operators $E_\tau : X \to X$ such that

$$E_\tau x \to x, \quad x \in X, \quad \tau \to 0. \tag{6.337}$$

If X is a "suitable function space", we are able to represent E_τ by a *mollifier* e_τ via the equation

$$E_\tau(x)(s) = \langle e_\tau(s, \cdot), x \rangle_X. \tag{6.338}$$

Instead of $x^\dagger$ we now look for $E_\tau x^\dagger$ for some suitable $\tau > 0$, thereby assuming that e_τ has a representation

$$A^* v_s^\tau = e_\tau(s, \cdot) \tag{6.339}$$

with $v_s^\tau \in Y$.

Mollification involving the pseudoinverse. If $Ax^\dagger = y$, we can compute $E_\tau x^\dagger$ as follows:

$$(E_\tau x^\dagger)(s) = \langle e_\tau(s, \cdot), x^\dagger \rangle_X = \langle A^* v_s^\tau, x^\dagger \rangle_X = \langle v_s^\tau, Ax^\dagger \rangle_Y = \langle v_s^\tau, y \rangle_Y, \tag{6.340}$$

i.e.,

$$(E_\tau x^\dagger)(s) = \langle v_s^\tau, y \rangle_Y. \tag{6.341}$$

Hence, the problem of solving (6.336) reduces to that of solving (6.339), which is also ill-posed as soon as (6.336) is. However, the right side of (6.339) (which is actually a family of equations depending on the parameter s) is usually given analytically, since the mollifier e_τ is chosen. Hence, there is no (or much less) error in the data of (6.339), and this equations can be solved (by regularization) much better than (6.336). As soon as an approximation for v_s^τ has been computed, it can be used to solve (6.336) for any right-hand side y via (6.341). If we define the operator $S_\tau : Y \to X$ via the estimate

$$(S_\tau y)(s) = \langle v_s^\tau, y \rangle, \tag{6.342}$$

then, by (6.341), this operator maps the right-hand side of (6.336) to mollified solutions. This motivates the term *approximate inverse* of A used for S_τ, also for the more general case that (6.353) is not solvable. In this case, (6.353) is replaced by

$$\|A^* v_s^\tau - e_\tau(s, \cdot)\|_X \to \min, \tag{6.343}$$

which, is equivalent to

$$AA^* v_s^\tau = Ae_\tau(s, \cdot). \tag{6.344}$$

Note that one needs the requirement that (6.344) is solvable. The function v_s^τ is called *reconstruction kernel*; uniqueness can be enforced by solving (6.344) in the best-approximate sense, i.e., by selecting the solution of (6.344) with minimal norm: $v_s^\tau = (A^*)^\dagger e_\tau(s,\cdot)$. Assume, for simplicity, that $\mathcal{R}(A)$ is dense in Y, so that $(AA^*)^{-1}$ exist. Then we have with v_s^γ defined by (6.343):

$$\begin{aligned} S_\tau y &= \langle (AA^*)^{-1}Ae_\tau(s,\cdot), y\rangle_Y = \langle e_\tau(s,\cdot), A^*(AA^*)^{-1}y\rangle_X \\ &= \langle e_\tau(s,\cdot), (A^*A)^\dagger A^* y\rangle_X = (E_\tau x^\dagger)(s), \end{aligned} \tag{6.345}$$

i.e., $S_\tau y$ is the mollified version of the best-approximate solution of (6.341). This justifies (6.343).

Let A be compact with singular system $\{\sigma_n; x_n, y_n\}$. Let a regularization method realized in standard way, i.e.,

$$x_\alpha = \sum_{n=1}^\infty \sigma_n F_\alpha(\sigma_n^2)\langle y, y_n\rangle_Y x_n. \tag{6.346}$$

If we assume that X and Y are suitable function spaces, then (6.346) can be written as

$$x_\alpha(s) = \langle v_s^\tau, y\rangle_Y \tag{6.347}$$

with

$$v_s^\tau(t) = \sum_{n=1}^\infty \sigma_n F_\alpha(\sigma_n^2) y_n(t) y_n(s). \tag{6.348}$$

Now, v_s^τ can be written in the form (6.344) with

$$e_\tau(s,t) = \sum_{n=1}^\infty \sigma_n^2 F_\alpha(\sigma_n^2) y_n(s) y_n(t). \tag{6.349}$$

Hence, x_α can be considered as a mollified solution $E_\tau x^\dagger$ with E_τ given (in the sense of (6.337)) by the mollifier (6.349), so that linear regularization methods can also be viewed as mollifier methods.

The choice of the mollifier, of course, depends on what one wants to achieve; frequently used choices are integral kernels (such as Haar kernel, sinc kernel, etc.). Note again that each of these mollifiers can be applied only to equations, so that (6.344) admits a solution.

Space multiscale mollification. Let $\{\tau_j\}_{j\in\mathbb{N}}$ be a positive, strictly monotonically decreasing sequence satisfying $\lim_{j\to\infty}\tau_j = 0$. We start from $E_{\tau_j}: X \to X$ with $E_{\tau_j}x \to x$ such that

$$\underbrace{AE_{\tau_j}}_{=A_{\tau_j}} x \to Ax = y, \quad x \in X, \quad \tau_j \to 0. \tag{6.350}$$

If X is a "suitable function space", we are led to represent $A_{\tau_j} = AE_{\tau_j}$ by a *mollifier* a_{τ_j} via the equation

$$A_{\tau_j}(x)(s) = \langle \underbrace{Ae_{\tau_j}}_{=a_{\tau_j}}(s,\cdot), x\rangle_X = \langle \underbrace{AA^*v_s^{\tau_j}}_{=a_{\tau_j}}(s,\cdot), x\rangle_X, \tag{6.351}$$

i.e.,

$$A_{\tau_j}(x)(s) = \langle a_{\tau_j}(s,\cdot), x\rangle_X, \tag{6.352}$$

thereby assuming that a_{τ_j} has a representation

$$A^* w_s^{\tau_j} = a_{\tau_j}(s,\cdot) = AA^* v_s^{\tau_j} \tag{6.353}$$

with $w_s^{\tau_j} \in Y$. So, corresponding to the sequence $\{\tau_j\}_{j\in\mathbb{N}}$, there is a scale-dependent sequence $\{a_{\tau_j}\}_{j\in\mathbb{N}}$ called the *mollifier scaling function.* The *mollifier wavelets* $\{(wa)_{\tau_j})\}_{j\in\mathbb{N}}$, which are to be understood as the difference between two consecutive mollifier scaling functions

$$(wa)_{\tau_j} = a_{\tau_{j+1}} - a_{\tau_j}, \qquad j \in \mathbb{N}. \tag{6.354}$$

Now, let $J \in \mathbb{N}$ be any scale. We consider the mollifier scaling function a_{τ_J} to scale J and the corresponding mollifier wavelet functions $(wa)_{\tau_j}$, $j = J_0, \ldots, J$, $J_0 < J$, given by (6.354). Then an elementary calculation shows that

$$a_{\tau_J} = a_{\tau_{J_0}} + \sum_{j=J_0}^{J-1} (wa)_{\tau_j}. \tag{6.355}$$

The lowpass filtered version A_{τ_j} of A to the scale j is given by

$$A_{\tau_j}(x)(s) = \langle a_{\tau_j}(s,\cdot), x\rangle_X, \tag{6.356}$$

while the bandpass filtered version $(WA)_{\tau_j}$ to the scale j is given by

$$(WA)_{\tau_j}(x)(s) = \langle (wa)_{\tau_j}(s,\cdot), x\rangle_X. \tag{6.357}$$

The *mollifier multiscale process* has the following structure:

Let $A_{\tau_{J_0}}(x)(s)$ be the initial lowpass filter of $A(x)(s)$ to the scale J_0. Let $(WA)_{\tau_j}$, $j = J_0, \ldots, J-1$, be the successive detailed information $(WA)_{\tau_j}$, $j = J_0, \ldots, J-1$, given by (6.357). Then the following mollifier reconstruction formula is valid:

$$\lim_{J\to\infty} \left| A(x)(s) - \left(A_{\tau_{J_0}}(x)(s) + \sum_{j=1}^{J-1} (WA)_{\tau_{J_0+j}}(x)(s) \right) \right| = 0. \tag{6.358}$$

The multiscale procedure derived here may be illustrated by the following computational scheme:

$$\begin{array}{ccccccccc} (WA)_{\tau_{J_0}} & & & & (WA)_{\tau_{J_0+1}} & & & & \\ & \searrow & & & & \searrow & & & \\ A_{\tau_{J_0}} & \longrightarrow & + & \longrightarrow & A_{\tau_{J_0+1}} & \longrightarrow & + & \longrightarrow & A_{\tau_{J_0+2}} \quad \cdots . \end{array}$$

This scheme represents a "tree algorithm" for approximating the operator A.

Inversion via multiscale mollifier decorrelation. To explain the basic idea of *mollifier inversion* we use (6.356) and (6.357) in a multiscale decomposition scheme: The approach presented here is based on the fact that, for a sufficiently large scale index J, the mollifier $A_{\tau_J}(x)(s)$ may be understood as an approximation of $A(x)(s)$ with negligible error, i.e.,

$$A(x)(s) \simeq A_{\tau_J}(x)(s). \tag{6.359}$$

Moreover, we assume that $\alpha_i^{N_J} = A(x)(s_i) \simeq A_{\tau_J}(x)(s_i^{N_J}), i = 1, \ldots, N_J$, are known. Our goal is to determine x from

$$(6.360) \qquad A_{\tau_J}(x)(s_i^{N_J}) = \langle a_{\tau_J}(s_i^{N_J}, \cdot), x\rangle_X \simeq \sum_{k=1}^{M_J} \beta_k^{M_J}\ a_{\tau_J}(s_i^{N_J}, q_k^{M_J})\ x(q_k^{M_J}),$$

where $\beta_k^{M_J}$ are known weight coefficients. Hence, $\beta_k^{M_J}\ x(q_k^{M_J})$, $k = 1, \ldots, M_J$, can be determined by solving a linear system of the type

$$(6.361) \qquad \alpha_i^{N_J} = \sum_{k=1}^{M_J} \beta_k^{M_J}\ a_{\tau_J}(s_i^{N_J}, q_k^{M_J})\ x(q_k^{M_J}), \quad i = 1, \ldots, N_J,$$

from the known values $A(x)(s_i^{N_J}), i = 1, \ldots, N_J$. Since the weights $\beta_k^{M_J}$ are known, the values $x(q_k^{M_J})$ are immediately obtainable from (6.361), so that $x(s)$ follows from

$$(6.362) \qquad x(s) \simeq E_{\tau_J}(x)(s) = x_{\tau_J}(x)(s) \simeq \sum_{k=1}^{M_J} \beta_k^{M_J}\ e_{\tau_J}(s, q_k^{M_J})\ x(q_k^{M_J}).$$

Even more, if we use the same constituents of (6.362), for $j = J_0, \ldots, J-1$,

$$(6.363) \qquad E_{\tau_j}(x)(s) \simeq \sum_{k=1}^{M_J} \beta_k^{M_J}\ e_{\tau_j}(s, q_k^{M_J})\ x(q_k^{M_J}),$$

$$(6.364) \qquad (WE)_{\tau_j}(x)(s) \simeq \sum_{k=1}^{M_J} \beta_k^{M_J}\ (we)_{\tau_j}(s, q_k^{M_J})\ x(q_k^{M_J}),$$

with

$$(6.365) \qquad (we)_{\tau_j} = e_{\tau_{j+1}} - e_{\tau_j}, \quad j = J_0, \ldots, J-1,$$

then the formulas (6.363), (6.364) enable us to perform a multiscale decorrelation analysis of x.

The multiscale mollifier inversion as indicated by (6.362) and (6.363), (6.364) is an extremely useful methodology for applying potential methods in geologic exploration. As a matter of fact, gravity and magnetic field inversion leading to density distribution and material magnetization, respectively, may be described both by inverse problems of 3D potential theory (cf. Section 4.2). Both inverse gravimetry and magnetometry may be formulated by a Fredholm integral equation of the first kind (see Section 7.2). The essential mollifier idea is to replace occurring monopole/dipole kernels in the volume integral representations of Earth's gravitational/magnetic field, respectively, by associated mollifier regularizations $a_{\tau_j}(\cdot, \cdot)$ which lead back to Haar-type kernels $e_{\tau_j}(\cdot, \cdot)$. In doing so, both the associated wavelet kernels $(wa)_{\tau_j}(\cdot, \cdot)$ and $(we)_{\tau_j}(\cdot, \cdot)$, respectively, show *local supports* which become smaller with increasing scale j. In more detail, the applicability of mollifiers in the proposed multiscale framework concerned with geoexploration is characterized as follows: The gravimetric/magnetometric mollifier approach of Chapter 7 (see also W. Freeden [**270**], C. Blick et al. [**100**]) is interested in associated sequences of wavelet kernels, respectively, which show decreasing space-localized supports with increasing scale for the simultaneous multiscale approximation of both gravitational/magnetic potential and mass density/magnetization distribution, so that numerically economic and efficient implementations become applicable. Moreover, the convolutions with the inverse wavelet kernels simultaneously allow to model bandpass filtered information of the potential as well as the density/magnetization distribution, so that both signatures admit a certain decorrelation into detail information (see Chapter 7). The substantial idea is that, if the wavelet kernels of the gravitational/magnetic potential are constructed in close geophysical relevance, the detail information of the density/magnetization distribution reflects the structure of the geological layers.

All in all, the proposed multiscale mollifier inversion methods in gravimetry/magnetometry provide 3D-separations of local geological features by specifying the transitions of the geologic strata.

6.12. Backus-Gilbert method

The Backus-Gilbert method (cf. [**54**–**56**]) treats moment problems of the type

$$\langle x, k_i \rangle_X = \beta_i, \quad i = 1, \ldots, n, \tag{6.366}$$

with given elements $k_i \in X$, for example $X = L^2(\mathcal{G})$, $\mathcal{G} \subset \mathbb{R}^q$ measurable bounded region, so that the system (6.366) takes the form

$$\int_{\mathcal{G}} k_i(t)\, x(t)\, dt = \beta_i, \quad i \in \{1, \ldots, n\}, \tag{6.367}$$

which can be thought of as resulting from discretizing an integral equation of the first kind

$$\int_{\mathcal{G}} k(s,t)\, x(t)\, dt = \beta(s) \tag{6.368}$$

by collocating at points $s_1, \ldots, s_n$, so that $k_i(t) = k(s_i, t)$, $\beta_i = \beta(s_i)$.

With $Ax = (\langle x, k_1 \rangle_X, \ldots, \langle x, k_n \rangle_X)^T$, $\beta = (\beta_1, \ldots, \beta_n)^T$, (6.366) can be written in the form $Ax = \beta$ with $X = L^2(\mathcal{G})$, $Y = \mathbb{R}^n$. In the Backus-Gilbert method, one looks for an approximate inverse $S : \mathbb{R}^n \to L^2(\mathcal{G})$ for A by defining

$$Sy = \sum_{i=1}^{n} y_i v_i \tag{6.369}$$

with functions $v_i \in L^2(\mathcal{G})$ to be determined by the following procedure: Since

$$SAx = \sum_{i=1}^{n} \langle x, k_i \rangle_X v_i = \langle x, \sum_{i=1}^{n} k_i v_i \rangle_X, \tag{6.370}$$

i.e., for the concrete case (6.367)

$$(SAx)(s) = \int_{\mathcal{G}} x(t) \left[\sum_{i=1}^{n} k_i(t) v_i(s) \right] dt, \tag{6.371}$$

one should aim at determining the functions v_i such that $\sum_{i=1}^{n} k_i(t) v_i(s)$ is "close" to $\delta(s-t)$:

$$\sum_{i=1}^{n} k_i(t) v_i(s) \approx \delta(s-t). \tag{6.372}$$

The question is how to formalize the requirement (6.372). In the classical approach to the Backus-Gilbert method [**54**–**56**], this is done by minimizing, for any fixed $s \in \mathcal{G}$ and some chosen number τ, the functional

$$(v_1, \ldots, v_n) \mapsto \int_{\mathcal{G}} |s-t|^{2\tau} \left| \sum_{k=1}^{n} k_i(t) v_i \right|^2 dt. \tag{6.373}$$

Under the normalization constraint

$$\int_{\mathcal{G}} \sum_{i=1}^{n} k_i(t) v_i \, dt = 1, \tag{6.374}$$

we then take $v_i(s)$ by definition equal to v_i. The constraint (6.374) just says that for $x = 1, (SAx)(s) = x(s)$. The parameter τ (in [**54**], $\tau = 1$) determines the concrete method.

The common feature between mollification and the Backus-Gilbert method is the following: In both cases, an approximate inverse (determined by v_s^ρ or by the $v_i(s)$) is determined independently from the right hand side of the equation, which can then be used to explicitly represent an approximate solution via (6.347) or via (6.369). By use of Lagrange multipliers, the Backus-Gilbert basis functions $v_1, \ldots, v_n$ can be determined pointwise from the linear system

$$\begin{pmatrix} G(s) & w \\ w^T & 0 \end{pmatrix} \begin{pmatrix} v \\ \lambda \end{pmatrix} = \begin{pmatrix} 0 \\ 1 \end{pmatrix}, \quad s \in \mathcal{G}, \tag{6.375}$$

with

$$(G(s))_{ij} = \int_{\mathcal{G}} |s-t|^{2\tau} k_i(t) k_j(t) \, dt, \quad i, j \in \{1, \ldots, n\}, \tag{6.376}$$

$$w_i = \int_{\mathcal{G}} k_i(t) \, dt, \qquad i \in \{1, \ldots, n\}. \tag{6.377}$$

Note that the matrix of this system depends on s while in the corresponding system (6.344) for mollifier methods, s enters only in the right-hand side.

6.13. Stochastic regularization methods

We let X be a Hilbert space. Our statistical approach starts from an equation of type (see, e.g., [**547**])

$$Ax = y + \eta \tag{6.378}$$

where x, y, η are considered as values of jointly distributed random variables.

Random variables. On the probability space Ω equipped with the probability measure P we understand $\xi : \Omega \to X$ as Hilbert space-valued random variable. If X is a function space, then we denote by ξ a *stochastic process*. For $x \in X$ we obtain by $\xi_x = \langle x, \xi \rangle$ a real-valued random variable. Hence, for a complete orthogonal system $\{x_i\}$ in X, we are able to introduce by $\xi^{(i)} = \xi_{x_i} = \langle x_i, \xi \rangle$ an infinite number of jointly distributed random variables.

In what follows we suppose that ξ has a *vanishing expectation value*, i.e., we have

$$E[\langle x, \xi \rangle] = 0 \tag{6.379}$$

for all $x \in X$ (note that (6.379) does not mean any restriction, if (6.379) is violated we are allowed to go over to the random variable $\tilde{\xi} = \xi - E[\xi]$). Furthermore, we assume that the random variable has a finite second moment, so that

(i)

$$E[|\langle x, \xi\rangle|^2] < \infty \quad \text{for all} \quad x \in X, \tag{6.380}$$

for all $x \in X$

(ii) the expectation value is continuous at x.

Then it follows that $E[\langle x, \xi\rangle\langle \xi, z\rangle]$ is a continuous, symmetric, nonnegative bilinear form on X, hence, there exists a linear, continuous, selfadjoint, nonnegative operator $R_{\xi\xi} : X \to X$ satisfying

$$\langle R_{\xi\xi} x, z\rangle = Cov(x, z) = E[\langle x, \xi\rangle\langle \xi, z\rangle], \tag{6.381}$$

$R_{\xi\xi}$ is called *covariance operator.*

Example. Let $\mathcal{G}$ be a regular region. Assume that $X = L^2(\mathcal{G})$ and $\xi(\alpha)$ is defined for $\alpha \in \mathcal{G}$. Then we are able to identify the covariance operator with the covariance function resulting in the *autocovariance function* given by

$$R_{\xi\xi}(\alpha, \beta) = Cov(\xi(\alpha), \xi(\beta)) = E[\xi(\alpha) \cdot \xi(\beta)]. \tag{6.382}$$

Application to an $L^2(\mathcal{G})$-function x yields the identity

$$R_{\xi,\xi} x(\alpha) = \int_{\mathcal{G}} R_{\xi\xi}(\alpha, \beta)\, x(\beta)\, d\beta. \tag{6.383}$$

As *white noise* we denote the Gaussian process characterized by

$$R_{\xi\xi}(\alpha, \beta) = \eta^2\, \delta(\alpha - \beta), \tag{6.384}$$

i.e.,

$$R_{\xi\xi} = \eta^2 I. \tag{6.385}$$

Let ξ, η be jointly distributed random variables with $\xi : \Omega \to X$, $\zeta : \Omega \to Y$. Then we define the *cross-covariance operator* by $R_{\xi\zeta} : Y \to X$ as follows:

$$\langle R_{\xi\zeta} y, x\rangle = E\left[\langle y, \zeta\rangle_Y \langle x, \xi\rangle_X\right]. \tag{6.386}$$

Best linear estimator and Tikhonov-Phillips regularization. Next we have a look at the equation

$$A\xi = \beta + \zeta, \quad \xi : \Omega \to X, \quad \beta, \zeta : \Omega \to Y. \tag{6.387}$$

We assume that the inverse A^{-1} exists. The problem is to estimate ξ under the knowledge of β and ζ. To this end, we suppose that $E[\xi] = E[\beta] = 0$ such that $R_{\xi\zeta} = 0$ (i.e., ξ and ζ are uncorrelated) and $R_{\zeta\zeta}^{-1}$ exists. It is not difficult to show that (see, e.g., [**547**])

$$R_{\beta\beta} = AR_{\xi\xi}A^* + R_{\zeta\zeta}, \tag{6.388}$$

$$R_{\xi\beta} = R_{\xi\xi}A^*. \tag{6.389}$$

In order to realize a least-squares estimation we have to consider a *linear estimator* of ξ, i.e., a random variable $\xi_L = L\beta$, where $L : Y \to X$ is a linear and continuous operator such that $x_L = Ly$ is a solution of the operator equation. Central in our considerations is the following statement, that is standard in statistical geoscience:

If $R_{\zeta\zeta}$ is assumed to have a continuous inverse, then

$$L = R_{\xi\beta}R_{\beta\beta}^{-1} = R_{\xi\xi}A^*(AR_{\xi\xi}A^* + R_{\zeta\zeta})^{-1} \tag{6.390}$$

minimizes the functional $E\left[|\langle x, \xi - L\beta\rangle|^2\right]$ for all x.

Under the simplifying assumptions

$$R_{\xi\xi} = I, \quad R_{\zeta\zeta} = \eta^2 I$$

we obtain for (6.390)

$$L = A^*(AA^* + \eta^2 I)^{-1} = (A^*A + \eta^2 I)^{-1}A^*. \tag{6.391}$$

In other words, in similarity to the Tikhonov-Phillips regularization, we are led to normal equations in order to determine the best linear estimator.

For arbitrary covariance operators we obtain $x = Ly$ by minimizing

$$\langle R_{\zeta\zeta}^{-1}(Ax - y), Ax - y\rangle + \langle R_{\xi\xi}^{-1}x, x\rangle = |||Ax - y|||^2 + \|x\| \tag{6.392}$$

with

$$|||y||| = \langle R_{\zeta\zeta}^{-1}y, y\rangle \tag{6.393}$$

and

$$\|x\| = \langle R_{\xi\xi}^{-1}x, x\rangle\ . \tag{6.394}$$

Observing these facts we are finally able to come to the following conclusion:

The best linear estimator is a special Tikhonov-Phillips regularization method, or, *the Tikhonov-Phillips regularization method is a special linear estimator.*

The stochastic approach also allows an interpretation of Bayes-estimation as Tikhonov-Phillips regularization (for further details the reader is referred, e.g., to K.-R. Koch [**501**] and the references therein).

6.14. Nonlinear ill-posed problems

Linear models for nonlinear relationships usually describe the reality in a very limited way. Thus it is required to work with nonlinear concepts. As a consequence, we next design some concepts for describing and solving nonlinear inverse problems (thereby following A. Rieder [**723**] in larger parts, however, mentioning some recent achievements). Because of the diversity of nonlinear phenomena, we should not expect a closed theory. Although research in the field of nonlinear inverse problems is still very much in development, a trend to the concrete problem solvers has been emerging over the last decade. As a consequence, stability and convergence analyses are mainly based on the specific properties of the individual case. Nevertheless, we focus our considerations on nonlinear inverse problems to possible aspects of a unifying theory, to provide a solid knowledge of which is quite helpful in the discussion of a concrete situation. In particular our interest is to show which specific difficulties nonlinearities cause and how they can be handled. Note that the extension of the regularization methodology to the nonlinear is not obvious, since neither a singular value decomposition can be carried out nor an adjoint of a nonlinear operator can be defined adequately. In today's scientific situation, this observation immediately leads, in addition to the Tikhonov-Phillips regularization, to Newton-type iterative methods for the regularization of nonlinear problems.

Local ill-posedness. We start from an equation of the form

$$\Phi(f) = g, \tag{6.395}$$

where $\Phi : \mathcal{D}(\Phi) \subset X \to Y$ is a nonlinear continuous operator with definition space $\mathcal{D}(\Phi)$.

For the following considerations we assume that Equation (6.395) is solvable, so there exists an $f^\dagger \in \mathcal{D}(\Phi)$ with $\Phi(f^\dagger) = g$.

Under what circumstances is the problem (6.395) considered to be ill-posed?

- If $f^\dagger$ is not an isolated solution of (6.395), i.e., for all $r > 0$ there exists an $f^r \in \mathbb{B}_r(f^\dagger) \cup \mathcal{D}(\Phi)$ with $\Phi(f^r) = g$. In this case, $f^\dagger$ cannot be uniquely reconstructed locally from the data.
- If $f^\dagger$ does not depend continuously on g, a nonreflected reconstruction from perturbed data leads to useless results.

The following characterization specifies the local character of the ill-posedness of nonlinear equations and covers both cases mentioned above:

The nonlinear operator equation (6.395) on a Banach space is called *locally ill-posed in* $f^\dagger \in \mathcal{D}(\Phi)$, if for every $r > 0$ there is a sequence $\{f_k^r\}_{k\in\mathbb{N}} \subset \mathbb{B}_r(f^\dagger) \cap \mathcal{D}(\Phi)$ which does not converge to $f^\dagger$, but whose image sequence $\{\Phi(f_k^r)\}_{k\in\mathbb{N}}$ converges to $\Phi(f^\dagger)$:

$$\lim_{k\to\infty} \|\Phi(f_k^r) - \Phi(f^\dagger)\| = 0, \quad \text{but} \quad f_k^r \not\to \infty. \tag{6.396}$$

Otherwise, (6.395) is locally well posed in $f^\dagger \in \mathcal{D}(\Phi)$, i.e., there exists an $r > 0$ such that, for all sequences $\{f_k^r\}_{k\in\mathbb{N}} \subset \mathbb{B}_r(f^\dagger) \cap \mathcal{D}(\Phi)$,

$$\lim_{k\to\infty} \|\Phi(f_k^r) - \Phi(f^\dagger)\|_Y = 0 \tag{6.397}$$

implies

$$\lim_{k\to\infty} \|f_k^r - f^\dagger\| = 0. \tag{6.398}$$

Clearly, in a nonisolated point of the inverse image of g, the equation (6.395) should be locally ill-posed.

Basic features of linear ill-posed problems. We have characterized the ill-posedness of linear operator equations in Hilbert spaces earlier: The problem $Ax = y$, $A \in \mathfrak{L}(X,Y)$, is ill-posed if and only if the image $\mathcal{R}(A)$ is not closed in Y. Linear problems in Hilbert spaces allow a simple, global characterization of the ill-posedness (after M.Z. Nashed [**625**]). For an operator $A \in \mathfrak{L}(X,Y)$ we have the equivalence relations:

- $Ax = y$ is ill-posed,
- $\mathcal{R}(A)$ is not closed in Y,
- $A^\dagger$ is discontinuous.

In particular, linear equations generated by compact operators are always ill-posed.

The linear equation $Ax = y$ is either locally well-posed in all points of X or locally ill-posed in all points. It is locally ill-posed if A is not injective or if the image $\mathcal{R}(A)$ is closed.

The question now is how this kind of ill-posedness is in relationship to nonlinear obligations? As an immediate answer we are led to the following context:

Characterization of nonlinear ill-posed problems. In the nonlinear case, there is no criterion as clear as in the linear case informing about local ill-posedness. Nonetheless, it should be outlined that the compactness of the nonlinear operator essentially implies local ill-posedness (for more details see, e.g., [**216**], [**441**], see also [**442**]).

The definition of compactness of a linear operator can be applied without modification to nonlinear operators $\Phi : \mathcal{D}(\Phi) \subset X \to Y$ is compact if and only if the image $\Phi(U)$ of a bounded set $U \subset \mathcal{D}(\Phi)$ is relatively compact in Y. The compactness of nonlinear operators, however, does not imply their continuity which is why the following definition becomes necessary:

An operator $\Phi : \mathcal{D}(\Phi) \subset X \to Y$ between normed spaces is called *completely continuous* if it is compact and continuous (note that linear compact operators are always completely continuous).

Furthermore, we need another property of a nonlinear operator:

An operator $\Phi : \mathcal{D}(\Phi) \subset X \to Y$ between Hilbert spaces is called *weakly sequentially closed* if for all sequences $\{x_n\}_{n\in\mathbb{N}} \subset \mathcal{D}(\Phi)$ holds (note that the weak convergence of the sequences $x_n \rightharpoonup x$ in X and $\Phi(x_n) \rightharpoonup y$ in Y implies $x \in \mathcal{D}(\Phi)$ as well as $\Phi(x) = y$).

Completely continuous and weakly sequentially closed operators have a useful property (see [**216**]).

A completely continuous and weakly sequentially closed operator $\Phi : \mathcal{D}(\Phi) \subset X \to Y$ transforms the weakly convergent sequence $\{x_n\}_{n\in\mathbb{N}} \subset \mathcal{D}(\Phi)$ into the strongly convergent sequence $\{\Phi(x-n)\}_{n\in\mathbb{N}}$, more precisely: $x_n \rightharpoonup x$ implies $\lim_{n\to\infty} \|\Phi(x_n) - \Phi(x)\| = 0$.

The next result is also taken from H.W. Engl et al. [**216**]. However, we formulate it in the version of B. Hofmann [**441**] (see also [**442**]).

Let $\Phi : \mathcal{D}(\Phi) \subset X \to Y$ be completely continuous and weakly sequentially closed. In addition, let X be separable and infinite-dimensional. Then the equation (6.395) *is locally ill-posed in $f^\dagger \in int(\mathcal{D}(\Phi))$.*

Next we are interested in the relation between the local ill-posedness of the nonlinear equation (6.395) in $f^\dagger$ and the ill-posedness of its linearization $\Phi'(f^\dagger)f = y$. This relation has been studied in detail by B. Hofmann [**441**] and O. Scherzer [**744**]. We adopt two results (their proofs can be found, e.g., in [**723**]):

Let X and Y be Hilbert spaces. Assume that the mapping $\Phi : \mathcal{D}(\Phi) \subset X \to Y$ is Fréchet-differentiable with locally Lipschitz-continuous derivative in $f^\dagger \in int(\mathcal{D}(\Phi))$, i.e. there is a constant $L > 0$ and a radius $r_L > 0$ with

$$\|\Phi'(f) - \Phi'(f^\dagger)\| \leq L\|f - f^\dagger\|_X \quad \text{for all} \quad f \in \mathbb{B}_{r_L}(f^\dagger) \subset \mathcal{D}(\Phi). \tag{6.399}$$

Then we are confronted with the following situation: If the nonlinear equation (6.395) *is locally ill-posed in* $f^\dagger$, *then the linearization* $\Phi'(f^\dagger)$ *is locally ill-posed in any* $f \in X$, *i.e., the image of* $\Phi'(f^\dagger)$ *is not closed in* Y *or* $\Phi'(f^\dagger)$ *is not injective.*

Unfortunately, this result cannot be reversed (cf. the counterexample in A. Rieder [**723**]). The local ill-posedness of (6.395) in $f^\dagger$ cannot be deduced from the local ill-posedness of the linearization $\Phi'(f^\dagger)$.

Thus the question arises why is the connection between the local ill-posedness of a nonlinear equation in $f^\dagger$ and the local ill-posedness of its linearization in $f^\dagger$ so casual in general? The reason is the Taylor expansion

$$\Phi(f) - \Phi(f^\dagger) = \Phi'(f^\dagger)(f - f^\dagger) + E(f, f^\dagger). \tag{6.400}$$

It does not adequately describe the local behavior of an ill-posed problem. Although the residual $E(f, f^\dagger)$ behaves like $o(\|f - f^\dagger\|)$ for $f \to f^\dagger$, it need not be small compared to the residual $\Phi(f) - \Phi(f^\dagger)$ or to the linearized residual $\Phi'(f^\dagger)(f - f^\dagger)$. As a matter of fact, if Φ is completely continuous, then $\Phi'(f^\dagger)$ is compact (see, e.g., [**881**]). Then $\Phi' f^\dagger)(f - f^\dagger)$ can be significantly smaller than $E(f, f^\dagger)$. Similarly, $\Phi(f) - (f^\dagger)$ can be much smaller than $f - f^\dagger$ or than $E(f, f^\dagger)$.

This situation, however, changes when the linearization error is controlled by the nonlinear residual, e.g., when Φ satisfies the following condition (6.401):

There is a value $r > 0$ *and a number* $\omega > 0$ *such that*

$$\|E(v, w)\| = \|\Phi(v) - \Phi(w) - \Phi'(w)(v - w)\| \leq \omega \|\Phi(v) - \Phi(w)\| \quad \textit{for all} \quad v, w \in \mathbb{B}_r(f^\dagger) \subset \mathcal{D}(\Phi). \tag{6.401}$$

In fact, the condition (6.401) implies for $\omega < 1$ the inequality:

$$1 - \omega \leq \frac{\|\Phi'(w)(v - w)\|}{\|\Phi(v) - \Phi(w\|)} \leq 1 + \omega. \tag{6.402}$$

In other words, nonlinear and linearized residual control each other.

Assume that the continuous operator $\Phi : \mathcal{D}(\Phi) \subset X \to X$ *between Hilbert spaces satisfies the condition* (6.401) *with* $\omega < 1$.

The problem (6.395) *is locally well-posed in* $f^\dagger \in \mathcal{D}(\Phi)$ *if and only if* $\Phi'(f^\dagger)$ *is injective and* $\mathbb{R}(\Phi'(f^\dagger))$ *is close (i.e , the problem* (6.395) *is locally well-posed if and only if the linearization of* (6.395) *is locally well-posed.)*

The problem (6.395) *is locally ill-posed in* $f^\dagger \in \mathcal{D}(\Phi)$ *if and only if* $\Phi'(f^\dagger)$ *has a nontrivial null space or* $R(\Phi'(f^\dagger))$ *is not closed.*

The equivalences in the last statement are also valid under a weaker condition. It is sufficient to assume (6.402), where $1 - \omega$ and $1 + \omega$ are replaced by arbitrary positive constants.

In the convergence analysis of Newton-type iterative methods for the stable solution of nonlinear ill-posed equations (as discussed later), we need a further constraint on the nonlinearity of Φ:

Let $\Phi : \mathcal{D}(\Phi) \subset X \to Y$ be an operator such that there exists a mapping $Q : X \times X \to L(Y)$ satisfying

$$\Phi'(v) = Q(v,w)\Phi'(w) \quad \text{and} \quad \|I - Q(v,w)\| \le C_Q \|v - w\| \quad \text{for all} \quad v, w \in \mathbb{B}_r(f^\dagger) \subset \mathcal{D}(\Phi). \tag{6.403}$$

The condition (6.403) enables us to formulate the following result:

Let X, Y be Banach spaces. Suppose that the operator $\Phi : \mathcal{D}(\Phi) \subset X \to Y$ satisfies (6.403). *Then we have*

(i) $\mathcal{N}(\Phi'(v)) = \mathcal{N}(\Phi'(w))$ *holds true for all* $v, w \in \mathbb{B}_r(f^\dagger)$.

(ii) *If $\mathcal{N}(\Phi'(f^\dagger))$ is nontrivial, then $f^\dagger \in int(\mathcal{D}(\Phi))$ cannot be an isolated solution of* (6.395).

(iii) *If $C_Q\, r < 1$, then* (6.403) *implies the estimate* (6.401) *with* $\omega = \frac{C_Q r}{1 - C_Q r}$. *For $C_Q r < \frac{1}{2}$, $\omega < 1$.*

Next we transfer some terms from the linear theory to the nonlinear context, thereby assuming that the nonlinear mapping $\Phi : \mathcal{D}(\Phi) \subset X \to Y$ is continuous.

(i) *Given $g \in \mathcal{R}(\Phi)$ and $f^* \in X$. $f^\dagger \in \mathcal{D}(\Phi)$ satisfying the minimal property*

$$\|f^\dagger - f^*\| = \min\{\|f - f^*\| : \Phi(f) = g\} \tag{6.404}$$

is called f^*-minimum norm solution with respect to g (note that, for linear operators, the minimum norm solution is uniquely determined).

(ii) *A regularization procedure for Φ is a pair $(\{R_t\}_{t>0}, \gamma)$ consisting of a family of continuous operators $R_t : X \times Y \to X$ and a mapping $\gamma : (0,\infty) \times Y \to (0,\infty)$, so that*

$$\sup\{\|f^\dagger - R_{\gamma(\varepsilon, g^\varepsilon)}(f^*, g^\varepsilon)\| : g^\varepsilon \in Y, \|g - g^\varepsilon\| \le \varepsilon\} \ \text{for}\ \varepsilon \to 0, \tag{6.405}$$

where $f^\dagger$ is an f^*-minimum norm solution with respect to $g \in \mathcal{R}(\Phi)$.

(iii) *Let $f^\dagger$ be a f^*-minimum norm solution with respect to $g \in \mathcal{R}(\Phi)$, and let Φ be Frechet-differentiable in $f^\dagger$. Moreover, let*

$$f^\dagger - f^* = |\Phi'(f^\dagger)|^\mu\, w \quad \text{for} \quad w \in X \quad \text{and} \quad \mu > 0. \tag{6.406}$$

Then $(\{R_t\}_{t>0}, \gamma)$ is called order-optimal if, for $\varepsilon > 0$ sufficiently small, there exists a positive constant C_μ such that

$$\sup\left\{\|f^\dagger - R_{\gamma(\varepsilon, g^\varepsilon)}(f^*, g^\varepsilon)\| : g^\varepsilon \in Y, \|g - g^\varepsilon\| \le \varepsilon\right\} \le C_\mu\, \|w\|^{\frac{1}{\mu+1}}\, \varepsilon^{\frac{p}{p+1}}. \tag{6.407}$$

Under certain circumstances, the f^*-minimum norm solution is locally unique:

Let $\Phi : \mathcal{D}(\Phi) \subset X \to Y$ be Fréchet-differentiable in $f^\dagger$. For $g = \Phi(f\dagger)$ we let $\mathbb{U}(g) = \{f \in \mathcal{D}(\Phi) | \Phi(f) = g\}$. Suppose that (6.401) *holds in $\mathbb{B}_r(f^\dagger)$, and let there be an $f^* \in X$ with $f^\dagger - f^* \in \mathcal{N}(\Phi'(f^\dagger))^\perp$.*

Then $f^\dagger$ is the unique f^*-minimum norm solution in $\mathbb{B}_r(f^\dagger)$ with respect to g:

$$\|f^\dagger - f^*\| < \|f - f^*\| \quad \text{for all} \quad f \in \mathbb{U}(g) \cap \mathbb{B}_r(f^\dagger) \backslash \{f^\dagger\}. \tag{6.408}$$

In fact, the property (6.406) implies $f^\dagger - f^* \in \mathcal{N}(\Phi'(f^\dagger))^\perp$. Therefore $f^\dagger$ must be a locally unique f^*-minimum norm solution with respect to $g = \Phi(f^\dagger)$.

Nonlinear Tikhonov-Phillips regularization. The Tikhonov-Phillips regularization can be easily formulated for nonlinear operator equations. To this end we recall the minimal property of the Tikhonov-Phillips regularized solution f_t^ε for linear problems with $A \in L(X, Y)$:

$$f_t^\varepsilon = \operatorname{argmin}\{\|Af - g^\varepsilon\|^2 + t\,\|f\|^2 \ : f \in \mathcal{N}(A)^\perp\}, \quad t > 0, \tag{6.409}$$

For the stable solution of (6.395) under the assumption of disturbed data $g^\varepsilon \in Y$ our purpose is to minimize the functional

$$J_t(f) = J_t(f, g_t^\varepsilon) := \|\Phi(f) - g^\varepsilon\|^2 + t\,\|f - f^*\|^2, \quad t > 0, \tag{6.410}$$

over $\mathcal{D}(\Phi)$. f^* determines which solution is approximated by a minimizing element of J_t.

Our procedure is reasonable; because the minimization problem allows a solution (cf. T.I. Seidmann [**760**]).

Let $\Phi : \mathcal{D}(\Phi) \subset X \to Y$ be a weakly sequentially closed mapping between Hilbert spaces. Suppose that $t > 0, f^* \in X$, and $g \in Y$. Then there exists an $f_t^\varepsilon \in \mathcal{D}(\Phi)$ with

$$J_t(f_t^\varepsilon) = \min\{J_t(f) : f \in \mathcal{D}(\Phi)\}. \tag{6.411}$$

Now, let f_t^ε denote a minimizing element of J_t. We ask ourselves to what extent f_t^ε is suitable for regularization. To this end (cf. T.I. Seidmann, C.R. Vogel [**761**], H.W. Engl et al. [**216**]), the stability of f_t^ε and g^ε, respectively, should be investigated first.

Let $\{g^{\varepsilon_k}\}_{k\in\mathbb{N}} \subset Y$ be a sequence converging to g^ε. Let $\{f_t^{\varepsilon_k}\}_{k\in\mathbb{N}}$ denote an associated sequence of minimizing elements of $J_t(\cdot, g^{\varepsilon_k})$.

Then there exists a convergent subsequence of $\{f_t^{\eta_k}\}_{k\in\mathbb{N}}$ and every convergent subsequence converges against a minimizing element f_t^ε of $J_t(\cdot, g^\varepsilon)$. Moreover, if f_t^ε is unique, the sequence $\{f_t^{\varepsilon_k}\}_{k\in\mathbb{N}}$ converges to f_t^ε.

The nonlinear Thikhonov-Phillips regularization $\{R_t^\eta\}_{T>0}$, $R_t^\eta : X \times Y \to X$, $\eta \geq 0$, is defined by $R_t^\eta(f^*, g^\varepsilon) := f_t^{\varepsilon,\eta}$, where $f_t^{\varepsilon,\eta}$ is determined by

$$J_t(f_t^{\varepsilon,\eta}, g^\varepsilon) \leq \min\{J_t(f, g^\varepsilon) : f \in \mathcal{D}(\Phi)\} + \eta.$$

With the parameter η we model errors which inevitably occur in a numerical minimization process of the functional. The existence of $f_t^{\varepsilon,\eta}$ is guaranteed.

Let $g \in \mathcal{R}(\Phi)$ and let $f^\dagger$ be a f^*-minimum norm solution of (6.395). The family $\{g_\varepsilon\}_{\varepsilon>0}$ satisfies $\|g - g^\varepsilon\| \leq \eta$ and the perturbation $\eta(t)$ behaves like $o(t)$ for $t \to 0$. Let us choose $\gamma : (0,\infty) \to (0,\infty)$ so that

$$\gamma(\varepsilon) \to 0 \quad \textit{as well as} \quad \frac{\varepsilon^2}{\gamma(\varepsilon)} \to 0 \quad \textit{for} \quad \varepsilon \to 0, \tag{6.412}$$

then $\{R_{\gamma(\varepsilon)}^{\eta(\gamma(\varepsilon))}(f^*, g^\varepsilon)\}_{\varepsilon>0}$ possesses a convergent subsequence. The limit of any convergent subsequence is a f^*-minimum norm solution of (6.395). If the f^*-minimum norm solution is uniquely determined, then we even have

$$\lim_{\varepsilon\to 0} \|R_{\gamma(\varepsilon)}^{\eta(\gamma(\varepsilon))}(f^*, g^\varepsilon) - f^\dagger\| = 0, \tag{6.413}$$

i.e., $(\{R_t^{\eta(t)}\}_{t>0,\gamma})$ *is a regularization method.*

Next we are concerned with the order optimality of the nonlinear Tikhonov-Phillips regularization (as obtained by H.W. Engl et al. [**216**], H.W. Engl et al. [**215**]).

Let $\Phi : \mathcal{D}(\Phi) \subset X \to Y$ *a weakly sequentially closed mapping between Hilbert spaces. Further, let* Φ *be Fréchet-differentiable in* $\mathbb{B}_\rho(f^\dagger) \subset \mathcal{D}(\Phi)$ *for a value* $\rho > 0$ *with a Lipschitz-continuous derivative in* $f^\dagger$*:*

$$\|\Phi'(f) - \Phi'(f^\dagger)\| \le L\|f - f^\dagger\| \quad \textit{for all} \quad f \in \mathbb{B}_\rho(f^\dagger). \tag{6.414}$$

Furthermore, let there be a $\mu \in [1,2]$ *for* $f^* \in X$ *such that* $f^\dagger - f^* \in \mathcal{R}(|\Phi'(f^\dagger)|^\mu)$*. In fact, this assumption guarantees the existence of an element* $w \in Y$ *with*

$$f^\dagger - f^* = \Phi'(f^\dagger)^*\, w. \tag{6.415}$$

The family $\{g^\varepsilon\}_{\varepsilon>0} \subset Y$ *satisfies* $\|g - g^\varepsilon\| \le \varepsilon$*, where* $g = \Phi(f^\dagger)$*, and the perturbation* $\eta(t)$ *behaves like* $O(t^2)$ *for* $t \to 0$*. Assume that the a priori paramter* $\gamma : (0,\infty) \to (0,\infty)$ *satisfy*

$$c_\gamma \varepsilon^{\frac{2}{\mu+1}} \le \gamma(\varepsilon) \le C_\gamma \varepsilon^{\frac{2}{\mu+1}} \tag{6.416}$$

with constants $0 < c_\gamma \le C_\gamma$*.*

If $L\|w\| < 1$ *and* $\rho > 2\|f^\dagger - f^*\|$*, then we have*

$$\|R_{\gamma(\varepsilon)}^{\eta(\gamma(\varepsilon))}(f^*, g^\varepsilon) - f^\dagger\|_X = O\left(\varepsilon^{\frac{\mu}{\mu+1}}\right) \quad \varepsilon \to 0. \tag{6.417}$$

Let us make some remarks about the aforementioned result (6.417):

- The weak sequential closedness of Φ plays only an indirect role in the proof: it guarantees the existence and stability of $\{f_t^{\varepsilon,\eta}\}_{t>0}$ for $\eta \ge 0$.
- While (6.395) may have several f^*-minimum norm solutions, only one of them can satisfy the above conditions.
- For $\mu = 1$ we are able to deduce the convergence order

$$\|\Phi\left(R_{\gamma(\varepsilon)}^{\eta(\gamma(\varepsilon))}(f^*, g^\varepsilon)\right) - g\| = O(\varepsilon) \quad \text{for} \quad \varepsilon \to 0 \tag{6.418}$$

  of the nonlinear residual.
- If we change, for $\mu = 1$, the a priori strategy γ by letting

$$\tilde{c}_\gamma \frac{\varepsilon}{|w|} \le \gamma(\varepsilon) \le \tilde{C}_r \frac{\varepsilon}{|w|}, \tag{6.419}$$

  then $(\{R_t^{\eta(t)}\}_{t>0,\gamma})$ is order-optimal provided η behaves like $O(\|w\|^2 t^2)$.
- A posteriori parameter strategies for the nonlinear Tikhonov Phillips method study O. Scherzer [**744**] as well as O. Scherzer et al. [**746**]. In order to prove order optimality, these contributions require conditions on the nonlinearity of type (6.401) and (6.403), respectively.

Newton's iterative method. In its well-known formulation, Newton's iteration (see, e.g., J.M. Ortega, W.C. Rheinboldt [**667**] and most textbooks in numerical analysis)

$$x_{k+1} = x_k - J(x_k)^{-1} f(x_k), \quad k = 0, 1, \ldots \tag{6.420}$$

is the ubiquitous method for solving systems of nonlinear equations in the form of $f(x) = 0$, where $J(x)$ is the Jacobian of the mapping f at x. It is well documented that Newton's iteration quadratically converges to any isolated solution under natural conditions such as the mapping is smooth and the initial iterate is near a solution at which the Jacobian is invertible.

Extending Newton's iteration beyond (6.420) by replacing the inverse with a certain kind of generalized inverses traces back to Gauss for solving least-squares solutions of overdetermined systems by the Gauss-Newton iteration with an assumption that the Jacobian is injective. For systems with rank-deficient Jacobians, A. Ben-Israel [**76**] is the first to propose using the Moore-Penrose inverse to generalize Newton's iteration as

$$x_{k+1} - x_k - J(x_k)^{\dagger} f(x_k), \quad k = 0, 1, \ldots \tag{6.421}$$

M.T. Chu [**151**] seems to be the first to prove the local convergence of the iteration (6.421) with minimal assumptions for underdetermined systems with surjective Jacobians. J.-P. Dedieu, M.-H. Kim [**180**] show that the iteration (6.421) locally quadratically converges to a nonisolated solution under the assumption that the Jacobian has a constant deficient rank in a neighborhood of the initial iterate. M.Z. Nashed, X. Chen [**630**] propose a Newton-like iteration by replacing the inverse with a certain outer inverse of the Jacobian. Moreover, they prove a quadratic convergence to a stationary point under a certain set of conditions. X. Chen et al. [**145**] along with Y. Levin, A. Ben-Israel [**535**] follow up with similar convergence results toward stationary points using outer inverses. Recent work by Z. Zeng [**882**] is suitable for any rank of the Jacobian at the zero. The iteration converges quadratically to a stationary point that is guaranteed to be a zero if the initial iterate is near a "semiregular zero" (for details see [**882**]).

Newton-type iterative regularization. The question is if Newton's methods are also suitable for the regularization of ill-posed equations? A positive answer can be found below:

For the numerical solution of (6.395) with perturbed data g^{ε} we consider the iteration rule

$$f_{n+1}^{\varepsilon} = f_n^{\varepsilon} + s_n^{\varepsilon}, \quad n \in \mathbb{N}_0, \tag{6.422}$$

with initial value $f_0 = f_0^{\varepsilon} \in \mathcal{D}(\Phi)$. The iteration step s_n^{ε} must be determined. For this we try to come as close as possible to the exact iteration step

$$s_n^{\varepsilon} = f^{\dagger} - f_n^{\varepsilon}, \tag{6.423}$$

here, $f^{\dagger}$ is a solution of (6.395). Note that, if s_n^{ε} and s_n^{e} coincide, then we have

$$f_{n+1}^{\varepsilon} = f^{\dagger}. \tag{6.424}$$

How can an approximation to s_n^{e} be obtained? If Φ is Fréchet-differentiable, the exact iteration step solves the linear system

$$\Phi'(f_n^{\varepsilon}) s_n^{e} = g - \Phi(f_n^{\varepsilon}) - E(f^{\dagger}, f_n^{\varepsilon}) = b_n, \tag{6.425}$$

where $E(v, w) = \Phi(v) - \Phi(w) - \Phi'(w)(v - w)$ is the linearization error. Unfortunately, we do not know the right-hand side b_n, but only a noisy version

$$b_n^\delta := g^\varepsilon - \Phi(f_n^\varepsilon) \quad \text{with} \quad \|b_n^\delta - b_n\| \le \varepsilon + \|E(f^\dagger, f_n^\varepsilon)\|. \tag{6.426}$$

The noise in b_n^δ has two origins, one is the noise in g^ε and the other is the linearization error.

We therefore determine s_n^ε as the solution of

$$\Phi'(f_n^\varepsilon)\ s = b_n^\delta. \tag{6.427}$$

Here, we have to take into account the fact that the ill-posedness of (6.395) inherits to (6.427). (This is the case, for example, when Φ and hence $\Phi'(f_n^\varepsilon)$ are completely continuous.) Since s_n^ε is determined from a linearization of (6.395) into f_n^ε, the iteration (6.422) is of Newton type. Depending on how approximations to s_n^e are obtained from (6.427), different procedures result.

To avoid error amplification, as in iterative methods for linear problems, the iteration (6.422) must be stopped in time. For this purpose, we employ the discrepancy principle: Accept as the solution of (6.395) that iteration $f_{n^*}^\varepsilon$ for which

$$\|\Phi(f_{n^*}^\varepsilon) - g^\varepsilon\| \le \tau\varepsilon < \|\Phi(f_n^\varepsilon) - g^\varepsilon\|, \quad n = 0, 1, \ldots, n^* - 1, \tag{6.428}$$

where the constant $\tau > 1$ is chosen fixed.

In the sequel, we assume $\Phi : \mathcal{D}(\Phi) \subset X \to Y$ to be a continuously differentiable mapping between Hilbert spaces X and Y. Moreover, let $f^\dagger \in \mathcal{D}(\Phi), g = \Phi(f^\dagger)$, $g^\varepsilon \in Y$ with $\|g - g^\varepsilon\| < \varepsilon, A = \Phi'(f^\dagger)$ and $A_n = \Phi'(f_n^\varepsilon)$.

F. Hettlich, W. Rundell [**428**] go one step further. Instead of the linearization (6.427), they use a quadratic approximation to Φ in f_n^ε to determine the iteration step. Another approach is proposed by U. Tautenhahn [**815**] by formulating the ill-posed nonlinear problem (6.395) as a well-posed fixed point problem.

Nonlinear Landweber method. The nonlinear Landweber method estimates by applying an undamped Landwerber step with initial value 0 to (6.427), i.e., we set

$$s_n^\varepsilon = A_n^* b_n^\delta \tag{6.429}$$

and

$$f_{n+1}^\varepsilon = f_n^\varepsilon + A_n^*(g^\varepsilon - \Phi(f_n^\varepsilon)), \quad n \in \mathbb{N}_0, \tag{6.430}$$

respectively. Moreover, assume that the operators A_n are scaled by $\|A_n\| \le 1$ for all $n \in \mathbb{N}_0$.

M. Hanke et al. [**399**] have analyzed the nonlinear Landweber method obtaining the following two results:

Let (6.428) *hold with* $\omega < 1/2$ *and* $r > 0$. *If* f_0 *is started sufficiently close to* $f^\dagger$, *then the nonlinear Landweber iteration* (6.430) *together with* (6.428) *is well-defined and convergent for* $\tau > 2(1 + \omega)/(1 - 2\omega) > 2$:

$$\{f_0, \ldots, f_{n^*}^\varepsilon\} \subset \mathbb{B}_r(f^\dagger) \quad \textit{as well as} \quad f_{n^*}^\varepsilon \to \tilde{f} \quad \textit{for} \quad \varepsilon \to 0,$$

where $\tilde{f} \in \mathbb{B}_r(f^\dagger)$ *also solves* (6.395).

Instead of (6.401), *assume the stronger condition* (6.403) *in* $\mathbb{B}_r(f^\dagger)$ *with* $C_Q\, r < 1/3$. *(This implies* (6.401) *in* $\mathbb{B}_{2r}(f_0)$ *with* $\omega < 1/2$*). Suppose that the initial value* $f_0 \in \mathbb{B}_r(f^\dagger)$ *satisfies* $f^\dagger - f_0 = |A\|^\mu\, w$ *for a* $w \in X$ *and a* $\mu \in [0,1]$. *If* $\|w\|$ *is sufficiently small, we have*

$$\|f^\varepsilon_{n^*} - f^\dagger\| \le C\;\; \varepsilon^{\frac{\mu}{\mu+1}}\;\|w\|^{\frac{1}{\mu+1}}. \tag{6.431}$$

Thus, the nonlinear Landweber method together with the discrepancy principle is an order-optimal regularization method.

Like the linear case, the nonlinear Landweber method converges very slowly. The numerical effort required to reach the stopping criterion (6.428) is, in general, considerable. A. Neubauer [**651**] investigated the convergence of the nonlinear Landweber method in Hilbert scales taking into account discretization effects. To speed up the nonlinear Landweber method, O. Scherzer [**745**] suggests a multilevel variant.

Gauss-Newton-type regularization. To compute s^ε_n, we apply a regularization procedure to (6.427) generated by a regularizing filter $\{F_t\}_{t>0}$, i.e.,

$$s^\varepsilon_n = F_{t_n}(A^*_nA_n)A^*_nb^\delta_n + (I + F_{t_n}(A^*_nA_n)A^*_nA_n)(f_0 - f^\varepsilon_n). \tag{6.432}$$

The second summand introduces the initial value $f_0 - f^\varepsilon_n$ into the regularization of (6.427). Altogether we obtain the iteration

$$f^\varepsilon_{n+1} = f_0 + F_{t_n}(A^*_nA_n)A^*_n(b^\delta_n - A_n(f_0 - f^\varepsilon_n)), \quad n \in \mathbb{N}_0, \tag{6.433}$$

which may be interpreted as a regularized version of a *Gauss-Newton iteration* (see, e.g., P. Deuflhard, A. Hohmann [**185**]).

Example. $F_t(\lambda) = (\lambda + t)^{-1}$: Using the example of the Tikhonov-Phillips filter $F_t(\lambda) = (\lambda+t)^{-1}$, we are able to calculate s^ε_n as the unique solution of the equation

$$(A^*_nA_n + t_nI)s^\varepsilon_n = A^*_nb^\delta_n + t_n(f_0 - f^\varepsilon_n) \tag{6.434}$$

An equivalent characterization of the iteration step is given by

$$s^\varepsilon_n = \operatorname{argmin}\{\|A_ns - b^\delta_n\|^2_Y + t_n\|s - (f_0 - f^\varepsilon_n)\|^2 : s \in X\}_v. \tag{6.435}$$

Example $F_t(\lambda) = \sum_{j=0}^{1/t-1}(1-\lambda)^j$. Iterative methods can also be used to determine s^ε_n. Thus, the choice of $F_t(\lambda) = \sum_{j=0}^{1/t-1}(1-\lambda)^j$ for $t^{-1} \in \mathbb{N}$ together with scaling $\|A_n\| \le 1$ the application of t_n^{-1} steps of the undamped Landweber method to (6.427) with initial value $f_0 - f^\varepsilon_n$:

$$\begin{aligned} s_{n,0} &= f_0 - f^\varepsilon_n, \\ s_{n,k+1} &= s_{n,k} + A^*_n(b^\delta_n - A_ns_{n,k}), \quad k = 0, 1, \dots t_n^{-1} - 1, \\ s^\varepsilon_n &= s_{n,t_n^{-1}}. \end{aligned} \tag{6.436}$$

The Gauss-Newton iteration (6.433) is terminated by the adapted discrepancy principle (6.437): stop with the iterated $f^\varepsilon_{n^*}$ if

$$\begin{aligned} &\max\left\{\|\Phi(f^\varepsilon_{n^*-1}) - g^\varepsilon\|, \|b^\delta_{n^*-1} - A_{n^*-1}s^\varepsilon_{n^*-1}\|\right\} \le \tau\varepsilon \\ &\quad < \max\left\{\|\Phi(f^\varepsilon_k) - g^\varepsilon\|, \|b^\delta_k - A_ks^\varepsilon_k\|\right\}, \quad k = 0, 1, \dots, n^* - 2. \end{aligned} \tag{6.437}$$

In the generality presented here the next statement was proved by B. Blaschke (-Kaltenbacher) [**486**]. Partial results were obtained earlier by A.B. Bakushinskii [**67**] and by B. Blaschke et al. [**98**].

Let $\{F_t\}_{t>0}$ be a regularizing filter with qualification $\mu_0 > 1$. Assume that

$$\sup\{|F_t(\lambda)| : 0 \le \lambda \le \overline{\lambda}\} = O(t^{-1}), \tag{6.438}$$

where $\overline{\lambda}$ is an upper bound for $\|\Phi'(\cdot)\|^2$ over $\mathbb{B}_\rho(f_0) \subset \mathcal{D}(\Phi)$. Further allow $f^\dagger \in \mathbb{B}_\rho(f_0)$ to represent

$$f^\dagger - f_0 = |\Phi'(f^\dagger)|^\mu \, w$$

for a $\mu \in [0, \overline{\mu}]$, $\overline{\mu} = \min\{1, \mu_0 - 1\}$, and an element $w \in X$. Let (6.403) *hold in $\mathbb{B}_{2\rho}(f_0)$ with $C_Q\ \rho < 1/2$. Let the sequence $\{t_n\}_{n\in\mathbb{N}_0}$ in* (6.433) *be a strictly monotonically decreasing zero sequence with $0 < \vartheta \le t_{n+1}/t_n$.*

If ρ and $\|w\|$ are sufficiently small and $\tau > 1$ is sufficiently large, then the Gauss-Newton iteration (6.433) *with the stopping rule* (6.437) *is well-defined, i.e. $\{f_0, \dots, f^\varepsilon_{n^*}\} \subset \mathbb{B}_\rho(f^\dagger)$, and we have*

$$\|f^\dagger - f^\varepsilon_{n^*}\| = O\left(\varepsilon^{\frac{\mu}{\mu+1}}\right) \quad \textit{for} \quad \varepsilon \to 0. \tag{6.439}$$

In a series of papers [**700**], [**701**], [**702**], Qi-Nian analyzed various variants and extensions of the Gauss-Newton iteration (6.433). P. Deuflhard et al. [**184**] transferred the affine-invariant theory of convergence of Newton's method for well-posed problems to the Gauss-Newton iteration.

Inexact Newton method. From the point of view of numerical efficiency, the Gauss-Newton iteration (6.433) has two weaknesses.

- Because of the convergence of $\{f^\varepsilon_{n*}\}_{\varepsilon>0}$ to $f^\dagger$, the sequence of Newton steps $\{s^\varepsilon_{n*}\}_{\varepsilon>0}$ must converge to zero for $\varepsilon \to 0$. Therefore, the initial value $f_0 - f^\varepsilon_{n*}$, which in general does not converge to zero for $\varepsilon \to 0$, is not very suitable for computing s^ε_{n*}. The initial value of zero seems to be more appropriate.
- The strength of the regularization of (6.427) for computing s^ε_n is fixed a priori by the default of $\{t_n\}$. A dynamic choice of $\{t_n\}$, on the other hand, allows us to take into account insights gained in the course of the iteration, thus allowing to increase the speed of convergence.

To meet both objections, A. Rieder [**723**] takes up the idea of inexact Newton procedures (due to C.T. Kelley [**492**]) and proposes an alternative iteration procedure, which, however, is not pursued further here.

6.15. Methodologies and dilemmas

Substantially we are able to distinguish *three aspects of regularization: (i) strategies of resolution and reconstruction, (ii) regularization-approximation schemes, (ii) regularization algorithms.* One of the purposes of our work is to dramatize this delineation with reference to specific methods and results.

The *strategy of resolution and reconstruction of ill-posed problems* involves one or more of the following intuitive ideas:

- change the notion of what is meant by a solution (e.g., ε-approximate solution: $\|Au - y\| \leq \varepsilon$, where $\varepsilon > 0$ is prescribed; quasi-solution: $\|Au - y\| \leq \|Ax - y\|$ for all $x \in \mathcal{M}$, a prescribed subset of the domain of A; least-squares solution of minimal norm, etc.),
- modify the operator equation or the problem itself,
- change the spaces and/or topologies,
- specify the type of involved noise.

The philosophy of resolution leads to the use of algebraic methods versus function space methods, statistical versus deterministic approaches, etc. By a regularization-approximation scheme we refer to a variety of methods such as Tikhonov's regularization, projection and filter methods, mollifier methods, multiscale methods, iterative approximation, etc., that can be applied to ill-posed problems. These schemes turn into algorithms once a resolution strategy can be effectively implemented. Unfortunately, this approach requires the determination of a suitable value of a certain parameter associated with the scheme (e.g., regularization parameter, mesh size, dimension of the subspace in the projection scheme, specification of the adequate level of a scale space, etc.). Even in their restriction to concrete obligations, this is not a trivial problem since it involves a trade-off between accuracy and numerical stability, a situation that does not usually arise in well-posed problems (guaranteeing, in accordance with the famous Hadamard approach, existence, uniqueness, and continuous dependence on the data).

By a *regularization-approximation scheme* we refer to a variety of methods such as Tikhonov's regularization, projection methods, multiscale methods, mollifier techniques, iterative approximation, etc., that can be applied to ill-posed problems. These schemes turn into algorithms once a resolution strategy can be effectively implemented. Unfortunately, this requires a determination of a suitable value of a certain parameter associated with the scheme (e.g., regularization parameter, mesh size, dimension of subspace in the projection scheme, specification of the level of a scale space, classification of noise, etc.). As already mentioned, this is not a trivial problem since it involves a trade-off between accuracy and numerical stability.

The *numerical analysis of all ill-posed problem* ultimately involves solutions of finite-dimensional problems in order to obtain numerical approximations. This fact often entails a *two-stage regularization.* One first may "regularize" the problem in function spaces and then apply numerical methods to approximate the solution of a well-posed problem (or a family of such problems). On the other hand, one may "discretize" or approximate the IPP by finite dimensional problems and then resolve numerical instability of these problems by methods of numerical Linear Algebra that are suitable for discrete ill-posed problems. Two alternate routes are represented by the following diagram:

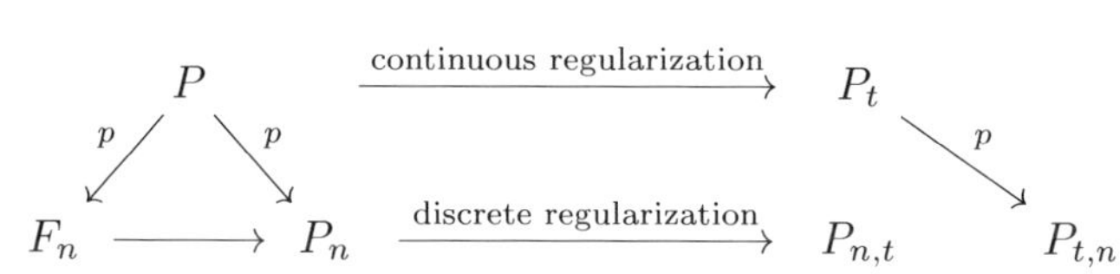

Here, P is a given ill-posed problem and P_t is a "regularized" version of P by use of a certain regularization scheme in a certain function space. F_n is a problem in a finite dimensional function space and P_n is a finite dimensional algebraic problem, both obtained by approximation of P; whereas $P_{n,t}$ denotes a "regularization" of the (numerically unstable) problem P_n and $P_{t,n}$ is the numerical approximation of the problem P_t, which is numerically stable for t not too small. In the diagram continuous regularization, discrete regularization are distinguished, and "p" denotes a generic "projection" (discretization, projection method, etc.).

At this stage, after having some knowledge of most of the regularization methods, we mention some procedures which provide concrete realizations of these schemes (see, e.g., M.Z. Nashed [**625**]):

$P \to P_t$	Tikhonov's regularization, quasi-reversibility methods, replacement of the IPP by a stable minimization problem depending on a parameter, iterative methods in function spaces;
$P \to F_n$	truncated singular value decompositions (TSVD), truncated series expansion, moment discretization, projection methods, (multiscale) regularizer operators in function spaces;
$P \to P_n$	finite difference method with collocation, reduction of F_n to algebraic equations;
$P_n \to P_{n,t}$	decomposition methods or regularization for linear algebraic equations, TSVD for matrices;
$P_\gamma \to P_{t,n}$	various numerical methods for solving well-posed problems, e.g., discretization, projection methods, multiscale procedures, etc.

Returning to the general scheme, we let $x, x_t, x_n, x_{t,n}, x_{n,t}$ denote, respectively, the "solution" (classical or least-square of minimal norm) of the problems $P, P_t, P_n, P_{t,n}, P_{n,t}$ in the absence of contamination in the data, and let $x^\varepsilon, x_t^\varepsilon$, etc., denote the corresponding solution when the data are contaminated (y is replaced by y^ε, where $\|y^\varepsilon - y\| \leq \varepsilon$ for some $\varepsilon > 0$). If we assume that the various *regularizer schemes* are convergent (e.g., $x_t \to x$ as $t \to 0$, $x_{n,t} \to x_n$ as $t \to 0$ for each fixed $n, x_n \to x$ as $n \to \infty$, etc.) in the absence of error, then estimates similar to the standard type studied earlier may be used to calculate an "approximate solution" x_ε as before. For example,

$$\|x_{t,n}^\varepsilon - x\| \leq \|x_{t,n}^\varepsilon - x_t^\varepsilon\| + \|x_t^\varepsilon - x_t\| + \|x_t - x\|. \tag{6.440}$$

Here, $\|x_{t,n}^\varepsilon - x_t^\varepsilon\|_X$ is an *approximation error estimate* that provides a rate of convergence of the approximation scheme for the well-posed problem P_t for a fixed t. $\|x_t^\varepsilon - x_t\|$ is an *estimate for the contamination error* which can be estimated if the robustness of P_t is known and $\|x_t - x\|$ is a *regularization error*. Similarly,

$$\|x_{n,t}^\varepsilon - x\| \leq \|x_{n,t}^\varepsilon - x_n^t\| + \|x_n^\varepsilon - x_n\| + \|x_n - x\|. \tag{6.441}$$

Note, however, that now $\|x_{n,t}^\varepsilon - x_n^t\|$ is an *error in the regularization of the problem* P_n, $\|x_n^\varepsilon - x_n\|$ is an *error due to the propagation of contamination into the discrete system*, etc.

The *dilemmas and methodologies of mathematical and numerical analysis of IPP* involve the following facts and observations:

(i) For most regularization and approximation schemes

$$\lim_{t\to 0} x_t = x \tag{6.442}$$

and

$$\lim_{n\to\infty} x_n = x \tag{6.443}$$

in the absence of contamination. For some approximation (projection) schemes, nonconvergence can take place even without contamination.

(ii) Neither the double limit

$$\lim_{\substack{t\to 0,\\ n\to\infty}} x^{\varepsilon}_{t,n}, \tag{6.444}$$

nor the iterated limits [$\lim_{n\to\infty}\lim_{t\to 0}$ and $\lim_{t\to 0}\lim_{n\to\infty}$] of $x^{\varepsilon}_{t,n}$ and $x^{\varepsilon}_{n,t}$ exist. In fact, $\|x^{\varepsilon}_{t,n}\|_X$ and $\|x^{\varepsilon}_{n,t}\|_X$ blow up as $n\to\infty$ and $t\to 0$. It should be emphasized that this *blow-up is intrinsically inherent in all IPP's regardless of any regularization-approximation scheme.* Thus, the best one can achieve for a numerical resolution of IPP is to minimize the error: $\|x - x^{\varepsilon}_{t,n}\|$ or $\|x - x^{\varepsilon}_{n,t}\|$, and to find "paths" along which $x_{t,n}$ and $x_{n,t}$ converge to x as $\varepsilon\to 0$.

(iii) The alternative routes diagram is *noncommutative*, in general. It is not always clear which path along the diagram is more effective. One has to compare the minimum errors of $\|x - x^{\varepsilon}_{t,n}\|_X$ and $\|x - x^{\varepsilon}_{n,t}\|_X$ and to take into consideration the computational complexity of the two paths. For some simple schemes, the diagram is commutative (e.g., Tikhonov regularization and TSVD for a compact operator commute).

(iv) Sharp resolutions of an ill-posed problem (i.e., an optimal compromise between accuracy and numerical stability) hing upon "optimal parameter choice criteria" (for t and n). Analytic criteria for this choice are often not available due to the lack of sharp rates of convergence in the preceding estimates, except for Tikhonov regularization or related methods based on simple variational principles. Often the parameter is chosen by an interactive computational scheme, based on rough analytic estimates.

Part 4

Regularization examples

CHAPTER 7

Regularization methodologies in geotechnology

In practice, the question arises, which regularization method seems to be suitable for the inverse problem to be treated? Usually, it is found out that with the straightforward application of exclusively one of the regularization methods, which we have outlined in the last chapters, the requirements of an acceptable physically meaningful approximation and a sufficiently stable "solution" cannot be satisfied. As a matter of fact, we have to adapt a regularization procedure to the specific physical and practical needs of the problem under consideration, so that mathematical modifications and extensions of the regularization methods presented earlier are indispensable. Often it is particularly helpful to combine conceptual ingredients of different regularization concepts and to connect them to a consistent whole, which is better adapted to the physical reality and the practical necessities of the concrete problem. This strategy will be demonstrated by three selected examples in geoengineering.

7.1. Potential theoretic auxiliaries

For the convenience of the reader we first recapitulate ordinary Cartesian and spherical nomenclature: Only if necessary we write $x_{(q)}$ instead of x to point out that x is an element of $\mathbb{R}^q$. The canonical orthonormal system in $\mathbb{R}^q$ is denoted by $\varepsilon^1, \ldots, \varepsilon^q$. Any element $x \in \mathbb{R}^q$ may be represented in Cartesian coordinates $x_i = x \cdot \varepsilon^i, i = 1, \ldots, q$, by $x = \sum_{i=1}^q (x \cdot \varepsilon^i)\, \varepsilon^i = \sum_{i=1}^q x_i\, \varepsilon^i$. In coordinates the *inner (scalar) product* of two elements $x, y \in \mathbb{R}^q$ is given by $x \cdot y = x^T y = \sum_{i=1}^q x_i y_i$. Clearly, $x^2 = |x|^2 = x \cdot x = x^T x$, $x \in \mathbb{R}^q$, i.e., *the Euclidean norm* in $\mathbb{R}^q$ is given by $|x| = \sqrt{x \cdot x} = \sqrt{x^T x}, x \in \mathbb{R}^q$.

The unit sphere in $\mathbb{R}^q$ is denoted by $\mathbb{S}^{q-1}$: $\mathbb{S}^{q-1} = \{x \in \mathbb{R}^q : |x| = 1\}$. Each $x \in \mathbb{R}^q$, $x = (x_1, \ldots, x_q)^{\mathrm{T}}$, $|x| \neq 0$, admits a representation of the form $x = r\xi$, $r = |x|$, $\xi = (\xi_1, \ldots, \xi_q)^{\mathrm{T}}$, where $\xi \in \mathbb{S}^{q-1}$ is the uniquely determined (unit) vector of x. Using the canonical orthonormal basis $\varepsilon^1, \ldots, \varepsilon^q$ in $\mathbb{R}^q$ (more accurately, $\varepsilon^1_{(q)}, \ldots, \varepsilon^q_{(q)}$ in $\mathbb{R}^q$) we are able to write $\xi_{(q)} \in \mathbb{S}^{q-1}$, $q \geq 3$, in the form

$$\xi_{(q)} = t\varepsilon^q + \sqrt{1-t^2}\, \xi_{(q-1)},\ t \in [-1,1],\ \xi_{(q-1)} \in \mathbb{S}^{q-2},\ t = \cos(\theta), \tag{7.1}$$

$$\xi_{(2)} = (\cos\varphi, \sin\varphi)^T,\ \xi_{(2)} \in \mathbb{S}^1,\ \varphi \in [0, 2\pi). \tag{7.2}$$

We have $\|\mathbb{S}^{q-1}\| = \int_{\mathbb{S}^{q-1}} dS_{(q-1)} = 2\ \frac{\pi^{\frac{q}{2}}}{\Gamma(\frac{q}{2})}$ (note that, for $q = 1$, $\|\mathbb{S}^0\| = 2$, is the "area" of the interval $\mathbb{S}^0 = [-1, 1]$). The *sphere in $\mathbb{R}^q$ with radius R around $y \in \mathbb{R}^q$* is denoted by $\mathbb{S}^{q-1}_R(y) = \{x \in \mathbb{R}^q : |x - y| = R\}$, and $\mathbb{S}^{q-1}_R$ is the sphere with radius R around 0 (i.e., $\mathbb{S}^{q-1}_R = \mathbb{S}^{q-1}_R(0)$). Furthermore, $\mathbb{B}^q_R(y)$ denotes *the ball in*

$\mathbb{R}^q$ *with center* $y \in \mathbb{R}^q$ *and radius* R: $\mathbb{B}^q_R(y) = \{x \in \mathbb{R}^q : |x - y| < R\}$. We simply write $\mathbb{B}^q_R$ for the ball with radius R around the origin 0. Moreover, we let $\mathbb{B}^q_1 = \mathbb{B}^q$.

By means of polar coordinates $x_{(q)} = r\xi_{(q)}$, $r = |x_{(q)}|$, $\xi_{(q)} \in \mathbb{S}^{q-1}$, the *gradient* ∇ in $\mathbb{R}^q$ can be represented in the form

$$\nabla_{x_{(q)}} = \xi_{(q)} \frac{\partial}{\partial r} + \frac{1}{r} \nabla^*_{\xi_{(q)}}, \tag{7.3}$$

where ∇^* is the *surface gradient* on $\mathbb{S}^{q-1}$. Moreover, in terms of spherical coordinates the *Laplace operator (Laplacian)* $\Delta = \nabla \cdot \nabla$ in $\mathbb{R}^q$ has the representation

$$\Delta_{x_{(q)}} = r^{1-q} \frac{\partial}{\partial r} r^{q-1} \frac{\partial}{\partial r} + \frac{1}{r^2} \Delta^*_{\xi_{(q)}}, \tag{7.4}$$

where Δ^* describes the *(Laplace-)Beltrami-operator* of the unit sphere $\mathbb{S}^{q-1}$ recursively given by

$$\Delta^*_{\xi_{(q)}} = (1 - t^2) \left(\frac{\partial}{\partial t} \right)^2 - (q-1) t \frac{\partial}{\partial t} + \frac{1}{1 - t^2} \Delta^*_{\xi_{(q-1)}}, \tag{7.5}$$

$$\Delta^*_{\xi_{(2)}} = \left(\frac{\partial}{\partial \varphi} \right)^2. \tag{7.6}$$

In analogy to the cartesian nomenclature, $\Delta^* = \nabla^* \cdot \nabla^*$, where $\nabla^* \cdot$ is the *surface divergence* on $\mathbb{S}^{q-1}$ (for more notational details see, e.g., [**326**]).

By a region $\mathcal{G} \subset \mathbb{R}^q$ we mean an open and connected set in $\mathbb{R}^q$. A bounded region $\mathcal{G} \subset \mathbb{R}^q$ is called a *regular region in* $\mathbb{R}^q$, if it is an open and connected set $\mathcal{G} \subset \mathbb{R}^q$, $q \geq 2$, for which

(i) its boundary $\partial \mathcal{G}$ constitutes an orientable, piecewise smooth Lipschitzian manifold of dimension $q - 1$,

(ii) the origin $0 \in \mathbb{R}^q$ is contained in $\mathcal{G}$,

(iii) $\mathcal{G}$ uniquely divides $\mathbb{R}^q$ into the "inner space" $\mathcal{G}$ and the "outer space" $\mathcal{G}^c = \mathbb{R}^q \backslash \overline{\mathcal{G}}$, $\overline{\mathcal{G}} = \mathcal{G} \cup \partial \mathcal{G}$

(note that, as geoscientifically relevant 3D regular regions, we may choose the (potato) Earth's body or parts of it, the interior of geoequipotential surfaces such as the geoid, telluroid, etc., but also ball, ellipsoid, cube, polyhedral bodies, etc. are included in accordance with the definition). The *volume of a regular region* $\mathcal{G} \subset \mathbb{R}^q$ is written by $\|\mathcal{G}\| = \int_{\mathcal{G}} dx$, where dx (or more accurately, $dx_{(q)}$) is the volume element. The *area of the boundary* $\partial \mathcal{G}$ *of a regular region* $\mathcal{G} \subset \mathbb{R}^q$ is given by $\|\partial \mathcal{G}\| = \int_{\partial \mathcal{G}} dS$, where dS (or more accurately, $dS_{(q-1)}$) is the surface element.

Next we start with the introduction of harmonic functions in regions $\mathcal{G}$ of Euclidean space $\mathbb{R}^q, q \geq 2$. For more details about potential theory, the reader is referred, e.g., to O.D. Kellogg [**493**], N. Aronszajn, T.M. Creese, L.J. Lipkin [**48**], E. Martensen [**558**], L.L. Helms [**424**], W. Walter [**853**], W. Freeden, C. Gerhards [**274**], and the references therein:

$U \in C^{(2)}(\mathcal{G})$ is called a *harmonic function* (sometimes also called a *harmonic*) in a region $\mathcal{G} \subset \mathbb{R}^q$, if it satisfies the Laplace equation

$$\Delta U(x) = \sum_{i=1}^{q} \left(\frac{\partial}{\partial x_i} \right)^2 U(x_1, \ldots, x_q) = 0 \ , \ x = (x_1, \ldots, x_q)^T \in \mathcal{G}. \tag{7.7}$$

$U \in C^{(2m)}(\mathcal{G})$, $m \in \mathbb{N}$, is called a *polyharmonic function of degree* m in $\mathcal{G} \subset \mathbb{R}^q$, if

$$\Delta^m U(x) = 0, \quad x \in \mathcal{G}. \tag{7.8}$$

A regular region $\mathcal{G} \subset \mathbb{R}^q$ allows the *Gauss Theorem*

$$\int_{\mathcal{G}} \nabla \cdot f(x)\, dx = \int_{\partial\mathcal{G}} f(x) \cdot \nu(x)\, dS(x) \tag{7.9}$$

for all continuously differentiable vector fields f on $\overline{\mathcal{G}}$ (throughout this work, ν is the (unit) normal field on $\partial\mathcal{G}$ directed to the exterior of $\overline{\mathcal{G}}$). By letting $f = \nabla F, F \in C^{(2)}(\overline{\mathcal{G}})$, we obtain from (7.9)

$$\int_{\mathcal{G}} \Delta F(x)\, dx = \int_{\partial\mathcal{G}} \frac{\partial F}{\partial \nu}(x)\, dS(x), \tag{7.10}$$

where $\frac{\partial}{\partial \nu}$ denotes the derivative in the direction of the outer (unit) normal field ν. Consequently, for all functions $F \in C^{(2)}(\overline{\mathcal{G}})$ satisfying the Laplace equation $\Delta F = 0$ in $\mathcal{G}$, we have

$$\int_{\partial\mathcal{G}} \frac{\partial F}{\partial \nu}(x)\, dS(x) = 0. \tag{7.11}$$

For all vector fields $f = F\ \nabla G$, $F \in C^{(1)}(\overline{\mathcal{G}})$, $G \in C^{(2)}(\overline{\mathcal{G}})$, we obtain from the Gauss Theorem the following Green theorems:

First Green Theorem. Suppose that $\mathcal{G} \subset \mathbb{R}^q$ is a regular region. For $F \in C^{(1)}(\overline{\mathcal{G}}), G \in C^{(2)}(\overline{\mathcal{G}})$ we have

$$\int_{\mathcal{G}} \{F(x)\Delta G(x) + \nabla F(x) \cdot \nabla G(x)\}\, dx = \int_{\partial\mathcal{G}} F(x) \frac{\partial G}{\partial \nu}(x)\, dS(x).$$

Taking $f = F\nabla G - G\nabla F$ with $F, G \in C^{(2)}(\overline{\mathcal{G}})$ we obtain

Second Green Theorem. Suppose that $\mathcal{G} \subset \mathbb{R}^q$ is a regular region. For $F, G \in C^{(2)}(\overline{\mathcal{G}})$ we have

$$\int_{\mathcal{G}} \{G(x)\Delta F(x) - F(x)\Delta G(x)\}\ dx = \int_{\partial\mathcal{G}} \left\{ G(x)\frac{\partial F}{\partial \nu}(x) - F(x)\frac{\partial G}{\partial \nu}(x) \right\}\ dS(x).$$

Finally, we mention an extension of the Second Green Theorem to Helmholtz operators $\Delta + \lambda, \lambda \in \mathbb{R}$, i.e., a transition to the metaharmonic context:

Extended Second Green Theorem. For any regular region $\mathcal{G} \subset \mathbb{R}^q$, $q \geq 2$, and for $F \in C^{(2m)}(\overline{\mathcal{G}})$, $m \in \mathbb{N}$, we have

$$\begin{aligned} \int_{\mathcal{G}} G(x)(\Delta + \lambda)^m F(x)\, dx &= \int_{\mathcal{G}} F(x)(\Delta + \lambda)^m G(x)\, dx \\ &+ \sum_{r=0}^{m-1} \int_{\partial\mathcal{G}} \Big(\frac{\partial}{\partial \nu}(\Delta + \lambda)^r F(x)\Big)\Big((\Delta + \lambda)^{m-(r+1)} G(x)\Big) dS(x) \\ &- \sum_{r=0}^{m-1} \int_{\partial\mathcal{G}} \Big((\Delta + \lambda)^r F(x)\Big)\Big(\frac{\partial}{\partial \nu}(\Delta + \lambda)^{m-(r+1)} G(x)\Big)\ dS(x). \end{aligned} \tag{7.12}$$

Harmonics. Next a survey about the theory of homogeneous harmonic polynomials and spherical harmonics of dimension q is given including the addition theorem, the Funk-Hecke formula, and the closure and completeness theorems (cf. C. Müller [**605, 607**], E.M. Stein, G. Weiss [**790**]). Our approach is self-contained. It consistently contains all essential properties (without proofs). Evidently - in the form as presented here - most of the material is known from the literature (see, e.g., W. Freeden [**252**], W. Freeden, C. Gerhards [**274**], W. Freeden, M. Gutting [**277**], W. Freeden, M. Schreiner [**321, 326**] and the references therein).

Let $Hom_n(\mathbb{R}^q)$ consist of all polynomials $H_n(q;\cdot)$ in q variables which are homogeneous of degree n (i.e., $H_n(q;\tau x) = \tau^n H_n(q;x)$ for all $\tau \in \mathbb{R}$ and all $x \in \mathbb{R}^q$). If $H_n(q;\cdot)$ is a member of the class $Hom_n(\mathbb{R}^q)$, then there exist complex numbers $C_\alpha = C_{\alpha_1,\ldots,\alpha_q}$ such that $H_n(q;x) = \sum_{[\alpha]=n} C_\alpha x^\alpha$. In Cartesian coordinates $x = (x_1,\ldots,x_q)^T$,

$$H_n(q;x_1,\ldots,x_q) = \sum_{\alpha_1+\ldots+\alpha_q=n} C_{\alpha_1,\ldots,\alpha_q} x_1^{\alpha_1}\ldots x_q^{\alpha_q}. \tag{7.13}$$

It is obvious that the set of monomials $x \mapsto x^\alpha$, $[\alpha] = n$, $x \in \mathbb{R}^q$, is a basis of the space $Hom_n(\mathbb{R}^q)$. The number of such monomials is precisely the number of ways a q–tuple can be chosen so that we have $[\alpha] = n$, i.e., the number of ways selecting $q-1$ elements out of a collection of $n+q-1$. This means that the dimension $\dim(Hom_n(\mathbb{R}^q))$ of $Hom_n(\mathbb{R}^q)$ is equal to

$$M(q;n) = \dim(Hom_n(\mathbb{R}^q)) = \begin{pmatrix} n+q-1 \\ q-1 \end{pmatrix} = \begin{pmatrix} n+q-1 \\ n \end{pmatrix} = \frac{\Gamma(n+q)}{\Gamma(q)\Gamma(n+1)}. \tag{7.14}$$

Let $H_n(q;\nabla_x)$ be the differential operator associated to $H_n(q;x)$ (i.e., replace x^α formally by $(\nabla_x)^\alpha$ in the expression of $H_n(q;x)$): $H_n(q;\nabla_x) = \sum_{[\alpha]=n} C_\alpha (\nabla_x)^\alpha$. If such an operator is applied to a homogeneous polynomial $U_n(q;\cdot)$ of the same degree $U_n(q;x) = \sum_{[\beta]=n} D_\beta x^\beta$, we obtain as result a number

$$H_n(q;\nabla_x)\overline{U_n(q;x)} = \sum_{[\alpha]=n} C_\alpha \overline{D_\alpha}\ \alpha!. \tag{7.15}$$

Clearly, we have $H_n(q;\nabla_x)\overline{U_n(q;x)} = \overline{U_n(q;\nabla_x)\overline{H_n(q;x)}}$, $H_n(q;\nabla_x)\overline{H_n(q;x)} \geq 0$. This enables us to introduce an inner product $(\cdot,\cdot)_{Hom_n(\mathbb{R}^q)}$ on the space $Hom_n(\mathbb{R}^q)$ by letting (cf. [**790**])

$$\langle H_n(q;\cdot), U_n(q;\cdot)\rangle_{Hom_n(\mathbb{R}^q)} = H_n(q;\nabla_x)\overline{U_n(q;x)}. \tag{7.16}$$

The space $Hom_n(\mathbb{R}^q)$ equipped with the inner product $\langle\cdot,\cdot\rangle_{Hom_n(\mathbb{R}^q)}$ is a finite-dimensional Hilbert space. The set of monomials $\{x \mapsto (\alpha!)^{-\frac{1}{2}} x^\alpha : [\alpha] = n\}$ forms an orthonormal system in the space $Hom_n(\mathbb{R}^q)$. For each $H_n(q;\cdot) \in Hom_n(\mathbb{R}^q)$ it follows in connection with the binomial theorem that

$$H_n(q;x) = \Big(H_n(q;\nabla_y)\Big)\frac{(x\cdot y)^n}{n!} = \frac{1}{n!}(x\cdot\nabla_y)^\alpha\ \overline{H_n(q;y)}. \tag{7.17}$$

This shows that $Hom_n(\mathbb{R}^q)$ equipped with the inner product $\langle\cdot,\cdot\rangle_{Hom_n(\mathbb{R}^q)}$ is an *$M(q;n)$-dimensional Hilbert space with the reproducing kernel*

$$K_{Hom_n(\mathbb{R}^q)}(x,y) = \frac{(x\cdot y)^n}{\Gamma(n+1)},\ x,y \in \mathbb{R}^q, \tag{7.18}$$

i.e., $K_{Hom_n(\mathbb{R}^q)}(x,\cdot) \in Hom_n(\mathbb{R}^q)$ for each $x \in \mathbb{R}^q$ and

$$\langle K_{Hom_n(\mathbb{R}^q)}(\cdot,x), H_n(q;\cdot)\rangle_{Hom_n(\mathbb{R}^q)} = H_n(q;x) \tag{7.19}$$

for each $x \in \mathbb{R}^q$ and each $H_n(q;\cdot) \in Hom_n(\mathbb{R}^q)$.

Let $H_{n,1}(q;\cdot),\ldots,H_{n,M(q;n)}(q;\cdot)$ be an orthonormal system in $Hom_n(\mathbb{R}^q)$ with respect to $\langle\cdot,\cdot\rangle_{Hom_n(\mathbb{R}^q)}$. Then each $H_n(q;\cdot) \in Hom_n(\mathbb{R}^q)$ can be represented in the form

$$H_n(q;\cdot) = \sum_{j=1}^{M(q;n)} \Big\langle H_n(q;\cdot), H_{n,j}(q;\cdot)\Big\rangle_{Hom_n(\mathbb{R}^q)} H_{n,j}(q;\cdot), \tag{7.20}$$

where

$$K_{Hom_n(\mathbb{R}^q)}(x,y) = \sum_{j=1}^{M(q;n)} H_{n,j}(q;x)\,\overline{H_{n,j}(q;y)},\ \ x,y \in \mathbb{R}^q. \tag{7.21}$$

In other words, we are led to the *addition theorem for homogeneous polynomials.*

Let $H_{n,1}(q;\cdot),\ldots,H_{n,M(q;n)}(q;\cdot)$ be an orthonormal system in $Hom_n(\mathbb{R}^q)$ with respect to $\langle\cdot,\cdot\rangle_{Hom_n(\mathbb{R}^q)}$. Then, for all $x,y \in \mathbb{R}^q$,

$$\frac{(x\cdot y)^n}{n!} = \sum_{j=1}^{M(q;n)} H_{n,j}(q;x)\,\overline{H_{n,j}(q;y)}. \tag{7.22}$$

Suppose that there are given $M(q;n)$ points $x_1,\ldots,x_{M(q;n)} \in \mathbb{R}^q$ such that

$$\sum_{k=1}^{M(q;n)} b_k \overline{H_{n,j}(q;x_k)} = \Big\langle H_n(q;\cdot), H_{n,j}(q;\cdot)\Big\rangle_{Hom_n(\mathbb{R}^q)}, \tag{7.23}$$

for $H_n(q;\cdot) \in Hom_n(\mathbb{R}^q)$ and $j = 1,\ldots,M(q;n)$. Then

$$H_n(q;x) = \sum_{k=1}^{M(q;n)} \sum_{j=1}^{M(q;n)} b_k\,\overline{H_{n,j}(q;x_k)}\,H_{n,j}(q;x),\ \ x \in \mathbb{R}^q. \tag{7.24}$$

This yields

$$H_n(q;x) = \sum_{k=1}^{M(q;n)} b_k\,\frac{(x\cdot x_k)^n}{n!},\ \ x \in \mathbb{R}^q. \tag{7.25}$$

Let $Harm_n(\mathbb{R}^q)$ be the class of all polynomials in $Hom_n(\mathbb{R}^q)$ that are harmonic:

$$Harm_n(\mathbb{R}^q) = \{H_n(q;\cdot) \in Hom_n(\mathbb{R}^q) :\ \Delta_x H_n(q;x) = 0,\ x \in \mathbb{R}^q\}. \tag{7.26}$$

For $n < 2$, of course, all homogeneous polynomials are harmonic. Assume that n is an integer with $n \geq 2$. Let $H_{n-2}(q;\cdot)$ be a homogeneous polynomial of degree $n-2$, i.e., $H_{n-2}(q;\cdot) \in Hom_{n-2}(\mathbb{R}^q)$. Then, for each homogeneous harmonic polynomial $K_n(q;\cdot)$, we have

$$\langle |\cdot|^2 H_{n-2}(q;\cdot), K_n(q,\cdot)\rangle_{Hom_n(\mathbb{R}^q)} = H_{n-2}(\nabla_x)\Delta_x \overline{K_n(q;x)} = 0. \tag{7.27}$$

This means $|\cdot|^2 H_{n-2}(q;\cdot)$ is orthogonal to $K_n(q;\cdot)$ in the sense of the inner product $\langle\cdot,\cdot\rangle_{Hom_n(\mathbb{R}^q)}$. Conversely, suppose that $K_n(q;\cdot) \in Harm_n(\mathbb{R}^q)$ is orthogonal to all L_n of the form

$$L_n(x) = |x|^2 H_{n-2}(q;x),\ H_{n-2}(q;\cdot) \in Hom_{n-2}(\mathbb{R}^q),\ x \in \mathbb{R}^q. \tag{7.28}$$

Then it follows that

$$0 = \left\langle |\cdot|^2 H_{n-2}(q;\cdot), K_n(q;\cdot)\right\rangle_{Hom_n(\mathbb{R}^q)} = \left\langle H_{n-2}(q;\cdot), \Delta K_n(q;\cdot)\right\rangle_{Hom_{n-2}(\mathbb{R}^q)} \tag{7.29}$$

for all $H_{n-2}(q;\cdot) \in Hom_{n-2}(\mathbb{R}^q)$. This is true if $\Delta K_n(q;\cdot) = 0$; i.e, $K_n(q;\cdot)$ is a homogeneous harmonic polynomial. So, $Hom_n(\mathbb{R}^q)$, $n \geq 2$, is the orthogonal direct sum of $Harm_n(\mathbb{R}^q)$ and $Harm_n^{\perp}(\mathbb{R}^q)$, where $Harm_n^{\perp}(\mathbb{R}^q) = |\cdot|^2 Hom_{n-2}(\mathbb{R}^q)$. Furthermore, the dimension $N(q;n)$ of $Harm_n(\mathbb{R}^q)$ can be determined as follows:

$$N(q;n) = \dim(Harm_n(\mathbb{R}^q)) = \dim(Hom_n(\mathbb{R}^q)) - \dim(Hom_{n-2}(\mathbb{R}^q)), \tag{7.30}$$

such that

$$N(q;n) = \frac{\Gamma(n+q)}{\Gamma(q)\Gamma(n+1)} - \frac{\Gamma(n+q-2)}{\Gamma(q)\Gamma(n-1)} = \frac{(2n+q-2)\Gamma(n+q-2)}{\Gamma(n+1)\Gamma(q-1)} = O\left(n^{q-2}\right).$$

Each homogeneous polynomial of degree n can be uniquely decomposed in the form

$$H_n(q;x) = \sum_{j=0}^{\lfloor \frac{n}{2} \rfloor} |x|^{2j} K_{n-2j}(q;x),\ x \in \mathbb{R}^q, \tag{7.31}$$

$K_{n-2j}(q;\cdot) \in Harm_{n-2j}(\mathbb{R}^q)$, where $\lfloor \frac{n}{2} \rfloor$ is the largest integer which is less or equal than $\frac{n}{2}$.

We are now interested in explicitly giving the projection $P_{Harm_n(\mathbb{R}^q)} H_n(q;\cdot)$ of a homogeneous polynomial $H_n(q;\cdot)$ to $Harm_n(\mathbb{R}^q)$. We restrict ourselves to dimensions $q \geq 3$. Induction states that

$$\begin{aligned}(y\cdot\nabla_x)^n &\frac{1}{|x|^{q-2}} \\ &= \frac{(-1)^n}{|x|^{q+2n-2}} \frac{1}{\Gamma\left(\frac{q-2}{2}\right)} \left(\sum_{m=0}^{\lfloor \frac{n}{2} \rfloor} \frac{(-1)^m \Gamma\left(\frac{q-2}{2}+n-m\right)}{\Gamma(m+1)} 2^{n-2m} |x|^{2m} \Delta^m \right) (y\cdot x)^n\end{aligned} \tag{7.32}$$

is valid for every $y \in \mathbb{R}^q$. Now, as we have seen in (7.25), every $H_n(q;\cdot) \in Hom_n(\mathbb{R}^q)$ can be represented for (suitable) $x_1, \ldots, x_{M(q;n)}$ as linear combination in terms of the functions $x \mapsto (x_i \cdot x)^n$, $x \in \mathbb{R}^q$, $i = 1, \ldots, M(q,n)$. Thus, we are led to the following conclusion: For all $H_n \in Hom_n(\mathbb{R}^q)$, $q \geq 3$, and all $x \in \mathbb{R}^q$, $|x| \neq 0$, we have

$$\begin{aligned}H_n(q;\nabla_x) &\frac{1}{|x|^{q-2}} \\ &= \frac{(-1)^n}{|x|^{q+2n-2}} \frac{2^n}{\Gamma\left(\frac{q-2}{2}\right)} \left(\sum_{m=0}^{\lfloor \frac{n}{2} \rfloor} \frac{(-1)^m \Gamma\left(\frac{q-2}{2}+n-m\right)}{\Gamma(m+1)} 2^{-2m} |x|^{2m} \Delta^m \right) H_n(q;x).\end{aligned} \tag{7.33}$$

Using the decomposition $H_n(q;x) = K_n(q;x) + |x|^2 H_{n-2}(q;x),\ x \in \mathbb{R}^q$, with $K_n(q;\cdot) \in Harm_n(\mathbb{R}^q),\ H_{n-2}(q;\cdot) \in Hom_{n-2}(\mathbb{R}^q)$, we find for $|x| \neq 0$

$$\begin{aligned}(7.34)\qquad \Big(H_n(q;\nabla_x)\Big)\frac{1}{|x|^{q-2}} &= K_n(q;\nabla_x)\frac{1}{|x|^{q-2}} \\ &= (-1)^n |x|^{2-2n-q} \frac{2^n \Gamma\left(\frac{q-2+2n}{2}\right)}{\Gamma\left(\frac{q-2}{2}\right)} K_n(q;x).\end{aligned}$$

This shows that, for $H_n(q;\cdot) \in Hom_n(\mathbb{R}^q)$,
(7.35)

$$P_{Harm_n(\mathbb{R}^q)}(H_n(q;\cdot)) = \sum_{m=0}^{\lfloor \frac{n}{2} \rfloor} (-1)^m \frac{2^{-2m}\Gamma\left(\frac{q-2}{2}+n-m\right)}{\Gamma\left(\frac{q-2}{2}+n\right)\Gamma(m+1)} |x|^{2m} \Delta^m H_n(q;x).$$

Observing the identity

$$(7.36)\qquad \Delta_x^k \frac{(x\cdot y)^n}{\Gamma(n+1)} = \frac{1}{\Gamma(n-2k+1)} |y|^{2k} (x\cdot y)^{n-2k},\ k \geq 0,$$

we get for $|x| \neq 0,\ |y| \neq 0$

$$\begin{aligned}(7.37)\quad & P_{Harm_n(\mathbb{R}^q)}\left(\frac{(x\cdot y)^n}{\Gamma(n+1)}\right) \\ &= \sum_{m=0}^{\lfloor \frac{n}{2} \rfloor} \frac{(-1)^m 2^{-2m}\Gamma\left(\frac{q-2}{2}+n-m\right)}{\Gamma\left(\frac{q}{2}+n-1\right)\Gamma(m+1)\Gamma(n-2m+1)} |x|^{2m}|y|^{2m}(x\cdot y)^{n-2m}.\end{aligned}$$

Suppose that $\{K_{n,j}(q;\cdot)\}_{j=1,\ldots,N(q;n)}$ is an orthonormal system in $Hom_n(\mathbb{R}^q)$ with respect to $\langle\cdot,\cdot\rangle_{Hom_n(\mathbb{R}^q)}$. Let $\{L_{n,j}(q;\cdot)\}_{j=1,\ldots,M(q;n)-N(q;n)}$ be an orthonormal system in $Harm_n^{\perp}(\mathbb{R}^q)$. Then the union of both systems forms an orthonormal system in $Hom_n(\mathbb{R}^q)$. Therefore we are able to deduce that

$$(7.38)\qquad P_{Harm_n(\mathbb{R}^q)}\left(\frac{(x\cdot y)^n}{\Gamma(n+1)}\right) = \sum_{j=1}^{N(q;n)} K_{n,j}(q;x)\,\overline{K_{n,j}(q;y)}.$$

By comparison of (7.37) and (7.38) we obtain the *addition theorem of homogeneous harmonic polynomials in* $\mathbb{R}^q$.

Let $\{K_{n,j}(q;\cdot)\}_{j=1,\ldots,N(q;n)}$ *be an orthonormal system in the space* $Harm_n(\mathbb{R}^q)$ *with respect to inner product* $\langle\cdot,\cdot\rangle_{Hom_n(\mathbb{R}^q)}$. *Then, for all* $x,y \in \mathbb{R}^q,\ q \geq 3,\ x = |x|\xi,\ y = |y|\eta;\ \xi,\eta \in \mathbb{S}^{q-1}$, *we have*

$$(7.39)\qquad \sum_{j=1}^{N(q;n)} K_{n,j}(q;x)\,\overline{K_{n,j}(q;y)} = \frac{\Gamma\left(\frac{q}{2}\right)|x|^n|y|^n}{\Gamma\left(\frac{q}{2}+n\right)2^n} N(q;n) P_n(q,\xi\cdot\eta),$$

where we have used the abbreviation
(7.40)

$$P_n(q;t) = \frac{\Gamma(n+1)\Gamma(q-2)}{\Gamma(n+q-2)\Gamma\left(\frac{q-2}{2}\right)} \sum_{m=0}^{\lfloor \frac{n}{2} \rfloor} \frac{(-1)^m\Gamma\left(\frac{q-2}{2}+n-m\right)}{\Gamma(m+1)\Gamma(n-2m+1)} (2t)^{n-2m},\ t \in [-1,1].$$

Next we discuss the important problem of how, for any pair of elements $H_n(q;\cdot) \in Harm_n(\mathbb{R}^q), K_n(q;\cdot) \in Harm_n(\mathbb{R}^q)$, the inner product $\langle\cdot,\cdot\rangle_{Hom_n(\mathbb{R}^q)}$ is related to the (usual) inner product $\langle\cdot,\cdot\rangle_{L^2(\mathbb{S}^{q-1})}$. In more detail, for $H_m(q;\cdot) \in Harm_m(\mathbb{R}^q)$, $K_n(q;\cdot) \in Harm_n(\mathbb{R}^q)$

$$\big\langle H_m(q;\cdot), K_n(q;\cdot)\big\rangle_{L^2(\mathbb{S}^{q-1})} = \frac{\delta_{mn}}{\mu_n(q)} \big\langle H_n(q;\cdot), K_n(q;\cdot)\big\rangle_{Hom_n(\mathbb{R}^q)}, \tag{7.41}$$

where $\mu_n(q)$ is given by

$$\mu_n(q) = \frac{2^n\Gamma\left(\frac{q}{2}+n\right)}{\|\mathbb{S}^{q-1}\|\Gamma\left(\frac{q}{2}\right)}. \tag{7.42}$$

We are led to the following reformulation of the *addition theorem for homogeneous harmonic polynomials.*

Let $K_{n,j}(q;\cdot) \in Harm_n(\mathbb{R}^q)$, $j = 1,\dots,N(q;n)$, be an $L^2(\mathbb{S}^{q-1})$-orthonormal system, i.e.,

$$\int_{\mathbb{S}^{q-1}} K_{n,j}(q;x)\, \overline{K_{r,s}(q;x)}\, dS_{(q-1)}(x) = \delta_{nr}\delta_{js}. \tag{7.43}$$

Then, for $x, y \in \mathbb{R}^q, x = |x|\xi,\ y = |y|\eta,\ \xi,\eta \in \mathbb{S}^{q-1}$, we have

$$\sum_{j=1}^{N(q;n)} K_{n,j}(q;x)\, \overline{K_{n,j}(q;y)} = \frac{N(q;n)}{\|\mathbb{S}^{q-1}\|} |x|^n |y|^n P_n(q, \xi\cdot\eta), \tag{7.44}$$

where $P_n(q;\cdot)$ is defined by (7.40).

To any orthonormal system $\{K_{n,j}(q;\cdot)\}_{j=1,\dots,N(q;n)}$ with respect to the inner product $\langle\cdot,\cdot\rangle_{Hom_n(\mathbb{R}^q)}$, there corresponds the system

$$\left\{\sqrt{\mu_n(q)}\; K_{n,j}(q;\cdot)\right\}_{j=1,\dots,N(q;n)}$$

orthogonal with respect to $\langle\cdot,\cdot\rangle_{L^2(\mathbb{S}^{q-1})}$, and vice versa.

W. Freeden, R. Reuter [**309**], W. Freeden, M. Schreiner [**321, 326**] use the $Hom_n(\mathbb{R}^q)$-topology to generate homogeneous harmonics polynomials exactly in integer mode.

Legendre polynomials. The function $P_n(q;\cdot) : t \mapsto P_n(q;t),\ t \in [-1,+1],\ n = 0,1,\dots,$ occuring in the addition theorem for homogeneous harmonic polynomials of degree n in q-dimension (see (7.40)) is known as the *Legendre polynomial of degree n and dimension q*. It is easily seen that $P_n(q;\cdot)$ satisfies the following properties:

(i) $P_n(q;\cdot)$ is a polynomial of degree n, i.e.,

$$P_n(q;t) \;=\; \alpha_n(q)\, t^n \;+\; \dots\;, \tag{7.45}$$

where

$$\alpha_n(q) \;=\; \frac{2^n\Gamma(n+\frac{q}{2})}{n!\Gamma(\frac{q}{2})N(q;n))} \;=\; \frac{n!}{2^n}\frac{\Gamma(\frac{q-1}{2})}{\Gamma(\frac{2n+q-1}{2})}\binom{2n+q-3}{n}. \tag{7.46}$$

(ii) $P_n(q;1) = 1$,

(iii) $\int_{-1}^{+1} P_n(q;t)P_m(q;t)(1-t^2)^{\frac{q-3}{2}}\, dt = 0, \quad n \neq m.$

$P_n(q;\cdot)$ is uniquely determined by the properties (i), (ii), (iii).

Applying the Cauchy-Schwarz inequality to the addition theorem, we obtain for $x, y \in \mathbb{R}^q$

$$\begin{aligned} \frac{N(q;n)}{\|\mathbb{S}^{q-1}\|}|x|^n|y|^n\,|P_n(q;\xi\cdot\eta)| &= \left|\sum_{j=1}^{N(q;n)} K_{n,j}(q;x)K_{n,j}(q;y)\right| \\ &\le \sqrt{\sum_{j=1}^{N(q;n)} |K_{n,j}(q;x)|^2}\sqrt{\sum_{j=1}^{N(q;n)} |K_{n,j}(q;y)|^2} \\ &= \frac{N(q;n)}{\|\mathbb{S}^{(q-1)}\|}|x|^n|y|^n P_n(q;1). \end{aligned} \tag{7.47}$$

Consequently, we have

$$|P_n(q;t)| \le P_n(q;1) = 1, \quad t \in [-1,+1]. \tag{7.48}$$

By elementary calculations we find

$$\left((1-t^2)\left(\frac{d}{dt}\right)^2 - (q-1)t\frac{d}{dt} + n(n+q-2)\right)P_n(q;t) = 0. \tag{7.49}$$

More concretely, $P_n(q;\cdot)$ is the only twice continuously differentiable eigenfunction of the " Legendre operator" given by

$$\mathrm{L}_t = (1-t^2)\left(\frac{d}{dt}\right)^2 - (q-1)t\frac{d}{dt} \tag{7.50}$$

on the interval $[-1,+1]$, corresponding to the eigenvalues $\mathrm{L}^\wedge(n)$ $(= \mathrm{L}^\wedge_{[-1,1]}(n)) = n(n+q-2), n = 0,1,\ldots$, that is bounded on $[-1,+1]$ with $P_n(q;1) = 1$.

The Legendre polynomial can be expressed by the so-called *Rodrigues formula* in the form

$$P_n(q;t) = (-1)^n \frac{\Gamma\left(\frac{q-1}{2}\right)}{2^n\Gamma\left(\frac{q-1}{2}+n\right)}(1-t^2)^{\frac{3-q}{2}}\left(\frac{d}{dt}\right)^n (1-t^2)^{n+\frac{q-3}{2}}. \tag{7.51}$$

The polynomial $P_n(q;\cdot)$ has exactly n different zeros in $(-1,+1)$.

Legendre polynomials satisfy a large number of recursion formulas. We mention only the following relations:

(i)

$$(q-1)P_n'(q;t) = n(n+q-2)P_{n-1}(q+2;t), \tag{7.52}$$

(ii)

$$(n+q-2)P_{n+1}(q;t) - (2n+q-2)tP_n(q;t) + nP_{n-1}(q;t) = 0, \tag{7.53}$$

(iii)

$$(2n+q-2)(1-t^2)P_n(q+2;t) = (q-1)\left(P_{n-1}(q;t) - P_{n+1}(q;t)\right), \tag{7.54}$$

(iv)

$$\left(\frac{d}{dt}\right)^l P_n(q;t) = \frac{2^l\Gamma\left(l+\frac{q}{2}\right)}{\Gamma\left(\frac{q}{2}\right)}\frac{N(q+2l;n-l)}{N(q;n)}P_{n-l}(q+2l;t). \tag{7.55}$$

In connection with the Rodrigues formula (7.51) we get by partial integration the *Rodrigues rule*, which is a helpful tool in the theory of Bessel functions (see Section 8.3): Let F be n times continuously differentiable on $[-1,+1]$. Then we have

(7.56)
$$\int_{-1}^{+1} F(t)P_n(q;t)(1-t^2)^{\frac{q-3}{2}}\, dt = \frac{\Gamma\left(\frac{q-1}{2}\right)}{2^n\Gamma\left(\frac{q-1}{2}+n\right)} \int_{-1}^{+1} (1-t^2)^{n+\frac{q-3}{2}} F^{(n)}(t)\, dt.$$

For $t \in [-1,1]$ and $|r|<1$, $q \geq 3$ we have (see, e.g., W. Freeden, M. Gutting [**277**])

(7.57)
$$\sum_{n=0}^{\infty} \frac{\Gamma(n+q-2)}{\Gamma(n+1)\Gamma(q-2)} P_n(q;t) r^n = \frac{1}{(1+r^2-2rt)^{\frac{q-2}{2}}}.$$

The two-dimensional analogue of (7.57) is given under obvious assumptions

(7.58)
$$\frac{1}{2}\ln\left(1-2rt+r^2\right) = -\sum_{n=1}^{\infty} \frac{1}{n} P_n(2;t)\, r^n.$$

As a special case we obtain with $x=|x|\xi$, $y=|y|\eta$, $\xi,\eta \in \mathbb{S}^{q-1}$, $q \geq 3$,

(7.59)
$$\frac{1}{|x-y|^{q-2}} = \frac{1}{|y|\left(1+\left(\frac{|x|}{|y|}\right)^2 - 2\frac{|x|}{|y|}\xi\cdot\eta\right)^{\frac{q-2}{2}}}$$
$$= \frac{1}{|y|}\sum_{n=0}^{\infty} \frac{\Gamma(n+q-2)}{\Gamma(n+1)\Gamma(q-2)} P_n(q;\xi\cdot\eta)\left(\frac{|x|}{|y|}\right)^n$$

provided that $|x|<|y|$. Moreover, we know that

(7.60)
$$|x-y|^{2-q} = \sum_{n=0}^{\infty} \frac{(-1)^n}{n!} |x|^n (\xi\cdot\nabla_y)^n |y|^{2-q}.$$

By comparison we therefore obtain the so-called *Maxwell representation*

(7.61)
$$(\xi\cdot\nabla_y)^n |y|^{2-q} = (-1)^n \frac{\Gamma(n+q-2)}{\Gamma(q-2)} \frac{P_n(q;\xi\cdot\eta)}{|y|^{n+q-2}}.$$

As $y \mapsto |y|^{2-q}$, $|y| \neq 0$, is (apart from a multiplicative constant) the fundamental solution of the Laplace equation in q dimensions (see (7.89)), Maxwell's representation tells that the Legendre polynomials may be obtained by repeated differentiations of the kernel function $|x-y|^{2-q}$ in the direction of ξ. The potential on the right side may be regarded as the potential of a pole of order n with the axis ξ at the origin.

The Legendre polynomials $P_n(q;\cdot)$, $q \geq 3$, as defined in our approach by (7.40) coincide (apart from a multiplicative constant) with Gegenbauer polynomials. More precisely, we have (see, e.g., [**554**])

(7.62)
$$\Gamma(q-2+n)P_n(q;t) = \Gamma(q-2)\Gamma(n+1)C_n^{\left(\frac{q-2}{2}\right)}(t), \quad t\in[-1,1],$$

where

$$C_n^{(\alpha)}(t) = \frac{1}{\Gamma(\alpha)} \sum_{k=0}^{\lfloor \frac{n}{2} \rfloor} (-1)^k \frac{\Gamma(\alpha+n-k)}{k!\,(n-2k)!}\,(2t)^{n-k}, \quad t \in [-1,1]. \tag{7.63}$$

Spherical harmonics. Let $H_n(q;\cdot)$ be a homogeneous harmonic polynomial of degree n in $\mathbb{R}^q$, i.e., $H_n(q;\cdot) \in Harm_n(\mathbb{R}^q)$. Its decomposition into radial and angular parts is straightforward: $H_n(q;\cdot)(x) = r^n\, Y_n(q;\xi), x = r\xi, r = |x|, \xi \in \mathbb{S}^{q-1}$. The restriction $Y_n(q;\cdot) = H_n(q;\cdot)|\mathbb{S}^{q-1}$ is called a *spherical harmonic of degree n and dimension q*. The set of all spherical harmonics of degree n and dimension q, i.e., the set of all restrictions $Y_n(q;\cdot) = H_n(q;\cdot)|\mathbb{S}^{q-1}$ with $H_n(q;\cdot) \in Harm_n(\mathbb{R}^q)$, is denoted by $Harm_n(\mathbb{S}^{q-1})$. In what follows we simply write $Harm_n$ instead of $Harm_n(\mathbb{R}^q)$ (or $Harm_n(\mathbb{S}^{q-1})$) if no confusion is likely to arise. Our purpose is to list the essential properties of functions in $Harm_n$, i.e., spherical harmonics of degree n and dimension q. For more details on spherical harmonics the reader is referred to C. Müller [**607**], W. Freeden [**254**], W. Freeden, M. Gutting [**277**], W. Freeden, M. Schreiner [**321**]. The last book also contains a vectorial as well as tensorial approach in Euclidean space $\mathbb{R}^3$.

The linear space $Harm_n$ is of dimension $N(q;n) = \dim(Harm_n)$. We already know that the spherical harmonics of different degrees are orthogonal in the sense of the $L^2(\mathbb{S}^{q-1})$-inner product

$$\langle Y_n(q;\cdot), Y_m(q;\cdot)\rangle_{L^2(\mathbb{S}^{q-1})} = \int_{\mathbb{S}^{q-1}} Y_n(q;\xi)\, \overline{Y_m(q;\xi)}\, dS_{(q-1)}(\xi) = 0, \quad n \neq m. \tag{7.64}$$

From the addition theorem of homogeneous harmonic polynomials it is an easy task to obtain the *addition theorem for spherical harmonics.*

Let $\{Y_{n,j}(q;\cdot)\}_{j=1,\dots,N(q;n)}$ be an $L^2(\mathbb{S}^{q-1})$-orthonormal system in $Harm_n$. Then, for any pair $(\xi,\eta) \in \mathbb{S}^{q-1} \times \mathbb{S}^{q-1}$, we have

$$\sum_{j=1}^{N(q;n)} Y_{n,j}(q;\xi)\, \overline{Y_{n,j}(q;\eta)} = \frac{N(q;n)}{\|\mathbb{S}^{q-1}\|} P_n(q;\xi\cdot\eta). \tag{7.65}$$

An immediate consequence is that

$$Y_n(q;\xi) = \frac{N(q;n)}{\|\mathbb{S}^{q-1}\|} \int_{\mathbb{S}^{q-1}} P_n(q;\xi\cdot\eta) Y_n(q;\eta)\, dS_{(q-1)}(\eta) \tag{7.66}$$

holds for all $\xi \in \mathbb{S}^{q-1}$ and all $Y_n(q;\cdot) \in Harm_n$, i.e., $K_{Harm_n}(q;\cdot)$ defined by

$$K_{Harm_n}(q;\xi\cdot\eta) = \frac{N(q;n)}{\|\mathbb{S}^{q-1}\|}\, P_n(q;\xi\cdot\eta), \quad \xi,\eta \in \mathbb{S}^{q-1}, \tag{7.67}$$

is the reproducing kernel of the $N(q;n)$–dimensional space $Harm_n$.

Within our notation *the two-dimensional analogue of the addition theorem* in terms of the circle functions "*circular harmonics*" $H_{n,j}(2;\cdot) : \mathbb{R}^2 \to \mathbb{R}$, $j = 1,2$, given by

$$H_{n,1}(2;x_{(2)}) = H_{n,1}(2;x_1,x_2) = \frac{1}{\sqrt{\pi}}\,\Im\big((x_2+ix_1)^n\big) = \frac{(-1)^{n+1}}{\sqrt{\pi}}|x|^n\cos(n\varphi),$$

$$H_{n,2}(2;x_{(2)}) = H_{n,2}(2;x_1,x_2) = \frac{1}{\sqrt{\pi}}\,\Re\big((x_2+ix_1)^n\big) = \frac{(-1)^{n+1}}{\sqrt{\pi}}|x|^n\sin(n\varphi),$$

$x_{(2)} \in \mathbb{R}^2$, $x_{(2)} = (x_1,x_2)^T, x_1 = r\cos\varphi$, $x_2 = r\sin\varphi$, $r = |x_{(2)}| = \sqrt{x_1^2+x_2^2}$, $0 \le \varphi < 2\pi$ is given by

$$\sum_{j=1}^{2} H_{n,j}\left(2;x_{(2)}\right)\overline{H_{n,j}\left(2;y_{(2)}\right)} = \frac{|x_{(2)}|^n|y_{(2)}|^n}{\pi}P_n\left(2;\xi_{(2)}\cdot\eta_{(2)}\right). \tag{7.68}$$

Next we are interested in the defining properties of the Legendre polynomial in $\mathbb{R}^q$. To this end we first show that the only function in $Harm_n$ that is invariant under orthogonal transformations having one point fixed can be characterized by the Legendre polynomial.

Let $H_n(q;\cdot)$ be a homogeneous harmonic polynomial of degree n satisfying the following properties:

(i) $H_n(q;tx) = H_n(q;x)$ for all orthogonal transformations t with $\det(t) = 1$ and $t\epsilon^q = \varepsilon^q$,

(ii) $H_n(q;\varepsilon^q) = 1$.

Then $H_n(q;\cdot)$ is uniquely determined, and we have with $x = r\xi$, $\xi \in \mathbb{S}^{q-1}$, $\xi = t\varepsilon^q + \sqrt{1-t^2}\,\xi_{(q-1)}$, $\xi_{(q-1)} \in \mathbb{S}^{q-2}$,

$$H_n(q;\xi) = r^nP_n(q;\xi\cdot\varepsilon^q) = r^nP_n(q,t), \tag{7.69}$$

where

$$P_n(q;t) = n!\,\Gamma\left(\frac{q-1}{2}\right)\sum_{l=0}^{\lfloor\frac{n}{2}\rfloor}\left(-\frac{1}{4}\right)^l\frac{(1-t^2)^l\,t^{n-2l}}{l!\,(n-2l)!\,\Gamma\left(l+\frac{q-1}{2}\right)}. \tag{7.70}$$

Funk-Hecke formula. We now come to the *Funk-Hecke formula* (see H. Funk [**338**], E. Hecke [**420**] for early versions, and C. Müller [**600, 605**] for the approach presented here).

Suppose that G is of class $L^1([-1,1])$. Then, for all $(\xi,\eta) \in \mathbb{S}^{q-1}\times\mathbb{S}^{q-1}$ and all $n \in \mathbb{N}_0$,

$$\int_{\mathbb{S}^{q-1}} G(\xi\cdot\zeta)P_n(q;\zeta\cdot\eta)\,dS_{(q-1)}(\zeta) = G^\wedge(n)\,P_n(q;\xi\cdot\eta), \tag{7.71}$$

where

$$G^\wedge(n) = G^\wedge_{\mathbb{S}^{q-1}}(n) = \|\mathbb{S}^{q-2}\|\int_{-1}^{+1} G(s)P_n(q;s)(1-s^2)^{\frac{q-3}{2}}\,ds. \tag{7.72}$$

From (7.71) we find by multiplication with $Y_n(q;\eta)$ and integration with regard to η the following *variant of the Funk-Hecke formula.*

Let G be of class $L^1([-1,+1])$. Then, for $Y_n(q;\cdot) \in \mathrm{Harm}_n$,

$$\int_{\mathbb{S}^{q-1}} G(\xi\cdot\eta)Y_n(q;\eta)\,dS_{(q-1)}(\eta) = G^\wedge(n)\,Y_n(q;\xi). \tag{7.73}$$

Laplace representation. Next we are interested in the Laplace representation of spherical harmonics. To this end we start from the integral (cf. C. Müller [**607**])

$$L_n(q;x) = \frac{1}{\|\mathbb{S}^{q-2}\|} \int_{\mathbb{S}^{q-2}} \left(x \cdot \varepsilon^q + ix_{(q-1)} \cdot \zeta_{(q-1)}\right)^n \, dS_{(q-2)}(\zeta_{(q-1)}), \tag{7.74}$$

where $x \in \mathbb{R}^q$ is of the form $x = (x \cdot \varepsilon^q)\varepsilon^q + x_{(q-1)}$, $x_{(q-1)} \in \mathbb{R}^{q-1}$. $L_n(q;\cdot)$ is a homogeneous harmonic polynomial in $\mathbb{R}^q$ of degree n. In addition, for all orthogonal transformations $\mathbf{t}$ with $\det(\mathbf{t}) = 1$ and $\mathbf{t}\epsilon^q = \varepsilon^q$, we have $L_n(q;\mathbf{t}x) = L_n(q;x)$ and $L_n(q;\varepsilon^q) = 1$. This tells us that

$$L_n(q;x) = r^n L_n(q;\xi) = r^n P_n(q;t), \tag{7.75}$$

where, as usual, $x = r\xi, \xi \in \mathbb{S}^{q-1}$ $\xi = t\varepsilon^q + \sqrt{1-t^2}\xi_{(q-1)}$, $\xi_{(q-2)} \in \mathbb{S}^{q-2}$. Hence, we have

$$P_n(q;t) = \frac{1}{\|\mathbb{S}^{q-2}\|} \int_{\mathbb{S}^{q-2}} \left(t + i\sqrt{1-t^2}\xi_{(q-1)} \cdot \zeta_{(q-1)}\right)^n \, dS_{(q-2)}(\zeta_{(q-1)}). \tag{7.76}$$

Equivalently, we have

$$P_n(q;t) = \frac{\|\mathbb{S}^{q-3}\|}{\|\mathbb{S}^{q-2}\|} \int_{-1}^{1} \left(t + is\sqrt{1-t^2}\right)^n (1-s^2)^{\frac{q-4}{2}} \, ds. \tag{7.77}$$

for $q \geq 3$, $n \geq 0$, and $t \in [-1,1]$.

This leads the following estimates for the Legendre polynomial (see W. Freeden, M. Gutting [**277**]): *For $q \geq 3$ the Legendre polynomials $P_n(q;\cdot)$ satisfy the inequalities*

$$|P_n(q;t)| \leq 1, \ n \geq 0, \ t \in [-1,1], \tag{7.78}$$

$$|P_n(q;t)| \leq \frac{\Gamma(\frac{q-1}{2})}{\sqrt{\pi}} \left(\frac{4}{n(1-t^2)}\right)^{\frac{q-2}{2}}, \ n \geq 1, \ t \in (-1,1). \tag{7.79}$$

The estimates remain valid for $q = 2$.

Bernstein kernel. In order to show the closure and completeness of the spherical harmonics in $L^2(\mathbb{S}^{q-1})$, we consider the *Bernstein kernel of degree n* (see [**232**], [**277**] for the vectorial and tensorial context in $\mathbb{R}^3$)

$$B_n(\xi \cdot \eta) = \left(\frac{1+\xi \cdot \eta}{2}\right)^n, \ \xi, \eta \in \mathbb{S}^{q-1}, \quad n \in \mathbb{N}_0. \tag{7.80}$$

Let F be of class $C^{(0)}(\mathbb{S}^{q-1})$. Then

$$\lim_{n\to\infty} \sup_{\xi\in\mathbb{S}^{q-1}} \left| \left(\frac{1}{4\pi}\right)^{\frac{q-1}{2}} \frac{\Gamma(n+q-1)}{\Gamma\left(n+\frac{q-1}{2}\right)} \int_{\mathbb{S}^{q-1}} \left(\frac{1+\xi \cdot \eta}{2}\right)^n F(\eta) \, dS_{(q-1)}(\eta) - F(\xi) \right| = 0. \tag{7.81}$$

It is not difficult to see that

$$\left(\frac{1}{4\pi}\right)^{\frac{q-1}{2}} \frac{\Gamma(n+q-1)}{\Gamma\left(n+\frac{q-1}{2}\right)} \int_{-1}^{+1} \left(\frac{1+t}{2}\right)^n P_k(q;t)(1-t^2)^{\frac{q-3}{2}} \, dt = \frac{\beta_n^k(q)}{\|\mathbb{S}^{q-2}\|}, \tag{7.82}$$

where we have used the abbreviation

$$\beta_n^k(q) = \frac{n!}{(n-k)!} \frac{(n+q-2)!}{(n+k+q-2)!}. \tag{7.83}$$

Note that

$$\beta_n^k(q) < \beta_{n+1}^k(q) \tag{7.84}$$

and

$$\lim_{n\to\infty} \beta_n^k(q) = 1. \tag{7.85}$$

Hence,

$$\left(\frac{1}{4\pi}\right)^{\frac{q-1}{2}} \frac{\Gamma(n+q-1)}{\Gamma\left(n+\frac{q-1}{2}\right)} \left(\frac{1+\xi\cdot\eta}{2}\right)^n = \sum_{k=0}^{n} \beta_n^k(q) \frac{N(q;k)}{\|\mathbb{S}^{q-1}\|} P_k(q;\xi\cdot\eta). \tag{7.86}$$

Closure and completeness. Summarizing our results we finally obtain in connection with the addition theorem for spherical harmonics:

For all $F \in C^{(0)}(\mathbb{S}^{q-1})$

$$\lim_{n\to\infty} \sup_{\xi\in\mathbb{S}^{q-1}} \left| F(\xi) - \sum_{k=0}^{n} \sum_{j=1}^{N(q;k)} \beta_n^k(q) \int_{\mathbb{S}^{q-1}} F(\eta)\overline{Y_{k,j}(q;\eta)}\, dS_{(q-1)}(\eta)\; Y_{k,j}(q;\xi) \right| = 0.$$

This limit relation enables us to prove the closure of the system of spherical harmonics.

The system $\{Y_{n,j}(q;\cdot)\}_{\substack{n=0,1,\ldots \\ j=1,\ldots,N(q,n)}}$ is closed in the space $L^2(\mathbb{S}^{q-1})$ with respect to $\|\cdot\|_{L^2(\mathbb{S}^{q-1})}$.

Fundamental solution for the Laplace operator in $\mathbb{R}^q$. Let $x \in \mathcal{G}$ be fixed, where $\mathcal{G}$ is a region in $\mathbb{R}^q$. We are looking for a harmonic function U in $\mathcal{G}\setminus\{x\}$ such that $U(x) = F(|x-y|), x \in \mathcal{G}\setminus\{x\}$; i.e., U depends only on the mutual distance of x and y. From the identity

$$\left(\frac{\partial}{\partial x_i}\right)^2 F(|x-y|) = F''(|x-y|)\frac{(x_i-y_i)^2}{|x-y|^2} + F'(|x-y|)\left(\frac{1}{|x-y|} - \frac{(x_i-y_i)^2}{|x-y|^3}\right) \tag{7.87}$$

we easily obtain $\Delta_x F(|x-y|) = F''(|x-y|) + \frac{q-1}{|x-y|}F'(|x-y|) = 0$. In other words, $F(|x-y|)$ can be written in the form

$$F(|x-y|) = \begin{cases} C_1 \ln(|x-y|) + C_2 & , \quad q=2 \\ C_1|x-y|^{2-q} + C_2 & , \quad q\geq 3 \end{cases} \tag{7.88}$$

with some constants C_1, C_2. By convention, the function F_q (in the nomenclature of Chapter 11, more explicitly, $F_q^{(1)}$) given by

$$x \mapsto F_q(|x-y|) = \begin{cases} -\frac{1}{2\pi}\ln(|x-y|) & , \quad q=2 \\ \frac{|x-y|^{2-q}}{(q-2)\|\mathbb{S}^{q-1}\|} & , \quad q\geq 3 \end{cases} \tag{7.89}$$

is called the *fundamental solution in $\mathbb{R}^q$ for the Laplace operator* Δ. Note that the one-dimensional fundamental solution for the Laplacian, i.e., the operator of the second order derivative, is given by the continuous function

$$x \mapsto F_1(|x-y|) = -\frac{1}{2}|x-y| = -\frac{1}{2}(x-y)\ \mathrm{sign}(x-y), \quad x,y\in\mathbb{R}. \tag{7.90}$$

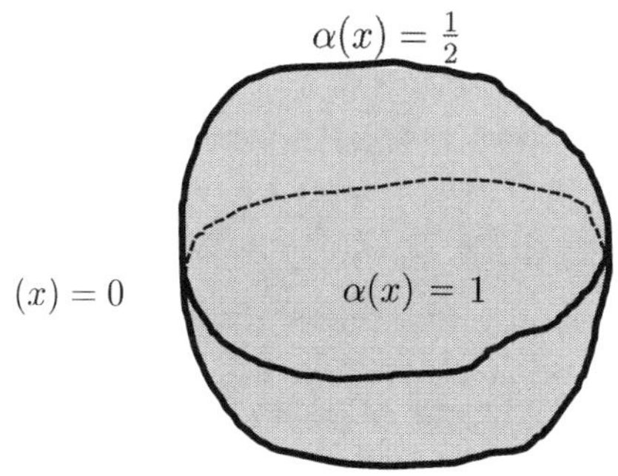

FIGURE 1. Solid angle subtended at $x \in \mathbb{R}^3$ by the surface $\partial\mathcal{G}$ of a regular region $\mathcal{G}$ with "smooth boundary".

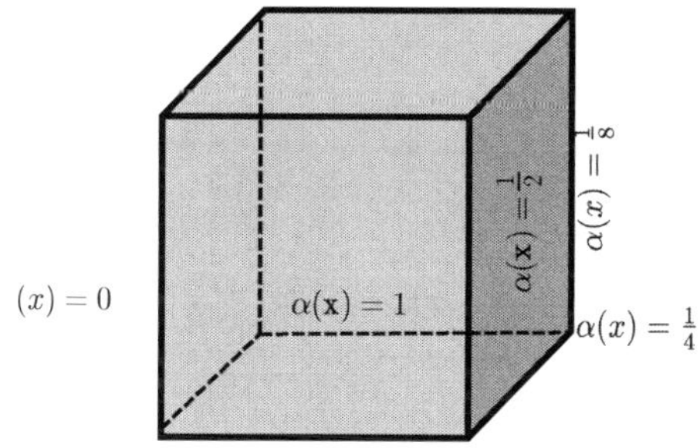

FIGURE 2. Solid angle subtended at $x \in \mathbb{R}^3$ by the surface $\partial\mathcal{G}$ of the "nonsmooth" cube $\mathcal{G} = (-1,1)^3$.

Its continuity is a powerful feature with nice applications in the one-dimensional lattice point and sampling theory.

The fundamental solution for the Laplace operator leads to the *Third Green Theorem in* $\mathbb{R}^q$ (see, e.g., [**493**], [**48**]):

Let $\mathcal{G}$ be a regular region in $\mathbb{R}^q$. Suppose that $U : \overline{\mathcal{G}} \to \mathbb{R}$ is twice continuously differentiable, i.e., $U \in C^{(2)}\left(\overline{\mathcal{G}}\right)$. Then we have

$$\int_{\partial\mathcal{G}} \left\{ F_q(|x-y|) \left(\frac{\partial U}{\partial \nu}(y) \right) - U(y) \left(\frac{\partial}{\partial \nu_y} F_q(|x-y|) \right) \right\} dS(y) - \int_{\mathcal{G}} F_q(|x-y|)\, \Delta U(y)\, dy = \alpha(x)\, U(x), \tag{7.91}$$

where $\alpha(x)$, $x \in \mathbb{R}^q$, is the solid angle at x subtended by the surface $\partial\mathcal{G}$.

The integral

$$\alpha(x) = -\int_{\partial\mathcal{G}} \frac{\partial}{\partial \nu_y} F_q(|x-y|)\, dS(y), \quad x \in \mathbb{R}^q. \tag{7.92}$$

is a measure for the "solid angle" $\alpha(x)$ (more explicitly, $\alpha_{\partial\mathcal{G}}(x)$) subtended by the boundary $\partial\mathcal{G}$ at the point $x \in \mathbb{R}^q$ (cf. Figures 1). Note that the concept of the solid angle is also valid for regular regions $\mathcal{G}$ such as cube, simplex, polyhedron, more concretely, to regular regions with solid angle $\alpha(x)$ at $x \in \mathbb{R}^q$ subtended by the surface $\partial\mathcal{G}$. For example, for the cube $\mathcal{G} = (-1,1)^3 \subset \mathbb{R}^3$ we have (i) $\alpha(g) = 1$ if g is located in the open cube $\mathcal{G}$, (ii) $\alpha(g) = \frac{1}{2}$ if g is located on one of the six faces of the boundary $\partial\mathcal{G}$ of the cube $\mathcal{G}$ but not on an edge or in a vertex, (iii) $\alpha(g) = \frac{1}{4}$ if g is located on one of the eight edges of $\partial\mathcal{G}$ but not in a vertex, (iv) $\alpha(g) = \frac{1}{8}$ if g is located in one of the eight vertices of $\partial\mathcal{G}$ (cf. Figure 2).

Finally we mention the *Poisson differential equation*, which is a classical result in potential theory (see, for example, S.G. Michlin [**584, 585**] or W. Freeden [**270**] for an alternative proof in $\mathbb{R}^3$ using mollifier techniques, that can be used in inverse gravimetry).

Let F be of class $C^{(0)}\left(\overline{\mathcal{G}}\right)$, $\overline{\mathcal{G}} = \mathcal{G} \cup \partial\mathcal{G} \subset \mathbb{R}^q$. Then $U : \overline{\mathcal{G}} \to \mathbb{R}$ given by

$$U(x) = \int_{\mathcal{G}} F(y)\ F_q(|x-y|)\ dy \tag{7.93}$$

is of class $C^{(1)}\left(\overline{\mathcal{G}}\right)$, and we have

$$\nabla U(x) = \int_{\mathcal{G}} F(y)\ \nabla_x F_q(|x-y|)\ dy. \tag{7.94}$$

If F is bounded in $\overline{\mathcal{G}} = \mathcal{G} \cup \partial\mathcal{G}$ satisfying a Lipschitz-condition in the neighborhood of the point $x \in \mathcal{G}$ (more precisely, $|F(y) - F(z)| \leq C_F|y-z|$ for all y, z in the neighborhood of the point $x \in \mathcal{G}$), then U as given by (7.93) is twice continuously differentiable in $x \in \mathcal{G}$, and we have

$$\Delta U(x) = -\ F(x). \tag{7.95}$$

Potential spaces and Runge-Walsh context. In what follows we restrict ourselves to the three-dimensional Euclidean space $\mathbb{R}^3$:

Let $\mathcal{B} \subset \mathbb{R}^3$ be a regular region (such as the Earth's interior). The function spaces $C^{(2)}(\mathcal{B})$, $C^{(2)}\left(\overline{\mathcal{B}}\right)$, $C^{(2)}(\mathcal{B}^c)$, $C^{(2)}\left(\overline{\mathcal{B}^c}\right)$, $\mathcal{B}^c = \mathbb{R}^3 \backslash \overline{\mathcal{B}}$ for twice continuously differentiable scalar fields, $c^{(2)}(\mathcal{B})$,$c^{(2)}\left(\overline{\mathcal{B}}\right)$,$c^{(2)}(\mathcal{B}^c)$, $c^{(2)}\left(\overline{\mathcal{B}^c}\right)$ for twice continuously differentiable vector fields, $\mathbf{c}^{(2)}(\mathcal{B})$, $\mathbf{c}^{(2)}\left(\overline{\mathcal{B}}\right)$, $\mathbf{c}^{(2)}(\mathcal{B}^c)$, $\mathbf{c}^{(2)}\left(\overline{\mathcal{B}^c}\right)$ for twice continuously differentiable tensor fields are defined in canonical way (see, e.g., [**326**] for more details).

Scalar potential spaces. We introduce $Pot\,(\mathcal{B})$ as the space of scalar potentials harmonic in $\mathcal{B}$:

$$Pot(\mathcal{B}) = \{F \in C^{(2)}(\mathcal{B}) :\ \Delta F = 0 \text{ in } \mathcal{B}\}. \tag{7.96}$$

$Pot^{(0)}(\overline{\mathcal{B}})$ is the space of continuous functions $F : \overline{\mathcal{B}} \to \mathbb{R}$, whose restrictions $F|\mathcal{B}$ are members of $Pot\,(\mathcal{B})$. In brief (but not quite mathematically accurate),

$$Pot^{(0)}(\overline{\mathcal{B}}) = Pot(\mathcal{B}) \cap C^{(0)}(\overline{\mathcal{B}}), \tag{7.97}$$

and, in similar way,

$$Pot^{(k)}(\overline{\mathcal{B}}) = Pot(\mathcal{B}) \cap C^{(k)}(\overline{\mathcal{B}}), \quad 0 \leq k \leq \infty. \tag{7.98}$$

Let U be of class $Pot^{(0)}\left(\overline{\mathcal{B}}\right)$. The *maximum/minimum principle* states

$$\sup_{x \in \overline{\mathcal{B}}} |U(x)| \leq \sup_{x \in \partial\mathcal{B}} |U(x)|. \tag{7.99}$$

Correspondingly, we define $Pot(\mathcal{B}^c), \mathcal{B}^c = \mathbb{R}^3 \backslash \mathcal{B}$, as the space of scalar potentials harmonic in $\mathcal{B}^c$ and regular at infinity:

(7.100)
$$Pot(\mathcal{B}^c) = \{F \in C^{(2)}(\mathcal{B}^c) : \Delta F = 0 \text{ in } \mathcal{B}^c, |F(x)| = O\left(|x|^{-1}\right), \quad |x| \to \infty\}.$$

$Pot^{(k)}(\overline{\mathcal{B}^c})$ is the space of k-times continuously differentiable functions $F : \overline{\mathcal{B}^c} \to \mathbb{R}$, whose restrictions $F|\mathcal{B}^c$ are members of $Pot(\mathcal{B}^c)$. In brief,

(7.101)
$$Pot^{(k)}(\overline{\mathcal{B}^c}) = Pot(\mathcal{B}^c) \cap C^{(k)}(\overline{\mathcal{B}^c}), \quad 0 \leq k \leq \infty.$$

Let U be of class $Pot^{(0)}\left(\overline{\mathcal{B}^c}\right)$. The *maximum/minimum principle* states

(7.102)
$$\sup_{x \in \overline{\mathcal{B}^c}} |U(x)| \leq \sup_{x \in \partial\mathcal{B}} |U(x)|.$$

For later use we introduce vectorial and tensorial counterparts of $Pot(\overline{\mathcal{B}}^c)$:

Vectorial potential spaces. $pot(\mathcal{B}^c)$ is the space of vectorial potentials given by

(7.103)
$$pot(\mathcal{B}^c) = \{f \in c^{(1)}(\mathcal{B}^c) : \nabla \cdot f = 0, \quad \nabla \wedge f = 0 \text{ in } \mathcal{B}^c, \\ |f(x)| = O\left(|x|^{-2}\right), |x| \to \infty\}.$$

We let

(7.104)
$$pot^{(k)}(\overline{\mathcal{B}^c}) = pot(\mathcal{B}^c) \cap c^{(k)}(\overline{\mathcal{B}^c}), \quad 0 \leq k \leq \infty.$$

Tensorial potential spaces. Similarly, the space $\mathbf{pot}\,(\mathcal{B}^c)$ of tensorial potentials is given by

(7.105)
$$\mathbf{pot}(\mathcal{B}^c) = \{\mathbf{f} \in \mathbf{c}^{(1)}\,(\mathcal{B}^c) : \nabla \cdot \mathbf{f} = 0, \quad \nabla \wedge \mathbf{f} = 0 \text{ in } \mathcal{B}^c, \\ |\mathbf{f}(x)| = O\left(|x|^{-3}\right), |x| \to \infty\}.$$

We set

(7.106)
$$\mathbf{pot}^{(k)}(\overline{\mathcal{B}^c}) = \mathbf{pot}(\mathcal{B}^c) \cap \mathbf{c}^{(k)}(\overline{\mathcal{B}^c}), \quad 0 \leq k \leq \infty$$

Outer harmonics in Euclidean space $\mathbb{R}^3$. Consider the sphere $\mathbb{S}^2_R \subset \mathbb{R}^3$ around the origin with radius $R > 0$. By virtue of the isomorphism $\mathbb{S}^2 \ni \xi \mapsto R\xi \in \mathbb{S}^2_R$, we assume functions $F : \mathbb{S}^2_R \to \mathbb{R}$ to be defined on $\mathbb{S}^2$. It is clear that the function spaces defined on $\mathbb{S}^2$ admit their natural generalizations as spaces of functions defined on $\mathbb{S}^2_R$. We have, for example, $C^{(\infty)}(\mathbb{S}^2_R)$, $L^p(\mathbb{S}^2_R)$, etc. Obviously, an $L^2(\mathbb{S}^2)$-orthonormal system of spherical harmonics forms an orthogonal system on $\mathbb{S}^2_R$ (with respect to $\langle\cdot,\cdot\rangle_{L^2(\mathbb{S}^2_R)}$). More explicitly, we have

(7.107)
$$\langle Y_{n,k}(3;\cdot), Y_{p,q}(3;\cdot)\rangle_{L^2(\mathbb{S}^2_R)} = \int_{\mathbb{S}^2_R} Y_{n,k}\left(3; \frac{x}{|x|}\right) Y_{p,q}\left(3; \frac{x}{|x|}\right) dS(x) = R^2 \delta_{n,p}\delta_{k,q}.$$

With the relationship $\xi \leftrightarrow R\xi$, the *surface gradient* $\nabla^{*;R}$ and the *Beltrami operator* $\Delta^{*;R}$ on $\mathbb{S}^2_R$, respectively, have the representation $\nabla^{*;R} = (1/R)\nabla^{*;1} = (1/R)\nabla^*$, $\Delta^{*;R} = (1/R^2)\Delta^{*;1} = (1/R^2)\Delta^*$, where ∇^*, Δ^* are the surface gradient and the Beltrami operator of the unit sphere $\mathbb{S}^2$. We have $\Delta^{*;R}Y_n(3;\cdot) =$

$(1/R^2)(\Delta^*)^\wedge(n)Y_n(3;\cdot)$ for $Y_n(3;\cdot)\in\mathrm{Harm}_n(\mathbb{S}^2)$. We introduce $\{Y^R_{n,k}(3;\cdot)\}_{\substack{n=0,1,\ldots\\k=1,\ldots,2n+1}}$ by letting

$$Y^R_{n,k}(3;x) = \frac{1}{R}Y_{n,k}\left(3;\frac{x}{|x|}\right), \quad x\in\mathbb{S}^2_R \tag{7.108}$$

Due to (7.107), the system $\{Y^R_{n,k}(3;\cdot)\}_{\substack{n=0,1,\ldots\\k=1,\ldots,2n+1}}$ is an orthonormal basis in $L^2(\mathbb{S}^2_R)$:

$$L^2(\mathbb{S}^2_R) = \overline{\operatorname{span}_{\substack{n=0,1,\ldots,\\k=1,\ldots,2n+1}}(Y^R_{n,k}(3;\cdot))}^{\|\cdot\|_{L^2(\mathbb{S}^2_R)}} \tag{7.109}$$

The system $\{H^R_{n,k}(3;\cdot)\}_{\substack{n=0,1,\ldots\\k=1,\ldots,2n+1}}$ of *outer harmonics* $H^R_{n,k}(3;\cdot)$ *of degree* n *and order* k defined by

$$H^R_{n,k}(3;x) = \left(\frac{R}{|x|}\right)^{n+1} Y^R_{n,k}(3;x), \qquad x\in\mathbb{R}^3\backslash\{0\}, \tag{7.110}$$

satisfies the following properties:

- $H^R_{n,k}(3;\cdot)$ is of class $C^{(\infty)}(\mathbb{R}^3\backslash\{0\})$,
- $H^R_{n,k}(3;\cdot)$ satisfies the Laplace equation in $\mathbb{R}^3\backslash\{0\}$,
- $H^R_n(3;\cdot)$ is regular at infinity, i.e.,

$$\left|H^R_{n,k}(3;x)\right| = O\left(\frac{1}{|x|}\right), \quad |x|\to\infty, \tag{7.111}$$

- $H^R_{n,k}(3;\cdot)\Big|\mathbb{S}^2_R = Y^R_{n,k}(3;\cdot) = \frac{1}{R}Y_{n,k}(3;\cdot)$,
- $\left\langle H^R_{n,k}(3;\cdot), H^R_{p,q}(3;\cdot)\right\rangle_{L^2(\mathbb{S}^2_R)} = \delta_{n,p}\delta_{k,q}$.

The representation of outer harmonics on spheres of different "altitudes" $\gamma > R > 0$ is characterized in accordance with (7.110) as follows:

$$H^R_{n,k}(3;\cdot) = \left(\frac{R}{\gamma}\right)^n H^\gamma_{n,k}(3;\cdot). \tag{7.112}$$

Locally uniform and uniform approximation by outer harmonics. From W. Freeden [**254**] we borrow the following *closure theorems in terms of outer harmonics*:

Let $\mathcal{B}\subset\mathbb{R}^3$ be a regular region such that $R<\inf_{x\in\partial\mathcal{B}}|x|$ ("Runge-Bjerhammar condition"). Then the following statements are valid for outer harmonics $H^R_{n,j} = H^R_{n,j}(3;\cdot)$:

- $\left\{H^R_{n,j}\Big|\partial\mathcal{B}\right\}_{\substack{n=0,1,\ldots\\j=1,\ldots,2n+1}}$ *is closed in* $L^2(\partial\mathcal{B})$.
- $\left\{H^R_{n,j}\Big|\partial\mathcal{B}\right\}_{\substack{n=0,1,\ldots\\j=1,\ldots,2n+1}}$ *is closed in* $C^{(0)}(\partial\mathcal{B})$.

Approximation methods. The second statement concerned with *uniform approximation* (see (7.124)) is more difficult to handle, seen from the view point of constructive approximation. The first statement

$$L^2(\partial\mathcal{B}) = \overline{\left\{ H^R_{n,j}\big|\partial\mathcal{B} \right\}_{\substack{n=0,1,\ldots \\ j=1,\ldots,2n+1}}}^{\|\cdot\|_{L^2(\partial\mathcal{B})}} \tag{7.113}$$

leads to two different manifestations of *locally uniform approximation*:

$C^{(0)}(\partial\mathcal{B})$-locally uniform concept. Let $\mathcal{B} \subset \mathbb{R}^3$ be a regular region such that $R < \inf_{x\in\partial\mathcal{B}} |x|$. For given $F \in C^{(0)}(\partial\mathcal{B})$, let U be the potential of class $Pot^{(0)}(\overline{\mathcal{B}^c})$ with $U|\partial\mathcal{B} = F$. Then, for every $k \in \mathbb{N}_0$ and all points $x \in \mathcal{K} \subset \mathcal{B}^c$ with $\mathcal{K}$ not-necessarily compact, but $\operatorname{dist}(\overline{\mathcal{K}}, \partial\mathcal{B}) > 0$, there exist an integer m (dependent on ε) and a set of coefficients $a_{0,1}, \ldots, a_{m,1}, \ldots, a_{m,2m+1}$ such that

$$\left(\int_{\partial\mathcal{B}} \left| F(x) - \sum_{n=0}^{m} \sum_{j=1}^{2n+1} a_{n,j} H^R_{n,j}(x) \right|^2 dS(x) \right)^{\frac{1}{2}} \leq \varepsilon \tag{7.114}$$

and

$$\sup_{x\in\overline{\mathcal{K}}} \left| \left(\nabla^{(k)} U\right)(x) - \sum_{n=0}^{m} \sum_{j=1}^{2n+1} a_{n,j} \left(\nabla^{(k)} H^R_{n,j}\right)(x) \right| \leq C\varepsilon \tag{7.115}$$

hold for all $k \in \mathbb{N}_0$, where $C > 0$ depends on $\mathcal{B}$. In other words (see also [**260**] for numerical applicability), the L^2-approximation in terms of outer harmonics on $\partial\mathcal{B}$ implies the uniform approximation (in the ordinary sense) on each subset $\overline{\mathcal{K}}$ with positive distance to $\partial\mathcal{B}$ (in particular, to each compact subset $\overline{\mathcal{K}}$ with positive distance to $\partial\mathcal{B}$ such that an approximation is guaranteed in locally compact topology). Unfortunately, (7.114) and (7.115) are non-constructive, since further information about the upper limit of m and the choice of the coefficients of the approximating linear combination is needed. In order to derive a more constructive approximation theorem the system of potential values can be orthonormalized on $\partial\mathcal{B}$. As a result we obtain a *"(generalized) Fourier expansion" (orthogonal Fourier approximation)* that shows locally uniform approximation.

Let $\mathcal{B} \subset \mathbb{R}^3$ be a regular region such that $R < \inf_{x\in\partial\mathcal{B}} |x|$. For given $F \in C^{(0)}(\partial\mathcal{B})$, let U be the solution of the exterior problem $U \in Pot^{(0)}(\overline{\mathcal{B}^c}), U|\partial\mathcal{B} = F$. Corresponding to the countably infinite sequence $\{H^R_{n,j}\}_{n=0,1,\ldots,j=,\ldots,2n+1}$ there exists a system $\{H_{n,j}(\partial\mathcal{B};\cdot)\}_{n=0,1,\ldots,j=1,\ldots,2n+1}$ in the space $\mathrm{Pot}^{(0)}(\mathbb{R}^3\backslash\mathbb{B}^3_R)$ such that its restriction $\{H_{n,j}(\partial\mathcal{B};\cdot)|\partial\mathcal{B}\}_{n=0,1,\ldots,j=,\ldots,2n+1}$ is orthonormal in the sense that

$$\int_{\partial\mathcal{B}} H_{n,j}(\partial\mathcal{B};y) H_{l,k}(\partial\mathcal{B};y)\, dS(y) = \delta_{n,l}\, \delta_{j,k}. \tag{7.116}$$

Consequently, U is representable in the form

$$U = \sum_{n=0}^{\infty} \sum_{j=1}^{2n+1} \left(\int_{\partial\mathcal{B}} F(y) H_{n,j}(\partial\mathcal{B};y)\, dS(y) \right) H_{n,j}(\partial\mathcal{B};\cdot). \tag{7.117}$$

Moreover, for every $k \in \mathbb{N}_0$ *and all points* $x \in \mathcal{K}$, $\mathcal{K} \Subset \mathcal{B}^c$, *i.e.,* $\mathcal{K} \subset \mathcal{B}^c$ *with* $\operatorname{dist}(\overline{\mathcal{K}}, \partial\mathcal{B}) > 0$, *each partial sum* $U^{(m)}, m \in \mathbb{N}_0$, *given by*

$$U^{(m)}(x) = \sum_{n=0}^{m} \sum_{j=1}^{2n+1} \left(\int_{\partial\mathcal{B}} F(y) H_{n,j}(\partial\mathcal{B}; y)\, dS(y) \right) H_{n,j}(\partial\mathcal{B}; x) \tag{7.118}$$

admits the estimate

$$\sup_{x \in \overline{\mathcal{K}}} \left| \left(\nabla^{(k)} U \right)(x) - \left(\nabla^{(k)} U^{(m)} \right)(x) \right| \tag{7.119}$$

$$\leq C \left(\int_{\partial\mathcal{B}} |F(y)|^2 \, dS(y) - \sum_{n=0}^{m} \sum_{j=1}^{2n+1} \left| \int_{\partial\mathcal{B}} F(y) H_{n,j}(\partial\mathcal{B}; y)\, dS(y) \right|^2 \right)^{\frac{1}{2}}.$$

Note that the orthonormalization procedure can be performed (e.g., by the Gram-Schmidt orthonormalization process) once and for all when the boundary surface $\partial\mathcal{G}$ of a regular region $\mathcal{B}$ is specified. In the same way, inner boundary-value problems can handled by Fourier expansions (orthogonal expansions) in terms of inner harmonics. Furthermore, locally uniform approximation by Fourier expansions can be obtained not only for (the multipole system of) outer harmonics, but also for certain mass point systems (monopoles) and related kernel representations. The details can be found, e.g., in [**260**], [**274**].

$L^2(\partial\mathcal{B})$-locally uniform concept. To every $F \in L^2(\partial\mathcal{B})$, there exists a uniquely determined potential U of the form (7.117). This leads to the definition of the Hilbert space $H_0(\overline{\mathcal{B}^c})$ of harmonic functions in $\mathcal{B}^c$ with square-integrable boundary values by letting

(7.120)

$$H_0(\overline{\mathcal{B}^c}) = \left\{ U = \sum_{n=0}^{\infty} \sum_{j=1}^{2n+1} \left(\int_{\partial\mathcal{B}} F(y)\; H_{n,j}(\partial\mathcal{B}; y)\, dS(y) \right) H_{n,j}(\partial\mathcal{B}; \cdot) : F \in L^2(\partial\mathcal{B}) \right\}.$$

$H_0(\overline{\mathcal{B}^c})$ equipped with the inner product

$$\langle U, \tilde{U} \rangle_{H_0(\overline{\mathcal{B}^c})} = \sum_{n=0}^{\infty} \sum_{j=1}^{2n+1} \int_{\partial\mathcal{B}} F(y) H_{n,j}(\partial\mathcal{B}; y)\, dS(y) \int_{\partial\mathcal{B}} \tilde{F}(y) H_{n,j}(\partial\mathcal{B}; y)\, dS(y),$$

with

$$U|\partial\mathcal{B} = F, \qquad \tilde{U}|\partial\mathcal{B} = \tilde{F}, \qquad F, \tilde{F} \in L^2(\partial\mathcal{B}), \tag{7.121}$$

is a Hilbert space (note that, for $H_0(\overline{\mathcal{B}^c})$, we are allowed to identify the norm $\|\cdot\|_{H_0(\overline{\mathcal{B}^c})}$ with the $\|\cdot\|_{L^2(\partial\mathcal{B})}$-norm).

In the theory of harmonic functions related to regular regions, a result first motivated by C. Runge [**739**] in one-dimensional complex analysis and later generalized by J.L. Walsh [**852**] and I.N. Vekua [**835**] to potential theory in three-dimensional Euclidean analysis is of basic interest. For our gravitationally relevant obligations, we use the *Runge concept* in the formulation following [**254**] (see also [**51**]). In fact, we are able to verify the following *Runge-Walsh variant*

$$\overline{H_0(\overline{\mathcal{G}^c})|\overline{\mathcal{B}^c}}^{\|\cdot\|_{C^{(0)}(\overline{\mathcal{B}^c})}} = Pot^{(0)}\left(\overline{\mathcal{B}^c}\right) \tag{7.122}$$

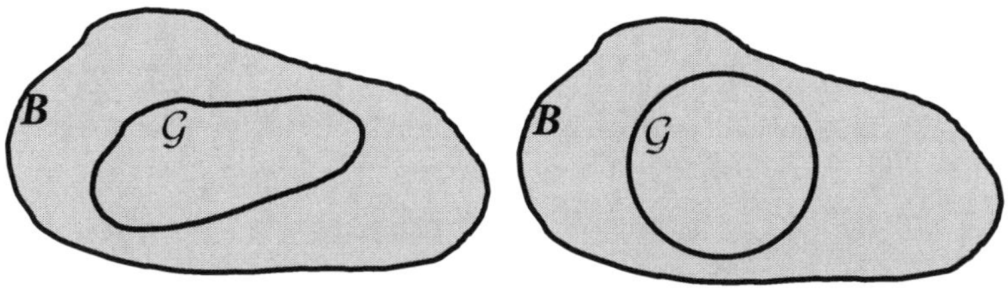

FIGURE 3. Runge regions (outer space situation): Any regular region $\mathcal{G} \subset \mathcal{B}$ (left) and "Runge-Bjerhammar ball" $\mathcal{G} = \mathbb{B}_R^3 \subset \mathcal{B}$ (right).

for arbitrary regular regions $\mathcal{G} \Subset \mathcal{B}$, i.e., $\mathcal{G} \subset \mathcal{B}$ with $\operatorname{dist}(\overline{\mathcal{G}}, \partial\mathcal{B}) > 0$ (cf. Figure 3, left). Hence, for every $\varepsilon > 0$ and $V \in Pot^{(0)}\left(\overline{\mathcal{B}^c}\right)$, there exists a potential $\tilde{V} \in H_0(\overline{\mathcal{G}^c})$, such that $\sup_{x\in\overline{\mathcal{B}^c}} |V(x) - \tilde{V}(x)| \leq \varepsilon$ (see also [**51**, **266**, **274**] for more details). In particular, under the "Runge-Bjerhammar condition" $R < \inf_{x\in\partial\mathcal{B}} |x|$, we are led with $\mathcal{G} = \mathbb{B}_R^3$ (cf. Figure 3, right) to

$$\overline{H_0(\overline{(\mathbb{B}_R^3)^c})|\overline{\mathcal{B}^c}}^{\|C^{(0)}\left(\overline{\mathcal{B}^c}\right)\|} = Pot^{(0)}\left(\overline{\mathcal{B}^c}\right). \tag{7.123}$$

$C^{(0)}(\partial\mathcal{B})$*-uniform concept.* Next we come to the closure of $\left\{H_{n,j}^R \middle| \partial\mathcal{B}\right\}_{\substack{n=0,1,\ldots \\ j=1,\ldots,2n+1}}$ in $C^{(0)}(\partial\mathcal{B})$, i.e.,

$$C^{(0)}(\partial\mathcal{B}) = \overline{\left\{H_{n,j}^R \middle| \partial\mathcal{B}\right\}_{\substack{n=0,1,\ldots \\ j=1,\ldots,2n+1}}}^{\|C^{(0)}(\partial\mathcal{B})\|}. \tag{7.124}$$

For any given function $F \in C^{(0)}(\partial\mathcal{B})$ and arbitrary $\varepsilon > 0$ there exists a linear combination $H_m = \sum_{n=0}^{m} \sum_{j=1}^{2n+1} a_{n,j} H_{n,j}^R$ such that $\|F - H_m\|_{C^{(0)}(\partial\mathcal{B})} \leq \varepsilon$. In connection with the maximum/minimum principle this leads to the result (see, e.g., [**852**], [**254**])

$$Pot^{(0)}(\overline{\mathcal{B}^c}) = \overline{\operatorname{span}_{\substack{n=0,1,\ldots; \\ m=1,\ldots,2n+1}} H_{n,m}^R|\overline{\mathcal{B}^c}}^{\|\cdot\|_{C^{(0)}(\overline{\mathcal{B}^c})}}. \tag{7.125}$$

Unfortunately, a constructive procedure of determining best approximate coefficients $a_{n,j}$ in the $C^{(0)}(\partial\mathcal{B})$-topology seems to be unknown. Therefore, it is a canonical replacement to introduce harmonic splines (characterized by variational principles) in reproducing Hilbert subspaces such as $H_0(\overline{(\mathbb{B}_R^3)^c})|\overline{\mathcal{B}^c}$, so that the spline method can be regarded as an immediate extension of the method of Fourier expansions to spaces with reproducing structure, hence, providing coefficients optimal in a Sobolev-like norm (see, e.g., [**284**]).

Finally it should be remarked that the approximation of boundary values and potential by the method of (generalized) Fourier expansion in terms of outer harmonics is achieved by superposition of functions with " oscillating character". The oscillations grow in number, but they decrease in size with increasing truncation order. The oscillating character of the Fourier expansions remains true (cf. [**260**]), if other trial bases are used (for example, monopoles, multipoles, certain kernel function representations such as Abel-Poisson kernels). Thus, Fourier expansions provide least squares approximation by successive oscillations, which become larger and larger in number, but smaller and smaller in amplitude. It is therefore not (as already A. Sommerfeld [**787**] had pointed out) a technique of "osculating character" (as, for example, interpolation or smoothing in reproducing kernel Hilbert spaces by harmonic splines as proposed, e.g., by [**257, 262, 284**], [**773**]).

7.2. Relevant inverse potential problems

Today, when working with the Earth's gravitational and magnetic potential for practical purposes, we are confronted with different types of inverse problems dependent on certain measurement procedures, for which the following examples may be regarded as exemplary prototypes for a large class of ill-posed inverse problems occurring in Applied Mathematics:

(i) *Satellite gravitational gradiometry (SGG)*, e.g., for geoidal determination, uses gravitational accelerations obtained from a gradiometer on board of a satellite. SGG mathematically leads to a surface integral equation which forms a convolution of the Poisson kernel with second derivatives of the gravitational Hesse tensor.

(ii) *Inverse gravimetry (IG)*, e.g., for geologic exploration, uses (intensities of) gravitational velocities obtained from gravimeters (usually on the Earth's surface, but also in boreholes for some areas). Following Newton's law, IG mathematically leads to a volumetric integral equation, which forms a convolution of the scalar-valued Earth's density distribution with the monopole kernel, i.e., the reciprocal distance between the integration variable and the computation point.

(iii) *Inverse magnetometry (IM)*, e.g., for geologic exploration as well as determination of ground water levels in mining areas, uses (intensities of) magnetic inductions obtained from magnetometers. Following Biot-Savart's law, IM mathematically leads to a volumetric integral equation, which forms a convolution of the vector-valued Earth's magnetization distribution with the dipole kernel, i.e., the gradient of the reciprocal distance between the integration variable and the computation point.

SSG on the one side, and IG as well as IM on the other side constitute inverse problems in form of a convolution Fredholm integral equation of first kind. As already mentioned (see also Section 2), SGG is characterized by a surface integral equation, while IG and IM lead to a volumetric integral expression. As a consequence, they are distinct not only in their mathematical form, but also in their inverse nature, hence, in their character of the ill-posedness. Nonetheless, in the opinion of the authors, both inverse problems have one property in common, they are "too ill-posed" to allow a successful application of a standard regularization technique for practical obligations. So, our decision here is to propose a regularization strategy, adequate under the specific observation of geoscientific aspects and practical needs. The results are quite different regularization concepts, where the regularization techniques earlier explained in this work take over the role of "component idea-giving moments" in the overall realization.

Satellite gravitational gradiometry (SGG). Let the closure $\overline{\mathcal{B}}$ of the regular region $\mathcal{B}$ in $\mathbb{R}^3$, designate the Earth's body including the Earth's surface $\partial\mathcal{B}$.

"Upward continuation" assigns the square-integrable function $F = V|\mathbb{S}^2_R, R < \inf_{x\in\partial\mathcal{B}} |x|$, of the gravitational potential $V \in H_0(\mathbb{R}^3\backslash\mathbb{B}^3_R)$ (cf. (7.120)) the corresponding gravitational tensor values $\mathbf{v} = (\nabla \otimes \nabla)\ V$ in the outer space $\mathcal{B}^c = \mathbb{R}^3\backslash\overline{\mathcal{B}}$ (e.g., on a satellite orbit $\Gamma \subset \mathbb{R}^3\backslash\overline{\mathcal{B}}$ with $\operatorname{dist}(\Gamma, \partial\mathcal{B}) > 0$).

In a preparatory spherically reflected nomenclature we are led to the mapping

$$\mathbf{\Lambda} : F \to \mathbf{v} = (\nabla \otimes \nabla)\ V = \mathbf{\Lambda}(F), \tag{7.126}$$

that assigns the function $F = V|\mathbb{S}^2_R$, $V \in H_0(\mathbb{R}^3\backslash\mathbb{B}^3_R)$, the tensor $\mathbf{v} = (\nabla\otimes\nabla)\ V = \mathbf{\Lambda}(F)$, $\mathbf{v} \in \mathbf{pot}(\mathcal{B}^c)$, via the surface integral equation of the first kind

$$\mathbf{v}(x) = \mathbf{\Lambda}(F)(x) = \int_{\mathbb{S}^2_R} (\nabla_x \otimes \nabla_x)\ \underbrace{\frac{1}{4\pi R}\frac{|x|^2 - R^2}{|x-y|^2}}_{Poisson\ kernel}\ F(y)\ dS(y), \tag{7.127}$$

where R indicates a height satisfying $R < \inf_{x\in\partial\mathcal{B}} |x|$. The result of the mapping $\mathbf{\Lambda}$ is a harmonic tensor field $\mathbf{v}$ in the outer space of the sphere $\mathbb{S}^2_R$, i.e., $\Delta\mathbf{v}(x) = 0,\ x \in \mathbb{R}^3\backslash\overline{\mathbb{B}^3_R}$, such that $F = V|\mathbb{S}^2_R$. The space $\mathbf{\Lambda}(L^2(\mathbb{S}^2_R))$ forms a tensor Hilbert space of Sobolev type (see, e.g., [**266, 302**]). For any R, γ with $R < \inf_{x\in\partial\mathcal{B}} |x| < \gamma \leq \inf_{x\in\Gamma} |x|$ we see that $\mathbf{\Lambda}(L^2(\mathbb{S}^2_R))|\mathbb{S}^2_\gamma$ is a strict subspace of the space $\mathbf{l}^2(\mathbb{S}^2_\gamma)$ of square-integrable tensor-fields on the sphere $\mathbb{S}^2_\gamma$.

Ill-posedness of SGG. The *SGG "downward continuation"* in its simplest form is to determine the square-integrable function F on the spherical (Earth's) surface $\mathbb{S}^2_R$ from the knowledge of the values of the tensorial function $(\nabla_x \otimes \nabla_x)V(x) = \mathbf{v}(x)$, $V \in H_0(\mathbb{R}^3\backslash\mathbb{B}^3_R)$, on a spherical satellite orbit $\mathbb{S}^2_\gamma$, i.e.,

$$\mathbf{v}(x) = \mathbf{\Lambda}(F)(x) = \int_{\mathbb{S}^2_R} \left((\nabla_x \otimes \nabla_x) \frac{1}{4\pi R}\frac{|x|^2 - R^2}{|x-y|^3} \right) \underbrace{F(y)}_{=V(y)}\ dS(y), \quad x \in \mathbb{S}^2_\gamma. \tag{7.128}$$

This spherically reflected formulation of SGG already shows that we are confronted with an ill-posed problem in the sense of Hadamard's classification (see, e.g., [**302**], [**322**], [**307**]):

- Because of the fact $\mathbf{\Lambda}(L^2(\mathbb{S}^2_R))|\mathbb{S}^2_\gamma \subsetneqq \mathbf{l}^2(\mathbb{S}^2_\gamma)$ the problem of "downward continuation" is not solvable for all tensor functions in $\mathbf{l}^2(\mathbb{S}^2_\gamma)$. "Downward continuation" is only feasible, if the integral equation $\mathbf{\Lambda}(F) = \mathbf{v}$ has a right side, which is a restriction of a harmonic tensor field $\mathbf{v}$ to $\mathbb{S}^2_\gamma$. In practice, the existence problem of a solution, however, is not of relevance, since to model F we have to start from discretely given tensor values of $\mathbf{v}$ on the satellite orbit.
- The operator $\mathbf{\Lambda}$ represents a bounded mapping of the space $L^2(\mathbb{S}^2_R)$ into the Sobolev space $\mathbf{\Lambda}(L^2(\mathbb{S}^2_R))|\mathbb{S}^2_\gamma \subsetneqq \mathbf{l}^2(\mathbb{S}^2_\gamma)$. It even has a uniquely defined inverse on $\mathbf{\Lambda}(L^2(\mathbb{S}^2_R))|\mathbb{S}^2_\gamma$. Unfortunately, according to the laws of functional analysis, this inverse is not bounded, i.e., the inversion process cannot be carried out stably.

Altogether, SGG turns out to be an exponentially ill-posed tensorial inverse problem which inextricably needs regularization because of the unboundedness of the existing inverse. In fact, it should be mentioned that two practice-relevant ways play an essential role in our SGG-methodologies, namely the "frequency based" and "space-based" approach. The frequency based spectral approach, because of the exponential ill-posedness, uses as essential tools tensor outer harmonics for the "downward continuation" of the gravitational potential. For the space-based approach, because of the seriously ill-posed SGG-structure, there are good reasons to restrict a regularization procedure to a multiscale mollifier-type methodology within a Runge-Walsh reflected potential theoretic framework (cf. (7.122)), so that geometrically realistic assumptions of the Earth's body and the satellite orbit can be retained.

The Runge-Walsh background for our regularization methodologies shows that the assumption of a spherical satellite orbit $\mathbb{S}^2_\gamma$ is not mathematically mandatory for the SGG-problem (cf. [**306**]). "Downward continuation" can also be understood more generally as the inverse problem of determining $F = V|\mathbb{S}^2_R$ from the integral equation

(7.129)

$$\mathbf{v}(x) = \mathbf{\Lambda}(F)(x) = \int_{\mathbb{S}^2_R} \left((\nabla_x \otimes \nabla_x) \frac{1}{4\pi R} \frac{|x|^2 - R^2}{|x-y|^3} \right) F(y)\, dS(y), \quad x \in \mathbb{R}^3 \backslash \overline{\mathcal{B}},$$

instead of (7.128). As a consequence, discrete potential values from the entire outer space of $\partial\mathcal{B}$ may find their way as input data into the inverse problem, so that we are able to accept real satellite positions of Γ (instead of a spherical orbit).

As we saw earlier, the Runge-Bjerhammar sphere $\mathbb{S}^2_R$ is a sphere lying entirely inside the "real" Earth $\overline{\mathcal{B}}$, i.e., $R < \inf_{x\in\partial\mathcal{B}} |x|$. According to the Runge-Walsh context (cf. (7.122)) we may assume that the (real) gravitational potential of the Earth $\overline{\mathcal{B}}$ as a member of $Pot^{(0)}(\overline{\mathcal{B}^c})$ can be approximated arbitrarily close by a restriction $V|\overline{\mathcal{B}^c}$ of a potential $V \in H_0(\mathbb{R}^3 \backslash \mathbb{B}^3_R)$ (in $\|\cdot\|_{C^{(0)}(\overline{\mathcal{B}^c})}$-norm).

Inverse gravimetry (IG). Our geomathematical point of departure in the case of IG is Poisson's differential equation, which is using the Laplace operator Δ for a regular region $\mathcal{B} \subset \mathbb{R}^3$, $\rho = -\Delta V$ in $\mathcal{B}$, where according to the Newtonian theory, V is the gravitational geopotential to the (sufficiently smooth) density distribution ρ in $\overline{\mathcal{B}}$. The general solution of the Poisson equation can be calculated with the help of the fundamental solution (monopole kernel) of the Laplace operator Δ in $\mathbb{R}^3$

$$F_3^{(1)}(|x-y|) = \frac{1}{4\pi} \frac{1}{|x-y|}, \quad x, y \in \mathbb{R}^3,\ x \neq y. \tag{7.130}$$

by the (volumetric) Newtonian integral representation

$$V(x) = S(\rho)(x) = \int_{\mathcal{B}} \underbrace{F_3^{(1)}(|x-y|)}_{monopole\ kernel}\ \rho(y)\, dy, \quad x \in \mathbb{R}^3. \tag{7.131}$$

For a regular region $\mathcal{B} \subset \mathbb{R}^3$, IG is the problem of determining the scalar-valued density function $\rho \in L^2(\mathcal{B})$ from (information of) the gravitational potential V in accordance with the integral equation (7.131).

We denote the image of $X = L^2(\mathcal{B})$ under the operator S by $Y = S(L^2(\mathcal{B}))$, i.e., Y is characterized by

$$Y = \{V : V = S(\rho) = \int_{\mathcal{B}} F_3^{(1)}(|\cdot - y|)\, \rho(y)\, dy,\ \rho \in L^2(\mathcal{B})\}. \tag{7.132}$$

Furthermore, for any subset $\mathcal{G} \subset \mathbb{R}^3$, we introduce the operator

$$S_{\mathcal{G}} : X = L^2(\mathcal{B}) \to Y|\mathcal{G} = S_{\mathcal{G}}(L^2(\mathcal{B})) \tag{7.133}$$

(more accurately, $S_{\mathcal{G}}^{\mathcal{B}}$), so that $Y|\mathcal{G}$ consists of all functions $S_{\mathcal{G}}(\rho)$ given by

$$\mathcal{G} \ni x \mapsto S_{\mathcal{G}}(\rho)(x) = \int_{\mathcal{B}} F_3^{(1)}(|x-y|)\,\rho(y)\,dy,\ \rho \in L^2(\mathcal{B}) \tag{7.134}$$

(note that S may be formally understood as $S_{\mathbb{R}^3}$ in our nomenclature). Clearly, $Y|\mathcal{G}$ forms a set of harmonic functions in $\mathcal{G}$, provided that $\mathcal{G}$ is a subset of $\mathcal{B}^c = \mathbb{R}^3 \backslash \overline{\mathcal{B}}$.

Unfortunately we are confronted with the difficulty that, contrary to $L^2(\partial\mathcal{B})$, the class $L^2(\mathcal{B})$ of square-integrable functions on a regular region $\mathcal{B}$ is not obtainable only by the L^2-completion of a harmonic function system. As a matter of fact, we have to take into account a so-called *"anharmonic function system"* (see, e.g., [**577**], [**284**]). Therefore, IG demands some preparatory knowledge about essential keystones of Newtonian potential theory:

Decomposition of the Hilbert space $L^2(\mathcal{B})$. Let $\mathcal{B} \subset \mathbb{R}^3$ be a regular region. $\mathcal{F} \in \mathfrak{D}'(\mathcal{B}) = \left(C_c^{(\infty)}(\mathcal{G})\right)'$, $\mathfrak{D}'(\mathcal{B})$ dual space of $\mathfrak{D}(\mathcal{B}) = C_c^{(\infty)}(\mathcal{G})$, is called *distributionally harmonic* if and only if $\Delta\mathcal{F} = 0$. The set of all regular harmonic $L^2(\mathcal{G})$-distributions in $\mathfrak{D}'(\mathcal{B})$ is denoted by $DistPot(\mathcal{B})$. The space $DistPot(\mathcal{B})$ apparently represents a generalization of the set $Pot(\mathcal{B})$ of harmonic functions in $\mathcal{B}$. Indeed (see, e.g., [**175**]), the set $DistPot(\mathcal{B})$ of all regular harmonic $L^2(\mathcal{B})$-distributions is a closed subspace of $L^2(\mathcal{B})$. It is known (see, e.g., [**854**]) that $Pot(\mathcal{B}) \subset DistPot(\mathcal{B})$. Even more, the so-called Weyl's Lemma tells us that $\int_{\mathcal{B}} F(x)\,\Delta\Phi(x)\,dx = 0$ for all $\Phi \in \mathfrak{D}(\mathcal{B})$ implies $F \in Pot(\mathcal{B})$, i.e., $DistPot(\mathcal{B}) \subset Pot(\mathcal{B})$. In consequence, we are led to the following remarkable result: Let $\mathcal{B} \subset \mathbb{R}^3$ be a regular region. Then $DistPot(\mathcal{B}) = Pot(\mathcal{B})$. A well-known result of functional analysis (see, e.g., [**175**]) implies the following decomposition of $L^2(\mathcal{B})$: Let $\mathcal{B} \subset \mathbb{R}^3$ be a regular region. Then

$$L^2(\mathcal{B}) = \overline{Pot(\mathcal{B})}^{\|\cdot\|_{L^2(\mathcal{B})}} \oplus \left(\overline{Pot(\mathcal{B})}^{\|\cdot\|_{L^2(\mathcal{B})}}\right)^{\perp}. \tag{7.135}$$

The members of the class $AnPot(\mathcal{B})$ defined by $AnPot(\mathcal{B}) = \left(\overline{Pot(\mathcal{B})}^{\|\cdot\|_{L^2(\mathcal{G})}}\right)^{\perp}$ are called *anharmonic functions* in $\mathcal{B}$ (see, e.g., [**577**]). Summarizing our results we therefore arrive at the following *decomposition theorem*: Let $\mathcal{B} \subset \mathbb{R}^3$ be a regular region. Then

$$L^2(\mathcal{B}) = Pot(\mathcal{B}) \oplus AnPot(\mathcal{B}). \tag{7.136}$$

Ill-posedness of IG. After these preparations it immediately follows that IG in its desired "explorational formulation" to determine ρ via the integral equation

$$\mathcal{B}^c = \mathbb{R}^3 \backslash \overline{\mathcal{B}} \ni x \mapsto V(x) = S_{\mathcal{B}^c}(\rho)(x) = \int_{\mathcal{B}} G(\Delta;|x-y|)\,\rho(y)\,dy,\ \rho \in L^2(\mathcal{B}), \tag{7.137}$$

i.e., for input data exclusively from $\mathcal{B}^c$, violates all criteria of Hadamard's classification (see, e.g., W. Freeden, M.Z. Nashed [**291**]):

- A solution of IG exists only if $V = S_{\mathcal{B}^c}(\rho)$ belongs to the space $Y|\mathcal{B}^c$. However, it should be pointed out that this restriction does not cause any numerical difficulty, since, in practice, the information of V is only finite-dimensional.
- The most serious problem is the nonuniqueness of the solution: If ρ is a member of class $L^2(\mathcal{B})$, then $S_{\mathcal{B}^c} : L^2(\mathcal{B}) \to Y|\mathcal{B}^c$ given by (7.137) defines a linear operator such that $S_{\mathcal{B}^c}(\rho)$ is continuous in $\overline{\mathcal{B}^c}$, harmonic in $\mathcal{B}^c$, and regular at infinity. The operator $S_{\mathcal{B}^c}$ as defined by (7.137) is surjective, but it is not injective. Indeed, the null space (kernel) of $S_{\mathcal{B}^c}$, i.e., $\mathcal{N}(S_{\mathcal{B}^c}) = AnHarm(\mathcal{B})$ consists of all functions in $L^2(\mathcal{B})$ that are orthogonal to harmonic functions in $\mathcal{B}$. In other words, $\mathcal{N}(S_{\mathcal{B}^c})$ constitutes the anharmonic functions in $\mathcal{B}$.
- The operator $S_{\mathcal{B}^c}$ represents an injective mapping which, however, has a discontinuous inverse.

The most critical problem for practical purposes is that no uniqueness of an interior geologic structure can be obtained from only terrestrial data. So, the problem (7.137) is not properly suited for the geological practice. As a matter of fact, any regularization method (i.e., also the mollifier methodology as discussed later in this work) additionally requires certain preliminary information, preferably from earlier borehole activities in the vicinity of the area $\mathcal{B}$ under investigation (see, e.g., W. Freeden [**270**] and the literature therein) to obtain a useful resolution of the geological strata. Moreover, it should be noted that, for explorational tasks, instead of the scalar gravitational potential V, the (measurement of the) intensity $|v|$ of the geopotential field $v = \nabla V$ is a more realistic gravimetric observational quantity, which leads over to a nonlinear IG-framework.

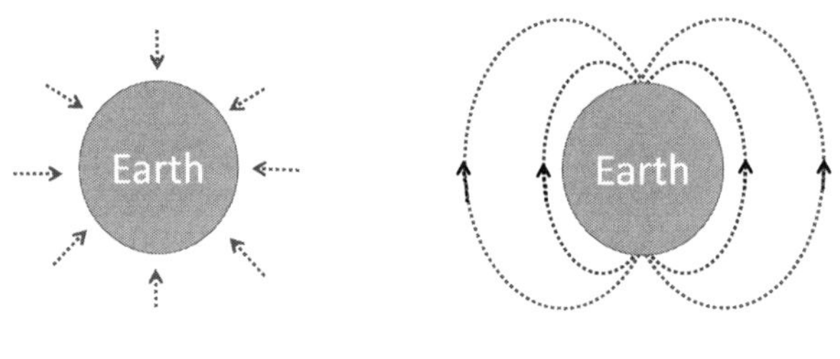

FIGURE 4. The gravitational field lines are monopolar, i.e., they point to the center of mass of the Earth (left), while the magnetic field lines are dipolar (right, see publications in geoscience, e.g., [**59**], [**118**], [**451**], [**661**] for more precise illustrations).

Inverse magnetometry (IM). Magnetometry, being another potential method, has a lot in common with gravimetry and, hence, they are often measured and interpreted together (see, e..g., [**797**], [**95**], [**274**]). Gravitational and magnetic exploration techniques are both passive in that they exploit naturally existing fields of the Earth. The measured quantities are integrated effects of the subsurface and need to be decorrelated in order to offer interpretability (cf. [**272**]). Both problems may be formulated by a Fredholm volumetric integral equation of the first kind. They lack well-posedness, from which the nonuniqueness is the most dramatic feature (see, e.g., [**58**], [**742**]). A regularization procedure similar to that of gravimetry is also feasible, since all criteria of Hadamard's classification are violated for IM similarly to inverse gravimetry. However, there are also significant differences. There are no magnetic monopoles (in contrast to gravitation) and, hence, dipoles are principal constituents for practical purposes. Magnetic poles can be repulsive or attractive, they are not always attractive as in the gravitational case (cf. Figure 4).

The *dipole magnetic induction* is the gradient field of a dipole potential for a distribution of magnetization m

$$
\begin{aligned}
(7.138)\qquad b(x) &= \nabla_x \int_{\mathcal{B}} \underbrace{\nabla_y F_3^{(1)}(|x-y|)}_{\textit{dipole kernel}} \cdot m(y)\, dy \\
&= -\nabla_x \frac{1}{4\pi} \int_{\mathcal{B}} \frac{(x-y)\cdot m(y)}{|x-y|^3}\, dy, \quad x \in \mathbb{R}^3.
\end{aligned}
$$

For points $x \in \mathbb{R}^3 \backslash \overline{\mathcal{B}}$ we have

$$
\begin{aligned}
(7.139)\qquad b(x) &= -\frac{1}{4\pi} \int_{\mathcal{B}} \nabla_x \frac{m(y)\cdot(x-y)}{|x-y|^3}\, dy \\
&= -\frac{1}{4\pi} \int_{\mathcal{B}} \left(\frac{m(y)}{|x-y|^3} - 3\frac{(x-y)\cdot m(y)}{|x-y|^5}(x-y) \right) dy.
\end{aligned}
$$

The magnetic field b is divergence-free, so that, for all points $x \in \mathbb{R}^3 \backslash \overline{\mathcal{B}}$, we have $\nabla \cdot b(x) = 0, x \in \mathbb{R}^3 \backslash \overline{\mathcal{B}}$. Obviously, inverse magnetometry written in analogy to inverse gravimetry (see, e.g., [**270**]) may be seen as the determination of the vectorial magnetization m using the integral equation (7.138) corresponding to the dipole magnetic induction b produced by susceptible material in $\overline{\mathcal{B}}$.

Macroscopic magnetostatics is characterized by a relationship between the magnetic induction b and the magnetic field h in the form

$$(7.140)\qquad h = b - m,$$

which is known (see, e.g., [**100**]) to be equivalent to a relationship between m and h (or b). In fact, for a large number of materials, there is an approximate linear relationship between m and h. If an object made of magnetically linear and additionally isotropic material is placed in a magnetic field produced by a distant electromagnet, the magnetization $m(x)$ at a source point x may be assumed to be the product of the *magnetic susceptibility* $\chi(x)$ and the magnetic field $h(x)$:

$$(7.141)\qquad m(x) = \chi(x)\, h(x).$$

The magnetic susceptibility is called *paramagnetic*, if $\chi > 0$; the magnetic susceptibility is called *diamagnetic*, if $\chi < 0$. Usually, χ is quite small for paramagnetic and diamagnetic materials. A linear relationship between h and m also implies a linear relationship between b and h. In fact, the magnetic dipole induction field b at the same source point x is given by $b(x) = h(x) + m(x)$. Consequently we are allowed to express (7.141) in terms of the susceptibility χ by substituting (7.141) into (7.140) so that $b(x) = (1+\chi(x))\, h(x) = \mu(x)\, h(x)$, where *the relative permeability* $\mu(x)$ is defined by $\mu(x) = 1 + \chi(x)$.

Summarizing we are led to the following conclusions: Magnetization is a vector quantity defined as the magnetic dipole moment per unit volume. The magnetization, m, at any location within a magnetic body $\overline{\mathcal{B}}$ is related to the field h at that location through the magnetic susceptibility, χ, by $m = \chi\ h$. In other words, the more susceptible a material, the more magnetized it becomes when placed in an inducing field. Another magnetic quantity is the dipole magnetic field, b, which is related to the field h through the magnetic permeability, μ, by $b = \mu\ h$. Permeability is related to susceptibility by $\mu(x) = 1 + \chi(x)$. In vacuum, we have $\mu = 1$ and in air, we have $\mu \simeq 1$. There essentially are two characterizations, whereby matter may acquire a macroscopic magnetic-moment distribution (by macroscopic we mean averaged over a large number of atoms), namely paramagnetic for $\chi > 0$ and diamagnetic for $\chi < 0$. In magnetic prospection, the quantity measured is b (or some functionals of it), although we are often interested in the field h. However, since usually $\mu \simeq 1$ we treat b as if it were h.

Inverse magnetometry in Born approximation. If the body's magnetization is parallel to the inducing field with magnitude and direction proportional to the susceptibility χ, we are led over to inverse magnetometry in so-called *Born approximation* for the integral equation (cf. [**100**])

$$b_{Born}(x) = \nabla_x \int_{\mathcal{B}} \nabla_y F_3^{(1)}(|x-y|) \cdot \underbrace{m(y)}_{=\chi(y)h(y)} \ dy. \tag{7.142}$$

The objective now is to render a feasible subsurface distribution of the susceptibility χ, that may have given rise to the anomalous fields in the magnetic survey data. This also is an inverse problem of ill-posed nature. In practice, one often works with input data in a certain direction l, i.e., $l \cdot b$. If the data are acquired in free space (i.e., air), then $\mu \simeq 1$. Thus we treat $l \cdot h$ as if it were $l \cdot b$. Therefore, in connection with (7.2), we are confronted with the inverse problem of determining the susceptibility χ from scalar data of $l \cdot b$:

$$l(x) \cdot b(x) \simeq (l(x) \cdot \nabla_x) \int_{\mathcal{B}} \left(\nabla_y F_3^{(1)}(|x-y|) \cdot \ b(y)\right) \chi(y)\ dy. \tag{7.143}$$

7.3. Satellite gravitational gradiometry (SGG)

The initial situation for the SGG-problem can be recapitulated briefly (see, e.g., R. Rummel et al. [**732–735, 737, 738**], W. Freeden et al. [**286, 302, 305, 321, 323, 324**]) as follows (cf. Figure 5):

Let the regular region $\mathcal{B}$ denote the Earth's interior. Suppose that satellite data

$$\mathbf{v} = (\nabla \otimes \nabla)\, V, \quad V \in H_0(\overline{\mathcal{B}^c}), \tag{7.144}$$

from the Earth's gravitational potential V are known on the "real satellite orbit" $\Gamma \subset \mathcal{B}^c$ with $\operatorname{dist}(\Gamma, \partial\mathcal{B}) > 0$.. *We look for the solution of the satellite gravitational gradiometry problem (SGG-problem) of determining the "terrestrial potential" $V|\partial\mathcal{B} \in L^2(\partial\mathcal{B})$ from (discrete data of) $\mathbf{v} = (\nabla \otimes \nabla)\, V$, on the real satellite orbit Γ.*

Seen from potential theoretic context, SGG is a nonstandard problem of potential theory (cf. [**274**]). The reasons are obvious:

- Tensorial SGG-data (or a scalar manifestation of them such as the second radial derivative on the orbit as input data) do not form the standard equipment of potential theory (such as, e.g., Dirichlet's or Neumann's boundary data). Thus, uniqueness cannot be deduced in the way as known (see, e.g., [**653**], [**493**], [**424**], [**95**], [**853**], [**274**]) for boundary data in classical boundary value problems (such as Dirichlet's or Neumann's boundary value problem). Nevertheless, "upward continuation" via Dirichlet's problem as the inverse to "downward continuation" plays an important role in the mathematical treatment of SGG.
- "Downward continuation" by SGG as the inverse of "upward continuation" (cf. [**735, 736**], [**302, 305, 306**], [**321, 323**], [**312**], [**307**]) leads to an ill-posed problem, since the data are not given on the boundary of the domain of interest, i.e., on the Earth's surface $\partial\mathcal{B}$, but on locations along the orbit Γ. So, SGG-data have its natural limit, what is reflected the exponential damping of the frequency parts (i.e., the orthogonal coefficients) of the gravitational potential with increasing satellite heights.

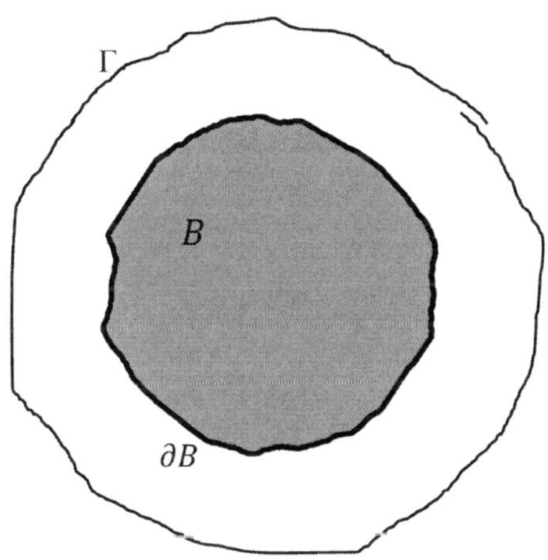

FIGURE 5. The (real) SGG geometric situation as interplay between the Earth's surface $\partial\mathcal{B}$ and the orbit Γ (sectional illustration). The ESA-satellite GOCE (cf. [**736**]) had been launched into its orbit with some additional height so as to perform the complex commissioning activities. In fact, the orbit was left to decay naturally due to atmospheric drag from the initial height of 278 km to the desired height of about 250 km.

In what follows the regularization process of SGG is realized in a twofold sense, namely in spectral form using tensor outer harmonics and in spatial form using mollifier tensor Abel-Poisson kernels:

Spectral ("frequency based") regularization. We first recapitulate the introduction of tensor spherical harmonics as proposed by W. Freeden, M. Schreiner [**324**]: By $\mathbf{l}^2(\mathbb{S}^2)$ we denote the Hilbert space of square-integrable tensor fields $\mathbf{f}:\mathbb{S}^2\to\mathbb{R}^3\otimes\mathbb{R}^3$ with the inner product

$$\langle \mathbf{f},\mathbf{g}\rangle_{\mathbf{l}^2(\mathbb{S}^2)} = \int_{\mathbb{S}^2} \mathbf{f}(\xi)\cdot\mathbf{g}(\xi)\ dS(\xi), \quad \mathbf{f},\mathbf{g}\in\mathbf{l}^2(\mathbb{S}^2), \tag{7.145}$$

and associated norm $\|\cdot\|_{\mathbf{l}^2(\mathbb{S}^2)}$ (note that the space $\mathbf{l}^2(\mathbb{S}^2)$ is the completion of $\mathbf{c}^{(0)}(\mathbb{S}^2)$ with respect to the norm $\|\cdot\|_{\mathbf{l}^2(\mathbb{S}^2)}$).

We introduce the operators $\mathbf{o}^{(i,k)} : C^{(\infty)}(\mathbb{S}^2) \to \mathbf{c}^{(\infty)}(\mathbb{S}^2)$, $i,k = 1,2,3$, which transform scalar functions into tensor fields:

$$\mathbf{o}^{(1,1)}F(\xi) = \xi \otimes \xi F(\xi), \tag{7.146}$$

$$\mathbf{o}^{(1,2)}F(\xi) = \xi \otimes \nabla^*_\xi F(\xi), \tag{7.147}$$

$$\mathbf{o}^{(1,3)}F(\xi) = \xi \otimes L^*_\xi F(\xi), \tag{7.148}$$

$$\mathbf{o}^{(2,1)}F(\xi) = \nabla^*_\xi F(\xi) \otimes \xi, \tag{7.149}$$

$$\mathbf{o}^{(2,2)}F(\xi) = \mathbf{i}_{\tan}(\xi)F(\xi), \tag{7.150}$$

$$\mathbf{o}^{(2,3)}F(\xi) = \left(\nabla^*_\xi \otimes \nabla^*_\xi - L^*_\xi \otimes L^*_\xi\right) F(\xi) + 2\nabla^*_\xi F(\xi) \otimes \xi, \tag{7.151}$$

$$\mathbf{o}^{(3,1)}F(\xi) = L^*_\xi F(\xi) \otimes \xi, \tag{7.152}$$

$$\mathbf{o}^{(3,2)}F(\xi) = \left(\nabla^*_\xi \otimes L^*_\xi + L^*_\xi \otimes \nabla^*_\xi\right) F(\xi) + 2L^*_\xi F(\xi) \otimes \xi, \tag{7.153}$$

$$\mathbf{o}^{(3,3)}F(\xi) = \mathbf{j}_{\tan}(\xi)F(\xi),\ \xi \in \mathbb{S}^2, \tag{7.154}$$

$F \in C^{(2)}(\mathbb{S}^2)$. The tensors $\mathbf{i}_{\tan} = \mathbf{i} - \xi \otimes \xi$ and $\mathbf{j}_{\tan} = \xi \wedge \mathbf{i}$ are the surface identity tensor and the surface rotation tensor, respectively. The adjoint operators $O^{(i,k)}$ to $\mathbf{o}^{(i,k)}$ satisfying

$$\langle \mathbf{o}^{(i,k)}F, \mathbf{f}\rangle_{\mathbf{l}^2(\mathbb{S}^2)} = \langle F, O^{(i,k)}\mathbf{f}\rangle_{L^2(\mathbb{S}^2)} \tag{7.155}$$

for $F \in C^{(2)}(\mathbb{S}^2)$ and $\mathbf{f} \in \mathbf{c}^{(2)}(\mathbb{S}^2)$ are given by

$$O^{(1,1)}\mathbf{f}(\xi) = \xi^T\mathbf{f}(\xi)\xi, \tag{7.156}$$

$$O^{(1,2)}\mathbf{f}(\xi) = -\nabla^*_\xi \cdot p_{\tan}\left(\xi^T\mathbf{f}(\xi)\right), \tag{7.157}$$

$$O^{(1,3)}\mathbf{f}(\xi) = -L^*_\xi \cdot p_{\tan}\left(\xi^T\mathbf{f}(\xi)\right), \tag{7.158}$$

$$O^{(2,1)}\mathbf{f}(\xi) = -\nabla^*_\xi \cdot p_{\tan}\left(\mathbf{f}(\xi)\xi\right), \tag{7.159}$$

$$O^{(2,2)}\mathbf{f}(\xi) = \mathbf{i}_{\tan}(\xi) \cdot \mathbf{f}(\xi), \tag{7.160}$$

$$\begin{aligned} O^{(2,3)}\mathbf{f}(\xi) = {} & \nabla^*_\xi \cdot p_{\tan}\left(\nabla^*_\xi \cdot \mathbf{p}_{\tan,*}\mathbf{f}(\xi)\right) - L^*_\xi \cdot p_{\tan}\left(L^*_\xi \cdot \mathbf{p}_{\tan,*}\mathbf{f}(\xi)\right) \\ & - 2\nabla^*_\xi \cdot p_{\tan}\left(\mathbf{f}(\xi)\xi\right), \end{aligned} \tag{7.161}$$

$$O^{(3,1)}\mathbf{f}(\xi) = -L^*_\xi \cdot p_{\tan}\left(\mathbf{f}(\xi)\xi\right), \tag{7.162}$$

$$\begin{aligned} O^{(3,2)}\mathbf{f}(\xi) = {} & L^*_\xi \cdot p_{\tan}\left(\nabla^*_\xi \cdot \mathbf{p}_{\tan}\mathbf{f}(\xi)\right) + \nabla^*_\xi \cdot p_{\tan}\left(L^*_\xi \cdot \mathbf{p}_{\tan}\mathbf{f}(\xi)\right) \\ & - 2L^*_\xi \cdot p_{\tan}\left(\mathbf{f}(\xi)\xi\right), \end{aligned} \tag{7.163}$$

$$O^{(3,3)}\mathbf{f}(\xi) = \mathbf{j}_{\tan}(\xi) \cdot \mathbf{f}(\xi), \tag{7.164}$$

where

$$p_{\tan}f(\xi) = f(\xi) - (\xi \cdot f(\xi))\xi,\ \xi \in \mathbb{S}^2, \tag{7.165}$$

$$\mathbf{p}_{\tan,*}\mathbf{f}(\xi) = \mathbf{f}(\xi) - \xi \otimes ((\mathbf{f}(\xi))^T\xi),\ \xi \in \mathbb{S}^2. \tag{7.166}$$

With the help of the operators $\mathbf{o}^{(i,k)}$ we are able to define a set of *tensor spherical harmonics* $\{\mathbf{y}^{(i,k)}_{n,m}\}_{i,k=1,2,3;\ n=0_{ik},\ldots;\ m=1,\ldots,2n+1}$ by

$$\mathbf{y}^{(i,k)}_{n,m} = \left(\mu^{(i,k)}_n\right)^{-1/2} \mathbf{o}^{(i,k)}Y_{n,m}, \tag{7.167}$$

where the normalization constants $\mu_n^{(i,k)}$ are given by

(7.168)
$$\mu_n^{(i,k)} = \begin{cases} 1, & (i,k) = (1,1), \\ 2, & (i,k) \in \{(2,2),(3,3)\}, \\ n(n+1), & (i,k) \in \{(1,2),(1,3),(2,1),(3,1)\}, \\ 2n(n+1)(n(n+1)-2), & (i,k) \in \{(2,3),(3,2)\}. \end{cases}$$

For simplicity, we use the abbreviation

(7.169)
$$0_{ik} = \begin{cases} 0, & (i,k) \in \{(1,1),(2,2),(3,3)\}, \\ 1, & (i,k) \in \{(1,2),(1,3),(2,1),(3,1)\}, \\ 2, & (i,k) \in \{(2,3),(3,2)\}. \end{cases}$$

By $\mathbf{harm}_n^{(i,k)}(\mathbb{S}^2)$ we denote the space of all tensor spherical harmonics of degree n and kind (i,k). By $\mathbf{harm}_n(\mathbb{S}^2)$ we denote the space of all tensor spherical harmonics of degree n.

Let $\{\mathbf{y}_{n,m}^{(i,k)}\}_{m=1,\ldots,2n+1}$ be an $\mathbf{l}^2(\mathbb{S}^2)$-orthonormal basis of $\mathbf{harm}_n^{(i,k)}(\mathbb{S}^2)$. Then, for any pair $(\xi,\eta) \in \mathbb{S}^2 \times \mathbb{S}^2$, the *addition theorem for tensor spherical harmonics*

(7.170)
$$\sum_{m=1}^{2n+1} \mathbf{y}_{n,m}^{(i,k)}(\xi) \otimes \mathbf{y}_{n,m}^{(p,q)}(\eta) = \frac{2n+1}{4\pi} \mathbf{P}_n^{(i,k,p,q)}(\xi,\eta),$$

is valid for $i,k,p,q \in \{1,2,3\}$, where $\mathbf{P}_n^{(i,k,p,q)} : \mathbb{S}^2 \times \mathbb{S}^2 \to \mathbb{R}^3 \otimes \mathbb{R}^3 \otimes \mathbb{R}^3 \otimes \mathbb{R}^3$ denote the *Legendre tensors of degree n* defined by

(7.171)
$$\mathbf{P}_n^{(i,k,p,q)} = \left(\mu_n^{(i,k)}\right)^{-1/2} \left(\mu_n^{(p,q)}\right)^{-1/2} \mathbf{o}_\xi^{(i,k)} \mathbf{o}_\eta^{(p,q)} P_n(\xi \cdot \eta), \quad \xi,\eta \in \mathbb{S}^2.$$

Note that, for sufficiently smooth tensor fields $\mathbf{f} : \mathbb{S}^2 \to \mathbb{R}^3 \otimes \mathbb{R}^3$ of the form

(7.172)
$$\mathbf{f}(\xi) = \sum_{i,k=1}^{3} F_{i,k}(\xi) \varepsilon^i \otimes \varepsilon^k, \quad x \in \mathbb{S}^2,$$

we set

(7.173)
$$\mathbf{o}_\xi^{(p,q)} \mathbf{f}(\xi) = \sum_{i,k=1}^{3} \left(\mathbf{o}_\xi^{(p,q)} F_{i,k}(\xi)\right) \otimes \varepsilon^i \otimes \varepsilon^k.$$

Explicit representations of the Legendre tensors $\mathbf{P}_n^{(i,k,p,q)}$ can be found in W. Freeden, M. Schreiner [**321, 326**]. The system $\{\mathbf{y}_{n,m}^{(i,k)}\}$ of tensor spherical harmonics was introduced (cf. [**321**]) by concentrating on the fact that the Helmholtz decomposition into normal and tangential tensor fields is fulfilled (cf. [**321, 326**], [**659**]). However, one disadvantage of the set of Helmholtz tensor spherical harmonics is that these functions introduced above are not eigenfunctions of the (scalar) Beltrami operator. Nonetheless, this property leads us to define so-called Hardy-Hodge tensor spherical harmonics and, subsequently, outer harmonics in such a way that they fulfill the Laplace equation in the outer space. This is the reason (cf. [**321, 326**]) why we introduce another set of operators $\tilde{\mathbf{o}}^{(i,k)} : C^{(\infty)}(\mathbb{S}^2) \to \mathbf{c}^{(\infty)}(\mathbb{S}^2)$, $i,k = 1,2,3$, based on the operators $\mathbf{o}^{(i,k)}$ by

$$
\begin{pmatrix} \tilde{\mathbf{o}}^{(1,1)}Y_n \\ \tilde{\mathbf{o}}^{(1,2)}Y_n \\ \tilde{\mathbf{o}}^{(2,1)}Y_n \\ \tilde{\mathbf{o}}^{(2,2)}Y_n \\ \tilde{\mathbf{o}}^{(3,3)}Y_n \end{pmatrix} = \mathbf{a}_D \begin{pmatrix} Y_n \\ Y_n \\ Y_n \\ Y_n \\ Y_n \end{pmatrix}, \quad \text{and} \quad \begin{pmatrix} \tilde{\mathbf{o}}^{(1,3)}Y_n \\ \tilde{\mathbf{o}}^{(2,3)}Y_n \\ \tilde{\mathbf{o}}^{(3,1)}Y_n \\ \tilde{\mathbf{o}}^{(3,2)}Y_n \end{pmatrix} = \mathbf{b}_D \begin{pmatrix} Y_n \\ Y_n \\ Y_n \\ Y_n \end{pmatrix}, \tag{7.174}
$$

where the matrix operators $\mathbf{a}_D$ and $\mathbf{b}_D$ are defined by

$$
\mathbf{a}_D = \begin{pmatrix} \mathbf{o}^{(1,1)}(D+1)(D+2) & -\mathbf{o}^{(1,2)}(D+2) & -\mathbf{o}^{(2,1)}(D+2) & -\frac{1}{2}o^{(2,2)}(D+2)(D+1) & \frac{1}{2}\mathbf{o}^{(2,3)} \\ \mathbf{o}^{(1,1)}D^2 & \mathbf{o}^{(1,2)}D & -\mathbf{o}^{(2,1)}(D-1) & -\frac{1}{2}\mathbf{o}^{(2,2)}D(D-1) & -\frac{1}{2}\mathbf{o}^{(2,3)} \\ \mathbf{o}^{(1,1)}(D+1)^2 & -\mathbf{o}^{(1,2)}(D+1) & \mathbf{o}^{(2,1)}(D+2) & \frac{1}{2}\mathbf{o}^{(2,2)}(D+2)(D+1) & -\frac{1}{2}\mathbf{o}^{(2,3)} \\ \mathbf{o}^{(1,1)}D(D-1) & \mathbf{o}^{(1,2)}(D-1) & \mathbf{o}^{(2,1)}(D-1) & -\frac{1}{2}\mathbf{o}^{(2,2)}D(D-1) & \frac{1}{2}\mathbf{o}^{(2,3)} \\ 0 & 0 & \mathbf{o}^{(2,1)} & -\frac{1}{2}\mathbf{o}^{(2,2)}D(D+1) & -\frac{1}{2}\mathbf{o}^{(2,3)} \end{pmatrix} \tag{7.175}
$$

and

$$
\mathbf{b}_D = \begin{pmatrix} \mathbf{o}^{(1,3)}(D+1) & \mathbf{o}^{(3,1)} & -\frac{1}{2}\mathbf{o}^{(3,2)} & -\frac{1}{2}\mathbf{o}^{(3,3)}D(D+1) \\ \mathbf{o}^{(1,3)}D & -\mathbf{o}^{(3,1)} & \frac{1}{2}\mathbf{o}^{(3,2)} & \frac{1}{2}\mathbf{o}^{(3,3)}D(D+1) \\ 0 & \mathbf{o}^{(3,1)}(D+2) & -\frac{1}{2}\mathbf{o}^{(3,2)} & \frac{1}{2}\mathbf{o}^{(3,3)}(D+2)(D+1) \\ 0 & \mathbf{o}^{(3,1)}(D-1) & \frac{1}{2}\mathbf{o}^{(3,2)} & -\frac{1}{2}\mathbf{o}^{(3,3)}D(D-1) \end{pmatrix}, \tag{7.176}
$$

and D is the (pseudodifferential) operator $D = (-\Delta + \frac{1}{4})^{1/2} - \frac{1}{2}$ of order 1 characterized by $DY_n = D^\wedge(n)Y_n = nY_n$ for all $Y_n \in \mathrm{Harm}_n(\mathbb{S}^2)$. The adjoint operators $\tilde{O}^{(i,k)} : \mathbf{c}^{(\infty)}(\mathbb{S}^2) \to C^{(\infty)}(\mathbb{S}^2)$, $i,k = 1,2,3$, to the operators $\tilde{\mathbf{o}}^{(i,k)}$ satisfying the equation

$$
\langle \tilde{\mathbf{o}}^{(i,k)}G, \mathbf{f} \rangle_{\mathbf{l}^2(\mathbb{S}^2)} = \langle G, \tilde{O}^{(i,k)}\mathbf{f} \rangle_{L^2(\mathbb{S}^2)}, \tag{7.177}
$$

$\mathbf{f} \in \mathbf{c}^{(\infty)}(\mathbb{S}^2)$, $G \in C^{(\infty)}(\mathbb{S}^2)$, are obviously given by

$$
\begin{pmatrix} \tilde{O}^{(1,1)}Y_n \\ \tilde{O}^{(1,2)}Y_n \\ \tilde{O}^{(2,1)}Y_n \\ \tilde{O}^{(2,2)}Y_n \\ \tilde{O}^{(3,3)}Y_n \end{pmatrix} = \mathbf{a}_D \begin{pmatrix} Y_n \\ Y_n \\ Y_n \\ Y_n \\ Y_n \end{pmatrix}, \quad \text{and} \quad \begin{pmatrix} \tilde{O}^{(1,3)}Y_n \\ \tilde{O}^{(2,3)}Y_n \\ \tilde{O}^{(3,1)}Y_n \\ \tilde{O}^{(3,2)}Y_n \end{pmatrix} = \mathbf{b}_D \begin{pmatrix} Y_n \\ Y_n \\ Y_n \\ Y_n \end{pmatrix}. \tag{7.178}
$$

After these preliminaries we are now in a position to introduce the tensor spherical harmonics $\tilde{\mathbf{y}}_{n,m}^{(i,k)} = \left(\tilde{\mu}_n^{(i,k)}\right)^{-1/2} \tilde{\mathbf{o}}^{(i,k)} Y_{n,m}$, $n = \tilde{0}_{ik}, \ldots; m = 1, \ldots, 2n+1$, with

$$
\tilde{0}_{ik} = \begin{cases} 0, & (i,k) \in \{(1,1),(2,1),(3,1)\}, \\ 1, & (i,k) \in \{(1,2),(1,3),(2,3),(3,3)\}, \\ 2, & (i,k) \in \{(2,2),(3,2)\}, \end{cases} \tag{7.179}
$$

and normalization constants

$$\tilde{\mu}_n^{(1,1)} = (n+2)(n+1)(2n+3)(2n+1), \tag{7.180}$$

$$\tilde{\mu}_n^{(1,2)} = 3n^4, \tag{7.181}$$

$$\tilde{\mu}_n^{(1,3)} = n(n+1)^2(2n+1), \tag{7.182}$$

$$\tilde{\mu}_n^{(2,1)} = (n+1)^2(2n+3)(2n+1), \tag{7.183}$$

$$\tilde{\mu}_n^{(2,2)} = n(n-1)(2n+1)(2n-1), \tag{7.184}$$

$$\tilde{\mu}_n^{(2,3)} = n^2(n+1)^2, \tag{7.185}$$

$$\tilde{\mu}_n^{(3,1)} = n^2(n+1)(2n+1), \tag{7.186}$$

$$\tilde{\mu}_n^{(3,2)} = n(n+1)^2(2n+1), \tag{7.187}$$

$$\tilde{\mu}_n^{(3,3)} = n^2(n-1)(2n+1). \tag{7.188}$$

In contrary to the system (7.167), each member of the system $\{\tilde{\mathbf{y}}_{n,m}^{(i,k)}\}$ is an eigenfunction of the Beltrami operator. More explicitly, we have the following result: Assume that $\{Y_{n,m}\}_{n=0,1,\ldots;\ m=1,\ldots,2n+1}$ is an $L^2(\mathbb{S}^2)$-orthonormal set of scalar spherical harmonics. Then, the set

$$\left\{\tilde{\mathbf{y}}_{n,m}^{(i,k)}\right\}_{i,k=1,2,3;\ n=\tilde{0}_{ik},\ldots;\ m=1,\ldots,2n+1}, \tag{7.189}$$

forms an $\mathbf{l}^2(\mathbb{S}^2)$-orthonormal set of tensor spherical harmonics which is closed in $\mathbf{c}^{(0)}(\mathbb{S}^2)$ and $\mathbf{l}^2(\mathbb{S}^2)$ with respect to $\|\cdot\|_{\mathbf{c}^{(0)}(\mathbb{S}^2)}$ and $\|\cdot\|_{\mathbf{l}^2(\mathbb{S}^2)}$, respectively, and complete in $\mathbf{l}^2(\mathbb{S}^2)$ with respect to $(\cdot,\cdot)_{\mathbf{l}^2(\mathbb{S}^2)}$. Furthermore, we are able to verify that

$$\Delta_\xi^* \tilde{\mathbf{y}}_{n,m}^{(1,1)} = -(n+2)(n+3)\tilde{\mathbf{y}}_{n,m}^{(1,1)}, \tag{7.190}$$

$$\Delta_\xi^* \tilde{\mathbf{y}}_{n,m}^{(1,2)} = -n(n+1)\tilde{\mathbf{y}}_{n,m}^{(1,2)}, \tag{7.191}$$

$$\Delta_\xi^* \tilde{\mathbf{y}}_{n,m}^{(1,3)} = -(n+1)(n+2)\tilde{\mathbf{y}}_{n,m}^{(1,3)}, \tag{7.192}$$

$$\Delta_\xi^* \tilde{\mathbf{y}}_{n,m}^{(2,1)} = -n(n+1)\tilde{\mathbf{y}}_{n,m}^{(2,1)}, \tag{7.193}$$

$$\Delta_\xi^* \tilde{\mathbf{y}}_{n,m}^{(2,2)} = -(n-1)(n-2)\tilde{\mathbf{y}}_{n,m}^{(2,2)}, \tag{7.194}$$

$$\Delta_\xi^* \tilde{\mathbf{y}}_{n,m}^{(2,3)} = -n(n-1)\tilde{\mathbf{y}}_{n,m}^{(2,3)}, \tag{7.195}$$

$$\Delta_\xi^* \tilde{\mathbf{y}}_{n,m}^{(3,1)} = -(n+1)(n+2)\tilde{\mathbf{y}}_{n,m}^{(3,1)}, \tag{7.196}$$

$$\Delta_\xi^* \tilde{\mathbf{y}}_{n,m}^{(3,2)} = -n(n-1)\tilde{\mathbf{y}}_{n,m}^{(3,2)}, \tag{7.197}$$

$$\Delta_\xi^* \tilde{\mathbf{y}}_{n,m}^{(3,3)} = -n(n+1)\tilde{\mathbf{y}}_{n,m}^{(3,3)}, \tag{7.198}$$

where the application of the Beltrami operator is understood component-by-component.

Because of the completeness of the tensor spherical harmonics (see the proof in [**277**]) every tensor field $\mathbf{f} \in \mathbf{l}^2(\mathbb{S}^2)$ can be written as an orthogonal (Fourier) series

$$\mathbf{f} = \sum_{i,k=1}^{3} \sum_{n=\tilde{0}_{ik}}^{\infty} \sum_{m=1}^{2n+1} \mathbf{f}^{(i,k)\wedge_{\mathbf{l}^2(\mathbb{S}^2)}}(n,m)\ \tilde{\mathbf{y}}_{n,m}^{(i,k)} \tag{7.199}$$

(in $\|\cdot\|_{\mathbf{l}^2(\mathbb{S}^2)}$–sense), where the Fourier coefficients are given by

$$\mathbf{f}^{(i,k)\wedge_{\mathbf{l}^2(\mathbb{S}^2)}}(n,m) = \langle \mathbf{f}, \tilde{\mathbf{y}}^{(i,k)}_{n,m}\rangle_{\mathbf{l}^2(\mathbb{S}^2)} = \int_{\mathbb{S}^2} \mathbf{f}(\xi)\cdot\tilde{\mathbf{y}}^{(i,k)}_{n,m}(\xi)\, dS(\xi). \tag{7.200}$$

Tensor outer harmonics. We introduce a class of *tensor outer harmonics* (cf. [**276**], [**321, 326**]) based on the definition of tensor spherical harmonics:

$$\mathbf{h}^{R;(1,1)}_{n,m}(x) = \frac{1}{R}\left(\frac{R}{|x|}\right)^{n+3} \tilde{\mathbf{y}}^{(1,1)}_{n,m}\left(\frac{x}{|x|}\right), \tag{7.201}$$

$$\mathbf{h}^{R;(1,2)}_{n,m}(x) = \frac{1}{R}\left(\frac{R}{|x|}\right)^{n+1} \tilde{\mathbf{y}}^{(1,2)}_{n,m}\left(\frac{x}{|x|}\right), \tag{7.202}$$

$$\mathbf{h}^{R;(1,3)}_{n,m}(x) = \frac{1}{R}\left(\frac{R}{|x|}\right)^{n+2} \tilde{\mathbf{y}}^{(1,3)}_{n,m}\left(\frac{x}{|x|}\right), \tag{7.203}$$

$$\mathbf{h}^{R;(2,1)}_{n,m}(x) = \frac{1}{R}\left(\frac{R}{|x|}\right)^{n+1} \tilde{\mathbf{y}}^{(2,1)}_{n,m}\left(\frac{x}{|x|}\right), \tag{7.204}$$

$$\mathbf{h}^{R;(2,2)}_{n,m}(x) = \frac{1}{R}\left(\frac{R}{|x|}\right)^{n-1} \tilde{\mathbf{y}}^{(2,2)}_{n,m}\left(\frac{x}{|x|}\right), \tag{7.205}$$

$$\mathbf{h}^{R;(2,3)}_{n,m}(x) = \frac{1}{R}\left(\frac{R}{|x|}\right)^{n} \tilde{\mathbf{y}}^{(2,3)}_{n,m}\left(\frac{x}{|x|}\right), \tag{7.206}$$

$$\mathbf{h}^{R;(3,1)}_{n,m}(x) = \frac{1}{R}\left(\frac{R}{|x|}\right)^{n+2} \tilde{\mathbf{y}}^{(3,1)}_{n,m}\left(\frac{x}{|x|}\right), \tag{7.207}$$

$$\mathbf{h}^{R;(3,2)}_{n,m}(x) = \frac{1}{R}\left(\frac{R}{|x|}\right)^{n} \tilde{\mathbf{y}}^{(3,2)}_{n,m}\left(\frac{x}{|x|}\right), \tag{7.208}$$

$$\mathbf{h}^{R;(3,3)}_{n,m}(x) = \frac{1}{R}\left(\frac{R}{|x|}\right)^{n+1} \tilde{\mathbf{y}}^{(3,3)}_{n,m}\left(\frac{x}{|x|}\right), \tag{7.209}$$

$x \in \mathbb{R}^3\backslash\mathbb{B}^3_R$, $n = \tilde{0}_{ik}, \ldots$, $m = 1, \ldots, 2n+1$. It is not difficult to show that the following properties are satisfied:

- $\mathbf{h}^{R;(i,k)}_{n,m}$ is of class $\mathbf{c}^{(\infty)}(\mathbb{R}^3\backslash\overline{\mathbb{B}^3_R})$,
- $\Delta_x \mathbf{h}^{R;(i,k)}_{n,m}(x) = 0$, $x \in \mathbb{R}^3\backslash\overline{\mathbb{B}^3_R}$, i.e., the component functions of $\mathbf{h}^{R;(i,k)}_{n,m}$ fulfill the Laplace equation,
- $\mathbf{h}^{R;(i,k)}_{n,m}|\mathbb{S}^2_R = \frac{1}{R}\tilde{\mathbf{y}}^{(i,k)}_{n,m}$,
- $\langle \mathbf{h}^{R;(i,k)}_{n,m}, \mathbf{h}^{R;(p,q)}_{l,s}\rangle_{\mathbf{l}^2(\mathbb{S}^2_R)} = \int_{\mathbb{S}^2_R} \mathbf{h}^{R;(i,k)}_{n,m}(x)\cdot\mathbf{h}^{R;(p,q)}_{l,s}(x)\, dS(x) = \delta_{i,p}\delta_{k,q}\delta_{n,l}\delta_{m,s}$, where $\mathbf{l}^2(\mathbb{S}^2_R)$ is the space of square-integrable tensor fields on $\mathbb{S}^2_R$,
- $|\mathbf{h}^{R;(i,k)}_{n,m}(x)| = O\left(|x|^{-1}\right)$, $|x| \to \infty$.

It must be emphasized that the spherically reflected formulation of the tensorial SGG-problem exclusively uses the tensor outer harmonics of kind $(1,1)$ specified by Eq. (7.201).

Runge-Walsh concept. In our approach (cf. Figure 6) we come back to the *Runge-Walsh concept* in the formulation [**254**]. The Runge-Walsh approximation property justifies the approximation of the Earth's gravitational potential on and outside the regular surface by a linear combination of scalar outer harmonics, i.e., by harmonic "trial functions of polynomial nature" showing a "harmonicity domain" $\mathcal{B}^c$

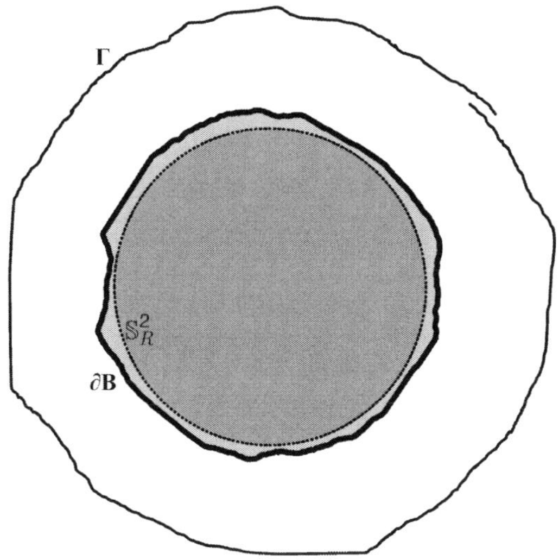

FIGURE 6. The spherically reflected Runge-Bjerhammar situation (for more details see [**254**] and [**51**]).

with $\mathbb{R}^3 \backslash \overline{\mathbb{B}_R^3} \supset \overline{\mathcal{B}^c}$. It should be remarked that the same property holds true, for example, for outer ellipsoidal harmonics outside a surface Ξ located totally in $\mathcal{B}$ with $\operatorname{dist}(\partial\mathcal{B}, \Xi) > 0$. However, for reasons of numerical economy and efficiency we restrict ourselves to outer spherical harmonics and Runge-Bjerhammar spheres. Moreover, the Runge-Walsh property can be verified for certain systems of monopoles and multipoles inside $\mathcal{B}$ (cf. [**254**], [**284**]), certain kernel representations as well as harmonic splines as well as scaling and wavelet functions (cf. [**266**], [**274**], [**284**]). From the Ph.D.-thesis H. Nutz [**659**] we know that

$$\mathbf{pot}^{(0)}(\overline{\mathcal{B}^c}) = \overline{\operatorname{span}_{\substack{n=0,1,\ldots;\\ m=1,\ldots,2n+1}} (\mathbf{h}_{n,m}^{R,(1,1)})|\overline{\mathcal{B}^c}}^{\|\cdot\|_{\mathbf{c}^{(0)}(\overline{\mathcal{B}^c})}}. \tag{7.210}$$

Again, it should be remarked that the same property holds true for other trial systems, such as tensor outer ellipsoidal harmonics outside an internal ellipsoid, fundamental systems of Hesse tensors of certain monopoles inside $\mathcal{B}$, certain tensor kernel representations as well as tensor harmonic splines and wavelets (see [**302**], [**305**]).

From the scalar Runge-Walsh concept (7.125) it can be deduced additionally (see, e.g., [**254**], [**274**], [**284**]) that, for an arbitrarily small number $\varepsilon > 0$ and any given $\tilde{V} \in Pot^{(0)}(\overline{\mathcal{B}^c})$, there exists a potential $V \in H_0(\mathbb{R}^3 \backslash \overline{\mathbb{B}_R^3})$ with $R < \inf_{x\in\partial\mathcal{B}} |x|$ such that

$$\sup_{x\in\overline{\mathcal{B}^c}} |\tilde{V}(x) - V(x)| \leq \varepsilon \tag{7.211}$$

and

$$\sup_{x\in\Gamma} \left| \underbrace{(\nabla\otimes\nabla)\tilde{V}(x)}_{=\tilde{\mathbf{v}}(x)} - \underbrace{(\nabla\otimes\nabla)V(x))}_{=\mathbf{v}(x)} \right| \leq C\,\varepsilon \tag{7.212}$$

for some positive constant C (depending on the geometries of $\partial\mathcal{B}$ and Γ). The calamity of evaluating the gravitational potential $\tilde{V}$ (which we consider as the Earth's gravitational potential) by a potential V in terms of outer harmonics is that these basis functions are globally supported. This is the reason why they do not show any space localization but ideal frequency (momentum) localization (for a more detailed description see [**297**]). This property makes outer harmonics laborous to use for high resolution modeling at local scale. As a matter of fact, the uncertainty principle (see, e.g., [**297**] and the references therein) leads to the conclusion that outer harmonics are well suited to resolve low-frequency problems at global scale, i.e., to represent trend approximations. As a well promising compromise, "sum conglomerates" of outer harmonics, i.e., so-called (outer) harmonic kernel functions, can be handled (see [**297**] offering a limited, but appropriately balanced range of frequency as well as space localization). Even better, we can construct families of kernels which control the increase of space localization at the cost of the decrease of frequency localization by specifying a scale parameter. This leads, at least for local purposes, to the multiscale philosophy by scaling and wavelet functions in space as well as frequency context.

Summarizing our considerations we are led to the following conclusion: A viable way to model SGG may be based on the Runge property that there exists a potential V outside a Runge- Bjerhammer sphere $\mathbb{S}^2_R$ inside the Earth in ε-accuracy ($\varepsilon > 0$, arbitrarily small) to the Earth's gravitational potential $\tilde{V}$ such that

$$\sup_{x\in\Gamma} \left| \underbrace{(\nabla\otimes\nabla)\tilde{V}(x)}_{=\tilde{\mathbf{v}}(x)} - \underbrace{(\nabla\otimes\nabla)V(x))}_{=\mathbf{v}(x)} \right| = \sup_{x\in\Gamma} |\tilde{\mathbf{v}}(x) - \mathbf{v}(x)| \leq C\,\varepsilon. \tag{7.213}$$

As a consequence of our excursion to the Runge-Walsh theory, the relation between the tensorial measurements $\tilde{\mathbf{v}}$, i.e., the gradiometer orbital data of $\tilde{V}$ on Γ and the auxiliary potential V outside the sphere $\mathbb{S}^2_R$ may be explained for points x on the satellite orbit Γ as follows (cf. [**307**]):

(in frequency-based approach) by a (pseudo)differential equation of the form

$$\sum_{n=0}^{\infty}\sum_{m=1}^{2n+1} V^{\wedge_{L^2(\mathbb{S}^2_R)}}(n,m)\sqrt{\tilde{\mu}_n^{(1,1)}}\ \mathbf{h}^{R;(1,1)}_{n,m} = (\nabla\otimes\nabla)V(x) = \mathbf{v}(x) \approx \tilde{\mathbf{v}}(x), \tag{7.214}$$

where

$$V^{\wedge_{L^2(\mathbb{S}^2_R)}}(n,m) = \int_{\mathbb{S}^2_R} V(x)\ H^R_{n,m}(x)\ dS(x).$$

(in space-based approach) by a linear Fredholm integral equation of the first kind (7.215)

$$\int_{\mathbb{S}^2_R} \underbrace{(\nabla_x\otimes\nabla_x)\,\frac{1}{4\pi R}\,\frac{|x|^2-R^2}{|x-y|^3}}_{=\mathbf{k}_R(x,y)}\ V(y)\ \ dS(y) = \ (\nabla\otimes\nabla)V(x) = \mathbf{v}(x) \approx \tilde{\mathbf{v}}(x),$$

(note that " $\approx$ " means "approximately equal", so that the input data $\tilde{\mathbf{v}}(x)$ may be replaced by $\mathbf{v}(x)$ in pointwise sense). In other words, SGG in frequency as well as space domain is appropriately modeled in the spherically reflected nomenclature of V, thereby taking as input data the tensor measurements $\tilde{\mathbf{v}}(x)$ of the (actual) Earth's potential $\tilde{V}$ on the orbit Γ.

SGG-uniqueness. Some words about the uniqueness of spherically reflected SGG should be made: Keeping in mind that any solution of the SGG problem in tensor spherical framework can be expressed as a series of outer harmonics due to the Runge-Walsh approximation property and observing the closure and completeness of the spherical harmonics in the space of square-integrable functions on spheres, we are able to conclude that the SGG problem is uniquely solvable (up to some low order spherical harmonics) involving the $O^{(1,1)}$, $O^{(1,2)}$, $O^{(2,1)}$, $O^{(2,2)}$, and $O^{(2,3)}$ components.

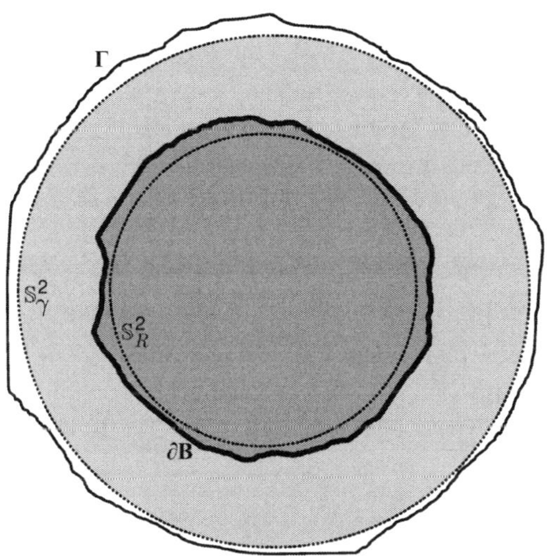

FIGURE 7. The geometric situation of satellite gravitational gradiometry as discussed in our frequency framework.

More concretely, we are able to formulate the following results: Let V be of class $Pot^{(0)}(\mathbb{R}^3 \backslash \mathbb{B}^3_\gamma)$, with $\gamma < \inf_{x\in\Gamma} |x|$ (cf. Figure 7) . Then the following statements are valid:

(1) $O^{(i,k)} \nabla \otimes \nabla V|\mathbb{S}^2_\gamma = 0$ if $(i,k) \in \{(1,3),(3,1),(3,2),(3,3)\}$,
(2) $O^{(i,k)} \nabla \otimes \nabla V|\mathbb{S}^2_\gamma = 0$ for $(i,k) \in \{(1,1),(2,2)\}$ if and only if $V|\mathbb{S}^2_\gamma = 0$,
(3) $O^{(i,k)} \nabla \otimes \nabla V|\mathbb{S}^2_\gamma = 0$ for $(i,k) \in \{(1,2),(2,1)\}$ if and only if $V|\mathbb{S}^2_\gamma$ is constant,
(4) $O^{(2,3)} \nabla \otimes \nabla V|\mathbb{S}^2_\gamma = 0$ if and only if $V|\mathbb{S}^2_\gamma$ is linear combination of spherical harmonics of degree 0 and 1.

The above SGG-conditions (see also [**276**]) give detailed information, which tensor components of the Hesse-tensor are suitable to guarantee the uniqueness of the SGG-problem. In fact, for a potential of class $Pot^{(0)}(\mathbb{R}^3 \backslash \mathbb{B}^3_\gamma)$ with vanishing spherical harmonic moments of degree 0 and 1 such as the Earth's disturbing potential (see, e.g., [**423**], [**824**]), uniqueness is assured for all listed cases.

SGG in frequency-based Sobolev space framework. The considerations below use some scalar pre-work to be found in [**266**], [**305**]. Our goal is to formulate the SGG-problem in terms of a pseudodifferential operator equation involving suitable Sobolev reference spaces, where the scalar case merely serves as a preparation of the tensorial context, which will be studied subsequently:

Scalar case. We let $\mathcal{A}$ be the linear space consisting of all sequences $\{A_n\}_{n\in\mathbb{N}_0}$ of real numbers $A_n \neq 0$, $n \in \mathbb{N}_0$: $\mathcal{A} = \{\{A_n\} : A_n \in \mathbb{R},\, A_n \neq 0,\, n \in \mathbb{N}_0\}$. We consider the set $\mathcal{E} = \mathcal{E}(\{A_n\}; \mathbb{R}^3\backslash\mathbb{B}^3_R)$ given by

$$\mathcal{E} = \Big\{ F \in Pot^{(\infty)}(\mathbb{R}^3\backslash\mathbb{B}^3_R) \ : \ \sum_{n=0}^{\infty}\sum_{m=1}^{2n+1} A_n^2 \left(F^{\wedge_{L^2(\mathbb{S}^2_R)}}(n,m)\right)^2 < \infty \Big\}, \tag{7.216}$$

where

$$F^{\wedge_{L^2(\mathbb{S}^2_R)}}(n,m) = \int_{\mathbb{S}^2_R} F(y)\, H^R_{n,m}(y)\, dS(y). \tag{7.217}$$

Note that we assume that R is chosen in accordance to the Runge-Walsh concept (cf. Figure 7). On $\mathcal{E}$ we introduce an inner product $\langle\cdot,\cdot\rangle_{H(\{A_n\};\mathbb{R}^3\backslash\mathbb{B}^3_R)}$ by

$$\langle F, G\rangle_{H(\{A_n\};\mathbb{R}^3\backslash\mathbb{B}^3_R)} = \sum_{n=0}^{\infty}\sum_{m=1}^{2n+1} A_n^2\, F^{\wedge_{L^2(\mathbb{S}^2_R)}}(n,m)\, G^{\wedge_{L^2(\mathbb{S}^2_R)}}(n,m), \quad F, G \in \mathcal{E}. \tag{7.218}$$

The associated norm is given by $\|F\|_{H(\{A_n\};\mathbb{R}^3\backslash\mathbb{B}^3_R)} = \left(\langle F, F\rangle_{H(\{A_n\};\mathbb{R}^3\backslash\mathbb{B}^3_R)}\right)^{1/2}$.

The *(scalar) Sobolev space* H $(= H(\{A_n\}; \mathbb{R}^3\backslash\mathbb{B}^3_R))$ is the completion of $\mathcal{E}$ under the norm $\|\cdot\|_{H(\{A_n\};\mathbb{R}^3\backslash\mathbb{B}^3_R)}$: $H(\{A_n\}; \mathbb{R}^3\backslash\mathbb{B}^3_R) = \overline{\mathcal{E}}^{\|\cdot\|_{H(\{A_n\};\mathbb{R}^3\backslash\mathbb{B}^3_R)}}$. H equipped with the inner product $\langle\cdot,\cdot\rangle_H$ is a separable Hilbert space. The system $\{H^{*\{A_n\}}_{n,m}(R;\cdot)\}$ given by $H^{*\{A_n\}}_{n,m}(R;x) = A_n^{-1} H^R_{n,m}(x), x \in \overline{\mathbb{R}^3\backslash\mathbb{B}^3_R}$, is a Hilbert basis (we also simply write $H^*_{n,m}(R;\cdot)$ instead of $H^{*\{A_n\}}_{n,m}(R;\cdot)$ if no confusion is likely to arise). Any function $F \in H$ can be expanded as a Fourier series in terms of the basis $H^*_{n,m}(R;\cdot)$:

$$F = \sum_{n=0}^{\infty}\sum_{m=1}^{2n+1} F^{\wedge_H}(n,m)\, H^{*\{A_n\}}_{n,m}(R;\cdot), \tag{7.219}$$

where $F^{\wedge_H}(n,m) = \langle F, H^*_{n,m}(R;\cdot)\rangle_H$ (note that, if no confusion is likely to arise, we will also use the notation $F^{\wedge_H}(n,m) = F^{\wedge_{H(\{A_n\};\mathbb{R}^3\backslash\mathbb{B}^3_R)}}(n,m)$). Furhermore, according to its construction, the space H has the reproducing kernel function $K(\cdot,\cdot) : \mathbb{R}^3\backslash\mathbb{B}^3_R \times \mathbb{R}^3\backslash\mathbb{B}^3_R \to \mathbb{R}$ given by

$$K(x,y) = \sum_{n=0}^{\infty}\sum_{m=1}^{2n+1} H^{*\{A_n\}}_{n,m}(R;x)\, H^{*\{A_n\}}_{n,m}(R;y), \quad x, y \in \mathbb{R}^3\backslash\mathbb{B}^3_R. \tag{7.220}$$

Our goal next is the introduction of a class of scalar Sobolev spaces based on the spherical symbol $\{(\Delta^{*;R})^\wedge(n)\}_{n\in\mathbb{N}_0}$ of the Beltrami operator $\Delta^{*;R}$ related to the sphere $\mathbb{S}^2_R$. We observe that

$$\Delta^{*;R} Y_{n,m} = \frac{1}{R^2}\Delta^* Y_{n,m} = -\frac{1}{R^2} n(n+1) Y_{n,m}, \quad n \in \mathbb{N}_0;\ m = 1,\ldots,2n+1. \tag{7.221}$$

In particular, we have $\Delta^{*;R} Y_{0,1} = 0$, which requires a shift by a nonzero constant, for example $\frac{1}{4R^2}$, to obtain invertibility. As a consequence we formally obtain

$$\left(-\Delta^{*;R} + \frac{1}{4R^2}\right)^{s/2} Y_{n,m} = \left(\frac{n+1/2}{R}\right)^{s} Y_{n,m} \tag{7.222}$$

and

$$\left(\left(-\Delta^{*;R} + \frac{1}{4R^2}\right)^{s/2} F\right)^{\wedge} (n,m) = \left(\frac{n+1/2}{R}\right)^{s} F^{\wedge}(n,m), \tag{7.223}$$

$n \in \mathbb{N}_0$, $m = 1, \ldots, 2n+1$. For $s \in \mathbb{R}$ we let

$$H_s = H\left(\left\{\left(\frac{n+1/2}{R}\right)^{s}\right\}; \mathbb{R}^3 \backslash \mathbb{B}_R^3\right), \tag{7.224}$$

such that the norm in H_s fulfills $\|F\|_{H_s} = \|(-\Delta_x^{*;R} + \frac{1}{4R^2})^{s/2} F\|_{L^2(\mathbb{S}_R^2)}$. For H_0 we identify the norm $\|\cdot\|_{H_0}$ with the $\|\cdot\|_{L^2(\mathbb{S}_R^2)}$-norm. In fact, the space H_0 may be understood as the space of all solutions of the Dirichlet boundary value problem in $\mathbb{R}^3 \backslash \mathbb{B}_R^3$ corresponding to $L^2(\mathbb{S}_R^2)$-boundary values (as known from (7.120) for $\mathcal{B} = \mathbb{B}_R^3$). Furthermore, if $t < s$, then we have $\|F\|_{H_t} \leq \|F\|_{H_s}$ and $H_s \subset H_t$.

In order to formulate some results (cf. [**266**], [**276**]) about the convergence of an expansion in terms of outer harmonics to a function in ordinary pointwise sense (Sobolev Lemma) we need the concept of *summable sequences* $\{A_n\}_{n=0,1,\ldots} \in \mathcal{A}$, i.e., $\{A_n\}_{n=0,1,\ldots} \in \mathcal{A}$ satisfying

$$\sum_{n=0}^{\infty} \frac{2n+1}{A_n^2} < \infty. \tag{7.225}$$

Sobolev lemma. Assume that the sequences $\{A_n\}_{n\in\mathbb{N}_0}, \{B_n\}_{n\in\mathbb{N}_0} \in \mathcal{A}$ are given in such a way that $\{B_n^{-1} A_n\}_{n\in\mathbb{N}_0}$, $B_n \neq 0$, is summable. Then, any F of class $H\left(\{B_n^{-1} A_n\}; \mathbb{R}^3 \backslash \mathbb{B}_R^3\right)$ corresponds to a potential of class $Pot^{(0)}(\mathbb{R}^3 \backslash \mathbb{B}_R^3)$. The Sobolev Lemma states that in the case of summability of the sequence $\{B_n^{-1} A_n\}_{n\in\mathbb{N}_0}$, the Fourier series in terms of the basis functions $H_{n,m}^* \in H\left(\{B_n^{-1} A_n\}; \mathbb{R}^3 \backslash \mathbb{B}_R^3\right)$ is continuous on the boundary $\mathbb{S}_R^2$. In particular, we have the following statement: If $F \in H_s(\mathbb{R}^3 \backslash \mathbb{B}_R^3)$ with $s > 1$, then F corresponds to a function of class $Pot^{(0)}(\mathbb{R}^3 \backslash \mathbb{B}_R^3)$.

For any F in $L^2(\mathbb{S}_R^2)$, there exists one and only one "harmonic continuation" $V \in H(\{A_n\}; \mathbb{R}^3 \backslash \mathbb{B}_R^3)$, which is given by

$$V(x) = \sum_{n=0}^{\infty} \sum_{m=1}^{2n+1} A_n^2 \; F^{\wedge_{L^2(\mathbb{S}_R^2)}}(n,m) \; H_{n,m}^*(R;x), \quad x \in \mathbb{R}^3 \backslash \mathbb{B}_R^3, \tag{7.226}$$

where

$$F^{\wedge_{L^2(\mathbb{S}_R^2)}}(n,m) = \int_{\mathbb{S}_R^2} F(y) \; H_{n,m}^R(y) \; dS(y) = \frac{1}{A_n^2} V^{\wedge_H}(n,m), \tag{7.227}$$

$n = 1, 2, \ldots$; $m = 1, \ldots, 2n+1$.

Tensorial case. In order to introduce Sobolev spaces for tensor fields we start from

$$\mathbf{e} = \left\{ \mathbf{f} \in \mathbf{pot}^{(\infty)}(\mathbb{R}^3 \backslash \mathbb{B}^3_R) : \sum_{n=0}^{\infty} \sum_{m=1}^{2n+1} A_n^2 \, (\mathbf{f}^{\wedge_{\mathbf{l}^2(\mathbb{S}^2_R)}}(n,m))^2 < \infty \right\}, \tag{7.228}$$

where $\mathbf{f}^{\wedge_{\mathbf{l}^2(\mathbb{S}^2_R)}}(n,m) = \int_{\mathbb{S}^2_R} \mathbf{f}(y) \cdot \mathbf{h}^{R;(1,1)}_{n,m}(y) \, dS(y)$. Equipped with the inner product

$$\langle \mathbf{f}, \mathbf{g} \rangle_{\mathbf{h}(\{A_n\};\mathbb{R}^3 \backslash \mathbb{B}^3_R)} = \sum_{n=0}^{\infty} \sum_{m=1}^{2n+1} A_n^2 \, \mathbf{f}^{\wedge_{\mathbf{l}^2(\mathbb{S}^2_R)}}(n,m) \, \mathbf{g}^{\wedge_{\mathbf{l}^2(\mathbb{S}^2_R)}}(n,m), \tag{7.229}$$

$\mathbf{f}, \mathbf{g} \in \mathbf{e}$, the space $\mathbf{e}$ becomes a pre-Hilbert space. We define the Sobolev space $\mathbf{h} = \mathbf{h}(\{A_n\}; \mathbb{R}^3 \backslash \mathbb{B}^3_R)$ as the completion of $\mathbf{e}$ under the norm $\| \cdot \|_{\mathbf{h}(\{A_n\};\mathbb{R}^3 \backslash \mathbb{B}^3_R)}$, which denotes the norm associated to $\langle \cdot, \cdot \rangle_{\mathbf{h}(\mathbb{R}^3 \backslash \mathbb{B}^3_R)}$: $\mathbf{h} = \overline{\mathbf{e}}^{\| \cdot \|_{\mathbf{h}(\{A_n\};\mathbb{R}^3 \backslash \mathbb{B}^3_R)}}$. If no confusion is likely to arise we also use $\mathbf{h}$ instead of $\mathbf{h}(\{A_n\}; \mathbb{R}^3 \backslash \mathbb{B}^3_R))$. The space $\mathbf{h}$ equipped with the inner product $\langle \cdot, \cdot \rangle_{\mathbf{h}(\{A_n\};\mathbb{R}^3 \backslash \mathbb{B}^3_R)}$ is a separable Hilbert space. The system $\{\mathbf{h}^{*\{A_n\}}_{n,m}(R; \cdot)\}_{n=0,1,\ldots,m=1,\ldots,2n+1}$, given by $\mathbf{h}^{*\{A_n\}}_{n,m}(R; x) = A_n^{-1} \mathbf{h}^{R;(1,1)}_{n,m}(x)$, $x \in \mathbb{R}^3 \backslash \mathbb{B}^3_R$, represents an $\mathbf{h}$-orthonormal Hilbert basis in $\mathbf{h}$ (note that we sometimes write $\mathbf{h}^*_{n,m}(R; \cdot)$ instead of $\mathbf{h}^{*\{A_n\}}_{n,m}(R; \cdot)$). As a consequence we can expand a function $\mathbf{f} \in \mathbf{h}$ as a Fourier series in terms of the basis $\mathbf{h}^{*\{A_n\}}_{n,m}(R; \cdot)$

$$\mathbf{f} = \sum_{n=0}^{\infty} \sum_{m=1}^{2n+1} \mathbf{f}^{\wedge_{\mathbf{h}(\{A_n\};\mathbb{R}^3 \backslash \mathbb{B}^3_R)}}(n,m) \, \mathbf{h}^{*\{A_n\}}_{n,m}(R; \cdot), \tag{7.230}$$

where $\mathbf{f}^{\wedge_{\mathbf{h}}}(n,m) = \mathbf{f}^{\wedge_{\mathbf{h}(\{A_n\};\, \mathbb{R}^3 \backslash \mathbb{B}^3_R)}}(n,m) = \langle \mathbf{f}, \mathbf{h}^{*\{A_n\}}_{n,m}(R; \cdot) \rangle_{\mathbf{h}}$. The Hilbert space $\mathbf{h}$ possesses the reproducing kernel $\mathbf{K}(\cdot,\cdot) : \mathbb{R}^3 \backslash \mathbb{B}^3_R \otimes \mathbb{R}^3 \backslash \mathbb{B}^3_R \to \mathbb{R}^3 \otimes \mathbb{R}^3 \otimes \mathbb{R}^3 \otimes \mathbb{R}^3$ given by

$$\mathbf{K}(x,y) = \sum_{n=0}^{\infty} \sum_{m=1}^{2n+1} \mathbf{h}^{*\{A_n\}}_{n,m}(x) \otimes \mathbf{h}^{*\{A_n\}}_{n,m}(y), \tag{7.231}$$

$x, y \in \mathbb{R}^3 \backslash \mathbb{B}^3_R$, i.e.,

- for all $x \in \mathbb{R}^3 \backslash \mathbb{B}^3_R$, $\tilde{O}^{(1,1)}_R \mathbf{K}_{\mathbf{h}}(\cdot, x) \in \mathbf{h}$, where the operator $\tilde{O}^{(1,1)}_R$ is the extension of the adjoint operator of $\tilde{\mathbf{o}}^{(1,1)}_R$ to tensor fields of rank four,
- $\tilde{O}^{(1,1)}_R \mathbf{f}(x) = \left\langle \tilde{O}^{(1,1)}_R \mathbf{K}_{\overline{\mathbf{h}(\mathbb{S}^{2,\mathrm{ext}}_R)}}(\cdot, x), \mathbf{f} \right\rangle_{\mathbf{h}}$ for $\mathbf{f} \in \mathbf{h}$ and $x \in \mathbb{R}^3 \backslash \mathbb{B}^3_R$.

Finally, in analogy to the scalar case, we set $\mathbf{h}_s = \mathbf{h}\left(\left\{ \left(\frac{n+1/2}{R} \right)^s \right\}; \mathbb{R}^3 \backslash \mathbb{B}^3_R \right)$.

Tensorial Sobolev lemma. Let the sequences $\{A_n\}_{n=0,1,\ldots}$, $\{B_n\}_{n \in \mathbb{N}_0} \in \mathcal{A}$ be given such that $\{B_n^{-1} A_n\}_{n=0,1,\ldots} \in \mathcal{A}$ is summable. Then each $\mathbf{f} \in \mathbf{h}\left(\{B_n^{-1} A_n\}; \mathbb{R}^3 \backslash \mathbb{B}^3_R\right)$ corresponds to a function of class $\mathbf{pot}^{(0)}(\mathbb{R}^3 \backslash \mathbb{B}^3_R)$. For any $\mathbf{f} \in \mathbf{l}^2(\mathbb{S}^2_R)$, there exists one and only one tensorial "harmonic upward continuation" $\mathbf{v} \in \mathbf{h}$ of the form

$$\mathbf{v}(x) = \sum_{n=0}^{\infty} \sum_{m=1}^{2n+1} A_n^2 \, \mathbf{f}^{\wedge_{\mathbf{l}^2(\mathbb{S}^2_R)}}(n,m) \, \mathbf{h}^*_{n,m}(R; x), \quad x \in \mathbb{R}^3 \backslash \mathbb{B}^3_R, \tag{7.232}$$

where

$$\mathbf{f}^{\wedge_{\mathbf{l}^2(\mathbb{S}^2_R)}}(n,m) = \int_{\mathbb{S}^2_R} \mathbf{f}(y) \cdot \mathbf{h}^{R;(1,1)}_{n,m}(y) \, S(y) = \frac{1}{A_n^2} \mathbf{v}^{\wedge_{\mathbf{h}}}(n,m). \tag{7.233}$$

Pseudodifferential operators. Since we know, at least in the spherically reflected context, which conditions guarantee the uniqueness of a SGG-solution we can turn to the question how to find a solution from tensorial data and what we mean with a spectral solution, when we have to take into account the ill-posedness. This leads to analyze the SGG-problem step by step by use of the concept of pseudodifferential operators (as introduced by [**811**], [**276**] for the scalar case).

We shortly introduce the framework of scalar as well as tensorial pseudodifferential operators (PDO's) in a parallel way:

Scalar pseudodifferential operators. Let $\{\Lambda^\wedge(n)\}_{n=0,1,\ldots}$ be a sequence of real numbers. The operator $\Lambda : H(\mathbb{R}^3\backslash\mathbb{B}^3_R) \to H(\mathbb{R}^3\backslash\mathbb{B}^3_\tau)$, $\tau \geq R > 0$, defined by

$$\Lambda F = \sum_{n=0}^{\infty}\sum_{m=1}^{2n+1} \Lambda^\wedge(n)\; F^{\wedge_{H(\mathbb{R}^3\backslash\mathbb{B}^3_R))}}(n,m)\; H^*_{n,m}(\tau;\cdot), \tag{7.234}$$

is called a *scalar pseudodifferential operator of order t* if

$$\lim_{n\to\infty} \frac{|\Lambda^\wedge(n)|}{\left(n+\frac{1}{2}\right)^t} = \text{const} \neq 0 \tag{7.235}$$

holds true for some $t \in \mathbb{R}$. If the limit

$$\lim_{n\to\infty} \frac{|\Lambda^\wedge(n)|}{\left(n+\frac{1}{2}\right)^t} = 0 \tag{7.236}$$

holds true for all $t \in \mathbb{R}$, the operator Λ is called a *scalar pseudodifferential operator of exponential order.* The sequence $\{\Lambda^\wedge(n)\}_{n=0,1,\ldots}$ is called the *symbol of the scalar PDO* Λ.

Tensorial pseudodifferential operators. Let $\{\mathbf{\Lambda}^\wedge(n)\}_{n=0,1,\ldots}$ be a sequence of real numbers. The operator $\mathbf{\Lambda} : H(\mathbb{R}^3\backslash\mathbb{B}^3_R) \to \mathbf{h}(\mathbb{R}^3\backslash\mathbb{B}^3_\tau)$, $\tau \geq R > 0$, defined by

$$\mathbf{\Lambda} F = \sum_{n=0}^{\infty}\sum_{m=1}^{2n+1} \mathbf{\Lambda}^\wedge(n)\; F^{\wedge_{H(\mathbb{R}^3\backslash\mathbb{B}^3_R))}}(n,m)\; \mathbf{h}^*_{n,m}(\tau;\cdot), \tag{7.237}$$

is called a *tensorial pseudodifferential operator of order t* if

$$\lim_{n\to\infty} \frac{|\mathbf{\Lambda}^\wedge(n)|}{\left(n+\frac{1}{2}\right)^t} = \text{const} \neq 0 \tag{7.238}$$

holds true for some $t \in \mathbb{R}$. If the limit

$$\lim_{n\to\infty} \frac{|\mathbf{\Lambda}^\wedge(n)|}{\left(n+\frac{1}{2}\right)^t} = 0 \tag{7.239}$$

holds true for all $t \in \mathbb{R}$, the operator $\mathbf{\Lambda}$ is called a *tensorial pseudodifferential operator of exponential order.* The sequence $\{\mathbf{\Lambda}^\wedge(n)\}_{n=0,1,\ldots}$ is called the *symbol of the tensorial PDO* $\mathbf{\Lambda}$.

Within the context of pseudodifferential operators we are able to formulate the SGG-problem (for more details on pseudodifferential operators the reader should consult [**811**], [**266**], [**276**], [**274**] and the references therein).

Upward/downward continuation. Let $\mathbb{S}^2_R$ be a Runge-Bjerhammar sphere inside the real Earth $\mathcal{B}$, i.e., $R < \inf_{x\in\partial\mathcal{B}} |x|$. Furthermore, let γ be smaller than the lowest possible altitude of the satellite, i.e., $\gamma < \inf_{x\in\Gamma} |x|$ (cf. Figure 8). Consider a potential of class $H(\{A_n\};\mathbb{R}^3\backslash\mathbb{B}^3_R)$ of the form

$$F = \sum_{n=0}^{\infty}\sum_{m=1}^{2n+1} F^{\wedge_{H(\{A_n\};\mathbb{R}^3\backslash\mathbb{B}^3_R)}}(n,m)\; H^{*\{A_n\}}_{n,m}(R;\cdot). \tag{7.240}$$

The *upward continuation operator* $\Lambda^{R,\gamma}_{up}$ associates to $F \in H(\{A_n\};\mathbb{R}^3\backslash\mathbb{B}^3_R)$ the solution $\Lambda^{R,\gamma}_{up}F$ of the Dirichlet problem $\Lambda^{R,\gamma}_{up}F \in Pot^{(0)}(\mathbb{R}^3\backslash\mathbb{B}^3_\gamma)$ corresponding to the boundary values $(\Lambda^{R,\gamma}_{up}F)|\mathbb{S}^2_\gamma = F|\mathbb{S}^2_\gamma$. The upward continuation operator $\Lambda^{R,\gamma}_{up}$ possesses the associated symbol $(\Lambda^{R,\gamma}_{up})^\wedge(n) = (R/\gamma)^n$. The inverse of $\Lambda^{R,\gamma}_{up}$ is called the *downward continuation operator*, so that $\Lambda^{R,\gamma}_{down} = (\Lambda^{R,\gamma}_{up})^{-1}$. It brings down the gravitational potential at height γ to the height R and has the associated symbol $(\Lambda^{R,\gamma}_{down})^\wedge(n) = (\gamma/R)^n$. It is obvious, that the upward continuation is well-posed, whereas the downward continuation leads to an ill-posed problem.

Pseudodifferential operator of the Hesse tensor. Let us consider the operator of the Hesse tensor $\nabla\otimes\nabla : H(\mathbb{R}^3\backslash\mathbb{B}^3_R) \to \mathbf{h}(\mathbb{R}^3\backslash\mathbb{B}^3_R)$, which associates to $F \in H(\{A_n\};\mathbb{R}^3\backslash\mathbb{B}^3_R)$ the solution $x \mapsto \mathbf{\Lambda F(x)} = (\nabla\otimes\nabla)\mathbf{F(x)}, \mathbf{x} \in \mathbb{R}^\mathbf{3}\backslash\mathbb{B}^\mathbf{3}_\mathbf{R}$, of the Dirichlet problem corresponding to the boundary values $(\mathbf{\Lambda F})|\mathbb{S}^\mathbf{2}_\mathbf{R} = (\nabla\otimes\nabla)\mathbf{F}|\mathbb{S}^\mathbf{2}_\mathbf{R}$. This is an operator of order 2 given by the symbol

$$\mathbf{\Lambda^\wedge(n,m)} = \frac{\sqrt{\mathbf{(n+2)(n+1)(2n+3)(2n+1)}}}{\mathbf{R^2}}, \quad \mathbf{n = 0,1,\ldots, m = 1,\ldots,2n+1.}$$

Tensorial SGG-operator. The SGG operator $\mathbf{\Lambda} : H(\mathbb{R}^3\backslash\mathbb{B}^3_R) \to \mathbf{h}(\mathbb{R}^3\backslash\mathbb{B}^3_\gamma)$ expressed in terms of outer harmonics is given by

(7.241)

$$\mathbf{\Lambda} H^*_{n,m}(R;x) = \frac{\sqrt{(n+2)(n+1)(2n+3)(2n+1)}}{\gamma^2}\left(\frac{R}{\gamma}\right)^n \mathbf{h}^*_{n,m}(\gamma;x), \quad x\in\mathbb{R}^3\backslash\mathbb{B}^3_\gamma.$$

The symbol of this operator is given by

$$\begin{aligned}&\mathbf{\Lambda}^\wedge(n,m)\\ &= \left(\frac{R}{\gamma}\right)^n \frac{\sqrt{(n+2)(n+1)(2n+3)(2n+1)}}{\gamma^2}, \quad n=0,1,\ldots, m=1,\ldots,2n+1.\end{aligned}$$

The SGG-operator applied to $F \in H(\mathbb{R}^3\backslash\mathbb{B}^3_R)$ is representable as an outer harmonic series expansion

$$\mathbf{\Lambda} F(x) = \sum_{n=0}^{\infty}\sum_{m=1}^{2n+1} \mathbf{\Lambda}^\wedge(n,m) F^{\wedge_{H(\mathbb{R}^3\backslash\mathbb{B}^3_R)}}(n,m)\; \mathbf{h}^*_{n,m}(\gamma;x). \tag{7.242}$$

Up to now, we assumed spherical geometry in connection with spherical harmonics, i.e., we presented spherical harmonics which are adequate for the common spherical approach, where the reference surface is supposed to be a sphere and the data points are given on a spherical satellite orbit. Next, we make the first steps to a Runge-Walsh concept which enables the application of arbitrary geometries. The basic idea underlying the Runge-Walsh concept is to specify two spheres as illustrated in Figure 7, thereby leading back to spherical basis functions and using the specific properties of outer harmonics. So, our frequency based strategy to handle the tensorial (inverse) SGG-problem is as follows: Starting from the data given on the real satellite orbit Γ (which is not necessarily required here to be a closed surface) we pull down the tensorial information to a sphere $\mathbb{S}^2_\gamma$ of radius γ with $\mathrm{dist}(\mathbb{S}^2_\gamma, \Gamma) > 0$. By virtue of "downward continuation" from $\mathbb{S}^2_\gamma$ to the sphere $\mathbb{S}^2_R$ inside the Earth $\mathcal{B}$ such that $\mathrm{dist}(\mathbb{S}^2_R, \partial\mathcal{B}) > 0$ we are able to express the solution in terms of scalar outer harmonics, i.e., we obtain a representation of the gravitational potential on the real Earth's surface $\partial\mathcal{B}$ using data on the real orbit Γ. The key idea of the solution process of the SGG-problem of determining the Earth's gravitational potential on the Earth's surface from orbital values $\mathbf{v}$ of the Hesse tensor on Γ (cf. Figure 7) is to use the "Runge-Walsh framework", that there exists a potential outside the Runge-Bjerhammar sphere $\mathbb{S}^2_R$ inside the Earth $\mathcal{B}$ with ε-accuracy ($\varepsilon > 0$, but arbitrarily small) to the real gravitational potential outside and on the Earth's surface $\partial\mathcal{B}$ under consistent approximation of the Hesse tensors of both potentials on the satellite orbit Γ.

Multiscale frequency regularization. Tensor spherical harmonics of type (1,1) allow us to express the Hesse tensor applied to solid (outer) harmonics in the form

$$(7.243)\quad (\nabla\otimes\nabla)\, H^R_{n,m} = \sqrt{\tilde{\mu}_n^{(1,1)}}\ \mathbf{h}^{R;(1,1)}_{n,m}, \quad \tilde{\mu}_n^{(1,1)} = (n+2)(n+2)(2n-3)(2n-1).$$

This leads to the *tensor-isotropic SGG-pseudodifferential equation* (cf. [**326**])

$$\mathbf{\Lambda}(V) = \sum_{n=0}^{\infty}\sum_{m=1}^{2n+1} V^{\wedge_{L^2(\mathbb{S}^2_R)}}(n,m)\sqrt{\tilde{\mu}_n^{(1,1)}}\ \mathbf{h}^{R;(1,1)}_{n,m} \ = \ (\nabla\otimes\nabla)V = \mathbf{v},$$

as spectral (frequency) representation for the inversion of the SGG-integral equation $\Lambda V = \mathbf{v}$. Equivalently, we have a representation as convolution

$$(7.244)\qquad \mathbf{\Lambda}(V) = \int_{\mathbb{S}^2_R} \mathbf{K}^{\mathbf{\Lambda}}_R(\cdot, y)\ V(y)\ dS(y) \ = \ (\nabla\otimes\nabla)V = \mathbf{v},$$

where the tensorial kernel $\mathbf{K}^{\mathbf{\Lambda}}_R(\cdot,\cdot)$ is given by

$$(7.245)\qquad \mathbf{K}^{\mathbf{\Lambda}}_R(x,y) \ = \ \sum_{n=0}^{\infty}\sum_{m=1}^{2n+1}\sqrt{\tilde{\mu}_n^{(1,1)}}\ \ H^R_{n,m}(x)\ \mathbf{h}^{R;(1,1)}_{n,m}(y)$$

in spectral way. By the completeness of the system $\{H^R_{n,m}\}$ this enables us to conclude in the framework $\mathbf{l}^2(\mathbb{S}^2_R)$ of square-integrable tensor fields on $\mathbb{S}^2_R$ that

$$(7.246)\qquad \begin{aligned} \langle(\nabla\otimes\nabla)V, \mathbf{h}^{R;(1,1)}_{p,q}\rangle_{\mathbf{L}^2(\mathbb{S}^2_R)} &= \int_{\mathbb{S}^2_R}(\nabla\otimes\nabla)V(y)\cdot\mathbf{h}^{R;(1,1)}_{p,q}(y)\ dS(y) \\ &= V^{\wedge_{L^2(\mathbb{S}^2_R)}}(p,q)\sqrt{\tilde{\mu}_n^{(1,1)}}. \end{aligned}$$

Consequently, we obtain the following expansion for the potential V in $\mathbb{R}^3\backslash\overline{\mathbb{B}^3_R}$

$$(7.247)\qquad \begin{aligned} V(x) &= \mathbf{\Lambda}^{-1}(\nabla\otimes\nabla V)(x) \\ &= \sum_{n=0}^{\infty}\sum_{m=1}^{2n+1}\langle\underbrace{\nabla\otimes\nabla V}_{=\mathbf{v}}, \mathbf{h}^{R;(1,1)}_{n,m}\rangle_{\mathbf{l}^2(\mathbb{S}^2_R)}(\tilde{\mu}_n^{(1,1)})^{-1/2} H^R_{n,m}(x), \end{aligned}$$

i.e.,

$$V(x) = \mathbf{\Lambda}^{-1}\mathbf{v}(x) = \int_{\mathbb{S}^2_R} \mathbf{K}_R^{\mathbf{\Lambda}^{-1}}(x,y)\cdot \mathbf{v}(y)\, dS(y) \tag{7.248}$$

with

$$\mathbf{K}_R^{\mathbf{\Lambda}^{-1}}(x,y) = \sum_{n=0}^{\infty}\sum_{m=1}^{2n+1} (\tilde{\mu}_n^{(1,1)})^{-1/2}\, H_{n,m}^R(x)\, \mathbf{h}_{n,m}^{R;(1,1)}(y). \tag{7.249}$$

This formula expresses the gravitational potential V in terms of the tensor $\mathbf{v}$ on locations x of the satellite orbit Γ. $\mathbf{\Lambda}$ is a linear bounded injective compact operator (see, e.g., [**266**] for more details) so it follows from functional analysis that the SGG-problem is ill-posed because of the unboundedness of $\mathbf{\Lambda}^{-1}$. Hence, the SGG-problem needs regularization. As described earlier, a *regularization strategy for the SGG-problem* is a sequence $\{R_j\}_{j\in\mathbb{N}_0}$ of linear bounded pseudodifferential operators R_j so that $\lim_{j\to\infty} R_j\mathbf{\Lambda}(V) = V$ in the outer space of $\mathbb{S}^2_R$, i.e., the operators $R_j\Lambda$ converge in pointwise sense to the identity operator. In principle, all aforementioned regularization methods are applicable to SGG. In what follows, however, we restrict ourselves on two essential *SGG-multiscale regularization strategies.*

Multiscale Tikhonov regularization strategy. This method makes use of the (non-bandlimited) isotropic Tikhonov-kernels (scaling functions) Φ_j, $j = 0,1,\ldots$, given by

$$\Phi_j(x,y) = \sum_{n=0}^{\infty}\sum_{m=1}^{2n+1} (\Phi_j)^\wedge(n) H_{n,m}^R(x) H_{n,m}^R(y) \tag{7.250}$$

with

$$(\Phi_j)^\wedge(n) = \frac{(\mathbf{\Lambda}^\wedge(n))^2}{(\mathbf{\Lambda}^\wedge(n))^2 + \mu_j^2}, \qquad n = 0,1,\ldots,\ j = 0,1,\ldots, \tag{7.251}$$

where $\{\mu_j\}$, $j = 0,1,\ldots$, is a sequence of real numbers satisfying $\lim_{j\to\infty}\mu_j = 0$. Then, the operators $R_j = \Phi_j * \mathbf{\Lambda}^{-1}$ constitute a multiscale regularization strategy. More concretely,

$$R_j\mathbf{\Lambda}(V)(x) = \sum_{n=0}^{\infty}\sum_{m=1}^{2n+1} (\Phi_j)^\wedge(n) \langle \underbrace{\nabla\otimes\nabla V}_{=\mathbf{v}}, \mathbf{h}_{n,m}^{R;(1,1)}\rangle_{\mathbf{l}^2(\mathbb{S}^2_R)} (\tilde{\mu}_n^{(1,1)})^{-1/2} H_{n,m}^R(x),$$

i.e.,

$$R_j\mathbf{\Lambda}(V)(x) = \int_{\mathbb{S}^2_R}\int_{\mathbb{S}^2_R} \Phi_j(x,z)\, \mathbf{K}_R^{\mathbf{\Lambda}^{-1}}(z,y)\, dS(z)\cdot(\nabla\otimes\nabla)V(y)\, dS(y).$$

Multiscale truncated singular value regularization strategy. The point of departure is a one-dimensional function $\varphi_0 : [0,\infty) \to \mathbb{R}$ with the following properties (cf. [**297**]): (1) $\varphi_0(0) = 1$, (2) φ_0 is monotonically decreasing, (3) φ_0 is continuous at 0. (4) $\varphi_0 : [0,\infty) \to \mathbb{R}$ has a local support, i.e., $\operatorname{supp}\varphi_0 \subset [0,1]$. Accordingly we are led to the isotropic scaling functions Φ_j, $j = 0,1,\ldots$, given by

$$\Phi_j(x,y) = \sum_{n=0}^{\infty}\sum_{m=1}^{2n+1} \varphi_0(2^{-j}n)\, H_{n,m}^R(x) H_{n,m}^R(y), \tag{7.252}$$

where

$$0 \leq (\Phi_j)^\wedge(n) = \varphi_0(2^{-j}n) \leq 1, \qquad n = 0, 1, \ldots \tag{7.253}$$

and

$$\lim_{j\to\infty} (\Phi_j)^\wedge(n) = \lim_{j\to\infty} \varphi_0(2^{-j}n) = 1, \qquad n = 0, 1, \ldots . \tag{7.254}$$

In accordance with our construction, the compact support of φ_0 implies that only finitely many $(\Phi_j)^\wedge(n)$ are different from 0. Obviously, it follows that $\operatorname{supp}\varphi_j \subset [0, 2^j]$. So, the bandlimited kernels Φ_j, $j = 0, 1, \ldots$, defined via a generator φ_0 satisfying the properties (1)–(4) as stated above define operators $R_j = \Phi_j * \mathbf{\Lambda}^{-1}$, which constitute a regularization strategy in the following sense: (1) R_j is bounded, (2) the limit relation $\lim_{j\to\infty} R_j \mathbf{\Lambda}(V) = V$ holds true in the outer space of $\mathbb{S}^2_R$.

SGG in space-based framework. As already pointed out, the SGG-problem can be modeled by a tensorial Fredholm integral equation of the first kind in space domain. In what follows we are interested in a SGG modeling method using exclusively arguments in space domain thereby involving Poisson kernels, where the geometric situation is illustrated by Figure 8.

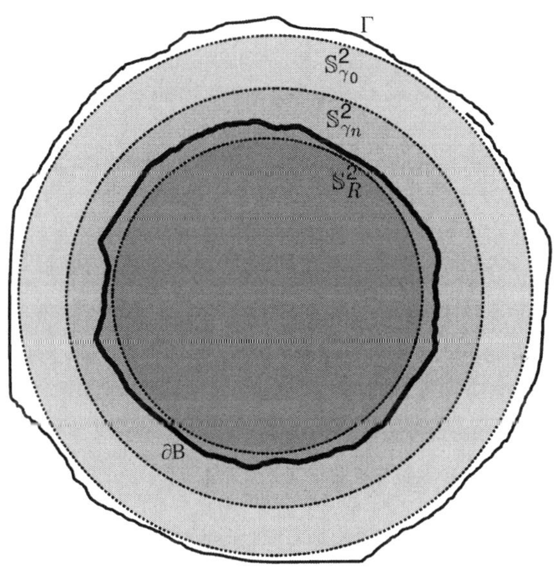

FIGURE 8. The geometric situation of satellite gravitational gradiometry as discussed in "downward continuation" multiscale space regularization procedure.

Fredholm integral equation. In the spherically reflected Runge-Bjerhammar framework ($R < \inf_{x\in\partial\mathcal{B}} |x|$) of a potential V approximating arbitrarily close the Earth's external gravitational potential $\tilde{V}$, the SGG-integral equation

$$\int_{\mathbb{S}^2_R} V(y) \underbrace{\nabla_x \otimes \nabla_x \frac{1}{4\pi R} \frac{|x|^2 - R^2}{|x-y|^3}}_{=\mathbf{k}_R(x,y)} \, dS(y) = (\nabla \otimes \nabla)V(x) = \mathbf{v}(x) \approx \tilde{\mathbf{v}}(x), \tag{7.255}$$

is valid for all locations x along the satellite orbit Γ, where

$$\mathbf{v}(x) = (\nabla \otimes \nabla)V(x) \approx (\nabla \otimes \nabla)\tilde{V}(x), \quad x \in \Gamma, \tag{7.256}$$

are the known tensorial input data, $\mathbf{k}_R(\cdot,\cdot)$ is as usual the Hesse-tensor of the *Poisson kernel*,

$$K_R(x,y) = \frac{1}{4\pi R}\frac{|x|^2 - R^2}{|x-y|^3}, \quad x \neq y, \tag{7.257}$$

and, in accordance with our construction, the scalar function $F = V|\partial\mathcal{B}$, i.e., the potential on the (actual) Earth's surface $\partial\mathcal{B}$ obtained by "upward continuation" via the integral formula

$$V|\partial\mathcal{B} = \int_{\mathbb{S}^2_R} V(y)\; K_R(\cdot,y)\; dS(y)|\partial\mathcal{B} \tag{7.258}$$

represents the desired SGG solution to the known Hesse tensor field $\mathbf{v}(x), x \in \Gamma$, given by

$$\mathbf{v}(x) = \int_{\mathbb{S}^2_R} V(y)\; \mathbf{k}_R(x,y)\; dS(y), \quad x \in \Gamma \tag{7.259}$$

(note that we do not require that Γ is a closed surface in Euclidean space $\mathbb{R}^3$).

Preparatory framework. By approximate integration over the sphere $\mathbb{S}^2_R$ (see, e.g., [**278**] for appropriate rules) we are led to a cubature formula of the form

$$\mathbf{v}(x) = \int_{\mathbb{S}^2_R} V(y)\; \mathbf{k}_R(x,y)\; dS(y) \approx \sum_{k=1}^{N_R} \underbrace{w_k^R V(y_k^R)}_{=:a_k^R} \mathbf{k}_R(x,y_k^R), \quad x \in \Gamma, \tag{7.260}$$

with appropriately given weights $w_k^R \in \mathbb{R}, k = 1,\ldots,N_R$, and nodes $y_k^R \in \mathbb{S}^2_R$, $k = 1,\ldots,N_R$ ("$\approx$" means that the the continuous sum, i.e., the integral, on the left side is approximated by an associated discrete cubature formula on the right side). In order to determine the coefficients $a_k^R \in \mathbb{R}, k = 1,\ldots,N_R$, we assume that tensor values $\mathbf{v}(x_l), l = 1,\ldots,L_R$, are known along the orbit Γ. Thus, SGG is reduced to a discrete problem, and a (discrete) solution can be obtained, e.g., by interpolatory requirements (in case of error-free data) or by smoothing and/or adjustment procedures (in case of error-affected data)

$$\mathbf{v}(x_l) \approx \sum_{k=1}^{N_R} a_k^R\; \mathbf{k}_R(x_l,y_k^R), \quad l = 1,\ldots,L_R. \tag{7.261}$$

Once the coefficients of the linear system, i.e., $a_k^R = w_k^R\; V(y_k^R), \quad k = 1,\ldots,N_R$ are available (note that the weights $w_k^R, k = 1,\ldots,N_R$, are known from the integration rule), the Runge-Bjerhammar potential V can be obtained from its discretization (7.260) as follows:

$$\begin{aligned} V(x) &= \int_{\mathbb{S}^2_R} V(y)\; K_R(x,y)\; dS(y) \\ &\approx V^R(x) = \sum_{k=1}^{N_R} \underbrace{w_k^R V(y_k^R)}_{=a_k^R} K_R(x,y_k^R), \quad x \in \mathbb{R}^3 \backslash \overline{\mathbb{B}^3_R},\; y_k^R \in \mathbb{S}^2_R. \end{aligned} \tag{7.262}$$

As a consequence, the restriction $V^R|\partial\mathcal{B}$ of V^R represents the desired approximate SGG-solution on the Earth's surface $\partial\mathcal{B}$ in space domain. Unfortunately, it should be noted that the linear system (7.261) with all tensorial input data situated in the "far field" Γ of the Runge-Bjerhammar sphere $\mathbb{S}^2_R$ represents a serious obstacle to establish a discrete SGG-solution V^R. In fact, it may be expected that the coefficient matrix of the linear system (7.261) tends to be ill-conditioned at least for larger integers N_R and L_R. This is the reason, why we propose a twofold regularization procedure, namely a "downward continuation" multiscale regularization-strategy to take advantage from the decorrelation property of multiscale structures and to suppress numerical instabilities as far as possible.

Multiscale mollifier framework. Let us consider a monotonically decreasing sequence $\{\gamma_n\}_{n=0,1,\ldots}$ of real numbers γ_n, i.e., $\gamma_n > \gamma_{n+1}$, $n = 0,1,\ldots$, satisfying the conditions (cf. Figure 8)

- the initial value γ_0 is equal to γ, i.e., we let $\gamma_0 = \gamma$,
- the limit of the sequence $\{\gamma_n\}_{n=0,1,\ldots}$ is equal to R, i.e., the radius R of the Runge-Bjerhammar sphere, i.e., $\lim_{n\to\infty} \gamma_n = R$.

In what follows, we identify the members of the sequence $\{\gamma_n\}_{n=0,1,\ldots}$ with the radii of spheres around the origin to act as the "scales" in an Poisson kernel filter-reflected low- and bandpass approach thereby observing that

$$K_{\gamma_n}(x,y) = \frac{1}{4\pi\gamma_n}\frac{|x|^2 - {\gamma_n}^2}{|x-y|^3}, \quad x,y \in \mathbb{S}^2_{\gamma_n},\ x \neq y, \tag{7.263}$$

shows increasing space localization with increasing "scale" n, i,e., with decreasing radii γ_n (for more details about the space localization of Poisson kernels see, e.g., [**297**] and the references therein).

Initial step: discrete Poisson lowpass filtering. Classical potential theory (see, e.g., [**493**]) tells us that the restriction $V|\mathbb{R}^3\backslash\overline{\mathbb{B}^3_{\gamma_0}}$ of the potential V can be represented in the form

$$V(x) = V|\mathbb{R}^3\backslash\mathbb{B}^3_{\gamma_0}(x) = \int_{\mathbb{S}^2_{\gamma_0}} V(y)\ K_{\gamma_0}(x,y)\ dS(y), \quad x \in \mathbb{R}^3\backslash\mathbb{B}^3_{\gamma_0}. \tag{7.264}$$

By approximate integration over the sphere $\mathbb{S}^2_{\gamma_0}$ we are led to a *cubature formula* V_{γ_0} approximating the integral representation (7.264) of V on $\mathbb{R}^3\backslash\mathbb{B}^3_{\gamma_0}$ in the form

$$V|\mathbb{R}^3\backslash\mathbb{B}^3_{\gamma_0}(x) \approx V_{\gamma_0}(x) = \sum_{k=1}^{N_{\gamma_0}} \underbrace{w_k^{\gamma_0} V(y_k^{\gamma_0})}_{=a_k^{\gamma_0}} K_{\gamma_0}(x,y_k^{\gamma_0}), \quad x \in \mathbb{R}^3\backslash\mathbb{B}^3_{\gamma_0}, \tag{7.265}$$

with appropriately given weights $w_k^{\gamma_0} \in \mathbb{R}, k = 1,\ldots,N_{\gamma_0}$, and prescribed nodes $y_k^{\gamma_0} \in \mathbb{S}^2_{\gamma_0}$, $k = 1,\ldots,N_{\gamma_0}$. Hence, we obtain by forming the Hesse operator

$$\mathbf{v}(x) \approx \mathbf{v}_{\gamma_0}(x) = (\nabla_x \otimes \nabla_x)V_{\gamma_0}(x) = \sum_{k=1}^{N_{\gamma_0}} \underbrace{w_k^{\gamma_0} V(y_k^{\gamma_0})}_{=a_k^{\gamma_0}} \mathbf{k}_{\gamma_0}(x,y_k^{\gamma_0}), \quad x \in \Gamma. \tag{7.266}$$

In order to determine the coefficients $a_k^{\gamma_0} \in \mathbb{R}$, $k = 1,\ldots,N_{\gamma_0}$, we assume in accordance with the observational SGG-situation that the tensor values $\mathbf{v}(x_l^{\gamma_0}), l =$

$1, \ldots, L_{\gamma_0}$, are known along the orbit Γ. In other words, the problem of determining $V|_{\mathbb{R}^3\backslash\mathbb{B}^3_{\gamma_0}}$ by a cubature based approximation V_{γ_0} is reduced to the discrete problem

$$(7.267) \qquad \mathbf{v}(x_l^{\gamma_0}) \approx \mathbf{v}_{\gamma_0}(x_l^{\gamma_0}) = \sum_{k=1}^{N_{\gamma_0}} a_k^{\gamma_0}\ \mathbf{k}_{\gamma_0}(x_l^{\gamma_0}, y_k^{\gamma_0}), \quad l = 1, \ldots, L_{\gamma_0}.$$

Once the linear system (7.267) is solved for the input data $\mathbf{v}(x_l^{\gamma_0}), l = 1, \ldots, L_{\gamma_0}$, so that the coefficients $a_k^{\gamma_0}, k = 1, \ldots, N_{\gamma_0}$, are available (once more, note that $w_k^{\gamma_0}$ is known from the integration rule), the potential V_{γ_0} as a discrete version to $V|\mathbb{R}^3\backslash\mathbb{B}^3_{\gamma_0}$ can be obtained from

$$(7.268) \qquad V_{\gamma_0}(x) = \sum_{k=1}^{N_{\gamma_0}} \underbrace{w_k^{\gamma_0} V(y_k^{\gamma_0})}_{=a_k^{\gamma_0}} K_{\gamma_0}(x, y_k^{\gamma_0}), \quad x \in \mathbb{R}^3\backslash\mathbb{B}^3_{\gamma_0}.$$

Subsequently, (regularization by) "downward continuation" yields the "0-scale" potential $V^R_{\gamma_0}$ in $\mathbb{R}^3\backslash\overline{\mathbb{B}^3_R}$ of the form

$$(7.269) \qquad V^R_{\gamma_0}(x) = \sum_{k=1}^{N_R} w_k^R\ V_{\gamma_0}\left(\gamma_0 \frac{y_k^R}{R}\right) K_R(x, y_k^R), \quad x \in \mathbb{R}^3\backslash\overline{\mathbb{B}^3_R},\ y_k^R \in \mathbb{S}^2_R,$$

hence, the restriction $V^R_{\gamma_0}|\partial\mathcal{B}$ of the potential $V^R_{\gamma_0}$ represents the "0-scale" SGG-lowpass solution on the (real) Earth's surface $\partial\mathcal{B}$, where the weights $w_k^R \in \mathbb{R}, k = 1, \ldots, N_R$, and nodes $y_k^R \in \mathbb{S}^2_R$, $k = 1, \ldots, N_R$, in (7.269) can be taken from (7.262) or another appropriate cubature formula. It sould be noted that, for reasons of comparability, we use the same cubature formula for all forthcoming "downward continued" n-scale" SGG-lowpass solutions on the Runge-Bjerhammar sphere $\mathbb{R}^3\backslash\overline{\mathbb{B}^3_R}$.

Subsequent steps: discrete Poisson lowpass and bandpass filtering. The potential $V|_{\mathbb{R}^3\backslash\overline{\mathbb{B}^3_{\gamma_1}}}$ admits the representation

$$(7.270) \qquad V(x) = V|\mathbb{R}^3\backslash\mathbb{B}^3_{\gamma_1}(x) = \int_{\mathbb{S}^2_{\gamma_1}} V(y)\ K_{\gamma_1}(x, y)\ dS(y), \quad x \in \mathbb{R}^3\backslash\mathbb{B}^3_{\gamma_1}.$$

By approximate integration over the sphere $\mathbb{S}^2_{\gamma_1}$ we are able to deduce a *cubature formula* V_{γ_1} of the integral representation (7.270) of $V|\mathbb{R}^3\backslash\mathbb{B}^3_{\gamma_1}$ in the form

$$(7.271) \qquad V(x) = V|\mathbb{R}^3\backslash\mathbb{B}^3_{\gamma_1}(x) \approx V_{\gamma_1}(x) = \sum_{k=1}^{N_{\gamma_1}} b_k^{\gamma_1}\ K_{\gamma_1}(x, y_k^{\gamma_1}), \quad x \in \mathbb{R}^3\backslash\mathbb{B}^3_{\gamma_1},$$

with $N_{\gamma_1} > N_{\gamma_0}$ and prescribed nodes $y_k^{\gamma_1} \in \mathbb{S}^2_{\gamma_1}$, $k = 1, \ldots, N_{\gamma_1}$, such that the weights satisfy the properties $b_k^{\gamma_1} = a_k^{\gamma_0}, k = 1, \ldots, N_{\gamma_0}$. Hence, we obtain by forming the Hesse operator

$$(7.272) \qquad \mathbf{v}(x) \approx \mathbf{v}_{\gamma_1}(x) = (\nabla_x \otimes \nabla_x)V_{\gamma_1}(x) = \sum_{k=1}^{N_{\gamma_1}} b_k^{\gamma_1}\ \mathbf{k}_{\gamma_1}(x, y_k^{\gamma_1}), \quad x \in \Gamma.$$

We consider the "wavelet potential" W_{γ_0} given by

$$(7.273) \qquad W_{\gamma_0} = V_{\gamma_1}|\mathbb{R}^3\backslash\mathbb{B}^3_{\gamma_0} - V_{\gamma_0}.$$

W_{γ_0} allows the discretization in the form

$$
\begin{aligned}
(7.274)\qquad W_{\gamma_0}(x) &= \sum_{k=1}^{N_{\gamma_1}} b_k^{\gamma_1} K_{\gamma_1}(x, y_k^{\gamma_1}) - \sum_{k=1}^{N_{\gamma_0}} a_k^{\gamma_0} K_{\gamma_0}(x, y_k^{\gamma_0}) \\
&= \sum_{k=N_{\gamma_0}+1}^{N_{\gamma_1}} b_k^{\gamma_1} K_{\gamma_1}(x, y_k^{\gamma_1}) \\
&\quad + \sum_{k=1}^{N_{\gamma_0}} a_k^{\gamma_0} \left(K_{\gamma_1}(x, y_k^{\gamma_1}) - K_{\gamma_0}(x, y_k^{\gamma_0})\right), \quad x \in \mathbb{R}^3 \backslash \mathbb{B}^3_{\gamma_0}.
\end{aligned}
$$

We use the canonical notation $\mathbf{w}_{\gamma_0} = (\nabla \otimes \nabla) W_{\gamma_0}$. Then, in connection with (7.274), the SGG-equations

$$
(7.275)\qquad \mathbf{w}_{\gamma_0}(x_l^{\gamma_1}) = \underbrace{\mathbf{v}_{\gamma_1}(x_l^{\gamma_1})}_{\approx \mathbf{v}(x_l^{\gamma_1})} - \mathbf{v}_{\gamma_0}(x_l^{\gamma_1}), \quad x_l^{\gamma_1} \in \Gamma, \ l = L_{\gamma_0} + 1, \ldots, L_{\gamma_1},
$$

lead to the linear system

$$
\begin{aligned}
(7.276)\qquad \mathbf{w}_{\gamma_0}(x_l^{\gamma_1}) &= \sum_{k=N_{\gamma_0}+1}^{N_{\gamma_1}} b_k^{\gamma_1} \ \mathbf{k}_{\gamma_1}(x_l^{\gamma_1}, y_k^{\gamma_1}) \\
&\quad + \sum_{k=1}^{N_{\gamma_0}} a_k^{\gamma_0} \ \left(\mathbf{k}_{\gamma_1}(x_l^{\gamma_1}, y_k^{\gamma_1}) - \mathbf{k}_{\gamma_0}((x_l^{\gamma_1}, y_k^{\gamma_0})\right)
\end{aligned}
$$

in the unknowns $b_k^{\gamma_1}$, $k = N_{\gamma_0} + 1, \ldots, N_{\gamma_1}$. Equivalently, we obtain

$$
(7.277)\qquad \sum_{k=N_{\gamma_0}+1}^{N_{\gamma_1}} b_k^{\gamma_1} \ \mathbf{k}_{\gamma_1}(x_l^{\gamma_1}, y_k^{\gamma_1}) \approx \mathbf{v}(x_l^{\gamma_1}) - \sum_{k=1}^{N_{\gamma_0}} a_k^{\gamma_0} \ \mathbf{k}_{\gamma_1}(x_l^{\gamma_1}, y_k^{\gamma_1}),
$$

$l = L_{\gamma_0} + 1, \ldots, L_{\gamma_1}$, where $a_k^{\gamma_0}, k = 1, \ldots, N_{\gamma_0}$, are known (note that (i) the points $x_l^{\gamma_1} \in \Gamma$, $l = L_{\gamma_0} + 1, \ldots, L_{\gamma_1}$, should be chosen in an appropriate way to guarantee an improvement provided by the bandpass potential W_{γ_0} of scale 0 to the whole multiscale approach and (ii) the choice $y_k^{\gamma_1} = y_k^{\gamma_0}$, $k = 1, \ldots, N_{\gamma_0}$, is not excluded). Once the coefficients $b_k^{\gamma_1}$, $k = N_{\gamma_0} + 1, \ldots, N_{\gamma_1}$ are calculated from (7.277), the potential V_{γ_1} is known from (7.271). The potential V_{γ_1} may be approximately regarded as the sum of the lowpass potential V_{γ_0} and the bandpass potential W_{γ_0} of scale 0 in $\mathbb{R}^3 \backslash \mathbb{B}^3_{\gamma_0}$, leading to the "1-scale" SGG-lowpass solution in $\mathbb{R}^3 \backslash \mathbb{B}^3_{\gamma_1}$.

"Downward continuation" yields the "1-scale" potential $V_{\gamma_1}^R$ in $\mathbb{R}^3 \backslash \mathbb{B}^3_R$ of the form

$$
(7.278)\qquad V_{\gamma_1}^R(x) := \sum_{k=1}^{N_R} w_k^R \ V_{\gamma_1}\left(\gamma_1 \frac{y_k^R}{R}\right) K_R(x, y_k^R), \quad x \in \mathbb{R}^3 \backslash \mathbb{B}^3_R, \ y_k^R \in \mathbb{S}^2_R,
$$

so that the restriction $V_{\gamma_1}^R | \partial\mathcal{B}$ of the potential $V_{\gamma_1}^R$ represents the "1-scale" SGG-lowpass solution on the (real) Earth's surface $\partial\mathcal{B}$ (as pointed out earlier, for reasons of comparability, we use the same cubature formula for the "downward continued" 1-scale" SGG-lowpass solution $V_{\gamma_1}^R$ as for the "downward continued" 0-scale" SGG-lowpass solution $V_{\gamma_0}^R$ on $\mathbb{R}^3 \backslash \mathbb{B}^3_R$).

Now, generally, assume that "downward continued" k-scale" SGG-lowpass solutions $V^R_{\gamma_k}$, $k = 1, \ldots n$, are calculated on $\mathbb{R}^3 \backslash \mathbb{B}^3_R$ in the indicated way. Our purpose is to determine $V^R_{\gamma_{n+1}}$:

The potential $V|\mathbb{R}^3 \backslash \mathbb{B}^3_{\gamma_{n+1}}$ admits the representation

(7.279)
$$V(x) = V|\mathbb{R}^3 \backslash \mathbb{B}^3_{\gamma_{n+1}}(x) = \int_{\mathbb{S}^2_{\gamma_{n+1}}} V(y)\ K_{\gamma_{n+1}}(x,y)\ dS(y), \quad x \in \mathbb{R}^3 \backslash \mathbb{B}^3_{\gamma_{n+1}},$$

Therefore, we obtain by applying the Hesse operator

(7.280)
$$\mathbf{v}(x) \approx \mathbf{v}_{\gamma_{n+1}}(x) = (\nabla_x \otimes \nabla_x) V_{\gamma_{n+1}}(x) = \sum_{k=1}^{N_{\gamma_1}} b_k^{\gamma_{n+1}}\ \mathbf{k}_{\gamma_{n+1}}(x, y_k^{\gamma_{n+1}}), \quad x \in \Gamma.$$

By approximate integration over the sphere $\mathbb{S}^2_{\gamma_{n+1}}$ we are able to deduce a cubature formula $V_{\gamma_{n+1}}$ of the integral representation (7.279) of $V|\mathbb{R}^3 \backslash \mathbb{B}^3_{\gamma_{n+1}}$ in the form

$$V_{\gamma_{n+1}}(x) = \sum_{k=1}^{N_{\gamma_{n+1}}} b_k^{\gamma_{n+1}}\ K_{\gamma_{n+1}}(x, y_k^{\gamma_{n+1}}), \quad x \in \mathbb{R}^3 \backslash \mathbb{B}^3_{\gamma_{n+1}}, \tag{7.281}$$

with $N_{\gamma_{n+1}} > N_{\gamma_n}$ and given nodes $y_k^{\gamma_{n+1}} \in \mathbb{S}^2_{\gamma_{n+1}}$, $k = 1, \ldots, N_{\gamma_{n+1}}$, such that

$$b_k^{\gamma_{n+1}} = b_k^{\gamma_n}, \quad k = 1, \ldots, N_{\gamma_n}. \tag{7.282}$$

We consider the "wavelet potential" W_{γ_n} given by

$$W_{\gamma_n} = V_{\gamma_{n+1}}|\mathbb{R}^3 \backslash \mathbb{B}^3_{\gamma_n} - V_{\gamma_n}. \tag{7.283}$$

Because of (7.282), W_{γ_n} allows the discretization in the form

$$\begin{aligned} W_{\gamma_n}(x) &= \sum_{k=1}^{N_{\gamma_{n+1}}} b_k^{\gamma_{n+1}} K_{\gamma_{n+1}}(x, y_k^{\gamma_{n+1}}) - \sum_{k=1}^{N_{\gamma_n}} b_k^{\gamma_n} K_{\gamma_n}(x, y_k^{\gamma_n}) \\ &= \sum_{k=N_{\gamma_n}+1}^{N_{\gamma_{n+1}}} b_k^{\gamma_{n+1}} K_{\gamma_{n+1}}(x, y_k^{\gamma_{n+1}}) \\ &\quad + \sum_{k=1}^{N_{\gamma_n}} b_k^{\gamma_n} \left(K_{\gamma_{n+1}}(x, y_k^{\gamma_{n+1}}) - K_{\gamma_n}(x, y_k^{\gamma_n}) \right), x \in \mathbb{R}^3 \backslash \mathbb{B}^3_{\gamma_n}, \end{aligned} \tag{7.284}$$

where the coefficients $b_k^{\gamma_n}, k = 1, \ldots, N_{\gamma_n}$, are known. By definition we set $\mathbf{w}_{\gamma_n} = (\nabla \otimes \nabla) W_{\gamma_n}$. Then, in connection with (7.284), the SGG-equations

$$\mathbf{w}_{\gamma_n}(x_l^{\gamma_{n+1}}) \approx \mathbf{v}(x_l^{\gamma_{n+1}}) - \mathbf{v}_{\gamma_n}(x_l^{\gamma_{n+1}}), \quad x_l^{\gamma_{n+1}} \in \Gamma, \ l = L_{\gamma_n} + 1, \ldots, L_{\gamma_{n+1}}, \tag{7.285}$$

lead to the linear system

$$\begin{aligned} \mathbf{w}_{\gamma_n}(x_l^{\gamma_{n+1}}) &= \mathbf{v}_{\gamma_{n+1}}(x_l^{\gamma_{n+1}}) - \mathbf{v}_{\gamma_n}(x_l^{\gamma_{n+1}}) \\ &= \sum_{k=N_{\gamma_n}+1}^{N_{\gamma_{n+1}}} b_k^{\gamma_{n+1}}\ \mathbf{k}_{\gamma_{n+1}}(x_l^{\gamma_{n+1}}, y_k^{\gamma_{n+1}}) \\ &\quad + \sum_{k=1}^{N_{\gamma_n}} b_k^{\gamma_n} \left(\mathbf{k}_{\gamma_{n+1}}(x_l^{\gamma_{n+1}}, y_k^{\gamma_{n+1}}) - \mathbf{k}_{\gamma_n}(x_l^{\gamma_{n+1}}, y_k^{\gamma_n}) \right) \end{aligned} \tag{7.286}$$

in the unknowns $b_k^{\gamma_{n+1}}, k = N_{\gamma_n}+1, \dots, N_{\gamma_{n+1}}$, i.e.,

$$\sum_{k=N_{\gamma_0}+1}^{N_{\gamma_{n+1}}} b_k^{\gamma_{n+1}}\, \mathbf{k}_{\gamma_{n+1}}(x_l^{\gamma_1}, y_k^{\gamma_{n+1}}) \approx \mathbf{v}(x_l^{\gamma_{n+1}}) - \sum_{k=1}^{N_{\gamma_n}} b_k^{\gamma_n}\, \mathbf{k}_{\gamma_{n+1}}(x_l^{\gamma_{n+1}}, y_k^{\gamma_{n+1}}), \tag{7.287}$$

$l = L_{\gamma_n}+1, \dots, L_{\gamma_{n+1}}$ (again, $x_l^{\gamma_{n+1}} \in \Gamma$, $l = L_{\gamma_n}+1, \dots, L_{\gamma_{n+1}}$ should be chosen in an appropriate way to guarantee an improvement by the bandpass potential W_{γ_n}). Once the coefficients $b_k^{\gamma_{n+1}}, k = N_{\gamma_n}+1, \dots, N_{\gamma_{n+1}}$ are calculated from (7.287), the potential $V_{\gamma_{n+1}}$ is known from (7.271). The potential $V_{\gamma_{n+1}}$ may be regarded as the sum of the lowpass potential V_{γ_n} and the bandpass potential W_{γ_n} of scale n in $\mathbb{R}^3 \backslash \mathbb{B}^3_{\gamma_n}$, and $V_{\gamma_{n+1}}$ represents the "$(n+1)$-scale" SGG-lowpass solution in $\mathbb{R}^3 \backslash \mathbb{B}^3_{\gamma_{n+1}}$.

"Downward continuation" leads to the "$(n+1)$-scale" potential $V^R_{\gamma_{n+1}}$ in $\mathbb{R}^3 \backslash \mathbb{B}^3_R$ of the form

$$V^R_{\gamma_{n+1}}(x) = \sum_{k=1}^{N_R} w_k^R\, V_{\gamma_{n+1}}\left(\gamma_{n+1}\frac{y_k^R}{R}\right) K_R(x, y_k^R), \quad x \in \mathbb{R}^3 \backslash \mathbb{B}^3_R,\ y_k^R \in \mathbb{S}^2_R, \tag{7.288}$$

so that the restriction $V^R_{\gamma_{n+1}}|\partial\mathcal{B}$ of $V^R_{\gamma_{n+1}}$ represents the "$(n+1)$-scale" SGG-lowpass solution on the (real) Earth's surface $\partial\mathcal{B}$.

Continuing our wavelet process we obtain "downward continued" k-scale" SGG-lowpass solutions $V^R_{\gamma_k}$, $k = 0, 1, \dots$, calculated on $\mathbb{R}^3 \backslash \mathbb{B}^3_R$, hence, available on the real Earth's surface $\partial\mathcal{B}$ and its exterior space just by restriction. We successively obtain lowpass and bandpass SGG-discretizations of the Earth's gravitational potential from a tensorial (error-free) SGG-dataset, thereby basing our multiscale procedure exclusively on operations in space domain (note that the error-affected case can be handled scale-by-scale by obvious smoothing and/or adjustment manipulations as already mentioned above).

Altogether, we are led to the following *conclusions*: Multiscale techniques using regularizing wavelets as constructed here, indeed, represent an outstanding methodology by its particular ability to exhaust all specific features of the "portion of the signal" by a "zooming-in" process. In fact, different powerful techniques for regularization are at the disposal of the analyst in frequency as well as space domain from globally via regionally up to locally reflected scales. All these facts lead to the conclusion that multiresolution methods are superior to any other SGG-solution strategy. As an immediate consequence, this work on multiscale regularization may be rated without any doubt as a well-promising and far-reaching methodology in future SGG-research. However, there is also no doubt that each method in approximation theory and numerics has its own aim and character. In fact, it is the essence of any numerical realization that it becomes optimal only with respect to especially specified criteria. In multiscale regularization we have to reflect the different stages of space/frequency localization so that the modeling process can be performed under the localization requirements necessary and sufficient for relevant geoscientific interpretation. A positive decision in one direction often amounts to a negative implication in an other direction. Ultimately, because of the uncertainty principle, it is impossible to provide a "cure-all methodology". This "sine qua non" ingredient in any mathematical approximation method is also the reason, why we are indispensibly led to two essential calamities in the "step by step" SGG-approximation by scaling and wavelet potentials as proposed here:

(1) The choice of the regularization scaling function is an important problem in SGG. Mathematically, all wavelet regularization strategies are equivalent, however, the right computational compromise between mathematical rigor and geodetic/geophysical relevance in respect to the geometry of the orbit, data width, accuracy of the data, and occurring noise level and characteristics is a task for future research, which should not be underestimated.

(2) Today, the amount of amplification for the gravitational potential is not yet suitably known on the orbit as an a priori state. As a consequence, a missing stopping strategy in the multiresolution regularization caused by manual input usually produces huge errors in the potential at the Earth's surface even from extremely small errors in the measurements. Thus there is a strong need for a SGG-stopping strategy in the near future.

A loss of information in SGG-modeling is unavoidable. An algorithm establishing an approximate solution for the inverse SGG-problem has to reflect as good as possible the intention of the geoscientific applicant. Gravitational a priori information today for the characterization of the right scaling and wavelet potentials for the SGG-solution process as well as a mathematically validated stopping strategy in multiscale regularization is an important challenge for future work.

Finally, it should be remarked that the SGG-regularization methods were successfully applied to the data of the ESA satellite GOCE by the Geomathematics Group, Department of Mathematics, University of Kaiserslautern, Rhineland-Palatinate, Germany, where numerical realizations by spectral multiscale regularization (see the context of Section 6.10) are given by use of the (scalar) second radial derivatives $\frac{x}{|x|} \cdot (\nabla_x \otimes \nabla_x) V(x) \frac{x}{|x|} = \left(\frac{x}{|x|} \cdot \nabla_x\right) \left(\frac{x}{|x|} \cdot \nabla_x\right) V(x)$ instead of Hesse-tensor data $(\nabla_x \otimes \nabla_x) V(x)$ of the geopotential V on satellite orbits Γ (see, e.g., [**313**]).

7.4. Inverse gravimetry (IG)

The classical approach to inverse gravimetry in geophysics and geodesy (see, e.g., [**737**], [**577–579, 582**]) is of spectral nature. More concretely, the Newton volume integral is expanded in terms of inner/outer harmonics, and the orthogonal coefficients of the density distribution are extracted from associated orthogonal coefficients of the gravitational potential. This procedure is very helpful for global Earth's obligations, but it is not adequate for local explorational purposes (as intended here). Nevertheless, spectral global solutions (as proposed by W. Freeden, M.Z. Nashed [**296**]) are of certain value, since they may be used as initial trend approximations for local resolution approaches.

Sparse space-based methods. Following the concept of distributions (i.e., generalized functions) the Dirac delta may be regarded as a kind of limit of a sequence of functions having a tall spike at the origin. The approximating functions of the sequence are thus "approximate" or "mollifier" Dirac functions. In what follows, we adopt the concept of Dirac sequences for 3D-volume approximation in geoexploration. However, for reasons of economic and efficient numerics, we are only interested in simply structured, locally supported Dirac sequences. More explicitly, the Dirac sequences are constituted by radial basis functions, so that their local support is just a ball. In doing so, we introduce Dirac mollifier scaling functions in such a way that their corresponding Newton mollifier potentials immediately become obvious and elementarily caculable.

A sequence $\{K_\tau\}_{\tau>0}$ of 1D-radially τ-dependent functions $r \mapsto K_\tau(r)$, $r \geq 0$, is called a *3D-Dirac mollifier scaling function* if it satisfies the following conditions:

(i) $K_\tau(r) = 0, \quad r > \tau,$

(ii) $K_\tau(r) \geq 0, \quad r \geq 0,$

(iii) $K_\tau|(0,\tau]$ is arbitrarily often differentiable,

(iv) $\{K_\tau\}_{\tau>0}$ satisfies $\lim_{\tau\to 0}\int_\delta^\infty K_\tau(r)\, dr = 0$ for every $\delta > 0$,

(v) $\{K_\tau\}_{\tau>0}$ satisfies the condition

$$\int_{\mathbb{R}^3} K_\tau(|y|)\, dy = \int_{\substack{|y|\leq\tau \\ y\in\mathbb{R}^3}} K_\tau(|y|)\, dy = 4\pi \int_0^\tau K_\tau(r)\, r^2\, dr = 1. \tag{7.289}$$

Let $\mathcal{B}$ be a regular region in Euclidean space $\mathbb{R}^3$. Let ρ be a continuous function on $\overline{\mathcal{B}}$. Then, the family $\{S_{K_\tau}\}_{\tau>0}$, of operators $S_{K_\tau} : \rho \mapsto S_{K_\tau}(\rho)$ given by

$$S_{K_\tau}(\rho)(x) = \int_{\substack{|x-y|\leq\tau \\ y\in\overline{\mathcal{B}}}} K_\tau(|x-y|)\, \rho(y)\, dy$$

constitutes an *approximating identity* of ρ in the already known sense, i.e., for each $x \in \overline{\mathcal{B}}$, the integral

$$S_{K_\tau}(\rho)(x) - \alpha(x)\rho(x) = \int_{\substack{|x-y|\leq\tau \\ y\in\overline{\mathcal{B}}}} K_\tau(|x-y|)\, (\rho(y) - \rho(x))\, dy \tag{7.290}$$

satisfies the limit relation

$$\lim_{\tau\to 0} S_{K_\tau}(\rho)(x) - \alpha_{\partial\mathcal{B}}(x)\rho(x) = 0 \tag{7.291}$$

(see, e.g., [**274**]), where $\alpha_{\partial\mathcal{B}}(x)$ is the solid angle subtended by the surface $\partial\mathcal{B}$ at the point $x \in \mathbb{R}^3$ (which justifies the name approximating identity for $\{S_{K_\tau}\}_{\tau>0}$ for points $x \in \mathcal{B}$, where $\alpha_{\partial\mathcal{B}}(x) = 1$). Formally, an approximate identity $\{S_{K_\tau}\}_{\tau>0}$ in $\mathcal{B}$ involving the sequence $\{K_\tau\}_{\tau>0}$ is constructed in such a way that the kernels $\{K_\tau\}_{\tau>0}$ converge to the Dirac function(al):

(7.292)

$$\lim_{\tau\to 0} \rho_{K_\tau}(x) - \rho(x) = \lim_{\tau\to 0} \int_{\substack{|x-y|\leq\tau \\ y\in\overline{\mathcal{B}}}} (K_\tau(|x-y|) - \delta(|x-y|))\, \rho(y)\, dy = 0, \quad x \in \mathcal{B}.$$

This is the reason why for decreasing "scales" τ the scale continuous "Dirac mollifier scaling functions $\{K_\tau\}_{\tau\geq 0}$ show an increasingly strong "spatial localization". The (formally defined) *scale-continuous "Dirac mollifier wavelet functions"*

$$(WK)_\tau = -\tau \frac{d}{d\tau} K_\tau \tag{7.293}$$

also show an increasingly strong spatial localization. Our interest for practical applicabilty, however, is in *scale-discrete "Dirac mollifier wavelet functions"*. To this end, we consider a positive, monotonically decreasing sequence $\{\tau_j\}_{j\in\mathbb{N}}$ of scaling parameters $\tau_j > 0$ satisfying $\lim_{j\to\infty} \tau_j = 0$. It is not difficult to verify that

$$\begin{aligned} (WK)_{\tau_j}(|x-y|) &= \int_{\tau_{j+1}}^{\tau_j} (WK)_\tau(|x-y|)\, \frac{1}{\tau}\, d\tau \\ &= \int_{\tau_{j+1}}^{\tau_j} -\tau \left(\frac{d}{d\tau} K_\tau(|x-y|)\right) \frac{1}{\tau}\, d\tau \\ &= K_{\tau_{j+1}}(|x-y|) - K_{\tau_j}(|x-y|). \end{aligned} \tag{7.294}$$

This leads to a family $\{W_{K_{\tau_j}}\}_{\tau_j>0}$ of operators $W_{K_{\tau_j}} : \rho \mapsto W_{K_{\tau_j}}(\rho)$ corresponding to $\{S_{K_{\tau_j}}\}_{\tau_j>0}$ defined by

$$W_{K_{\tau_j}}(\rho)(x) = \int_{\mathbb{R}^3} \underbrace{(WK)_{\tau_j}(|x-y|)}_{=K_{\tau_{j+1}}(|x-y|)-K_{\tau_j}(|x-y|)} \rho(y)\, dy, \tag{7.295}$$

where the scale-discrete *Dirac mollifier wavelet function* $\{(WK)_{\tau_j}\}_{j\in\mathbb{N}}$ is given by

$$(WK)_{\tau_j} = K_{\tau_{j+1}} - K_{\tau_j}, \quad j \in \mathbb{N}. \tag{7.296}$$

The mollifier scaling function $K_{\tau_{j+1}}$ to scale $j+1$ is the sum of the mollifier scaling function K_{τ_j} to the scale j and the corresponding mollifier wavelet function $(WK)_{\tau_j}$ to the scale j. Eq. (7.296) allows us to express the wavelet filtering

$$W_{K_{\tau_j}}(\rho)(x) = \int_{\substack{|x-y|\le\tau_j \\ y\in\overline{\mathcal{B}}}} (WK)_{\tau_j}(|x-y|)\, \rho(y)\, dy \tag{7.297}$$

as the difference between two successive filterings

$$\int_{\substack{|x-y|\le\tau_{j+1} \\ y\in\overline{\mathcal{B}}}} (K_{\tau_{j+1}}(|x-y|)\, \rho(y)\, dy - \int_{\substack{|x-y|\le\tau_j \\ y\in\overline{\mathcal{B}}}} (K_{\tau_j}(|x-y|)\, \rho(y)\, dy, \tag{7.298}$$

i.e., analogously to (7.296), but in operator nomenclature, we have

$$W_{K_{\tau_j}}(\rho) = S_{K_{\tau_{j+1}}}(\rho) - S_{K_{\tau_j}}(\rho). \tag{7.299}$$

Observing (7.299) we can control the *"detail information"*, which although in the integral

$$S_{K_{\tau_{j+1}}}(\rho)(x) = \int_{\substack{|x-y|\le\tau_{j+1} \\ y\in\overline{\mathcal{B}}}} (K_{\tau_{j+1}}(|x-y|)\, \rho(y)\, dy, \tag{7.300}$$

but not in the integral

$$S_{K_{\tau_j}}(\rho)(x) = \int_{\substack{|x-y|\le\tau_j \\ y\in\overline{\mathcal{B}}}} (K_{\tau_j}(|x-y|)\, \rho(y)\, dy \tag{7.301}$$

is included. The control of detail information is the key to our desired mollifier reflected decorrelation methodology. If we continue our scaling equation (7.296), the discrete wavelets $(WK)_{\tau_j}$, $j = 0, \ldots, J-1$, obey the relation

$$K_{\tau_J} = K_{\tau_{J_0}} + \sum_{j=J_0}^{J-1} (WK)_{\tau_j}. \tag{7.302}$$

For continuous ρ in $\overline{\mathcal{B}}$, the scaling equation

$$S_{K_{\tau_J}}(\rho) = S_{K_{\tau_{J_0}}}(\rho) + \sum_{j=J_0}^{J-1} W_{K_{\tau_j}}(\rho) \tag{7.303}$$

and the resulting limit relation (approximate identity)

$$\lim_{J\to\infty} S_{K_{\tau_J}}(\rho) = S_{K_{\tau_{J_0}}}(\rho) + \lim_{J\to\infty} \sum_{j=J_0}^{J-1} W_{K_{\tau_j}}(\rho) = \alpha_{\partial\mathcal{B}}\ \rho \tag{7.304}$$

hold true.

Once more, of particular interest in practice is that the mollifier scaling functions K_{τ_j} have local supports $\mathrm{supp}(K_{\tau_j}(|x-\cdot|)) = \overline{\mathbb{B}^3_{\tau_j}(x)} \cap \overline{\mathcal{B}}$, $x \in \overline{\mathcal{B}}$, which shrink with increasing scales. Simultaneously, the maxima of the functions $K_{\tau_j}(|x-\cdot|)$, $x \in \overline{\mathcal{B}}$, increase with increasing scales.

Multiscale Newton mollifier context. We pursue the idea of deriving a mollifier regularization of the Newton kernel, i.e., the fundamental solution of the negative Laplace operator. More explicitly, for a given Dirac mollifier scaling function $\{K_\tau\}_{\tau>0}$, we are looking for a family of *Newton mollifier scaling functions* $\left(F_3^{(1)}\right)_{K_\tau}(|x-y|)$ with the property

$$-\Delta_x \left(F_3^{(1)}\right)_{K_\tau}(|x-y|) \;=\; K_\tau(|x-y|), \quad x, y \in \mathbb{R}^3. \tag{7.305}$$

Since all kernels are radially dependent (i.e., dependent on $r = |x-y|$, $x, y \in \mathbb{R}^3$), we can express the Eq. (7.305) in 1D-nomenclature using the radial representation of the Laplace operator

$$-\frac{1}{r^2}\frac{\partial}{\partial r} r^2 \frac{\partial}{\partial r}\left(F_3^{(1)}\right)_{K_\tau}(r) \;=\; -\frac{1}{r}\frac{\partial^2}{\partial r^2}\left(r\left(F_3^{(1)}\right)_{K_\tau}(r)\right) = \; K_\tau(r). \tag{7.306}$$

For a Dirac mollifier scaling function $\{K_\tau\}_{\tau>0}$, we are therefore able to obtain formally a kernel $\left(F_3^{(1)}\right)_{K_\tau}(r)$ according to the formula $-\int \frac{1}{r^2}\left(\int r^2 K_\tau(r)\, dr\right) dr$.

Examples. First we mention the probably simplest example, viz. the Haar kernel of degree 0.

- The *Haar kernel of degree* 0 (simply called *Haar kernel*) is given by

$$K_\tau(r) = H_\tau^0(r) \;=\; \begin{cases} \frac{1}{\|\mathbb{B}^3_\tau\|}, & r < \tau, \\ 0, & r \geq \tau, \end{cases} \tag{7.307}$$

with $\|\mathbb{B}^3_\tau\| = \frac{4}{3}\pi\tau^3$ and $\left(F_3^{(1)}\right)_{H_\tau^0}(r)$ is given by

$$\left(F_3^{(1)}\right)_{H_\tau^0}(r) = \begin{cases} \frac{1}{4\pi r}, & \tau \le r \\ \frac{1}{8\pi\tau}(3 - \frac{1}{\tau^2}r^2), & r < \tau, \end{cases} \tag{7.308}$$

i.e.,

$$\left(F_3^{(1)}\right)_{H_\tau^0}(r) = \begin{cases} \frac{1}{4\pi r}, & \tau \le r, \\ \frac{1}{4\pi r} - \frac{(\tau-r)^2\ (r+2\tau)}{8\pi\tau_j{}^3 r}, & 0 < r < \tau, \\ \frac{3}{8\pi\tau}, & r = 0. \end{cases} \tag{7.309}$$

The critical point of the Haar kernel of degree 0 is the discontinuity, i.e., the ordinary Haar function $H_\tau = H_\tau^0$ satisfies the equation

$$-\Delta_x \left(F_3^{(1)}\right)_{H_\tau^0}(|x-y|) = H_\tau^0(|x-y|) = \begin{cases} 0, & |x-y| \ge \tau, \\ \frac{3}{4\pi\tau^3}, & |x-y| < \tau. \end{cases} \tag{7.310}$$

In what follows we are therefore interested in smoothed Haar kernel variants: For $r > 0$, an ordinary 1D-Taylor expansion of the degree $n+1$ yields

$$F_3^{(1)}(r) = \frac{1}{4\pi r} = \frac{1}{4\pi} \sum_{k=0}^{n+1} \frac{(-1)^k}{r^{k+1}}(r-\tau)^k + O((r-\tau)^{n+2}). \tag{7.311}$$

The sum on the right side of (7.311) allows the following reformulation

$$\begin{aligned} F_3^{(1)}(r) &= \frac{1}{4\pi} \sum_{k=0}^{n+1} \frac{(-1)^k}{r^{k+1}}(r-\tau)^k + O((r-\tau)^{n+2}) \\ &= \frac{1}{4\pi r} - \frac{((n+1)r + 2\tau)(r-\tau)^{n+2}}{8\pi r\tau^{n+3}} + O((r-\tau)^{n+2}). \end{aligned} \tag{7.312}$$

This leads to a mollification of $F_3^{(1)}(r)$ in the form

$$\left(F_3^{(1)}\right)_{K_\tau}(r) = \begin{cases} \frac{1}{4\pi r}, & \tau \le r, \\ \frac{1}{4\pi r} - \frac{(\tau-r)^{n+2}\ (n+1)r\ +2\tau)}{8\pi\tau^{n+3} r}, & 0 < r < \tau, \\ \frac{n+3}{8\pi\tau}, & r = 0. \end{cases} \tag{7.313}$$

An elementary calculation shows that, for $r \in [0,\tau)$, the kernel $r \mapsto \left(F_3^{(1)}\right)_{K_\tau}(r)$ is a polynomial of degree $n+2$. More explicitly, we have

(7.314)

$$\begin{aligned} &\left(F_3^{(1)}\right)_{K_\tau}(r) \\ &= \frac{1}{8\pi\tau^{n+3}} \sum_{k=0}^{n+1} \left(\binom{n+2}{k}(n+1)(-1)^{k+1} + 2(-1)^k \binom{n+2}{k+1} \right) \tau^{n+2-k} r^k \\ &\quad - \frac{(n+1)(-r)^{n+2}}{8\pi\tau^{n+3}}. \end{aligned}$$

Especially, for $n = 0$, we arrive at the kernels known from (7.307) and (7.309). Moreover, for $\tau \leq r$, it follows that

$$-\frac{1}{r}\frac{\partial^2}{\partial r^2}\left(r\ \left(F_3^{(1)}\right)_{K_\tau}(r)\right) = -\frac{1}{r}\frac{\partial^2}{\partial r^2}\left(r\frac{1}{4\pi r}\right) = 0, \tag{7.315}$$

whereas, for $\tau \geq r > 0$ we have

$$\begin{aligned} -\frac{1}{r}\frac{\partial^2}{\partial r^2}\left(r\ \left(F_3^{(1)}\right)_{K_\tau}(r)\right) &= -\frac{1}{r}\frac{\partial^2}{\partial r^2}\left(r\frac{1}{4\pi r} - \frac{((n+1)r+2\tau)(r-\tau)^{n+2}}{8\pi r\tau^{n+3}}\right) \\ &= \frac{1}{r}\frac{\partial}{\partial r}\left(\frac{(\tau-r)^{n+1}(n+3)((n+1)r+\tau)}{8\pi\tau^3}\right) \\ &= \frac{(n+1)(n+2)(n+3)}{8\pi}\ \frac{(\tau-r)^n}{\tau^{n+3}}. \end{aligned} \tag{7.316}$$

In other words, we end up with the following kernel variants: The *Haar kernel of degree n* is given by

$$K_\tau(r) = H_\tau^n(r) \ = \begin{cases} \frac{(n+1)(n+2)(n+3)}{8\pi}\ \frac{(\tau-r)^n}{\tau^{n+3}}, & r < \tau, \\ 0, & r \geq \tau. \end{cases} \tag{7.317}$$

The *negative Laplace-antiderivative of the Haar kernel of degree n* is given by

$$\left(F_3^{(1)}\right)_{H_\tau^n}(r) = \begin{cases} \frac{1}{4\pi r}, & \tau \leq r, \\ \frac{1}{4\pi r} - \frac{(\tau-r)^{n+2}\ (n+1)r\ +2\tau)}{8\pi\tau^{n+3}r}, & 0 < r < \tau, \\ \frac{n+3}{8\pi\tau_j}, & r = 0 \end{cases}$$

or

$$\begin{aligned} &\left(F_3^{(1)}\right)_{H_\tau^n}(r) \\ &= \begin{cases} \frac{1}{4\pi r}, & \tau \leq r, \\ \frac{1}{8\pi\tau^{n+3}}\sum_{k=0}^{n+1}\left(\binom{n+2}{k}(n+1)(-1)^{k+1} + 2(-1)^k\binom{n+2}{k+1}\right)\tau^{n+2-k}r^k \\ \quad -\frac{(n+1)(-r)^{n+2}}{8\pi\tau^{n+3}}, & 0 \leq r < \tau. \end{cases} \end{aligned}$$

Finally, the following properties central in our considerations should be mentioned:

- The smoothed Haar kernels H_τ^n satisfy the desired integral relation

$$\int_{\mathbb{R}^3}(-\Delta_y(\left(F_3^{(1)}\right)_{H_\tau^n}(|x-y|))\ dy = \int_{\mathbb{R}^3} H_\tau^n(|x-y|)\ dy \ = \ 1. \tag{7.318}$$

- Let $\mathcal{B}$ be a regular region. Assume that ρ is continuous in $\overline{\mathcal{B}}$. Then the limit relation

$$\lim_{\substack{\tau\to 0\\ \tau>0}}\int_{\mathcal{B}} H_\tau^n(|x-y|)\ \rho(y)\ dy \ = \ \alpha_{\partial\mathcal{B}}(x)\ \rho(x) \tag{7.319}$$

  holds true for $x \in \mathbb{R}^3$, where $\alpha_{\partial\mathcal{B}}(x)$ is the solid angle with respect to the surface $\partial\mathcal{B}$ at $x \in \mathbb{R}^3$.
- The function $\left(F_3^{(1)}\right)_{H_\tau^n}$ is $(n+1)$-times continuously differentiable.

- Let $\mathcal{B}$ be a regular region. Assume that ρ is continuous in $\overline{\mathcal{B}}$. Then the limit relation

$$\sup_{x\in\mathcal{B}} \left| \int_{\mathcal{B}} \left(F_1^3(|x-y|) - \left(F_3^{(1)}\right)_{H_\tau^n} (|x-y|) \right) dy \right| = O(\tau^2) \tag{7.320}$$

holds true as τ tends to zero (note that we omit analogous results in L^2-topology).

Approximate mollifier solution. In analogy to our previous considerations for the Haar kernel of degree 0 we use the operators

$$S_{\left(F_3^{(1)}\right)_{K_\tau}}(\rho) = V_{K_\tau} = \int_{\mathcal{B}} \left(F_3^{(1)}\right)_{K_\tau} (|\cdot - z|)\, \rho(z)\, dz, \ \rho \in L^2(\mathcal{B}), \tag{7.321}$$

and

$$S_{K_\tau}(\rho) = \rho_{K_\tau} = \int_{\mathcal{B}} K_\tau(|\cdot - z|)\, \rho(z)\, dz, \ \rho \in L^2(\mathcal{B}). \tag{7.322}$$

We denote the image of $X = L^2(\mathcal{B})$ under the operator $S_{\left(F_3^{(1)}\right)_{K_\tau}}$ by $Y_{K_\tau} = S_{\left(F_3^{(1)}\right)_{K_\tau}}(L^2(\mathcal{B}))$. So, instead of discussing the Newton integral equation

$$S(\rho) = V = \int_{\mathcal{B}} F_3^{(1)}(|\cdot - y|)\, \rho(y)\, dy, \ \rho \in L^2(\mathcal{B}), \tag{7.323}$$

we consider the mollifier Newton integral equation

$$S_{\left(F_3^{(1)}\right)_{K_\tau}}(\rho) = V_{K_\tau} = \int_{\mathcal{B}} \left(F_3^{(1)}\right)_{K_\tau} (|\cdot - z|)\, \rho(z)\, dz, \ \rho \in L^2(\mathcal{B}), \quad \tau > 0. \tag{7.324}$$

For a chosen Dirac mollifier scaling function $\{K_\tau\}_{\tau>0}$, the Newton mollifier scaling and wavelet functions can be treated completely analogous to the Dirac mollifier case. Therefore, we only give a short outline: We consider, for a positive, monotonically decreasing sequence $\{\tau_j\}_{j\in\mathbb{N}}$ of scale parameters with $\lim_{j\to\infty} \tau_j = 0$ (e.g., the dyadic sequence $\tau_j = C \cdot 2^{-j}$, $j \in \mathbb{N}$, with a constant C, which is chosen to match the size of the test area $\mathcal{B}$), the sequence of Newton mollifier scaling functions $\{\left(F_3^{(1)}\right)_{K_{\tau_j}}(\cdot)\}_{j\in\mathbb{N}}$, satisfying the property

$$-\Delta_x \left(F_3^{(1)}\right)_{K_{\tau_j}} (|x-y|) = K_{\tau_j}(|x-y|), \quad x, y \in \mathbb{R}^3. \tag{7.325}$$

The family of Newton mollifier wavelet functions $\{\left(WF_3^{(1)}\right)_{K_{\tau_j}}(\cdot)\}_{j\in\mathbb{N}}$ is defined via the (scale-discrete) difference equation

$$\left(WF_3^{(1)}\right)_{K_{\tau_j}} (|x-y|) = \left(F_3^{(1)}\right)_{K_{\tau_{j+1}}} (|x-y|) - \left(F_3^{(1)}\right)_{K_{\tau_j}} (|x-y|), \quad j \in \mathbb{N}_0, \tag{7.326}$$

$x, y \in \mathbb{R}^3$. For the "Newton mollifier potentials" corresponding to $\{K_{\tau_j}\}_{j\in\mathbb{N}_0}$

$$V_{K_{\tau_j}}(x) = S_{\left(F_3^{(1)}\right)_{K_{\tau_j}}}(\rho)(x) = \int_{\mathcal{B}} \left(F_3^{(1)}\right)_{K_{\tau_j}} (|x-y|)\, \rho(y)\, dy \tag{7.327}$$

and

$$(WV)_{K_{\tau_j}}(x) = W_{\left(F_3^{(1)}\right)_{K_{\tau_j}}}(\rho)(x) = \int_{\mathcal{B}} \left(WF_3^{(1)}\right)_{K_{\tau_j}} (|x-y|)\, \rho(y)\, dy \tag{7.328}$$

we have the reconstruction formula

$$V_{K_{\tau_J}}(x) = V_{K_{\tau_{J_0}}}(x) + \sum_{j=J_0}^{J-1} (WV)_{K_{\tau_j}}(x), \tag{7.329}$$

i.e., in operator nomenclature

$$S_{\left(F_3^{(1)}\right)_{K_{\tau_J}}}(\rho) = S_{\left(F_3^{(1)}\right)_{K_{\tau_{J_0}}}}(\rho) + \sum_{j=J_0}^{J-1} W_{\left(F_3^{(1)}\right)_{K_{\tau_j}}}(\rho), \tag{7.330}$$

so that

$$\lim_{j\to\infty} \sup_{x\in\overline{\mathcal{B}}} \left| V(x) - \left(V_{K_{\tau_{J_0}}}(x) + \sum_{j=J_0}^{J-1} (WV)_{K_{\tau_j}}(x) \right) \right| = 0. \tag{7.331}$$

It should be noted that each wavelet function of $\{\left(WF_3^{(1)}\right)_{K_{\tau_j}}(\cdot)\}_{j\in\mathbb{N}_0}$ has a local support, and this makes the decorrelation numerics efficient and economical. Clearly, the Newton mollifier scaling and wavelet functions $\left(F_3^{(1)}\right)_{K_{\tau_j}}$ and $\left(WF_3^{(1)}\right)_{K_{\tau_j}}$ generate lowpass or bandpass filtered mollifier versions $S_{\left(F_3^{(1)}\right)_{K_{\tau_J}}}$ and $W_{\left(F_3^{(1)}\right)_{K_{\tau_j}}}$ to the Newton potential V, respectively. Via the negative Laplace derivative we simultaneously obtain lowpass or bandpass filtered versions of the density distribution ρ in the form

$$\begin{aligned} \rho_{K_{\tau_j}}(x) &= -\int_{\mathcal{B}} \Delta_x \left(F_3^{(1)}\right)_{K_{\tau_j}} (|x-y|)\, \rho(y)\, dy \\ &= \int_{\mathcal{B}} K_{\tau_j}(|x-y|)\, \rho(y)\, dy \end{aligned} \tag{7.332}$$

and

$$\begin{aligned} (W\rho)_{K_{\tau_j}}(x) &= \rho_{K_{\tau_{j+1}}}(x) - \rho_{K_{\tau_j}}(x) \\ &= -\int_{\mathcal{B}} \Delta_x \left(WF_3^{(1)}\right)_{K_{\tau_j}} (|x-y|)\, \rho(y)\, dy\ . \end{aligned} \tag{7.333}$$

Mollifier wavelet inversion. After these preparations we first devote our attention to the basic idea already known from the Haar case: For sufficiently large scale J the Newton mollifier potential $V_{K_{\tau_J}}$ can be understood as an approximation of the Newton potential V with negligible error, i.e.,

$$V(x) \simeq V_{K_{\tau_J}}(x) = \int_{\mathcal{B}} \left(F_3^{(1)}\right)_{K_{\tau_J}} (|x-y|)\, \rho(y)\, dy, \quad x\in\mathbb{R}^3. \tag{7.334}$$

An appropriate cubature formula over $\overline{\mathcal{B}}$ yields the representation

$$V(x) \simeq \sum_{i=1}^{N_J} \left(F_3^{(1)}\right)_{K_{\tau_J}} \left(|x-y_i^{N_J}|\right)\, w_i^{N_J}\, \rho(y_i^{N_J}), \tag{7.335}$$

where $w_i^{N_J}\in\mathbb{R}$, $y_i^{N_J}\in\overline{\mathcal{G}}$, $i=1,\ldots,N_J$, are the known weights and nodes, respectively. Since the coefficients $a_i^{N_J} = w_i^{N_J}\,\rho\left(y_i^{N_J}\right), i=1,\ldots,N_J$, are unknown, they must be calculated by solving a linear system of equations of the type

$$V\left(x_k^{M_J}\right) = \sum_{i=1}^{N_J} \left(F_3^{(1)}\right)_{K_{\tau_J}} \left(\Delta; |x_k^{M_J} - y_i^{N_J}|\right) a_i^{N_J}, \quad k = 1, \ldots, M_J, \tag{7.336}$$

from the known gravitational values $V\left(x_k^{M_J}\right)$ at the positions $x_k^{M_J} \in \overline{\mathcal{B}}$, $k = 1, \ldots, M_J$. Since the integration weights $w_i^{N_J}$ are known, one obtains the density values $\rho\left(y_i^{N_J}\right)$ immediately from (7.4), so that the density ρ is calculated from

$$\rho(x) \simeq \rho_{K_{\tau_J}}(x) = \sum_{i=1}^{N_J} K_{\tau_j}\left(|x - y_i^{N_J}|\right) w_i^{N_J} \rho(y_i^{N_J}), \quad x \in \mathcal{B}. \tag{7.337}$$

Simultaneously, by use of the same cubature formula, the wavelet inversion of the gravimetric problem leads to

$$\rho_{K_{\tau_j}}(x) = \sum_{i=1}^{N_J} \underbrace{(-\Delta_x \left(F_3^{(1)}\right)_{K_{\tau_j}} \left(|x - y_i^{N_J}|\right))}_{=K_{\tau_j}\left(|x-y_i^{N_J}|\right)} a_i^{N_J}, \quad j = J_0, \ldots, J, \tag{7.338}$$

and

$$(W\rho)_{K_{\tau_j}}(x) = \sum_{i=1}^{N_J} \underbrace{(-\Delta_x \left(WF_3^{(1)}\right)_{K_{\tau_j}} \left(|x - y_i^{N_J}|\right)}_{=(WK)_{\tau_j}\left(|x-y_i^{N_J}|\right)} a_i^{N_J} \quad j = J_0+1, \ldots, J. \tag{7.339}$$

In other words, a mollifier decorrelation analysis of the mass density distribution ρ from given potential data is established for all scale-levels. However, we are confronted with several difficulties in the described mollifier wavelet inversion process:

- (*Cubature*) In order to apply an appropriate cubature formula (7.335) over regular regions $\mathcal{B} \subset \mathbb{R}^3$ such as Euler summation based lattice point equidistributions (cf. [**278**]) in the sense of Weyl, multivariate Romberg cubature [**274**], etc., we need a certain amount of nodes $y_i^{N_J} \in \overline{\mathcal{G}}$, $i = 1, \ldots, N_J$. Since the total number N_J of nodes and the number M_J of given potential values for the numerics should be in reasonable proportion, there is usually a lack of applicable data material, so that a best approximate cubature formula is a useful, but very time-consuming tool.
- (*Uniqueness*) As already pointed out, the lack of uniqueness of inverse gravimetry means that an inversion with exclusively terrestrial material (especially for deeper structures) is not effective. Pre-information and/or additional information from inside (for example, from earlier borehole activities) is urgently required to guarantee a "certain a priori constraint for the uniqueness".
- (*Pre-Information*) Even if the number of data and suitable pre-information are secured, the critical point of the inversion process described is the solution of the linear system (7.336). Since the kernel $(F_3^{(1)})_{K_{\tau_J}}(|\cdot - \cdot|)$ does not have a local support, an economic numeric is hardly feasible. Stability problems in the aforementioned solution process are therefore unavoidable.

Note that the solution of the linear system (7.336) is the more stable in numerical sense and realistic from practical point of view, the more internal data information is available. In practice, however, a high amount of internal data information is often a wishful thinking. This fact describes the crux of the problem with which terrestrial/internal inverse gravimetry is confronted.

Despite the problems in the mollifier inversion process described above, there are some ways to avoid and/or weak the calamities, which are the subject of the following hints:

Inversion using pre-information. In order to counter the difficulties of the non-uniqueness of the solution and to avoid the full-sized coefficient matrix of the linear system (7.337), we can assume that from terrestrial (in combination with available internal) preliminary information the potential V can be associated with a known lowpass filtered potential $V_{K_{\tau_{J_0}}}$, $J_0 < J$. In fact, we are led to discuss two choices to introduce $V_{K_{\tau_{J_0}}}$:

First variant. We consider a Newton potential with a mollified density distribution

$$V_{K_{\tau_{J_0}}}(x) = \int_{\mathcal{B}} F_3^{(1)}(|x-y|)\, \rho_{K_{\tau_{J_0}}}(y)\, dy, \tag{7.340}$$

where

$$\rho_{K_{\tau_{J_0}}}(y) = \int_{\mathcal{B}} K_{\tau_{J_0}}(|y-z|)\, \rho(z)\, dz. \tag{7.341}$$

This leads to a convolution of the density distribution ρ with a filtered Newton kernel in the form

$$\begin{aligned} V_{K_{\tau_{J_0}}}(x) &= \int_{\mathcal{B}} \int_{\mathcal{B}} F_3^{(1)}(|x-y|)\, K_{\tau_{J_0}}(|y-z|)\, dy\, \rho(z)\, dz \\ &= \int_{\mathcal{B}} \underbrace{\int_{\mathbb{B}^3_{\tau_{J_0}}(z)\cap\mathcal{B}} F_3^{(1)}(|x-y|)\, K_{\tau_{J_0}}(|y-z|)\, dy}_{=\Gamma_{K_{\tau_{J_0}}}(|x-z|)}\, \rho(z)\, dz. \end{aligned} \tag{7.342}$$

Note that the 1D term $\Gamma_{K_{\tau_{J_0}}}(r)$, $r = |x-z|$, may be calculated in an a priori step and stored elsewhere. An appropriate integration formula leads to the sum expression

$$\begin{aligned} & V(x) - V_{K_{\tau_{J_0}}}(x) \\ &\simeq V_{K_{\tau_J}}(x) - V_{K_{\tau_{J_0}}}(x) \\ &= \int_{\mathcal{B}} \left(\Gamma_{K_{\tau_J}}(|x-y|) - \Gamma_{K_{\tau_{J_0}}}(|x-y|)\right)\, \rho(y)\, dy \\ &\simeq \sum_{i=1}^{N_J} \left(\Gamma_{K_{\tau_J}}(|x-y_i^{N_J}|) - \Gamma_{K_{\tau_{J_0}}}(|x-y_i^{N_J}|)\right)\, w_i^{N_J}\, \rho\left(y_i^{N_J}\right), \end{aligned} \tag{7.343}$$

where $w_i^{N_J} \in \mathbb{R}$, $y_i^{N_J} \in \mathbb{B}_{\tau_{J_0}}(x)$, $i = 1, \ldots, N_J$, are the known weights and nodes, respectively. The coefficients $a_i^{N_J} = w_i^{N_J}\, \rho\left(y_i^{N_J}\right), i = 1, \ldots, N_J$, are calculated from the known values $V\left(x_k^{M_J}\right) - V_{K_{\tau_{J_0}}}\left(x_k^{M_J}\right), x_k^{M_J} \in \mathbb{R}^3, k = 1, \ldots, M_J$, via the linear system

$$
\begin{aligned}
(7.344) \qquad & V(x_k^{M_J}) - V_{K_{\tau_{J_0}}}(x_k^{M_J}) \\
&= \sum_{i=1}^{N_J} \left(\Gamma_{K_{\tau_J}}(|x - y_i^{N_J}|) - \Gamma_{K_{\tau_{J_0}}}(|x - y_i^{N_J}|) \right) \; w_i^{N_J} \; \rho\left(y_i^{N_J}\right),
\end{aligned}
$$

$k = 1, \ldots, M_J$.

Second variant. We consider a potential with a mollified Newton kernel

$$
(7.345) \qquad V_{K_{\tau_{J_0}}}(x) \; = \; \int_{\mathcal{B}} \left(F_3^{(1)}\right)_{K_{\tau_{J_0}}} (|x-y|) \; \rho(y) \; dy.
$$

In this case, for a sufficiently large J and for all $x \in \mathbb{R}^3$, we have

$$
\begin{aligned}
(7.346) \qquad & V(x) - V_{K_{\tau_{J_0}}}(x) \\
&\simeq V_{K_{\tau_J}}(x) - V_{K_{\tau_{J_0}}}(x) \\
&= \int_{\mathcal{B}} \left(\left(F_3^{(1)}\right)_{K_{\tau_J}} (|x-y|) - \left(F_3^{(1)}\right)_{K_{\tau_{J_0}}} (|x-y|) \right) \; \rho(y) \; dy.
\end{aligned}
$$

The difference kernel occurring in the integral in (7.346) has a local support. More explicitly, for all $x \in \mathbb{R}^3$ with $|x - y| \geq \tau_{J_0}$, we have

$$
(7.347) \qquad \left(F_3^{(1)}\right)_{K_{\tau_J}} (|x-y|) - \left(F_3^{(1)}\right)_{K_{\tau_{J_0}}} (|x-y|) \; = \; 0.
$$

This property is of great significance for our numerical intention. With a suitable *cubature formula over* $\overline{\mathcal{B}}$ we obtain

$$
\begin{aligned}
(7.348) \quad & V(x) - V_{K_{\tau_{J_0}}}(x) \\
&\simeq \sum_{i=1}^{N_J} \left(\left(F_3^{(1)}\right)_{K_{\tau_J}} \left(|x - y_i^{N_J}|\right) - \left(F_3^{(1)}\right)_{K_{\tau_{J_0}}} \left(|x - y_i^{N_J}|\right) \right) \; w_i^{N_J} \; \rho\left(y_i^{N_J}\right),
\end{aligned}
$$

where $w_i^{N_J} \in \mathbb{R}$, $y_i^{N_J} \in \mathbb{B}_{\tau_{J_0}}(x)$, $i = 1, \ldots, N_J$, are the known weights and nodes, respectively. The coefficients $a_i^{N_J} = w_i^{N_J} \; \rho\left(y_i^{N_J}\right), i = 1, \ldots, N_J$, are calculated from the known values $V\left(x_k^{M_J}\right) - V_{K_{\tau_{J_0}}}\left(x_k^{M_J}\right)$, $x_k^{M_J} \in \mathbb{R}^3$, $k = 1, \ldots, M_J$, via the linear system of equations

$$
\begin{aligned}
(7.349) \quad & V(x_k^{M_J}) - V_{K_{\tau_{J_0}}}(x_k^{M_J}) \\
&= \sum_{i=1}^{N_J} \left(\left(F_3^{(1)}\right)_{K_{\tau_J}} \left(|x_k^{M_J} - y_i^{N_J}|\right) - \left(F_3^{(1)}\right)_{K_{\tau_{J_0}}} \left(|x_k^{M_J} - y_i^{N_J}|\right) \right) \; a_i^{N_J},
\end{aligned}
$$

$k = 1, \ldots, M_J$, which numerically permits the property (7.347).

In other words, the efficiency and economy in the solution process of the linear system (7.347) depends essentially on the distribution and density of the data points $x_k^{M_J}$, $k = 1, \ldots, M_J$, for the available potential values $V(x_k^{M_J})$, $k = 1, \ldots, M_J$:

- If the potential values are available for data points $x_k^{M_J}$, $k = 1, \ldots, M_J$, in "appropriately distributed" density over $\overline{\mathcal{B}}$, the economy is the higher the more pre-information is available due to the property (7.347), i.e., the greater J_0 can be chosen with $J_0 < J$.
- For exclusively given terrestrial potential values (i.e., $x_k^{M_J}$, $k = 1, \ldots, M_J$, are points of the intersection of the Earth's surface with the boundary $\partial\mathcal{B}$), however, the support of (7.347) causes a strong restriction, so that we cannot carry out deeper geological modeling due to the limited extension of the support. In fact, if one is interested in exploration at greater depths, the support has to be adapted, i.e., J_0 has to be chosen appropriately small, which of course affects the numerical economy.

Again, the typical calamity for IPP's of finding appropriate parameter strategies becomes obvious. All in all, by the second inversion variant we are confronted with the unfortunate situation that numerical effectiveness and modeling at greater depth are mutually exclusive.

Gravimetric RKHS context. In gravimetric practice, spline methods within a certain RKHS-framework are often helpful for determining interpolation and smoothing solutions and for completing data material by best approximations. In what follows we methodically orient ourselves towards [**257**], [**262**], [**291**] and [**299**].

The key to the application of spline methods in gravimetry is provided by the reproducing kernel Hilbert space structure of the reference space $Y = S(L^2(\mathcal{B}))$ with S given by

$$S(\rho)(x) = V(x) = \int_{\mathcal{B}} \underbrace{\frac{1}{4\pi} \frac{1}{|x-y|}}_{=F_3^{(1)}(|x-y|)} \rho(y)\, dy, \quad x \in \mathbb{R}^3, \tag{7.350}$$

for square-integrable functions ρ, i.e., ρ is a function of class $X = L^2(\mathcal{B})$. We remember the relationship

$$\begin{aligned} V(x) &= -\Delta_x \int_{\mathcal{B}} \underbrace{\frac{1}{4\pi} \frac{1}{|x-y|}}_{=F_3^{(1)}(|x-y|)} V(y)\, dy \\ &= -\Delta_x \int_{\mathcal{B}} \underbrace{\int_{\mathcal{B}} F_3^{(1)}(|x-y|)\, F_3^{(1)}(\Delta; |y-z|)\, dy}_{=K_Y(x,z)}\ \rho(z)\, dz. \end{aligned} \tag{7.351}$$

The inner product $\langle \cdot, \cdot \rangle_Y$ for Newton potentials $V', V'' \in Y$ of the form

$$V' = \int_{\mathcal{B}} F_3^{(1)}(|\cdot - y|)\, \rho'(y)\, dy, \qquad V'' = \int_{\mathcal{B}} F_3^{(1)}(|\cdot - y|)\, \rho''(y)\, dy \tag{7.352}$$

is given by

$$
\begin{aligned}
&\langle V', V''\rangle_{Y=S(L^2(\mathcal{B}))} \\
&\quad = \int_{\mathcal{B}} \left(-\Delta_x \int_{\mathcal{B}} F_3^{(1)}(|x-y|)\rho'(y)\, dy\right)\left(-\Delta_x \int_{\mathcal{B}} F_3^{(1)}(|x-y|)\rho''(y)\, dy\right)\, dx \\
&\quad = \int_{\mathcal{B}} \rho'(x)\, \rho''(x)\, dx \\
&\quad = \langle \rho', \rho''\rangle_{X=L^2(\mathcal{B})}
\end{aligned}
$$

In other words, the inner product may be understood as an extension from the Hilbert space of the density functions $\rho', \rho'' \in X = L^2(\mathcal{B})$ to the space of Newton volume potentials $V', V'' \in Y$ in the form $\langle V', V''\rangle_Y = \langle \rho', \rho''\rangle_X$.

We consider the kernel

$$
(x,z) \mapsto K_Y(x,z) = \int_{\mathcal{B}} F_3^{(1)}(|x-y|)\, F_3^{(1)}(|y-z|)\, dy, \quad x, z \in \mathbb{R}^3. \tag{7.353}
$$

For each $z \in \mathbb{R}^3$, $K_Y(\cdot, z)$ is an element of Y, and for each $x \in \mathbb{R}^3$, $K_Y(x, \cdot)$ is an element of Y. For each $x \in \mathbb{R}^3$ and each $V \in Y$ we can therefore form the scalar product $\langle K_Y(x,\cdot), V\rangle_Y$. It follows that

$$
\begin{aligned}
(K_Y(x,\cdot), V)_Y &= \left\langle \int_{\mathcal{B}} F_3^{(1)}(|x-y|)\, F_3^{(1)}(|y-\cdot|)\, dy, \int_{\mathcal{B}} F_3^{(1)}(|\cdot - y|)\rho(y)\, dy)\right\rangle_Y \\
&= \int_{\mathcal{B}} F_3^{(1)}(|x-z|)\, \rho(z)\, dz \\
&= V(x). \qquad (7.354)
\end{aligned}
$$

Consequently, the kernel (7.353) has the *reproducing property*

$$
\langle K_Y(x,\cdot), V\rangle_Y = V(x), \quad x \in \mathbb{R}^3, \tag{7.355}
$$

which is of great importance for an efficient setup of a spline theory.

Spline interpolation. Our goal is to find a suitable approximation $S_N^V \in Y$ of the potential V (in the $\|\cdot\|_Y$- sense) under the interpolatory conditions

$$
S_N^V(x_i) = V(x_i), \quad i = 1, \ldots, N. \tag{7.356}
$$

In the space Y, a value $V(x)$ can be expressed in a point $x \in \mathbb{R}^3$ by an evaluation functional $\mathcal{E}_x : V \mapsto \mathcal{E}_x(V) = V(x), V \in Y$, for which the boundedness condition

$$
|\mathcal{E}_x(V)| = |V(x)| \leq \sqrt{K_Y(x,x)}\, \|V\|_Y \tag{7.357}
$$

is valid. The given values $V(x_i)$, $x_i \in \mathbb{R}^3$, $i = 1, \ldots, N$, in the points of X_N are assumed to be associated to a linearly independent system of (bounded) evaluation functionals $\mathcal{E}_{x_1}, \ldots, \mathcal{E}_{x_N}$ on Y. Our aim is to find a *minimum norm interpolating function* $S_N^V \in Y$ as a linear combination of the functions $\mathcal{E}_{x_i} K_Y(\cdot,\cdot) = K_Y(x_i, \cdot)$, $i = 1\ldots, N$, which corresponds exactly to the projection of the geopotential V on the N-dimensional linear subspace spanned by the system $\mathcal{E}_{x_i} K_Y(\cdot,\cdot)$, $i = 1, \ldots, N$ (see, e.g., [**257**]).

In the following we outline an explicit way to establish the explicit representation of $S_N^V \in Y$: The space $\mathcal{S}_Y(\mathcal{E}_{x_1}, \ldots, \mathcal{E}_{x_N})$ of all linear combinations

$$S = \sum_{i=1}^{N} a_i \mathcal{E}_{x_i} K_Y(\cdot,\cdot) = \sum_{i=1}^{N} a_i K_Y(x_i,\cdot), \ a_i \in \mathbb{R}, \tag{7.358}$$

is called the *spline space in* Y *with respect to* $\mathcal{E}_{x_1}, \ldots, \mathcal{E}_{x_N}$. Clearly, $\mathcal{S}_Y(\mathcal{E}_{x_1}, \ldots, \mathcal{E}_{x_N})$ is a N-dimensional subspace of Y. More than that, there is one and only one spline $S \in \mathcal{S}_Y(\mathcal{E}_{x_1}, \ldots, \mathcal{E}_{x_N})$, which has the interpolatory property $S(x_i) = V(x_i)$, $i = 1, \ldots, N$. Note that the coefficient matrix of the linear system of equations

$$\sum_{i=1}^{N} a_i^N K_Y(x_i, x_j) = V(x_j), \quad j = 1, \ldots, N,$$

is regular as Gram matrix of a linearly independent system $\{K_Y(x_i,\cdot)\}_{i=1,\ldots,N}$. We denote the uniquely determined interpolating spline by S_N^V.

Let S be a spline function of class $\mathcal{S}_Y(\mathcal{E}_{x_1}, \ldots, \mathcal{E}_{x_N})$ of the form (7.358). *Then, for every $P \in Y$, the spline formula*

$$\langle S, P\rangle_Y = \sum_{i=1}^{N} a_i \mathcal{E}_{x_i}(P) = \sum_{i=1}^{N} a_i P(x_i) \tag{7.359}$$

holds true (which is an immediate consequence of the reproducing kernel structure in Y).

Using the spline formula (7.359) we can easily guarantee the following property of the interpolating spline S_N^V (see, e.g., [**257**]):

Suppose that there are known the values $V(x_i), i = 1, \ldots, N$, from $V \in Y$. The minimum norm interpolation problem for determining the function S_N^V via the variational condition

$$\|S_N^V\|_Y = \min_{\substack{P \in Y, \\ P(x_i) = V(x_i), i=1,\ldots,N}} \|P\|_Y$$

is well-posed in the sense that its solution S_N^V exists, is unique, and depends continuously on the data $V(x_1), \ldots, V(x_N)$. The uniquely determined solution S_N^V is explicitly given by

$$S_N^V(x) = \sum_{i=1}^{N} a_i^N K_Y(x_i, x), \quad x \in \mathbb{R}^3,$$

where the coefficients $a_1^N, \ldots, a_N^N$ result as the solution of the linear system (7.359). *and the reproducing kernel of the Hilbert space Y is given by* (7.353).

The construction principle is designed so that the interpolating spline has the minimum Y-energy norm among all interpolators. This interpretation is reflected by one-dimensional cubic spline interpolation, where the interpolating spline indeed shows minimal "curvature-energy" (in the linearized sense) (see, e.g., [**375**]). Thus, according to our construction, it may be expected that the interpolating spline S_N^V is largely free from oscillations. From the spline approach above we simultaneously obtain an approximation of the density distribution ρ_N^V as a linear combination of the functions $x \mapsto -\Delta_x \int_{\mathcal{G}} F_3^{(1)}(|x-z|)\, F_3^{(1)}(|z-x_i|)\ dz$, $i=1,\ldots,N$, which is harmonic in $\mathcal{B}\backslash\{x_i : i=1,\ldots,N\}$:

$$\rho_N^V(x) = \sum_{i=1}^{N} a_i^N F_3^{(1)}(x-x_i). \tag{7.360}$$

Alternative approach. It should be remarked that the coefficients a_i^N, $i=1,\ldots,N$, of the spline result (7.360) for the density distribution can be obtained in an alternative way by inserting

$$\rho_N(y) = \sum_{i=1}^{N} a_i^N\ F_3^{(1)}(|y-x_i|) \tag{7.361}$$

into the Newton integral (7.131) and then going on to the interpolation conditions

$$\begin{aligned} V(x_k) &= \int_{\mathcal{B}} F_3^{(1)}(|x-y|)\ \rho_N(y)\ dy \\ &= \sum_{i=1}^{N} a_i^N \int_{\mathcal{B}} F_3^{(1)}(|x_i-y|)\ F_3^{(1)}(|x_k-y|)\ dy, \end{aligned} \tag{7.362}$$

$k=1,\ldots,N$.

The *critical point with spline interpolation*, as presented here, is an adequate solution of the linear system. In what follows we will comment this problem in more detail:

- The choice of the cubature formula to be used for establishing the coefficient matrix strongly depends on the geometry of the region $\mathcal{B}$. Euler summation based lattice point equidistributions in the sense of Weyl, Romberg cubature, best approximate cubature formulas (see, e.g., [**278**]), etc. and the references therein) are appropriately applicable tools for approximate integration over a regular region $\mathcal{B}$. Nonetheless, the computational work should not be underestimated. In particular, equidistributions (in the sense of Weyl) provide simply structured cubature formulas, which, however, show a slow speed of convergence. Thus we are often confronted with the calamity that a simply structured cubature needs a huge number of integration nodes, which can hardly be expected from the observational side in exploration practice. Thus, a best approximate integration formula corresponding to a limited number of nodes forms a way out of this situation, but these formulas are very time consuming. So, on the one side we save time by applying the reproducing structure, on the other side the coefficients of the linear equation system have to be generated by numerical integration.
- The coefficient matrix $(K_Y(x_i,x_j))_{i,j=1,\ldots,N}$ is full-sized. An advantage is the symmetry and positive definiteness of the coefficient matrix, that enables, e.g., Cholesky's decomposition method. Numerical instability results from a growing "correlation" between the "constituting kernel functions" within the interpolation process. This fact can be seen by looking at the *correlation coefficient* C_{corr} (see [**312**]):

$$C_{\text{corr}} = \frac{\langle K_Y(x_k,\cdot),\ K_Y(x_i,\cdot)\rangle_Y}{\|\ K_Y(x_k,\cdot)\ \|_Y \|\ K_Y(x_i,\cdot)\ \|_Y} = \frac{K_Y(x_k,x_i)}{\sqrt{K_Y(x_k,x_k)K_Y(x_i,x_i)}}. \tag{7.363}$$

It follows that $C_{\rm corr}$ tends towards 1 if x_k and x_i become "closer together". The numerical difficulties initiated W. Freeden, M.Z. Nashed [**291**, **296**] to replace the reproducing kernel by Gaussian bell function representations. The goal was to enable exponential sum reduction and to apply algorithms based on Prony's method (for more details about Prony's method, e.g., the following references should be mentioned: [**28**], [**689**], [**717**]). Another procedure is to introduce locally supported mollifier spline wavelets (see [**291**, **296**], [**312**]).

Mollifier spline wavelet methods. Let $\{K_\tau\}_{\tau>0}$ be a Dirac mollifier scaling function. We designate by $Y_{K_\tau} = S_{\left(F_3^{(1)}\right)_{K_\tau}}(L^2(\mathcal{B}))$, $\tau > 0$, the space of all *Newton mollified integrals* $S_{G_{K_\tau}}(\rho)$ given by

$$S_{\left(F_3^{(1)}\right)_{K_\tau}}(\rho)(x) = \int_{\mathcal{B}} \left(F_3^{(1)}\right)_{K_\tau}(|x-y|)\, \rho(y)\, dy, \quad \rho \in L^2(\mathcal{B}). \tag{7.364}$$

For $\tau > 0$, an associated mollifier version of the kernel

$$K_Y(x,y) = \int_{\mathcal{B}} F_3^{(1)}(|x-z|)\, F_3^{(1)}(|y-z|)\, dz \tag{7.365}$$

is given by

$$K_{Y_{K_\tau}}(x,y) = \int_{\mathcal{B}} \left(F_3^{(1)}\right)_{K_\tau}(|x-z|) \left(F_3^{(1)}\right)_{K_\tau}(|z-y|)\, dz, \quad x,y \in \mathbb{R}^3. \tag{7.366}$$

Note that, from (7.366), we see that

$$\begin{aligned} &\sum_{k=1}^{N}\sum_{i=1}^{N} a_k\, a_i \int_{\mathcal{B}} \left(F_3^{(1)}\right)_{K_\tau}(|x_k-z|) \left(F_3^{(1)}\right)_{K_\tau}(|z-x_i|)\, dz \\ &\qquad = \int_{\mathcal{B}} \left| \sum_{k=1}^{N} a_k \left(F_3^{(1)}\right)_{K_\tau}(|x_k-z|) \right|^2 dz > 0. \end{aligned} \tag{7.367}$$

Moreover, the matrix

$$\left(\int_{\mathcal{B}} \left(F_3^{(1)}\right)_{K_\tau}(|x_k-z|) \left(F_3^{(1)}\right)_{K_\tau}(|z-x_i|)\, dz \right)_{k,i=1,\ldots,N} \tag{7.368}$$

is of Gram nature, so that it is non-negative definite and regular, provided that the system $\{\left(F_3^{(1)}\right)_{K_\tau}(|x_i - \cdot|)\}_{i=1,\ldots,N}$ is linearly independent. In other words, the image $Y_{K_\tau} = S_{\left(F_3^{(1)}\right)_{K_\tau}}(L^2(\mathcal{B})), \tau > 0$, defines a Hilbert space $(Y_{K_\tau}, \langle\cdot,\cdot\rangle_{Y_{K_\tau}})$ possessing (7.366) as the reproducing kernel. The image space $Y_{K_\tau} = S_{\left(F_3^{(1)}\right)_{K_\tau}}(L^2(\mathcal{B}))$ of $X = L^2(\mathcal{B})$ under the operator $S_{\left(F_3^{(1)}\right)_{K_\tau}}$ given by (7.364) is a reproducing kernel Hilbert space possessing the reproducing kernel (7.366) i.e.,

$$\langle K_{Y_{K_\tau}}(x,\cdot), V_{K_\tau} \rangle_{Y_{K_\tau}} = V_{K_\tau}(x), \quad x \in \mathbb{R}^3. \tag{7.369}$$

In the space $Y_{K_\tau} = S_{\left(F_3^{(1)}\right)_{K_\tau}}(L^2(\mathcal{B}))$, minimum norm (spline) interpolation may be performed in the already known way:

Let $\{x_1, \ldots, x_N\} \subset \mathbb{R}^3$ be a point system, such that the evaluation functionals $\mathcal{E}_{x_1}, \ldots, \mathcal{E}_{x_N}$ are linearly independent. Then, within the set

$$\mathcal{I}^V_{\mathcal{E}_{x_1},\ldots,\mathcal{E}_{x_N}} = \{U \in Y_{K_\tau} : \mathcal{E}_{x_i}[U] = \mathcal{E}_{x_i}[V] = V(x_i),\ i = 1, \ldots, N\}, \tag{7.370}$$

the minimum norm interpolation problem of finding S^V_N that satisfies

$$\|S^V_N\|_{Y_{K_\tau}} = \inf_{U \in \mathcal{I}^V_{\mathcal{E}_{x_1},\ldots,\mathcal{E}_{x_N}}} \|U\|_{Y_{K_\tau}} \tag{7.371}$$

is well-posed, i.e., its solution exists, is unique and depends continuously on the data $V(x_1), \ldots, V(x_N)$. The uniquely determined solution S^V_N is given in the explicit form

$$S^V_N(x) = \sum_{i=1}^N a_i^N \mathcal{E}_{x_i} K_{Y_{K_\tau}}(\cdot, x), \quad x \in \mathbb{R}^3, \tag{7.372}$$

where the coefficients $a_1^N, \ldots, a_N^N$ are determined by solving the linear system of equations

$$\sum_{i=1}^N a_i^N \mathcal{E}_{x_j} \mathcal{E}_{x_i} K_{Y_{K_\tau}}(\cdot, \cdot) = V(x_j), \quad j = 1, \ldots, N. \tag{7.373}$$

Consequently, we obtain an approximation of the density distribution ρ^V_N as a linear combination of the functions

$$x \mapsto \int_{\mathcal{B}} K_\tau(|x - z|) \left(F_3^{(1)}\right)_{K_\tau} (|z - x_i|)\ dz,\ i = 1, \ldots, N, \tag{7.374}$$

which are <u>not</u> harmonic:

$$\rho^V_N(x) = \sum_{i=1}^N a_i^N \int_{\mathcal{B}} K_\tau(|x - z|) \left(F_3^{(1)}\right)_{K_\tau} (|z - x_i|)\ dz. \tag{7.375}$$

Moreover, for purposes of decorrelation, we are able to take advantage of the sparse character of the resulting wavelets to reduce considerably the computational effort in the solution process. Finally it should be remarked that, for sufficiently small $\tau > 0$, the integrals (7.374) are "close to" $x \mapsto F_3^{(1)}(|z - x_i|)$, $i = 1, \ldots, N$, so that we end up with the already known harmonic approximation in the limit case:

$$\rho^V_N(x) = \sum_{i=1}^N a_i^N F_3^{(1)}(|z - x_i|). \tag{7.376}$$

In practice, a spline approach by means of evaluation functionals as presented until now is unrealistic. For practical purposes other observables than potential values have to come into play such as gravity anomalies, gravity disturbances, deflections of the vertical, satellite-to-satellite (SST)-data, satellite gravitational gradiometry (SGG)-data (cf. [**270**]). These observables, however, involve derivatives applied to the gravitational potential. As long as a derivative is taken at a point of $\mathcal{B}^c$, applied to a regular Newton volume integral, the linear functionals representing these observables (cf. [**262**], [**305**]) are bounded on the reproducing kernel Hilbert space $Y = S(L^2(\mathcal{B}))$. However, in the case of terrestrial observables, a derivative in a point on the boundary $\partial\mathcal{B}$ does not allow necessarily to guarantee the boundedness on $Y = S(L^2(\mathcal{B}))$. The same difficulty happens for derivatives at points inside $\mathcal{B}$. As a consequence, spline methods have to go over to "weakened" spline versions to guarantee the boundedness of measurable gravitational observables. In fact, the occurrence of derivatives leads us to document at least two "weakened" spline variants, namely

- the mollifier spline approach involving Y_{K_τ} as reproducing kernel Hilbert space,
- the spline Hilbert space approach corresponding to a Runge-Bjerhammar ball $\mathbb{B}^3_R$ inside $\mathcal{B}$, i.e., R is chosen such that $R < \inf_{x\in\partial\mathcal{B}} |x|$. In the last case, terrestrial observables relative to $\mathcal{B}$ become inner observables relative to $(\mathbb{B}^3_R)^c$, so that we are able to use original spline theory, but now outside the Runge-Bjerhammar ball (note that the Runge-Bjerhammar ball $\mathbb{B}^3_R$ may be replaced by any regular region $\mathcal{G} \subset \mathcal{B}$ (see Fig. 3, left) with positive distance to $\partial\mathcal{B}$ (see [**254**], [**51**]), however, for reasons of computational simplicity, the Runge-Bjerhammar ball $\mathbb{B}^3_R$ is usually used, for example, in Physical Geodesy (see the sectional illustration of Figure 3, right),

Spline solutions in terms of Gaussian sums resulting in reduced exponential sums. We are interested in spline approximation as an inversion procedure based on evaluation functionals. As, in spline approximation, one needs to evaluate reproducing kernel sum expressions when calculating the full-sized spline matrix, we have to find a proper replacement of

$$K_{A[L^2(\mathcal{G})]}(x,y) = \frac{1}{(4\pi)^2}\int_{\mathcal{G}} \frac{1}{|x-z||y-z|}\, dz, \;\; x, y, \in \mathbb{R}^3, \tag{7.377}$$

that can be evaluated in economic and efficient way. Our goal is to substitute the kernel of a monopole $|x-z|^{-1}$ by a linear combination of Gaussians, i.e.,

$$\frac{1}{|z-x|} \approx \sum_{m=1}^{M} \omega_m e^{-\alpha_m |x-z|^2}, \tag{7.378}$$

so that

$$\begin{aligned} K_{A[L^2(\mathcal{G})]}(x,y) &= \frac{1}{(4\pi)^2}\int_{\mathcal{G}} \frac{1}{|x-z||y-z|}\, dz \\ \simeq K^M_{A[L^2(\mathcal{G})]}(x,y) &= \frac{1}{(4\pi)^2}\sum_{m=1}^{M}\sum_{n=1}^{M} \omega_m\omega_n \int_{\mathcal{G}} e^{-\alpha_m|x-z|^2} e^{-\alpha_n|y-z|^2}\, dz \end{aligned} \tag{7.379}$$

(note that the right side of (7.379) allows the separation into Cartesian coordinates; thus, multi-dimensional integrals over cuboids can be handled iteratively, just by one-dimensional integration. Moreover, (7.379) is explicitly caculable using the Funk-Hecke formula of spherical harmonics in the case of a ball (cf. [**122**])). The critical point concerning (7.378) is to specify the coefficients α_m, ω_m, $m = 1, \ldots, M$, and the integer M. Different approaches have been proposed in the literature: In W. Hackbusch [**388**], the approximation is attacked by a Newton-type optimization procedure. In W. Hackbusch et al. [**389**], a Remez algorithm exploits specific properties of a certain error functional (not discussed here in more detail). Fast multipole methods (see, e.g., [**147**], [**373**]) also provide tools of great numerical significance. Our approach here parallels the concepts presented by G. Beylkin, L., Monzón [**82**], [**83**]. This concept starts with an initial approximation obtained by the appropriate discretization of an integral expression of $|z-x|^{-1}$. Afterwards, in order to reduce the number M of terms of the Gaussian sum on the right side of (7.378), an algorithm is applied based on Prony's method.

An advantage is that we are able to develop our results for the one-dimensional fuction $r \mapsto r^{-1}$, $r \in [\delta, 1]$, with some sufficiently small $\delta > 0$:

$$\frac{1}{r} \simeq \sum_{m=1}^{M} \omega_m e^{-\alpha_m r^2}, \quad r \in [\delta, 1] \tag{7.380}$$

(note that we can transform any closed interval to this type of interval so that we can restrict ourselves, without loss of generality to the interval $[\delta, 1]$).

Our interest is in guaranteeing that

$$\left| \frac{1}{r} - \sum_{m=1}^{M} \omega_m e^{-\alpha_m r^2} \right| \leq \frac{\varepsilon_{rel}}{r}, \quad r \in [\delta, 1], \tag{7.381}$$

where $\varepsilon_{rel} > 0$ is (a bound of) the *relative error of the approximation* (note that the relative error states, up to which significant digit our approximations are correct; in fact, if we compare the absolute error with the relative error via the relation $\varepsilon_{rel} = \max_{r\in[\delta,1]}(\varepsilon_{abs}\, r)$, it becomes clear that the relative error will become larger than the absolute error under our assumptions).

Initial approximation. We start from the integral representation

$$\frac{1}{r} = \frac{2}{\sqrt{\pi}} \int_{-\infty}^{\infty} e^{-r^2 e^{2s} + s}\, ds, \tag{7.382}$$

that easily follows from

$$\frac{2}{\sqrt{\pi}} \int_a^b e^{-r^2 e^{2s} + s}\, ds = \frac{2}{\sqrt{\pi}} \int_{e^a}^{e^b} e^{-(rt)^2}\, dt = \frac{1}{r} \left(\operatorname{erf}(re^b) - \operatorname{erf}(re^a) \right) \tag{7.383}$$

by applying the limit relations $\lim_{b\to\infty} \operatorname{erf}(re^b) = 1, \lim_{a\to-\infty} \operatorname{erf}(re^a) = 0$, where erf is the *error function* given by

$$\operatorname{erf}(s) = \frac{2}{\sqrt{\pi}} \int_0^s \exp(-t^2)\, dt, \ s \geq 0. \tag{7.384}$$

Next we apply the composite *trapezoidal rule* (see, e.g., [**278**]) to discretize the integral (7.382) in order to deduce an initial approximation of the form (7.380). In more detail, if a, b are given such that $-\infty < a < 0 < b < \infty$ and $M + 1$ is the total number of grid points in the trapezoidal rule, then using the abbreviations $s_m = a + mh, m = 0, \ldots, M$ and $h = \frac{b-a}{M}$ we are led to coefficients ω_m and α_m of the Gaussian sum (7.378) in the form $\omega_m = \frac{2h}{\sqrt{\pi}} e^{s_m}, \quad m = 0, \ldots, M$, and $\alpha_m = e^{2s_m}, \quad m = 0, \ldots, M$. So, our initial approximation is of the form

$$\frac{1}{r} \simeq \frac{2h}{\sqrt{\pi}} \sum_{m=0}^{M} e^{s_m} \exp(-e^{2s_m} r^2), \quad r \in [\delta, 1]. \tag{7.385}$$

Now, we are prepared to deal with the question of how to choose a, b, and M. First it is readily seen that

$$\left| 1 - r \int_a^b e^{-r^2 e^{2s} + s}\, ds \right| = r \int_b^{\infty} e^{-r^2 e^{2s} + s}\, ds + r \int_{-\infty}^{a} e^{-r^2 e^{2s} + s}\, ds. \tag{7.386}$$

Our desire is to estimate the occurring integrals such that

$$(7.387)\qquad r\int_{-\infty}^{a} e^{-r^2e^{2s}+s}\,ds < \frac{1}{4}\varepsilon_{rel}, \qquad r\int_{b}^{\infty} e^{-r^2e^{2s}+s}\,ds < \frac{1}{4}\varepsilon_{rel},$$

for all $r \in [\delta, 1]$. As a consequence, the error of the truncated integral in the limits from a to b is at most $\frac{1}{2}\varepsilon_{rel}$, i.e.,

$$(7.388)\qquad \left|1 - r\int_{a}^{b} e^{-r^2e^{2s}+s}\,ds\right| < \frac{1}{2}\varepsilon_{rel}.$$

Observing the special structure of the definite integral (7.383) we are able to conclude that

$$(7.389)\qquad r\int_{-\infty}^{a} e^{-r^2e^{2s}+s}\,ds = |\,\mathrm{erf}(re^a)| \leq \mathrm{erf}(e^a)$$

and

$$(7.390)\qquad r\int_{b}^{\infty} e^{-r^2e^{2s}+s}\,ds = |1 - \mathrm{erf}(re^b)| \leq 1 \quad \mathrm{erf}(\delta e^b)$$

for all $r \in [\delta, 1]$. Combining these results we finally obtain for a and b the following representations: $a = \ln\left(\mathrm{erf}^{-1}\left(\frac{1}{4}\varepsilon_{rel}\right)\right)$, $b = \ln\left(\frac{1}{\delta}\,\mathrm{erf}^{-1}\left(1 - \frac{1}{4}\varepsilon_{rel}\right)\right)$. All in all, the error of the truncated integral is assured to have a bound less than $\frac{1}{2}\varepsilon_{rel}$.

Since the composite trapezoidal rule converges to the integral of the order $O(M^{-2})$, it is evident that the Gaussian sum approximation becomes better the larger M is chosen. In fact, the substitution of r^{-1} by a finite sum of exponentials can be theoretically based on the following result due to [**82**]:

For any δ with $0 < \delta \leq 1$ and $0 < \varepsilon \leq \frac{1}{2}$, there exist positive numbers α_m and ω_m, such that

$$(7.391)\qquad \left|\frac{1}{r} - \sum_{m=1}^{M} \omega_m \exp(\ \alpha_m r^2)\right| < \frac{1}{r}\varepsilon, \quad r \in [\delta, 1],$$

with

$$(7.392)\qquad M = \log\varepsilon^{-1}(c_0 + c_1\log\varepsilon^{-1} + c_2\log\delta^{-1}),$$

where $c_k, k = 0, 1, 2,$ are constants. For fixed δ, we have

$$(7.393)\qquad M = O(\log^2(\varepsilon^{-1})).$$

However, this result does not provide us with the required coefficients of the Gaussian sum. Therefore, in the sequel, we are required to handle a proper determination of M in numerical way. More concretely, for integration limits a and b fixed (as specified above) we are interested in M, so that the stipulated relative error ε_{rel}, in view of (7.385), can be fulfilled, by considering the expression

$$(7.394)\qquad \varepsilon_{rel} \leq \varepsilon(M) = \max_{r\in[\delta,1]}\left|1 - \frac{2h}{\sqrt{h}}\sum_{m=0}^{M} e^{s_m}\exp(-e^{2s_m}r^2)\right|,$$

where h and s_m are known from the trapezoidal rule setting. Detailed studies can be found in the thesis [**122**] (note that a relative error ε_{rel} of 10^{-14} implying $8.993 \cdot 10^{-15}$ for $\varepsilon(M)$ with $M = 402$ already means, that the occurring error is at the second last significant digit in terms of double precision).

Assuming that we have N data points for spline approximation we need exactly $\frac{N(N-1)}{2}$ evaluations of the reproducing kernel (7.377). In total, we have to sum up $M^2 \frac{N(N-1)}{2}$ summands, where we already have taken into account, that the spline interpolation matrix is symmetric and the matrix diagonal is explicitly calculable. Since we are not able to reduce the number of data points N - even worse, we are required to increase the number of data points N in order to get a better approximation - we are interested in reducing the number M of summands in the Gaussian sum. In order to achieve this goal, we are able to use an algorithm proposed in [**83**], which is based on Prony's method. As a matter of fact, this algorithm provides a significant reduction of the number of terms, if the coefficients α_m in a Gaussian sum are relatively small, without losing considerable accuracy.

Sum reduction. We start our considerations from a Gaussian sum with small exponents, i.e., we let $E_{M_0}(r) = \sum_{m=1}^{M_0} p_m e^{-\beta_m r^2}$, where M_0 is large. Our purpose is to seek another Gaussian sum E_N possessing $N \ll M_0$ terms to approximate $E_{M_0}(r)$ in the given interval $[\delta, 1]$. Since the values β_m are supposed to be small, we are allowed to use the Taylor expansion of the exponential function up to a degree $2N-1$, so that

$$E_{M_0}(r) \approx \sum_{m=1}^{M_0} p_m \sum_{k=0}^{2N-1} \frac{(-\beta_m r^2)^k}{k!} = \sum_{k=0}^{2N-1} \underbrace{\left(\sum_{m=1}^{M_0} p_m (-\beta_m)^k \right)}_{=:h_k} \frac{r^{2k}}{k!}. \tag{7.395}$$

Note that h_k is a fast decaying sequence, such that $N \ll M_0$ is resonable. Now, we express the values h_k in the form

$$h_k = \sum_{m=1}^{N} \omega_m (-\alpha_m)^k, \quad k = 0, \dots 2N-1, \tag{7.396}$$

where $\omega_m, \alpha_m \in \mathbb{R}$, $m = 1, \dots, N$. Our goal (cf. [**83**]) is to determine the unknown coefficients ω_m, α_m using *Prony's method.* In more detail, the Prony method aims at finding the coefficients of the polynomial

$$Q(x) = \prod_{m=1}^{N} (x + \alpha_m) = \sum_{k=1}^{N} q_k x^k, \tag{7.397}$$

where $q_N = 1$. Note that

$$\begin{aligned} \sum_{l=0}^{N} h_{k+l} q_l &= \sum_{l=0}^{N} \left(\sum_{m=1}^{N} \omega_m (-\alpha_m)^{k+l} \right) q_l \\ &= \sum_{m=1}^{N} \omega_m (-\alpha_m)^k \underbrace{\sum_{l=0}^{N} q_l (-\alpha_m)^l}_{=0} = 0, \end{aligned} \tag{7.398}$$

since $-\alpha_1, \dots, -\alpha_N$ are the roots of the polynomial Q. As a consequence, we obtain $\sum_{l=0}^{N-1} h_{k+l} q_l = -h_{N+k} = b_k, k = 0, \dots, N-1$. Thus, the vector $q = (q_0, \dots, q_{N-1})^T$ is the solution of the vectorial system $Hq = b$, where $b = (b_0, \dots, b_{N-1})^T$ and $H = \{h_{k+l}\}_{k,l=0,\dots,N-1}$ is a $N \times N$ Hankel matrix. Solving this system in the unknowns q we obtain the coefficients of the polynomial Q, so that we are able to determine the roots α_m (if needed by a numerical method). Afterwards, we solve the Vandermonde system (7.396) and obtain the coefficients ω_m, $m = 1, \ldots, 2N-1$. By observing these preliminaries we are able to set up the desired Gaussian sum E_N to approximate E_{M_0}, i.e.,

$$(7.399) \qquad E_{M_0}(r) = \sum_{m=1}^{M_0} p_m e^{-\beta_m r^2} = \sum_{k=0}^{\infty} h_k \frac{r^{2k}}{k!}$$
$$\simeq \sum_{k=0}^{\infty} \sum_{m=1}^{N} w_m (-\alpha_m)^k \frac{r^{2k}}{k!} = \sum_{m=1}^{N} \omega_m e^{-\alpha_m r^2} = E_N(r).$$

Next we exemplarily present the effectiveness of the reduction procedure by applying it to the initial approximation studied earlier in this section. We start from the initial approximation (7.385) and split it at M_0 so that

$$\sum_{m=1}^{M} \underbrace{\frac{2h}{\sqrt{\pi}} e^{s_m}}_{\omega_m} \exp\left(- \underbrace{e^{2s_m}}_{=\alpha_m} r^2 \right) = \sum_{m=1}^{M_0} \omega_m \exp\left(-\alpha_m r^2\right) + \sum_{m=M_0+1}^{M} \omega_m \exp(-\alpha_m r^2).$$

The question is how to choose M_0 and afterwards N? We recall that the coefficients α_m have to be rather small. The decisive point here is the truncated Taylor expansion of $\exp(x)$ which provides a good approximation, e.g., on a closed interval in $[0, 1]$. This is the reason why we choose M_0 to be the maximum of all m satisfying $s_m < 0, m = 1, \ldots, M_0$. Then, we have $\alpha_m < 1$ for all $m = 1, \ldots, M_0$, numerical tests documented in the thesis [**122**]), have shown that the choice of $N = 6$ leads to good results).

If we assume the reference region $\mathcal{G}$ especially to be of spherical shape, e.g., $\mathcal{G}$ is the ball $\mathbb{B}_R(0)$ of radius R around the origin, the reproducing kernel is available in explicit form, so that the computation time for numerical purposes can be kept short (for details see [**294**]).

Spline smoothing. In case of erroneous data, spline interpolation must be replaced by a *spline smoothing* technique in $Y = \mathcal{S}(L^2(\mathcal{B}))$. It turns out that the "smoothing spline" minimizes a functional that controls the fitting of an approximating function, under statistically oriented conditions: For a given point system $\{x_1, \ldots, x_N\}$ the problem of adapting a smooth function to a given amount of data $(x_1, \tilde{\gamma}_1), \ldots, (x_N, \tilde{\gamma}_N)$ is usually realized by looking for a potential $P \in Y$ so that the functional

$$(7.400) \qquad \sigma_{\beta,\delta}(P) = \sum_{k=1}^{N} \left(\frac{\mathcal{E}_{x_k} P - \tilde{\gamma}_k}{\beta_k} \right)^2 + \delta \, \|P\|_Y^2$$

is minimized in Y, where $\beta = (\beta_1, \ldots, \beta_N)^T \in \mathbb{R}^N$ are predetermined positive weights and $\delta \geq 0$ is any parameter. If possible, the values β_k, $k = 1, \ldots, N$, should be adjusted to the standard deviations of the measured data values $\tilde{\gamma}_k$, $k = 1, \ldots, N$. The parameter δ provides a measure of smoothness. A small value of δ associates confidence in the measured values, while a large value of δ does the opposite. If we formally select $\delta = 0$, smoothing will result in least-squares approximation. If we have $\mathcal{E}_{x_k} P = P(x_k) = \gamma_k = \tilde{\gamma}_k$, $k = 1, \ldots, N$, and $\delta \neq 0$, we are back in spline interpolation.

Let δ, $\beta_1, \ldots, \beta_N$ be given positive constants. Suppose that $(x_k, \tilde{\gamma}_k)$, $k = 1, \ldots, N$, are given data points. Then there exists a uniquely determined spline $S \in \mathcal{S}_Y(\mathcal{E}_{x_1}, \ldots, \mathcal{E}_{x_N})$, so that the inequality

$$\sigma_{\beta,\delta}(S) \leq \sigma_{\beta,\delta}(P)$$

applies to all $P \in Y$ (with equality if and only if $P = S$). S is uniquely determined by the linear system of equations

$$\mathcal{E}_{x_k} S + \delta\, \beta_k^2\, a_k = \tilde{\gamma}_k, \quad k = 1, \ldots, N.$$

The proof is standard, it can be found, for example, in [**257**]. Obviously, the condition of the coefficient matrix of the linear system improves because of more diagonal dominance with an increasing smoothing parameter δ. Our explanations imply that simultaneously with the spline approximation S for the potential V

$$V(x) \simeq S(x) = \sum_{i=1}^{N} a_i \int_{\mathcal{B}} F_3^{(1)}(|x-y|)\, F_3^{(1)}(|y-x_i|)\, dy, \quad x \in \mathcal{B}, \tag{7.401}$$

the resulting approximation ρ_S for the density ρ in $\mathcal{B}\backslash\{x_1, \ldots x_N\}$ is given by

$$\rho(x) \simeq \rho_S(x) = -\sum_{i=1}^{N} a_i F_3^{(1)}(|x-x_i|). \tag{7.402}$$

In other words, following this approach, the density distribution ρ has a harmonic representation.

For the sake of simplicity, we were so far concerned with evaluation functionals $\mathcal{E}_{x_i}$, $i = 1, \ldots, N$, in smoothing methodology. However, all arguments remain valid also for general linear bounded functionals $\mathcal{A}_i, i = 1, \ldots, N$ on Y, which characterize gravitational observables (as, e.g., specified in [**270**]), where $\mathcal{A}_i V$, $i = 1, \ldots, N$, denote the data values available by observation and/or measurement. To perform a smoothing procedure for known gravitational quantities $\mathcal{A}_i V, i = 1, \ldots, N$, we use the functional (7.400), which describes the "trade-off" between confidence in the potential data and smoothness of the approximating spline: More concretely, our consideration starts from the assumption for the potential $V \in Y = S(L^2(\mathcal{B}))$ that we are confronted as a result of observation and measurement with a set of real numbers $\mathcal{A}_i(V) = \tilde{\alpha}_i$, $i = 1, \ldots, p$; $\mathcal{L}_j(V) = \lambda_j$, $j = 1, \ldots, q$, corresponding to a system of $(p+q)$ bounded functionals ("observables") $\mathcal{A}_1, \ldots, \mathcal{A}_p$; $\mathcal{L}_1, \ldots, \mathcal{L}_q$ on Y.

Combined spline interpolation and smoothing. Following a contribution by W. Freeden, B. Witte [**331**] we are interested in an approximation of the potential $V \in Y = S(L^2(\mathcal{B}))$ by a spline S, which is "smooth" in the Y-norm and satisfies the following properties:

$$\begin{aligned} \mathcal{A}_i(S) &\text{ is "close" to } \mathcal{A}_i(V) = \tilde{\alpha}_i, \quad i = 1, \ldots, p, \\ \mathcal{L}_i(S) &\text{ is equal to } \mathcal{L}_i(V) = \lambda_i, \quad i = 1, \ldots, q. \end{aligned}$$

Suppose that $\mathcal{A}_1, \ldots, \mathcal{A}_p$ and $\mathcal{L}_1, \ldots, \mathcal{L}_q$ are systems of linear functionals bounded on $Y = S(L^2(\mathcal{G}))$. Let V be an (unknown) potential of class Y, such that the erroneous values α_k, $k = 1, \ldots, p$, for $\mathcal{A}_k(V)$, $k = 1, \ldots, p$, and the exact values λ_k, $k =$

$1, \dots, q$, for $\mathcal{L}_k(V)$, $k = 1, \dots, q$, are known. δ, $\beta_1, \dots, \beta_p$ are suitably given positive constants resulting from a priori statistical knowledge. Then, the coefficients

$$a \in \mathbb{R}^p, \quad a = (a_1, \dots, a_p)^T,$$
$$l \in \mathbb{R}^q, \quad l = (l_1, \dots, l_q)^T,$$

of the spline function $S \in Y$ of the form

$$S(x) = \sum_{i=1}^{p} a_i \mathcal{A}_i K_Y(\cdot, x) + \sum_{j=1}^{q} l_j \mathcal{L}_j K_Y(\cdot, x) \tag{7.403}$$

are uniquely determined by the linear equations

$$\mathcal{A}_i S + \delta\, \beta_i^2\, a_i = \tilde{\alpha}_i\ , \quad i = 1, \dots, p, \tag{7.404}$$
$$\mathcal{L}_j S = \lambda_j\ , \quad j = 1, \dots, q. \tag{7.405}$$

The spline $S \in Y$, which is uniquely determined by the equations (7.404) and (7.405), represents the only solution in Y that satisfies the inequality

$$\begin{aligned} \sigma_{\beta,\delta}(S) &= \sum_{i=1}^{p} \left(\frac{\mathcal{A}_i(S) - \tilde{\alpha}_i}{\beta_i} \right)^2 + \delta(S, S)_Y \\ \le \sigma_{\beta,\delta}(P) &= \sum_{i=1}^{p} \left(\frac{\mathcal{A}_i(P) - \tilde{\alpha}_i}{\beta_i} \right)^2 + \delta(P, P)_Y \end{aligned} \tag{7.406}$$

for all $P \in Y$ with $\mathcal{L}_j(P) = \lambda_j, \quad j = 1, \dots, q$.

Mollifier spline inversion. As already known, for a sufficiently large J, the reproducing kernel

$$K_Y(x, z) = \int_{\mathcal{B}} F_3^{(1)}(|x - y|)\ F_3^{(1)}(|y - z|)\ dy \tag{7.407}$$

can be replaced by the mollifier kernel

$$K_{Y_{\tau_J}}(x, z) = \int_{\mathcal{B}} \left(F_3^{(1)} \right)_{K_{\tau_J}} (|x - y|)\ \left(F_3^{(1)} \right)_{K_{\tau_J}} (|y - z|)\ dy. \tag{7.408}$$

While the right side of (7.407), for $x, z \in \overline{\mathcal{B}}$, represents an improper integral, the mollifier approximation (7.408) is a proper integral for all $x, z \in \mathbb{R}^3$. A spline potential $S \in Y$ approximating the potential $V \in Y$ with respect to $\mathcal{A}_i$, $i = 1, \dots, N$, of the form

$$V(x) \simeq S(x) = \sum_{i=1}^{N} a_i \mathcal{A}_i K_Y(\cdot, x) \tag{7.409}$$

can be replaced by the mollifier spline

$$V(x) \simeq S^N_{K_{\tau_J}}(x) = \sum_{i=1}^{N} a_i \mathcal{A}_i K_{Y_{K_{\tau_J}}}(\cdot, x), \quad x \in \mathcal{B}.$$

Applying the negative Laplace derivative we are able to calculate the corresponding mollifier density function in the (non-harmonic) representation (with $K_{\tau_J}(\cdot, \cdot)$ from,

e.g., (7.411))
(7.410)

$$\rho(x) \simeq \rho_{S^N_{K_{\tau_J}}}(x) = -\sum_{i=1}^{N} a_i \int_{\mathcal{B}} K_{\tau_J}(|x-y|)\, \mathcal{L}_i \left(F_3^{(1)}\right)_{K_{\tau_J}} (|y-\cdot|)\, dy, \quad x \in \mathcal{B}.$$

Of course, we can proceed to scales $j < J$ in the usual way, already explained earlier, thus obtaining decorrelations from spline approximants.

In contrast to the spline S, the mollifier spline $S_{K_{\tau_J}}$ can also be used for linear functionals $\mathcal{A}_i$ that are not bounded with respect to Y, such as higher order radial derivatives, oblique derivatives or negative Laplace derivatives in the interior of the test area. In particular, terrestrial gravimetric data and even gravimetric borehole data can be used as derivative functionals. This formally leads to the following *combined interpolation/smoothing mollifier approximation* adequately adapted to the goal of inverse gravimetry.

Let $\mathcal{A}_1, \ldots, \mathcal{A}_p$ be linear functionals representing the "observable-type of oblique derivative, gravitational (Bouguer) anomaly, etc. in terrestrial points $y_1, \ldots, y_p$ (see, e.g., [**270**] *for details). Let $\mathcal{L}_1, \ldots, \mathcal{L}_q$ be linear functionals representing the negative Laplace derivative in the points $x_1, \ldots, x_q$ of the interior $\mathcal{B}$. Let V be a potential for which measured values $\tilde{\alpha}_k$, $k = 1, \ldots, p$, i.e.,*

$$\mathcal{A}_k(V) = \tilde{\alpha}_k, \quad k = 1, \ldots, p, \tag{7.411}$$

and exact values $\rho(x_k) = \rho_k$, $k = 1, \ldots, q$, i.e.,

$$\mathcal{L}_k(V) = -\Delta V(x_k) = \rho_k, \; k = 1, \ldots, q, \tag{7.412}$$

are known. Suppose that $\delta, \beta_1, \ldots, \beta_p$ are suitably given (statistically motivated) positive constants.

Then the coefficients

$$a_{\tau_J} \in \mathbb{R}^p, \quad a_{\tau_J} = (a_1^{\tau_J}, \ldots, a_p^{\tau_J})^T, \tag{7.413}$$

$$l_{\tau_J} \in \mathbb{R}^q, \quad l_{\tau_J} = (l_1^{\tau_J}, \ldots, l_q^{\tau_J})^T, \tag{7.414}$$

of the mollifier spline

$$S_{K_{\tau_J}}(x) = \sum_{i=1}^{p} a_i^{\tau_J} \mathcal{A}_i K_{Y_{K_{\tau_J}}}(\cdot, x) + \sum_{j=1}^{q} l_j^{\tau_J} \mathcal{L}_j K_{Y_{K_{\tau_J}}}(\cdot, x), \tag{7.415}$$

obtained from the linear equations

$$\mathcal{A}_i S_{K_{\tau_J}} + \delta\, \beta_i^2\, a_i^{\tau_J} = \tilde{\alpha}_i\,, \quad i = 1, \ldots, p, \tag{7.416}$$

$$\mathcal{L}_j S_{K_{\tau_J}} = \rho_j\,, \quad j = 1, \ldots, q, \tag{7.417}$$

provide a combined interpolation/smoothing approximation to erroneous terrestrial gravitational data and exactly known density values.

This is a remarkable result in the theory of inverse gravimetry, since it combines available values from potential and density distribution in canonical way.

Nonlinear optimization regularization. Next we are concerned with a realistic "pre-information scenario" (note that a smoothing procedure is unavoidable when handling measurements, since these data are usually affected with noise, i.e., they are subject to errors due to the measuring process and the measuring instruments, which must be compensated): We suppose that the gravitational vector ∇V (or a "Bouguer variant" of it, see, e.g., [**270**]) is modelable in the form

$$(7.418)\quad (W\nabla V)_{K_{\tau_{J_0}}}^{K_{\tau_J}}(x) = \int_{\mathcal{B}} \nabla_x \left(\left(WF_3^{(1)} \right)_{K_{\tau_{J_0}}}^{K_{\tau_J}} (|x-y|) \right) \rho(y)\, dy, \quad x \in \mathbb{R}^3,$$

where we have used the abbreviation

$$(7.419)\quad (\left(WF_3^{(1)} \right)_{K_{\tau_{J_0}}}^{K_{\tau_J}} (|x-y|) = \left(F_3^{(1)} \right)_{K_{\tau_J}} (|x-y|) - \left(F_3^{(1)} \right)_{K_{\tau_{J_0}}} (|x-y|).$$

In similarity to spline smoothing (see, e.g., [**331**]) we consider the functional

$$(7.420)\quad \sigma_{\beta,\delta}^{cont}(\rho) = \sum_{i=1}^{N} \left(\frac{|(W\nabla V)_{K_{\tau_{J_0}}}^{K_{\tau_J}}(x_i)| - |\nabla\tilde{V}(x_i)|}{\beta_i} \right)^2 + \delta \int_{\mathcal{B}} (\rho(y) - \tilde{\rho}(y))^2\, dy,$$

where the intensities

$$(7.421)\qquad |\nabla\tilde{V}(x_i)| = |\nabla_x \tilde{V}(x)|_{x=x_i}|, \qquad x_i \in \mathbb{R}^3,\ i = 1, \dots, N,$$

are given and $\tilde{\rho}$ designates an observed pre-information about the mass density distribution ρ in $\mathcal{B}$. In order to come to numerical results, we have to discretize $\sigma_{\beta,\delta}^{cont}(\rho)$ by applying an appropriate cubature formula leading to the functional $\sigma_{\beta,\delta}^{discr}(\rho_1, \dots, \rho_N)$ given by

$$(7.422)\quad \sigma_{\beta,\delta}^{discr}(\rho_1, \dots, \rho_N)$$
$$= \sum_{i=1}^{N} \frac{1}{\beta_i^2} \left(\left| \sum_{k=1}^{M} w_k \nabla_x \left(\left(WF_3^{(1)} \right)_{K_{\tau_{J_0}}}^{K_{\tau_J}} (|x-y_k|)|_{x=x_i} \right) \rho_k \right| - \left| \nabla\tilde{V}(x_i) \right| \right)^2$$
$$+ \delta \sum_{k=1}^{M} w_k \left| \rho_k - \tilde{\rho}(y_k) \right|^2.$$

Here, $w_k, y_k,\ k = 1, \dots, M$, are the known positive weights and the prescribed nodes of the cubature formula, respectively, and $\rho_k = \rho(y_k),\ k = 1, \dots, M, \eta$ are the unknown mass density values for points $y_k,\ k = 1, \dots, M$, in $\mathcal{B}$. In other words, the $\tilde{\rho}$-vector $(\tilde{\rho}(y_1), \dots, \tilde{\rho}(y_M))^T$ collects the known density pre-information of the unknown density value vector $(\rho(y_1), \dots, \rho(y_M))^T$, and $|\nabla\tilde{V}(x_i)|,\ i = 1, \dots, N$, characterizes the known intensities $|\nabla V(x_i)|,\ i = 1, \dots, N$, in the points $x_i \in \mathbb{R}^3$. The values $\beta_i,\ i = 1, \dots, N$, should be adjusted to the standard deviations of the observed data values $|\nabla\tilde{V}(x_i)|,\ i = 1, \dots, N$. The positive parameter δ serves for the smoothness control, i.e., the larger the choice of δ is, the stronger the influence of the a priori information $(\tilde{\rho}(y_1), \dots, \tilde{\rho}(y_M))^T$ on the unknown density values $(\rho(y_1), \dots, \rho(y_M))^T$ to be calculated.

In fact, it is not difficult to see that the expression $(\sigma_{\beta,\delta}^{discr}(\rho_1,\ldots,\rho_M))^2$ can be interpreted as the square of the $(N+M)$-dimensional vector

(7.423)

$$\begin{pmatrix} \frac{1}{\beta_1}\left|\sum_{k=1}^{M} w_k \nabla\left(\left(F_3^{(1)}\right)_{K_{\tau_J}}(|x_1-y_k|) - \left(F_3^{(1)}\right)_{K_{\tau_{J_0}}}(|x_1-y_k|)\right)\rho_k\right| - \left|\nabla\tilde{V}(x_1)\right| \\ \vdots \\ \frac{1}{\beta_N}\left|\sum_{k=1}^{M} w_k \nabla\left(\left(F_3^{(1)}\right)_{K_{\tau_J}}(|x_N-y_k|) - \left(F_3^{(1)}\right)_{K_{\tau_{J_0}}}(|x_N-y_k|)\right)\rho_k\right| - \left|\nabla\tilde{V}(x_N)\right| \\ \sqrt{\delta}\,\sqrt{w_1}\,|\rho_1-\tilde{\rho}(y_1)| \\ \vdots \\ \sqrt{\delta}\,\sqrt{w_M}\,|\rho_M-\tilde{\rho}(y_M)| \end{pmatrix}.$$

The unknown coefficients are to be determined in such a way, that the term $(\sigma_{\beta,\delta}^{discr}(\rho_1,\ldots,\rho_M))^2$ is minimized in the set of all M-tupel $(\rho_1,\ldots,\rho_M)^T$. It should be noted that the result of the minimization procedure is strongly dependent on the "smoothing parameter" δ. The mollifier wavelet

(7.424)

$$x \mapsto \left(WF_3^{(1)}\right)_{K_{\tau_{J_0}}}^{K_{\tau_J}}(|x-y|) = \left(F_3^{(1)}\right)_{K_{\tau_J}}(|x-y|) - \left(F_3^{(1)}\right)_{K_{\tau_{J_0}}}(|x-y|), \quad x \in \mathcal{B},$$

is known to possess a ball as support, the width of which is determined by the scale index J_0 and thus J_0 , as already explained, has a decisive influence on the desired exploration depth. In consistency with our model specification, the scale index $J > J_0$ should be chosen as large as possible. Clearly, the nonlinear minimization problem $(\sigma_{\beta,\delta}^{discr}(\rho_1,\ldots,\rho_M))^2 \to \min$ can be solved by a standard solver in numerics, which will not be discussed here.

Finally, it should be mentioned that the regularization techniques as presented here found their application to geothermal exploration during the last years in the German state Saarland, where huge anthropogenic impacts on mining prevents the use of reflection seismics (research project "SPE" (funding reference number: FKZ 0324061), PI Prof. Dr. W. Freeden, University of Kaiserslautern, Germany), corporate manager Prof. Dr. M. Bauer, CBM - Gesellschaft für Consulting, Business und Management mbH, Bexbach, Germany) funded by "Federal Ministry for Economic Affairs and Energy, Berlin" and the "Project Management Jülich", responsible project managers Dr. W. Monser, Dr. K. Schreiber.

7.5. Inverse magnetometry (IM)

Our magnetometric mollifier approach is interested in associated sequences of dipole and Dirac wavelet kernels, respectively, which show decreasing space-localized supports with increasing scale for the simultaneous multiscale approximation of both magnetometric potential and magnetization density distribution, so that numerically economic and efficient implementations become applicable. In fact, the convolutions with the inverse wavelet kernels simultaneously allow to model bandpass filtered information of the potential as well as the magnetization distribution, so that both signatures admit a certain decorrelation into detail information.

The substantial idea is that, if the wavelet kernels of the magnetometric potential are constructed in close geophysical relevance, the detail information of the magnetization distribution reflects the structure of the geological layers. The critical point, however, is the choice of the associated dipole potential and Dirac wavelet kernels. In principle, an infinite number of kernel variants can be taken for purposes of approximation.

Nonetheless, this work offers a reasonable compromise. It will be shown that an advantageous connection of our magnetometric interest with appropriate tools of constructive approximation is given by the probably simplest type of a space-localizing Dirac mollifier, namely the Haar kernels. As a matter of fact, a mollifier sequence of Haar-type kernels immediately leads to a counterpart of a dipole mollifier sequence by forming taylorized versions of the dipole potential kernel, so that mollifier divergences of the vectorial dipole integral via the (negative) Laplace operator in 3D Euclidean space, indeed, reduce to Haar-type kernels after inversion. As a consequence, the simultaneous approach using associated dipole potential and Dirac wavelets opens a cornucopia of perspectives (see [**270**], [**100**] and the references therein), for instance, signature decorrelation, noise cancellation, data compression, etc. The disadvantage of the classical Haar kernel, however, is its discontinuity. This often leads to numerical instabilities, especially in a marginal strip of around the boundary of a bounded exploration area. This phenomenon can be avoided similarly as in inverse gravimetry, at least to some extend, by, e.g., smoothed variants of the Haar kernels.

The dipole magnetic induction for a distribution of magnetization m is given by

$$b(x) = \nabla_x \int_{\mathcal{B}} \underbrace{\nabla_y F_3^{(1)}(|x-y|)}_{=\gamma(x,y)} \cdot m(y)\, dy \tag{7.425}$$

The replacement of the singular kernel $\gamma(x,y)$ by the regular (mollifier) Haar kernel $\gamma_{H_\tau^n}(x,y)$ enables an approximation in the form

$$b_{H_\tau^n}(x) = \nabla_x \int_{\mathcal{B}} \underbrace{\nabla_y \left(F_3^{(1)}\right)_{H_\tau^n}(|x-y|)}_{=\gamma_{H_\tau^n}(x,y)} \cdot m(y)\, dy \tag{7.426}$$

where the mollifier vectorial dipole kernel $\gamma_{H_\tau^n}(x,y)$, $x,y \in \mathbb{R}^3$, possesses the explicit representation

$$\gamma_{H_\tau^n}(x,y) = \begin{cases} -\frac{x-y}{4\pi|x-y|^3}, & \tau_j \le |x-y|, \\ -\frac{x-y}{4\pi|x-y|^3} - \frac{(x-y)(\tau_j-|x-y|)^{n+2}(n+1)}{8\pi|x-y|^2{\tau_j}^{n+3}} & \\ \quad + \frac{(x-y)(\tau_j-|x-y|)^{n+2}(2\tau_j+|x-y|(n+1))}{8\pi|x-y|^3{\tau_j}^{n+3}} & \\ \quad + \frac{(x-y)(\tau_j-|x-y|)^{n+1}(n+2)(8\pi\tau_j+|x-y|(n+1))}{8\pi|x-y|^2{\tau_j}^{n+3}}, & 0 < |x-y| < \tau_j, \\ 0, & |x-y| = 0, \end{cases}$$

The tensorial mollifier scaling function $\mathfrak{g}_{H_\tau^n}(x,y) = \nabla_x \gamma_{H_\tau^n}(x,y),\ x,y \in \mathbb{R}^3$, is given by

(7.427)

$$\mathfrak{g}_{H_\tau^n}(x,y) = \begin{cases} (x-y)(x-y)^T \frac{3}{4|x-y|^5\pi} - \mathbf{i}\frac{1}{4\pi|x-y|^3}, & \tau_{j_j} \le |x-y|, \\ -(x-y)(x-y)^T \left(\frac{6{\tau_j}^3(\tau_j-|x-y|)^n - 6{\tau_j}^{n+3}}{8|x-y|^5{\tau_j}^{n+3}\pi} \right. & \\ \quad + \frac{3|x-y|^3 n^2(\tau_j-|x-y|)^n}{8|x-y|^5{\tau_j}^{n+3}\pi} & \\ \quad + \frac{|x-y|^3 n^3(\tau_j-|x-y|)^n + 2|x-y|^3 n(\tau_j-|x-y|)^n}{8|x-y|^5{\tau_j}^{n+3}\pi} & \\ \quad + \frac{6|x-y|n{\tau_j}^2(\tau_j-|x-y|)^n + 3|x-y|^2 n\tau_j(\tau_j-|x-y|)^n}{8|x-y|^5{\tau_j}^{n+3}\pi} & \\ \quad \left. + \frac{3|x-y|^2 n^2\tau_j(\tau_j-|x-y|)^n}{8|x-y|^5{\tau_j}^{n+3}\pi} \right) & \\ +\mathbf{i}\left(\frac{(\tau_j-|x-y|)^{n+2}(|x-y|+2\tau_j+|x-y|n)}{8\pi|x-y|^3\tau_j n+3} - \frac{1}{4\pi|x-y|^3} \right. & \\ \quad - \frac{(\tau_j-|x-y|)^{n+2}(n+1)}{8\pi|x-y|^2{\tau_j}^{n+3}} & \\ \quad \left. + \frac{(\tau_j-|x-y|)^{n+1}(n+2)(|x-y|+2\tau_j+|x-y|n)}{8\pi|x-y|^2{\tau_j}^{n+3}} \right), & 0 < |x-y| < \tau_j, \\ -\frac{n^3+6n^2+11n+6}{24\pi{\tau_j}^3}, & 0 = |x-y|, \end{cases}$$

where $\mathbf{i}$ denotes the unit matrix.

Multiscale mollifier approximation. For numerical purposes, the continuous mollifier parameter τ will be replaced by a strictly monotonically decreasing sequence τ_j, $j \in \mathbb{N}_0$, with $\lim_{j\to\infty} \tau_j = 0$, and filtering has to be performed with corresponding τ_j-low- and bandpass mollifier convolutions. The *scale-discrete vectorial wavelet kernels* $(W\gamma)_{H_{\tau_j}^n}(x,y),\ x,y \in \mathbb{R}^3$, given by

$$(W\gamma)_{H_{\tau_j}^n}(x,y) = \gamma_{H_{\tau_{j+1}}^n}(x,y) - \gamma_{H_{\tau_j}^n}(x,y), \tag{7.428}$$

are used for the geological interpretation by multiscale mollifier decorrelation. Later, they also serve as a basis for the intended inversion process to deduce the magnetization distribution from geomagnetic field data.

There are several ways to obtain mollifier approximations of the magnetic induction, from which we study two approaches in more detail:

Tensor-based variant. As *j-scale lowpass filtered magnetic dipole induction* we obtain

$$b_{H_{\tau_j}^n}(x) = \int_{\mathcal{B}} \mathfrak{g}_{H_{\tau_j}^n}(x,y)\, m(y)\, dy, \tag{7.429}$$

and the *j-scale bandpass filtered magnetic dipole induction* is given by

$$(Wb)_{H_{\tau_j}^n} = b_{H_{\tau_{j+1}}^n} - b_{H_{\tau_j}^n}, \tag{7.430}$$

i.e.,

$$(Wb)_{H_{\tau_j}^n}(x) = \int_{\mathcal{B}} \underbrace{(\mathfrak{g}_{H_{\tau_{j+1}}^n}(x,y) - \mathfrak{g}_{H_{\tau_j}^n}(x,y))}_{=(W\mathfrak{g})_{H_{\tau_j}^n}(x,y)}\, m(y)\, dy. \tag{7.431}$$

For integers j with $J_0 \leq j \leq J-1$, this allows a decorrelation of $b_{H^n_{\tau_J}}$ via the identity

$$b_{H^n_{\tau_J}}(x) = b_{H^n_{\tau_{J_0}}}(x) + \sum_{j=J_0}^{J-1} (Wb)_{H^n_{\tau_j}} . \tag{7.432}$$

For sufficiently large J, we have

$$b(x) \simeq b_{H^n_{\tau_J}}(x) = \int_{\mathcal{B}} \mathfrak{g}_{H^n_{\tau_J}}(x,y)\, m(y)\, dy, \quad x \in \mathbb{R}^3. \tag{7.433}$$

In order to establish an inversion process we obtain by using a suitable cubature formula over $\overline{\mathcal{B}}$,

$$b(x) \simeq b_{H^n_{\tau_J}}(x) \simeq \sum_{i=1}^{N_J} \mathfrak{g}_{H^n_{\tau_J}}(x, y_i^{N_J})\ \underbrace{w_i^{N_J}\ m(y_i^{N_J})}_{=a_i^{N_J}}, \quad x \in \mathbb{R}^3, \tag{7.434}$$

and

$$(Wb)_{H^n_{\tau_J}}(x) \simeq \sum_{i=1}^{N_J} (W\mathfrak{g})_{H^n_{\tau_J}}(x, y_i^{N_J})\ \underbrace{w_i^{N_J}\ m(y_i^{N_J})}_{=a_i^{N_J}}, \quad x \in \mathbb{R}^3, \tag{7.435}$$

where $w_i^{N_J} \in \mathbb{R}$ and $y_i^{N_J} \in \mathbb{R}^3, i = 1, \ldots, N_J$, are the known weights and nodes, respectively. Hence, if a finite set of vector values of the field b is known at certain locations, the approximate formula (7.434) provides a linear system in the unknown coefficients $a_i^{N_J}$, $i = 1, \ldots, N_J$, and, consequently, $m(y_i^{N_J})$, $i = 1, \ldots, N_J$.

Scalar-based variant. The field b is convolved against the scalar mollifiers $H^n_{\tau_j}$ leading to

$$b_{H^n_{\tau_j}}(x) = b_{H^n_{\tau_j}}(x) = \int_{\mathcal{B}} H^n_{\tau_j}(|x-y|)\ b(y)\ dy \tag{7.436}$$

hence,

$$b_{H^n_{\tau_j}}(x) = \nabla_x \int_{\mathcal{B}} \int_{\mathcal{B}} H^n_{\tau_j}(|x-y|) \left(\nabla_y \frac{1}{4\pi} \frac{1}{|z-y|} \right) dy \cdot m(z)\ dz, \tag{7.437}$$

where, for $x, z \in \mathcal{B}$, the integrals

$$\iota_{H^n_{\tau_j}}(x,z) \ = \ \int_{\mathcal{B}} H^n_{\tau_j}(|x-y|) \underbrace{\left(\nabla_y \frac{1}{4\pi} \frac{1}{|z-y|} \right)}_{=\gamma(z,y)} dy \tag{7.438}$$

can be calculated in an a priori step and stored elsewhere for later use.

Application of a suitable cubature formula yields

$$b_{H^n_{\tau_J}}(x) \simeq \nabla_x \sum_{i=1}^{N_J} \iota_{H^n_{\tau_J}}(x, y_i^{N_J}) \cdot \underbrace{w_i^{N_J} m(y_i^{N_J})}_{=a_i^{N_J}} \tag{7.439}$$

Therefore, if a finite set of vector values of b is known, the cubature formula (7.439) provides a linear system in the vectorial unknows $a_i^{N_J}, i = 1, \ldots, N_J$, and consequently the vector values $m(z_i^{N_J})$.

For $j = J_0, \ldots, J-1$, the constituting ingredients of a multi-scale decorrelation are obtainable from

$$(7.440)\quad b_{H^n_{\tau_j}}(x) = \nabla_x \sum_{i=1}^{N_J} \int_{\mathcal{B}} H^n_{\tau_j}(|x-y|) \left(\nabla_y \frac{1}{4\pi} \frac{1}{|y - y_i^{N_J}|} \right) dy \cdot \underbrace{w_i^{N_J} m(y_i^{N_J})}_{=a_i^{N_J}}.$$

and

$$(7.441)\quad (Wb)_{H^n_{\tau_j}}(x) = \nabla_x \sum_{i=1}^{N_J} \int_{\mathcal{B}} (WH)^n_{\tau_j}(|x-y|) \left(\nabla_y \frac{1}{4\pi} \frac{1}{|y - y_i^{N_J}|} \right) dy \cdot \underbrace{w_i^{N_J} m(y_i^{N_J})}_{=a_i^{N_J}}.$$

Mollifier wavelet inversion using pre-information. In order to overcome the difficulties of the non-uniqueness of the solution and to avoid a full-sized coefficient matrix of the linear equation system, we can assume (as in the case of gravimetry [**270**]) that from terrestrial in combination with available internal preliminary information, the field b can be associated with a known lowpass filtered field $b_{H^n_{\tau_{J_0}}}$, $J_0 < J$. In fact, we are led to discuss two choices to handle pre-information by a mollifier Haar field $b_{H^n_{\tau_{J_0}}}$:

Tensor-based variant. We assume that

$$(7.442)\quad b_{H^n_{\tau_{J_0}}}(x) = \int_{\mathcal{B}} \mathfrak{g}_{H^n_{\tau_{J_0}}}(x,y)\, m(y)\, dy.$$

is known. Then, for sufficiently large J we obtain the following expression

$$(7.443)\quad b(x) - b_{H^n_{\tau_{J_0}}}(x) \simeq b_{H^n_{\tau_J}}(x) - b_{H^n_{\tau_{J_0}}}(x) = \int_{\mathcal{B}} \left(\mathfrak{g}_{H^n_{\tau_J}}(x,y) - \mathfrak{g}_{H^n_{\tau_{J_0}}}(x,y) \right) m(y)\, dy.$$

The difference kernel occurring in the integral in (7.443) has a local support. More explicitly, for all $x \in \mathbb{R}^3$ with $|x-y| \geq \tau_{J_0}$, we have $\mathfrak{g}_{H^n_{\tau_J}}(x,y) - \mathfrak{g}_{H^n_{\tau_{J_0}}}(x,y) = 0$. In connection with a suitable cubature formula over $\overline{\mathcal{B}}$ we obtain

$$(7.444)\quad b(x) - b_{H^n_{\tau_{J_0}}}(x) \simeq b_{H^n_{\tau_J}}(x) - b_{H^n_{\tau_{J_0}}}(x) \simeq \sum_{i=1}^{N_J} \left(\mathfrak{g}_{H^n_{\tau_J}}(x, y_i^{N_J}) - \mathfrak{g}_{H^n_{\tau_{J_0}}}(x, y_i^{N_J}) \right) \underbrace{w_i^{N_J}\, m(y_i^{N_J})}_{=a_i^{N_J}},$$

where $w_i^{N_J} \in \mathbb{R}$ and $y_i^{N_J} \in \mathbb{R}^3, i = 1, \ldots, N_J$, are the known weights and nodes, respectively. The vectorial coefficients $a_i^{N_J} \in \mathbb{R}^3, a_i^{N_J} = w_i^{N_J}\, m(y_i^{N_J}), i = 1, \ldots, N_J$, are calculated by using the known values $b(x_k^{L_J}) - b_{H^n_{\tau_{J_0}}}(x_k^{L_J}), x_k^{L_J} \in \mathbb{R}^3, k = 1, \ldots, L_J$, from the linear system of equations

$$(7.445)\quad b(x_k^{L_J}) - b_{H^n_{\tau_{J_0}}}(x_k^{L_J}) = \sum_{i=1}^{N_J} \left(\mathfrak{g}_{H^n_{\tau_J}}(x_k^{L_J}, y_i^{N_J}) - \mathfrak{g}_{H^n_{\tau_{J_0}}}(x_k^{L_J}, y_i^{N_J}) \right) a_i^{N_J}, \quad k = 1, \ldots, L_J.$$

The solvability of the linear system (7.445) depends essentially on the distribution and density of the data points $x_k^{L_J}$, $k = 1, \dots, L_J$, for the dipole potential $B(x_k^{L_J})$, $k = 1, \dots, L_J$, in relation to the prescribed knots $y_k^{N_J}$, $k = 1, \dots, N_J$, occurring in the cubature formula:

- If the dipole potential values are available for data points $x_k^{L_J}$, $k = 1, \dots, L_J$, in "appropriately distributed" width over $\overline{\mathcal{B}}$, the economy of the solvability is the higher the greater J_0 can be chosen with $J_0 < J$.
- For exclusively available terrestrial values (i.e., $x_k^{L_J}$, $k = 1, \dots, L_J$, are points of the intersection of the Earth's surface with the boundary $\partial\mathcal{B}$), however, the situation $J_0 < J$, J_0 large, means a strong restriction. In fact, if one is interested in exploration at greater depths, the support has to be adapted, i.e., J_0 has to be chosen appropriately small, which of course affects the numerical economy.

The behavior of the vectorial magnetic induction in relation to vectorial magnetization distribution is similar to that of the scalar gravitational potential in relation to the scalar density distribution. As a consequence, our above approach using pre-information associated with a scale $J_0 < J$ is certainly extremely helpful numerically, however, numerical effectiveness and modeling at greater depths are mutually exclusive for terrestrial data sets. For computational purposes we have to find an appropriate compromise dependent on the available data system.

Scalar-based variant. We consider the mollified field

$$b_{H^n_{\tau_{J_0}}}(x) = \nabla_x \int_{\mathcal{B}} \underbrace{\nabla_y \left(F_3^{(1)}\right)(|x-y|)}_{=\gamma(x,y)} \cdot m_{H^n_{\tau_{J_0}}}(y)\, dy, \tag{7.446}$$

with

$$m_{H^n_{\tau_{J_0}}}(y) = \int_{\mathcal{B}} H^n_{\tau_{J_0}}(|y-z|)\, m(z)\, dy = \int_{\substack{|y-z|\le\tau_{J_0} \\ z\in\mathcal{B}}} H^n_{\tau_{J_0}}(|y-z|)\, m(z)\, dz. \tag{7.447}$$

This leads to a convolution of the vectorial magnetization m with a filtered dipole kernel in the form

$$b_{H^n_{\tau_{J_0}}}(x) = \nabla_x \int_{\mathcal{B}} \int_{\mathcal{B}} H^n_{\tau_{J_0}}(|y-z|)\, \gamma(x,y)\, dy \;\cdot\; m(z)\, dz. \tag{7.448}$$

For sufficiently large J and $J_0 < J$ we obtain the following expression

$$\begin{aligned} b(x) - b_{H^n_{\tau_{J_0}}}(x) &\simeq b_{H^n_{\tau_J}}(x) - b_{H^n_{\tau_{J_0}}}(x) \\ &= \nabla_x \int_{\mathcal{B}} \int_{\mathcal{B}} H^n_{\tau_J}(|y-z|)\gamma(x,y)\, dy \cdot m(z)\, dz \\ &\quad - \nabla_x \int_{\mathcal{B}} \int_{\mathcal{B}} H^n_{\tau_{J_0}}(|y-z|)\gamma(x,y)\, dy \cdot m(z)\, dz. \end{aligned} \tag{7.449}$$

The vector kernels $\iota_{H^n_{\tau_j}}(x,z) = \int_{\mathcal{B}} H^n_{\tau_j}(|y-z|)\, \gamma(x,y)\, dy$, $j \in \{J_0, J\}$, can be evaluated in an a priori step. Thus, the difference kernel occurring in the integral on the right side of (7.449) represents an alternative that should be kept in mind for numerical purposes.

Mollifier Born optimization. It should be mentioned that the magnetization distribution inside $\mathcal{B}$ can be obtained for our scalar-based approach in already known way via linear equations generated by appropriate cubature formulas (this can be realized in a similar way as in the gravimetric case; for more details see [**100**]). In our approach here we only discuss the similarities to the mollifier Born context, i.e., we are concerned with a "pre-information methodology" involving susceptibility instead of magnetization distribution. To this end we suppose that the field b is modelable in the form

$$(Wb)_{H_{\tau^n_{J_0}}}^{H_{\tau^n_J}}(x) = \int_{\mathcal{B}} \nabla_x \left((W\gamma)_{H_{\tau^n_{J_0}}}^{H_{\tau^n_J}}(x,y) \right) \cdot b(y)\, \chi(y)\, dy, \quad x \in \mathbb{R}^3, \tag{7.450}$$

where

$$(W\gamma)_{H_{\tau^n_{J_0}}}^{H_{\tau^n_J}}(x,y) = \gamma_{H_{\tau^n_J}}(x,y) - \gamma_{H_{\tau^n_{J_0}}}(x,y). \tag{7.451}$$

We consider the "smoothing functional"

$$\sigma^{\mathrm{cont}}_{\beta,\delta}(\chi) = \sum_{i=1}^{N} \left(\frac{|(Wb)_{H_{\tau^n_{J_0}}}^{H_{\tau^n_J}}(x_i)| - |\tilde{b}(x_i)|}{\beta_i} \right)^2 + \delta \int_{\mathcal{B}} (\chi(y) - \tilde{\chi}(y))^2\, dy, \tag{7.452}$$

where the observed intensities of the field b, i.e., $|\tilde{b}(x_i)|, x_i \in \mathbb{R}^3, i = 1, \ldots, N$, are given and $\tilde{\chi}$ is an observed pre-information about the susceptibility χ in $\mathcal{B}$. We discretize $\sigma^{\mathrm{cont}}_{\beta,\delta}(\chi)$ by applying an appropriate cubature formula leading to the functional $\sigma^{\mathrm{discr}}_{\beta,\delta}(\chi_1, \ldots, \chi_N)$ given by

(7.453)

$$\begin{aligned}
&\sigma^{\mathrm{discr}}_{\beta,\delta}(\chi_1, \ldots, \chi_N) \\
&\quad = \sum_{i=1}^{N} \frac{1}{\beta_i^2} \left(\left| \sum_{k=1}^{L} w_k \left(\nabla_x (W\gamma)_{H_{\tau^n_{J_0}}}^{H_{\tau^n_J}}(x, y_k)|_{x=x_i} \cdot b(y_k) \right) \chi_k \right| - \left| \tilde{b}(x_i) \right| \right)^2 \\
&\qquad + \delta \sum_{k=1}^{L} w_k \left| \chi_k - \tilde{\chi}(y_k) \right|^2 .
\end{aligned}$$

Here, $w_k, y_k,\ k = 1, \ldots, L$, are the known positive weights and the prescribed nodes of the cubature formula, respectively, and $\chi_k = \chi(y_k), k = 1, \ldots, L$, are the unknown susceptibility values for the points $y_k,\ k = 1, \ldots, L$, in $\mathcal{G}$. In other words, $(\tilde{\chi}(y_1), \ldots, \tilde{\chi}(y_L))^T$ stands for the known pre-information of the unknown susceptibility vector $(\chi(y_1), \ldots, \chi(y_L))^T$, and $|\tilde{b}(x_i)|,\ i = 1, \ldots, N$, characterizes the known (observed) intensities of the secondary field vectors $|b(x_i)|$, $i = 1, \ldots, N$, in the points $x_i \in \mathbb{R}^3$. The values $\beta_i,\ i = 1, \ldots, N$, should be adjusted to the standard deviations of the observed data values $|\tilde{b}(x_i)|$, $i = 1, \ldots, N$. The positive parameter δ serves for the smoothness control, i.e., the larger the choice of δ is, the stronger the influence of the a priori information $(\tilde{\chi}(y_1), \ldots, \tilde{\chi}(y_L))^T$ on the unknown values $(\chi(y_1), \ldots, \chi(y_L))^T$ to be calculated. In fact, it is not hard to verify that the expression $(\sigma^{\mathrm{discr}}_{\beta,\delta}(\chi_1, \ldots, \chi_N))^2$ can be identified as the square of the $(N + L)$-dimensional vector

$$
\begin{pmatrix}
\frac{1}{\beta_1}\left|\sum_{k=1}^{L} w_k \left(\nabla_x (W\gamma)_{H_{\tau^n_{J_0}}}^{H_{\tau^n_J}}(x, y_k)|_{x=x_1} \cdot b(y_k)\right)\chi_k\right| - \left|\tilde{b}(x_1)\right| \\
\vdots \\
\frac{1}{\beta_N}\left|\sum_{k=1}^{L} w_k \left(\nabla_x (W\gamma)_{H_{\tau^n_{J_0}}}^{H_{\tau^n_J}}(x, y_k)|_{x=x_N} \cdot b(y_k)\right)\chi_k\right| - \left|\tilde{b}(x_N)\right| \\
\sqrt{\delta}\,\sqrt{w_1}\,|\chi_1 - \tilde{\chi}(y_1)| \\
\vdots \\
\sqrt{\delta}\,\sqrt{w_L}\,|\chi_L - \tilde{\chi}(y_L)|
\end{pmatrix}. \tag{7.454}
$$

The unknown coefficients are determined in such a way, that $(\sigma^{\mathrm{discr}}_{\beta,\delta}(\chi_1, \ldots, \chi_L)^2$ is minimized in the set of all M-tupel $(\chi_1, \ldots, \chi_L)^T$. It should be stressed that the result of the minimization procedure is dependent on the "smoothing parameter" δ. The mollifier wavelet

$$
x \mapsto (W\gamma)_{H_{\tau^n_{J_0}}}^{H_{\tau J}}(x, y) = \gamma_{H_{\tau^n_J}}(x, y) - \gamma_{H_{\tau^n_{J_0}}}(x, y), \quad x \in \mathcal{B}, \tag{7.455}
$$

is known to possess a ball as support, the width of which is determined by the scale index J_0 and thus, as already explained, has a decisive influence on the desired exploration depth. In consistency with our model specification, the scale index $J > J_0$ should be chosen as large as possible. The nonlinear minimization problem $(\sigma^{\mathrm{discr}}_{\beta,\delta}(\chi_1, \ldots, \chi_L))^2 \to \min$ can be solved by standard solvers in numerics.

Finally, it should be remarked that the IM regularization techniques as presented here found their application to geothermal exploration during the last years in the German state Saarland (research project SYSEXPL (FKZ 03EE4002A&B), PI Prof. Dr. W. Freeden, University of Kaiserslautern, Germany), corporate manager Prof. Dr. M. Bauer, CBM - Gesellschaft für Consulting, Business und Management mbH, Bexbach, Germany) funded by "Federal Ministry for Economic Affairs and Energy, Berlin" and the "Project Management Jülich", responsible project managers Dr. V. Monser, Dr. K. Schreiber.

Part 5

Sampling tools

CHAPTER 8

Lattice point and special function theoretic auxiliaries

We recapitulate some results of the geometric and analytic theory of numbers (see, e.g., E. Landau [**520**], E. Hlawka [**436–438**], G.H. Hardy, E.M. Wright [**416**], F. Fricker [**335**], E. Hlawka, et al. [**439**], E. Krätzel [**510, 511**] for more details). Our intent is to provide some key material as a basic reference for our later work.

8.1. Geometric number theory

Lattices. Let $g_1, \ldots, g_q$ be linearly independent vectors in the q-dimensional Euclidean space $\mathbb{R}^q$. The set Λ (more precisely, $\Lambda_{(q)}$) of all points

$$g = n_1 g_1 + \ldots + n_q g_q, \quad n_i \in \mathbb{Z},\ i = 1, \ldots, q, \tag{8.1}$$

is called a *lattice in* $\mathbb{R}^q$ *with basis* $g_1, \ldots, g_q$ (see Figure 1 for a 2D example). Clearly, the vectors $\varepsilon^1, \ldots, \varepsilon^q$ form a basis of the unit lattice $\mathbb{Z}^q$.

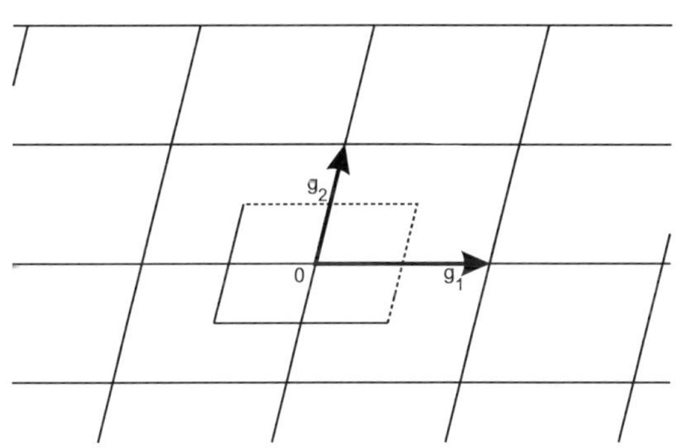

FIGURE 1. Two-dimensional lattice and its fundamental cell $\mathcal{F}_\Lambda$.

The half-open parallelotope $\mathcal{F}_\Lambda$ consisting of the points $x \in \mathbb{R}^q$ with

$$x = t_1 g_1 + \ldots + t_q g_q, \quad -\frac{1}{2} \leq t_i < \frac{1}{2},\ i = 1, \ldots, q, \tag{8.2}$$

is called the *fundamental cell of the lattice* Λ (cf. Figure 1). Obviously, there are infinitely many cells of Λ reflecting the Λ-periodicity. $\mathcal{F}_\Lambda$, as specified by (8.2), is both simple and appropriate for our purposes. From Linear Algebra (see, e.g., [**175**]) it is well known that the volume $\|\mathcal{F}_\Lambda\|$ of the fundamental cell $\mathcal{F}_\Lambda$ is equal to the quantity

$$\|\mathcal{F}_\Lambda\| = \int_{\mathcal{F}_\Lambda} dx = \sqrt{\det\left((g_i \cdot g_j)_{i,j=1,\ldots,q}\right)}. \tag{8.3}$$

For each lattice point $g \in \Lambda$ we have $\|\mathcal{F}_\Lambda\| = \|\mathcal{F}_\Lambda + \{g\}\| = \|\mathcal{F}_\Lambda - \{g\}\|$. Because of $(\mathcal{F}_\Lambda + \{g\}) \cap (\mathcal{F}_\Lambda + \{g'\}) = \emptyset$ for $g \neq g'$, $g, g' \in \Lambda$, we have

$$\mathbb{R}^q = \bigcup_{g \in \Lambda} (\mathcal{F}_\Lambda + \{g\}) = \bigcup_{g \in \Lambda} (\mathcal{F}_\Lambda - \{g\}) . \tag{8.4}$$

Inverse lattices. Since the vectors $g_1, \ldots, g_q$ are assumed to be linearly independent, there exists a system of vectors $h_j, j = 1, \ldots, q$, in $\mathbb{R}^q$ such that

$$h_j \cdot g_i = \delta_{i,j} = \begin{cases} 0 & , \quad i \neq j, \\ 1 & , \quad i = j, \end{cases} \tag{8.5}$$

($\delta_{i,j}$ is the Kronecker symbol). In more detail (see, e.g., J.W.S. Cassels [**141**], C.G. Lekkerkerker [**532**]), for $i, j = 1, \ldots, q$, we let $g_i \cdot g_j = \Gamma_{ij}$. The scalars Γ^{ij}, $i, j = 1, \ldots, q$, are defined by $\sum_{j=1}^q \Gamma^{ij} \Gamma_{jk} = \delta_{i,k}$. The vectors h_j, $j = 1, \ldots, q$, given by $h_j = \sum_{k=1}^q \Gamma^{jk} g_k, j = 1, \ldots, q$, satisfy the equations

$$h_j \cdot g_i = \sum_{k=1}^q \Gamma^{jk} g_k \cdot g_i = \sum_{k=1}^q \Gamma^{jk} \Gamma_{ki} = \delta_{j,i}, \quad i, j = 1, \ldots, q. \tag{8.6}$$

Moreover, we have

$$h_i \cdot h_j = \sum_{k=1}^q \Gamma^{ik} g_k \cdot \sum_{l=1}^q \Gamma^{jl} g_l = \sum_{l=1}^q \Gamma^{jl} \sum_{k=1}^q \Gamma^{ik} \Gamma_{kl} = \Gamma^{ji}.$$

The lattice with basis $h_1, \ldots, h_q$ given by (8.1) is called the *inverse (or dual) lattice* Λ^{-1} to Λ. The inverse lattice Λ^{-1} consists of all vectors $h \in \mathbb{R}^q$ such that the inner product $h \cdot g$ is an integer for all $g \in \Lambda$. Obviously, $\Lambda = (\Lambda^{-1})^{-1}$. Moreover, for the fundamental cell $\mathcal{F}_{\Lambda^{-1}}$ of the inverse lattice Λ^{-1} we have $\|\mathcal{F}_{\Lambda^{-1}}\| = \|\mathcal{F}_\Lambda\|^{-1}$.

Let $\Lambda = \tau \mathbb{Z}^q$ be the lattice generated by the "dilated basis" $\tau\varepsilon^1, \ldots, \tau\varepsilon^q$, where τ is a positive number and $\varepsilon^1, \ldots, \varepsilon^q$, as usual, forms the canonical orthonormal basis in $\mathbb{R}^q$. Then, the volume of the fundamental cell of $\tau\mathbb{Z}^q$ is $\|\mathcal{F}_\Lambda\| = \tau^q$. The generating vectors of the inverse lattice Λ^{-1} are $\tau^{-1}\varepsilon^1, \ldots, \tau^{-1}\varepsilon^q$. The volume of the fundamental cell of the inverse lattice is given by $\|\mathcal{F}_{\Lambda^{-1}}\| = \tau^{-q} = \|\mathcal{F}_\Lambda\|^{-1}$. In particular, for $\tau = 1$, i.e., the unit lattice $\Lambda = \mathbb{Z}^q$, we have $\Lambda^{-1} = \mathbb{Z}^q = \Lambda$ such that $\|\mathcal{F}_{\Lambda^{-1}}\| = 1 = \|\mathcal{F}_\Lambda\|$. This fact has been used permanently in our one-dimensional theory above, and it is always obvious throughout this work whenever $\Lambda = \mathbb{Z}^q$ (for more details on lattices see, e.g., [**141**], [**532**]).

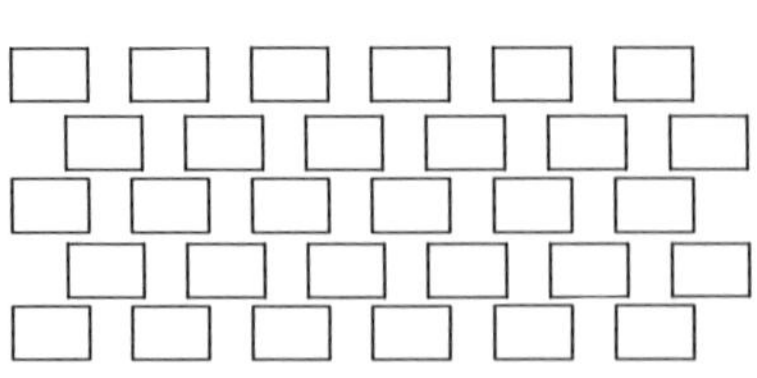

FIGURE 2. A two-dimensional figure lattice associated to a rectangle.

Basic results of the geometry of numbers. If $\Lambda \subset \mathbb{R}^q$ is a lattice and $x \in \mathbb{R}^q$, then $\Lambda + \{x\} = \{g + x \in \mathbb{R}^q : g \in \Lambda\} \subset \mathbb{R}^q$ is called the *(point) lattice based at* x. If $\Lambda \subset \mathbb{R}^q$ is a lattice and $\mathcal{M} \subset \mathbb{R}^q$ is a set, then the sum $\mathcal{M} + \Lambda$ is called the *figure lattice* associated to $\mathcal{M}$ by Λ in $\mathbb{R}^q$ (see Figure 2 for rectangles in $\mathbb{R}^2$). If $\mathcal{M}$ is a ball around the origin 0, then the figure lattice $\mathcal{M} + \Lambda$ is called a *ball lattice.*

For a set $\mathcal{M} \subset \mathbb{R}^q$ we define *the characteristic function* $\mathcal{X}_{\mathcal{M}}$ by

$$\mathcal{X}_{\mathcal{M}}(x) = \begin{cases} 1 & , \quad x \in \mathcal{M} \\ 0 & , \quad x \notin \mathcal{M}. \end{cases} \tag{8.7}$$

A point $x \in \mathbb{R}^q$ belongs to the figure lattice $\mathcal{M}+\Lambda$ if and only if $\sum_{g\in\Lambda} \mathcal{X}_{\mathcal{M}}(x+g) \geq 1$. The point $x \in \mathbb{R}^q$ does not belong to the figure lattice $\mathcal{M} + \Lambda$ if and only if $\sum_{g\in\Lambda} \mathcal{X}_{\mathcal{M}}(x + g) = 0$.

FIGURE 3. An example of a two-dimensional covering lattice.

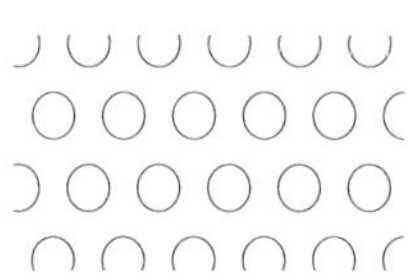

FIGURE 4. An example of a two dimensional filling lattice.

Covering and filling lattices. If for all points from $\mathbb{R}^q$ the formula

$$\sum_{g\in\Lambda} \mathcal{X}_{\mathcal{M}}(x + g) \geq 1 \tag{8.8}$$

is valid, then we call Λ a *covering lattice* for $\mathcal{M}$ (see Figure 3) because in this case $\mathbb{R}^q = \mathcal{M} + \Lambda$. If, on the other hand, for all points $x \in \mathbb{R}^q$, the formula

$$\sum_{g\in\Lambda} \mathcal{X}_{\mathcal{M}}(x + g) \leq 1 \tag{8.9}$$

holds true, the lattice Λ is called a *filling lattice* for $\mathcal{M}$ (see Figure 4). In this case an arbitrary point $x \in \mathbb{R}^q$ can belong to at most one of the sets $\mathcal{M} + \{g\}$ with $g \in \Lambda$.

Blichfeldt's and Minkowski's lattice point theorems. If we restrict ourselves to the characteristic function $F = \mathcal{X}_{\mathcal{M}}$, where $\mathcal{M}$ is a bounded measurable set, the volume $\|\mathcal{M}\|$ of $\mathcal{M}$ is given by $\int_{\mathbb{R}^q} \mathcal{X}_{\mathcal{M}}(x)\, dx = \|\mathcal{M}\|$. The relation $\sum_{g\in\Lambda} \mathcal{X}_{\mathcal{M}}(x+g) \geq \frac{\|\mathcal{M}\|}{\|\mathcal{F}_\Lambda\|}$ says that at least $\lceil \|\mathcal{M}\| \;/\; \|\mathcal{F}_\Lambda\| \rceil$ points of the form $x + g$, $g \in \Lambda$, lie in the set $\mathcal{M}$. This leads to *Blichfeldt's theorem*:

For each lattice Λ and each nonempty, bounded, measurable set $\mathcal{M}$ there exists a point $x \in \mathbb{R}^q$ such that the number N of lattice points of the lattice $\Lambda + \{x\}$ based at $x \in \mathbb{R}^q$ lying inside the set $\mathcal{M}$ satisfies the relation $N \geq \frac{\|\mathcal{M}\|}{\|\mathcal{F}_\Lambda\|}$.

In Blichfeldt's theorem little is assumed about the set $\mathcal{M}$. In particular, the point $x \in \mathbb{R}^q$ serving as the origin for the lattice $\Lambda + \{x\}$ remains completely unspecified. If one wishes to avoid this lack of information, additional assumptions on the set $\mathcal{M}$ are necessary.

A set $\mathcal{M}$ is said to be *symmetric* with respect to a point $z \in \mathbb{R}^q$ if $x = z + a$ belonging to the set implies $x' = z - a$ also does. The point z is called the center of symmetry of this set. A set $\mathcal{M}$ is said to be *star-shaped* with respect to a point z if $x = z + a$ belonging to the set implies that $x_t = z + ta$ also does for all $t \in [0, 1]$. A set $\mathcal{M}$ is called *convex* if it is star-shaped with respect to each of its points. A criterion for convex sets is as follows: A set $\mathcal{M}$ is convex if and only if, for each two points x and z belonging to it, so does the line segment consisting of all points $y_t = x + t(z - x)$, $t \in [0, 1]$, too. With these preparations we are able to formulate *Minkowski's lattice point theorem* that is a classical result in the geometry of numbers.

If x is the center of symmetry of the symmetric, bounded, and convex region $\mathcal{G} \subset \mathbb{R}^q$, and $\Lambda \subset \mathbb{R}^q$ is a lattice with $\|\mathcal{G}\| > 2^q \ \|\mathcal{F}_\Lambda\|$, then $\mathcal{G}$ contains at least one lattice point $z = x + g$, distinct from $x \in \mathbb{R}^q, g \in \Lambda$, from the lattice $\Lambda + \{x\}$ based at x.

For the cube $\mathcal{G} = \left\{x + \sum_{i=1}^q t_i \varepsilon^i \in \mathbb{R}^q : t_i \in (-1, 1), \ \ i = 1, \ldots, q\right\}$, we are able to deduce that the condition $\|\mathcal{G}\| > 2^q \ \|\mathcal{F}_\Lambda\|$ cannot be weakened for Λ being the lattice $\mathbb{Z}^q$. However, we are able to formulate a variant of Minkowski's theorem showing the less restrictive inequality $\|\mathcal{G}\| \geq 2^q \ \|\mathcal{F}_\Lambda\|$.

If x denotes the center of symmetry of the symmetric, compact, convex set $\mathcal{K}$, and if $\Lambda \subset \mathbb{R}^q$ is a lattice with $\|\mathcal{K}\| \geq 2^q \ \|\mathcal{F}_\Lambda\|$, then $\mathcal{K}$ contains at least one lattice point $y = x + g$ distinct from $x \in \mathbb{R}^q, g \in \Lambda$, from the lattice $\Lambda + \{x\}$ based at x.

For more details the reader is referred, e.g., to L.J. Mordell [**593, 594**], E. Hlawka [**437**], E. Hlawka, J. Schoißengeier, R. Taschner [**439**], and the references therein. The (standard) proof of Minkowski's theorem is based on Mordell's work (cf. [**593**]). Later on, we give a different proof of Minkowski's Theorem completely based on an analytic proposition, namely a multi-dimensional variant of the Poisson summation formula. This concept is motivated by the two-dimensional approach presented in C. Müller [**604**].

Lattice points inside circles. The branch of geometric lattice point theory concerned with lattice point summation inside balls has a long history, which reaches back to L. Euler [**221, 222**] and C.F. Gauss [**344, 354**]. Enlightening accounts of the developments within this theory are, e.g., due to G.H. Hardy [**410**], E. Landau [**520**], A. Walfisz [**849–851**], to mention just a few. We recapitulate some results on the number of lattice points inside circles $\mathbb{S}^1_N$ of radii $N > \frac{\sqrt{2}}{2}$ around the origin 0; more specifically, we deal with closed disks $\overline{\mathbb{B}^2_N}$ of radii $N > \frac{\sqrt{2}}{2}$ (for more background material the reader is referred, e.g., to the monographs F. Fricker [**335**],

E. Krätzel [**510**], a survey paper is A. Ivic et al. [**460**] (moreover, we refer to W. Freeden [**268**], W. Freeden, M.Z. Nashed [**296, 298**]).

The problem of determining the total number of lattice points of $\mathbb{Z}^2$ inside and on a circle with radius N, i.e., the determination of the quantity

$$\sharp_{\mathbb{Z}^2}\left(\overline{\mathbb{B}_N^2}\right) = \sharp\left\{(n_1,n_2)^T \in \mathbb{Z}^2 : n_1^2+n_2^2 \le N^2\right\} = \sum_{\substack{n_1^2+n_2^2\le N^2 \\ (n_1,n_2)^T\in\mathbb{Z}^2}} 1 \tag{8.10}$$

is very old. In fact, for the sum (8.10) C. F. Gauß [**344**] found a simple, but efficient method for its estimation: Associate to every square the Northwest edge as lattice point. The union of all squares with lattice points inside $\overline{\mathbb{B}_N^2}$ defines a polyhedral set $\mathbb{P}_N^2$ with area $\|\mathbb{P}_N^2\| = \sharp_{\mathbb{Z}^2}\left(\overline{\mathbb{B}_N^2}\right)$ (cf. Figure 5). Since the diagonal of each

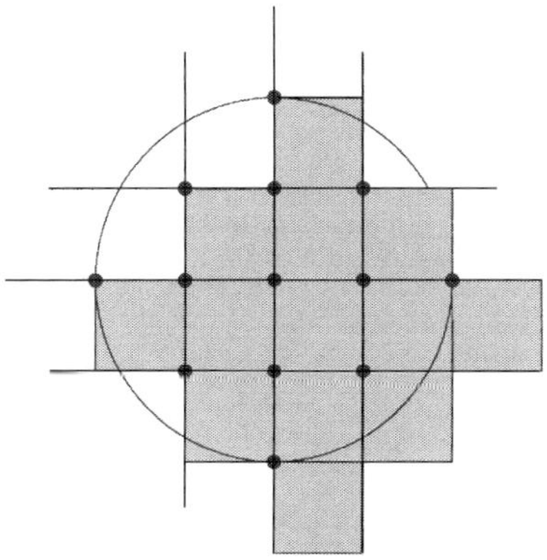

FIGURE 5. The polyhedral set $\mathbb{P}_N^2$.

square is $\sqrt{2}$, the geometry of Figure 5 shows that

$$\pi\left(N-\frac{\sqrt{2}}{2}\right)^2 \le \sharp_{\mathbb{Z}^2}\left(\overline{\mathbb{B}_N^2}\right) \le \pi\left(N+\frac{\sqrt{2}}{2}\right)^2. \tag{8.11}$$

Therefore, $\sharp_{\mathbb{Z}^2}\left(\overline{\mathbb{B}_N^2}\right) - \pi N^2$ after division by N is bounded for $N\to\infty$, which is usually written by use of the O-symbol as

$$\sharp_{\mathbb{Z}^2}\left(\overline{\mathbb{B}_N^2}\right) = \pi N^2 + O(N). \tag{8.12}$$

So, the number of lattice points in $\overline{\mathbb{B}_N^2}$ is equal to the area of that circle plus a remainder of the order of the boundary. In particular, we are led to a representation of the number π:

$$\lim_{N\to\infty} \frac{\sharp_{\mathbb{Z}^2}\left(\overline{\mathbb{B}_N^2}\right)}{N^2} = \pi. \tag{8.13}$$

Lattice points inside balls. Just as in the planar case, the problem of counting the number of points $(n_1, \ldots, n_q)^T$ with all the n_i integers inside spheres $\mathbb{S}_N^{q-1}$ of radii N centered at the origin 0 may be discussed. More concretely, by $\sharp_{\mathbb{Z}^q}\left(\overline{\mathbb{B}_N^q}\right) = \sum_{\substack{|g|\leq N\\ g\in\mathbb{Z}^q}} 1$ we denote the *total number of lattice points* g of the lattice $\mathbb{Z}^q$ such that $|g| \leq N$; i.e., g is a member of $\overline{\mathbb{B}_N^q}$. Following the idea of C.F. Gauss [**344**] we consider the associated polyhedron

$$\mathbb{P}_N^q = \bigcup_{\substack{g\in\overline{\mathbb{B}_N^q}\\ g\in\mathbb{Z}^q}} (\mathcal{F}_{\mathbb{Z}^q} + \{g\}), \tag{8.14}$$

where, as usual, $\mathcal{F}_{\mathbb{Z}^q} + \{g\}$ is the translate of $\mathcal{F}_{\mathbb{Z}^q}$ by the lattice point $g \in \mathbb{Z}^q$. Clearly, the volume $\|\mathbb{P}_N^q\|$ of $\mathbb{P}_N^q$ is equal to $\sharp_{\mathbb{Z}^q}\left(\overline{\mathbb{B}_N^q}\right)$. There exists a constant d such that the estimate $|x| \leq d$ holds true for all $x \in \mathcal{F}_{\mathbb{Z}^q}$ (for example, $d = \frac{\sqrt{q}}{2}$), so that for $\mathbb{P}_N^q$ the following inclusion is valid with $N > d$: $\overline{\mathbb{B}_{N-d}^q} \subset \mathbb{P}_N^q \subset \overline{\mathbb{B}_{N+d}^q}$. This inclusion leads to the following inequality

$$\left\|\overline{\mathbb{B}_{N-d}^q}\right\| \leq \underbrace{\sharp_{\mathbb{Z}^q}\left(\overline{\mathbb{B}_N^q}\right)}_{=\|\mathbb{P}_N^q\|} \leq \left\|\overline{\mathbb{B}_{N+d}^q}\right\|. \tag{8.15}$$

In accordance with the Gaussian approach we therefore obtain the following fundamental result in the theory of lattice points inside spheres:

$$\sharp_{\mathbb{Z}^q}\left(\overline{\mathbb{B}_N^q}\right) = \frac{\pi^{\frac{q}{2}}}{\Gamma(\frac{q}{2}+1)} N^q + O\left(N^{q-1}\right), \quad N \to \infty.$$

Lattice point problems. The problem of determining (see, e.g., E. Landau [**520**]) the value

$$\alpha_q = \inf\left\{\gamma : \sharp_{\mathbb{Z}^q}\left(\overline{\mathbb{B}_{\sqrt{N}}^q}\right) = \left\|\overline{\mathbb{B}_{\sqrt{N}}^q}\right\| + O(N^\gamma)\right\} \tag{8.16}$$

is known as the *lattice point problem for the sphere* (for $q = 2$, *lattice point problem for the circle*).

2 D lattice point problem. From the Gaussian concept we obtain $\alpha_2 \leq \frac{1}{2}$. Considerable improvements became rather long and made use of some of the more heavy machinery in analysis (see Table 1 for some results). *Hardy's conjecture* [**411**] tells us that

$$\sharp_{\mathbb{Z}^2}\left(\overline{\mathbb{B}_{\sqrt{N}}^2}\right) - \pi N = O\left(N^{\frac{1}{4}+\varepsilon}\right) \tag{8.17}$$

for every $\varepsilon > 0$. This conjecture seems to be still a challenge for future work (see, e.g., F. Fricker [**334**], and the references therein, however, when replacing the lattice points by "small lattice discs" the 2D "lattice circle problem" turns out to be a straightforward exercise (see W. Freeden [**268**])).

3 D lattice point problem. For the case $q = 3$, we already know that

$$\sharp_{\mathbb{Z}^3}\left(\overline{\mathbb{B}_{\sqrt{N}}^3}\right) - \left\|\overline{\mathbb{B}_{\sqrt{N}}^3}\right\| = O(N). \tag{8.18}$$

In other words, for the 3D *lattice point problem*

$$\alpha_3 = \inf\left\{\gamma : \sharp_{\mathbb{Z}^3}\left(\overline{\mathbb{B}_{\sqrt{N}}^3}\right) - \left\|\overline{\mathbb{B}_{\sqrt{N}}^3}\right\| = O(N^\gamma)\right\} \tag{8.19}$$

TABLE 1. Improvements in the Upper Limit for the Circle Problem: Incomplete incremental improvements are listed in characterizing the upper limit for the circle problem.

$\frac{1}{3}$	G. Voronoi (1903)
$\frac{37}{112}$	J.E. Littlewood, A. Walfisz (1924)
$\frac{27}{82}$	J.G. van der Corput (1928)
$\frac{12}{37}$	J.-R. Chen (1963)
$\frac{139}{429}$	G. Kolesnik (1985)
$\frac{131}{416}$	M.N. Huxley (2003)

it is known that $\alpha_3 \leq 1$. In order to improve this result great efforts must be made (similarly to the two-dimensional case). E. Landau [**520**] proved $\alpha_3 \leq \frac{3}{4}$. I.M. Vinogradov [**839**] with $\alpha_3 \leq \frac{11}{16}$ and O.M. Fomenko [**237**] with $\alpha_3 \leq \frac{701}{1020}$ gave essential improvements; more recent progress is due to F. Chamizo, H. Iwaniec [**142**] with $\alpha_3 \leq \frac{29}{44}$ and D. R. Heath-Brown [**419**] with $\alpha_3 \leq \frac{21}{32}$. Again, the proofs of the improvements are rather long and technical.

4 D lattice point problem. For *the dimension* $q = 4$, the classical result due to Lagrange (that each positive integer can be expressed as the sum of at most four integral squares) helps us to verify an estimate (for more details see, e.g., F. Fricker [**334, 335**], E. Krätzel [**511**]) of the form

$$\sharp_{\mathbb{Z}^4}\left(\overline{\mathbb{B}^4_{\sqrt{N}}}\right) - \left\|\overline{\mathbb{B}^4_{\sqrt{N}}}\right\| = O\Big(N\ln(N)\Big), \tag{8.20}$$

and it is known that the ln-term in (8.20) cannot be omitted. In fact, A. Walfisz [**851**] used some deep analysis to show an estimate of the form

$$\sharp_{\mathbb{Z}^4}\left(\overline{\mathbb{B}^4_{\sqrt{N}}}\right) - \left\|\overline{\mathbb{B}^4_{\sqrt{N}}}\right\| = O\left(N(\ln(N))^{\frac{2}{3}}\right). \tag{8.21}$$

In other words, we are confronted with the situation that $\alpha_4 = 1$, however,

$$\sharp_{\mathbb{Z}^4}\left(\overline{\mathbb{B}^4_{\sqrt{N}}}\right) - \left\|\overline{\mathbb{B}^4_{\sqrt{N}}}\right\| \neq o\Big(N\Big). \tag{8.22}$$

qD *lattice point problem* ($q \geq 5$). For *higher dimensions* $q \geq 5$, the situation of the *lattice point problem for the sphere* changes drastically (see, e.g., E. Grosswald [**383**] for more details). The asymptotic growth of the number of points has been known for a long time. In fact, we have the following relations, leaving no room for progress.

For $q \geq 5$,

$$\alpha_q = \frac{q-2}{2}, \tag{8.23}$$

and

$$\sharp_{\mathbb{Z}^q}\left(\overline{\mathbb{B}^q_{\sqrt{N}}}\right) - \left\|\overline{\mathbb{B}^q_{\sqrt{N}}}\right\| = O\left(N^{\alpha_q}\right). \tag{8.24}$$

So, the impression that the lattice point problem for higher-dimensional spheres becomes more and more problematic is wrong. The essential difficulties arising in two- and three-dimensional cases do not occur for dimensions $q \geq 5$. The dimension $q = 4$ is an intermediate case. Once more, the reason seems to be that the decomposition of a positive integer as a sum of integral squares for dimensions $q \geq 5$ is not as "pathologic" as, for instance, for $q = 2$ and $q = 3$. Later on, we shall see that the O-term in (8.24) may be represented by an alternating series in term of Bessel functions. This expansion may be considered as a pre-step to the development of cardinal series in Shannon sampling.

8.2. Periodization and Poisson summation formula

Let Λ be a lattice in $\mathbb{R}^q$. The functions $\Phi_h,\ h \in \Lambda^{-1}$, defined by

$$\Phi_h(x) = \frac{1}{\sqrt{\|\mathcal{F}_\Lambda\|}}\, e(h \cdot x) = \frac{1}{\sqrt{\|\mathcal{F}_\Lambda\|}}\, e^{2\pi i h \cdot x}, \quad x \in \mathbb{R}^q, \tag{8.25}$$

are Λ-*periodic*, i.e., $\Phi_h(x+g) = \Phi_h(x)$ for all $x \in \mathbb{R}^q$ and all $g \in \Lambda$.

Spaces of periodic functions. The space of all $F \in C^{(m)}(\mathbb{R}^q)$ that are Λ-periodic is denoted by $C^{(m)}_\Lambda(\mathbb{R}^q)$, $0 \leq m \leq \infty$. $L^p_\Lambda(\mathbb{R}^q)$, $1 \leq p < \infty$, is the space of all $F : \mathbb{R}^q \to \mathbb{C}$ that are Λ-periodic and Lebesgue-measurable on $\mathcal{F}_\Lambda$ with

$$\|F\|_{L^p_\Lambda(\mathbb{R}^q)} = \left(\int_{\mathcal{F}_\Lambda} |F(x)|^p \, dx\right)^{\frac{1}{p}} < \infty. \tag{8.26}$$

Clearly, $C^{(0)}_\Lambda(\mathbb{R}^q) \subset L^p_\Lambda(\mathbb{R}^q)$. As is well known, $L^2_\Lambda(\mathbb{R}^q)$ is the completion of $C^{(0)}_\Lambda(\mathbb{R}^q)$ with respect to the norm $\|\cdot\|_{L^2_\Lambda(\mathbb{R}^q)}$:

$$L^2_\Lambda(\mathbb{R}^q) = \overline{C^{(0)}_\Lambda(\mathbb{R}^q)}^{\|\cdot\|_{L^2_\Lambda(\mathbb{R}^q)}}. \tag{8.27}$$

The system $\{\Phi_h\}_{h\in\Lambda^{-1}}$ of multi-dimensional "periodic polynomials" is orthonormal with respect to the $L^2_\Lambda(\mathbb{R}^q)$-inner product:

$$(\Phi_h, \Phi_{h'})_{L^2_\Lambda(\mathbb{R}^q)} = \int_{\mathcal{F}_\Lambda} \Phi_h(x)\, \overline{\Phi_{h'}(x)}\, dx = \delta_{h,h'}. \tag{8.28}$$

Spectrum of Δ. An easy calculation yields $(\Delta_x + \Delta^\wedge(h))\,\Phi_h(x) = 0, \Delta^\wedge(h) = 4\pi^2 h^2,\ h \in \Lambda^{-1}$. We shall say that λ is an *eigenvalue of the lattice* Λ with respect to the operator Δ if there is a nontrivial solution U of the differential equation $(\Delta + \lambda)U = 0$ satisfying the "boundary condition of periodicity" $U(x+g) = U(x)$ for all $g \in \Lambda$. The function U is called an *eigenfunction of the lattice* Λ with respect to the eigenvalue λ and the operator Δ. It is easy to conclude (see, e.g., [**433**]) that the functions Φ_h are the only eigenfunctions. Furthermore, the values $\Delta^\wedge(h) = 4\pi^2 h^2,\ h \in \Lambda^{-1}$, are the only eigenvalues of Δ with respect to the lattice Λ (note that we usually write $\Delta^\wedge(h)$ instead of $\Delta^\wedge_\Lambda(h)$ if no confusion is likely to arise). The set of all eigenvalues $\Delta^\wedge(h)$ with respect to Δ is the *spectrum* $\mathrm{Spect}_\Delta(\Lambda)$:

$$\mathrm{Spect}_\Delta(\Lambda) = \left\{\Delta^\wedge(h) : \ \Delta^\wedge(h) = 4\pi^2 h^2,\ h \in \Lambda^{-1}\right\}. \tag{8.29}$$

Fourier expansions. The system $\{\Phi_h\}_{h\in\Lambda^{-1}}$ is a closed and complete system in $(C^{(0)}_\Lambda(\mathbb{R}^q); \|\cdot\|_{L^2_\Lambda(\mathbb{R}^q)})$ as well as in $(L^2_\Lambda(\mathbb{R}^q); \|\cdot\|_{L^2_\Lambda(\mathbb{R}^q)})$. The fundamental result in Fourier analysis is that each $F \in L^2_\Lambda(\mathbb{R}^q)$ can be represented by its (orthogonal) *Fourier series*

$$\lim_{N\to\infty} \left\| F - \sum_{\substack{|h|\le N \\ h\in\Lambda^{-1}}} F^\wedge_{\mathcal{F}_\Lambda}(h)\Phi_h \right\|_{L^2_\Lambda(\mathbb{R}^q)} = 0, \tag{8.30}$$

where the *Fourier coefficients* $F^\wedge_{\mathcal{F}_\Lambda}(h)$ of F are given by

$$F^\wedge_{\mathcal{F}_\Lambda}(h) = \int_{\mathcal{F}_\Lambda} F(x)\ \overline{\Phi_h(x)}\ dx, \quad h \in \Lambda^{-1}. \tag{8.31}$$

The *Parseval identity* tells us that, for each $F \in L^2_\Lambda(\mathbb{R}^q)$,

$$\int_{\mathcal{F}_\Lambda} |F(x)|^2\ dx = \sum_{h\in\Lambda^{-1}} \left|F^\wedge_{\mathcal{F}_\Lambda}(h)\right|^2. \tag{8.32}$$

Let Λ be an arbitrary lattice in $\mathbb{R}^q$. Suppose that an (appropriate) function F is given in Euclidean space $\mathbb{R}^q$. The question is what is its periodic analog? Our purpose is to show that there are at least two ways of obtaining a periodization of a function under consideration:

The "*first periodization*" is straightforward:

$$x \mapsto S(x) = \sum_{g\in\Lambda} F(x+g). \tag{8.33}$$

Since this (formal) sum S is extended over all points of the lattice Λ it is obviously periodic, i.e.,

$$S(x+g') = \sum_{g\in\Lambda} F(x+g+g') = \sum_{g\in\Lambda} F(x+g) = S(x) \tag{8.34}$$

for all $g' \in \Lambda$.

The "*second periodization*" is based on the Fourier expansion

$$x \mapsto \sum_{h\in\Lambda^{-1}} S^\wedge_{\mathcal{F}_\Lambda}(h)\ \Phi_h(x), \tag{8.35}$$

where

$$S^\wedge_{\mathcal{F}_\Lambda}(h) = \int_{\mathcal{F}_\Lambda} S(x)\ \overline{\Phi_h(x)}\ dx,\ x \in \mathbb{R}^q. \tag{8.36}$$

The Fourier coefficients $S^{\wedge}_{\mathcal{F}_\Lambda}(h)$, $h \in \Lambda^{-1}$, of the series (8.35) are expressible in the form

$$S^{\wedge}_{\mathcal{F}_\Lambda}(h) = \int_{\mathcal{F}_\Lambda} \sum_{g\in\Lambda} F(x+g)\, \overline{\Phi_h(x)}\, dx = \sum_{g\in\Lambda} \int_{\mathcal{F}_\Lambda + \{g\}} F(x)\, \overline{\Phi_h(x)}\, dx, \tag{8.37}$$

where, as usual, $\mathcal{F}_\Lambda + \{g\}$ denotes the translate of $\mathcal{F}_\Lambda$ containing the lattice point $g \in \Lambda$. Since the fundamental cell $\mathcal{F}_\Lambda$ of the lattice Λ defines translates $\mathcal{F}_\Lambda + \{g\}$, $g \in \Lambda$, that are mutually disjoint, their union is the whole Euclidean space $\mathbb{R}^q$. This observation implies the identity

$$\int_{\mathcal{F}_\Lambda} \sum_{g\in\Lambda} F(x+g)\, \overline{\Phi_h(x)}\, dx = \int_{\mathbb{R}^q} F(x)\, \overline{\Phi_h(x)}\, dx = \frac{1}{\sqrt{\|\mathcal{F}_\Lambda\|}} F^{\wedge}_{\mathbb{R}^q}(h). \tag{8.38}$$

The formal conclusion (see, e.g., E.M. Stein, G. Weiss [**790**]) is that the first and second periodization are identical. Hence, we are led to the *Poisson summation formula for the Euclidean space* $\mathbb{R}^q$ and arbitrary lattices in $\mathbb{R}^q$

$$\sum_{g\in\Lambda} F(x+g) = \frac{1}{\|\mathcal{F}_\Lambda\|} \sum_{h\in\Lambda^{-1}} \int_{\mathbb{R}^q} F(y)\, e^{-2\pi i h\cdot y}\, dy\, e^{2\pi i h\cdot x}. \tag{8.39}$$

The identity (8.39) may be rigorously understood in many ways. The statement below taken from E.M. Stein, G. Weiss [**790**] was a first attempt to validate the Poisson summation formula for the standard lattice $\mathbb{Z}^q$ in Euclidean space $\mathbb{R}^q$, however, under conditions implying the absolute convergence of all occurring series involved in the process of periodization.

Suppose that the functions F and $F^{\wedge}_{\mathbb{R}^q}$ satisfy the conditions

$$F(x) = O\left(\frac{1}{|x|^{q+\varepsilon}}\right), \quad |x| \to \infty,\ \varepsilon > 0, \tag{8.40}$$

$$F^{\wedge}_{\mathbb{R}^q}(x) = O\left(\frac{1}{|x|^{q+\varepsilon}}\right), \quad |x| \to \infty,\ \varepsilon > 0, \tag{8.41}$$

so that the Fourier inversion formula is valid. Then, F and $F^{\wedge}_{\mathbb{R}^q}$ are continuous in $\mathbb{R}^q$ such that

$$\sum_{g\in\mathbb{Z}^q} F(x+g) = \sum_{h\in\mathbb{Z}^q} \int_{\mathbb{R}^q} F(y)\, e^{-2\pi i h\cdot y}\, dy\, e^{2\pi i h\cdot x}. \tag{8.42}$$

In particular, its classical version is as follows:

$$\sum_{g\in\mathbb{Z}^q} F(g) = \sum_{h\in\mathbb{Z}^q} \int_{\mathbb{R}^q} F(y)\, e^{-2\pi i h\cdot y}\, dy. \tag{8.43}$$

The four series in the last two identities converge absolutely.

An extremely unfortunate consequence for the purposes in our work is the continuity of the function F as well as its Fourier transform $F^{\wedge}_{\mathbb{R}^q}$. So, weaker (sufficient) conditions for the validity of the Poisson summation formula in Euclidean spaces $\mathbb{R}^q$ are needed in lattice point as well as sampling problems such that the absolute convergence is not implied for the series on the left side of (8.42), (8.43). As an additional result, the summation process becomes significant for establishing sufficient convergence criteria for multi-dimensional alternating sums (see W. Freeden [**268**] for more details) occurring in lattice point as well as sampling theory.

8.3. Metaharmonic theory and Bessel function asymptotics

First we deal with the decomposition of certain metaharmonic functions into radial and angular parts. In particular, we are interested in the (radial) cylinder functions of dimension q under the particular aspect of their applicability in our lattice point and sampling theory. So, special emphasis is laid on asymptotic expansions of entire (integral) solutions of the Helmholtz equation (cf. C. Müller [**600, 606**], see also W. Freeden [**268**]).

$U \in C^{(2)}(\mathcal{G})$ is called a *metaharmonic function* (occasionally also called *a metaharmonic*) with respect to the Helmholtz operator $\Delta + \lambda$, $\lambda \in \mathbb{R}$, in a region $\mathcal{G} \subset \mathbb{R}^q$ if it satisfies the Helmholtz equation

$$(\Delta + \lambda)\, U(x) = 0\ ,\ x \in \mathcal{G}. \tag{8.44}$$

$U \in C^{(2m)}(\mathcal{G})$, $m \in \mathbb{N}$, is called a *polymetaharmonic function of degree m* in $\mathcal{G}$ if

$$(\Delta + \lambda)^m\, U(x) = 0, \quad x \in \mathcal{G}. \tag{8.45}$$

Bessel functions. By a coordinate transformation $(\Delta + \lambda)\, U = 0$, $\lambda \in \mathbb{R}\backslash\{0\}$, can be reduced to $(\Delta + 1)U = 0$ or $(\Delta - 1)U = 0$, respectively. The best known solutions of these Helmholtz equations are $x \mapsto e^{ix\cdot\eta}$ or $x \mapsto e^{x\cdot\eta}$, $x \in \mathbb{R}^q$, $\eta \in \mathbb{S}^{q-1}$, respectively. Since the Helmholtz operators $\Delta \pm 1$ are linear, more general solutions can be obtained by superposition. This is the basic idea of (regular) Bessel and Hankel functions (for more details the reader is referred to, e.g., G.N. Watson [**857**], C. Müller [**600, 606, 607**], W. Freeden, M. Gutting [**277**], and the references therein).

The point of departure for our work is the entire solution $U_n : \mathbb{R}^q \to \mathbb{C}$ of the Helmholtz equation $\Delta U_n + U_n = 0$ in $\mathbb{R}^q$ of the form

$$U_n(x) = \frac{i^{-n}}{\|\mathbb{S}^{q-1}\|} \int_{\mathbb{S}^{q-1}} e^{ix\cdot\eta}\ Y_n(q;\eta)\ dS_{(q-1)}(\eta), \tag{8.46}$$

where $Y_n(q;\cdot)$ is a member of the space Harm_n of all spherical harmonics of degree n and dimension q. In terms of (standard) polar coordinates $x = r\xi, r = |x|$, $\xi \in \mathbb{S}^{q-1}$, the Funk-Hecke formula yields the decomposition

$$U_n(r\xi) = i^{-n} \frac{\|\mathbb{S}^{q-2}\|}{\|\mathbb{S}^{q-1}\|} \int_{-1}^{1} e^{irt} P_n(q;t)(1-t^2)^{\frac{q-3}{2}}\ dt\ \ Y_n(q;\xi). \tag{8.47}$$

In other words, a separation of the variables into a radial and an angular part is achieved in the form $U_n(x) = J_n(q;r)\ Y_n(q;\xi)$, $x = r\xi$, $\xi \in \mathbb{S}^{q-1}$. The function $J_n(q;\cdot)$ given by

$$J_n(q;r) = i^{-n} \frac{\|\mathbb{S}^{q-2}\|}{\|\mathbb{S}^{q-1}\|} \int_{-1}^{1} e^{irt} P_n(q;t)(1-t^2)^{\frac{q-3}{2}}\ dt, \quad r \geq 0, \tag{8.48}$$

is called the *Bessel function of order n and dimension q* (cf. C. Müller [**607**]).

The Bessel functions can be seen as the orthogonal (i.e., Fourier) coefficients of an orthogonal series in terms of Legendre polynomials.

For $x \in \mathbb{R}^q$, $x = r\xi$, $r = |x|$, $\xi \in \mathbb{S}^{q-1}$, *and* $\eta \in \mathbb{S}^{q-1}$ *we have*

$$e^{ix\cdot\eta} = e^{ir\xi\cdot\eta} = \sum_{n=0}^{\infty} i^n N(q;n) J_n(q;r) P_n(q;\xi\cdot\eta), \tag{8.49}$$

where, for each $r \in [0,\infty)$, *the series converges absolutely and uniformly with respect to* $\xi \cdot \eta$.

We consider, for $x, y \in \mathbb{R}^q$, the integral

$$J_0(q;|x-y|) = \frac{1}{\|\mathbb{S}^{q-1}\|} \int_{\mathbb{S}^{q-1}} e^{ix\cdot\zeta}\, e^{-iy\cdot\zeta}\, dS_{(q-1)}(\zeta). \tag{8.50}$$

The Parseval identity for the (orthogonal system of) Legendre polynomials shows that

$$\sum_{n=0}^{\infty} (N(q;n))^2 J_n(q;|x|) J_n(q;|y|) \int_{\mathbb{S}^{q-1}} P_n(q;\xi\cdot\zeta) P_n(q;\eta\cdot\zeta)\, dS_{(q-1)}(\zeta)$$

$$= \|\mathbb{S}^{q-1}\| \sum_{n=0}^{\infty} N(q;n) J_n(q;|x|) J_n(q;|y|) P_n(q;\xi\cdot\eta). \tag{8.51}$$

This gives the *addition theorem of Bessel functions*:

For $x, y \in \mathbb{R}^q$ *with* $x = |x|\xi, y = |y|\eta$, $\xi, \eta \in \mathbb{S}^{q-1}$, *we have*

$$J_0(q;|x-y|) = \sum_{n=0}^{\infty} N(q;n) J_n(q;|x|) J_n(q;|y|) P_n(q;\xi\cdot\eta). \tag{8.52}$$

For $r \in [0,\infty)$, *the Bessel function* $J_n(q;\cdot)$ *permits the representation by the power series*

$$J_n(q;r) = \Gamma\left(\frac{q}{2}\right)\left(\frac{r}{2}\right)^n \sum_{k=0}^{\infty} \frac{(-1)^k}{k!} \frac{(\frac{r}{2})^{2k}}{\Gamma\left(n+k+\frac{q}{2}\right)}. \tag{8.53}$$

Since the spherical harmonics of degree n and dimension q are the eigenfunctions of the Beltrami operator corresponding to the eigenvalues $(\Delta^*)^\wedge(n) = n(n+q-2)$, $n = 0, 1, \ldots$, i.e., $(\Delta^* + (\Delta^*)^\wedge(n))Y_n(q;\cdot) = 0$, $Y_n(q;\cdot) \in Harm_n$, it follows from the Helmholtz equation that

$$\left(\underbrace{\frac{1}{r^{q-1}}\frac{\partial}{\partial r} r^{q-1}\frac{\partial}{\partial r} + \frac{1}{r^2}\Delta^*_{(q)}}_{=\Delta_{(q)}} + 1\right) J_n(q;r) Y_n(q;\xi) = 0. \tag{8.54}$$

This shows the following differential equation for the Bessel functions

$$J_n''(q;r) + \frac{q-1}{r} J_n'(q;r) + \left(1 - \frac{n(n+q-2)}{r^2}\right) J_n(q;r) = 0. \tag{8.55}$$

In connection with the recursion relation for the Legendre polynomial

$$(n+q-2)P_{n+1}(q;t) = (2n+q-2)tP_n(q;t) - nP_{n-1}(q;t) \tag{8.56}$$

we are able to deduce the recursion relation

$$(n+q-2)J_{n+1}(q;r) + (2n+q-2)J_n'(q;r) = nJ_{n-1}(q;r). \tag{8.57}$$

Even more, we are able to verify the following recursion relations:

$$J_{n-1}(q;r) + J_{n+1}(q;r) = \frac{2n+q-2}{r} J_n(q;r), \tag{8.58}$$

$$J_{n-1}(q;r) - J_{n+1}(q;r) = 2J_n'(q;r) + \frac{q-2}{r} J_n(q;r), \tag{8.59}$$

$$J_{n+1}(q;r) = -r^n \frac{d}{dr}\left(r^{-n} J_n(q;r)\right), \tag{8.60}$$

$$J_{n-1}(q;r) = r^{2-q-n} \frac{d}{dr}\left(r^{n+q-2} J_n(q;r)\right). \tag{8.61}$$

Let $H_n(q;\cdot)$ be a homogeneous harmonic polynomial of the form $H_n(q;x) = r^n Y_n(q;\xi)$, $x \in \mathbb{R}^q$, $x = r\xi$, $r = |x|$, $x \in \mathbb{S}^{q-1}$. Then it is not hard to see that

$$H_n(q;\nabla_x)\, e^{ix\cdot\eta} = i^n e^{ix\cdot\eta}\, H_n(q;\eta), \quad \eta \in \mathbb{S}^{q-1}. \tag{8.62}$$

In connection with

$$J_0(q;|x|) = \frac{1}{\|\mathbb{S}^{q-1}\|} \int_{\mathbb{S}^{q-1}} e^{ix\cdot\eta}\, dS_{(q-1)}(\eta) \tag{8.63}$$

this yields

$$H_n(q;\nabla_x) J_0(q;|x|) = \frac{i^n}{\|\mathbb{S}^{q-1}\|} \int_{\mathbb{S}^{q-1}} e^{ix\cdot\eta} Y_n(q;\eta)\, dS_{(q-1)}(\eta) \tag{8.64}$$

such that

$$H_n(q;\nabla_x) J_0(q;|x|) = (-1)^n J_n(q;r) H_n(q;\xi). \tag{8.65}$$

In other words, Bessel functions of any order may be obtained by differentiation of the Bessel function $J_0(q;\cdot)$ of order 0.

From the power series of the Bessel functions we are immediately able to verify the following asymptotic relation:

For fixed n we have

$$J_n(q;r) = \Gamma\left(\frac{q}{2}\right)\left(\frac{r}{2}\right)^n + O(r^{n+2}), \quad r \to 0. \tag{8.66}$$

All in all, the relevant properties of Bessel functions for our later work can be summarized as follows (cf. C. Müller [**607**]):

The Bessel function $J_n(q;\cdot)$ satisfies the following relations:

(i) *For $r > 0$, $\xi, \eta \in \mathbb{S}^{q-1}$,*

$$e^{r\xi\cdot\eta} = \sum_{n=0}^{\infty} N(q;n)\; J_n(q;r)\; P_n(q,\xi\cdot\eta),$$

(ii) *For $x = |x|\xi$, $y = |y|\eta$*

$$J_0(q;|x-y|) = \sum_{n=0}^{\infty} N(q;n)\; J_n(q;|x|)\; J_n(q;|y|)\; P_n(q;\xi\cdot\eta),$$

(iii) *$J_n(q;\cdot)$ is a solution of the differential equation*

$$J_n''(q;r) + \frac{q-1}{r} J_n'(q;r) + \left(1 - \frac{n(n+q-2)}{r^2}\right) J_n(q;r) = 0,$$

(iv) *$J_n(q;\cdot)$ satisfies the recursion relation*

$$(n+q-2)J_{n+1}(q;r)+(2n+q-2)J_n'(q;r)-nJ_{n-1}(q;r)=0,$$

(v)

$$J_{n-1}(q;r)=r^{2-q-n}\frac{d}{dr}\left(r^{n+q-2}J_n(q;r)\right),$$

(vi)

$$\lim_{n\to\infty}\frac{J_n(q;r)}{\frac{\Gamma(\frac{q}{2})(\frac{r}{2})^n}{\Gamma(n+\frac{q}{2})}}=1.$$

(vii) *For homogeneous harmonic polynomials $H_n(q;\cdot)$ of degree n*

$$H_n(q;\nabla_x)J_0(q;|x|)=(-1)^nJ_n(q;|x|)H_n(q;\xi),$$

(viii) *For n fixed and $r\to 0$,*

$$J_n(q;r)=\Gamma\left(\frac{q}{2}\right)\left(\frac{r}{2}\right)^n+O(r^{n+2}).$$

Notational convention. The traditional notation, which is mostly used in the literature, is based on the observation that the Bessel functions of integral order n and dimension q are expressible as Bessel functions of order $n+\frac{q-2}{2}$ and dimension 2. We discuss this aspect in more detail: Letting, as usual (see G.N. Watson [**857**]),

$$J_\nu(r)=J_\nu(2;r)=\left(\frac{r}{2}\right)^\nu\sum_{k=0}^{\infty}\frac{(-1)^k}{k!}\frac{\left(\frac{r}{2}\right)^{2k}}{\Gamma(\nu+k+1)},\quad \nu\geq 0, \tag{8.67}$$

and observing

$$J_n(q;r)=\left(\frac{r}{2}\right)^n\Gamma\left(\frac{q}{2}\right)\sum_{k=0}^{\infty}\frac{(-1)^k}{k!}\frac{\left(\frac{r}{2}\right)^{2k}}{\Gamma(n+k+\frac{q}{2})},\quad n\in\mathbb{N}_0, \tag{8.68}$$

we are led by comparison to the relation

$$J_n(q;r)=\Gamma\left(\frac{q}{2}\right)\left(\frac{r}{2}\right)^{\frac{2-q}{2}}J_{n+\frac{q-2}{2}}(r),\quad n\in\mathbb{N}_0. \tag{8.69}$$

In particular, we have

$$J_0(3;r)=\sqrt{\frac{\pi}{2}}\frac{J_{\frac{1}{2}}(2;r)}{\sqrt{r}}=\frac{J_1(1;r)}{r}. \tag{8.70}$$

Furthermore we see that

$$J_0(3;r)=\frac{1}{2}\sqrt{\pi}\sum_{k=0}^{\infty}\frac{(-1)^kr^{2k}}{2^{2k}\Gamma(k+1)\Gamma\left(k+\frac{3}{2}\right)}=\sum_{k=0}^{\infty}\frac{(-1)^kr^{2k}}{(2k+1)!}=\frac{\sin(r)}{r}. \tag{8.71}$$

Hence, the sinc-function is expressible in the context of Bessel functions

$$J_0(3;r)=\sqrt{\frac{\pi}{2}}\frac{J_{\frac{1}{2}}(2;r)}{\sqrt{r}}=\frac{J_1(1;r)}{r}=\frac{\sin(r)}{r}=\operatorname{sinc}(r). \tag{8.72}$$

So, the notational framework of (8.72) prepares the role of the 1D sinc-kernel function taken in the multivariate context of this work, when number theoretically reflected sampling methodologies come into play. Finally it should be remarked (cf. C. Müller [**607**]), that the relations (8.67), (8.68), and (8.69) also allow us to understand $J_\nu(q;\cdot)$ for nonnegative real values ν by

$$J_\nu(q;r) = \Gamma\left(\frac{q}{2}\right)\left(\frac{r}{2}\right)^{\frac{2-q}{2}} J_{\nu+\frac{q-2}{2}}(r). \tag{8.73}$$

For later use in our lattice point theoretical as well as sampling context we mention the following integral representation:

$$\begin{aligned}\int_{\substack{|x|\leq R\\ x\in\mathbb{R}^q}} e^{-2\pi i w\cdot x}\,dx &= \|\mathbb{S}^{q-1}\|\int_0^R r^{q-1} J_0(q;2\pi r|w|)\,dr \\ &= \|\mathbb{S}^{q-1}\|\;R^q\,\frac{J_1(q;2\pi|w|R)}{2\pi|w|R},\quad R>0,\quad w\in\mathbb{R}^q.\end{aligned} \tag{8.74}$$

Equivalently we have in Watson's nomenclature

$$\int_{\substack{|x|\leq R\\ x\in\mathbb{R}^q}} e^{-2\pi i w\cdot x}\,dx \;=\; R^{\frac{q}{2}}\,\frac{J_{\frac{q}{2}}(2;2\pi\;|w|R)}{|w|^{\frac{q}{2}}},\quad R>0,\quad w\in\mathbb{R}^q \tag{8.75}$$

(note that, in accordance with our notation (cf. (8.67)), $J_{\frac{q}{2}}(\cdot) = J_{\frac{q}{2}}(2;\cdot)$).

Hankel functions. The Bessel functions are regular at 0. We now discuss a pair of functions which are regular at infinity.

For $r>0$, the functions $H_n^{(1)}(q;\cdot)$, $H_n^{(2)}(q;\cdot)$ defined by

$$H_n^{(1)}(q;r) = -2i^{-n}\frac{\|\mathbb{S}^{q-2}\|}{\|\mathbb{S}^{q-1}\|}\int_{1+0i}^{1+\infty i} e^{irt}P_n(q;t)(1-t^2)^{\frac{q-3}{2}}\,dt, \tag{8.76}$$

$$H_n^{(2)}(q;r) = 2i^{-n}\frac{\|\mathbb{S}^{q-2}\|}{\|\mathbb{S}^{q-1}\|}\int_{-1+0i}^{-1+\infty i} e^{irt}P_n(q;t)(1-t^2)^{\frac{q-3}{2}}\,dt \tag{8.77}$$

are called *Hankel functions of the first and second kind of order n and dimension q*, respectively.

From the Definition of the Hankel functions it immediately follows (cf. C. Müller [**607**]) that $H_n^{(2)}(q;r) = \overline{H_n^{(1)}(q;r)}$.

For n fixed and $r\to\infty$ we have

$$H_n^{(1)}(q;r) = \frac{\Gamma(\frac{q}{2})}{\sqrt{\pi}}\left(\frac{2}{r}\right)^{\frac{q-1}{2}} i^{-n-\frac{q-1}{2}}e^{ir} + O\left(r^{-\frac{q+1}{2}}\right), \tag{8.78}$$

$$H_n^{(2)}(q;r) = \frac{\Gamma(\frac{q}{2})}{\sqrt{\pi}}\left(\frac{2}{r}\right)^{\frac{q-1}{2}} i^{n+\frac{q-1}{2}}e^{-ir} + O\left(r^{-\frac{q+1}{2}}\right). \tag{8.79}$$

Observing the identity

$$J_n(q;r) = \frac{1}{2}\left(H_n^{(1)}(q;r) + H_n^{(2)}(q;r)\right) \tag{8.80}$$

we obtain the following asymptotic expansion:

For n fixed and $r\to\infty$,

$$J_n(q;r) = \frac{\Gamma(\frac{q}{2})}{\sqrt{\pi}}\left(\frac{2}{r}\right)^{\frac{q-1}{2}}\cos\left(n\frac{\pi}{2} + (q-1)\frac{\pi}{4} - r\right) + O\left(r^{-\frac{q+1}{2}}\right). \tag{8.81}$$

Based on techniques due to G. N. Watson [**857**] we are led to the following the O-term estimate:

For $n \in \mathbb{N}_0, m \in \mathbb{N}$ fixed and $r \to \infty$

$$J_n(q;r) = \frac{\Gamma(\frac{q}{2})}{2^{\frac{3-q}{2}}\sqrt{\pi}} r^{\frac{1-q}{2}}$$

$$\times \left(e^{i(r - \frac{\pi}{2}(n+\frac{q-2}{2}) - \frac{\pi}{4})} \sum_{l=0}^{m-1} \frac{(n+\frac{q-2}{2}, l)}{(-2ir)^l} + e^{-i(r-\frac{\pi}{2}(n+\frac{q-2}{2})-\frac{\pi}{4})} \sum_{l=0}^{m-1} \frac{(n+\frac{q-2}{2}, l)}{(2ir)^l} \right)$$

$$+ O\left(r^{-(m+\frac{q-1}{2})} \right),$$

where

$$\left(n + \frac{q-2}{2}, 0 \right) = 1$$

and

$$\left(n + \frac{q-2}{2}, l \right) = \frac{1}{l!} \left(n(n+q-2) + \frac{(q-3)(q-1)}{4} \right)$$

$$\cdots\cdot \left(n(n+q-2) + \frac{(q+2l-3)(q-2l-1)}{4} \right),$$

whereas, for $n \to \infty$ and fixed $r > 0$,

$$\lim_{n\to\infty} \frac{H_n^{(1)}(q;r)}{\frac{i}{\pi}\Gamma\left(\frac{q}{2}\right)\Gamma\left(n+\frac{q}{2}-1\right)\left(\frac{2}{r}\right)^{n+q-2}} = 1.$$

Kelvin functions. For $r > 0$, the function $K_n(q;\cdot)$ given by

$$K_n(q;r) = \int_1^\infty e^{-rt} P_n(q;t)(t^2-1)^{\frac{q-3}{2}}\, dt \tag{8.82}$$

is called the *Kelvin function (or modified Hankel function) of order n and dimension* q (cf. C. Müller [**607**]).

It follows that

$$H_n^{(1)}(q;ir) = 2\frac{\|\mathbb{S}^{q-2}\|}{\|\mathbb{S}^{q-1}\|} i^{1-q-n} K_n(q;r), \tag{8.83}$$

$$H_n^{(2)}(q;-ir) = 2\frac{\|\mathbb{S}^{q-2}\|}{\|\mathbb{S}^{q-1}\|} i^{n+q-1} K_n(q;r). \tag{8.84}$$

Moreover, we find for $r > 0$

$$K_{n-1}(q;r) - K_{n+1}(q;r) = -\frac{2n+q-2}{r} K_n(q;r) \tag{8.85}$$

and

$$nK_{n-1}(q;r) + (n+q-2)K_{n+1}(q;r) = (2n+q-2)K_n'(q;r), \tag{8.86}$$

which again follows from the recursion relation for the Legendre polynomial.

The Hankel functions as well as the Kelvin functions are not defined at the origin, but they have characteristic singularities there.

For $r \to 0$ and $n+q-3 \geq 0$ we have

$$K_n(q;r) = \frac{\Gamma(q-1)}{2^{q-1}\Gamma\left(\frac{q}{2}\right)}\Gamma\left(n+\frac{q-2}{2}\right)\left(\frac{2}{r}\right)^{n+q-2} + O\left(r^{-n-q+4}\right). \tag{8.87}$$

The function $x \mapsto K_0(q;|x|)$, $x \in \mathbb{R}^q\backslash\{0\}$, shows the same singularity behavior at the origin as the fundamental solution $x \mapsto F_q^{(1)}(|x|)$, $x \in \mathbb{R}^q\backslash\{0\}$, for the Laplace operator in $\mathbb{R}^q$. This fact is of particular significance for the characterization of the singularity behavior of the multi-dimensional lattice (Euler-Green) function establishing the multi-dimensional Euler-type summation formula.

For $r \to \infty$ and $n \geq 0$ fixed we obtain

(8.88)

$$K_n(q;r) = \int_1^\infty e^{-rt}P_n(q;t)(t^2-1)^{\frac{q-3}{2}}\,dt = e^{-r}\left(2^{\frac{q-3}{2}}\frac{\Gamma\left(\frac{q-1}{2}\right)}{r^{\frac{q-1}{2}}} + O(r^{-\frac{q+1}{2}})\right).$$

This leads to the following asymptotic relations:

For $r \to \infty$ and $n \geq 0$ fixed,

$$K_n(q;r) = \frac{e^{-r}}{2}\Gamma\left(\frac{q-1}{2}\right)\left(\left(\frac{2}{r}\right)^{\frac{q-1}{2}} + O\left(r^{-\frac{q+1}{2}}\right)\right). \tag{8.89}$$

The Kelvin function $K_n(q;\cdot)$ satisfies the following relations:

(i) *For $x = |x|\xi$, $y = |y|\eta$,*

$$K_0(q;|x-y|) = \sum_{n=0}^{\infty} N(q;n)I_n(q;|x|)K_n(q;|y|)P_n(q;\xi\cdot\eta),$$

(ii) *$K_n(q;\cdot)$ is a solution of the differential equation*

$$K_n''(q;r) - \frac{q-1}{r}K_n'(q;r) - \left(1+\frac{n(n+q-2)}{r^2}\right)K_n(q;r) = 0,$$

(iii) *$K_n(q;\cdot)$ satisfies the recursion relations*

$$(n+q-2)K_{n+1}(q;r) + (2n+q-2)K_n'(q;r) + nK_{n-1}(q;r) = 0,$$

$$K_{n-1}(q;r) - K_{n+1}(q;r) = -\frac{2n+q-2}{r}K_n(q;r),$$

(iv)

$$(-1)^m r^{-n-m}K_{n+m}(q;r) = \left(\frac{1}{r}\frac{d}{dr}\right)^m\left(K_n(q;r)\,r^{-n}\right),$$

$$(-1)^m r^{n-m+q-2}K_{n-m}(q;r) = \left(\frac{1}{r}\frac{d}{dr}\right)^m\left(r^{n+q-2}K_n(q;r)\right),$$

(v)

$$\lim_{n\to\infty}\frac{2\pi\,K_n(q;r)}{\Gamma\left(\frac{q-1}{2}\right)\left(\frac{2}{r}\right)^{n+q-2}\Gamma\left(n+\frac{q}{2}-1\right)} = 1,$$

(vi) *For $n+q-3>0$ fixed and $r\to 0$,*

$$K_n(q;r)=\frac{\Gamma(q-1)}{2^{q-1}\Gamma\left(\frac{q}{2}\right)}\Gamma\left(n+\frac{q-2}{2}\right)\left(\frac{2}{r}\right)^{n+q-2}+O(r^{-n-q+4})$$

and, for $n=0$, $q=2$ and $r\to 0$,

$$K_0(q;r) = -\ln(r)+O(1).$$

(vii) *For $r\to\infty$ and $n\geq 0$ fixed*

$$K_n(q;r)=\frac{e^{-r}}{2}\Gamma\left(\frac{q-1}{2}\right)\left(\left(\frac{2}{r}\right)^{\frac{q-1}{2}}+O\left(r^{-\frac{q+1}{2}}\right)\right).$$

In particular,

$$K_0(3;r)=\frac{e^{-r}}{r},\quad K_n(3;r)=r^n\left(\frac{1}{r}\frac{d}{dr}\right)^n\frac{e^{-r}}{r}. \tag{8.90}$$

Notational convention. It is interesting to relate the Kelvin function of dimension q to their counterparts of dimension 2. For real values ν we have (see G.N. Watson [**857**], W. Magnus, W, F. Oberhettinger, R.P. Soni [**554**])

$$K_\nu(r)=K_\nu(2;r)=\frac{\sqrt{\pi}}{\Gamma\left(\nu+\frac{1}{2}\right)}\left(\frac{r}{2}\right)^\nu\int_1^\infty e^{-rt}(t^2-1)^{\nu-\frac{1}{2}}\,dt. \tag{8.91}$$

By comparison we find

$$K_n(q;r)=\frac{\Gamma\left(\frac{q-1}{2}\right)}{\sqrt{\pi}}\left(\frac{r}{2}\right)^{\frac{2-q}{2}}K_{n+\frac{q-2}{2}}(r). \tag{8.92}$$

Classification for the Helmholtz equation. The solutions of the Helmholtz equation can be subdivided into three classes depending on their domain of definition:

- solutions for the "inner space" of a fixed ball $\mathbb{B}^q_R$, $R>0$,
- solutions for the "outer space" of a fixed ball $\mathbb{B}^q_R$, $R>0$,
- "entire" solutions valid for the whole Euclidean space $\mathbb{R}^q$.

From the literature about the Helmholtz equation(cf. C. Müller [**606**] for the 3D-case) we borrow the following expansion theorems for inner and/or outer space solutions involving Bessel and Hankel functions.

Expansion for the inner space of a ball. Let U be a function of class $C^{(2)}\left(\mathbb{B}^q_R\right)$ such that $\Delta U+U=0$ in $\mathbb{B}^q_R$. Then there exists a sequence $\{Y_n(q;\cdot)\}_{n=0,1,\ldots}$ of spherical harmonics such that

$$U(x)=\sum_{n=0}^\infty J_n(q;r)Y_n(q;\xi), \tag{8.93}$$

$x=r\xi$, $\xi\in\mathbb{S}^{q-1}$, where the series is absolutely and uniformly convergent for all $x\in\overline{\mathbb{B}^q_r}$ with $r=|x|<R$. Conversely, if (8.93) holds uniformly for all $x\in\overline{\mathbb{B}^q_r}$ with $r=|x|<R$, then U is of class $C^{(2)}(\mathbb{B}^q_R)$ satisfying $\Delta U+U=0$ in $\mathbb{B}^q_R$.

Expansion for the outer space of a ball. Let U be of class $C^{(2)}\left(\mathbb{R}^q\backslash\overline{\mathbb{B}^q_R}\right)$ such that $\Delta U+U=0$ in $\mathbb{R}^q\backslash\overline{\mathbb{B}^q_R}$. Then there exist sequences $\{Y_n^{(i)}(q;\cdot)\}_{n=0,1,\ldots}$, $i=1,2$, of spherical harmonics such that

$$U(x)=\sum_{n=0}^{\infty}\left\{H_n^{(1)}(q;r)Y_n^{(1)}(q;\xi)+H_n^{(2)}(q;r)Y_n^{(2)}(q;\xi)\right\}, \tag{8.94}$$

$x=r\xi$, $\xi\in\mathbb{S}^{q-1}$, where the series (8.94) is absolutely and uniformly convergent for all $x\in\overline{\mathbb{B}^q_r}$ with $r=|x|>R$. Conversely, if (8.94) holds uniformly for all $x\in\mathbb{R}^q\backslash\overline{\mathbb{B}^q_r}$ with $r=|x|>R$, then U is of class $C^{(2)}\left(\mathbb{R}^q\backslash\overline{\mathbb{B}^q_R}\right)$ satisfying $\Delta U+U=0$ in $\mathbb{R}^q\backslash\overline{\mathbb{B}^q_R}$.

Multivariate Bessel-type integral asymptotics. The study of the lattice point and sampling series needs to know more about the asymptotics of the metaharmonic context determining their convergence behavior. To this end we especially start from a finite sum of the form

$$U^{(N)}(x)=(2\pi)^{\frac{1-q}{2}}\left\|\mathbb{S}^{q-1}\right\|\sum_{n=0}^{N}i^nJ_n(q;r)Y_n(q;\xi) \tag{8.95}$$

with $x=r\xi$, $r=|x|$, $\xi\in\mathbb{S}^{q-1}$, $Y_n(q;\cdot)\in Harm_n(\mathbb{S}^{q-1})$, and $N\in\mathbb{N}_0$, satisfies the Helmholtz wave equation $\Delta U^{(N)}(x)+U^{(N)}(x)=0$, $x\in\mathbb{R}^q$; i.e., $U^{(N)}$ is an entire solution of the Helmholtz equation in $\mathbb{R}^q$. Remembering the asymptotic expansion of the Bessel function

$$J_n(q;r)=\frac{\Gamma\left(\frac{q}{2}\right)}{\sqrt{\pi}}\left(\frac{2}{r}\right)^{\frac{q-1}{2}}\cos\left(n\frac{\pi}{2}+(q-1)\frac{\pi}{4}-r\right)+O\left(r^{-\frac{q+1}{2}}\right) \tag{8.96}$$

we obtain

$$\begin{aligned} &r^{\frac{q-1}{2}}U^{(N)}(r\xi)\\ &\quad=e^{i\left(r-(q-1)\frac{\pi}{4}\right)}F^{(N)}(q;\xi)+e^{-i\left(r-(q-1)\frac{\pi}{4}\right)}F^{(N)}(q;-\xi)+o(1), \end{aligned}\tag{8.97}$$

where

$$F^{(N)}(q;\xi)=\sum_{n=0}^{N}Y_n(q;\xi),\quad F^{(N)}(q;-\xi)=\sum_{n=0}^{N}(-1)^nY_n(q;\xi). \tag{8.98}$$

Using the estimate (cf. [**857**], [**554**])

$$\begin{aligned} J_n(q;r)=&\frac{\Gamma\left(\frac{q}{2}\right)}{\sqrt{\pi}2^{\frac{3-q}{2}}}r^{\frac{1-q}{2}}\\ &\times\left(e^{i\left(r-\frac{\pi}{2}\left(n+\frac{q-2}{2}\right)-\frac{\pi}{4}\right)}\sum_{l=0}^{m-1}\frac{(n+\frac{q-2}{2},l)}{(-2ir)^l}+e^{-i\left(r-\frac{\pi}{2}\left(n+\frac{q-2}{2}\right)-\frac{\pi}{4}\right)}\sum_{l=0}^{m-1}\frac{(n+\frac{q-2}{2},l)}{(-2ir)^l}\right)\\ &+O\left(\frac{1}{r^{m+\frac{q-1}{2}}}\right),\quad m\geq 1, \end{aligned}$$

with (see (8.3) and (8.82))

$$\left(n+\frac{q-2}{2},0\right)=1 \tag{8.99}$$

and

(8.100)
$$\left(n + \frac{q-2}{2},\ l\right) = \frac{1}{l!}\left(n(n+q-2) + \frac{(q-1)(q-3)}{4}\right) \cdot \ldots \cdot \left(n(n+q-2) + \frac{(q+2l-3)(q-2l-1)}{4}\right)$$

we are led to introduce the "operators"

$$O(q;0) = 1, \tag{8.101}$$

(8.102)
$$O(q;l) = \frac{1}{l!}\left(\frac{(q-1)(q-3)}{4} - \Delta^*\right) \cdot \ldots \cdot \left(\frac{(q+2l-3)(q-2l-1)}{4} - \Delta^*\right).$$

In terms of (8.101) and (8.102) we obtain the asymptotic relation

(8.103)
$$\begin{aligned} r^{\frac{q-1}{2}} U^{(N)}(r\xi) &= (2\pi)^{\frac{1-q}{2}} \left\|\mathbb{S}^{q-1}\right\| \sum_{n=0}^{N} i^n J_n(q;r) Y_n(q;\xi) \\ &= i^{\frac{1-q}{2}} e^{ir} \sum_{l=0}^{m-1} \left(\frac{1}{-2\pi i r}\right)^l O(q;l) F^{(N)}(q;\xi) \\ &\quad + i^{\frac{q-1}{2}} e^{-ir} \sum_{l=0}^{m-1} \left(\frac{1}{2\pi i r}\right)^l O(q;l) F^{(N)}(q;-\xi) \\ &\quad + o\left(r^{1-m}\right), \quad r \to \infty. \end{aligned}$$

Asymptotic expansions of the type (8.97) and (8.103), respectively, are characteristic for entire solutions of the Helmholtz wave equation. Unfortunately, when the attempt is made to formulate an asymptotic expansion of type (8.103) for $N \to \infty$, we are confronted with serious problems of the convergence for the right side of (8.103). Nevertheless, a helpful result of interest for our purpose is due to C. Müller (see, e.g., [**600**]):

Let F be of class $C^{(k)}(\mathbb{S}^{q-1})$, $k \geq \frac{q}{2} + 1$. Then, uniformly with respect to all $\xi \in \mathbb{S}^{q-1}$, we have for $r \to \infty$

(8.104)
$$\begin{aligned} \left(\frac{r}{2\pi}\right)^{\frac{q-1}{2}} \int_{\mathbb{S}^{q-1}} F(\eta) e^{ir\xi\cdot\eta}\, dS_{(q-1)}(\eta) \\ = i^{\frac{1-q}{2}} e^{ir} F(\xi) + i^{\frac{q-1}{2}} e^{-ir} F(-\xi) \ + \ o(1). \end{aligned}$$

Furthermore, in an analogous way as in the theory of Bessel functions, the last estimate admits a canonical extension.

Let F be of class $C^{(k)}(\mathbb{S}^{q-1})$ with $k \geq \frac{q}{2} + 2m - 1, m \in \mathbb{N}$. Then

$$\left(\frac{r}{2\pi}\right)^{\frac{q-1}{2}} \int_{\mathbb{S}^{q-1}} e^{ir\xi\cdot\eta} F(\eta)\, dS_{(q-1)}(\eta) \tag{8.105}$$

$$= i^{\frac{1-q}{2}} e^{ir} \sum_{l=0}^{m-1} \left(\frac{1}{-2ir}\right)^{l} O(q;l) F(\xi)$$

$$+ \; i^{\frac{q-1}{2}} e^{-ir} \sum_{l=0}^{m-1} \left(\frac{1}{2ir}\right)^{l} O(q;l) F(-\xi)$$

$$+ \; o\left(r^{1-m}\right),$$

where

$$O(q;0) = 1, \tag{8.106}$$

$$O(q;l) = \frac{1}{l!} \prod_{j=1}^{l} \left(\frac{(q+2j-3)(q-2j-1)}{4} - \Delta^*\right), \; l \in \mathbb{N}, \tag{8.107}$$

and Δ^* is the Beltrami operator on the unit sphere $\mathbb{S}^{q-1}$ (note that, for the constant function $F = 1$, (8.105) leads back to (8.104)).

Part 6

Sampling methodologies

CHAPTER 9

Sampling over continuously connected pointsets

Early engineering approaches to univariate sampling were known in the first half of the last century and used basic Fourier theory. In principle, all mathematical facets to be needed for 1D Shannon sampling theorem were already available to L.J. Mordell [**590–592**] and G.H. Hardy [**413**], but their interest was mainly in number theoretic problems. Following the historic explanations by P.L. Butzer and his group [**123, 128**], bandlimited sampling variants were given by E.T. Whittaker [**862, 863, 865**] and K. Ogura [**660**]. In later years it became known that the classical sampling theorem had been presented before C.E. Shannon [**764, 765**] to the Russian communication community by V.A. Kotel'nikov [**508**]. In a more implicit verbal form, it had also been described in the German literature by H. Raabe [**705**] (for more details concerning the history see H.D. Lüke [**552**]). As a consequence of the discovery of the several independent introductions of the sampling theorem, people started to refer to the theorem by including the names of the aforementioned authors, resulting in such phrases as the "Whittaker-Kotel'nikov-Shannon (WKS) sampling theorem": It essentially shows that, for a given *sample rate*, perfect reconstruction can be guaranteed for any *bandwidth* smaller than the sample rate (see also H. Nyquist [**657**], [**658**]) and the bandwidth is reflected by the choice of an adopted sinc-like kernel of type (1.13).

9.1. Univariate Fourier theory based Shannon sampling

Using 1D lattice nomenclature we are able to recapitulate the classical Shannon sampling expansion within the following framework: Consider the 1D lattice Λ given by $\Lambda = \sigma\mathbb{Z} = \{g = \sigma n : n \in \mathbb{Z}, \sigma > 0\}$. Let $F^{\wedge}_{\mathcal{F}_\Lambda}$ be the Fourier transform (cf. (1.1)), given by

$$F^{\wedge}_{\mathcal{F}_\Lambda}(x) = \int_{\mathcal{F}_\Lambda} F(z)\, e^{-2\pi i x\cdot z}\, dz, \quad x \in \mathbb{R}, \tag{9.1}$$

for some $F \in L^2(\mathcal{F}_\Lambda), \mathcal{F}_\Lambda = [-\frac{\sigma}{2}, \frac{\sigma}{2})$, extended Λ-periodically to the real line $\mathbb{R}$, i.e., F is assumed to be of class $L^2_\Lambda(\mathbb{R})$.

Fourier expansion based Shannon sampling. We expand the extension F in a series of the form

$$F(x) = \sum_{h\in\Lambda^{-1}} c_h\, \Phi_h(x), \quad \Phi_h(x) = \frac{1}{\sqrt{\|\mathcal{F}_\Lambda\|}}\, e^{2\pi i h\cdot x}, \tag{9.2}$$

where $\Lambda^{-1} = \sigma^{-1}\mathbb{Z}$ is the inverse lattice of Λ, $\|\mathcal{F}_\Lambda\| = \sigma$ is the length of the fundamental cell $\mathcal{F}_\Lambda$, and $\Sigma_{h\in\Lambda^{-1}} \ldots$ means that the sum is extended over all h of the inverse lattice Λ^{-1}. We know that the system $\{\Phi_h\}_{h\in\Lambda^{-1}}$ is orthonormal with respect to the $L^2(\mathcal{F}_\Lambda)$-inner product (cf. (8.25)). The coefficients c_h are given in the form (cf. (1.18))

$$c_h = \int_{\mathcal{F}_\Lambda} F(t)\,\overline{\Phi_h(t)}\,dt = \frac{1}{\sqrt{\|\mathcal{F}_\Lambda\|}} \int_{\mathcal{F}_\Lambda} F(t)\,e^{-2\pi i h\cdot t}\,dt. \tag{9.3}$$

Comparing (9.1) and (9.3) we are led to the representation

$$c_h = \frac{1}{\sqrt{\|\mathcal{F}_\Lambda\|}}\,F^{\wedge}_{\mathcal{F}_\Lambda}(h). \tag{9.4}$$

Substituting (9.3) into (9.1) we obtain the identity

$$F^{\wedge}_{\mathcal{F}_\Lambda}(t) = \int_{\mathcal{F}_\Lambda} \sum_{h\in\Lambda^{-1}} c_h \frac{1}{\sqrt{\|\mathcal{F}_\Lambda\|}} e^{2\pi i h\cdot x}\,e^{-2\pi i x\cdot t}\,dx, \quad t\in\mathbb{R},3 \tag{9.5}$$

such that, by interchanging the order of integration and summation, we arrive at

$$F^{\wedge}_{\mathcal{F}_\Lambda}(t) = \sum_{h\in\Lambda^{-1}} F^{\wedge}_{\mathcal{F}_\Lambda}(h) \frac{1}{\|\mathcal{F}_\Lambda\|} \int_{\mathcal{F}_\Lambda} e^{2\pi i x\cdot(h-t)}\,dx. \tag{9.6}$$

The identity (9.6) immediately leads to the celebrated classical expansion of a bandlimited signal

$$F^{\wedge}_{\mathcal{F}_{\sigma\mathbb{Z}}}(t) = \sum_{h\in\frac{1}{\sigma}\mathbb{Z}} F^{\wedge}_{\mathcal{F}_{\sigma\mathbb{Z}}}(h)\,\frac{\sin(2\pi(h-t)\frac{\sigma}{2})}{2\pi(h-t)\frac{\sigma}{2}}. \tag{9.7}$$

We equivalently find that

$$F^{\wedge}_{\mathcal{F}_{\sigma\mathbb{Z}}}(t) = \sum_{n\in\mathbb{Z}} F^{\wedge}_{\mathcal{F}_{\sigma\mathbb{Z}}}\left(\frac{n}{\sigma}\right)\operatorname{sinc}\left(\pi\sigma(\frac{n}{\sigma}-t)\right). \tag{9.8}$$

Altogether, in the Fourier theoretic terminology as used here, Shannon's sampling result (9.7) can be rephrased as follows:

Any function $F^{\wedge}_{\mathcal{F}_\Lambda}$, $\Lambda=\sigma\mathbb{Z}$, bandlimited to the fundamental cell $\mathcal{F}_\Lambda = \left[-\frac{\sigma}{2},\frac{\sigma}{2}\right)$ of the lattice $\Lambda=\sigma\mathbb{Z}$, i.e., the Fourier integral (9.1) may be reconstructed from the sequence of samples

$$\{F^{\wedge}_{\mathcal{F}_\Lambda}(h)\}_{h\in\Lambda^{-1}} \tag{9.9}$$

via the "cardinal series"

$$F^{\wedge}_{\mathcal{F}_\Lambda}(t) = \sum_{h\in\Lambda^{-1}} F^{\wedge}_{\mathcal{F}_\Lambda}(h)\,\underbrace{\frac{1}{\|\mathcal{F}_\Lambda\|}\int_{\mathcal{F}_\Lambda} e^{2\pi i x\cdot(h-t)}\,dx}_{=K_{\mathcal{F}_\Lambda}(h-t)}, \quad t\in\mathbb{R}, \tag{9.10}$$

where the series is absolutely and uniformly convergent on any compact set of the real line $\mathbb{R}$. Moreover, for functions $F^{\wedge}_{\mathcal{F}_\Lambda}$ satisfying the Shannon sampling theorem, we have

$$\int_{\mathbb{R}} |F^{\wedge}_{\mathcal{F}_\Lambda}(t)|^2\,dt = \sum_{h\in\Lambda^{-1}} |F^{\wedge}_{\mathcal{F}_\Lambda}(h)|^2 < \infty. \tag{9.11}$$

Thus there is an isometric one-to-one correspondence between bandlimited functions $F^{\wedge}_{\mathcal{F}_\Lambda}$ in $L^2_\Lambda(\mathbb{R})$ and sequences $\{F^{\wedge}_{\mathcal{F}_\Lambda}(h)\}_{h\in\Lambda^{-1}}$ in the space of Λ^{-1}-summable sequences $\ell^2(\Lambda^{-1})$.

Obviously, the Fourier transform $F^{\wedge}_{\mathcal{F}_\Lambda}$ (considered as a one-dimensional signal) is an infinitely repeated replication of the samples at points h of the inverse lattice Λ^{-1}. The fundamental cell $\mathcal{F}_\Lambda = [-\frac{\sigma}{2}, \frac{\sigma}{2})$ of the lattice $\Lambda = \sigma\mathbb{Z}$, is called the *base band.*

Let us consider a function $F^{\wedge}_{\mathcal{F}_{\tau\mathbb{Z}}}$ (dependent on the time variable) of base band τ on the lattice $\Lambda = \sigma\mathbb{Z}$. Following H. Nyquist [**657, 658**], for a given *sample rate* σ, perfect reconstruction is guaranteed for any base band with $\tau \leq \sigma$. If the sample rate τ is too large, i.e., $\tau > \sigma$, the reconstruction exhibits imperfections known as *aliasing.* In the case of *undersampling,* i.e., $\tau > \sigma$, the contributions of higher frequencies will not be filtered out. Instead they occur in a frequency translated way, leading to the well known phenomenon of *aliasing* errors (see, e.g., [**99**]). In the case of *oversampling,* i.e., $\tau < \sigma$, the signal can be sampled, where "superfluous information" can be removed (see, e.g., M.Z. Nashed, Q. Sun [**636**]) or used in an appropriate way to accelerate the convergence of the series (see, e.g., R.J. Marks II [**561**]).

Engineering approach to Shannon sampling. For the rigorous justification of the engineering approach (as given by, e.g., A. Papoulis [**678, 679**], A.V. Oppenheim, R.W. Schäfer [**664**]) to sampling, it is helpful to recapitulate (see Chapter 2) that the definition of the Fourier transform $\mathcal{F}_\mathbb{R}$ is extendable from $\mathfrak{S}$ (the space of all tempered distributions) to $\mathfrak{S}'$ (the space of all continuous linear functionals on $\mathfrak{S}$). More concretely, mathematicians consider the sequence of sampled values $\{F^{\wedge}_{\mathcal{F}_\mathbb{Z}}(h)\}_{h\in\mathbb{Z}}$ as a member in the space $\ell^2(\mathbb{Z})$ of all square-summable sequences on $\mathbb{Z}$ (cf. M.Z. Nashed, G.G. Walter [**643**]). Electrical engineers like to continue to think of this sequence as a time signal. In essence, engineers view the Shannon sampling theorem in terms of an impulse train and a lowpass filter in the following way:

$$F^{\wedge}_{\mathcal{F}_\mathbb{Z}}(t) \xrightarrow{\text{sampler}} (F^{\wedge}_{\mathcal{F}_\mathbb{Z}})^*(t) \xrightarrow{\text{lowpass filter}} F^{\wedge}_{\mathcal{F}_\mathbb{Z}}(t). \tag{9.12}$$

For the *rigorous mathematical justification of the engineering approach to sampling* we have to work, so that the sampling map (i.e., the "sampled impulse train")

$$S(F^{\wedge}_{\mathcal{F}_\mathbb{Z}}) = (F^{\wedge}_{\mathcal{F}_\mathbb{Z}})^*(t) = \sum_{h\in\mathbb{Z}} F^{\wedge}_{\mathcal{F}_\mathbb{Z}}(h)\ \delta(t-h), \tag{9.13}$$

and the lowpass filter map

$$P(F^{\wedge}_{\mathcal{F}_\mathbb{Z}})^*(t) = \sum_{h\in\mathbb{Z}} F^{\wedge}_{\mathcal{F}_\mathbb{Z}}(h) P(\delta(h-t)) = \sum_{h\in\mathbb{Z}} F^{\wedge}_{\mathcal{F}_\mathbb{Z}}(h)\ \frac{\sin \pi(t-h)}{\pi(t-h)}. \tag{9.14}$$

are defined on appropriate spaces, and their composition PS represents the identity map.

The mathematical difficulty to characterize the framework, in which the operators S and P are sensefully given, is not resolved in the engineering approach. More concretely, we are confronted with the following situation:

- The sampler S takes $F^{\wedge}_{\mathcal{F}_\mathbb{Z}}$ into $(F^{\wedge}_{\mathcal{F}_\mathbb{Z}})^*$, which is out of the space of band-limited functions. Obviously, $(F^{\wedge}_{\mathcal{F}_\mathbb{Z}})^*$ is not a signal with a finite L^2-energy.
- The question is in what sense the impulse train (9.13) series does converge? Obviously, the series converges in the sense of tempered distributions to a distribution.

- The map P recovers $F^{\wedge}_{\mathcal{F}_{\mathbb{Z}}}$, at least formally, since we are allowed to write

(9.15) $$P(\delta(t-h)) = \int_{\mathcal{F}_{\mathbb{Z}}} e^{-2\pi i x \cdot t} e^{2\pi i x \cdot h} \, dx = \frac{\sin \pi(t-h)}{\pi(t-h)}.$$

  If P is continuous (in some topology), then we arrive at the cardinal series

(9.16) $$P(F^{\wedge}_{\mathcal{F}_{\mathbb{Z}}})^*(t) = \sum_{h\in\mathbb{Z}} F^{\wedge}_{\mathcal{F}_{\mathbb{Z}}}(h) P(\delta(h-t)) = \sum_{h\in\mathbb{Z}} F^{\wedge}_{\mathcal{F}_{\mathbb{Z}}}(h) \, \frac{\sin \pi(t-h)}{\pi(t-h)}.$$

  However, since $\mathfrak{S}'$ is not a Hilbert space, we do not know if P is a continuous operator.

- Still another difficulty arises! P is not well-defined on $\mathfrak{S}'$. As a matter of fact, P corresponds under the Fourier transform to the multiplication of the Fourier transform of a member G in $\mathfrak{S}'$ with the characteristic function of $\overline{\mathcal{F}_{\mathbb{Z}}}$:

(9.17) $$P(G)(t) = (\chi_{\overline{\mathcal{F}_{\mathbb{Z}}}} G^{\wedge}_{\mathbb{R}})^{\wedge}_{\mathbb{R}}(-t).$$

  Unfortunately the characteristic function is not a multiplier in $\mathfrak{S}'$. Hence we are led to restrict ourselves to a subspace of $\mathfrak{S}'$, in which $\chi_{\overline{\mathcal{F}_{\mathbb{Z}}}}$ is a multiplier.

These issues between the engineering approach to sampling and its mathematical justification (cf. [**392**]) have been resolved in M.Z. Nashed, G.G. Walter [**644**] and M.Z. Nashed, Q. Sun [**637**]. A rigorous proof of the engineering approach can be obtained by considering the sampling map S as an operator from the Paley-Wiener space $B^{\wedge}_{\mathcal{F}_{\mathbb{Z}}}$ (as introduced in (1.12)) to the Sobolev space $H^{-1}(\mathbb{R})$ and the fitering map P as an operator (actually an orthogonal projection) from $H^{-1}(\mathbb{R})$ to $B^{\wedge}_{\mathcal{F}_{\mathbb{Z}}}$:

(9.18) $$B^{\wedge}_{\mathcal{F}_{\mathbb{Z}}} \xrightarrow{S} H^{-1}(\mathbb{R}) \xrightarrow{P} B^{\wedge}_{\mathcal{F}_{\mathbb{Z}}}.$$

This proposition does not require that $F^{\wedge}_{\mathcal{F}_{\mathbb{Z}}}$ be bandlimited, but the next one in which we consider projection operators in the space $H^{-1}(\mathbb{R})$ does. Clearly, we have $B_{\mathcal{F}_{\mathbb{Z}}} \subset L^2(\mathbb{R}) \subset H^{-1}(\mathbb{R})$, $B_{\mathcal{F}_{\mathbb{Z}}}$ is closed in $H^{-1}(\mathbb{R})$ in the topology of $H^{-1}(\mathbb{R})$.

Note that the 1D series

(9.19) $$\sum_{g\in\mathbb{Z}} F^{\wedge}_{\mathcal{F}_{\mathbb{Z}}}(g) \, \frac{\sin \pi(t-g)}{\pi(t-g)}$$

converges in the sense of $L^2(\mathbb{R})$ (since $\{\mathrm{sinc}(\pi(t-g))\}_{g\in\mathbb{Z}}$ is an orthonormal sequence and $\{F^{\wedge}_{\mathcal{F}_{\mathbb{Z}}}(g)\}_{g\in\mathbb{Z}}$ is of class $\in \ell^2(\mathbb{Z})$) and, hence, in the sense of $H^{-1}(\mathbb{R})$.

- The characteristic functions are multipliers in the space of Fourier transforms of the elements in $H^r(\mathbb{R})$, which is known to be the image under the Fourier transform of $L^2_r(\mathbb{R})$, the space of square-integrable functions with respect to the measure $d\mu(x) = (1+|x|^2)^r \, dx$. So, we may consider $H^r(\mathbb{R})$ with values $r < -\frac{1}{2}$ (in fact, $H^{-1}(\mathbb{R})$ is canonically taken, since $\delta \in H^{-1}$).
- We are led to the sampling map $S(F^{\wedge}_{\mathcal{F}_{\mathbb{Z}}}) = (F^{\wedge}_{\mathcal{F}_{\mathbb{Z}}})^*$ (see (9.13)) as a map onto $H^{-1}(\mathbb{R})$. Then the partial sums of the "sampled impulse train"

(9.20) $$\sum_{|h|\le N} F^{\wedge}_{\mathcal{F}_{\mathbb{Z}}}(h) \, \delta(t-h)$$

belong to $H^{-1}(\mathbb{R})$ and converge in the norm of $H^{-1}(\mathbb{R})$ to $(F^{\wedge}_{\mathcal{F}_{\mathbb{Z}}})^*$. This proposition does not require the signal to be bandlimited.

- When we consider a projector onto the space $H^{-1}(\mathbb{R})$, $B_{\mathcal{F}_{\mathbb{Z}}} \subset L^2(\mathbb{R}) \subset H^{-1}(\mathbb{R})$, and $B_{\mathcal{F}_{\mathbb{Z}}}$ is closed in the topology of $H^{-1}(\mathbb{R})$. Hence, the orthogonal projection of $H^{-1}(\mathbb{R})$ onto $B_{\mathcal{F}_{\mathbb{Z}}}$ is well-defined. The reproducing kernel $K_{\mathcal{F}_{\mathbb{Z}}}$ of $B_{\mathcal{F}_{\mathbb{Z}}}$ enables us to compute this projection. In fact, (9.16) holds in the norm of $H^{-1}(\mathbb{R})$ (see [**643**]). Moreover, the series (9.16) converges to $P(F^{\wedge}_{\mathcal{F}_{\mathbb{Z}}})^*(t)$ by the continuity of the orthogonal projector and to $F^{\wedge}_{\mathcal{F}_{\mathbb{Z}}}(t)$ by the sampling theorem. Thus, we have $P(F^{\wedge}_{\mathcal{F}_{\mathbb{Z}}})^* = F^{\wedge}_{\mathcal{F}_{\mathbb{Z}}}$.

The above ideas provide not only the engineering arguments, but they also suggest important extensions to other RKHS (cf. [**643, 644**]).

Summarizing our considerations we obtain the following result: For purposes of Paley-Wiener space based sampling extensions we are able to interpret our procedure entirely within the space $H^{-1}(\mathbb{R})$. For $F^{\wedge}_{\mathcal{F}_{\mathbb{Z}}}$ a bandlimited function, i.e., an element of the Hilbert space $B_{\mathcal{F}_{\mathbb{Z}}}$, $S : F^{\wedge}_{\mathcal{F}_{\mathbb{Z}}} \mapsto SF^{\wedge}_{\mathcal{F}_{\mathbb{Z}}} = (F^{\wedge}_{\mathcal{F}_{\mathbb{Z}}})^*$ may be considered as a map from $H^{-1}(\mathbb{R})$ into $H^{-1}(\mathbb{R})$. The map $P : (F^{\wedge}_{\mathcal{F}_{\mathbb{Z}}})^* \mapsto F^{\wedge}_{\mathcal{F}_{\mathbb{Z}}}$ is an orthogonal projector of $H^{-1}(\mathbb{R})$ onto the closed subspace $B_{\mathcal{F}_{\mathbb{Z}}}$ of $H^{-1}(\mathbb{R})$. This leads to the following proposition:

Let

$$(F^{\wedge}_{\mathcal{F}_{\mathbb{Z}}})^*(t) = \sum_{n\in\mathbb{Z}} F^{\wedge}_{\mathcal{F}_{\mathbb{Z}}}(n)\ \delta(t-n) \tag{9.21}$$

be the sampled impulse train obtained by sampling a bandlimited function $F^{\wedge}_{\mathcal{F}_{\mathbb{Z}}}$. Then the partial sums of the series (9.21) *belong to $H^{-1}(\mathbb{R})$ and converge in the sense of $H^{-1}(\mathbb{R})$ to $(F^{\wedge}_{\mathcal{F}_{\mathbb{Z}}})^*$.*

Paley-Wiener RKHS properties. In Shannon sampling (cf. [**675**]) it is observed that the space of all square-integrable functions bandlimited to the fundamental cell of $\mathcal{F}_\Lambda$ of a lattice $\Lambda = \sigma\mathbb{Z}$ is a reproducing kernel Hilbert space on the real line, i.e., the Paley-Wiener space $B_{\mathcal{F}_\Lambda}$ with the reproducing kernel

$$K_{\mathcal{F}_\Lambda}(h-t) = \frac{1}{\|\mathcal{F}_\Lambda\|}\int_{\mathcal{F}_\Lambda} e^{2\pi i x\cdot(h-t)}\,dx = \frac{\sin(\pi\sigma(h-t))}{\pi\sigma(h-t)} = \mathrm{sinc}(\pi\sigma(h-t)),\ t\in\mathbb{R}, \tag{9.22}$$

although this fact is not used in the classical proof of the Shannon sampling theorem.

The Paley-Wiener space $B_{\mathcal{F}_\Lambda}$ has many interesting properties collected, for example, in [**629**]:

- The sequence

$$\{K_{\mathcal{F}_\Lambda}(\cdot - h)\}_{h\in\Lambda^{-1}} \tag{9.23}$$

  is an orthonormal basis for $B_{\mathcal{F}_\Lambda}$, i.e.,

$$\int_{\mathcal{F}_\Lambda} K_{\mathcal{F}_\Lambda}(t-h)\ \overline{K_{\mathcal{F}_\Lambda}(t-h')}\ dt = \delta_{h',h} = \begin{cases} 1, & h'=h, \\ 0, & h'\neq h. \end{cases} \tag{9.24}$$

- The sequence (9.23) shows the discrete orthogonality property:

$$K_{\mathcal{F}_\Lambda}(h'-h) = \delta_{h',h}. \tag{9.25}$$

- $F^{\wedge}_{\mathcal{F}_\Lambda}(\cdot - c) \in B_{\mathcal{F}_\Lambda}$ and $\|F^{\wedge}_{\mathcal{F}_\Lambda}(\cdot - c)\|_{L^2(\mathbb{R})} = \|F^{\wedge}_{\mathcal{F}_\Lambda}\|_{L^2(\mathbb{R})}$ for all $F^{\wedge}_{\mathcal{F}_\Lambda} \in B_{\mathcal{F}_\Lambda}$ and $c \in \mathbb{R}$, such that $B_{\mathcal{F}_\Lambda}$ is a *unitarily translation-invariant subspace of* $L^2(\mathbb{R})$.
- $B_{\mathcal{F}_\Lambda}$ is a shift-invariant subspace of $L^2(\mathbb{R})$

$$B_{\mathcal{F}_\Lambda} = \left\{ \sum_{h\in\Lambda} c(h)\; K_{\mathcal{F}_\Lambda}(\cdot - h) : \sum_{h\in\Lambda} |c(h)|^2 < \infty \right\}. \tag{9.26}$$

The *interpolatory background of bandlimited sampling* can be recapitulated briefly as follows: A mathematical way to interpolate samples $F^{\wedge}_{\mathcal{F}_\Lambda}(h'), h' \in \Lambda^{-1}$, $\Lambda^{-1} = \frac{1}{\sigma}\mathbb{Z}$, may be realized by use of sinc-functions (cf. Figure 1). The sequence (9.23) shows the discrete orthogonality property: $\operatorname{sinc}(\pi\sigma(h - h')) = \delta_{h,h'}$. As a consequence, $F^{\wedge}_{\mathcal{F}_\Lambda}$ given by the so-called *cardinal series*

$$F^{\wedge}_{\mathcal{F}_\Lambda}(t) = \sum_{h\in\Lambda^{-1}} F^{\wedge}_{\mathcal{F}_\Lambda}(h)\; \operatorname{sinc}(\pi\sigma\,(h - t)), \quad t \in \mathbb{R}, \tag{9.27}$$

satisfies the *interpolatory property*

$$F^{\wedge}_{\mathcal{F}_\Lambda}(t)|_{t=h'} = \sum_{h\in\Lambda^{-1}} F^{\wedge}_{\mathcal{F}_\Lambda}(h)\; \operatorname{sinc}(\pi\sigma\,(h - t))|_{t=h'} = F^{\wedge}_{\mathcal{F}_\Lambda}(h'), \quad h' \in \Lambda^{-1}. \tag{9.28}$$

9.2. Reproducing kernel induced sampling concepts

Next we come to ideas and concepts involving the essential use of functional analytic means: First some extensions of the classical Shannon sampling theorem are shown to have a setting in an associated RKHS. This was partially recognized by K. Yao [**872**], who already showed that there is a relation between the sampling theorem and the reproducing kernel in some special cases. Of course, within this context, the prototype example is the Shannon sampling theorem. We do not repeat Yao's work, but present some other examples from a slightly different point of view. M.Z. Nashed, G.G. Walter [**643**] have shown that general versions of sampling theorems hold true for certain RKHSs over "continuously connected pointsets". This approach also includes sampling identities related to other transforms than the Fourier transform (as realized in the classical Shannon theory). For example, Legendre, Sturm-Liouville, Jacobi, and Laguerre transforms are among the examples which may be investigated, as well as sampling using frames. Error analyses, such as truncation error, aliasing, jitter error, and amplitude error, are of significant interest. Second, in more recent papers, several authors (see, e.g., [**21**], [**23**], [**396**], [**636,637**]) approach RKHS sampling from the different directions and ask whether an RKHS exists for each sampling theorem. It turns out that the answer to this question is in the affirmative (under mild conditions to be satisfied for a sampling sequence). Indeed we are led to the remarkable statement that every sampling expansion induces an RKHS. In canonical continuation of the study of RKHSs over "continuously connected pointsets" (such as in 1D intervals), P. Jorgensen and his colleagues (cf. [**471–475,477–480**]) propose to go over to sampling settings in RKHSs over discretely given point sets. Central means of these concepts are point masses of finite norm. Moreover, sampling algorithms are in the focus, which go beyond interpolation.

Paley-Wiener space induced sampling. As we already saw (see, e.g., [**872**]), the Paley-Wiener Space $B_{\mathcal{F}_{\mathbb{Z}}}$ has a number of remarkable properties: It is a reproducing kernel Hilbert space with reproducing kernel known from (9.22), such that

$$F(t) = \int_{\mathbb{R}} K_{\mathcal{F}_{\mathbb{Z}}}(t-y)\ F(y)\ dy,, \quad t \in \mathbb{R},\ F \in B_{\mathcal{F}_{\mathbb{Z}}}. \tag{9.29}$$

$K_{\mathcal{F}_{\mathbb{Z}}}(t-y)$ has zeros at $t = h, y = h'$, where $h, h' \in \mathbb{Z}, h \neq h'$; and the functions

$$S_h = K_{\mathcal{F}_{\mathbb{Z}}}(\cdot - h), h \in \mathbb{Z}\ (= \mathbb{Z}^{-1}), \tag{9.30}$$

are the sampling constituents. In addition to the discrete orthogonality property $S_h(h') = \delta_{h,h'}$, these functions are also orthonormal in $L^2(\mathbb{R})$, i.e.,

$$\langle S_h, S_{h'} \rangle_{L^2(\mathbb{R})} = \int_{\mathbb{R}} S_h(t)\ S_{h'}(t)\ dt = \delta_{h,h'}. \tag{9.31}$$

In fact, the sequence $\{S_h\}_{h\in\mathbb{Z}}$ constitutes an orthonormal basis in $B_{\mathcal{F}_{\mathbb{Z}}}$. Thus it follows that

$$F = \sum_{h\in\mathbb{Z}} \langle F, S_h \rangle_{L^2(\mathbb{R})}\ S_h \tag{9.32}$$

for each $F \in B_{\mathcal{F}_{\mathbb{Z}}}$, with convergence in the sense of $B_{\mathcal{F}_{\mathbb{Z}}}$. Moreover, we have

$$\langle F, S_h \rangle_{L^2(\mathbb{R})} = \int_{\mathbb{R}} K_{\mathcal{F}_{\mathbb{Z}}}(t-h)\ F(t)\ dt = F(h), \tag{9.33}$$

and, hence, (9.32) is mainly a restatement of the sampling theorem. The reproducing kernel may be recovered from the functions S_h, since its expansion may be written in the form

$$K_{\mathcal{F}_{\mathbb{Z}}}(t-y) = \sum_{h\in\mathbb{Z}} S_h(y) S_h(t) = \sum_{h\in\mathbb{Z}} K_{\mathcal{F}_{\mathbb{Z}}}(t-h) K_{\mathcal{F}_{\mathbb{Z}}}(h-y), \quad t, y \in \mathbb{R}. \tag{9.34}$$

Nonequidistant sampling. The sampling sequence $\{S_h\}_{h\in\mathbb{Z}}$ as introduced by (9.30) shows the dual orthogonality property. However, it is known that sampling sequences need not to be orthogonal. *Nonequidistant Paley-Wiener sampling* (see [**675**]) is an example: Assume that $\{t_h\}_{h\in\mathbb{Z}}$ is a sequence of real numbers such that $t_{-h} = t_h$ with $t_0 = 0$ and $|t_h - h| \leq \frac{1}{4} - \varepsilon, h \in \mathbb{Z}$, for some $\varepsilon > 0$. We set

$$G(t) = t \prod_{\substack{h\neq 0\\ h\in\mathbb{Z}}} \left(1 - \frac{t}{t_h}\right). \tag{9.35}$$

G is an entire function of exponential type at most π (cf. [**643**]). Each $F \in B_{\mathcal{F}_{\mathbb{Z}}}$ satisfies

$$F(t) = \sum_{h\in\mathbb{Z}} F(t_h)\ \frac{G(t)}{G'(t_h)(t-h)} = \sum_{h\in\mathbb{Z}} F(t_h)\ S_G(t, t_h), \tag{9.36}$$

where we have used the abbreviation

$$S_G(t, t_h) = \frac{G(t)}{G'(t_h)(t-h)}. \tag{9.37}$$

Each function $S_G(\cdot, t_h)$ is a member of the space $B_{\mathcal{F}_\mathbb{Z}}$, but they are not necessarily orthogonal. Indeed, if $t_h = h$, then we obtain exactly S_h known from (9.30). However, we are able to verify the *biorthogonality* in the sense

$$\int_{\mathbb{R}} S_G(t, t_h)\ K_{\mathcal{F}_\mathbb{Z}}(t_{h'} - t)\ dt = S_G(t_{h'}, t_h) = \delta_{h',h}, \tag{9.38}$$

since $K_{\mathcal{F}_\mathbb{Z}}(\cdot - \cdot)$ is known to be the reproducing kernel of $B_{\mathcal{F}_\mathbb{Z}}$.

RKHS extensions of Shannon sampling. Let H be a $\mathbb{K}$-separable RKHS which is assumed to be a subspace of $L^2(\mathbb{R})$ closed in the topology of $H^{-1}(\mathbb{R})$. Let $K(\cdot, \cdot)$ be the reproducing kernel of H. Such kernels are known to be hermitian. Moreover, they may be expressed in the form

$$K(t, u) = \sum_{n \in \mathbb{N}} \overline{\Phi_n(t)}\ \Phi_n(u) \tag{9.39}$$

for some complete orthonormal system $\{\Phi_n\}_{n\in\mathbb{N}}$ (see, e.g., E. Hille [**434**], H.S. Shapiro [**766**]). However, the system $\{\Phi_n\}_{n\in\mathbb{N}}$ does not necessarily correspond to the sampling functions (as in the case of classical Shannon sampling). On the other side, this is not precluded and the Φ_n's may be of the form $\Phi_n(t) = c_n\ K(t_n, t)$ for some sequence $\{t_n\}_{n\in\mathbb{N}} \subset \mathbb{R}$ in which case we have

$$F(t) = \sum_{n \in \mathbb{N}} \int_{\mathbb{R}} F(s) K(t_n, s)\ ds\ \ c_n \Phi_n(t) = \sum_{n \in \mathbb{N}} c_n\ F(t_n)\ \Phi_n(t), \quad F \in H. \tag{9.40}$$

This expansion already is a *type of a sampling theorem.* The coefficients c_n may be calculated from the orthonormality of $\{\Phi_n\}_{n\in\mathbb{N}}$:

$$\delta_{n,m} = \int_{\mathbb{R}} c_n K(t_n, t)\ \overline{c_m K(t_m, t)}\ dt = c_n\ \overline{c_m}\ K(t_n, t_m). \tag{9.41}$$

Proposition. Let $K(\cdot, \cdot)$ be the reproducing kernel of the reproducing kernel Hilbert space H associated to (9.39)*. If the complete orthonormal system $\{\Phi_n\}_{n\in\mathbb{N}}$ satisfies $\Phi_n(t) = c_n\ K(t_n, t)$ for some sequences $\{c_n\}_{n\in\mathbb{N}}$ and $\{t_n\}_{n\in\mathbb{N}}$, then*

(i) $K(t_n, t_m) = 0,\ n \neq m, \quad n, m = 1, 2, \ldots,$
(ii) $|c_n|^{-2} = K(t_n, t_n),\ n \in \mathbb{N},$
(iii) *the sampling series* (9.40) *converges uniformly provided that there exists a positive constant C such that $K(t, t) \leq C < \infty$.*

In fact, the first two conditions (i), (ii) are obvious from (9.41): For $F \in H$, the conclusion (iii) follows from the Cauchy-Schwarz inequality

$$\begin{aligned} \left| F(t) - \sum_{\substack{|n| \leq N \\ n \in \mathbb{N}}} c_n F(t_n)\ \Phi_n(t) \right| &= \left| \int_{\mathbb{R}} K(t, u) \left(F(u) - \sum_{\substack{|n| \leq N \\ n \in \mathbb{N}}} c_n F(t_n)\ \Phi_n(u) \right) du \right| \\ &\leq \{K(t, t)\}^{1/2} \left\| F - \sum_{\substack{|n| \leq N \\ n \in \mathbb{N}}} c_n F(t_n)\ \Phi_n \right\|_H \end{aligned} \tag{9.42}$$

such that

$$\left| F(t) - \sum_{\substack{|n|\le N \\ n\in\mathbb{N}}} c_n F(t_n)\ \Phi_n(t) \right| = o(1),\ N \to \infty. \tag{9.43}$$

Converse proposition. A converse also holds true under the condition that the point sequence $\{t_n\}_{n\in\mathbb{N}}$ is a *zero sequence* in H, i.e., $K(t_n, t_m) = 0$, $n \neq m$ while $K(t_n, t_n) > 0$. Indeed, if $\{t_n\}_{n\in\mathbb{N}}$ is a zero sequence, then the functions

$$\Phi_n(t) = \frac{K(t_n, t)}{(K(t_n, t_n))^{1/2}}, \qquad n \in \mathbb{N}, \tag{9.44}$$

constitute an orthonormal sequence. If, in addition, $\{t_n\}_{n\in\mathbb{N}}$ is a *set of uniqueness* for H, i.e., $F(t_n) = 0$, $n \in \mathbb{N}$, implies $F = 0$, then the sequence $\{\Phi_n\}_{n\in\mathbb{N}}$ is complete in H (note that this explains the fact that any function F orthogonal to all the functions Φ_n satisfies

$$F(t_n) = \int_{\mathbb{R}} K(t_n, u)\ F(u)\ du = 0, \quad n \in \mathbb{N}, \tag{9.45}$$

and, hence, $F = 0$).

Modified proposition. If the sequence $\{\Phi_n\}_{n\in\mathbb{N}}$ in (9.40) is not complete in H, then we can form the subspace $H^\diamond$ consisting of its closed linear span. This space is again an RKHS with RK given by

$$K^\diamond(t, u) = \sum_{n\in\mathbb{N}} \frac{K(t, t_n) K(t_n, u)}{K(t_n, t_n)}. \tag{9.46}$$

It is clear that $K^\diamond(\cdot, \cdot)$ must be the (unique) reproducing kernel. Since $K^\diamond(t_n, t) = K(t_n, t), n \in \mathbb{N}$, the functions Φ_n have the form given. Since it is complete in $H^\diamond$, we obtain the modified proposition: Let $K(\cdot, \cdot)$ have a zero sequence $\{t_n\}_{n\in\mathbb{N}}$ of real numbers, i.e., $K(t_n, t_m) = 0, n \neq m$, $K(t_n, t_n) > 0$. Then there is an RKHS $H^\diamond$ with kernel given by (9.46) such that $\{\Phi_n\}_{n\in\mathbb{N}}$ given by $\Phi_n = c_n\ K^\diamond(t_n, t), n \in \mathbb{N}$, admits a sampling series (9.40) for all $F \in H^\diamond$.

Finally, we are led to some remarks:

- The assumption that $K(\cdot, \cdot)$ is a reproducing kernel of an RKHS is satisfied by all positive definite functions (see, e.g., N. Aronszajn [**47**]). Thus we need merely assume that the kernel is positive definite.
- Because of the continuity of $K(\cdot, \cdot)$, a zero sequence $\{t_n\}_{n\in\mathbb{N}}$ does not have a limit point. If it did, then this limit point would be a term in the sequence, say t_m, and a limit of a subsequence, say $s_j \to t_m$. However, we have

$$\int_{\mathbb{R}} |K(t, t_m)|^2\ dt = \lim_{s_j \to t_m} \int_{\mathbb{R}} K(t, s_j) K(t_m, t)\ dt = 0, \tag{9.47}$$

which contradicts the assumption that $K(t_m, t_m) > 0$.

More general sampling theorem. We formulate (cf. M.Z. Nashed, G.G. Walter [**643**]) a *sampling theorem* in accordance with Sobolev space oriented settings.

Let H be an RKHS which is a subspace of $L^2(\mathbb{R})$ closed in $H_{-1}(\mathbb{R})$ and closed under differentiation. Suppose that F is of class H. Let the continuous kernel $K(\cdot,\cdot)$ have the zero sequence $\{t_n\}_{n\in\mathbb{N}}$, which is additionally assumed to be a set of uniqueness for H. Let F satisfy the condition

$$\frac{F(t)}{K(t,t)} = O(|t|^{-2}), \tag{9.48}$$

and assume that $\{t_n\}_{n\in\mathbb{N}}$ is asymptotic to $\{n\}_{n\in\mathbb{N}}$ as $n\to\infty$ (this is a sufficient growth condition on t_n). Then the series

$$t \mapsto \sum_{n\in\mathbb{N}} F(t_n)\frac{\delta(t-t_n)}{K(t_n,t_n)},\ t\in\mathbb{R}, \tag{9.49}$$

converges in the sense of $H^{-1}(\mathbb{R})$, the orthogonal projection of (9.49) *onto H equals F, and*

$$F(t) = \sum_{n\in\mathbb{N}} F(t_n)\frac{K(t_n,t)}{K(t_n,t_n)} \tag{9.50}$$

uniformly on sets for which $K(t,t)$ is bounded.

M.Z. Nashed, G.G. Walter [**643**] remark that the hypotheses about the growth of $F(t)$ relative to $K(t,t)$ and the growth condition on $\{t_n\}_{n\in\mathbb{N}}$ can be replaced by a more general single condition on the growth of the sampled values. One possibility is to study a series given by

$$\sum_{n\in\mathbb{N}} \frac{F(t_n)}{K(t_n,t_n)} e^{2\pi i t_n\cdot\omega}. \tag{9.51}$$

Clearly, the sum (9.51) converges in the sense of $L^2([-A,A])$ for any $A>0$, since it converges uniformly for $\omega\in\mathbb{R}$ to a bounded function W. Hence,

$$\int_{\mathbb{R}} \underbrace{\left|\sum_{n\in\mathbb{N}} \frac{F(t_n)}{K(t_n,t_n)} e^{2\pi i t_n\cdot\omega} - W(\omega)\right|^2 \frac{1}{\omega^2+1}}_{=U(\omega)}\, d\omega = \left(\int_{-\infty}^{A} + \int_{-A}^{A} + \int_{A}^{\infty}\right) U(\omega)\, d\omega. \tag{9.52}$$

The middle integral on the right side of (9.52) can be made small because of the $L^2([-A,A])$-convergence, while the other two integrals are small for A large enough, e.g.,

$$\int_A^\infty U(\omega)\, d\omega \ \le\ \frac{C}{(A^2+1)^{1/4}}, \tag{9.53}$$

where C is a suitable positive constant. Summarizing our considerations we obtain the following proposition:

Let $F\in H$ satisfy the condition

$$\sum_{n\in\mathbb{N}} \left|\frac{F(t_n)}{K(t_n,t_n)}\right| < \infty. \tag{9.54}$$

Then (9.49) converges in the sense of $H^{-1}(\mathbb{R})$, and the conclusion (9.50) follows immediately.

Error analysis. We assume a *weak separation property of the zeros* $\{t_n\}_{n\in\mathbb{N}}$ of the RK $K(\cdot,\cdot)$ in H: There exists a number $\rho > 0$ such that $t_{n+1} - t_n \geq \rho\,|t_n|^r > 0$ for some $r \in (-1,1)$, where we have assumed that the points $\{t_n\}_{n\in\mathbb{N}}$ are arranged in increasing order. We already know that the sequence $\{t_n\}_{n\in\mathbb{N}}$ has no limit point, but they could diverge to ∞ arbitrarily slowly. In all of our already discussed examples this property is satisfied with $r = 0$ (note that the results mentioned here on error analysis are in the line with the corresponding results for the Paley-Wiener space (see, e.g., P.L. Butzer and his group [**123, 126, 128, 129**])).

As in the Shannon sampling theorem, there are several types of errors (see [**643**]): (1) truncation error, (2) aliasing error, (3) amplitude error, (4) jitter (or location) error, most of which are self-explanatory.

Truncation error. An expression for the truncation error is easily obtained. Indeed, for $F \in H$,

$$(9.55)\qquad e_T(N) = \left\| F - \sum_{\substack{n\leq N\\ n\in\mathbb{N}}} F(t_n)\frac{K(t_n,\cdot)}{K(t_n,t_n)}\right\|_H^2 = \sum_{\substack{n>N\\ n\in\mathbb{N}}} \frac{|F(t_n)|^2}{K(t_n,t_n)},$$

which must converge to zero by the orthogonality and completeness of $K_H(t_n,\cdot)$. However, this convergence may be very slow; it can be improved somewhat by considering $G(t) = (K(t,t))^{-1/2}F(t)$ instead of F. Then both $t \mapsto t^pG(t), t \in \mathbb{R}$, and $t \mapsto (t^p\ G(t))', t \in \mathbb{R}$, belong to $L^2(\mathbb{R})$ by hypothesis, and hence, their Fourier transforms do as well. That means that both $s \mapsto (G^{(p)})^\wedge(s)$ and $s \mapsto s(G^{(p)})^\wedge(s)$ belong to $L^2(\mathbb{R})$. Moreover, we have

$$(9.56)\quad \int_{\mathbb{R}} |(G^{(p)})^\wedge(s)|\, ds \leq \left\{\int_{\mathbb{R}} |(G^{(p)})^\wedge(s)|^2(s^2+1)\, ds\right\}^{1/2} \left\{\int_{\mathbb{R}} \frac{1}{s^2+1}\, ds\right\}^{1/2}$$

and so $(G^{(p)})^\wedge$ is of class $L^1(\mathbb{R})$ as well. Hence, by the Riemann-Lebesgue Lemma of Fourier theory (see (9.55)), $t^pG(t) \to 0$ as $|t| \to \infty$. This yields to the following result: Let the function $t \mapsto [K(t,t)]^{-1/2}\ t^pF(t), t \in \mathbb{R}$, constitute a member of class $H^1(\mathbb{R})$ for some integer $p > 0$. Then, as $|t| \to \infty$, $t^p[K(t,t)]^{-1/2}F(t) \to 0$. Hence, the error is dominated by $\sum_{\substack{n>N\\ n\in\mathbb{N}}} |t_n|^{-2p}$, and from $t_{n+1} - t_n \geq \rho\,|t_n|^r > 0$ it follows that $|t_n| \geq C|n|^{1/(1-r)}$, where C is a constant.

The truncation error (9.55) *for F satisfying the aforementioned hypothesis is for $p > (1-r)/2$,*

$$(9.57)\qquad e_T(N) = O\left(N^{-\frac{2p}{1-r}+1}\right).$$

Aliasing error. If $F \notin H$ but belongs to the Sobolev space $H^1(\mathbb{R})$, it is still possible to form the sampling series

$$(9.58)\qquad \sum_{n\in\mathbb{N}} F(t_n)\frac{K(t_n,\cdot)}{K(t_n,t_n)}.$$

This series converges to some element Q in H by Bessel's inequality. The aliasing error is the difference between F and Q, whose norm may be expressed as follows:

$$e_A = \|Q-F\|_H^2 = \|F\|_H^2 - \sum_{n\in\mathbb{N}} |\langle F,\Phi_n\rangle_H|^2 + \sum_{n\in\mathbb{N}} |\langle F,\Phi_n\rangle_H|^2 - \sum_{n\in\mathbb{N}} \frac{(F(t_n))^2}{K(t_n,t_n)}$$
$$= \operatorname{dist}^2(F,H) + \sum_{n\in\mathbb{N}} \frac{|Q(t_n)-F(t_n)|^2}{K(t_n,t_n)}.$$

Thus, the aliasing error shows two components, namely one arising from the distance $\operatorname{dist}(F,H)$ from F to the space H and the other from an amplitude error. For certain F somewhat better results are possible (cf. [**643**]).

Amplitude error. We assume that the amplitude error takes the form of a chopping error, i.e., the error is given by

$$\varepsilon_n = F(t_n) - F^*(t_n), \tag{9.59}$$

where $F^*(t_n)$ is the approximate value of $F(t_n)$ such that $|F^*(t_n)| \le |F(t_n)|$. We further assume that the error is uniformly bounded, i.e., $\varepsilon_n \le \varepsilon$, $n\in\mathbb{N}$. Then the total amplitude error is given by

$$\begin{aligned} e_A &= \left\| \sum_{n\in\mathbb{N}} (F(t_n)-F^*(t_n)) \frac{K(t_n,\cdot)}{K(t_n,t_n)} \right\|_H^2 \\ &= \sum_{n\in\mathbb{N}} \frac{\varepsilon_n^2}{K(t_n,t_n)} \\ &\le \sum_{\substack{n\le[\varepsilon^{-m}]\\ n\in\mathbb{N}}} \frac{\varepsilon_n^2}{K(t_n,t_n)} + \sum_{\substack{n>[\varepsilon^{-m}]\\ n\in\mathbb{N}}} \frac{\varepsilon_n^2}{K(t_n,t_n)}, \end{aligned} \tag{9.60}$$

where m is any positive number and $[x]$ is, as always, the "integer part" of x. By substituting the bound $\varepsilon_n \le \varepsilon$, $n\in\mathbb{N}$ into the first expression in (9.60) and using the fact that $\varepsilon_n^2 \le 2(F(t_n))^2$ we find that the error is bounded by

$$e_A \le \varepsilon^{-m+2} \sup_{n\in\mathbb{N}} \frac{1}{K(t_n,t_n)} + 2 \sum_{\substack{n>[\varepsilon^{-m}]\\ n\in\mathbb{N}}} \frac{(F(t_n))^2}{K(t_n,t_n)}. \tag{9.61}$$

By virtue of the bound (9.57) e_R satisfies the relation

$$e_A \le \frac{1}{e}\varepsilon^{-m+2} + O\left([\varepsilon^{-m}]^{1+\frac{2p}{r-1}}\right) \tag{9.62}$$

which is minimized by taking $-m+2 = -m(1+2p/(r-1))$ or $m = (1-r)/p$.

Let F satisfy the above hypothesis. Suppose that $K(t,t) \ge C > 0$. Assume that the chopping error satisfies $\varepsilon_n \le \varepsilon$, $n\in\mathbb{N}$. Then the total amplitude error is of the order

$$e_A = O(\varepsilon^{1+\frac{r-1}{p}}). \tag{9.63}$$

Jitter error. The location error or jitter error arises, when the location of the sampling points t_n are not exactly known. If t_n is replaced by $t_n + \delta_n$, then this error is given by

$$e_J = \left\| \sum_{n\in\mathbb{N}} (F(t_n) - F(t_n + \delta_n)) \frac{K(t_n, \cdot)}{K(t_n, t_n)} \right\|_H , \tag{9.64}$$

where $|\delta_n| \leq \delta$, and δ is a small positive number. In doing so we assume that $K(\cdot,\cdot)$ is a kernel of convolution type $K(t,u) = Z(t-u)$, such that $Z(0) \neq 0$. We use the fact that, since $F \in H^1(\mathbb{R})$,

$$F'(t) = \int_{\mathbb{R}} \frac{\partial K(t,u)}{\partial t} F(u)\, du = \int_{\mathbb{R}} K(t,u)\; F'(u)\, du \tag{9.65}$$

and, therefore, $F' \in H$. Hence, there is a t_n, τ_n exist with $t_n \leq \tau_n \leq t_n + \delta_n$,

$$F(t_n) - F(t_n + \delta_n) = F'(\tau_n)\, \delta_n = \delta_n \int_{\mathbb{R}} K(\tau_n, u)\; F'(u)\, du, \tag{9.66}$$

and the jitter error becomes

$$\begin{aligned} e_J = \sum_{n\in\mathbb{N}} \delta_n^2 \frac{[\int_{\mathbb{R}} K(\tau_n,u) F'(u)\, du]^2}{Z(0)} &\leq \sum_{\substack{n\leq[\delta^{-m}]\\ n\in\mathbb{N}}} \delta_n^2 \|F'\|^2_{L^2(\mathbb{R})} + \sum_{\substack{n>[\delta^{-m}]\\ n\in\mathbb{N}}} \{\ldots\} \\ &\leq \delta^{-m+2} \|F'\|^2_{L^2(\mathbb{R})} + O\left(\sum [\delta^{-m}]^{1+\frac{2p}{r-1}}\right) \end{aligned}$$

(note that, for $|n|$ sufficiently large, $|t_n + \delta_n|^{-p} \leq |t_n - \delta|^{-p}$).

Let $K(t,u)$ be of the convolution form $K(t,u) = Z(t-u)$ such that $Z(0) \neq 0$. Assume that F satisfies the above hypothesis. Then

$$e_J = O(\delta^{1+\frac{r-1}{p}}). \tag{9.67}$$

Nonorthogonal sampling. Let $K(\cdot,\cdot)$ be the reproducing kernel of H. Assume that $\{t_n\}_{n\in\mathbb{N}}$ is some sequence of points.

Finite-dimensional subspaces. We denote by $H_{1,\ldots,n}$ the subspace spanned by $K(t_1,\cdot),\ldots,K(t_n,\cdot)$. It is of course a reproducing kernel Hilbert space. We additionally require that $t_1,\ldots,t_n$ be chosen such that the matrix $K = [K(t_i,t_j)]_{i,j=1,\ldots,n}$ is nonsingular.

Let K given by (9.2) be nonsingular. Then the reproducing kernel $K_{H_{1,\ldots,n}}(t,u)$ of the space $H_{1,\ldots,n}$ is given by

$$K_{H_{1,\ldots,n}}(t,u) = [K(t_1,t)\ldots K(k_n,t)]\;\; K^{-1} \begin{bmatrix} K(t_1,u) \\ \vdots \\ K(t_n,u) \end{bmatrix} \tag{9.68}$$

and the biorthogonal sequence is given by $S_i, i = 1,2,\ldots,n$, where

$$\begin{bmatrix} S_1(t) \\ \vdots \\ S_n(t) \end{bmatrix} = K^{-1} \begin{bmatrix} K(t_1,t) \\ \vdots \\ K(t_n,t) \end{bmatrix} \tag{9.69}$$

An orthogonal basis of $H_{1,\dots,n}$ is given by

$$\begin{bmatrix} \Phi_1(t) \\ \vdots \\ \Phi_n(t) \end{bmatrix} = K^{-1/2} \begin{bmatrix} K(t_1,t) \\ \vdots \\ K(t_n,t) \end{bmatrix}. \tag{9.70}$$

The proof of each of the assertions is straightforward. We only remark that for (9.70) we need the positive definiteness of K, which follows from the positive definiteness of $K_{H_{1,\dots,n}}(\cdot,\cdot)$. In fact, $[S_i(t_j)]_{i,j=1,\dots,n} = K^{-1}K = I$, so that we are led to the result:

If $F \in H_{1,\dots,n}$, then the functions S_j given by (9.69) *satisfy the sampling property*

$$F = \sum_{\substack{j\le n \\ j\in\mathbb{N}}} F(t_j)\ S_j. \tag{9.71}$$

It should be remarked that the orthogonal complement of $H_{1,\dots,n}$ in H is also an RKHS (see., e.g., in [**766**]).

Infinite-dimensional subspaces spanned by frames. If we wish to extend the above results to an infinite sequence of points $\{t_n\}_{n\in\mathbb{N}}$ at which to sample $F \in H$, we need to worry about the convergence. An appropriate theory to use here is the theory of frames (see, e.g., R.M. Young [**874**]): We denote the closure in H of the linear span of $\{K(t_n,\cdot)\}_{n\in\mathbb{N}}$ by $H_{1,\dots,\infty}$, and require $\{K(t_n,\cdot)\}_{n\in\mathbb{N}}$ to constitude a frame in $H_{1,\dots,\infty}$, i.e., that there be constants C', C'' such that

$$C'\ \|F\|_H^2 \le \sum_{n\in\mathbb{N}} \left| \int_{\mathbb{R}} K(t,t_n)\ F(t)\ dt \right|^2 \le C''\ \|F\|_H^2 \tag{9.72}$$

for all $F \in H_{1,\dots,\infty}$. Obviously, for the special choice $F = K(t_i,\cdot), i \in \mathbb{N}$, we obtain

$$C\ K(t_i,t_i) \le \sum_{n\in\mathbb{N}} |K(t_n,t_i)|^2 \le C'\ K(t_i,t_i), \tag{9.73}$$

which gives a necessary condition for the inequality (9.72).

We form the (infinite) matrix $K = [K(t_i,t_j)]_{i,j\in\mathbb{N}}$, each of the rows or columns of (9.2) are square-summable, i.e., the rows or columns belong to $\ell^2(\mathbb{N})$ by (9.73). The matrix K is formally self-adjoint, i.e., $K^* = K$, and maps $\ell^2(\mathbb{N})$ into a space of sequences dominated by $K(t_n,t_n)$. We suppose further (cf. [**643**]) that

$$\| \sum_{n=m}^{p} a_n K(t_n,\cdot)\|_H \to 0, \quad p,m \to \infty, \tag{9.74}$$

provided that $\{a_n\}_{n\in\mathbb{N}}$ is a square-summable sequence, i.e., $\{a_n\}_{n\in\mathbb{N}} \in \ell^2(\mathbb{N})$. This assumption ensures that the matrix K maps the space of square-summable sequences into itself, i.e., $\ell^2(\mathbb{N})$ into $\ell^2(\mathbb{N})$. Hence, for $\{a_n\}_{n\in\mathbb{N}} \in \ell^2(\mathbb{N})$, the series $\sum_{n\in\mathbb{N}} a_n K(t_n,\cdot)$ converges to an element $F \in H_{1,\dots,\infty}$ with

$$\overline{F(t_n)} = \int_{\mathbb{R}} \overline{K(t_n,u)}\ F(u)\ du = \sum_{i=1}^{\infty} a_i K(t_i,t_n). \tag{9.75}$$

From (9.72) it follows that $\overline{F(t_n)}$ is square-summable and, hence, the range of the operator K given by the infinite matrix K is contained in $\ell^2(\mathbb{N})$. Moreover, again by (9.72), $F(t_n) = 0$ for all n implies $\|F\|_H = 0$, so that K is one-to-one.

Let K be the operator induced by the infinite matrix $[K(t_i, t_j)]_{i,j \in \mathbb{R}}$ and let $K(t_n, \cdot)$ satisfy (9.74). Then K constitudes a self-adjoint operator from $\ell^2(\mathbb{N})$ into $\ell^2(\mathbb{N})$ with a positive spectrum contained in the interval $[C', C'']$, where C' and C'' are given as in (9.72).

With these properties of K we can imitate the structure of the finite-dimensional case. The matrix K as an operator on $\ell^2(\mathbb{N})$ has an inverse K^{-1} which is also bounded. We denote by $k(t)$, $s(t)$, and $\phi(t)$ elements of $\ell^2(\mathbb{N})$ given as follows:

$$
\begin{aligned}
k(t) &= (K(t_1, t), \ldots, K(t_n, t), \ldots), \\
s(t) &= (S_1(t), \ldots, S_n(t), \ldots), \\
\phi(t) &= (\Phi_1(t), \ldots, \Phi_n(t), \ldots).
\end{aligned}
\tag{9.76}
$$

In this notation we obtain the following result (in close analogy to the finite-dimensional case):

Let $\{t_n\}_{n \in \mathbb{N}}$ be a sequence such that $K(t_n, t)$ satisfies (9.72) and (9.74) with $K(t,t) \leq C$. Then the reproducing kernel $K_{H_{1,\ldots,\infty}}(t,u)$ of $H_{1,\ldots,\infty}$ is given by

$$K_{H_{1,\ldots,\infty}}(t,u) = \langle k(t), K^{-1} k(u) \rangle_{\ell^2(\mathbb{N})}, \tag{9.77}$$

the biorthogonal sequence is given by

$$s(t) = K^{-1} k(t), \tag{9.78}$$

and the orthogonal basis of $H_{1,\ldots,\infty}$ is expressible by

$$\phi(t) = K^{-1/2} k(t). \tag{9.79}$$

As a consequence, the sampling theorem may be formulated in the form

$$F(t) = \left\langle \int_{\mathbb{R}} F(u) k(u)\, du, s(t) \right\rangle_{\ell^2(\mathbb{N})}, \qquad F \in H_{1,\ldots,\infty}. \tag{9.80}$$

Frame property. The biorthogonal sequence $\{S_n\}_{n \in \mathbb{N}}$ can also be obtained more directly from the theory of frames (cf. [**643**]): We suppose that (9.72) holds for all $F \in H_{1,\ldots,\infty}$, and let T be the map on H defined by

$$TF = \sum_{n \in \mathbb{N}} F(t_n) K_{H_{1,\ldots,\infty}}(t_n, \cdot). \tag{9.81}$$

Then, T is a bounded invertible, self-adjoint operator on H. If $\{K_{H_{1,\ldots,\infty}}(t_n, \cdot)\}_{n \in \mathbb{N}}$ is also exact, i.e., no proper subset is a frame, then $S_n = T^{-1} K_{H_{1,\ldots,\infty}}(t_n, \cdot)$ provides the biorthogonal sequence (cf. [**874**]). By operating in (9.81) with T^{-1} we obtain the sampling theorem

$$F = \sum_{n \in \mathbb{N}} F(t_n)\, S_n, \quad F \in H_{1,\ldots,\infty}, \tag{9.82}$$

and
(9.83)

$$K_{H_{1,\ldots,\infty}}(t,u) = \sum_{n\in\mathbb{N}} S_n(t) K_{H_{1,\ldots,\infty}}(t_n,u) = \sum_{n\in\mathbb{N}} T^{-1} K(t_n,t) K_{H_{1,\ldots,\infty}}(t_n,t).$$

This may be compared with (9.77) and (9.68).

Sampling reprostructure involving trigonometric polynomials. The sampling reprostructure involving the example of a finite-dimensional space of trigonometric polynomials leads to the well-known Dirichlet kernel (see, e.g., [**884**]) as reproducing kernel (for the corresponding orthogonal sampling system the reader is referred to [**643**]).

Sampling reprostructure involving orthogonal polynomials. We suppose that w is a weight function on $\mathbb{R}$ and let $P_n, n = 0,1,\ldots,N$, be the first $N+1$ normalized orthogonal polynomials (for example, for $w(t) = (1-t)^\alpha(1+t)^\beta \chi_{[-1,1]}(t)$ we have the Jacobi polynomials). The RKHS consists of all polynomials of degree at most N with the inner product

$$\langle F,G\rangle_{L^2_w(\mathbb{R})} = \int_{\mathbb{R}} w(t)\; F(t)\overline{G(t)}\; dt. \tag{9.84}$$

Its reproducing kernel is given by the Christoffel-Darboux formula (see, e.g., [**277**])

$$K(x,t) = \frac{\alpha_n}{\alpha_{N+1}} \frac{P_{N+1}(x)P_N(t) - P_N(x)P_{N+1}(t)}{x-t} = \sum_{n=0}^{N} P_n(x)P_n(t) \tag{9.85}$$

where α_n is the leading coefficient of P_n. We take $t_1,\ldots,t_{N+1}$ to be the (distinct) zero of P_{N+1}. Then $K(t_i,t_j) = 0$ for $i \neq j$ and $k(t_i,t_i) = \sum_{n=0}^N P_n^2(t_i) > 0$, since the zeros are interlaced. Thus, the sampling sequence is $\{K(t_i,\cdot)/K(t_i,t_i)\}_{i=1,\ldots,N+1}$, and the sampling theorem may be formulated as

$$F(t) = \sum_{i=1}^{N+1} F(t_i)\frac{K(t_i,t)}{K(t_i,t_i)} = \sum_{i=1}^{N+1} F(t_i)\frac{P_{N+1}(t)}{P'_{N+1}(t_i)(t-t_i)}, \tag{9.86}$$

as should be expected. The associated orthogonal sequence is given by
(9.87)

$$\Phi_i(t) = \frac{K(t_i,t)}{(K(t_i,t_i))^{1/2}} = \left(\frac{\alpha_N}{\alpha_{N+1}}\right)^{1/2} \frac{P_N(t_i)P_{N+1}(t)}{(P_N(t_i)P'_{N+1}(t_i))^{1/2}(t-t_i)},\; i = 1,\ldots,N+1,$$

each of which is a polynomial of degree N.

Kramer's sampling approach. H.P. Kramer [**512**] generalized the sampling theorem by replacing the trigonometric system by an arbitrary complete orthogonal system of the form $\{\Psi(x,t_n)\}_{n\in\mathbb{Z}}$, where $\Psi(\cdot,t) \in L^2(\mathcal{G})$ for $\mathcal{G} \subset \mathbb{R}$ being an interval. The sampling theorem was applied to convolution integrals F of the form

$$F(t) = \int_{\mathcal{G}} \Psi(x,t)\; \tilde{F}(x)\; dx, \quad \tilde{F} \in L^2(\mathcal{G}). \tag{9.88}$$

The sampling functions are given by

$$S_n(t) = \frac{\int_{\mathcal{G}} \Psi(x,t)\; \overline{\Psi(x,t_n)}\; dx}{\int_{\mathcal{G}} \Psi(x,t_n)\; \overline{\Psi(x,t_n)}\; dx}, \tag{9.89}$$

such that the sampling theorem is of the form

$$F(t) = \sum_{n=-\infty}^{\infty} F(t_n) S_n(t), \tag{9.90}$$

which follows easily from the orthogonal expansion of F.

We only consider the case, where the norm of $\Psi_n = \Psi(\cdot, t_n)$ is 1 (cf. [**643**]) (note that the general case can be attacked in a similar, though more complicated manner). Let H denote the space of all functions of the form (9.88), with the induced inner product, i.e.,

$$\langle \tilde{F}, \tilde{G} \rangle_{L^2(\mathcal{G})} = \int_{\mathcal{G}} \tilde{F}(x)\, \overline{\tilde{G}(x)}\, dx. \tag{9.91}$$

There is no ambiguity here since the transform (9.88) is one-to-one on the sequence $\{t_n\}_{n\in\mathbb{N}}$, hence, for all t. Since $\Psi(\cdot, t)$ is of class $L^2(\mathcal{G})$, its expansion

$$\Psi(\cdot, t) = \sum_{n\in\mathbb{N}} S_n(t)\ \Psi_n \tag{9.92}$$

converges in the sense of $L^2(\mathcal{G})$. We introduce the kernel $K(\cdot, \cdot)$ by

$$K(t, u) = \int_{\mathcal{G}} \overline{\Psi(x, t)}\ \Psi(x, u)\ dx. \tag{9.93}$$

Since $K(t, \cdot) \in H$ for each t, we find that

$$\langle F, K(t, \cdot) \rangle_H = \int_{\mathcal{G}} \Psi(x, t)\ \tilde{F}(x)\ dx = F(t), \tag{9.94}$$

i.e., $K(t, u)$ is the reproducing kernel. Clearly, we have

$$S_n(t) = K(t_n, t), \tag{9.95}$$

$$S_n(t_k) = \delta_{n,k}, \tag{9.96}$$

$$\langle S_n, S_m \rangle_H = \delta_{n,m}, \tag{9.97}$$

and

$$K(t, u) = \sum_{n\in\mathbb{N}} S_n(t)\ \overline{S_n(u)}. \tag{9.98}$$

Finally, it should be remarked that, by use of functional analytic means, we are able to deal with a variety of Hilbert and Banach spaces that admit sampling expansions with respect to univariate families $\{S_n\}_{n\in\mathbb{N}}$. In all cases, the sampling expansions (9.90) depend on the sampling points $\{t_n\}_{n\in\mathbb{N}}$, but not on the function F under discussion. Those function spaces arising in connection with sampling expansions (9.90) include certain Sobolev spaces, Wiener amalgam space, translation-invariant spaces, shift-invariant spaces, and spaces modeling signals with finite rate of innovation (for more details the reader is referred to, e.g., [**636, 637**] and the literature therein).

Sampling expansions. C.V. van der Mee et al. [**572**] base their sampling considerations on the following equivalent statements in a separable Hilbert space H (see also R.M. Young [**874**]):

- *The sequence $\{K(t_j,\cdot)\}_{j\in\mathbb{Z}}$ forms a Riesz basis in H.*
- *The sequence $\{K(t_j,\cdot)\}_{j\in\mathbb{Z}}$ is complete in H, and there exist positive constants C', C'' such that for arbitrary integers n,m and arbitrary positive scalars*

$$C' \sum_{s=-m}^{n} |c_s|^2 \leq \left\| \sum_{s=-m}^{n} c_s K(t_s,\cdot) \right\|_H^2 \leq C'' \sum_{s=-m}^{n} |c_s|^2. \tag{9.99}$$

- *The sequence $\{K(t_j,\cdot)\}_{j\in\mathbb{Z}}$ is complete in H, and its Gram matrix $[K(t_i,t_j)]_{i,j\in\mathbb{Z}}$ generates a bounded invertible operator on $\ell^2(\mathbb{Z})$.*

These statements immediately predetermine sampling variants:

Variant 1: Let $K(\cdot,\cdot)$ denote the the reproducing kernel of the RKHS H and $\langle\cdot,\cdot\rangle_H$ its inner product. Assume that $\{t_j\}_{j\in\mathbb{Z}}$ is an infinitely many distinct sampling point sequence in an interval $\mathcal{G}\subset\mathbb{R}$. Then the following statements are equivalent:

- *There exist positive constants C', C'' such that*

$$C' \, \|F\|_H \leq \left(\sum_{n\in\mathbb{Z}} |F(t_n)|^2 \right)^{\frac{1}{2}} \leq C'' \, \|F\|_H \tag{9.100}$$

holds for every $F\in H$, and no such relation holds for any proper subset of sampling points.

- *The sequence $\{K(t_j,\cdot)\}_{j\in\mathbb{Z}}$ forms a Riesz basis in H.*
- *The sequence $\{K(t_j,\cdot)\}_{j\in\mathbb{Z}}$ is complete in H, and its Gram matrix $[K(t_i,t_j)]_{i,j\in\mathbb{Z}}$ is bounded and strictly positive selfadjoint on $\ell^2(\mathbb{Z})$.*

If any of these conditions hold, we have the sampling expansion

$$F(t) = \sum_{j\in\mathbb{Z}} F(t_j)\, \frac{K(t_j,t)}{K(t_j,t_j)}. \tag{9.101}$$

In particular, if the three properties (i) $\mathcal{G}=\mathbb{R}$, (ii) the sampling points are equidistant (i.e., $t_j=\sigma j, j\in\mathbb{Z}$), (ii) the RKHS H is unitarily translation-invariant, i.e., for any $F\in H$ and $c\in\mathbb{R}$, we have $F(\cdot - c)\in H$ and $\|F(\cdot-c)\|_H = \|F\|_H$, are satisfied, the bi-infinite Gram matrix $[\kappa_{i,j}]_{i,j\in\mathbb{Z}} = [K(t_i,t_j)]_{i,j\in\mathbb{Z}}$ is a *Toeplitz matrix* (i.e., $\kappa_{i,j}=\kappa_{i-j}, i,j\in\mathbb{Z}$).

Variant 2: Let $K(\cdot,\cdot)$ denote the reproducing kernel of a unitarily translation-invariant Hilbert space H and $\langle\cdot,\cdot\rangle_H$ its inner product. Assume that $\{t_j\}_{j\in\mathbb{Z}}$ with $t_j=\sigma j, j\in\mathbb{Z}$, is an an equidistant sampling point sequence. Then the following statements are equivalent:

- *There exist positive constants C', C'' such that*

$$C' \, \|F\|_H \leq \left(\sum_{n\in\mathbb{Z}} |F(\sigma n)|^2 \right)^{\frac{1}{2}} \leq C'' \, \|F\|_H, \quad F\in H, \tag{9.102}$$

and no such relation holds for any proper subset of sampling points.

- *The sequence $\{K(\sigma j,\cdot)\}_{j\in\mathbb{Z}}$ is a Riesz basis in H.*

• *The sequence* $\{K(\sigma j, \cdot)\}_{j\in\mathbb{Z}}$ *is complete in* H, *and the Toeplitz matrix* $[\kappa_{i-j}]_{i,j\in\mathbb{Z}}$ *is bounded and strictly positive selfadjoint on* $\ell^2(\mathbb{Z})$.

If any of these conditions hold, we have the sampling expansion

$$F(t) \;=\; \sum_{j\in\mathbb{Z}} F(\sigma j)\, \frac{K(\sigma j, t)}{K(\sigma j, \sigma j)}. \tag{9.103}$$

9.3. Sampling sets in reproducing kernel spaces

Our next interest is in a general construction principle for the reproducing kernel Hilbert spaces which admit a discrete *sampling set* $\mathcal{S} \subset \mathcal{G}$. To this end we remember a few definitions and some preliminary results of functional analysis that will be needed in this context:

Sampling sets in reproducing kernel Hilbert spaces. A *frame* for a separable Hilbert space H is a sequence $\{F_n\}_{n\in\mathbb{Z}} \subset H$ such that there exist two positive constants C', C'' with the property that

$$C'\; \|F\|_H^2 \le \sum_{n\in\mathbb{Z}} |\langle F, F_n\rangle_H|^2 \le C''\; \|F\|_H^2, \quad F \in H. \tag{9.104}$$

The optimal constants (maximal for C' and minimal for C'') are called *frame bounds.* When $C' = C'' = 1$, $\{F_n\}_{n\in\mathbb{Z}}$ is called a *normalized tight frame.*

A system $\{F_n\}_{n\in\mathbb{Z}}$ is called a *Bessel sequence* if we only require the right side inequality of (9.104) to hold. Later, our considerations also need the concept of strongly disjoint or orthogonal frames as introduced in R. Balan [**63**] and D. Han, D. Larson [**394**]: Two Bessel sequences $\{F_n\}_{n\in\mathbb{Z}}$ and $\{G_n\}_{n\in\mathbb{Z}}$ are called *strongly disjoint* if

$$\sum_{n\in\mathbb{Z}} \langle F, F_n\rangle_H\; G_n = 0 \tag{9.105}$$

holds for all $F \in H$.

Let $\{F_n\}_{n\in\mathbb{Z}}$ be a frame for a Hilbert space H. The associated *analysis operator* T is the bounded linear operator defined by

$$TF = \sum_{n\in\mathbb{Z}} \langle F, F_n\rangle_H\; e_n, \tag{9.106}$$

where $\{e_n\}_{n\in\mathbb{Z}}$ is the standard orthonormal basis for $\ell^2(\mathbb{Z})$. It is easy to verify that the range space TH is closed and T is a boundedly invertible operator from H to TH. Moreover, we have $T^*e_n = F_n$ for each n. T^* is called the *synthesis operator.*

Suppose now that

$$A = T^*T \tag{9.107}$$

(this operator is known as the *frame operator* for the system $\{F_n\}_{n\in\mathbb{Z}}$). Then $\{A^{-1}F_n\}_{n\in\mathbb{Z}}$, which is called *standard dual frame*, provides the *reconstruction formula*

$$F = \sum_{n\in\mathbb{Z}} \langle F, A^{-1}F_n\rangle_H F_n = \sum_{n\in\mathbb{Z}} \langle F, F_n\rangle_H A^{-1}F_n, \quad x \in H, \tag{9.108}$$

where the convergence is in the H-Hilbert space norm.

Let H be a reproducing kernel Hilbert space of functions on $\mathcal{G}$. A *sampling set* is a countable subset $\mathcal{S}$ of $\mathcal{G}$ with the property that there exist two positive constants C' and C'' such that

$$C' \ \|F\|_H^2 \leq \sum_{\sigma \in \mathcal{S}} |F(\sigma)|^2 \leq C'' \ \|F\|_H^2, \quad F \in H. \tag{9.109}$$

In connection with (9.109) we are able to conclude (cf. [**396**]) that the evaluation linear functional sequence $\{K_\sigma : \sigma \in \mathcal{S}\}$, $K_\sigma = K(\cdot, \sigma), \sigma \in \mathcal{S}$, forms a frame for H, and that any function $F \in H$ can be reconstructed from its sampling on the set $\mathcal{S}$ by the following reconstruction formula

$$F = \sum_{\sigma \in \mathcal{S}} F(\sigma) \ P_\sigma, \quad F \in H, \tag{9.110}$$

where $\{P_\sigma : \sigma \in \mathcal{S}\}$ is a dual frame of $\{K_\sigma = K(\cdot, \sigma) : \sigma \in \mathcal{S}\}$. The series in (9.110) converges in the Hilbert space norm. Moreover, by (9.110), we obtain a reconstruction formula in a reproducing kernel Hilbert space H, that converges in the Hilbert space norm and also in pointwise sense.

Summarizing our results we obtain the following statement (see, e.g., [**637**]):

Assume that $\mathcal{S}$ is a sampling set for a reproducing kernel Hilbert space H on a (continuously connected) domain $\mathcal{G}$. Then there exists a frame $\{P_\sigma : \sigma \in \mathcal{S}\}$ for H such that

$$F = \sum_{\sigma \in \mathcal{S}} F(\sigma) \ P_\sigma, \quad F \in H, \tag{9.111}$$

where the convergence is both in norm and pointwise sense.

Sampling sets in reproducing kernel Banach spaces. A *reproducing kernel Banach space (RKBS)* is a Banach space B of functions on a set $\mathcal{G}$ such that the evaluation functions $F \to F(t)$ are continuous for each $t \in \mathcal{G}$.

A countable subset $\mathcal{S}$ of $\mathcal{G}$ is called a *p-sampling set* ($p \geq 1$) *for the Banach space* B, if there exists two positive constants C' and C'' such that

$$C' \ \|F\|_B \leq \left(\sum_{\sigma \in \mathcal{S}} |F(\sigma)|^p \right)^{1/p} \leq C'' \ \|F\|_B, \quad F \in B. \tag{9.112}$$

Clearly, p-sampling sets in reproducing kernel Banach spaces can be characterized by p-frames (cf. Section 9.2): Let B be a Banach space. Assume that B^* is its dual (note that the dual relation $P(F)$ will be still denoted by $\langle F, P \rangle_B$, where $F \in B$ and $P \in B^*$).

A countable set $\{P_k\}_{k \in \mathbb{Z}} \subset B^*$ is called a *p-frame,* if there exist positive constants C' and C'' such that

$$C' \ \|F\|_B \leq \left(\sum_{k \in \mathbb{Z}} |\langle F, P_k \rangle_B|^p \right)^{1/p} \leq C'' \ \|F\|_B, \quad F \in B, \tag{9.113}$$

i.e., the *analysis operator* T defined by $T : B \ni F \mapsto \{\langle F, P_k \rangle_B\} \in \ell^p(\mathbb{Z})$ is bounded from both above and below.

If $\{P_k\}_{k\in\mathbb{Z}}$ is p-frame for a Banach space B, and if, in addition, there exists a bounded linear operator $R : \ell^p(\mathbb{Z}) \to B$ such that $R\ T = I_B$, with I_B the identity operator on B, then $\{P_k\}_{k\in\mathbb{Z}}$ is called a *Banach frame with respect to the $\ell^p(\mathbb{Z})$-space*. In the case of a Banach frame we have the reconstruction formula

$$F = \sum_{k\in\mathbb{Z}} \langle F, P_k\rangle_B\ \tilde{P}_k, \quad F \in B, \tag{9.114}$$

where $\tilde{P}_k = R\ e_k$ with $\{e_k\}_{k\in\mathbb{Z}}$ is the standard basis of $\ell^p(\mathbb{Z})$.

9.4. Sampling induced reproducing kernel space framework

In a context, where sampling expansions are to be used, both

- the *sampling map* $F \mapsto \{F(t_n)\}_{n\in\mathbb{Z}}$
- the *inverse map* $\{F(t_n)\}_{n\in\mathbb{Z}} \mapsto F$

need to be continuous. Thus, the evaluation functional $t \mapsto F(t), t \in \mathcal{G}$, needs to be continuous for all F. Equivalently, the signal resides in an RKHS, even though this RKHS is not explicitly identified. So we are canonically led to the question:

How can we characterize all the reproducing kernel Hilbert spaces that allow the existence of a sampling set?

In what follows we characterize reproducing kernel Hilbert spaces on a set $\mathcal{G} \subset \mathbb{R}^q$ that have a prescribed discrete set $\mathcal{S}$ as a sampling set. The results generalize the corresponding work related to Riesz bases by M.Z. Nashed, G.G. Walter [**644**] and are related to the work by C.V. van der Mee, et al. [**572**] about sampling expansions. More concretely, they are based on a series of papers given by M.Z. Nashed and his colleagues during the last two decades (see, e.g., [**396, 636, 643, 644**]).

Reproducing kernel Hilbert space (RKHS) approach. Let $\mathcal{G}$ be a "continuously connected poinset", e.g., a region in $\mathbb{R}^q, q \geq 1$, and let $\mathcal{S}$ be a countable discrete subset of $\mathcal{G}$. Assume that $\{J_n\}_{n\in\mathbb{Z}}$ is a frame for a Hilbert space J, i.e., there exist two positive constants C', C'' with the property that

$$C'\ \|F\|_J^2 \leq \sum_{n\in\mathbb{Z}} |\langle F, J_n\rangle_J|^2 \leq C''\ \|F\|_J^2, \quad F \in J. \tag{9.115}$$

Suppose that $S_n : \mathcal{G} \to \mathbb{R}$ (or $\mathbb{C}$) forms a sequence $\{S_n\}_{n\in\mathbb{Z}}$ of functions satisfying the following two conditions:

(H1) $\{S_n(t)\}_{n\in\mathbb{Z}} \in \ell^2(\mathbb{Z})$ *for every* $t \in \mathcal{G}$,

(H2) $\{\eta_\sigma\}_{\sigma\in\mathcal{S}}$ *is a frame for* $\ell^2(\mathbb{Z})$, *where* η_σ *is given by* $\eta_\sigma = \{S_n(\sigma)\}_{n\in\mathbb{Z}}$.

We introduce an auxiliary function $\tilde{F}$ given by

$$\tilde{F}(t) \ = \ \sum_{n\in\mathbb{Z}} S_n(t)\ J_n, \quad t \in \mathcal{G}. \tag{9.116}$$

The function $\tilde{F}$ is well-defined, since $\{J_n\}_{n\in\mathbb{Z}}$ is a frame and $\sum_{n\in\mathbb{Z}} |S_n(t)|^2 < \infty$. Clearly, if $F = 0$ in J, then $\langle F, \tilde{F}(t)\rangle_J = 0$ for all $t \in \mathcal{G}$. Conversely, if $\langle F, \tilde{F}(t)\rangle_J = 0$ for all $t \in \mathcal{G}$, then the ℓ^2-sequence $\xi = \{\langle J_n, F\rangle_J\}_{n\in\mathbb{Z}}$ satisfies

$$\langle \xi, \eta_\sigma\rangle_{\ell^2(\mathbb{Z})} = \sum_{n\in\mathbb{Z}} S_n(\sigma)\ \langle F, J_n\rangle_J = 0, \quad \sigma \in \mathcal{S}. \tag{9.117}$$

This, together with the assumptions that $\{\eta_\sigma\}_{\sigma\in\mathcal{S}}$ is a frame for $\ell^2(\mathbb{Z})$ and that $\{J_n\}_{n\in\mathbb{Z}}$ is a frame for the Hilbert space J, enables us to verify that $F = 0$.

The function $\tilde{F}$ leads to a construction of a space H of functions on $\mathcal{G}$ by letting

$$H = \left\{ \langle F, \tilde{F}(t)\rangle_J : F \in J \right\}. \tag{9.118}$$

The linear map $J \ni F \mapsto \langle F, \tilde{F}(t)\rangle_J \in H$ is one-to-one and onto. So, H becomes a Hilbert space with the inner product on H defined by

$$\langle \Phi, \Psi\rangle_H = \langle F, G\rangle_J, \tag{9.119}$$

where $\Phi(t) = \langle F, \tilde{F}(t)\rangle_J$ and $\Psi(t) = \langle G, \tilde{F}(t)\rangle_J$, $t \in \mathcal{G}$. Moreover, the space H is a reproducing kernel Hilbert space. In fact, for $t \in \mathcal{G}$, we have

$$|\Phi(t)| = |\langle F, \tilde{F}(t)\rangle_J| \leq \|\tilde{F}(t)\|_J\ \|F\|_J = \|\tilde{F}(t)\|_J\ \|\Phi\|_H, \qquad \Phi \in H. \tag{9.120}$$

For the reproducing kernel Hilbert space H introduced in (9.118), we are able to formulate the following characterizations concerning its sampling sets (for the details of the proof of (i) and (ii) the reader is referred to D. Han et al. [**396**]):

(i) *$\mathcal{S}$ is a sampling set for the reproducing kernel Hilbert space H as indicated in* (9.118).

(ii) *Every reproducing kernel Hilbert space admitting a sampling set $\mathcal{S}$ can be constructed in the way as explained above.*

We call a sampling set $\mathcal{S}$ of H *exact* if the corresponding evaluation functional sequence forms a Riesz basis for H . Then the following proposition follows immediately from standard frame theory (see, e.g., D. Han, D. Larson [**394**]):

A sampling set $\mathcal{S}$ for a Hilbert space H of functions is exact, if and only if

$$\{\ \{\Phi(\sigma)\}_{\sigma\in\mathcal{S}} : \Phi \in H\} = \ell^2(\mathcal{S}). \tag{9.121}$$

In the construction principle proposed above, if we further require that both $\{J_n\}_{n\in\mathbb{Z}}$ and $\{\eta_\sigma\}_{\sigma\in\mathcal{S}}$ be Riesz bases for J and $\ell^2(\mathcal{S})$, respectively, then we are allowed to conclude that

$$\{\ \{\Phi(\sigma)\}_{\Phi\in H} : \sigma \in \mathcal{S}\} = T_\eta T_\Phi H = \ell^2(\mathcal{S}), \tag{9.122}$$

where T_Φ and T_η are the analysis operator for $\{\Phi(\sigma)\}_{\sigma\in\mathcal{S}}$ and $\{\eta_\sigma\}_{\sigma\in\mathcal{S}}$, respectively. Thus, following the aforementioned proposition, $\mathcal{S}$ is an exact sampling set for H. This leads to the following equivalent statements:

(i) *H is a reproducing kernel Hilbert space with an exact sampling set.*

(ii) *H can be constructed as in the indicated construction with $\{J_n\}_{n\in\mathbb{Z}}$ and $\{\eta_\sigma\}_{\sigma\in\mathcal{S}}$ being Riesz bases for H and $\ell^2(\mathbb{Z})$, respectively.*

In any nonexact sampling case, we always have an oversampling for the function space (cf. [**396**]).

Reproducing kernel Banach space (RKBS) approach. Unlike the Hilbert space case, the reconstruction operator does not necessarily exist for p-frames. In fact, it is known from O. Christensen, D. Stoeva [**148**], that there exists a p-frame $\{P_k\}_{k\in\mathbb{Z}}$ for a Banach space B, for which no family $\{\tilde{P}_k\}_{k\in\mathbb{Z}} \subset B$ satisfies $F = \sum_{k\in\mathbb{Z}} \langle F, P_k\rangle_B \tilde{P}_k$, $F \in B$. As a consequence, the existence of a reconstruction operator is assumed in the context of Banach frames [**21, 24, 139, 150, 381**]. Nonetheless, a reconstruction formula is available for some special p-frames. For instance, it was proved in P. Casazza et al. [**139**] that under certain natural conditions on the generators, a translation p-frame for a finite generated shift-invariant subspace is always a Banach frame.

In general, a p-frame $\{P_k\}_{k\in\mathbb{Z}}$ is a Banach frame if and only if the range space $T\mathcal{B}$ of the analysis operator T is complemented in $\ell^p(\mathbb{Z})$ (cf. O. Christensen, D. Stoeva [**150**]). Therefore, similarly to sampling in a Hilbert space, p-sampling for a Banach space is related to a Banach space frame and a p-frame in the space. In particular, for a Banach space B of functions on a set $\mathcal{G}$, a countable subset $\mathcal{S}$ of $\mathcal{G}$ is a p-sampling set for the space B, if and only if the evaluation functionals $F \to F(\sigma), \sigma \in \mathcal{S}$, form a p-frame for the space B.

Following D. Han et al. [**396**] we are able to conclude that a reconstruction formula always exists for a p-frame $\{F \to F(\sigma) : \sigma \in \mathcal{S}\}$ associated with a p-sampling set $\mathcal{S}$ on a reproducing kernel Banach space B:

Assume that $\mathcal{S}$ is a p-sampling set for a reproducing kernel Banach space B on a set $\mathcal{G}$. Then there exists a sequence of functions $\{S_{\sigma'}(\sigma)\}_{\sigma'\in\mathcal{S}}$ on $\mathcal{G}$ such that the following properties are satisfied:

(B1) $\{S_\sigma(t)\} \in \ell^q(\mathcal{S})$ *for every* $t \in \mathcal{G}$, *where* $\frac{1}{p} + \frac{1}{p'} = 1$.

(B2) $\{\eta_\sigma\}_{\sigma\in\mathcal{S}}$ *is a p-frame for* $T\ B$, *where* $\eta_\sigma = \{S_{\sigma'}(\sigma)\}_{\sigma'\in\mathcal{S}}$ *and* $T : B \to \ell^p(\mathcal{S})$ *is defined by* $TF = \{F(\sigma')\}_{\sigma'\in\mathcal{S}}$.

(B3) $F(t) = \sum_{\sigma\in\mathcal{S}} F(\sigma)\ S_\sigma(t)$, $t \in \mathcal{G}$, *where the convergence is in pointwise sense.*

Finally (cf. [**396**]), we are led to the general construction principle of reproducing kernel Banach spaces having $\mathcal{S}$ as a p-sampling set:

Let $\{Q_k\} \subset B^*$ be a p-frame for a Banach space B and $\mathcal{S}$ be a countable subset of $\mathcal{G}$. Suppose that $S_k, k \in \mathbb{Z}$, satisfy the conditions:

(B1) $\{S_k(t)\} \in \ell^q(\mathbb{Z})$ *for every* $t \in \mathcal{G}$, *where* $\frac{1}{p} + \frac{1}{p'} = 1$.

(B2) $\{\eta_\sigma\}_{\sigma\in\mathcal{S}}$ *is a p-frame for* $\ell^p(\mathbb{Z})$, *where* $\eta_\sigma = \{S_k(\sigma)\}_{k\in\mathbb{Z}}$.

Then

$$\tilde{B} = \left\{ \tilde{F} = \sum_{k\in\mathbb{Z}} \langle F, Q_k\rangle_B S_k(t) : \quad F \in B \right\} \tag{9.123}$$

is a reproducing kernel Banach space with $\mathcal{S}$ as a p-sampling set of $\tilde{B}$, where the norm of $\tilde{F} \in \tilde{B}$ with

$$\tilde{F} = \sum_{k\in\mathbb{Z}} \langle F, Q_k\rangle_B\ S_k, \quad F \in B, \tag{9.124}$$

is defined to be $\|F\|_B$.

9.5. Sampling in shift-invariant spaces

We begin with recapitulating known results about Shannon's sampling theorem, however, by formally replacing Λ by its inverse lattice Λ^{-1}. In doing so we obtain from (9.6) the Shannon sampling theorem in the form

$$F^{\wedge}_{\mathcal{F}_{\Lambda^{-1}}}(t) = \sum_{g\in\Lambda} F^{\wedge}_{\mathcal{F}_{\Lambda^{-1}}}(g) \underbrace{\|\mathcal{F}_{\Lambda}\| \int_{\mathcal{F}_{\Lambda^{-1}}} e^{-2\pi i x\cdot(t-g)}\, dx}_{=K_{\mathcal{F}_{\Lambda^{-1}}}(t-g)}. \tag{9.125}$$

Moreover, we know from (9.26) that the space $V(K_{\mathcal{F}_{\Lambda^{-1}}}(\cdot))$ defined by

$$V(K_{\mathcal{F}_{\Lambda^{-1}}}) = \left\{ \sum_{g\in\Lambda} c(g)\, K_{\mathcal{F}_{\Lambda^{-1}}}(\cdot - g) : \sum_{g\in\Lambda} |c(g)|^2 < \infty \right\}. \tag{9.126}$$

is a unitarily translation-invariant subspace of $L^2(\mathbb{R})$. In many cases in applications, however, functions that are obtained by measurements tend to have frequency components that decay for higher frequencies, so that these functions are not bandlimited in the strict sense. As a consequence, it will be advantageous to use nonbandlimited models that retain some of the simplicity and structure of Shannon's bandlimited model but are more amenable to numerical implementation and are more flexible for approximating real data. One such example are the shift-invariant spaces.

In 1977, A. Papoulis [**679**] extended the 1D-Shannon sampling theorem to so-called multichannel sampling, showing that a 1D bandlimited signal can be reconstructed from samples of response of a finite number of linear shift-invariant systems. This concept has been generalized to multidimensional signals (cf. [**116**]), too. Although Papoulis's approach is straightforward, it had been pointed out in [**115**] that potential instabilities arise. A temporal point of view on Papoulis's idea led to the framework of dynamical sampling started by A. Aldroubi et al. [**27**]. This work also points out the instabilities of the approach and some ways of dealing with them. Additional papers by A. Aldroubi and coauthors in this framework that should be mentioned are [**29**], [**30**], and [**31**].

A review of the main ideas of sampling in shift-invariant spaces is given by a palette of contributions, e.g., A. Aldroubi, K. Gröchenig [**21**], A. Aldroubi, M. Unser [**33**], M.Z. Nashed, G.G. Walter [**643**], M.Z. Nashed, Q. Sun [**637**], M. Unser [**827**]. Our approach below only gives a concise impression:

Uniform sampling in shift-invariant spaces. Sampling in shift-invariant spaces retains some of the basic constituents of the classical theory but generalizes the basic ingredient by replacing the sinc-function by an arbitrary generator function $H \in L^2(\mathbb{R})$. Thus, uniform sampling leads to a reference space of type

$$V(H) = \left\{ \sum_{g\in\Lambda} c_g\, H(\cdot - g) : \sum_{g\in\Lambda} |c_g|^2 < \infty \right\}. \tag{9.127}$$

This means that any function, which is continuously defined, is characterized by a sequence of coefficients c_g; this is the discrete signal representation that is used to do signal processing calculations. Note that the coefficients are not necessarily the samples of the signal, and that can be something quite different from the sinc-function. Indeed, our essential motivation is to discover functions that are simpler to handle numerically and have a much faster decay. So, for the spaces $V(H)$ to be well defined, some additional conditions on the generator H must be imposed.

The standard condition in wavelet theory is often stated in the Fourier (spectral) domain as

$$C' \leq \left\| \sum_{g\in\Lambda} |H_{\mathbb{R}}^{\wedge}(\cdot - g)|^2 \right\|_{L^2(\mathbb{R})} \leq C''. \tag{9.128}$$

This condition implies that $V(H)$ is a closed subspace of $L^2(\mathbb{R})$. Furthermore, the family of functions $\{H(\cdot - g)\}_{g\in\Lambda}$ constitutes a Riesz basis of $V(H)$, i.e., the image of an orthonormal basis under an invertible linear transformation (cf. I. Daubechies [**174**]). The theory of Riesz bases asserts the existence of a dual basis. Specifically, for any Riesz basis for $V(H)$ of the form $\{H(\cdot - g)\}_{g\in\Lambda}$, there exists a unique function H_{dual} such that $\{H_{\text{dual}}(\cdot - g)\}_{g\in\Lambda}$ is also a Riesz basis for $V(H)$, and the functions $H_{\text{dual}}(\cdot - g) \in V(H)$ form the unique dual basis determined by the biorthogonality condition

$$\int_{\mathbb{R}} H(x-g)\; H_{\text{dual}}(x-g')\; dx = \delta_{g,g'}, \quad g, g' \in \Lambda. \tag{9.129}$$

Since the dual generator H_{dual} belongs to $V(H)$, it can be expressed in the form

$$H_{\text{dual}} = \sum_{g\in\Lambda} c_g\; H(\cdot - g). \tag{9.130}$$

where the coefficients c_g are to be determined in such a way that $\{c_g\}_{g\in\Lambda}$ exists and belongs to $\ell^2(\Lambda)$.

A straightforward approach is the so-called *minimum error sampling*. This methodology enforces an optimal solution in the least-squares sense by the projection

$$\mathrm{P}_{V(H)}F = \sum_{g\in\Lambda} \int_{\mathbb{R}} F(x)\; H_{\text{dual}}(x-g)\; dx\; H(\cdot - g). \tag{9.131}$$

Indeed, the term (9.130) may be written in the Fourier domain as follows:

$$(H_{\text{dual}})_{\mathbb{R}}^{\wedge}(x) = H_{\mathbb{R}}^{\wedge}(x)\; B(e^{2\pi i x}) \tag{9.132}$$

where the term $B(e^{2\pi i x})$ has to be determined. From (9.129) it follows that

$$\sum_{g\in\Lambda} c_g \int_{\mathbb{R}} H(x)\; H(x-g')\; dx = \delta_{0,g'}. \tag{9.133}$$

By observing the biorthogonality constraint (9.129) and by solving (9.133) in the Fourier domain we obtain

$$B(e^{2\pi i x}) \left(\sum_{g\in\Lambda} |H_{\mathbb{R}}^{\wedge}(x-g)|^2 \right) = 1. \tag{9.134}$$

In connection with (9.132) we are therefore able to deduce that

$$(H_{\text{dual}})^{\wedge}_{\mathbb{R}}(x) = \frac{H^{\wedge}_{\mathbb{R}}(x)}{\sum_{g \in \Lambda} |H^{\wedge}_{\mathbb{R}}(x-g)|^2}. \tag{9.135}$$

It follows from (9.131) that

$$\mathrm{P}_{V(H)}F = \sum_{g \in \Lambda} \int_{\mathbb{R}} F(x) \frac{H^{\wedge}_{\mathbb{R}}(x-g)}{\sum_{g' \in \Lambda} |H^{\wedge}_{\mathbb{R}}(x-g')|^2} \, dx \; H(\cdot - g). \tag{9.136}$$

All in all, we realize that the validity of the above approach relies of the following three conditions (usually formulated for the lattice $\Lambda = \mathbb{Z}$, see, e.g., M. Unser [**827**]):

- $c = \{c_g\}_{g \in \Lambda}$ is square-summable, i.e.. $\{c_g\}_{g \in \Lambda} \in \ell^2(\Lambda)$, to guarantee the series convergence,
- $\{H(\cdot - g)\}_{g \in \Lambda}$ is a Riesz basis of $V(H)$, i.e., there exist positive constants C', C'' with $C' \leq C''$ such that

$$C' \;\; \|c\|_{\ell^2(\Lambda)} \leq \left\| \sum_{g \in \Lambda} c_g \; H(\cdot - g) \right\|^2_{\mathrm{L}^2(\mathbb{R})} \leq C'' \;\; \|c\|_{\ell^2(\Lambda)}, \tag{9.137}$$

  in order to ensure a unique representation. In approximation theory, we say that H has stable translates and is a stable generator (cf. [**466**], [**467**], [**468**]). Moreover, in the reference space $V(H)$, the Riesz basis property (9.137) is equivalent to the inequalities stated in (9.128) (cf. [**33**], [**21**]),
- $\{H(\cdot - g)\}_{g \in \Lambda}$ constitutes a *partition of unity*

$$\sum_{g \in \Lambda} H(x-g) = 1, \; x \in \mathbb{R}, \tag{9.138}$$

  so that a function can be represented as closely as possible, by specifying sufficiently small sampling intervals.

Other techniques for determining sequences $c = \{c_g\}_{g \in \Lambda}$ are known (for more details the reader is referred, e.g., to [**828**], [**827**], [**24**], [**637**]).

Nonuniform sampling in shift-invariant spaces. This methodology is relatively more recent. It leads to reference spaces

$$V(H) = \left\{ \sum_{g \in \Lambda} c_g \; H(\cdot - x_g) \; : \; \sum_{g \in \Lambda} |c_g|^2 < \infty, \quad \{x_g\}_{g \in \Lambda} \in \mathbb{R} \right\}. \tag{9.139}$$

The problems here to be solved can be characterized as follows: (1) for some generator H we have to specify the conditions on the sampling points $\{x_g\}_{g \in \Lambda} \in \mathbb{R}$ such that a signal can be uniquely and stably reconstructed from its samples, (2) practicable algorithms have to be designed for realizing the reconstruction.

Nonuniform sampling does not readily allow to derive closed form expressions for the coefficient sequence $\{c_g\}_{g \in \Lambda}$ corresponding to a set of sample points $\{x_g\}_{g \in \Lambda}$. Early contributions focussed on the perturbation of the lattice structure (see [**643**], [**149**]). A. Aldroubi, K. Gröchenig [**21**] resort to iterative algorithms (further methodologies are found in [**20**], [**19**], [**23**], [**22**], see also the references therein). In particular, the contribution [**2**], which is more general than [**22**], is worth mentioning in non-uniform sampling and perturbation techniques.

Sampling under nonbandlimitedness. The most famous of sampling theorems is undoubtedly the Shannon sampling theorem, which states that bandlimited signals can be recovered exactly from their discrete samples, under a sufficient sampling rate. The unanimous celebration has lead many scientists to start thinking about sampling theory exclusively in terms of bandlimitedness, which is nothing but a sufficient condition. The bandlimitedness of the Shannon sampling theorem is not a necessary condition. Unlimited sampling framework is described in [**94**]. In our approach (see Section 14.3), even the Shannon sampling theorem itself is shown to hold under nonbandlimitedness, as far as the functions to be sampled satisfy certain conditions at infinity.

In fact, sampling theorems can also be devised for nonbandlimited signals, as long as they possess *finitely many degrees of freedom.* This remarkable fact was brought to the attention of the signal processing community by M. Vetterli et al. [**837**] by introducing the finite rate of innovation (FRI) framework, concerned with the sampling of not-necessarily bandlimited signals in the form

$$F = \sum_{g\in\Lambda}\sum_{k=1}^{L} c_{g,k}\, H_k(\cdot - x_g), \tag{9.140}$$

where the finite sequence of functions $\{H_k\}_{k=1,\dots,L}$ is a priori given. In fact, signals of type (9.140) turn out to be totally specified by the set $\{c_{g,k}\}_{g\in\Lambda,k=1,\dots,L}$ of amplitudes and locations $\{x_g\}_{g\in\Lambda}$, which are called the *degrees of freedom.* Denoting, for some interval $[x_\alpha, x_\beta]$, by $C_F(x_\alpha, x_\beta)$ a function that counts the degrees of freedom in F over $[x_\alpha, x_\beta]$ we define the *rate of innovation* as follows: $\rho = \lim \frac{1}{N} C_F(-\frac{N}{2}, \frac{N}{2})$. Following [**837**], a signal with finite rate of innovation is understood to be a signal, whose parametric representation is given by (9.140) with a finite rate ρ (note that bandlimited signals can also be viewed as signals of finite rate of innovation).

A prototypical sparse example (cf. [**775**]) is the T-periodic *Dirac stream*

$$F = \sum_{g\in\Lambda}\sum_{k=1}^{L} c_k\, \delta(\cdot - t_k - Tg), \tag{9.141}$$

where $c_k \in \mathbb{Z}$ and $t_k \in [0, T]$. The Dirac stream (9.141) has $2L$ degrees of freedom $\{c_k, t_k\}_{k=1,\dots,L}$ per period T, yielding a finite rate of innovation of $\rho = \frac{2L}{T}$. Intuitively, any lossless sampling scheme for (9.141) must therefore have a sampling rate at least as large as the rate of innovation ρ, or it will be impossible to fix all the degrees of freedom. In [**101**], T. Blu et al. describe a sampling scheme achieving the second best sampling rate after the critical innovation rate, permitting to recover the signal innovations from the knowledge of $2L+1$ consecutive Fourier coefficients of F. Unfortunately, this scheme is very sensitive to noise perturbations in the collected samples. This is because the recovery of the innovations t_k relies on the resolution of a so-called annihilating equation, whose solvability requires the Toeplitz matrix built from the Fourier coefficients to be rank deficient. While this structural constraint is guaranteed to hold in the case of noiseless recovery of Dirac streams, it can break in the presence of noise (for correctives see [**101**] and the references therein).

All in all, the classical finite rate of innovation framework (9.141), as introduced by M. Vetterli et al. [**837**], aims at estimating the innovations $\{c_k, t_k\}_{k=1,\dots,L}$ of a T-periodic stream of Diracs. The locations t_k are first estimated by a nonlinear method, and then arranged into a Vandermonde system whose solution yields the Dirac amplitudes [**101**]. The recovery of the locations t_k relies on the annihilating equation, dating from Prony's work, which cancels out the Fourier series coefficients of F by convolving them with a particular filter, called the annihilating filter.

CHAPTER 10

Sampling over discretely given pointsets

Both the continuous and the discrete variant of reproducing kernel Hilbert spaces arise from a given positive definite kernel, i.e., a corresponding pre-Hilbert form and subsequently a Hilbert space completion. Nevertheless, a canonical question arises for both variants of reproducing kernel Hilbert spaces:

"What are the functions in the completion specifying a certain RKHS?"

It turns out that the Hilbert space completions are subtle:

Reproducing kernel Hilbert spaces over continuously connected areas (CRKHSs) are Hilbert spaces of functions, which are not always transparent from the reproducing kernel $K(\cdot,\cdot)$ itself. CRKHSs have been studied extensively since the pioneering paper by N. Aronszajn [**47**]. They further play an important role in the theory of partial differential operators (PDO), for example, as Green's functions of second order elliptic PDOs (see, e.g., [**650**], [**393**]. Other applications include engineering, physics, machine-learning theory [**514**], [**784**], [**166**], stochastic processes [**35**], [**36**], [**34**], [**37**], [**38**], numerical analysis, and more [**873**], [**703**], [**140**], [**876**], [**526**], [**838**], [**756**], [**421**], [**768**], [**747**]. Further examples of CRKHSs are the following: Hardy spaces or Bergman spaces (for complex domains), Sobolev spaces and Dirichlet spaces (see P.J. Davis [**175**], K.A. Okoudjou, R.S. Strichartz, E.K. Tuley [**662**], R.S. Strichartz [**800, 801**] for real domains, or for fractals), Cameron-Martin Hilbert spaces from Gaussian processes (in continuous time domain), bandlimited L^2-functions from signal analysis (see, e.g., [**298**] and the references therein), spline spaces in multivariate periodic spaces (see, e.g., [**263**]), in (geodetically reflected) spherical approximation and adjustment theory (cf. [**256–258, 297, 328**]), and solution of boundary value problems of potential theory, elasticity and electromagnetic theory (see, e.g., [**262, 274, 275, 282–284, 327**]). Altogether, the literature so far has focused on the theory of kernel functions defined on continuously connected domains, either domains in Euclidean space, or complex domains in one or more variables.

Reproducing kernel Hilbert spaces over discrete point distributions (DRKHSs) show certain applications, such that their features are quite different from those of their continuous counterparts. The results of the discrete approach are applicable, e.g., to the study of admissible distributions of discrete sample points for Brownian motion, for related Gaussian fields. Moreover, a DRKHS context can be discussed in terms of binomial coefficients. Examples of DRKHSs (see, e.g., P. Jorgensen, F. Tian [**478**]) are discrete versions of the Cameron-Martin Hilbert space and energy-Hilbert spaces corresponding to graph-Laplacians. Furthermore, there is a related discrete reproducing kernel notion on prescribed countably infinite discrete reference sets $\mathcal{S}$ called "relative." This means that increments have kernel representations: Consider functions F in H, but suppose instead that, for every pair of points x, y in $\mathcal{S}$, each of the difference values can be represented by a kernel from H. We then say that H is a discrete *relative reproducing kernel Hilbert space.* The "relative" variant is of more recent vintage, and it is used in the study of electrical networks (voltage differences), in analysis of Gaussian processes such as Gaussian fields and in geodetic levelling networks (see, e.g., K. Borre, P. Meissl [**110**] for an early concept).

The transition from continuous to discrete RHKS structure in the concept of "nonuniform sampling" may be characterized as follows: In the theory of sampling involving functional analysis, one studies Hilbert spaces consisting of signals, understood in a very general sense. One develops analytic tools and algorithms, allowing one to draw inference for an "entire"(or global) signal from partial information obtained from carefully chosen distributions of sample points. While the better known and classical sampling algorithms (e.g., Shannon) are based on interpolation, recent theories go beyond this. An early motivation is the work of H. Landau [**521**]. In this setting, it is possible to make precise the notion of "average sampling rates" in general configurations of sample points. The study [**478**] starts with the general system of positive definite kernels and their discrete RKHSs or relative RKHSs. With some use of metric geometry and of spectral theory for operators in a Hilbert space, this leads to sampling theorems for a host of nonuniform point configurations. The modern theory of nonuniform sampling is vast, and it specializes into a variety of subareas. For example, the following papers (and the literature cited there) will give an idea of the diversity of points of view: A. Aldroubi, C. Leonetti [**23**], A. Aldroubi, M. Unser [**33**] S. Kahn, R.M. Goodall, R. Dixon [**482**], Y. Yi Lin, L.D. Brown [**873**], L. Zhou, X. Li, F. Pan, [**883**].

10.1. Dirac mass

Let $\mathcal{S}$ be a countable and infinite discrete set, and $\mathcal{F}(\mathcal{S})$ the set of all finite subsets $\mathcal{A}$ of $\mathcal{S}$. Let $K : \mathcal{S} \times \mathcal{S} \to \mathbb{C}$ be positive definite in the sense of (2.80). Suppose that H is the associated reproducing kernel Hilbert space satisfying the reproducing property:

$$\langle F, K_x \rangle_H = F(x), \quad x \in \mathcal{S}, \quad F \in H. \tag{10.1}$$

For $\mathcal{A} \in \mathcal{F}(\mathcal{S})$ we understand by $H_{\mathcal{A}}$ the closed span of $\{K_x\}_{x \in \mathcal{A}} \subset H$, i.e.,

$$H_{\mathcal{A}} = \overline{\text{span}\{K_x : x \in \mathcal{A}\}}^{\|\cdot\|_H}. \tag{10.2}$$

$P_{\mathcal{A}}$ denotes the orthogonal projection of H onto $H_{\mathcal{A}}$. For $\mathcal{A} \in \mathcal{F}(\mathcal{B})$, we denote by $K_{\mathcal{A}}$ the $\sharp\mathcal{A} \times \sharp\mathcal{A}$ matrix $K_{\mathcal{A}} = [K(x,y)]_{(x,y) \in \mathcal{A} \times \mathcal{A}}$.

Central for our considerations (cf. [**478**]) is the following result: Let K : $\mathcal{S} \times \mathcal{S} \to \mathbb{C}$ be positive definite, and let H be the corresponding discrete RKHS. Let F be an arbitrary function on $\mathcal{S}$. Then F is a member of H, if and only if there is a constant $C(= C_F) < \infty$ such that, for every finitely supported function $\Xi : \mathcal{S} \to \mathbb{C}$, we have

$$\left| \sum_{x\in\mathcal{S}} \overline{\Xi(x)}\ F(x) \right|^2 \leq C \sum_{x,y\in\mathcal{S}} K(x,y)\ \overline{\Xi(x)}\ \Xi(y) \tag{10.3}$$

with a constant $C = C_F$ independent of Ξ.

Essential for the treatment of discrete reproducing kernel space methods is the definition of the *Dirac mass* at an element $x \in \mathcal{S}$:

A discrete RKHS H is said to have the *discrete mass* property, if $\delta_x \in H$ for all $x \in \mathcal{S}$, where

$$\delta_x(y) = \begin{cases} 1, & \text{if} \quad x = y \\ 0, & \text{if} \quad x \neq y \end{cases}, \tag{10.4}$$

is the *Dirac mass at* $x \in \mathcal{S}$.

Using the notation $(H_{\mathcal{A}})^\perp$ for the orthogonal complement

$$(H_{\mathcal{A}})^\perp = H \ominus H_{\mathcal{A}} = \{\Phi \in H : \langle F, \Phi\rangle_H = 0, F \in H_{\mathcal{A}}\} \tag{10.5}$$

we are able to verify (cf. (2.84)) that, for $x \in \mathcal{S}$,

$$\delta_x - \sum_{y\in\mathcal{A}} (K^{(\mathcal{A})})^{-1}\delta_x)(y)\ K_y \in H^\perp_{\mathcal{A}}. \tag{10.6}$$

This leads to the formulation of the following identities:

Suppose that $\mathcal{A} \subset \mathcal{F}(\mathcal{S})$, $x \in \mathcal{S}$. Assume that δ_x is a member H. Then

$$P_{\mathcal{A}}(\delta_x)(\cdot) = \sum_{y\in\mathcal{A}} (K^{(\mathcal{A})})^{-1}\delta_x)(y) K_y \tag{10.7}$$

and

$$(K^{(\mathcal{A})})^{-2}\delta_x)(x) - \|P_{\mathcal{A}}\delta_x\|^2_H. \tag{10.8}$$

If $\delta_x \in H$, then

$$\sup_{\mathcal{A}\in\mathcal{F}(\mathcal{S})} (K^{(\mathcal{A})})^{-1}\delta_x)(x) = \|\delta_x\|^2_H. \tag{10.9}$$

A slight abuse of notation should be indicated: One makes formally use of the expressions for $P_{\mathcal{A}}(\delta_x)$ even in the case, when δ_x might not be in H. For all finite $\mathcal{A}$, $P_{\mathcal{A}}(\delta_x)$ is of class H. However, for δ_x being a member in H, we must have an additional boundedness assumption to be satisfied.

Assume that $\mathcal{S}$ is countably infinite. Let $K : \mathcal{S} \times \mathcal{S} \to \mathbb{R}$ be positive definite. Let H be the corresponding DRKHS. Then, with $x' \in \mathcal{S}$ fixed, the following three conditions (1), (2), (3) *are equivalent (see* [**475**]*):*

(1) $\delta_{x'} \in H$,

(2) *there exists a constant $C_{x'} > 0$ such that for all $\mathcal{A} \in \mathcal{F}(\mathcal{S})$, the following estimate holds true:*

$$|\Xi(x')|^2 \leq C_{x'} \sum_{(x,y)\in\mathcal{A}\times\mathcal{A}} \overline{\Xi(x)}\ \Xi(y)\ K(x,y), \tag{10.10}$$

(3) *For* $\mathcal{A} \in \mathcal{F}(\mathcal{S})$,

$$K^{(\mathcal{A})} = [K(x,y)]_{(x,y)\in\mathcal{A}\times\mathcal{A}} \tag{10.11}$$

as a $\sharp\mathcal{A} \times \sharp\mathcal{A}$ *matrix satisfies*

$$\sup_{\mathcal{A}\in\mathcal{F}(\mathcal{S})} ((K^{(\mathcal{A})})^{-1}\delta_{x'})(x') < \infty. \tag{10.12}$$

According to S. Karlin, Z. Ziegler [**490**] we say that K is *strictly positive,* if and only if det $K^{(\mathcal{A})} > 0$ for all $\mathcal{A} \in \mathcal{F}(\mathcal{S})$.

Now we are prepared to formulate the following result:

Suppose $K : \mathcal{S} \times \mathcal{S} \to \mathbb{R}$ *is strictly positive. Set* $D_{\mathcal{A}} = \det\ K^{(\mathcal{A})}$. *If* $x \in \mathcal{S}$, *and* $\mathcal{A} \in \mathcal{F}(\mathcal{S})$, *denote by* $(K')^{(\mathcal{A})}$ *the minor in* $K^{(\mathcal{A})}$ *obtained by omitting row* x *and column* x, *then* $\delta_x \in H$ *holds if and only if*

$$\sup_{\mathcal{A}\in\mathcal{F}(\mathcal{S})} \frac{D'_{\mathcal{A}}}{D_{\mathcal{A}}} < \infty. \tag{10.13}$$

10.2. Infinite networks and relative Hilbert spaces

An important special case of relative RKHSs is that of infinite networks and/or graphs (see, e.g., K. Borre, P. Meissl [**110**] concerning geodetic networks). Here, one is interested in studying "large weighted graphs" (vertices $\mathcal{S}$, edges $\mathcal{E}$, and weights as functions assigned on the edges $\mathcal{E}$), and our motivation derives from learning, where "learning" is understood broadly to include learning of a suitable probability distribution, i.e., meaning learning from samples of training data. Other applications of an analysis of weighted graphs include statistical mechanics, such as infinite spin models, and large digital networks. In this respect it is natural to ask how to approach the analysis on "large" systems in optimal way by going over to infinite structures. In fact, it turns out that an analysis via infinite weighted graphs can be performed adequately (cf. [**478**]). This is so even if some of the questions in learning theory may, in fact, refer to only "large" finite graphs. Another reason (among others) for studying an infinite context is that *statistical features* in such an analysis are best predicted by the consideration of probability spaces corresponding to measures on infinite sample spaces. Moreover the latter are best designed from a consideration of infinite weighted graphs, as opposed to their finite counterparts.

An *infinite network* $(\mathcal{S}, \mathcal{E}, C)$ consists of the following ingredients:

- $\mathcal{S}$ forms a set of *vertices* with $\sharp\mathcal{S} = \aleph_0$, where $\aleph_0$ is the cardinality of the natural numbers,
- $\mathcal{E} \subset \mathcal{S} \times \mathcal{S}\backslash\{\text{diagonal}\}$ constitute the *edges,*
- $C : \mathcal{E} \to [0, \infty)$ is a fixed symmetric function representing *conductance.*

We assume that $(\mathcal{S}, \mathcal{E})$ is *connected,* i.e., for $x, y \in \mathcal{S}$, there exists $(x_i x_{i+1})_{i=0}^{n-1} \in \mathcal{E}$ such that $x_0 = x$, $x_n = y$.

Now, let $(\mathcal{S}, \mathcal{E}, C)$ be an infinite network. We denote by $H_{\mathcal{E}}$ the Hilbert space equipped, for $F, G \in H_{\mathcal{E}}$, with the inner product given by

$$\langle F, G\rangle_{H_{\mathcal{E}}} = \frac{1}{2} \sum_{(xy)\in\mathcal{E}} C_{xy} \left(\overline{F(x)} - \overline{F(y)}\right)(G(x) - G(y)), \tag{10.14}$$

i.e.,

$$H_{\mathcal{E}} = \left\{ F : \mathcal{S} \to \mathbb{C} : \ \|F\|^2_{H_{\mathcal{E}}} = \frac{1}{2} \sum_{(xy)\in\mathcal{E}} C_{xy} \, |F(x) - F(y)|^2 < \infty \right\}. \tag{10.15}$$

Since $(\mathcal{S}, \mathcal{E})$ is assumed to be connected, we are able to conclude that there exists, for any given pair $x, y \in \mathcal{S}$, an integer $n \in \mathbb{N}$ and $(x_i \; x_{i+1})_{i=0}^{n-1} \in \mathcal{E}$ such that $x_0 = x$ and $x_n = y$. Hence, for all $F \in H_{\mathcal{E}}$, we obtain by application of the Cauchy-Schwarz inequality

$$\begin{aligned} |F(y) - F(y)| &= \left| \sum_{i=0}^{n-1} F(x_{i+1}) - F(x_i) \right| \\ &= \left| \sum_{i=0}^{n-1} \frac{1}{\sqrt{C_{x_i x_{i+1}}}} \sqrt{C_{x_i x_{i+1}}} (F(x_{i+1}) - F(x_i)) \right| \\ &\leq \left(\sum_{i=0}^{n-1} \frac{1}{C_{x_i c_{i+1}}} \right)^{\frac{1}{2}} \left(\sum_{i=0}^{n-1} C_{x_i x_{i+1}} |F(x_{i+1}) - F(x_i)|^2 \right)^{\frac{1}{2}} \\ &\leq C \, \|F\|_{H_{\mathcal{E}}}, \end{aligned} \tag{10.16}$$

where C is a positive constant. Thus, the application of the Riesz Theorem of functional analysis to the Hilbert space $H_{\mathcal{E}}$ implies the existence of a member $v_{xy} \in H_E$ such that

$$F(x) - F(y) = \langle v_{xy}, F \rangle_{H_{\mathcal{E}}}, \quad F \in H_{\mathcal{E}}. \tag{10.17}$$

Summarizing our considerations (cf. [**471, 477**]) we therefore obtain the following two results:

(1) *$H_{\mathcal{E}}$ is a Hilbert space,*

(2) *For all $x, y \in \mathcal{S}$, there exists $v_{x,y} \in H_{\mathcal{E}}$ (called a dipole) such that*

$$F(x) - F(y) = \langle v_{xy}, F \rangle_{H_{\mathcal{E}}}, \quad F \in H_{\mathcal{E}}. \tag{10.18}$$

It is convenient to choose a *fixed base-point*, say $o \in V$, and set $v_x = v_{xo}$. In this case, (10.18) takes the form

$$F(x) - F(o) = \langle v_{xo}, F \rangle_{H_{\mathcal{E}}}, \quad F \in H_{\mathcal{E}}. \tag{10.19}$$

The Hilbert space $H_{\mathcal{E}}$ consists of "relative" constituents, thus, $H_{\mathcal{E}}$ is called a *relative RKHS*. The corresponding positive definite kernel is given as follows:

$$K(x, y) = \langle v_x, v_y \rangle_{H_{\mathcal{E}}}, \quad (x, y) \in \mathcal{S} \times \mathcal{S}. \tag{10.20}$$

All in all, a given infinite network $(\mathcal{S}, \mathcal{E}, C, H_{\mathcal{E}})$ is said to have a *finite point mass* if and only if

$$\delta_x \in H_{\mathcal{E}}, \quad x \in \mathcal{S}. \tag{10.21}$$

Obviously, the condition (10.21) will be automatic if, for all $x \in \mathcal{S}$,

$$\sharp\{y \in \mathcal{S} : (xy) \in \mathcal{E}\} < \infty, \tag{10.22}$$

but the finite-point mass case holds in many examples, where (10.22) is not assumed.

Assume that the condition (10.22) *is satisfied in* $(\mathcal{S}, \mathcal{E}, C, H_{\mathcal{E}})$. *For functions* F *on* $\mathcal{S}$, *set*

$$(\Delta_c F)(x) = \sum_{y \sim x} c_{xy}(F(x) - F(y)). \tag{10.23}$$

Suppose that $\{v_x\}_{x \in \mathcal{S} \setminus \{o\}}$ *is a system of dipoles (in the sense of* (10.18)*). We let*

$$K_x(y) = \langle v_y, v_x \rangle_H, \quad y \in \mathcal{S}. \tag{10.24}$$

Then a direct computation yields the identity

$$(\Delta_c K_x)(y) = \delta_{xy}, \quad x, y \in \mathcal{S} \setminus \{o\}. \tag{10.25}$$

10.3. Point masses of finite reproducing kernel Hilbert space norm

Next we turn to the general case of positive definite kernels and the case of (discrete) RKHSs as well as relative RKHSs, such that the point-mass condition (10.21) is satisfied. To this end, we consider a symmetric pair of operators (cf. [**478**]) associated with an RKHS having point masses of finite norm:

Symmetric pair of operators. Our purpose is to show that there is a canonically defined symmetric pair (A, B) of operators A, B characterized as follows: Assume that $\mathcal{S}$ is countably infinite. Let $K : \mathcal{S} \times \mathcal{S} \to \mathbb{C}$ (or $\mathbb{R}$) be positive definite. Let H be the corresponding RKHS. If (10.10) holds, i.e., H has the finite-mass property, then there exists a dual pair of operators (A, B) characterized by

$A : \ell^2(\mathcal{S}) \to H$, $\mathcal{D}(A) = \text{span}\{\delta_x\}$ is dense in $\ell^2(\mathcal{S})$, satisfying

$$A\delta_x = \delta_x \in H, \tag{10.26}$$

$B : H \to l^2(\mathcal{S})$, $\mathcal{D}(B) = \text{span}\{K_x\}$ is dense in H, satisfying

- in the case of a nonrelative DRKHS:

$$BK_x = \delta_x. \tag{10.27}$$

- in the case of a relative DRKHS:

$$BK_x = \delta_x - \delta_o. \tag{10.28}$$

In accordance with our nomenclature we set

$$\langle \xi, \eta \rangle_{\ell^2(\mathcal{S})} = \sum_{x \in \mathcal{S}} \overline{\xi(x)}\, \eta(x), \tag{10.29}$$

$$\|\xi\|^2_{\ell^2(\mathcal{S})} = \sum_{x \in \mathcal{S}} |\xi(x)|^2 < \infty, \tag{10.30}$$

$$\|K_x\|^2_H = K(x, x), \quad x \in \mathcal{S}. \tag{10.31}$$

It follows (cf. (10.27)- (10.28)) that

$$\langle A\delta_x, K_y \rangle_H = \langle \delta_x, BK_y \rangle_{\ell^2(\mathcal{S})}, \qquad x, y \in \mathcal{S}, \tag{10.32}$$

where

- in the case of a nonrelative DRKHS:

$$\langle \delta_x, K_y \rangle_H = \delta_{xy} = \langle \delta_x, \delta_y \rangle_{\ell^2(\mathcal{S})}, \tag{10.33}$$

- in the case of a relative DRKHS

$$\langle \delta_x, K_y \rangle_H = \delta_x(y) - \delta_x(o). \tag{10.34}$$

As an immediate consequence we obtain the following proposition:

The system (A, B) is a symmetric pair, i.e.,

$$\langle AU, V \rangle_H = \langle U, BV \rangle_{\ell^2(\mathcal{S})}, \quad U \in \mathcal{D}(A),\ V \in \mathcal{D}(B). \tag{10.35}$$

We use the following notation for the closure of linear operators $H_1 \xrightarrow{T} H_2$, where T has a dense domain in H_1 and the domain of T^* is assumed to be dense in H_2. The graph of T is given by

$$G(T) = \left\{ \begin{pmatrix} Q \\ TQ \end{pmatrix} : Q \in \mathrm{dom}(T) \right\} \subseteq \begin{pmatrix} H_1 \\ \oplus \\ H_2 \end{pmatrix}. \tag{10.36}$$

The $\begin{pmatrix} H_1 \\ \oplus \\ H_2 \end{pmatrix}$-closure of $G(T)$ is the graph of the "closure" of T, written by $\overline{T}$, in brief, $\overline{G(T)} = G(\overline{T})$. Therefore,

$$T^{**} = \overline{T}. \tag{10.37}$$

For the operators (A, B) as introduced above we are able to conclude the following results:

(1) *$A \subset B^*$ and $B \subset A^*$.*

(2) *$A^*\overline{A}$ and $B^*\overline{B}$ have dense domains and are selfadjoint, more concretely,* $\mathrm{dom}(A^*\overline{A})$ *is dense in $\ell^2(\mathcal{S})$, and $A^*\overline{A}$ is selfadjoint in $\ell^2(\mathcal{S})$,* $\mathrm{dom}(B^*\overline{B})$ *is dense in H, and $B^*\overline{B}$ is selfadjoint in H.*

(3) *From factorization methods in terms of unbounded operators (cf.* [**796**]*) it follows that*

$$\overline{A} = V(A^*\overline{A})^{\frac{1}{2}} = (\overline{A}A^*)^{\frac{1}{2}}V \tag{10.38}$$

$$\overline{B} = W(B^*\overline{B})^{\frac{1}{2}} = (\overline{B}B^*)^{\frac{1}{2}}W \tag{10.39}$$

with partial isometries $V : \ell^2(\mathcal{S}) \to H$, $W : H \to \ell^2(\mathcal{S})$, such that

$$V^*V = I_{\ell^2(\mathcal{S})} - \mathrm{Proj}\, ker(\overline{A}), \tag{10.40}$$

$$W^*W = I_H - \mathrm{Proj}\, ker(\overline{B}). \tag{10.41}$$

Next, we suppose that $(\mathcal{S}, K, H)$ satisfies the condition that $\delta_x \in H$, $x \in \mathcal{S}$. Let (A, B) be the associated symmetric pair from (10.35). We consider the spaces

$$\{Q \in H : (A^*Q)(x) = 0, \ \ x \in \mathcal{S}\} = \ker(A^*) \tag{10.42}$$

and

$$\overline{\mathrm{span}\{\delta_x : x \in \mathcal{S}\}}^{\|\cdot\|_H} = \{\delta_x : x \in \mathcal{S}\}^{\perp\perp}. \tag{10.43}$$

Since the term $\{\delta_x : x \in \mathcal{S}\}^{\perp\perp}$ refers to the inner product $\langle \cdot, \cdot \rangle_H$, we are able to show from the duality (10.35) that $\langle Q, \delta_x \rangle_H = 0$, $x \in \mathcal{S}$ holds if and only if

$(A^*Q)(x) = 0$, $x \in \mathcal{S}$. Moreover, if $\langle Q, \delta_x \rangle_H = 0$, $x \in \mathcal{S}$, then $Q \in \operatorname{dom}(A^*)$, and we have

$$(A^*Q)(x) = \langle \delta_x, A^*Q \rangle_{\ell^2(\mathcal{S})} = \langle A\delta_x, Q \rangle_H = \langle \delta_x, Q \rangle_H. \tag{10.44}$$

This leads to the following conclusion:

Let $(\mathcal{S}, K, H)$ satisfy the condition that $\delta_x \in H$, $x \in \mathcal{S}$. Let (A, B) be the associated symmetric pair satisfying (10.35)*. Then we have the orthogonal splitting*

$$H = \ker(A^*) \oplus \{\delta_x : x \in \mathcal{S}\}^{\perp\perp}. \tag{10.45}$$

We deal with the following significant questions: Assume that the point-mass property, i.e., $\delta_x \in H$, $x \in \mathcal{S}$, holds true. The question is how to compute the positive definite (p.d.) K with $\mathcal{S} \times \mathcal{S} \ni (x, y) \to K(x, y)$ and the induced positive definite (p.d.) kernel $\mathcal{S} \times \mathcal{S} \ni (x, y) \to \langle \delta_x, \delta_y \rangle_H$, where we are not assuming that $\delta_x \in \operatorname{dom}(\overline{B})$? In addition, we are confronted with the problems, how to compute the kernels $K(\cdot, \cdot)$ and $\langle \delta_x, \delta_y \rangle_H$? The answers are collected in the following way:

- *Let $K : \mathcal{S} \times \mathcal{S} \to \mathbb{C}$ (or $\mathbb{R}$) be positive definite, and assume that*

$$\delta_x \in H, \quad x \in \mathcal{S}, \tag{10.46}$$

i.e., H is a RKHS with point masses. Let (A, B) be the canonical symmetric pair. Then

$$K_x \in \operatorname{dom}(B^*\overline{B}), \quad x \in \mathcal{S}. \tag{10.47}$$

- *Let $(\mathcal{S}, K, H)$ such that $\delta_x \in H$, $x \in \mathcal{S}$.*
  1. *Then $AB : H \to H$ is a symmetric operator in the space H with dense domain* $\operatorname{span}\{K_x\} = \mathcal{D}(B)$*, and $AB\, K_x = \delta_x \in H$.*
  2. *Moreover, $B^*\overline{B}$ is a selfadjoint extension of AB (as an operator in H).*
- *Let $(\mathcal{S}, K, H)$ be given such that $\delta_x \in H$, $x \in \mathcal{S}$. Then $\delta_x \in \operatorname{dom}(A^*\overline{A})$, $x \in \mathcal{S}$, is equivalent to $\langle \delta_x, \delta_\cdot \rangle_H \in \ell^2(\mathcal{S})$, $x \in \mathcal{S}$, i.e.,*

$$\delta_x \in \operatorname{dom}(A^*\overline{A}), \quad x \in \mathcal{S} \tag{10.48}$$

$$\Updownarrow$$

$$\langle \delta_x, \delta_\cdot \rangle_H \in \ell^2(\mathcal{S}), \quad x \in \mathcal{S}. \tag{10.49}$$

- *Let $(\mathcal{S}, K, H)$ be given as above. Assume that $\delta_x \in H$, $x \in \mathcal{S}$. Then*

$$\langle \delta_x, \delta_y \rangle_H = \lim_{\mathcal{A}} (K^{(\mathcal{A})})^{-1}_{x,y}, \tag{10.50}$$

where the RKHS is the inductive limit over the filter of all finite subsets $\mathcal{A}$ of $\mathcal{S}$, and

$$K^{(\mathcal{A})} = [\langle K_x, K_y \rangle_H]_{(x,y) \in \mathcal{A} \times \mathcal{A}} = [K(x, y)]_{(x,y) \in \mathcal{A} \times \mathcal{A}}, \tag{10.51}$$

is a Gram matrix.

In fact, we obtain with (10.8)

$$
\begin{aligned}
\langle \delta_x, \delta_y \rangle_H &= \lim_A \langle \delta_x, P_A(\delta_y) \rangle_H \\
&= \lim_A \langle \delta_x, \sum_s (K^{(\mathcal{A})})^{-1}_{sy} K_s \rangle_H \\
&= \lim_A (K^{(\mathcal{A})})^{-1}_{xy},
\end{aligned} \tag{10.52}
$$

which proves the assertion (10.51).

10.4. Infinite square matrices

Let $(\mathcal{S}, K, H)$ be given in the above indicated way, assuming that $\delta_x \in H$, $x \in \mathcal{S}$. We consider the infinite $\mathcal{S} \times \mathcal{S}$ matrices D, K, and C, where D is the matrix determined by $D_{xy} = \langle \delta_x, \delta_y \rangle_H, (x,y) \in \mathcal{S} \times \mathcal{S}, D = [D_{xy}]_{(x,y) \in \mathcal{S} \times \mathcal{S}}$, $C = [C_{xy}]_{(x,y) \in \mathcal{S} \times \mathcal{S}}$ is given such that

$$
\delta_x = \sum_y C_{xy} K_y, \tag{10.53}
$$

and $K = [K_{xy}]_{(x,y) \in \mathcal{S} \times \mathcal{S}}$ is given by

$$
K_{xy} = K(x,y). \tag{10.54}
$$

It should be noted that, in the context of infinite matrices below for the infinite summations involved, we are making use of the limit concept of (10.52).

In terms of the identity matrix $I = [\delta_{xy}]_{(x,y) \in \mathcal{S} \times \mathcal{S}}$ we find $I = C\ K$. Forming the H-inner product of δ_x (as given in (10.53)) and K_z we obtain

$$
\delta_{xz} = \sum_{y \in \mathcal{S}} C_{xy} K_{yz} \Leftrightarrow I = C\ K. \tag{10.55}
$$

Observing $\langle \delta_x, K_z \rangle_H = \delta_{xz}$, we find

$$
\begin{aligned}
D_{xx'} &\overset{\text{by (10.4)}}{=} \langle \delta_x, \delta_{x'} \rangle_H \\
&\overset{\text{by (10.53)}}{=} \langle \sum_{y \in \mathcal{S}} C_{xy} K_y, \sum_{y' \in \mathcal{S}} C_{x',y} K_{y'} \rangle_H \\
&= \sum_y \sum_{y' \in \mathcal{S}} C_{xy} C_{x'y'} \langle K_y, K_{y'} \rangle_H \\
&= \sum_{y \in \mathcal{S}} \sum_{y' \in \mathcal{S}} C_{xy} K_{yy'} C_{x'y'} \\
&\overset{\text{by (10.54)}}{=} (CKC^T)_{xx'}
\end{aligned} \tag{10.56}
$$

which leads to the conclusion $D = CKC^T$.

In the consideration below we will be concerned with matrix algebra for "infinite square matrices." More precisely, we shall apply matrix algebra to pairs of matrices, where in each matrix factor, both rows and columns are indexed by the same given countable infinite set $\mathcal{S}$. We begin with

$$C = K^{-1}, \tag{10.57}$$

or equivalently,

$$(KC)_{xy} = \delta_{xz} \tag{10.58}$$

if and only if

$$\sum_z K_{xz} C_{zy} = \delta_{xy}. \tag{10.59}$$

Assuming that the given positive definite function K is real-valued we obtain $K = K^T$, since $K(x, y) = \langle K_x, K_y \rangle_H = \langle K_y, K_x \rangle_H = K(y, x)$. Hence, we finally have $D = C^T = (K^T)^{-1} = K^{-1}$.

10.5. Point masses and concrete models

We concretize our approach by the presentation of special examples:

Suppose that $\mathcal{S} \subset \mathcal{B} \subset \mathbb{R}^q$, where $\mathcal{S}$ is countable and discrete and $\mathcal{B}$ is a continuously connected domain. In this case, we obtain two kernels: K on $\mathcal{B} \times \mathcal{B}$, and $K^{(\mathcal{S})} = K|\mathcal{S} \times \mathcal{S}$ on $\mathcal{S} \times \mathcal{S}$ by restriction. If $x \in \mathcal{S}$, then $K_x^{(\mathcal{S})} = K(\cdot, x)$ is a function on $\mathcal{S}$, while $K_x = K(\cdot, x)$ is a function on $\mathcal{B}$. Clearly, the corresponding RKHSs are different, $H_\mathcal{S}$ versus H, where $H_\mathcal{S}$ is a RKHS of functions on $\mathcal{S}$, while H is a RKHS of functions on $\mathcal{B}$.

If $\mathcal{A} \in \mathcal{S}$ is a finite subset, and Ξ is a function on $\mathcal{A}$, then we have

$$\| \sum_{x \in \mathcal{A}} \Xi(x) K_x^{(\mathcal{S})} \|_{H_\mathcal{S}} = \| \sum_{x \in \mathcal{A}} \Xi(x) K_x \|_H. \tag{10.60}$$

As a consequence (cf. [**478**]) this implies the following result:

$H_\mathcal{S}$ is isometrically contained in H via $K_x^{(\mathcal{S})} \mapsto K_x, x \in \mathcal{S}$.

We discuss special cases of kernels $K : \mathcal{B} \times \mathcal{B} \to \mathbb{R}$ with restriction $K^{(\mathcal{S})} : \mathcal{S} \times \mathcal{S} \to \mathbb{R}$, where $\mathcal{S}$ is a countable discrete subset of $\mathcal{B}$. Indeed, δ_{x_i} has a different meaning with reference to the Hilbert spaces $H_\mathcal{S}$ and H, respectively:

- In the first case, it is simply δ_{x_i} given by

$$\delta_{x_i}(y) = \begin{cases} 1 & y = x_i \\ 0 & y \in \mathcal{S} \backslash \{x_i\}. \end{cases} \tag{10.61}$$

- In the second case, δ_{x_i} is a Schwartz distribution.

Note that we shall abuse a notation, writing δ_x in both cases.

Cameron-Martin Hilbert space. We consider the restriction from $[0,\infty)\times[0,\infty)$ to $\mathcal{S}\times\mathcal{S}$ of a special continuous positive definite kernel K on $[0,\infty)\times[0,\infty)$, namely

$$K(s,t) = s \wedge t = \min(s,t). \tag{10.62}$$

The resulting RKHS of the kernel $K(\cdot,\cdot)$ in (10.62) is the Cameron-Martin Hilbert space of functions F on $[0,\infty)$ with distributional derivative $F' \in L^2([0,\infty))$, equipped with the norm

$$\|F\|_H^2 = \int_0^\infty |F'(t)|^2\,dt < \infty. \tag{10.63}$$

In the Cameron-Martin Hilbert space framework of H, pick a discrete subset $\mathcal{S}\subset[0,\infty)$, then we know that the RKHS of the $\mathcal{S}\times\mathcal{S}$ restricted kernel, $K^{(\mathcal{S})}$, is isometrically embedded into H. More concretely, by setting $J^{(\mathcal{S})}(K_x^{(\mathcal{S})}) = K_x, \quad x\in\mathcal{S}$, we obtain an operator $J^{(\mathcal{S})}$ extending by "closed span" to an isometry $H_{\mathcal{S}} \xrightarrow{J^{(\mathcal{S})}} H$. It further follows that the range of $J^{(\mathcal{S})}$ may have an infinite co-dimension. In fact, $P_{\mathcal{S}} = J^{(\mathcal{S})}(J^{(\mathcal{S})})^*$ is the projection onto the range of $J^{(\mathcal{S})}$. The ortho-complement is as follows:

$$H\ominus H_{\mathcal{S}} = \{\Psi\in H : \Psi(x) = 0,\ x\in\mathcal{S}\}. \tag{10.64}$$

For K and $K^{(\mathcal{S})}$ given as in (10.5) we especially study the set $\mathcal{S}=\mathbb{N}$. Then, 1D generators of wavelet functions (cf. [**114**], [**694**]) provide nonzero functions Ψ on $[0,\infty)$, such that $\Psi\in H\ominus H_{\mathcal{S}}$. More precisely,

$$0 < \int_0^\infty |\Psi'(t)|^2\,dt < \infty, \tag{10.65}$$

where Ψ' is the distributional (weak) derivative. Furthermore,

$$\Psi(g) = 0, \quad g\in\mathbb{N}. \tag{10.66}$$

An explicit solution to (10.5), (10.65), (10.66) is given by $\Psi(t) = \text{sinc}\,(\pi t),\ t\in\mathbb{R}$. This enables us to generate an infinite-dimensional set of solutions.

Brownian motion. Consider the covariance function of the standard Brownian motion $B_t, t\in[0,\infty)$, i.e., a Gaussian process $\{B_t\}_{t\in[0,\infty)}$ with mean zero and covariance function

$$E(B_sB_t) = s\wedge t = \min(s,t). \tag{10.67}$$

We fix a set $\mathcal{S}$ by $\mathcal{S}:\ 0 < x_1 < x_2 < \ldots < x_i < x_{i+1} < \ldots$. Our purpose is to verify (see [**478**]) that the restriction to $\mathcal{S}\times\mathcal{S}$ for the ordered set(10.5) shows the discrete mass property: Let $H_{\mathcal{S}}$ be the RKHS corresponding to

$$K^{(\mathcal{S})}(x,y) = x\wedge y = \min(x,y), (x,y)\in\mathcal{S}\times\mathcal{S}. \tag{10.68}$$

For the ordered sets $\mathcal{A}_n = \{x_1, x_2, \ldots, x_n\}$ of finite subsets of $\mathcal{S}$ we consider the matrices

$$K^{(\mathcal{A}_n)} = \begin{bmatrix} x_1 & x_1 & x_1 & \ldots & x_1 \\ x_1 & x_2 & x_2 & \ldots & x_2 \\ x_1 & x_2 & x_3 & \ldots & x_3 \\ \ddots & \ddots & \ddots & \ddots & \ddots \\ x_1 & x_2 & x_3 & \ldots & x_n \end{bmatrix} = (x_i\wedge x_j)_{i,j=1}^n \tag{10.69}$$

Our goal is to guarantee (10.12) for K_S. To do so we have to compute all the determinants $D_n = \det(K^{(\mathcal{A}_n)})$. It is not hard to show (see [**478**]) that the matrix

$$\begin{bmatrix} x_1 & x_1 & x_1 & \dots & x_1 \\ x_1 & x_2 & x_2 & \dots & x_2 \\ x_1 & x_2 & x_3 & \dots & x_3 \\ \ddots & \ddots & \ddots & \ddots & \ddots \\ x_1 & x_2 & x_3 & \dots & x_n \end{bmatrix} \tag{10.70}$$

can be transformed by elementary manipulations in unitary equivalence to

$$\begin{bmatrix} x_1 & 0 & 0 & \dots & 0 \\ 0 & x_1 - x_2 & 0 & \dots & 0 \\ 0 & 0 & x_3 - x_2 & \dots & 0 \\ \vdots & \vdots & \vdots & \ddots & \vdots \\ 0 & \dots & 0 & \dots & x_n - x_{n-1} \end{bmatrix}. \tag{10.71}$$

Thus, by induction, we obtain

$$D_n = \det\left((x_i \wedge x_j)_{i,j=1}^n\right) = x_1(x_2 - x_1)(x_3 - x_2)\dots(x_n - x_{n-1}). \tag{10.72}$$

In connection with (10.8), this leads to

$$\|P_{\mathcal{A}_n}(\delta_{x_1})\|_{H_S}^2 = (K^{(\mathcal{A}_n)})^{-1}(\delta_{x_1})(x_1). \tag{10.73}$$

We introduce the abbreviation $\zeta_{(n)} = (K^{(\mathcal{A}_n)})^{-1}(\delta_{x_1})(\cdot)$. Then, from (10.73) it follows that

$$\|P_{\mathcal{A}_n}(\delta_{x_1})\|_{H_S}^2 = \zeta_{(n)}(x_1). \tag{10.74}$$

A direct computation shows the (1,1)-minor of the matrix $(K^{(\mathcal{A}_n)})^{-1}$ is given by

$$D'_{n-1} = \det((x_i \wedge x_j)_{i,j=2}^n) = x_2(x_3 - x_2)(x_4 - x_3)\dots(x_n - x_{n-1}). \tag{10.75}$$

Thus, straightforward calculation gives

$$\zeta_{(1)}(x_1) = \frac{1}{x_1}, \tag{10.76}$$

$$\zeta_{(2)}(x_1) = \frac{x_2}{x_1(x_2 - x_1)}, \tag{10.77}$$

$$\zeta_{(3)}(x_1) = \frac{x_2(x_3 - x_2)}{x_1(x_2 - x_1)(x_3 - x_2)} = \frac{x_2}{x_1(x_2 - x_1)}, \tag{10.78}$$

$$\zeta_{(4)}(x_1) = \frac{x_2(x_3 - x_2)(x_4 - x_3)}{x_1(x_2 - x_1)(x_3 - x_2)(x_4 - x_3)} = \frac{x_2}{x_1(x_2 - x_1)}. \tag{10.79}$$

$$\vdots$$

Altogether, we have

$$\zeta_{(1)}(x_1) = \frac{1}{x_1}, \tag{10.80}$$

$$\zeta_{(n)}(x_1)) = \frac{x_1}{x_1(x_2 - x_1)}, \quad n = 2, 3, \dots, \tag{10.81}$$

and

$$\|\delta_{x_1}\|_{H_S}^2 = \frac{x_2}{x_1(x_2 - x_1)}. \tag{10.82}$$

For all $n \geq 2$, we have

$$P_{\mathcal{A}_n}(\delta_{x_1}) = P_{\mathcal{A}_2}(\delta_{x_1}). \tag{10.83}$$

Therefore,

$$\delta_{x_1} \in H_{\mathcal{S}}^{(\mathcal{A}_2)} = span\{K_{x_1}^{(\mathcal{S})}, K_{x_2}^{(\mathcal{S})}\} \tag{10.84}$$

and

$$\delta_{x_1} = \zeta_{(2)}(x_1)K_{x_1}^{(\mathcal{G})} + \zeta_{(2)}(x_2)K_{x_2}^{(\mathcal{G})}. \tag{10.85}$$

In particular,

$$\zeta_{(2)}(x_1) = \frac{x_2}{x_1(x_2 - x_1)}, \tag{10.86}$$

$$\zeta_{(2)}(x_2) = -\frac{1}{x_2 - x_1} = \frac{1}{x_1 - x_2}, \tag{10.87}$$

and

$$\|\delta_{x_1}\|_{H_S}^2 = \frac{x_2}{x_1(x_2 - x_1)}. \tag{10.88}$$

We come back to the situation $\delta_{x_1} \in H_{\mathcal{S}}$, $\mathcal{S} = \{x_1 < x_2 < \ldots\} \subset [0, \infty)$ under the restriction (10.67), i.e., the covariance kernel of Brownian motion. We want to investigate the situation that $\delta_{x_i} \in H_{\mathcal{S}}$, even if $i > 1$. For that purpose we only need to modify the index notation as known from $\delta_{x_1} \in \mathcal{H}_{\mathcal{S}}$. The details are sketched as follows:

Fix $\mathcal{S} = \{x_i\}_{i=1}^{\infty}$, $x_1 < x_2 < \ldots$. Then it follows that

$$P_{\mathcal{A}_n}(\delta_{x_i}) = \begin{cases} 0 & \text{if} \quad n < i - 1 \\ \sum_{s=1}^{n} (K^{(\mathcal{A}_n)})^{-1}\delta_{x_i})(x_s)K_{x_s} & \text{if} \quad n \geq i \end{cases} \tag{10.89}$$

and

$$\|P_{\mathcal{A}_n}(\delta_{x_i})\|_H^2 = \begin{cases} 0, & \text{if} \quad n < i - 1, \\ \frac{1}{x_i - x_{i-1}}, & \text{if} \quad n = i, \\ \frac{x_{i+1} - x_{i-1}}{(x_i - x_{i-1})(x_{i+1} - x_i)}, & \text{if} \quad n > i. \end{cases} \tag{10.90}$$

This leads to the following conclusion:

$$\delta_{x_i} \in \operatorname{span}\{K_{x_{i-1}}^{(\mathcal{S})}, K_{x_i}^{(\mathcal{S})}, K_{x_{i+1}}^{(\mathcal{S})}\}, \tag{10.91}$$

and

$$\|\delta_{x_i}\|_H^2 = \frac{x_{i+1} - x_{i-1}}{(x_i - x_{i-1})(x_{i+1} - x_i)}. \tag{10.92}$$

Moreover, we obtain the following result:

If $\mathcal{S} \subset [0, \infty)$ is countable and $x_l \in \mathcal{S}$ is an accumulation point (from $\mathcal{S}$), then we have $\|\delta_{x_l}\|_{H_{\mathcal{S}}} = \infty$.

Example. We mention the special case $\mathcal{A}_4$ with $0 < x_1 < x_2 < x_3 < x_4$:

$$P_{\mathcal{A}_4}(\delta_{x_3}) = \sum_{y\in\mathcal{A}_4} \zeta^{(\mathcal{A}_4)}(y)K_y(\cdot) \tag{10.93}$$
$$\zeta^{(\mathcal{A}_4)} = (K^{(\mathcal{A}_4)})^{-1}\delta_{x_3}.$$

That is

$$\underbrace{\begin{bmatrix} x_1 & x_1 & x_1 & x_1 \\ x_1 & x_2 & x_2 & x_2 \\ x_1 & x_2 & x_3 & x_3 \\ x_1 & x_2 & x_3 & x_4 \end{bmatrix}}_{=\mathbf{K}^{(\mathcal{A}_4)}(x_i,x_j))_{i,j=1}^4} \begin{bmatrix} \zeta^{(\mathcal{A}_4)}(x_1) \\ \zeta^{(\mathcal{A}_4)}(x_2) \\ \zeta^{(\mathcal{A}_4)}(x_3) \\ \zeta^{(\mathcal{A}_4)}(x_4) \end{bmatrix} = \begin{bmatrix} 0 \\ 0 \\ 1 \\ 0 \end{bmatrix} \tag{10.94}$$

and

$$\begin{aligned} \zeta^{(\mathcal{A}_4)}(x_3) &= \frac{x_1(x_2-x_1)(x_4-x_2)}{x_1(x_2-x_1)(x_3-x_2)(x_4-x_3)} \\ &= \frac{x_4-x_2}{(x_3-x_2)(x_4-x_3)} \\ &= \|\delta_{x_3}\|_H^2. \end{aligned} \tag{10.95}$$

Sparse sample points. Let $\mathcal{S} = \{x_i\}_{i=1}^{\infty}$, where $x_i = \frac{i(i-1)}{2}$, $i \in \mathbb{N}$. Clearly, $x_{i+1} - x_i = i$, and so it is easy to see that

$$\|\delta_{x_i}\|_H^2 = \frac{x_{i+1}-x_{i-1}}{(x_i-x_{i-1})(x_{i+1}-x_i)} = \frac{2i-1}{(i-1)i} \longrightarrow 0,\ i\to\infty. \tag{10.96}$$

It follows that $\|\delta_{x_i}\|_H \to 0$ for $i\to\infty$, if the set $\mathcal{S} = \{x_i\}_{i=1}^{\infty} \subset [0,\infty)$ is sparse.

Nested set sequences. Suppose $\mathcal{S} = \cup_n \mathcal{A}_n$ with $\mathcal{A}_n \subset \mathcal{A}_{n+1}$. If $x_1 \in \mathcal{A}_n$, then we have

$$P_{\mathcal{A}_n}(\delta_{x_1}) = \sum_{y\in\mathcal{A}_n} \langle x_1, (K^{(\mathcal{A}_n)})^{-1}y\rangle_{\ell^2} K_y \tag{10.97}$$

and

$$\|P_{\mathcal{A}_n}(\delta_{x_1})\|_H^2 = \langle x_1, (K^{(\mathcal{A}_n)})^{-1}x_1\rangle_{\ell^2(\mathcal{S})} = (K^{(\mathcal{A}_n)})^{-1}\delta_{x_1})(x_1). \tag{10.98}$$

The expression $\|P_{\mathcal{A}_n}(\delta_{x_1})\|_{\mathcal{S}}^2$ is monotone in n, i.e.,

$$\|P_{\mathcal{A}_n}(\delta_{x_1})\|_H^2 \le \|P_{\mathcal{A}_{n+1}}(\delta_{x_1})\|_H^2 \le \cdots \le \|\delta_{x_1}\|_H^2, \tag{10.99}$$

such that

$$\sup_{n\in\mathbb{N}} \|P_{\mathcal{A}_n}(\delta_{x_1})\|_H^2 = \lim_{n\to\infty} \|P_{\mathcal{A}_n}(\delta_{x_1})\|_H^2 = \|\delta_{x_1}\|_H^2. \tag{10.100}$$

For other applications of reproducing kernel Hilbert spaces to the analysis of Gaussian processes the reader is referred to, e.g., [**474**].

Next we deal with the following question: Let $\mathcal{S} \subset \mathbb{R}$ be a countable discrete subset (for example, $\mathcal{S} = \mathbb{Z}$). The question is when does $K^{(\mathcal{S})} = K|\mathcal{S}\times\mathcal{S}$ show the discrete mass property? Positive answers, or not, will be discussed below for some examples in more detail:

Discrete Cameron-Martin RKHSs from restrictions. We let $\mathcal{B} = [0, \infty)$, and $K : \mathcal{B} \times \mathcal{B} \to \mathbb{R}$ with $K(x, y) = x \wedge y = \min(x, y)$, so that H represents the *Cameron-Martin Hilbert space*, consisting of functions $F \in L^2(\mathbb{R})$ such that

$$\int_0^\infty |F'(x)|^2 \, dx < \infty, \quad F(0) = 0. \tag{10.101}$$

Restriction 1: We choose $\mathcal{S} = \mathbb{N}_0 \subset \mathcal{B}$ and write, as usual, $H_{\mathcal{S}}$. The space $H_{\mathcal{S}}$ consists of all F satisfying

$$\sum_n |F(n) - F(n+1)|^2 < \infty. \tag{10.102}$$

We remember the discrete Laplacian $\Delta = \Delta_C$ (see (10.23)), i.e.,

$$(\Delta F)(x) = \sum_{y \sim x} C_{xy}(F(x) - F(y)), \tag{10.103}$$

with C the corresponding conductance, and apply Δ_C to functions F on $\mathcal{S} = \mathbb{N}_0 \subset \mathcal{B}$. Setting $C = 1$ we obtain

$$(\Delta F)(n) = 2F(n) - F(n-1) - F(n+1). \tag{10.104}$$

By vitrue of (10.25) we have $\Delta K_n = \delta_n$, and we arrive at the following result:

$$\delta_n = 2K_n - K_{n+1} - K_{n-1} \in H_{\mathcal{S}}. \tag{10.105}$$

Note that

$$\langle 2K_n - K_{n+1}, K_m \rangle_{H_{\mathcal{S}}} = \langle \delta_n, K_m \rangle_{H_{\mathcal{S}}} = \delta_{n,m}. \tag{10.106}$$

Restriction 2: Fix vertices $\mathcal{S} = \{x_i\}_{i=1}^\infty$,

$$0 < x_1 < x_2 < \ldots < x_i < x_{i+1} < \infty, \quad x_i \to \infty. \tag{10.107}$$

Assign the conductance as the reciprocal distance between two subsequent elements

$$C_{i,i+1} = C_{i+1,i} = \frac{1}{x_{i+1} - x_i}. \tag{10.108}$$

Now, in the RKHS-framework of

$$H_{\mathcal{S}} = \left\{ F : \mathcal{S} \to \mathbb{C} : \sum_{i=1}^\infty C_{i,i+1} \left| F(x_{i+1}) - F(x_i) \right|^2 < \infty \right\}. \tag{10.109}$$

application to the discrete Laplacian yields

$$(\Delta F)(x_i) = \left(\frac{1}{x_{i+1} - x_i} + \frac{1}{x_i - x_{i-1}} \right) F(x_i) \tag{10.110}$$

$$- \frac{1}{x_i - x_{i-1}} F(x_{i-1}) - \frac{1}{x_{i+1} - x_i} F(x_{i+1}). \tag{10.111}$$

Equivalently, we have

$$(\Delta F)(x_i) = (C_{i,i+1} + C_{i,i-1})F(x_i) - C_{i,i-1}F(x_{i-1}) - C_{i,i+1}F(x_{i+1}). \tag{10.112}$$

In connection with (10.112) it follows that $\Delta K_{x_i} = \delta_{x_i}$ where $K : \mathcal{B} \times \mathcal{B} \to \mathbb{R}$, $K(x, y) = x \wedge y = \min(x, y)$, provides the restriction from $[0, \infty) \times [0, \infty)$ to $\mathcal{S} \times \mathcal{S}$. So, we obtain

$$\delta_{x_i} = (C_{i,i+1} + C_{i,i-1})K_{x_i} - C_{i,i+1}K_{x_{i+1}} - C_{i,i-1}K_{x_{i-1}} \in H_{\mathcal{G}}, \tag{10.113}$$

so that the right side in the last equation is a finite sum.

Brownian bridge. We choose $\mathcal{B}$ as the interval $[0,1]$, and set

$$K_{\text{bri}}(s,t) = s \wedge t - st, \ (s,t) \in [0,1] \times [0,1]. \tag{10.114}$$

Then, (10.114) is the covariance function for the Brownian bridge $B_{bri}(t)$, i.e.,

$$B_{\text{bri}}(0) = B_{bri}(1) = 0 \tag{10.115}$$

$$B_{\text{bri}}(t) = (1-t)\ B\left(\frac{t}{1-t}\right), \quad 0 < t < 1; \tag{10.116}$$

where $B(t)$ is Brownian motion at t. The associated Cameron-Martin space H_{bri} is given by

$$H_{\text{bri}} = \{F : F' \in L^2([0,1]), F(0) = F(1) = 0\} \tag{10.117}$$

with

$$\|F\|^2_{H_{\text{bri}}} = \int_0^1 |F'(s)|^2\, ds < \infty. \tag{10.118}$$

If $\mathcal{S} = \{x_i\}_{i=1}^{\infty}$, $x_1 < x_2 < \ldots < 1$, is a discrete subset of $\mathcal{B}$, then we have for $\mathcal{A}_n \in \mathcal{F}(\mathcal{S})$, $\mathcal{A}_n = \{x_1, x_2, \ldots, x_n\}$,

$$K^{(\mathcal{A}_n)} = [K_{\text{bri}}(x_i, x_j)]_{i,j=1}^n, \tag{10.119}$$

so that (cf. (10.114)

$$\det\ K^{(\mathcal{A}_n)} = x_1(x_2 - x_1)\ldots(x_n - x_{n-1})(1 - x_n). \tag{10.120}$$

As a result, we obtain $\delta_{x_i} \in H_{\mathcal{S}}^{bri}$ for all i, and we have

$$\|\delta_{x_i}\|^2_{H_{\mathcal{S}}^{\text{bri}}} = \frac{x_{i+1} - x_{i-1}}{(x_{i+1} - x_i)(x_i - x_{i-1})}. \tag{10.121}$$

In fact,

$$\lim_{x_i \to 1} \|\delta_{x_i}\|^2_{H_{\mathcal{S}}^{\text{bri}}} = \infty. \tag{10.122}$$

Binomial RKHS. It is possible to associate a positive definite kernel to the standard binomial coefficients (cf. [**9**], [**478**]). Next we outline the properties ofthe resulting framework. Among the conclusions is that the point-masses have infinite RKHS H-norm.

Suppose that $\mathcal{S}$ is equal to the set $\mathbb{N}_0$. Let

$$K_{H_b}(x,y) = \sum_{n=0}^{x \wedge y} \binom{x}{n}\binom{y}{n}, \quad (x,y) \in \mathcal{S} \times \mathcal{S}, \tag{10.123}$$

where, as usual,

$$\binom{x}{n} = \frac{x(x-1)\ldots(x-n+1)}{n!} \tag{10.124}$$

denotes the standard binomial coefficient from the binomial expansion, so that H_b is the corresponding RKHS. Set

$$E_n(x) = \left\{ \begin{array}{ll} \binom{x}{n} & \text{if} \quad n \le x \\ 0 & \text{if} \quad n > x. \end{array} \right\} \tag{10.125}$$

Under these assumptions the following properties (cf. [**9**]) hold true:

(i) $E_n \in H_b, n \in \mathcal{S}$,

(ii) $\{E_n\}_{n\in\mathcal{S}}$ is an orthonormal basis in the Hilbert space H_b.

(iii) Set $\mathcal{A}_n = \{0, 1, 2, \dots, n\}$. Then the expansion

$$P_{\mathcal{A}_n} F = \sum_{k=0}^{n} \langle E_k, F\rangle_{H_b} E_k \tag{10.126}$$

is well defined for all functions for all functions $F : \mathcal{S} \to \mathbb{C}$.

(iv) For given $F : \mathcal{S} \to \mathbb{C}$ of the form (10.126)

$$F \in H_b \Leftrightarrow \sum_{k=0}^{\infty} |\langle E_k, F\rangle_{H_b}|^2 < \infty; \tag{10.127}$$

and, in this case, we have

$$\|F\|^2_{H_b} = \sum_{k=0}^{\infty} |\langle E_k, F\rangle_{H_b}|^2. \tag{10.128}$$

Fix $x_1 \in \mathcal{S}$, and let $F_1 = \delta_{x_1}$ with

$$F_1(y) = \begin{cases} 1 & \text{if } \ y = x_1 \\ 0 & \text{if } \ y \neq x_1, \end{cases} \tag{10.129}$$

then we have

$$\|P_{\mathcal{A}_n}(\delta_{x_1})\|^2_{H_b} = \sum_{k=x_1}^{n} \binom{k}{x_1}^2. \tag{10.130}$$

In [**9**], as auxiliary result, it is verified that, for all $m, n \in \mathcal{S}$, such that $m \leq n$, the identity

$$\delta_{mn} = \sum_{j=m}^{n} (-1)^{m+j} \binom{n}{j}\binom{j}{m}. \tag{10.131}$$

holds true. This result is an essential tool to invert the matrix $L^{(n)}$, that is simply a truncated Pascal triangle, arranged to fit into a lower triangular matrix:

$$L^{(n)} = \begin{bmatrix}
1 & 0 & 0 & 0 & \cdots & \cdots & 0 & \cdots & 0 & 0 \\
1 & 1 & 0 & 0 & \cdots & \cdots & 0 & \cdots & 0 & 0 \\
1 & 2 & 1 & 0 & & & \vdots & & \vdots & \vdots \\
1 & 3 & 3 & 1 & \ddots & & \vdots & & \vdots & \vdots \\
\vdots & \vdots & \vdots & \vdots & \ddots & & \vdots & & \vdots & \vdots \\
\vdots & \vdots & \vdots & \vdots & & 1 & 0 & & \vdots & \vdots \\
1 & \cdots & \binom{x}{y} & \binom{x}{y+1} & \cdots & * & 1 & \ddots & \vdots & \vdots \\
\vdots & \vdots & \vdots & \vdots & & & & \ddots & 0 & \vdots \\
\vdots & \vdots & \vdots & \vdots & & & & & 1 & 0 \\
1 & \vdots & \binom{n}{y} & \binom{n}{y+1} & \cdots & \cdots & \cdots & \cdots & n & 1
\end{bmatrix}. \tag{10.132}$$

In more detail, for all $n \in \mathbb{N}$, the inverse of the lower triangle matrix $L^{(n)}$ with the entries

$$(L^{(n)})_{xy} = \begin{cases} \binom{x}{y} & \text{if} \quad y \leq x \leq n \\ 0 & \text{if} \quad x < y \end{cases} \tag{10.133}$$

is given by

$$(L^{(n)})^{-1}_{xy} = \begin{cases} (-1)^{x-y}\binom{x}{y} & \text{if} \quad y \leq x \leq n \\ 0 & \text{if} \quad x < y. \end{cases} \tag{10.134}$$

All in all, we are able to formulate the following results:

Let H_b, K_{H_b}, be defined as indicated above with the lower triangle matrix $L^{(n)}$, $n \in \mathbb{N}$. We introduce the $(n+1)\times(n+1)$ matrix

$$K^{(n)} = [K_{H_b}(x,y)]_{(x,y)\in\mathcal{A}_n\times\mathcal{A}_n}. \tag{10.135}$$

(i) *Then $K^{(n)}$ is invertible with*

$$(K^{(n)})^{-1} = ((L^{(n)})^T)^{-1}(L^{(n)})^{-1}, \tag{10.136}$$

i.e., $(K^{(n)})^{-1}$ constitutes an (upper triangle) $\times$ (lower triangle) factorization.

(ii) *For the diagonal entries in the $(n+1)\times(n+1)$ matrix $(K^{(n)})^{-1}$ we have*

$$\langle x, (K^{(n)})^{-1}x\rangle_{l^2} = \sum_{k=x}^{n} \binom{k}{x}^2. \tag{10.137}$$

Thus we obtain

$$\|P_{\mathcal{A}_n}(\delta_{x_1})\|^2_{H_b} = \langle x_1, (K^{(n)})^{-1}x_1\rangle_{H_b} \tag{10.138}$$

for all $x_1 \in \mathcal{A}_n$, so that

$$\|P_{\mathcal{A}_n}(\delta_{x_1})\|^2_{H_b} = \sum_{k=x_1}^{n} \binom{k}{x_1}^2 \tag{10.139}$$

$$= 1 + \binom{x_1+1}{x_1}^2 + \binom{x_1+2}{x_1}^2 + \ldots + \binom{n}{x_1}^2. \tag{10.140}$$

Therefore we have

$$\|\delta_{x_1}\|^2_{H_b} = \sum_{k=x_1}^{\infty} \binom{k}{x_1}^2 = \infty. \tag{10.141}$$

The point-mass property in H_b is violated, i.e., no δ_x is in the RKHS H_b.

10.6. Point mass samples and interpolation

Let H be an RKHS generated by a positive definite kernel $(x,y) \mapsto K(x,y)$, $(x,y) \in \mathcal{S}\times\mathcal{S}$.

A discrete subset $\mathcal{S}$ of a continuously connected domain $\mathcal{G}$ is said to form *a set of point-mass samples*, if and only if the span of $\{K_x : x \in \mathcal{S}\}$ is dense in H, i.e.,

$$H \; = \; \overline{\operatorname{span}_{x\in\mathcal{S}} K(\cdot,x)}^{\|\cdot\|_H} \tag{10.142}$$

Let $\mathcal{G}, K$ and, thus, H be given, and assume that $\mathcal{S} \subset \mathcal{G}$ is a countable discrete subset. Then $\mathcal{S}$ forms a set of point-mass samples, if there exists $\varepsilon > 0$ such that

$$\sum_{x \in \mathcal{S}} |F(x)|^2 \geq \varepsilon \, \|F\|_H^2, \quad F \in H. \tag{10.143}$$

(In order to show that $\text{span}\{K_x : x \in \mathcal{S}\}$ is dense in H, we only have to verify that

$$0 = F(x) = \langle K_x, F \rangle_H, \quad x \in \mathcal{S}, \tag{10.144}$$

implies $F = 0$ in H. This conclusion, however, follows immediately from the estimate (10.143)).

Let $K : \mathcal{G} \times \mathcal{G} \to \mathbb{C}$ be positive definite, and let $\mathcal{S}$ be the corresponding RKHS. Let $\mathcal{S} \subset \mathcal{G}$ be a set of point-mass samples. Designate by $H^{(\mathcal{S})}$ the RKHS of the restricted kernel function $K^{(\mathcal{S})}$, defined by

$$K^{(\mathcal{S})}(x, y) = K(x, y), \quad (x, y) \in \mathcal{S} \times \mathcal{S}. \tag{10.145}$$

Then the assignment

$$W^{(\mathcal{S})} K_x^{(\mathcal{S})} = K_x, \quad x \in \mathcal{S} \tag{10.146}$$

extends by linearity and norm-closure to an isometry $W^{(\mathcal{S})}$ of $H^{\mathcal{S})}$ onto H.

Suppose now that

$$F = \sum_{x \in \mathcal{S}} C_x \; K_x \tag{10.147}$$

is a sum representation of F in H with uniquely determined coefficients $\{C_x\}_{x \in \mathcal{S}}$. For $y \in \mathcal{S}$, it follows that

$$\langle \delta_y, F \rangle_H = \sum_{x \in \mathcal{S}} C_x \underbrace{\langle \delta_y, K_x \rangle_H}_{= \delta_{yx}} = C_y \tag{10.148}$$

and

$$\langle \delta_y, F \rangle_H = \langle A\delta_y, F \rangle_H = \langle \delta_y, A^* F \rangle_{\ell^2} = (A^* F)(y), \quad y \in \mathcal{S}. \tag{10.149}$$

This leads to an *interpolation formula* within the context $\mathcal{G}, K, H, \mathcal{S}$ given above.

Let $\mathcal{S} \subset \mathcal{G}$ be a set of point-mass samples. Let (A, B) be the associated dual pair of operators as described by (10.26), (10.27), *and* (10.28). *Then, for all $F \in H$,*

$$F = \sum_{s \in \mathcal{S}} (A^* F)(s) K_s. \tag{10.150}$$

Equivalently,

$$F(x) = \sum_{s \in \mathcal{S}} (A^* F)(s) K(x, s), \quad x \in \mathcal{G}, \tag{10.151}$$

and the convergence in (10.150), (10.151) *is guaranteed if and only if*

$$\sum_{(s,t) \in \mathcal{S} \times \mathcal{S}} \overline{(A^* F)(s)} (A^* F)(t) K(s, t) < \infty \tag{10.152}$$

Moreover,

$$\|F\|_H^2 = \sum_{(s,t)\in S\times S} \overline{(A^*F)(s)}(A^*F)(t)K(s,t), \tag{10.153}$$

provided that (10.152) *holds true.*

It should be noted that there are many classical RKHSs $(K, H, \mathcal{G})$, which contain sets of point-mass samples, but nonetheless, the point-masses δ_x are not in H, i.e., $\|\delta_x\|_H = \infty$ for all $x \in \mathcal{G}$.

Shannon's sampling formula. Let $\mathcal{G}$ be the Euclidean space $\mathbb{R}$, and let

$$H = \left\{ F \in L^2(\mathbb{R}) : \operatorname{supp}(F_{\mathbb{R}}^{\wedge}) \subset \mathcal{F}_{\mathbb{Z}} = \left[-\frac{1}{2}, \frac{1}{2} \right) \right\}. \tag{10.154}$$

It is known that $(x, y) \mapsto K(x, y) = \operatorname{sinc}(\pi(x - y)), x, y \in \mathbb{R}$, represents a positive definite kernel turning H into an RKHS with the RK $K(\cdot,\cdot)$. Then, in the above nomenclature, $\{K_h : h \in \mathbb{Z}\}$ relates to a set $\mathcal{S} = \mathbb{Z}$ of point-mass samples. Moreover, as we know, $\{K_h\}_{h\in\mathbb{Z}}$ is an orthonormal basis (ONB) in H. As a consequence, we arrive at Shannon's sampling formula.

In addition to $\mathbb{Z}$, there are many other choices of discrete point-mass samples for H. For related functional analytic studies of sampling spaces corresponding to irregular distribution of sample-points see, e.g., [**472**], [**474**].

CHAPTER 11

Polyharmonic finite bandwidth sampling

Our point of departure is the introduction of the fundamental solutions for iterated Laplacians in $\mathbb{R}^q$: By induction on m we obtain (see N. Aronszajn et al. [**48**])

$$\Delta^m |x|^k = A_{m,k}|x|^{k-2m}, \tag{11.1}$$

where we have used the abbreviations

$$\begin{aligned} A_{0,1} &= 1, \\ A_{m,k} &= k(k-2)\ldots(k-2m+2)(k-2+q)(k-4+q)\ldots(k-2m+q). \end{aligned}$$

An elementary calculation yields in terms of the Gamma function

$$A_{m,k} = 2^{2m}\frac{\Gamma\left(\frac{k}{2}+1\right)\Gamma\left(\frac{k+q}{2}\right)}{\Gamma\left(\frac{k}{2}-m+1\right)\Gamma\left(\frac{k+q}{2}-m\right)} \tag{11.2}$$

except when k is even and $k \leq -2$ or $k+q$ is even and $k+q \leq 0$. Moreover,

$$\Delta\left(|x|^k \ln(|x|)\right) = k(k+q-2)|x|^{k-2}\ln(|x|) + \ (2k+q-2)|x|^{k-2}. \tag{11.3}$$

Again, by induction on m, we find that

$$\Delta^m\left(|x|^k \ln(|x|)\right) = A_{m,k}|x|^{k-2m}\ln(|x|) + |x|^{k-2m}\sum_{i=1}^{m} A_{i-1,k}A_{m-i,k-2i}(2k+q+2-4i).$$

By virtue of (11.2) the preceding formula can be transformed into the identity (see [**48**])

$$\Delta^m\left(|x|^k \ln(|x|)\right) = A_{m,k}|x|^{k-2m}\left(\ln(|x|) + \sum_{i=1}^{m}\left(\frac{1}{k-2i+2} + \frac{1}{k+q-2i}\right)\right). \tag{11.4}$$

For each operator Δ^l there corresponds a *characteristic function* $S_q^{(l)}$, $q \geq 2$, solving the equation $\Delta^l S_q^{(l)} = 0$ in $\mathbb{R}^q\backslash\{0\}$. To be more concrete, $S_q^{(l)}$ is given by

$$S_q^{(l)}(x) = \begin{cases} \frac{|x|^{2l-q}}{\gamma_{l-1}}, & q \text{ odd}, \quad l = 1, 2, \ldots, \\ \frac{|x|^{2l-q}}{\tilde{\gamma}_{l-1}}, & q \text{ even}, \quad l = 1, 2, \ldots, \frac{q}{2}-1, \\ \frac{-|x|^{2l-q}\ln(|x|)}{\tilde{\gamma}_{l-1}}, & q \text{ even}, \quad l = \frac{q}{2}, \frac{q}{2}+1, \ldots. \end{cases} \tag{11.5}$$

The constants γ_l and $\tilde{\gamma}_l$ are given (in the notation of (11.2)) as follows:

q odd:

$$\gamma_l = A_{l,2l+2-q} = 2^{2l} l! \frac{\Gamma\left(l+2-\frac{q}{2}\right)}{\Gamma\left(2-\frac{q}{2}\right)}, \tag{11.6}$$

$l = 0, 1, \ldots,$

q even:

$q = 2r \geq 4.$

$$\tilde{\gamma}_l = A_{l,2l+2-q} = (-1)^l 2^{2l} l! \frac{(r-2)!}{(r-l-2)!}, \tag{11.7}$$

$l = 0, \ldots, r-2,$

$$\tilde{\gamma}_l = (-1)^{r+1} 2^{2l-1} (r-2)!\; l!\; (l+1-r)!, \tag{11.8}$$

$l = r-1, r, \ldots,$

$q = 2$:

$$\tilde{\gamma}_l = 2^{2l} (l!)^2, \quad l \in \mathbb{N}_0. \tag{11.9}$$

For brevity we let

$$A_q = \begin{cases} \frac{1}{(q-2)\|\mathbb{S}^{q-1}\|} & , \quad q \neq 2, \\ \frac{1}{\|\mathbb{S}^{q-1}\|} & , \quad q = 2. \end{cases} \tag{11.10}$$

The *fundamental solution* $F_q^{(l)}(|\cdot|) : \mathbb{R}^q \backslash \{0\} \to \mathbb{R}$ *for the polyharmonic equation* $\Delta^l F_q^{(l)} = 0$ is defined by

(11.11)

$$F_q^{(l)}(|x|) = \begin{cases} A_q \frac{|x|^{2l-q}}{\gamma_{l-1}}, & q \text{ odd}, \quad l = 1, 2, \ldots, \\ A_q \frac{|x|^{2l-q}}{\tilde{\gamma}_{l-1}}, & q = 2r, \quad l = 1, \ldots, r-1 \\ A_q \left(\frac{-|x|^{2l-q} \ln(|x|)}{\tilde{\gamma}_{l-1}} + \frac{C_{l-\frac{q}{2}}}{\tilde{\gamma}_{l-1}} |x|^{2l-q} \right), & q = 2r, \quad l = r, r+1, \ldots, \end{cases}$$

where the constants C_l, $l = 0, 1, \ldots,$ are given by

$$C_0 = 0, \tag{11.12}$$

$$C_l = \sum_{k=1}^{l} \frac{1}{2k} + \sum_{k=r}^{l+r-1} \frac{1}{2k}, \quad l > 0. \tag{11.13}$$

In fact, with this choice of the constants C_l, it immediately follows that for all dimensions $q = 2, 3, \ldots$ and all degrees $l = 1, 2, \ldots$ we have

$$\Delta^m F_q^{(l)} = F_q^{(l-m)}, \quad m < l. \tag{11.14}$$

11.1. Multivariate integral formulas to boundary conditions

Next we formulate integral formulas that are canonical generalizations of the third Green theorem on regular regions $\mathcal{G}$ in $\mathbb{R}^q$ (cf.(11.15)). The theory of polyharmonic functions on $\mathcal{G}$ leads to formulas under prescribed conditions on the boundary $\partial\mathcal{G}$. These identities are used to establish multivariate spline functions on regular regions $\mathcal{G}$ under given boundary conditions.

By standard arguments we obtain from the extended second Green theorem the following extension of the third Green theorem (see [**48**]).

Let $\mathcal{G} \subset \mathbb{R}^q$, $q \geq 2$, be a regular region. Let $F : \overline{\mathcal{G}} \to \mathbb{R}$ be of the class $C^{(2m)}\left(\overline{\mathcal{G}}\right), \overline{\mathcal{G}} = \mathcal{G} \cup \partial\mathcal{G}$, $m \geq 1$. Then

$$\begin{aligned}\alpha(x)\, F(x) = &\sum_{k=0}^{m-1} \int_{\partial\mathcal{G}} F_q^{(k+1)}(|x-y|) \left(\frac{\partial}{\partial\nu_y}\Delta_y^k F(y)\right) dS(y) \\ &-\sum_{k=0}^{m-1} \int_{\partial\mathcal{G}} \left(\frac{\partial F_q^{(k+1)}}{\partial\nu_y}(|x-y|)\right) \Delta_y^k F(y)\, dS(y) \\ &-\int_{\mathcal{G}} F_q^{(m)}(|x-y|)\, \Delta_y^m F(y)\, dy,\end{aligned} \tag{11.15}$$

where $\alpha(x)$, $x \in \mathbb{R}^q$ (more accurately, $\alpha_{\partial\mathcal{G}}(x)$) is the solid angle at x subtended by the surface $\partial\mathcal{G}$ and $F_q^{(m)}$ is given by (11.11).

For $m = 1$, the identity (11.15) is the usual third Green theorem for the Laplacian. In what follows we only discuss the situation that x is a point inside the regular region $\mathcal{G} \subset \mathbb{R}^q$, $q \geq 2$. Then we are able to deduce from (11.15) that, for all $F \in C^{(2m)}(\overline{\mathcal{G}}), \overline{\mathcal{G}} = \mathcal{G} \cup \partial\mathcal{G}$,

$$\begin{aligned}F(x) = &-\int_{\mathcal{G}} \left(\Delta_y^m F_q^{(2m)}(|x-y|)\right) \left(\Delta_y^m F(y)\right) dy \\ &+\int_{\partial\mathcal{G}} \mathcal{B}_{(q)}^{(m)} \left[F(y), \Delta_y^m F_q^{(2m)}(|x-y|)\right] dS(y),\end{aligned} \tag{11.16}$$

where the boundary term is explicitly given by

$$\begin{aligned}&\int_{\partial\mathcal{G}} \mathcal{B}_{(q)}^{(m)} \left[F(y), \Delta_y^m F_q^{(2m)}(|x-y|)\right] dS(y) \\ &\quad= \sum_{k=0}^{m-1} \int_{\partial\mathcal{G}} \left(\Delta_y^{m-(k+1)} F_q^{(2m)}(|x-y|)\right) \left(\frac{\partial}{\partial\nu_y}\Delta_y^k F(y)\right) dS(y). \\ &\quad-\sum_{k=0}^{m-1} \int_{\partial\mathcal{G}} \left(\frac{\partial}{\partial\nu}\Delta_y^{m-(k+1)} F_q^{(2m)}(|x-y|)\right) \left(\Delta_y^k F(y)\right) dS(y).\end{aligned} \tag{11.17}$$

Easy to handle in practical applications are formulas for which the "boundary term" (11.17) vanishes. The simplest example should be mentioned: Suppose that $F \in C^{(2m)}(\overline{\mathcal{G}})$ satisfies the *homogeneous boundary conditions* (HBC)

$$\Delta^k F|\partial\mathcal{G} = 0, \tag{11.18}$$

$$\frac{\partial}{\partial\nu}\Delta^k F|\partial\mathcal{G} = 0, \tag{11.19}$$

$k = 0, \ldots, m-1$. Then it is clear that

$$\int_{\partial\mathcal{G}} \mathcal{B}_q^{(m)} \left[F(y), \Delta_y^m F_q^{(2m)}(|x-y|)\right] dS(y) = 0, \tag{11.20}$$

and we have

$$F(x) = -\int_{\mathcal{G}} F_q^{(m)}(|x-y|)\, \Delta_y^m F(y)\, dy, \tag{11.21}$$

11.2. Sampling to Dirichlet conditions

We are interested in an integral formula of the form (11.16), (11.17) corresponding to the prescribed Dirichlet boundary conditions

$$\Delta^k F|\partial\mathcal{G} = B_k, \quad k = 0, \ldots, m-1. \tag{11.22}$$

Integral formula to "Dirichlet conditions". For $x \in \mathcal{G}$, we introduce an auxiliary function $G(x,\cdot) \in C^{(4m)}(\overline{\mathcal{G}})$ as the solution of the polyharmonic boundary value problem

$$\Delta^{2m} G(x,\cdot)|\mathcal{G} = 0 \tag{11.23}$$

such that

$$\Delta^k G(x,\cdot)|\partial G = F_q^{(2m-k)}(|x-\cdot|)|\partial\mathcal{G}, \quad k = 0, \ldots, 2m-1 \tag{11.24}$$

(note that, under our assumptions, $F_q^{(2m-k)}(|x-\cdot|)|\partial\mathcal{G}, k = 0, \ldots 2m$, are infinitely often differentiable functions). Furthermore, it is not hard to see that the boundary value problem (11.23), (11.24) is equivalent to the following system of boundary value problems:

$$\Delta Q_1(x,\cdot)|\mathcal{G} = 0, \tag{11.25}$$

$$Q_1(x,\cdot)|\partial\mathcal{G} = \Delta^{2m-1} F_q^{(2m)}(|x-\cdot|)|\partial\mathcal{G}, \tag{11.26}$$

$$\vdots \qquad \vdots \qquad \vdots$$

$$\Delta Q_{2m-1}(x,\cdot)|\mathcal{G} = Q_{2m-2}(x,\cdot), \tag{11.27}$$

$$Q_{2m-1}(x,\cdot)|\partial\mathcal{G} = \Delta F_q^{(2m)}(|x-\cdot|)|\partial\mathcal{G}, \tag{11.28}$$

$$\Delta G(x,\cdot)|\mathcal{G} = Q_{2m-1}(x,\cdot), \tag{11.29}$$

$$G(x,\cdot)|\partial\mathcal{G} = F_q^{(2m)}(|x-\cdot|)|\partial\mathcal{G}. \tag{11.30}$$

Obviously, each of the problems can be solved separately. Since F is assumed to be of class $C^{(2m)}(\overline{\mathcal{G}})$, we obtain from the second Green theorem

$$\int_{\mathcal{G}} \{\underbrace{(\Delta_y^{2m} G(x,y))}_{=0} F(x) - (\Delta_y^m G(x,y))\Delta_y^m F(y)\}\, dy = \int_{\partial\mathcal{G}} \mathcal{B}_{(q)}^m \left[F(y), \Delta^m G(x,y)\right]\, dy$$

such that

$$\begin{aligned} 0 = & \int_{\mathcal{G}} \left(\Delta_y^m G(x,y)\right)\left(\Delta_y^m F(y)\right)\, dy \\ & + \sum_{k=0}^{m} \int_{\partial\mathcal{G}} \left(\Delta_y^{2m-(k+1)} G(x,y)\right)\left(\frac{\partial}{\partial\nu_y}\Delta_y^k F(y)\right)\, dS(y). \\ & - \sum_{k=0}^{m-1} \int_{\partial\mathcal{G}} \left(\frac{\partial}{\partial\nu_y}\Delta_y^{2m-(k+1)} G(x,y)\right)\left(\Delta_y^k F(y)\right)\, dS(y) \end{aligned} \tag{11.31}$$

This leads to the following result:

Let $\mathcal{G} \subset \mathbb{R}^q$ be a regular region. Suppose that $x \in \mathcal{G}$. Let B_k, $k = 0, \ldots m-1$, be given functions of the class $C^{(0)}(\overline{\mathcal{G}}), \overline{\mathcal{G}} = \mathcal{G} \cup \partial\mathcal{G}$. Let $F \in C^{(2m)}(\overline{\mathcal{G}})$, $\overline{\mathcal{G}} = \mathcal{G} \cup \mathcal{G}$, $m \geq 1$, satisfy the boundary conditions

$$\Delta^k F|\partial\mathcal{G} = B_k, \; k = 0, \ldots, m-1.$$

Furthermore, suppose that, for $x \in \mathcal{G}$, $G(x,\cdot) \in C^{(4m)}(\overline{\mathcal{G}})$ solves the boundary value problem

$$\Delta^{2m} G(x,\cdot)|\mathcal{G} = 0,$$
$$\Delta^{k} G(x,\cdot)|\partial\mathcal{G} = F_q^{(2m-k)}(|x-\cdot|)|\partial\mathcal{G}, \quad k = 0,\ldots,2m-1.$$

Then

$$F(x) = \int_{\mathcal{G}} \left(\Delta_y^m D^{(2m)}(x,y)\right)\left(\Delta_y^m F(y)\right)\, dy \tag{11.32}$$
$$+ \sum_{k=0}^{m-1} \int_{\partial\mathcal{G}} \underbrace{B_k(y)}_{=\Delta_y^k F(y)} \left(\frac{\partial}{\partial\nu_y}\Delta_y^{2m-(k+1)} D^{(2m)}(x,y)\right) dS(y),$$

where we have used the abbreviation

$$D^{(2m)}(x,y) = G(x,y) - F_q^{(2m)}(|x-y|). \tag{11.33}$$

The integral formula (11.32) is a reformulation of the identity (11.15) corresponding to given "Dirichlet boundary conditions" $\Delta^k F|\partial\mathcal{G}$, $k = 0,\ldots,m-1$, at the price of the additional introduction of the function $D^{(2m)}(x,\cdot)$ (dependent on $x \in \mathcal{G}$).

Generating homogeneous boundary conditions. The integral formula (11.32) can be used to generate homogeneous boundary conditions on $\partial\mathcal{G}$. If $k = 0,\ldots,m-1$, we introduce functions $W_k \in C^{(2m)}(\overline{\mathcal{G}})$, such that

$$\Delta^{k+1} W_k|\mathcal{G} = 0 \tag{11.34}$$

and

$$\Delta^k W_k|\partial\mathcal{G} = B_k, \tag{11.35}$$

$$W_k|\partial\mathcal{G} = \ldots = \Delta^{k-1} W_k|\partial\mathcal{G} = 0 \tag{11.36}$$

(note that, for $m = 1$, the condition (11.36) should be omitted). The boundary value problem (11.34), (11.35), and (11.36) is equivalent to

$$\Delta W_0|\mathcal{G} = 0, \tag{11.37}$$
$$W_0|\partial\mathcal{G} = B_0, \tag{11.38}$$

and, for $k = 1,\ldots,m-1$,

$$\Delta V_1|\mathcal{G} = 0, \tag{11.39}$$
$$V_1|\partial\mathcal{G} = B_k, \tag{11.40}$$
$$\Delta V_2|\mathcal{G} = V_1, \tag{11.41}$$
$$V_2|\partial\mathcal{G} = 0, \tag{11.42}$$
$$\vdots \qquad \vdots$$
$$\Delta V_k|\mathcal{G} = V_{k-1}, \tag{11.43}$$
$$V_k|\partial\mathcal{G} = 0, \tag{11.44}$$
$$\Delta W_k|\mathcal{G} = V_k, \tag{11.45}$$
$$W_k|\partial\mathcal{G} = 0, \tag{11.46}$$

where each of the problems can be treated separately. If we choose $F - \sum_{k=0}^{m-1} W_k$ instead of F in (11.32) we obtain the following integral formula (showing homogeneous boundary conditions).

Under the aforementioned assumptions

$$F(x) - \sum_{k=0}^{m-1} W_k(x) = \int_{\mathcal{G}} (\Delta_y^m D^{(2m)}(x,y))(\Delta_y^m F(y))\, dy. \tag{11.47}$$

The same procedure as described above for Dirichlet boundary conditions also applies to Neumann boundary conditions (the details are straightforward).

11.3. Finite bandwidth splines

The identity (11.47) can be used to introduce *splines on regular regions* $\mathcal{G} \subset \mathbb{R}^q$. We again restrict ourselves to the Dirichlet case (note that the Neumann case follows analogously):

Let $H_0^{(2m)}(\overline{\mathcal{G}})$ denote the space of all $F : \overline{\mathcal{G}} \to \mathbb{R}$ satisfying

$$F|\partial\mathcal{G} = \Delta F|\partial\mathcal{G} = \ldots = \Delta^{m-1} F|\partial\mathcal{G} = 0 \tag{11.48}$$

and

$$\Delta^m F \in L^2(\mathcal{G}). \tag{11.49}$$

$H_0^{(2m)}(\overline{\mathcal{G}})$ is endowed with the inner product

$$\langle F, G\rangle_{H_0^{(2m)}(\overline{\mathcal{G}})} = \int_{\mathcal{G}} (\Delta^m F(x))(\Delta^m G(x))\, dx, \tag{11.50}$$

such that, in connection with (11.32), we have

$$\Delta^k D^{(2m)}(x;\cdot) \mid \partial\mathcal{G} = 0 \tag{11.51}$$

for $k = 1, \ldots, 2m-1$. Applying these homogeneous boundary conditions we find

$$\left\langle D^{(2m)}(x,\cdot), F\right\rangle_{H_0^{(2m)}(\overline{\mathcal{G}})} = \int_{\mathcal{G}} \Delta_y^m D^{(2m)}(x,y)\ \Delta^m F(y)\, dy = F(x) \tag{11.52}$$

for all $F \in H_0^{(2m)}(\overline{\mathcal{G}})$ (note that $\|F\|_{H_0^{(2m)}(\overline{\mathcal{G}})} = 0$ is equivalent to $\Delta^m F \mid \overline{\mathcal{G}} = 0$ with $F \mid \partial\mathcal{G} = \Delta F \mid \partial\mathcal{G} = \ldots = \Delta^{m-1} F \mid \partial\mathcal{G} = 0$, hence, potential theory shows that $F = 0$).

$H_0^{(2m)}(\overline{\mathcal{G}})$ is a Sobolev-type space with the reproducing kernel $D^{(2m)}$ given by (11.33).

Consider now a finite set of points $x_1, \ldots, x_N \in \mathcal{G}$ such that

$$\det(D^{(2m)}(x_i, x_l))_{i,l=1,\ldots,N} \neq 0. \tag{11.53}$$

Then, for all choices $a = (a_1, \ldots, a_N)^T$ of coefficients $a_i \in \mathbb{R}$, $i = 1, \ldots, N$, and for all $F \in H_0^{(2m)}(\overline{\mathcal{G}})$,

$$\sum_{i=1}^N a_i\Big(F(x_i) - \sum_{k=0}^{m-1} W_k(x_i)\Big) = \int_{\mathcal{G}} \left(\Delta_y^m \sum_{i=1}^N a_i D^{(2m)}(x_i, y)\right) (\Delta^m F(y))\, dy. \tag{11.54}$$

The last identity may be interpreted as a *spline integration formula*, allowing the formulation of the following statement:

The spline function

$$S^{\alpha} = \sum_{i=1}^{N} a_i \, D^{(2m)}(x_i, \cdot) \tag{11.55}$$

uniquely determined by

$$S^{\alpha}(x_i) = \sum_{l=1}^{N} a_l D^{(2m)}(x_l, x_i) = \alpha_i, \quad i = 1, \ldots, N, \tag{11.56}$$

with

$$\alpha = (\alpha_1, \ldots, \alpha_N)^T, \quad \alpha_i = y_i - \sum_{k=0}^{m-1} W_k(x_i), \quad i = 1, \ldots, N, \tag{11.57}$$

satisfies the $H_0^{(2m)}(\mathcal{G})$*-integral formula*

$$\int_{\mathcal{G}} (\Delta^m F(y))^2 \; dy = \int_{\mathcal{G}} (\Delta^m S^{\alpha}(y))^2 \; dy + \int_{\mathcal{G}} (\Delta^m (F(y) - S^{\alpha}(y)))^2 \; dy \tag{11.58}$$

for all $F \in H_0^{(2m)}(\overline{\mathcal{G}})$ with $F(x_i) = y_i$, $i = 1, \ldots, N$. In fact, for all $F \in H_0^{(2m)}(\overline{\mathcal{G}})$ with $F(x_i) = y_i$, $i = 1, \ldots, N$, we see that

$$\int_{\mathcal{G}} \Delta^m S^{\alpha}(y) \; \Delta^m F(y) \; dy = \sum_{i=1}^{N} a_i \underbrace{(y_i - \sum_{k=0}^{m-1} W_k(x_i))}_{=\alpha_i}. \tag{11.59}$$

Moreover, we have

$$\int_{\mathcal{G}} \Delta^m S^{\alpha}(y) \; \Delta^m S^{\alpha}(y) \; dy = \sum_{i=1}^{N} a_i \underbrace{(y_i - \sum_{k=0}^{m-1} W_k(x_i))}_{=\alpha_i}. \tag{11.60}$$

Consequently it follows that

$$\begin{aligned} &\int_{\mathcal{G}} \Delta^m (F(y) - S^{\alpha}(y)) \; \Delta^m (F(y) - S^{\alpha}(y)) \; dy \\ &\quad = \int_{\mathcal{G}} (\Delta^m F(y))^2 \; dy + \int_{\mathcal{G}} (\Delta^m S^{\alpha}(y))^2 \; dy - 2 \sum_{i=1}^{N} a_i \alpha_i \\ &\quad = \int_{\mathcal{G}} (\Delta^m F(y))^2 \; dy - \int_{\mathcal{G}} (\Delta^m S^{\alpha}(y)) \; dy. \end{aligned} \tag{11.61}$$

This guarantees the minimum norm interpolation property (11.58) (note that spline smoothing and best approximation may be obtained in the standard way as, for example, described in [**256**], [**257**], and [**331**]).

The spline (minimum norm) interpolation problem

$$\left(\int_{\mathcal{G}} |\Delta_y^m S^{\alpha}(y)| dy \right)^{1/2} = \min_{\substack{F \in H_0^{(2m)}(\overline{\mathcal{G}}) \\ F(x_i) = y_i \\ i = 1, \ldots, N}} \left(\int_{\mathcal{G}} |\Delta_y^m F(y)|^2 \; dy \right)^{1/2} \tag{11.62}$$

is uniquely determined by

$$S^\alpha = \sum_{i=1}^{N} a_i \; D^{(2m)}(x_i, \cdot), \tag{11.63}$$

where

(11.64)

$$\begin{pmatrix} a_1 \\ \vdots \\ a_n \end{pmatrix} = \begin{pmatrix} D^{(2m)}(x_1,x_1) & \dots & D^{(2m)}(x_1,x_N) \\ \vdots & & \vdots \\ D^{(2m)}(x_N,x_1) & \dots & D^{(2m)}(x_N,x_N) \end{pmatrix}^{-1} \begin{pmatrix} y_1 - \sum\limits_{k=0}^{m-1} W_k(x_1) \\ \vdots \quad \vdots \\ y_N - \sum\limits_{k=0}^{m-1} W_k(x_N) \end{pmatrix}.$$

In accordance with our spline procedure above we are also able to introduce *spline Lagrangians in* $H_0^{(2m)}(\overline{\mathcal{G}})$ by setting

$$S^{\epsilon^k} = \sum_{l=1}^{N} a_l^k D^{(2m)}(\cdot, x_l), \;\; k = 1, \dots, N, \tag{11.65}$$

where the coefficients a_l^k satisfy the linear equations

$$\sum_{r=1}^{N} a_r^k D^{(2m)}(x_l, x_r) = \delta_{k,l}, \quad k, l = 1, \dots, N. \tag{11.66}$$

As a consequence, S^α can be reconstructed from its samples $S^\alpha(x_1), \dots, S^\alpha(x_N)$ by the $H_0^{(2m)}(\overline{\mathcal{G}})$*-spline sampling formula*

$$S^\alpha = \sum_{k=1}^{N} S^\alpha(x_k) \; S^{\epsilon^k}. \tag{11.67}$$

CHAPTER 12

Polyharmonic infinite bandwidth sampling

Asymptotic integral relations involving fundamental solutions of the iterated Laplace operator serve as the point of departure for the definition of polyharmonic splines with respect to a finite scattered data set in Euclidean spaces $\mathbb{R}^q$ (in historic language, *thin plate splines*). Finally, an approach to spline sampling on lattices in $\mathbb{R}^q$ is presented, where the term *“ polyharmonic cardinal spline”* is conventially used, when the set of knots forms a lattice.

12.1. Integral formulas over Euclidean spaces

For $q \geq 2$ and $x \in \mathbb{B}_N^q$, we obtain from (11.15)

$$F(x) = \sum_{l=0}^{m-1} \int_{\substack{|y|=N \\ y\in\mathbb{R}^q}} \left(\Delta_y^{2m-l-1} F_q^{(2m)}(|x-y|)\right) \left(\frac{\partial}{\partial\nu}\Delta^l F(y)\right) dS(y) \tag{12.1}$$
$$- \sum_{l=0}^{m-1} \int_{\substack{|y|=N \\ y\in\mathbb{R}^q}} \left(\frac{\partial}{\partial\nu_y}\Delta_y^{2m-l-1} F_q^{(2m)}(|x-y|)\right) \left(\Delta^l F(y)\right) dS(y)$$
$$- \int_{\substack{|y|\leq N \\ y\in\mathbb{R}^q}} \left(\Delta_y^{m} F_q^{(2m)}(|x-y|)\right) \left(\Delta^m F(y)\right)\, dy,$$

provided that F is $2m$ times continuously differentiable in balls $\overline{\mathbb{B}_N^q}$ around the origin with radius N. Our goal is to make a transition to infinity, obtaining $\mathbb{R}^q$ as limit of the balls $\mathbb{B}_N^q$ around the origin with radii N.

It is useful to list some properties of the *fundamental solution* $F_q^{(2m)}$ defined by (11.11) and to add some integral estimates involving this function:

- In the sense of the space of tempered distributions $\mathfrak{S}'(\mathbb{R}^q)$, $\Delta^{2m} F_q^{(2m)} = \delta$, where δ denotes the Dirac distribution.
- If $m > \frac{q}{4}, m \in \mathbb{N}$, then the kernel $y \mapsto F_q^{(2m)}(|x-y|)$ constitutes a continuous function in $\mathbb{R}^q$ for each (fixed) $x \in \mathbb{R}^q$.
- For $x \in \mathbb{R}^q$ fixed and $N \to \infty$, the asymptotic integral estimates

$$\int_{\substack{|y|=N \\ y\in\mathbb{R}^q}} \left|\Delta_y^{2m-l-1} F_q^{(2m)}(|x-y|)\right|\, dS(y) = O(N^{2l+1}), \quad N \to \infty, \tag{12.2}$$

and

$$\int_{\substack{|y|=N \\ y\in\mathbb{R}^q}} \left|\frac{\partial}{\partial\nu_y}\Delta_y^{2m-l-1} F_q^{(2m)}(|x-y|)\right|\, dS(y) = O(N^{2l+1}), \quad N \to \infty \tag{12.3}$$

hold true for $l = 0, \ldots, m-1$.

12.2. Asymptotically defined reproducing kernel Hilbert space

The asymptotic integral estimates (12.2) and (12.3) give rise to introduce the following function classes:

Class $A^{(m)}(\mathbb{R}^q)$*:* A function $F \in C^{2m}(\mathbb{R}^q)$ is said to be of class $A^{(m)}(\mathbb{R}^q)$, if, for $l = 0, \ldots, m-1$, the asymptotic relations

$$\Delta^l F(y) = o\left(|y|^{-(1+2l)}\right), \quad |y| \to \infty, \tag{12.4}$$

$$\nabla \Delta^l F(y) = o\left(|y|^{-(1+2l)}\right), \quad |y| \to \infty. \tag{12.5}$$

are valid (uniformly with respect to all directions $y/|y|$).

Class $B^{(m)}(\mathbb{R}^q)$*:* A function $F \in C^{(2m)}(\mathbb{R}^q)$ is said to be of class $B^{(m)}(\mathbb{R}^q)$, if there exists $\varepsilon > 0$ such that

$$\Delta^m F(y) = O\left(|y|^{-(2+2m+\varepsilon)}\right), \quad |y| \to \infty. \tag{12.6}$$

Assume, that $F \in C^{(2m)}(\mathbb{R}^q)$ is of class $A^{(m)}(\mathbb{R}^q)$. Then the surface integrals in (12.1) can be estimated as follows:

$$o\left(\sum_{l=0}^{m-1} N^{-(1+2l)} \int_{\substack{|y|=N \\ y\in\mathbb{R}}} \left|F_q^{(l+1)}(|x-y|)\right| + \left|\frac{\partial}{\partial \nu_y} F_q^{(l+1)}(|x-y|)\right| \, dS(y)\right). \tag{12.7}$$

Hence, in connection with the relations (12.2), (12.3), we find that

$$\sum_{l=0}^{m-1} \int_{\substack{|y|=N \\ y\in\mathbb{R}^q}} \left(\frac{\partial}{\partial \nu_y} F_q^{(l+1)}(|x-y|)\right) \left(\Delta^l F(y)\right) \, dS(y) = o(1), \quad N \to \infty, \tag{12.8}$$

and

$$\sum_{l=0}^{m-1} \int_{\substack{|y|=N \\ y\in\mathbb{R}^q}} \left(F_q^{(l+1)}(|x-y|)\right) \left(\frac{\partial}{\partial \nu} \Delta^l F(y)\right) \, dS(y) = o(1), \quad N \to \infty. \tag{12.9}$$

Furthermore, suppose that $F \in C^{(2m)}(\mathbb{R}^q)$ is of class $B^{(m)}(\mathbb{R}^q)$. Then it follows that

$$\begin{aligned} &\int_{N' \le |y| \le N} \left|F_q^{(m)}(|x-y|)\right| \left|\Delta_y^m F(y)\right| \, dy \\ &\qquad = O\left(\int_{N'}^{N} \frac{1}{(1+r)^{2(m+1)+\varepsilon}} \int_{|y|=r} \left|F_q^{(m)}(|y|)\right| \, dS(y) \, dr\right). \end{aligned} \tag{12.10}$$

Thus the integral

$$\int_{\mathbb{R}^q} \left(\Delta_y^m F_q^{(2m)}(|x-y|)\right) \left(\Delta^m F(y)\right) \, dy \tag{12.11}$$

is absolutely convergent.

Summarizing our results we arrive at the following limit relation:

Suppose that $F \in C^{(2m)}(\mathbb{R}^q)$, $m \in \mathbb{N}$, $q \geq 2$, *is of class* $A^{(m)}(\mathbb{R}^q) \cap B^{(m)}(\mathbb{R}^q)$: *Then, for* $x \in \mathbb{R}^q$, *the limit*

$$\begin{aligned}(12.12)\qquad &\int_{\mathbb{R}^q} (\Delta_y^m(-F_q^{(2m)}(|x-y|)))\ (\Delta^m F(y))\ dy \\ &= \lim_{N\to\infty} \int_{\substack{|y|\leq N\\ y\in\mathbb{R}^q}} (\Delta_y^m(-F_q^{(2m)}(|x-y|)))\ (\Delta^m F(y))\ dy\end{aligned}$$

exists, and we have the reproducing kernel formula

$$(12.13)\qquad F(x) = \int_{\mathbb{R}^q} (\Delta_y^m(-F_q^{(2m)}(|x-y|)))\ (\Delta^m F(y))\ dy.$$

12.3. Thin plate spline concept

For functions $F, G \in C^{(2m)}(\mathbb{R}^q)$, $m > \frac{q}{4}$, of class $A^{(m)}(\mathbb{R}^q) \cap B^{(m)}(\mathbb{R}^q)$, we introduce the inner product

$$(12.14)\qquad \langle F, G\rangle_{H^{(2m)}(\mathbb{R}^q)} = \int_{\mathbb{R}^q} (\Delta^m F(y))\ (\Delta^m G(y))\ dy.$$

We let

$$(12.15)\qquad H^{(2m)}(\mathbb{R}^q) \ = \ \overline{A^{(m)}(\mathbb{R}^q) \cap B^{(m)}(\mathbb{R}^q)}^{\|\cdot\|_{H^{(2m)}(\mathbb{R}^q)}}.$$

$H^{(2m)}(\mathbb{R}^q)$ is a Sobolev-like space possessing the reproducing kernel

$$(12.16)\qquad (x,y) \mapsto -F_q^{(2m)}(|x-y|), \quad x, y \in \mathbb{R}^q,$$

so that for all $F \in H^{(2m)}(\mathbb{R}^q), m > \frac{q}{4}$,

$$(12.17)\qquad F(x) = \int_{\mathbb{R}^q} (\Delta_y^m(-F_q^{(2m)}(|x-y|)))\ (\Delta^m F(y))\ dy.$$

Furthermore, we have

$$(12.18)\qquad F_q^{(2m)}(|x-z|) = \int_{\mathbb{R}^q} (\Delta_y^m(-F_q^{(2m)}|x-y|)))\ (\Delta_y^m(F_q^{(2m)}(|y-z|)))\ dy.$$

The integral formula (12.18) enables us to introduce *spline functions* $S : \mathbb{R}^q \to \mathbb{R}$ of the form

$$(12.19)\qquad S(x) = \sum_{i=1}^N a_i\ (-F_q^{(2m)}(|x_i - x|)), \quad x \in \mathbb{R}^q, \ \ a_1, \ldots, a_N \in \mathbb{R},$$

for a pointset $\{x_1, \ldots, x_N\} \subset \mathbb{R}^q$ satisfying

$$(12.20)\qquad \det\left(-F_q^{(2m)}(|x_i - x_j|)\right)_{i,j=1,\ldots N} \neq 0.$$

From (12.17) we immediately obtain for all $F \in H^{(2m)}(\mathbb{R}^q), m > \frac{q}{4}$, the *spline integration formula*

$$(12.21)\qquad \sum_{i=1}^N a_i F(x_i) = \int_{\mathbb{R}^q} (\Delta^m(S(y)))(\Delta^m F(y))\ dy.$$

Furthermore, we have

$$(12.22)\qquad \sum_{i=1}^N a_i S(x_i) = \int_{\mathbb{R}^q} (\Delta^m(S(y)))\ (\Delta^m S(y))\ dy.$$

This leads to the following characterization of *spline interpolation* in the Sobolev-like space $H^{(2m)}(\mathbb{R}^q)$:

Suppose that $m > \frac{q}{4}$. *Let* $\beta = (\beta_1, \ldots, \beta_N)^T$ *be* N *given real values. The spline function*

$$S^\beta = \sum_{i=1}^{N} a_i^\beta \; (-F_q^{(2m)}(|x_i - \cdot|)) \tag{12.23}$$

determined by

$$S^\beta(x_j) = \sum_{i=1}^{N} a_i^\beta(-F_q^{(2m)}(|x_j - x_i|)) = \beta_j, \quad j = 1, \ldots, N, \tag{12.24}$$

satisfies

$$\int_{\mathbb{R}^q} (\Delta^m F(y))^2 \; dy = \int_{\mathbb{R}^q} \left(\Delta^m S^\beta(y)\right)^2 \; dy + \int_{\mathbb{R}^q} \left(\Delta^m (F(y) - S^\beta(y))\right)^2 \; dy \tag{12.25}$$

for all $F \in H^{(2m)}(\mathbb{R}^q)$ *interpolating the data points* $(x_i, F(x_i)), F(x_i) = \beta_i$, $i = 1, \ldots, N$.

Finally we arrive at the following result in spline interpolation:

The spline interpolation problem

$$\left(\int_{\mathbb{R}^q} \left|\Delta^m S^\beta(y)\right|^2 \; dy \right)^{1/2} = \min_{\substack{F \in H^{(2m)}(\mathbb{R}^q) \\ F(x_i) = \beta_i, \\ i=1,\ldots,N}} \left(\int_{\mathbb{R}^q} \left|\Delta^m F(y)\right|^2 \; dy \right)^{1/2} \tag{12.26}$$

is uniquely determined by

$$S^\beta = \sum_{i=1}^{N} a_i^\beta \; (-F_q^{(2m)}(|x_i - \cdot|)), \tag{12.27}$$

where

$$\begin{pmatrix} a_1^\beta \\ \vdots \\ a_N^\beta \end{pmatrix} = \begin{pmatrix} -F_q^{(2m)}(|x_1 - x_1|) & \ldots & -F_q^{(2m)}(|x_1 - x_N|) \\ \vdots & & \vdots \\ -F_q^{(2m)}(|x_N - x_1|) & \ldots & -F_q^{(2m)}(|x_N - x_N|) \end{pmatrix}^{-1} \begin{pmatrix} \beta_1 \\ \vdots \\ \beta_N \end{pmatrix}. \tag{12.28}$$

At least for $q = 2, m = 1$, the integral $\int_{\mathbb{R}^q} |\Delta^m F(x)|^2 \; dx$ may be interpreted (under some linearizing assumptions) as the bending energy of a thin plate of infinite extent, F denoting here the deflection normal to the rest position (supposed, of course, to the plane).

It should be remarked that the idea of interpolating by use of fundamental solutions to powers of the Laplacian is not new. Although the earliest published work devoted to the subject seems to be R.L. Harder, R.N. Demarais [**409**], it is known that many mathematicians were aware of the idea and many consequences, either from the viewpoint of reproducing kernel Hilbert space theory or from the extension of the variational aspect of univariate spline theory to the multivariate case (see, e.g., J. Duchon [**195**], J. Meinguet [**573**]). Finally, it should be noted that spline sampling can be realized in various way (see, e.g., the monograph M.D. Buhmann [**119**] for further references).

12.4. Cardinal spline concept

I.J. Schoenberg's theory of univariate cardinal splines (cf. [**755**]) may be regarded as a development of certain properties of those functions F, which satisfy the differential equation $\Delta^m F = (\frac{d^2}{dx^2})^m F = 0$ on the complement of the integer lattice $\mathbb{Z}$ and show appropriate smoothness conditions on all of the Euclidean space $\mathbb{R}$. A canonical extension of this concept to the multivariate case has to consider in its structural manifestation (see W. Madych, S. Nelson [**553**] and also the literature therein) those functions F which satisfy the differential equations $\Delta^m F = 0$ on the complement of the integer lattice $\mathbb{Z}^q$ in $\mathbb{R}^q$ and show appropriate smoothness conditions on all of the Euclidean space $\mathbb{R}^q$.

In what follows, in a similar way, our approach considers the case where the differential operators are powers of the Laplacian. W. Madych, S. Nelson [**553**] are concerned with the basic properties of cardinal splines mainly together with the problems of existence and uniqueness of cardinal interpolation. The motivation for our work is the explanation of the variational framework based on the integral formula (12.18) over $\mathbb{R}^q$.

Let Λ be an arbitrary lattice in $\mathbb{R}^q$. Suppose that $F : \mathbb{R}^q \to \mathbb{R}$ is of class $H^{(2m)}(\mathbb{R}^q)$, $m > \frac{q}{4}$. Then we know the following integral formula

$$F(x) = \int_{\mathbb{R}^q} (\Delta^m(-F_q^{(2m)}(|x-y|)))(\Delta^m F(y))\, dy. \tag{12.29}$$

A sequence $\{a_h\}_{h\in\Lambda^{-1}}$ of real numbers $a_h, h \in \Lambda^{-1}$, is called an $H^{(2m)}(\mathbb{R}^q)$-*admissible sequence*, if $\sum_{h\in\Lambda^{-1}} a_h F_q^{(2m)}(|h-\cdot|)$ is a member of $H^{(2m)}(\mathbb{R}^q)$, i.e., the sequence $\{a_h\}_{h\in\Lambda^{-1}}$ satisfies the condition

$$\sum_{h\in\Lambda^{-1}} \sum_{h'\in\Lambda^{-1}} a_h\, a_{h'}\, F_q^{(2m)}(|h-h'|) < \infty. \tag{12.30}$$

Assume that $\{a_h\}_{h\in\Lambda^{-1}}$ is an $H^{(2m)}(\mathbb{R}^q)$-admissible sequence. Then, for every $F \in H^{(2m)}(\mathbb{R}^q)$,

$$\sum_{h\in\Lambda^{-1}} a_h F(h) = \int_{\mathbb{R}^q} \left(\Delta_y^m\left(-\sum_{h\in\Lambda^{-1}} a_h F_q^{(2m)}(|h-y|)\right)\right)(\Delta^m F(y))\, dy. \tag{12.31}$$

This result enables us to characterize *cardinal spline interpolation* for data points $(h, F(h))_{h\in\Lambda^{-1}}$ with $F \in H^{(2m)}(\mathbb{R}^q)$ and $F(h) = \beta_h, h \in \Lambda^{-1}$:

The cardinal spline function

$$S_m^\beta = \sum_{h\in\Lambda^{-1}} a_h^\beta(-F_q^{(2m)}(|h-\cdot|)) \tag{12.32}$$

satisfying

$$S_m^\beta(h') = \sum_{h\in\Lambda^{-1}} a_h^\beta(-F_q^{(2m)}(|h-h'|)) = \beta_{h'}, \quad h' \in \Lambda^{-1}, \tag{12.33}$$

satisfies the integral property

$$\int_{\mathbb{R}^q} (\Delta^m F(y))^2\, dy = \int_{\mathbb{R}^q} (\Delta^m S_m^\beta(y))^2\, dy + \int_{\mathbb{R}^q} (\Delta^m (F(y) - S_m^\beta(y)))^2\, dy, \tag{12.34}$$

for all $F \in H^{(2m)}(\mathbb{R}^q)$ interpolating the data points $(h', F(h') = \beta_{h'})_{h' \in \Lambda^{-1}}$.

Fundamental function of cardinal spline interpolation. For all $x \in \mathbb{R}^q \backslash \{0\}$, the Fourier transform $(F_q^{(2m)})^\wedge_{\mathbb{R}^q}$ (see, e.g., [**553**]) is given by

$$(F_q^{(2m)})^\wedge_{\mathbb{R}^q}(x) = \frac{1}{(-4\pi^2 x^2)^{2m}}. \tag{12.35}$$

We consider the the so-called *fundamental function of cardinal spline interpolation* L_{2m} with $4m > q$ defined via its Fourier transform

$$(L_{2m})^\wedge_{\mathbb{R}^q}(x) = \frac{1}{\|\mathcal{F}_\Lambda\|} \frac{(F_q^{(2m)})^\wedge_{\mathbb{R}^q}(x)}{\sum\limits_{g \in \Lambda} (F_q^{(2m)})^\wedge_{\mathbb{R}^q}(x-g)} = \frac{1}{\|\mathcal{F}_\Lambda\|} \frac{(-4\pi^2 x^2)^{-2m}}{\sum\limits_{g \in \Lambda} (-4\pi^2 |x-g|^2)^{-2m}}. \tag{12.36}$$

The function L_{2m} shows the following properties:

$$\sum_{g' \in \Lambda} (L_{2m})^\wedge_{\mathbb{R}^q}(x - g') = \frac{1}{\|\mathcal{F}_\Lambda\|} \sum_{g' \in \Lambda} \frac{(-4\pi^2 |x-g'|^2)^{-2m}}{\sum\limits_{g' \in \Lambda} (-4\pi^2 |x - g' - g|^2)^{-2m}} = \frac{1}{\|\mathcal{F}_\Lambda\|}. \tag{12.37}$$

and

$$\begin{aligned} (L_{2m})(h') &= \int_{\mathbb{R}^q} (L_{2m})^\wedge_{\mathbb{R}^q}(y)\, e^{2\pi i h' \cdot y}\, dy \\ &= \sum_{g \in \Lambda} \int_{\mathcal{F}_\Lambda} (L_{2m})^\wedge_{\mathbb{R}^q}(y-g)\, e^{2\pi i h' \cdot y}\, dy \\ &= \int_{\mathcal{F}_\Lambda} \underbrace{\sum_{g \in \Lambda} (L_{2m})^\wedge_{\mathbb{R}^q}(y-g)}_{=\frac{1}{\|\mathcal{F}_\Lambda\|}}\, e^{2\pi i h' \cdot y}\, dy \\ &= \delta_{h',0},\ h' \in \Lambda^{-1}. \end{aligned} \tag{12.38}$$

For our purposes it is helpful to represent the Fourier transform of L_m in the following way:

$$(L_{2m})^\wedge_{\mathbb{R}^q}(x) = (-4\pi^2 x^2)^{-2m} (Q_{2m})^\wedge_{\mathbb{R}^q}(x), \tag{12.39}$$

where

$$(Q_{2m})^\wedge_{\mathbb{R}^q}(x) = (-4\pi^2 x^2)^{2m} (L_{2m})^\wedge_{\mathbb{R}^q}(x) = \frac{1}{\|\mathcal{F}_\Lambda\|} \frac{1}{\sum\limits_{g \in \Lambda} (-4\pi^2 |x-g|^2)^{-2m}}. \tag{12.40}$$

The importance of $(Q_{2m})^\wedge_{\mathbb{R}^q}$ in this context is based on its Λ-periodicity. It enables us to expand $(Q_{2m})^\wedge_{\mathbb{R}^q}$ into the Fourier series

$$(Q_{2m})^\wedge_{\mathbb{R}^q} = \frac{1}{\|\mathcal{F}_\Lambda\|} \sum_{h \in \Lambda^{-1}} \int_{\mathcal{F}_\Lambda} (Q_{2m})^\wedge_{\mathbb{R}^q}(x)\, e^{-2\pi i h \cdot x}\, dx\, e^{2\pi i h \cdot x}, \tag{12.41}$$

where we know from W. Madych, S. Nelson [**553**], that there are positive constants A and A' such that

$$\left| \int_{\mathcal{F}_\Lambda} (Q_{2m})^\wedge_{\mathbb{R}^q}(x)\ e^{-2\pi i h\cdot x}\ dx \right| \leq A\ e^{-A'\ |h|}. \tag{12.42}$$

So, the series (12.41) is absolutely and uniformly convergent. Moreover, from the series (12.41) it follows that the Fourier transform of $(Q_{2m})^\wedge_{\mathbb{R}^q}$, i.e., Q_{2m} given by

$$Q_{2m}(x) \;=\; \int_{\mathbb{R}^q} (Q_{2m})^\wedge_{\mathbb{R}^q}(y)\ e^{2\pi i x\cdot y}\ dy, \tag{12.43}$$

is a distributional sum of constant multiples of translates of the delta function

$$Q_{2m}(x) \;=\; \sum_{h\in\Lambda^{-1}} \frac{1}{\sqrt{\|\mathcal{F}_\Lambda\|}} \int_{\mathcal{F}_\Lambda} (Q_{2m})^\wedge_{\mathbb{R}^q}(x)\ e^{-2\pi i h\cdot x}\ dx\ \delta(h-x). \tag{12.44}$$

In connection with (12.35) and (12.40) we therefore obtain

$$L_{2m}(x) = \sum_{h\in\Lambda^{-1}} ((Q_{2m})^\wedge_{\mathbb{R}^q})^\wedge_\Lambda(h)\ F_q^{(2m)}(|x-h|) = \int_{\mathbb{R}^q} Q_{2m}(x-y)\ F_q^{(2m)}(|y|)\ dy, \tag{12.45}$$

where

$$((Q_{2m})^\wedge_{\mathbb{R}^q})^\wedge_\Lambda(h) = \frac{1}{\sqrt{\|\mathcal{F}_\Lambda\|}} \int_{\mathcal{F}_\Lambda} (Q_{2m})^\wedge_{\mathbb{R}^q}(x)\ e^{-2\pi i h\cdot x}\ dx. \tag{12.46}$$

Cardinal spline sampling. L_{2m} is a cardinal polyharmonic spline such that

$$\sum_{h'\in\Lambda^{-1}} \beta_{h'} L_{2m}(h-h') = \beta_h, \quad h \in \Lambda^{-1}, \tag{12.47}$$

for data points $(h,\beta_h)_{h\in\Lambda^{-1}}, \beta_h = F(h), F \in H^{(2m)}(\mathbb{R}^q)$, so that the interpolating cardinal polyharmonic spline S_m^β in (12.34) has the unique sampling representation

$$S_m^\beta(x) = \sum_{h\in\Lambda^{-1}} \underbrace{S_m^\beta(h)}_{=\beta_h}\ L_{2m}(x-h). \tag{12.48}$$

12.5. Relationship to the multivariate Shannon kernel

We conclude this chapter with an extension of a univariate result due to I. J. Schoenberg [**755**] to our multivariate approach to cardinal splines: For $m > \frac{q}{4}$ we know from (12.36) that

$$\begin{aligned}(L_{2m})^\wedge_{\mathbb{R}^q}(x) &= \frac{(-4\pi^2 x^2)^{-2m}}{\|\mathcal{F}_\Lambda\| \sum_{g\in\Lambda} (-4\pi^2|x-g|^2)^{-2m}} \\ &= \left(\frac{1}{\|\mathcal{F}_\Lambda\|} + \frac{1}{\|\mathcal{F}_\Lambda\|} \sum_{\substack{g\in\Lambda\\ g\neq 0}} \left(\frac{|x|^2}{|x-g|^2} \right)^{2m} \right)^{-1}.\end{aligned} \tag{12.49}$$

Let the lattice Λ under consideration be generated by the vectors $g_1,\ldots,g_q \in \mathbb{R}^q$, so that any $g \in \Lambda$ different from the origin can be expressed as a linear combination $g = \sum_{i=1}^q n_i g_i, n_i \in \mathbb{Z}$, with at least one integer $n_i, i \in \{1,\ldots,q\}$, different from 0. Thus, for all $x \in \mathcal{F}_\Lambda$, we have $\frac{|x|^2}{|x-g|^2} < 1$.

In what follows we distinguish two cases:

Case: $x \in \mathcal{F}_\Lambda$. In connection with (12.49), this estimate implies the limit relation

$$\lim_{m\to\infty} (L_{2m})^{\wedge}_{\mathbb{R}^q}(x) = \frac{1}{\|\mathcal{F}_\Lambda\|}. \tag{12.50}$$

Case: $x \notin \mathcal{F}_\Lambda$. We remember

$$(L_{2m})^{\wedge}_{\mathbb{R}^q}(x) \;=\; (-4\pi^2 x^2)^{-2m}\,(Q_{2m})^{\wedge}_{\mathbb{R}^q}(x). \tag{12.51}$$

Furthermore, we have $(Q_{2m})^{\wedge}_{\mathbb{R}^q}(x+g') = (Q_{2m})^{\wedge}_{\mathbb{R}^q}(x)$ for all $x \in \mathbb{R}^q$ and $g' \in \Lambda$, i.e., $(Q_{2m})^{\wedge}_{\mathbb{R}^q}$ is Λ-periodic. Now, choose an arbitrary point $g'' \neq 0$ of the lattice Λ. Then the Λ-periodicity of $(Q_{2m})^{\wedge}_{\mathbb{R}^q}$ enables us to write

$$\begin{aligned}(L_{2m})^{\wedge}_{\mathbb{R}^q}(x) &= (-4\pi^2 x^2)^{-2m}(Q_{2m})^{\wedge}_{\mathbb{R}^q}(x-g'') \\ &= \left(\frac{|x-g''|^2}{|x|^2}\right)^{2m} (L_{2m})^{\wedge}_{\mathbb{R}^q}(x-g'').\end{aligned} \tag{12.52}$$

For $x \in \mathcal{F}_\Lambda \backslash \{g''\}$ we therefore obtain the limit relation

$$\lim_{m\to\infty} (L_{2m})^{\wedge}_{\mathbb{R}^q}(x) = 0. \tag{12.53}$$

Together with (12.50) and the dominated convergence theorem of analysis this implies that $(L_{2m})^{\wedge}_{\mathbb{R}^q}$ converges to the characteristic function of $\mathcal{F}_\Lambda$ multiplied with the factor $\frac{1}{\|\mathcal{F}_\Lambda\|}$. As a consequence, we obtain

$$\lim_{m\to\infty} \int_{\mathcal{G}} (L_{2m})^{\wedge}_{\mathbb{R}^q}(a)\, e^{2\pi i x\cdot a}\, da = \frac{1}{\|\mathcal{F}_\Lambda\|} \int_{\mathcal{F}_\Lambda \cap \mathcal{G}} e^{2\pi i x\cdot a}\, da \tag{12.54}$$

for $\mathcal{G}$ being a regular region in $\mathbb{R}^q$. By taking the Fourier transform over $\mathbb{R}^q$ we obtain a *relationship of L_m to the multivariate Shannon kernel*

$$\lim_{m\to\infty} L_{2m}(x) = \frac{1}{\|\mathcal{F}_\Lambda\|} \int_{\mathcal{F}_\Lambda} e^{2\pi i x\cdot a}\, da \tag{12.55}$$

uniformly for all points $x \in \mathbb{R}^q$.

CHAPTER 13

Polymetaharmonic finite bandwidth sampling

Our aim now is to demonstrate that the one-dimensional Shannon sampling theory may be regarded as particular extensions of one-dimensional Euler/Poisson summation formulas and that finite bandwidth Shannon sampling is strongly related to the lattice point context of Hardy-Landau identities known from number theory.

13.1. Univariate lattice point theory based Shannon sampling

We first explain the classical Euler summation formula based on the $\mathbb{Z}$-periodic lattice (i.e., Euler-Green) function $G(\Delta;\cdot)$ relative to the (one-dimensional) Laplace operator Δ, i.e., the second order derivative, in its equivalence to the standard Poisson summation formula (see L.J. Mordell [**590, 591**]).

Univariate Euler summation formula. Under the assumption of second order continuous derivatives of a function F on the interval $[0,n], n \in \mathbb{N}$, L. Euler (1707-1783) succeeded in finding the *summation formula*

$$
\begin{aligned}
(13.1) \qquad \sum_{\substack{g\in[0,n]\\ g\in\mathbb{Z}}} {}' F(g) &= \sum_{\substack{0<g<n\\ g\in\mathbb{Z}}} F(g) + \frac{1}{2}\left(F(0)+F(n)\right) \\
&= \int_0^n F(x)\,dx + \frac{1}{12}\left(F'(n)-F'(0)\right) + \int_0^n \underbrace{\left(-\frac{1}{2}B_2(x)\right)}_{=G(\Delta;x)} F''(x)\,dx,
\end{aligned}
$$

where B_2 given by

$$
B_2(x) = (x-\lfloor x\rfloor)^2 - (x-\lfloor x\rfloor) + \frac{1}{6} \tag{13.2}
$$

is known as the *"Bernoulli function of degree 2"* in today's nomenclature.

The particular structure of the summation formula captures the delicate details of the connection between integration, i.e., "continuous summation", and its various discretizations. Obviously, it subsumes and resolves the appropriate bridge between continuous and discrete summation within a single exposition. So, it provides the canonical bridge to modern sampling. At this stage, however, it should be pointed out that L. Euler himself used this interrelation between continuous and discrete sums only for estimating sums and series by virtue of integrals. It was actually C. MacLaurin (1698-1746), who discovered the summation formula (13.1) independently in 1742, to use it for the evaluation of integrals in terms of sums. Hence, C. MacLaurin's understanding seen from a view point of application is much closer to Shannon sampling in the sense of expressing a function in integral form by a cardinal series taken over discrete functional values at lattice points.

The essential idea of the Euler summation formula as used in our context is different from classical procedures, which can be found in the literature (see, e.g., [**416**], [**46**] and the references therein). Our approach is based on the physically motivated interpretation of the Bernoulli function B_2 occurring in the (one-dimensional) Euler summation formula by means of mathematical physics as the Green function $G(\Delta;\cdot)$ with respect to the (one-dimensional) Laplace operator $\Delta = \nabla \cdot \nabla$ corresponding to "boundary conditions" of $\mathbb{Z}$-periodicity. In doing so, the Bernoulli function, i.e., in the jargon of mathematical physics, the *Euler-Green function*, acts as the connecting tool to convert a differential equation involving the one-dimensional Laplace operator corresponding to periodic boundary conditions into an associated integral equation, i.e., the summation formula. More concretely, the $\mathbb{Z}$-periodic Euler-Green function $G(\Delta;\cdot)$ for the Laplace operator Δ is constructed so as to have the absolutely and uniformly convergent bilinear expansion

$$G(\Delta; x-y) = \sum_{\substack{\Delta^\wedge(h)\neq 0\\ h\in\mathbb{Z}}} \frac{e^{2\pi i h\cdot x} e^{-2\pi i h\cdot y}}{-\Delta^\wedge(h)}, \quad x,y\in\mathbb{R}, \tag{13.3}$$

where the sequence $\{\Delta^\wedge(h)\}_{h\in\mathbb{Z}}$ forms the spectrum $\mathrm{Spect}_\Delta(\mathbb{Z})$ of the Laplace operator Δ, i.e., $(\Delta + \Delta^\wedge(h))\ e^{2\pi i h\cdot x} = 0, x\in\mathbb{R}$, so that $\Delta^\wedge(h) = 4\pi^2 h^2,\ h\in\mathbb{Z}$. Observing the special values

$$G(\Delta;0) = G(\Delta;n) = \sum_{\substack{\Delta^\wedge(h)\neq 0\\ h\in\mathbb{Z}}} \frac{1}{-\Delta^\wedge(h)} = -\frac{1}{12} \tag{13.4}$$

and the explicit representation (13.2) of the Fourier series expansion (13.3) we are able to reformulate the Euler summation formula (13.1). Integration by parts yields

$$\begin{aligned}\sum_{\substack{g\in[0,n]\\ g\in\mathbb{Z}}} {}' F(g) &= \int_0^n F(x)\, dx + \sum_{\substack{h\neq 0\\ h\in\mathbb{Z}}} \left(\frac{\nabla F(n) - \nabla F(0)}{4\pi^2 h^2} - \int_0^n \frac{e^{2\pi i h\cdot x}}{4\pi^2 h^2}\, \Delta F(x)\, dx \right) \\ &= \int_0^n F(x)\, dx - \lim_{N\to\infty} \sum_{\substack{|h|\leq N\\ h\neq 0}} \frac{1}{2\pi i h} \int_0^n \nabla F(x)\, e^{2\pi i h\cdot x}\, dx,\end{aligned} \tag{13.5}$$

such that the classical 1D *Poisson summation formula* comes into play

$$\sum_{\substack{g\in[0,n]\\ g\in\mathbb{Z}}} {}' F(g) = \lim_{N\to\infty} \sum_{\substack{|h|\leq N\\ h\in\mathbb{Z}}} \int_0^n F(x)\, e^{2\pi i h\cdot x}\, dx. \tag{13.6}$$

In spite of their apparent dissimilarity, the Euler summation formula (13.1) and the Poisson summation formula (13.6) are equivalent for twice continuously differentiable functions on the interval $[0,n]$.

Seen from the mathematical point of view, the Euler summation formula (13.1) and the Poisson summation formula (13.6) are illustrations of how a generalization can lead to the solution of seemingly independent problems. The "building blocks" of the bridge between the two equivalent formulas (13.1) and (13.6) are the *defining constituents of the $\mathbb{Z}$-lattice (i.e., Euler-Green) function* $G(\Delta;\cdot)$, which can be characterized as follows:

(*Periodicity*) $G(\Delta;\cdot)$ is continuous in $\mathbb{R}$ and $\mathbb{Z}$-periodic

$$G(\Delta;x) = G(\Delta;x+g),\ x\in\mathbb{R},\ g\in\mathbb{Z}. \tag{13.7}$$

(*Differential equation*) $\Delta G(\Delta;\cdot)$ "coincides" apart from an additive constant with the Dirac function(al)

$$\Delta_x G(\Delta;x) = -1,\ x\in\mathbb{R}\backslash\mathbb{Z}. \tag{13.8}$$

(*Characteristic singularity*) $G(\Delta;\cdot)$ possesses the singularity of the 1D-fundamental solution of the Laplace operator

$$G(\Delta;x) - \frac{1}{2}\,|x| = O(1),\ x\to 0. \tag{13.9}$$

(*Normalization*). $G(\Delta;\cdot)$ integrated over a whole period interval of length 1 is assumed to be zero, i.e.,

$$\int_{\mathcal{J}_{\mathbb{Z}}} G(\Delta;x)\,dx = \int_{-\frac{1}{2}}^{\frac{1}{2}} G(\Delta;x)\,dx = 0. \tag{13.10}$$

The $\mathbb{Z}$-lattice function $G(\Delta;\cdot)$ is uniquely determined by the properties (13.7)-(13.10). More generally, for *arbitrary* intervals $[c,d]\subset\mathbb{R}$, $c<d$, and *arbitrary* twice

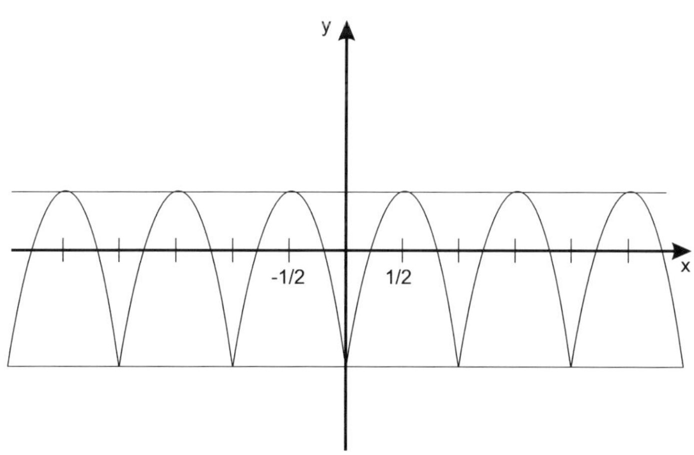

FIGURE 1. Graphical illustration of the Green function $G(\Delta;\cdot)$

continuously differentiable (weight) functions F on $[c,d]$, the constituents (13.7)-(13.10) of the Euler-Green function $G(\Delta;\cdot)$ enable us to guarantee the equivalence of the *Euler summation formula*

$$\sum_{\substack{g\in[c,d]\\ g\in\mathbb{Z}}}{}' \, F(g) = \int_c^d F(x)\, dx + \int_c^d G(\Delta;x)\, \Delta F(x)\, dx + \{F(x)\, (\nabla G(\Delta;x)) - G(\Delta;x)\, (\nabla F(x))\}\,\Big|_c^d \tag{13.11}$$

and the *Poisson summation formula*

$$\sum_{\substack{g\in[c,d]\\ g\in\mathbb{Z}}}{}' \, F(g) \quad = \quad \lim_{N\to\infty} \sum_{\substack{|h|\le N\\ h\in\mathbb{Z}}} \int_c^d F(x)\, e^{2\pi i h\cdot x}\, dx, \tag{13.12}$$

where we have used the convention

$$\sum_{\substack{g\in[c,d]\\ g\in\mathbb{Z}}}{}' \, F(g) = \sum_{\substack{c<g<d\\ g\in\mathbb{Z}}} F(g) \,+\, \frac{1}{2} \sum_{\substack{g=c,d\\ g\in\mathbb{Z}}} F(g), \tag{13.13}$$

and the second sum on the right side of (13.13) only occurs if c and/or d is an integer.

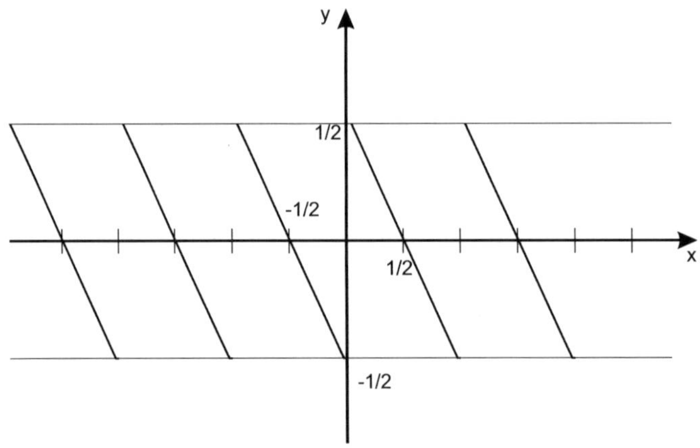

FIGURE 2. The derivative $\nabla G(\Delta;x)$ of the $\mathbb{Z}$-lattice (Euler-Green) function $G(\Delta;x)$.

Univariate Hardy-Landau identity. Of particular interest in lattice point theory is the special case of a constant weight function (i.e., $F = 1$), leading to the *one-dimensional "Hardy-Landau identity"* in the following equivalent representations:

$$\sum_{\substack{g\in[c,d]\\ g\in\mathbb{Z}}}{}' \, 1 \;=\; (d-c) \;+\; (\nabla G(\Delta;x))\,\Big|_c^d, \tag{13.14}$$

$$\sum_{\substack{g\in[c,d]\\ g\in\mathbb{Z}}}{}' \, 1 = (d-c) \,+\, \sum_{\substack{h\neq 0\\ h\in\mathbb{Z}}} \frac{1}{2\pi i h}\, e^{2\pi i h\cdot x}\Big|_c^d. \tag{13.15}$$

The formula (13.15) compares the number of $\mathbb{Z}$-lattice points inside an interval $[c,d]$ with the length $d-c$ of the interval $[c,d]$ under the explicit knowledge of the remainder term (usually called, the *discrepancy*) as an expression in terms of the derivative of the lattice function (see Figure 2) and equivalently by means of a convergent one-dimensional alternating series.

In the number-theoretic literature, the one-dimensional Hardy-Landau identity is usually formulated on an interval $[-R, R]$, $R > 0$, that is symmetric to the origin, using the nomenclature of Bessel functions. Under the particular choice $[-R, R]$ we have

(13.16)
$$\int_{-R}^{R} e^{2\pi i h\cdot x}\, dx = 2R\ \mathrm{sinc}(2\pi|h|R) = \frac{\sin(2\pi R|h|)}{\pi|h|} = \frac{J_1(1; 2\pi R|h|)}{\pi|h|}, \qquad R > 0,$$

so that

$$\sum_{\substack{|g|\le R\\ g\in\mathbb{Z}}}{}' \, 1 \; = \; 2R \; + \sum_{\substack{h\neq 0\\ h\in\mathbb{Z}}} \frac{J_1(1; 2\pi|h|R)}{\pi|h|} = \sum_{h\in\mathbb{Z}} \frac{J_1(1; 2\pi|h|R)}{\pi|h|}. \tag{13.17}$$

For positive $R < \frac{1}{2}$ we obtain a first sampling result

$$1 \; = \; \sum_{h\in\mathbb{Z}} 1 \; \frac{J_1(1; 2\pi|h|R)}{\pi|h|}. \tag{13.18}$$

The one-dimensional Poisson summation formula yields the identity

$$\lim_{N\to\infty} \sum_{\substack{|g|\le N\\ g\in\mathbb{Z}}} \frac{J_1(1; 2\pi R|h|)}{\pi|h|} = \sum_{h\in\mathbb{Z}} \int_{\mathbb{R}} \frac{J_1(1; 2\pi R|h|)}{\pi|h|}\, e^{-2\pi i h\cdot x}\, dx. \tag{13.19}$$

The (discontinuous) integral on the right side of (13.19) is explicitly known. By virtue of the Fourier inversion formula (see, e.g., [**857**]) we have

$$\lim_{N\to\infty} \int_{-N}^{N} \frac{J_1(1; 2\pi R|h|)}{\pi|h|}\, e^{-2\pi i h\cdot x}\, dx = \begin{cases} 1, & |h| < R,\\ \frac{1}{2}, & |h| = R,\\ 0, & |h| > R, \end{cases} \tag{13.20}$$

so that the one-dimensional Hardy-Landau identity allows an immediate interpretation as Poisson-type summation formula, too. In addition, the close relation between the Hardy-Landau summation and the metaharmonicity of the summands (cf. [**268**]) becomes obvious in the identity (13.15), since the function $e(h\cdot) = e^{2\pi i h\cdot}, h \in \mathbb{Z}$, satisfies *the 1 D Helmholtz equation* $(\Delta + \Delta^\wedge(h))\, e(h\cdot x) = 0, x \in \mathbb{R}$, corresponding to the "wave number" $\lambda = \Delta^\wedge(h) = 4\pi^2 h^2, h \in \mathbb{Z}$; i.e., $e(h\cdot)$, $h \in \mathbb{Z}$, is *metaharmonic* in $\mathbb{R}$.

As a consequence, the Euler and Poisson summation formulas offer two significant perspectives, which turn out to be of decisive importance in lattice point sampling

- to specify convergence criteria for alternating infinite series and to verify limits and asymptotic relations of infinite lattice point sums (see, e.g., W. Freeden [**268**] for more details).
- to evaluate integrals constituting bandlimited as well as nonbandlimited functions and to estimate the error, and to provide settings of constructive approximation (see, e.g., W. Freeden, M. Gutting [**278**] for more details on integration and numerical cubature).

Variants of the Euler summation formula. Our aim next is to introduce the lattice (i.e., Euler-Green) function $G(L+\lambda;\cdot)$ with respect or the operator

$$L+\lambda = \Delta + 4\pi i\ a\cdot\nabla + \lambda \tag{13.21}$$

and the "boundary condition" of Λ-periodicity, where $\lambda \in \mathbb{R}, a \in \mathbb{R}, \Lambda = \sigma\mathbb{Z}, \sigma > 0, \Lambda^{-1} = \sigma^{-1}\mathbb{Z}$, $\mathcal{F}_\Lambda = [-\frac{\sigma}{2}, \frac{\sigma}{2})$, $\|\mathcal{F}_\Lambda\| = \sigma$. To this end we notice that

$$(L + L^\wedge(h))\ e^{2\pi i h\cdot x} = 0,\ x \in \mathbb{R}, \tag{13.22}$$

where

$$L^\wedge(h) = 4\pi^2 h^2 - 4\pi i a\cdot 2\pi i h = 4\pi^2\left((h+a)^2 - a^2\right),\ h \in \Lambda^{-1}. \tag{13.23}$$

Keeping these facts in mind we are distributionally able to interpret the expression $L\ G(L;\cdot)$ as a modified Dirac function(al) δ_L given by

$$\delta(x) = \frac{1}{\|\mathcal{F}_\Lambda\|} \sum_{\substack{L^\wedge(h)=\lambda \\ h\in\Lambda^{-1}}} e^{2\pi i h\cdot x} + \delta_{L+\lambda}(x), \quad x \in \mathbb{R}, \tag{13.24}$$

where, as always, the sum $\sum_{L^\wedge(h)=\lambda, h\in\Lambda^{-1}} \ldots$ is assumed to be zero if there is no lattice point $h \in \Lambda$ satisfying $L^\wedge(h) = \lambda$. In this case, $\delta(x) = \delta_{L+\lambda}(x)$. Thus we are confronted with a striking difference between the case of eigenvalues and noneigenvalues of the operator L, which reflects the situation, which we have to expect for the structural context of Euler summation formulas.

The formal background provided by (13.24) motivates the following constituting properties of the 1D Λ-lattice (i.e., Euler-Green) function:

$G(L+\lambda;\cdot) : \mathbb{R}\backslash\Lambda \to \mathbb{C}$, $\lambda \in \mathbb{R}$ fixed, is called Λ-*lattice function (or Euler-Green function)* for the operator $L+\lambda, L = \Delta+4\pi i a\cdot\nabla,\ a \in \mathbb{R}$, if it satisfies the following properties:

(i) For all $x \in \mathbb{R}\setminus\Lambda$ and $g \in \Lambda$

$$G(L+\lambda; x+g) = G(L+\lambda; x). \tag{13.25}$$

(ii) For all $x \in \mathbb{R}\backslash\Lambda$

$$\overline{G(L+\lambda; x)} = G(L+\lambda; -x). \tag{13.26}$$

(iii) $G(L+\lambda;\cdot)$ is twice continuously differentiable for all $x \in \mathbb{R}\setminus\Lambda$.
For $\lambda \notin \mathrm{Spect}_L(\Lambda)$, i.e., $\lambda \neq 4\pi^2\left((h+a)^2 - a^2\right),\ h \in \Lambda^{-1}$,

$$(L+\lambda)G(L+\lambda; x) = 0. \tag{13.27}$$

For $\lambda \in \mathrm{Spect}_L(\Lambda)$, i.e., $\lambda = 4\pi^2\left((h+a)^2 - a^2\right)$

$$(L+\lambda)G(L+\lambda; x) = -\frac{1}{\|\mathcal{F}_\Lambda\|} \sum_{\substack{(L+\lambda)^\wedge(h)=0 \\ h\in\Lambda^{-1}}} e^{2\pi i h\cdot x}. \tag{13.28}$$

(iv) In a neighborhood of the origin

$$G(L;x) - \frac{1}{2}\,|x| = O(1),\ x \to 0, \tag{13.29}$$

(so that $\nabla G(L;x)$ shows a finite jump as singularity in the origin, hence, in all lattice points h of Λ because of condition (ii)).

(v) For all $h \in \Lambda^{-1}$ with $(L+\lambda)^\wedge(h) = 0$

$$\int_{\mathcal{F}_\Lambda} G(L+\lambda; x)\; e^{-2\pi i h\cdot x}\; dx = 0. \tag{13.30}$$

For all $h \in \Lambda^{-1}$ and $\lambda \in \mathbb{R}$ we readily see by a double integration by parts, that

$$\int_{\mathcal{F}_\Lambda} G(L+\lambda; x)\; e^{-2\pi i h\cdot x}\; dx = \begin{cases} 0, & (L+\lambda)^\wedge(h) = 0, \\ -\frac{1}{\|\mathcal{F}_\Lambda\|}\frac{1}{(L+\lambda)^\wedge(h)}, & (L+\lambda)^\wedge(h) \neq 0. \end{cases} \tag{13.31}$$

As a consequence, for each number $\lambda \in \mathbb{R}$, there exists one and only one Λ-lattice (i.e., Euler-Green) function $G(L+\lambda;\cdot)$ satisfying the constituting conditions (i) – (v) listed above (the proof follows from standard arguments of the Fourier theory). Moreover, it is easy to see that $G(L+\lambda;\cdot)$ has the absolutely and uniformly convergent bilinear orthogonal expansion

$$G(L+\lambda; x-y) \;=\; \sum_{\substack{(L+\lambda)^\wedge(h)\neq 0\\ h\in\Lambda^{-1}}} \frac{\Phi_h(x)\;\overline{\Phi_h(y)}}{-(L+\lambda)^\wedge(h)}, \quad x, y \in \mathbb{R}, \tag{13.32}$$

i.e.,

$$G(L+\lambda; x-y) \;=\; \frac{1}{\|\mathcal{F}_\Lambda\|} \sum_{\substack{4\pi^2\left((h+a)^2-a^2\right)\neq\lambda\\ h\in\Lambda^{-1}}} \frac{e^{2\pi i h\cdot(x-y)}}{\lambda - 4\pi^2\left((h+a)^2-a^2\right)}, \quad x, y \in \mathbb{R},$$

(note that $-(L+\lambda)^\wedge(h) = \lambda - L^\wedge(h) = \lambda - 4\pi^2\left((h+a)^2 - a^2\right)$).

The Λ-lattice function $G(L+\lambda;\cdot)$ for the operator $L+\lambda$ acts as the connecting tool to convert the differential equation involving the Helmholtz-type operator $L+\lambda$ corresponding to Λ-periodic "boundary conditions" into an associated integral equation, which turns out to be the counterpart of the Euler summation formula (13.1) in the particular manifestation for the differential operator $L+\lambda$.

Euler-type summation formula. Observing the defining properties of the Λ-lattice function $G(L+\lambda;\cdot)$ for the operator $L+\lambda$ we are immediately able to formulate an Euler summation formula analogously to the case of the Helmholtz operator $\Delta+\lambda$ (cf. [**268**]) by an ordinary twofold integration by parts.

Let Λ be an arbitrary lattice in $\mathbb{R}$. Let $(c,d) \subset \mathbb{R}, c < d$, be a finite interval. Let H be a twice continuously differentiable function on $[c,d]$. Then, for $\lambda \notin \mathrm{Spect}_L(\Lambda)$, we have

$$\begin{aligned} \sum_{\substack{g\in[c,d]\\ g\in\Lambda}}{}' H(g) \tag{13.33} \\ &= \int_{[c,d]} G(L+\lambda; x)\; (\overline{L}+\lambda) H(x)\; dx \\ &\quad + \left\{H(x)\,(\nabla_x + 4\pi i a)\, G(L+\lambda; x)\right\}\Big|_c^d \\ &\quad - \left\{G(L+\lambda; x)\nabla_x H(x)\right\}\Big|_c^d, \end{aligned}$$

while, for $\lambda \in \mathrm{Spect}_L(\Lambda)$, *we have*

$$\sum_{\substack{g\in[c,d]\\ g\in\Lambda}}{}' H(g) = \frac{1}{\|\mathcal{F}_\Lambda\|} \sum_{\substack{(L+\lambda)^\wedge(h)=0\\ h\in\Lambda^{-1}}} \int_{[c,d]} H(x)\, e^{2\pi i h\cdot x}\, dx + \int_{[c,d]} G(L+\lambda;x)\, (\overline{L}+\lambda) H(x)\, dx + \{H(x)\, (\nabla_x + 4\pi i a)\, G(L+\lambda;x)\}\big|_c^d - \{G(L+\lambda;x)\nabla_x H(x)\}\big|_c^d, \tag{13.34}$$

where a is an arbitrary real number. $\overline{L}$ *is the conjugate complex of L, i.e.,* $\overline{L} = \Delta - 4\pi i a\cdot\nabla$, $a \in \mathbb{R}$. *and*

$$\sum_{\substack{g\in[c,d]\\ g\in\Lambda}}{}' H(g) = \sum_{\substack{g\in(c,d)\\ g\in\Lambda}} H(g) + \frac{1}{2} \sum_{\substack{g=c,d\\ g\in\Lambda}} H(g). \tag{13.35}$$

Once more, $\sum_{L^\wedge(h)=\lambda, h\in\Lambda^{-1}} \cdots$ *means that the summation has to be extended over all* $h \in \Lambda^{-1}$ *satisfying* $L^\wedge(h) = \lambda$. *In the case, that there is no lattice point* $h \in \Lambda^{-1}$ *satisfying* $L^\wedge(h) = \lambda$, *we understand the sum to be zero.*

The Euler summation formulas (13.33) and (13.34) express a sum of functional values in terms of Fourier integrals over $[c, d]$ and associated boundary terms, thereby involving the derivatives of the function H up to the second order. The difference between the two cases (13.33) and (13.34) shows that "resonance phenomena" occur in the case of eigenvalues.

The summation formulas (13.33) and (13.34) lead to the following corollary:

Let Λ *be an arbitrary lattice in* $\mathbb{R}$. *Let* $(c, d) \subset \mathbb{R}, c < d$, *be a finite interval. Let H be a twice continuously differentiable function on* $[c, d]$. *Then we have*

$$\sum_{\substack{g\in[c,d]\\ g\in\Lambda}}{}' e^{2\pi i a\cdot g} H(g) = \frac{1}{\|\mathcal{F}_\Lambda\|} \sum_{\substack{(L+\lambda)^\wedge(h)=0\\ h\in\Lambda^{-1}}} \int_{[c,d]} e^{2\pi i a\cdot x} H(x)\, e^{2\pi i h\cdot x}\, dx + \int_{[c,d]} G(L+\lambda;x)\, (\overline{L}+\lambda)\, e^{2\pi i a\cdot x} H(x)\, dx + \{e^{2\pi i a\cdot x} H(x)\, (\nabla_x + 4\pi i a)\, G(L+\lambda;x)\}\big|_c^d - \{G(L+\lambda;x)\nabla_x e^{2\pi i a\cdot x} H(x)\}\big|_c^d. \tag{13.36}$$

The Euler-type summation formula (13.36) coincides with (13.11), i.e., the classical "resonance-type" Euler summation formula, by letting $\Lambda = \mathbb{Z}$, $\lambda = 0$, and $a = y = 0$.

Poisson-type summation formula over finite intervals. Next we choose the 1D-ball $\mathbb{B}^1_N(y) = (-N+y, N+y)$ as the reference interval in the identity (13.36). Furthermore, we replace the term $H(x)$ especially by $F^{\wedge}_{\mathbb{B}^1_R}(x-y), y \in \mathbb{R}$, with F being of class $C^{(2)}(\overline{\mathbb{B}^1_R})$, $R > 0$, where the Fourier transform $F^{\wedge}_{\mathbb{B}^1_R}$ is given by the integral

$$(13.37) \qquad F^{\wedge}_{\mathbb{B}^1_R}(x) = \int_{\mathbb{B}^1_R} F(z)\, e^{-2\pi i x\cdot z}\, dz = \int_{\substack{|z|\le R\\ z\in\mathbb{R}}} F(z)\, e^{-2\pi i x\cdot z}\, dz, \quad x \in \mathbb{R}.$$

We are led to the following summation formula:

For given $y \in \mathbb{R}$, let F be a twice continuously differentiable function on $\overline{\mathbb{B}^1_N(y)}$. Then we have

$$(13.38) \qquad \sum_{\substack{|g-y|\le N\\ g\in\Lambda}}{}' e^{2\pi i a\cdot g} F^{\wedge}_{\mathbb{B}^1_R}(g-y)$$

$$= \frac{1}{\|\mathcal{F}_\Lambda\|} \sum_{\substack{4\pi^2((h+a)^2-a^2)=\lambda\\ h\in\Lambda^{-1}}} e^{2\pi i y\cdot(h+a)} \int_{\substack{|x|\le N\\ x\in\mathbb{R}}} F^{\wedge}_{\mathbb{B}^1_R}(x)\, e^{2\pi i x\cdot(h+a)}\, dx$$

$$+ \int_{\substack{|x-y|\le N\\ x\in\mathbb{R}}} G(L+\lambda;x)\, (\overline{L}_x+\lambda)\, e^{2\pi i a\cdot x} F^{\wedge}_{\mathbb{B}^1_R}(x-y)\, dx$$

$$- \left\{G(L+\lambda;x)\, \nabla_x e^{2\pi i a\cdot x} F^{\wedge}_{\mathbb{B}^1_R}(x-y)\right\}\Big|_{-N+y}^{N+y}$$

$$+ \left\{ e^{2\pi i a\cdot x} F^{\wedge}_{\mathbb{B}^1_R}(x-y)\, (\nabla_x + 4\pi i a)\, G(L+\lambda;x)\right\}\Big|_{-N+y}^{N+y}.$$

Obviously, the identity (13.38) compares a finite sum of functional values in lattice points of Λ with a finite sum of Fourier transforms corresponding to lattice points $h \in \Lambda^{-1}$ with $4\pi^2((h+a)^2 - a^2) = \lambda$ under explicit knowledge of the remainder term.

Observing the elementary identities

$$(13.39) \qquad \Delta_x e^{2\pi i a\cdot x} F^{\wedge}_{\mathbb{B}^1_R}(x-y) = e^{2\pi i a\cdot x}\left(\Delta_x + 4\pi i a\cdot\nabla_x - 4\pi^2 a^2\right) F^{\wedge}_{\mathbb{B}^1_R}(x-y),$$

$$(13.40) \qquad a\cdot\nabla_x e^{2\pi i a\cdot x} F^{\wedge}_{\mathbb{B}^1_R}(x-y) = e^{2\pi i a\cdot x}\left(a\cdot\nabla_x + 2\pi i a^2\right) F^{\wedge}_{\mathbb{B}^1_R}(x-y),$$

and

$$(13.41) \quad (\Delta_x - 4\pi i a\cdot\nabla_x)\, e^{2\pi i a\cdot x} F^{\wedge}_{\mathbb{B}^1_R}(x-y) = e^{2\pi i a\cdot x}\left(\Delta_x + 4\pi^2 a^2\right) F^{\wedge}_{\mathbb{B}^1_R}(x-y),$$

we obtain

$$(13.42) \qquad \overline{L}_x\, e^{2\pi i a x} F^{\wedge}_{\mathbb{B}^1_R}(x-y) = e^{2\pi i a\cdot x}(\Delta_x + 4\pi^2 a^2) F^{\wedge}_{\mathbb{B}^1_R}(x-y).$$

Thus we arrive at the following Euler-type summation formula:

For $a, y \in \mathbb{R}$ and $N > 0$ we have

$$
\begin{aligned}
(13.43)\quad & \sum_{\substack{|g-y|\le N \\ g-y\in\Lambda}}{}' e^{2\pi i a\cdot g} F^{\wedge}_{\mathbb{B}^1_R}(g-y) \\
&= \frac{1}{\|\mathcal{F}_\Lambda\|} \sum_{\substack{4\pi^2((h+a)^2-a^2)=\lambda \\ h\in\Lambda^{-1}}} e^{2\pi i y\cdot(h+a)} \int_{\substack{|x|\le N \\ x\in\mathbb{R}}} F^{\wedge}_{\mathbb{B}^1_R}(x)\, e^{2\pi i x\cdot(h+a)}\, dx \\
&+ \int_{\substack{|x-y|\le N \\ x\in\mathbb{R}}} G(L+\lambda; x)\, e^{2\pi i a\cdot x} \left(\Delta_x + \lambda + 4\pi^2 a^2\right) F^{\wedge}_{\mathbb{B}^1_R}(x-y)\, dx \\
&+ \left\{ e^{2\pi i a\cdot x} G(L+\lambda;x) \nabla_x F^{\wedge}_{\mathbb{B}^1_R}(x-y) \right\}\Big|_{-N+y}^{N+y} \\
&- \left\{ F^{\wedge}_{\mathbb{B}^1_R}(x-y) e^{2\pi i a\cdot x} (\nabla_x + 4\pi i a) G(L+\lambda;x) \right\}\Big|_{-N+y}^{N-+y}.
\end{aligned}
$$

Now, for $x \neq y$, we get by a twofold integration by parts, i.e., the one-dimensional Green theorem,

$$
\begin{aligned}
(13.44)\quad F^{\wedge}_{\mathbb{B}^1_R}(x-y) = & -\frac{1}{4\pi^2(x-y)^2} \int_{\substack{|w|\le R \\ w\in\mathbb{R}}} \Delta_w F(w)\, e^{-2\pi i w\cdot(x-y)}\, dw \\
& -\frac{1}{4\pi^2(x-y)^2} \left\{ (-2\pi i(x-y))\, e^{-2\pi i w\cdot(x-y)} F(w) \right\}\Big|_{-R}^{R} \\
& +\frac{1}{4\pi^2(x-y)^2} \left\{ (\nabla_w F(w))\, e^{-2\pi i w\cdot(x-y)} \right\}\Big|_{-R}^{R}.
\end{aligned}
$$

So, we are confronted with the fact that the second term of the right side of (13.44) is of order $O(|x-y|^{-1})$, while the other terms of the right side of (13.44) are of order $O(|x-y|^{-2})$ for $|x-y| \to \infty$, because of the continuity properties of F and its derivatives (note that this observation is not critical in the univariate context, but it becomes problematic in higher dimensions, where methods of the stationary phase and/or asymptotic relations of the metaharmonic theory have to be applied (cf. [**268**])).

In connection with the continuity of $G(L+\lambda;\cdot)$ and the piecewise continuity and boundedness of its first derivative we obtain for all "wave numbers" λ

$$\sum_{\substack{|g-y|\leq N\\ g-y\in\Lambda}} e^{2\pi a\cdot g} F^{\wedge}_{\mathbb{B}^1_R}(g-y) \tag{13.45}$$
$$= \frac{1}{\|\mathcal{F}_\Lambda\|} \sum_{\substack{4\pi^2((a+h)^2-a^2)=\lambda\\ h\in\Lambda^{-1}}} e^{2\pi iy\cdot(a+h)} \int_{\substack{|x|\leq N\\ x\in\mathbb{R}}} F^{\wedge}_{\mathbb{B}^1_R}(x)\, e^{2\pi ix\cdot(a+h)}\, dx$$
$$+ \int_{\substack{|x-y|\leq N\\ x\in\mathbb{R}}} G(L+\lambda;x)\, e^{2\pi ia\cdot x} \left(\Delta_x+\lambda+4\pi^2a^2\right) F^{\wedge}_{\mathbb{B}^1_R}(x-y)\, dx$$
$$+\, o(1)$$

for $N\to\infty$. The key step of our considerations is as follows: Choosing the "wave number" in a special way, i.e., $\lambda=4\pi^2(R^2-a^2)$ we see that

$$\left(\Delta_x+\lambda+4\pi^2a^2\right) F^{\wedge}_{\mathbb{B}^1_R}(x-y) = \int_{\substack{|w|\leq R\\ w\in\mathbb{R}}} \underbrace{4\pi^2(R^2-w^2)F(z)}_{=\tilde{F}(\omega)}\, e^{-2\pi iw\cdot(x-y)}\, dw. \tag{13.46}$$

Therefore, because of the factor $4\pi^2(R^2-w^2)$, twofold integration by parts yields the asymptotic relation

$$(\Delta_x+\lambda+4\pi^2a^2)F^{\wedge}_{\mathbb{B}^1_R}(x-y) = O\left(|x-y|^{-2}\right),\quad |x-y|\to\infty. \tag{13.47}$$

Consequently, the indefinite integral

$$\lim_{N\to\infty} \int_{\substack{|x-y|\leq N\\ x\in\mathbb{R}}} G(L+\lambda;x)\, e^{2\pi ia\cdot x} \left(\Delta_x+\lambda+4\pi^2a^2\right) F^{\wedge}_{\mathbb{B}^1_R}(x-y)\, dx \tag{13.48}$$

is absolutely convergent for the special choice $\lambda=4\pi^2(R^2-a^2)$.

Considering the Fourier series expansion (13.32) with $\lambda=4\pi^2(R^2-a^2)$ we obtain

$$\int_{\substack{|x-y|\leq N\\ x\in\mathbb{R}}} G(L+\lambda;x)\, e^{2\pi ia\cdot x}\, F^{\wedge}_{\mathbb{B}^1_R}(x-y)\, dx$$
$$= \frac{1}{\|\mathcal{F}_\Lambda\|} \sum_{\substack{|h+a|\neq R\\ h\in\Lambda^{-1}}} \frac{e^{2\pi iy\cdot(a+h)}}{4\pi^2(R^2-(a+h)^2)} \int_{|z|\leq N} e^{2\pi iz\cdot(a+h)} \int_{\substack{|w|\leq R\\ w\in\mathbb{R}}} \tilde{F}(w)e^{-2\pi iw\cdot z}\, dw\, dz,$$

where we have introduced the abbreviation $\tilde{F}(w)=4\pi^2(R^2-w^2)F(w), w\in\mathbb{R}$. The application of the inversion formula of the one-dimensional Fourier theory (see, e.g.,

[**694**]) yields

$$\lim_{N\to\infty}\int_{|z|\le N} e^{2\pi i z\cdot(a+h)}\int_{\substack{|w|\le R\\ w\in\mathbb{R}}}\tilde{F}(w)e^{-2\pi i w\cdot z}\,dw\,dz \tag{13.49}$$

$$= \begin{cases} \tilde{F}(h+a), & |h+a|<R,\\ \frac{1}{2}\tilde{F}(h+a), & |h+a|=R,\\ 0, & |h+a|>R. \end{cases}$$

After some obvious manipulations we therefore obtain

$$\frac{1}{\|\mathcal{F}_\Lambda\|}\sum_{\substack{|a+h|\le R\\ h\in\Lambda^{-1}}}{}' e^{2\pi i y\cdot(a+h)}F(a+h) = \lim_{N\to\infty}\sum_{\substack{|g-y|\le N\\ g\in\Lambda}} e^{2\pi i a\cdot g}F^{\wedge}_{\mathbb{B}^1_R}(g-y). \tag{13.50}$$

Replacing Λ by its inverse lattice Λ^{-1} we obtain as a variant of the *one-dimensional Poisson summation formula*:

The lattice point identity

$$\sum_{\substack{|a+g|\le R\\ g\in\Lambda}}{}' F(a+g)\; e^{2\pi i y\cdot(a+g)} \tag{13.51}$$

$$= \lim_{N\to\infty}\frac{1}{\|\mathcal{F}_\Lambda\|}\sum_{\substack{|h-y|\le N\\ h\in\Lambda^{-1}}} e^{2\pi i a\cdot h}\int_{\substack{|x|\le R\\ x\in\mathbb{R}}} F(x)\; e^{-2\pi i x\cdot(h-y)}\,dx$$

holds true for arbitrary $R>0$ and all $a,y\in\mathbb{R}$, provided that F is a twice continuously differentiable function on the interval $\overline{\mathbb{B}^1_R}=[-R,R]$.

Fourier theoretic interpretation. On the one side, it is not difficult to see that

$$\sum_{\substack{|a+g+g'|\le R\\ g\in\Lambda}}{}' e^{2\pi i y\cdot(a+g+g')}F(a+g+g') = \sum_{\substack{|a+g|\le R\\ g\in\Lambda}}{}' e^{2\pi i y\cdot(a+g)}F(a+g) \tag{13.52}$$

holds for all $g'\in\Lambda$. Hence, for every $R>0$ and fixed $y\in\mathbb{R}$, the function

$$a\mapsto \sum_{\substack{|a+g|\le R\\ g\in\Lambda}}{}' e^{2\pi i y\cdot(a+g)}F(a+g),\quad a\in\mathbb{R} \tag{13.53}$$

is Λ-periodic, hence, the function

$$a\mapsto \lim_{N\to\infty}\frac{1}{\|\mathcal{F}_\Lambda\|}\sum_{\substack{|h-y|\le N\\ h\in\Lambda^{-1}}} e^{2\pi i a\cdot h}\int_{\substack{|x|\le R\\ x\in\mathbb{R}}} F(x)\; e^{-2\pi i x\cdot(h-y)}\,dx,\quad a\in\mathbb{R}, \tag{13.54}$$

is Λ-periodic. The formula (13.51) therefore means that the piecewise continuous function (13.53) is representable by its Fourier series (13.54) in the pointwise sense, provided that F is twice continuously differentiable on $\overline{\mathbb{B}^1_R}$. In other words, the Fourier coefficients of the function (13.53) are given by

$$
\begin{aligned}
(13.55)\qquad & \int_{\mathcal{F}_\Lambda} \sum_{\substack{|a+g|\le R\\ g\in\Lambda}}{}' e^{2\pi i y\cdot(a+g)} F(a+g)\, e^{-2\pi i a\cdot h}\, da \\
&= \int_{\mathcal{F}_\Lambda} \sum_{\substack{|a+g|\le R\\ g\in\Lambda}}{}' e^{2\pi i y\cdot(a+g)} F(a+g) e^{-2\pi i (a+g)\cdot h}\, da \\
&= \int_{\mathcal{F}_\Lambda} \sum_{\substack{|a+g|\le R\\ g\in\Lambda}}{}' F(a+g) e^{-2\pi i (a+g)\cdot(h-y)}\, da \\
&= \int_{\substack{|x|\le R\\ x\in\mathbb{R}}} F(x)\, e^{-2\pi i x\cdot(h-y)}\, dx
\end{aligned}
$$

for all $h \in \Lambda^{-1}$. On the other side, (13.51) can be rewritten in the following equivalent form:

The lattice point identity

$$
\begin{aligned}
(13.56)\qquad & \sum_{\substack{|a+h|\le R\\ h\in\Lambda^{-1}}}{}' F(a+h)\, \frac{1}{\|\mathcal{F}_\Lambda\|}\, e^{2\pi i y\cdot h} \\
&= \lim_{N\to\infty} \sum_{\substack{|g-y|\le N\\ g\in\Lambda}} e^{2\pi i a\cdot(g-y)} \int_{\substack{|x|\le R\\ x\in\mathbb{R}}} F(x)\, e^{-2\pi i x\cdot(g-y)}\, dx
\end{aligned}
$$

holds true for arbitrary $R > 0$ and all $a, y \in \mathbb{R}$, provided that F is a twice continuously differentiable function on the interval $\overline{\mathbb{B}^1_R}$.

For $a \in \mathbb{R}$, the function

$$
(13.57)\qquad y \mapsto \lim_{N\to\infty} \sum_{\substack{|g-y|\le N\\ g\in\Lambda}} e^{2\pi i a\cdot(g-y)} \int_{\substack{|x|\le R\\ x\in\mathbb{R}}} F(x)\, e^{-2\pi i x\cdot(g-y)}\, dx, \quad y\in\mathbb{R},
$$

are periodic with respect to the lattice Λ. The Fourier coefficients of the function (13.57) are given by

$$
\begin{aligned}
(13.58)\qquad & \int_{\mathcal{F}_\Lambda} \lim_{N\to\infty} \sum_{\substack{|g-y|\le N\\ g\in\Lambda}} e^{2\pi i a\cdot(g-y)} \int_{\substack{|x|\le R\\ x\in\mathbb{R}}} F(x)\, e^{-2\pi i x\cdot(g-y)}\, dx\; e^{-2\pi i y\cdot h}\, dy \\
&= \int_{\mathcal{F}_\Lambda} \lim_{N\to\infty} \sum_{\substack{|g-y|\le N\\ g\in\Lambda}} e^{2\pi i (a+h)\cdot(g-y)} \int_{\substack{|x|\le R\\ x\in\mathbb{R}}} F(x)\, e^{-2\pi i x\cdot(g-y)}\, dx\, dy \\
&= \int_{\mathbb{R}} e^{2\pi i (a+h)\cdot w} \int_{\substack{|x|\le R\\ x\in\mathbb{R}}} F(x)\, e^{-2\pi i x\cdot w}\, dx\, dw
\end{aligned}
$$

for all $h \in \Lambda^{-1}$, where the Fourier inversion formula (see, e.g., [**694**]) tells us that

(13.59)
$$\int_{\mathbb{R}} e^{2\pi i(a+h)\cdot w} \int_{\substack{|x|\le R\\ x\in\mathbb{R}}} F(x)\, e^{-2\pi i x\cdot w}\, dx\, dw = \begin{cases} F(a+h), & |a+h| < R,\\ \frac{1}{2}F(a+h), & |a+h| = R,\\ 0, & |a+h| > R. \end{cases}$$

The function (13.57) is representable by its Fourier series in the pointwise sense, provided that F is twice continuously differentiable on $\overline{\mathbb{B}^1_R}$.

Preparatory number theoretic interpretation. The formula (13.56) can be regarded as an extended one-dimensional realization of the famous Hardy-Landau identity. Indeed, for $a = y = 0$, and constant weight function $F = 1$, the identity (13.56) expresses the number of Λ–lattice points in $\overline{\mathbb{B}^1_R}$ in terms of an alternating series

(13.60)
$$\sum_{\substack{|g|\le R\\ g\in\Lambda}}{}' 1 = \frac{1}{\|\mathcal{F}_\Lambda\|} \lim_{N\to\infty} \sum_{\substack{|h|\le N\\ h\in\Lambda^{-1}}} \frac{J_1(1;2\pi|h|R)}{\pi|h|} = \frac{2R}{\|\mathcal{F}_\Lambda\|} \lim_{N\to\infty} \sum_{\substack{|h|\le N\\ h\in\Lambda^{-1}}} \operatorname{sinc}(2\pi|h|R).$$

Thus, the "Λ-*lattice discrepancy*" $\sum_{\substack{|g|\le R\\ g\in\Lambda}}{}' 1 - 2R\frac{1}{\|\mathcal{F}_\Lambda\|}$ allows the series expansion

$$\lim_{N\to\infty} \frac{1}{\|\mathcal{F}_\Lambda\|} \sum_{\substack{0<|h|\le N\\ h\in\Lambda^{-1}}} \frac{J_1(1;2\pi|h|R)}{\pi|h|}. \tag{13.61}$$

Moreover, it follows from (13.56) that

(13.62)
$$\sum_{\substack{|a+g|\le R\\ g\in\Lambda}}{}' e^{2\pi i y\cdot(a+g)} = \lim_{N\to\infty} \frac{1}{\|\mathcal{F}_\Lambda\|} \sum_{\substack{|h-y|\le N\\ h\in\Lambda^{-1}}} e^{2\pi i a\cdot h} \frac{J_1(1;2\pi|h-y|R)}{\pi|h-y|}, \quad y\in\mathbb{R},$$

holds true for all $a, y \in \mathbb{R}$. In particular, if $g = 0$ is the only lattice point in $\mathbb{B}^\wedge_R(a)$, we obtain the identity

$$e^{2\pi i y\cdot a} = \lim_{N\to\infty} \sum_{\substack{|h-y|\le N\\ h\in\Lambda^{-1}}} e^{2\pi i a\cdot h} \frac{1}{\|\mathcal{F}_\Lambda\|} \frac{J_1(1;2\pi|h-y|R)}{\pi|h-y|}, \quad y\in\mathbb{R}, \tag{13.63}$$

which can be regarded as one of the keystones for Papoulis' approach (cf. [**678**]) to one-dimensional sampling. It demonstrates that, for every point $y \in \mathbb{R}$, $e^{2\pi i y\cdot a}$ can be obtained from its samples $e^{2\pi i y\cdot h}, h \in \Lambda^{-1}$. This observation is another strong link to a close relationship of lattice point theory and sampling.

Finite bandwidth Shannon sampling. Next our purpose is to recognize the classical Shannon sampling theory in the Euler/Poisson-type summation reflected approach presented above. Our point of departure for realizing Shannon lattice point sampling actually starts from the Poisson-type summation formula

$$\sum_{\substack{|a+g|\leq R\\ g\in\Lambda}}{}' F(a+g)\, e^{2\pi i y\cdot(a+g)} \tag{13.64}$$

$$= \lim_{N\to\infty} \frac{1}{\|\mathcal{F}_\Lambda\|} \sum_{\substack{|h-y|\leq N\\ h\in\Lambda^{-1}}} e^{2\pi i a\cdot h} \int_{\substack{|x|\leq R\\ x\in\mathbb{R}}} F(x)\, e^{-2\pi i x\cdot(h-y)}\, dx$$

From (13.64) we are immediately led to

$$F^{\wedge}_{\mathbb{B}^1_R}(y) = \int_{\mathbb{B}^1_R} F(a)\, e^{-2\pi i y\cdot a}\, da = \int_{\mathcal{F}_\Lambda} \sum_{\substack{|a+g|\leq R\\ g\in\Lambda}}{}' F(a+g) e^{-2\pi i y\cdot(a+g)}\, da, \tag{13.65}$$

provided that $\overline{\mathbb{B}^1_R} \subset \overline{\mathcal{F}_\Lambda}$. So, it follows from (13.64) that

$$F^{\wedge}_{\mathbb{B}^1_R}(-y) = \lim_{N\to\infty} \sum_{\substack{|h-y|\leq N\\ h\in\Lambda^{-1}}} F^{\wedge}_{\mathbb{B}^1_R}(h-y) \frac{1}{\|\mathcal{F}_\Lambda\|} \frac{\sin(\pi h)}{\pi h}. \tag{13.66}$$

for all $\overline{\mathbb{B}^1_R} \subset \overline{\mathcal{F}_\Lambda}$. Consequently, going over to the usual Shannon context $\overline{\mathbb{B}^1_R} = \overline{\mathcal{F}_\Lambda}$, we obtain for the problem of reconstructing a bandlimited signal $F^{\wedge}_{\mathcal{F}_\Lambda}$ of the form

$$F^{\wedge}_{\mathcal{F}_\Lambda}(y) = \int_{\mathcal{F}_\Lambda} F(a)\, e^{-2\pi i a\cdot y}\, da, \quad F \in C^{(2)}\left(\overline{\mathcal{F}_\Lambda}\right), \quad y \in \mathbb{R}, \tag{13.67}$$

from (13.66) the sampling identity

$$F^{\wedge}_{\mathcal{F}_\Lambda}(-y) = \sum_{h\subset\Lambda^{-1}} F^{\wedge}_{\mathcal{F}_\Lambda}(h) \frac{1}{\|\mathcal{F}_\Lambda\|} \frac{\sin(\pi(h+y))}{\pi(h+y)}, \quad y \in \mathbb{R}. \tag{13.68}$$

Hence, by replacing $-y$ by y we obtain

$$F^{\wedge}_{\mathcal{F}_\Lambda}(y) = \sum_{h\in\Lambda^{-1}} F^{\wedge}_{\mathcal{F}_\Lambda}(h) \frac{1}{\|\mathcal{F}_\Lambda\|} \frac{\sin(\pi(h-y))}{\pi(h-y)}, \quad y \in \mathbb{R}. \tag{13.69}$$

The formula (13.69), indeed, is the classical *Shannon sampling formula.* It tells us that $F^{\wedge}_{\mathcal{F}_\Lambda}$, as defined by (13.67), can be recovered by its points on the lattice $\mathbb{Z}$.

In fact, it should be noted that the proof of our Shannon approach can be provided by a one-dimensional integration of a weighted Hardy-Landau-type lattice point identity. This observation can be documented also in the conventional Bessel function nomenclature of lattice point theory

$$F^{\wedge}_{\mathcal{F}_\Lambda}(y) = \sum_{h\in\Lambda^{-1}} F^{\wedge}_{\mathcal{F}_\Lambda}(h) \frac{1}{\|\mathcal{F}_\Lambda\|} \frac{J_1(1;\pi|h-y|)}{\pi|h-y|}, \quad y \in \mathbb{R}. \tag{13.70}$$

Summarizing in the lattice point nomenclature as used in this chapter we are able to rephrase Shannon's sampling result (13.68) as follows:

Let $\Lambda = \sigma\mathbb{Z}$, $\sigma > 0$, designate the lattice in $\mathbb{R}$ consisting of all points σg, $g \in \mathbb{Z}$. Moreover, denote by $\Lambda^{-1} = \frac{1}{\sigma}\mathbb{Z}$ the inverse lattice to Λ. Then, any function $F^{\wedge}_{\mathcal{F}_\Lambda}$

bandlimited to the fundamental cell $\mathcal{F}_\Lambda = \left[-\frac{\sigma}{2}, \frac{\sigma}{2}\right)$ of the lattice $\Lambda = \sigma\mathbb{Z}$, i.e., the Fourier transform over the finite interval $\mathcal{F}_\Lambda$

$$F^{\wedge}_{\mathcal{F}_\Lambda}(y) = \int_{\mathcal{F}_\Lambda} F(x) e^{-2\pi i x\cdot y}\, dx, \quad y \in \mathbb{R} \tag{13.71}$$

may be reconstructed from the sequence of samples $\{F^{\wedge}_{\mathcal{F}_\Lambda}(h)\}_{h\in\Lambda^{-1}}$ via its *cardinal series*

$$F^{\wedge}_{\mathcal{F}_\Lambda}(y) = \sum_{h\in\Lambda^{-1}} F^{\wedge}_{\mathcal{F}_\Lambda}(h) \underbrace{\frac{1}{\|\mathcal{F}_\Lambda\|}\int_{\mathcal{F}_\Lambda} e^{2\pi i x\cdot(h-y)}\, dx}_{=K_{\mathcal{F}_\Lambda}(h-y)}, \quad y \in \mathbb{R}, \tag{13.72}$$

where the symbol $\sum_{h\in\Lambda^{-1}} \ldots$, as always, means that the sum is extended over all lattice points h of the inverse lattice Λ^{-1}. Clearly, for $\Lambda = \sigma\mathbb{Z}$, $\sigma > 0$, we have $\|\mathcal{F}_{\Lambda^{-1}}\| = \frac{1}{\|\mathcal{F}_\Lambda\|} = \frac{1}{\sigma}$ and $K_{\mathcal{F}_\Lambda}(h-y)$ represents the kernel

$$K_{\mathcal{F}_\Lambda}(h-y) = \frac{1}{\|\mathcal{F}_\Lambda\|}\int_{\mathcal{F}_\Lambda} e^{2\pi i x\cdot(h-y)}\, dx = \frac{1}{\sigma}\frac{J_1(1;\pi|h-y|)}{\pi|h-y|}, \quad y\in\mathbb{R},\ h\in\Lambda^{-1}. \tag{13.73}$$

So, the Fourier transform $F^{\wedge}_{\mathcal{F}_\Lambda}$ (considered as a one-dimensional signal) is an infinitely repeated replication of the samples at points h of the inverse lattice Λ^{-1}. Combining (13.72) and (13.73) we equivalently obtain the well-known sampling formula

$$F^{\wedge}_{\mathcal{F}_{\sigma\mathbb{Z}}}(y) = \sum_{n=-\infty}^{\infty} F^{\wedge}_{\mathcal{F}_{\sigma\mathbb{Z}}}\left(\frac{n}{\sigma}\right)\frac{J_1\left(1;\pi\sigma|\frac{n}{\sigma}-y|\right)}{\pi\sigma|\frac{n}{\sigma}-y|}, \quad y\in\mathbb{R}, \tag{13.74}$$

such that, as announced earlier, Shannon sampling provides the bridge between continuous and discrete versions of a $\mathcal{F}_{\sigma\mathbb{Z}}$-bandlimited signal $F^{\wedge}_{\mathcal{F}_{\sigma\mathbb{Z}}}$.

Fourier transformed values based Shannon-type sampling (finite bandwidth). The univariate variant (13.51) of the *Poisson summation formula*, i.e., in our nomenclature, the *Poisson-type summation formula*, allows the following equivalent writing

$$\begin{aligned} &\sum_{\substack{|a+g|\le R\\ g\in\Lambda}}{}' F(a+g)\, e^{2\pi i y\cdot(a+g)} \\ &\qquad = \lim_{N\to\infty}\frac{1}{\|\mathcal{F}_\Lambda\|}\sum_{\substack{|h-y|\le N\\ h\in\Lambda^{-1}}} e^{2\pi i a\cdot h}\int_{\substack{|x|\le R\\ x\in\mathbb{R}}} F(x)\, e^{-2\pi i x\cdot(h-y)}\, dx. \end{aligned} \tag{13.75}$$

In fact, this leads to a a cardinal series resulting from the right side of (13.75), summing up functional values of the Fourier transform $F^{\wedge}_{\mathbb{B}^1_R}$ of $F\in C^{(2)}(\mathbb{B}^1_R)$, $R>0$, in y-shifted lattice points. Our goal is to go over from the Poisson-type summation formula (13.75) to a *Shannon-type sampling theorem*: Choosing $\Lambda=\sigma\mathbb{Z}$, $\sigma>0$, we obtain by integration of both sides of (13.75) over $\mathbb{B}^1_\tau$, $\tau>0$, the following Shannon-type sampling result (for the proof the reader is referred to the multivariate theory in this work (see also [**288**])):

Assume that Λ is the lattice $\sigma\mathbb{Z}$, $\sigma > 0$. Suppose that F is a member of the class $C^{(2)}(\overline{\mathbb{B}^1_\rho})$. Then, for $\mathbb{B}^1_\rho$, $\mathbb{B}^1_\tau$ with $\rho, \tau > 0$ arbitrary, we have

(13.76)

$$\begin{aligned}
&\sum_{\substack{(\mathcal{F}_{\sigma\mathbb{Z}}+\{g'\})\cap\mathbb{B}^1_\tau\neq\emptyset\\ g'\in\sigma\mathbb{Z}}} F^\wedge_{\mathbb{B}^1_\rho\cap\bigcup_{g\in\sigma\mathbb{Z}}(((\mathbb{B}^1_\tau\cap(\mathcal{F}_{\sigma\mathbb{Z}}+\{g'\}))-\{g'\})+\{g\})}(y)\\
&= \sum_{\substack{(\mathcal{F}_{\sigma\mathbb{Z}}+\{g'\})\cap\mathbb{B}^1_\tau\neq\emptyset\\ g'\in\sigma\mathbb{Z}}} \int_{\mathbb{B}^1_\rho\cap\bigcup_{g\in\Lambda}(((\mathbb{B}^1_\tau\cap(\mathcal{F}_{\sigma\mathbb{Z}}+\{g'\}))-\{g'\})+\{g\})} F(x)\, e^{-2\pi i y\cdot x}\, dx\\
&= \sum_{h\in\frac{1}{\sigma}\mathbb{Z}} \underbrace{\int_{\mathbb{B}^1_\rho} F(x)\, e^{-2\pi i h\cdot x}\, dx}_{=F^\wedge_{\mathbb{B}^1_\rho}(h)}\ \underbrace{\frac{1}{\|\mathcal{F}_{\sigma\mathbb{Z}}\|}\int_{\mathbb{B}^1_\tau} e^{2\pi i x\cdot(h-y)}\, dx}_{=K_{\mathbb{B}^1_\tau}(h-y)}.
\end{aligned}$$

Actually, the identity (13.76) provides an explicit quantification of the earlier mentioned “Nyquist-context”, i.e., they explicitly include all manifestations of over- and undersampling involving the intervals $\mathbb{B}^1_\rho, \mathbb{B}^1_\tau$ for the lattice $\Lambda = \sigma\mathbb{Z}, \sigma > 0$:

- *Oversampling* means that the signal can be sampled, where “superfluous information” can be removed (see, e.g., [**636**]) or used in an appropriate way to accelerate the convergence of the series (see, e.g., [**561**]).
- Otherwise, we speak of *undersampling*. In the case of undersampling we are confronted with the well-known phenomenon of *aliasing*, that the signal reconstructed from the samples is different from the original continuous signal, where the lattice point approach presented here provides the aliasing error in explicit form.

Some particular geometric (Nyquist-)configurations, i.e., special choices of intervals $\mathbb{B}^1_\rho, \mathbb{B}^1_\tau$ should be studied in more detail:

For $\overline{\mathbb{B}^1_\tau} = [-\frac{\sigma}{2}, \frac{\sigma}{2}] = \overline{\mathcal{F}_{\sigma\mathbb{Z}}}$ and $\mathbb{B}^1_\rho$ arbitrary, we obtain

(13.77)

$$\begin{aligned}
&\sum_{\substack{(\mathcal{F}_{\sigma\mathbb{Z}}+\{g'\})\cap\mathbb{B}^1_\tau\neq\emptyset\\ g'\in\sigma\mathbb{Z}}} F^\wedge_{\mathbb{B}^1_\rho\cap\bigcup_{g\in\sigma\mathbb{Z}}(((\mathbb{B}^1_\tau\cap(\mathcal{F}_{\sigma\mathbb{Z}}+\{g'\}))-\{g'\})+\{g\})}(y)\\
&= F^\wedge_{\mathbb{B}^1_\rho}(y)\\
&= \sum_{h\in\frac{1}{\sigma}\mathbb{Z}} F^\wedge_{\mathbb{B}^1_\rho}(h)\ K_{\mathbb{B}^1_\tau}(h-y).
\end{aligned}$$

We continue with $\overline{\mathbb{B}^1_\rho} = [-\frac{\sigma}{2}, \frac{\sigma}{2}]$ and $\mathbb{B}^1_\tau$ arbitrary. This yields the identity

$$\sum_{\substack{(\mathcal{F}_{\sigma\mathbb{Z}}+\{g'\})\cap\mathbb{B}^1_\tau\neq\emptyset \\ g'\in\sigma\mathbb{Z}}} F^\wedge_{\mathbb{B}^1_\rho\cap \bigcup_{g\in\sigma\mathbb{Z}}(((\mathbb{B}^1_\tau\cap(\mathcal{F}_{\sigma\mathbb{Z}}+\{g'\}))-\{g'\})+\{g\})}(y) \tag{13.78}$$
$$= \sum_{\substack{\mathcal{F}_{\sigma\mathbb{Z}}+\{g'\}\cap\mathbb{B}^1_\tau\neq\emptyset \\ g'\in\sigma\mathbb{Z}}} F^\wedge_{\mathbb{B}^1_\rho\cap((\mathbb{B}^1_\tau\cap(\mathcal{F}_\Lambda+\{g'\}))-\{g'\})}(y)$$
$$= \sum_{h\in\frac{1}{\sigma}\mathbb{Z}} F^\wedge_{\mathbb{B}^1_\rho}(h)\, K_{\mathbb{B}^1_\tau}(h-y).$$

For $\overline{\mathbb{B}^1_\rho}$, $\overline{\mathbb{B}^1_\tau} \subset [-\frac{\sigma}{2}, \frac{\sigma}{2}]$ we have

$$\sum_{\substack{(\mathcal{F}_{\sigma\mathbb{Z}}+\{g'\})\cap\mathbb{B}^1_\tau\neq\emptyset \\ g'\in\sigma\mathbb{Z}}} F^\wedge_{\mathbb{B}^1_\rho\cap \bigcup_{g\in\sigma\mathbb{Z}}(((\mathbb{B}^1_\rho\cap(\mathcal{F}_{\sigma\mathbb{Z}}+\{g'\}))-\{g'\})+\{g\})}(y) \tag{13.79}$$
$$= F^\wedge_{\mathbb{B}^1_\rho\cap\mathbb{B}^1_\tau}(y)$$
$$= \sum_{h\in\frac{1}{\sigma}\mathbb{Z}} F^\wedge_{\mathbb{B}^1_\rho}(h)\;\; K_{\mathbb{B}^1_\tau}(h-y).$$

For $\overline{\mathbb{B}^1_\rho} \subset \overline{\mathbb{B}^1_\tau} \subset [-\frac{\sigma}{2}, \frac{\sigma}{2}]$ we have

$$F^\wedge_{\mathbb{B}^1_\rho}(y) = \sum_{h\in\frac{1}{\sigma}\mathbb{Z}} F^\wedge_{\mathbb{B}^1_\rho}(h)\;\; K_{\mathbb{B}^1_\tau}(h-y), \tag{13.80}$$

whereas, for $\overline{\mathbb{B}^1_\tau} \subset \overline{\mathbb{B}^1_\rho} \subset [-\frac{\sigma}{2}, \frac{\sigma}{2}]$,

$$F^\wedge_{\mathbb{B}^1_\tau}(y) = \sum_{h\in\frac{1}{\sigma}\mathbb{Z}} F^\wedge_{\mathbb{B}^1_\rho}(h)\, K_{\mathbb{B}^1_\tau}(h-y). \tag{13.81}$$

In particular, for $\overline{\mathbb{B}^1_\rho} = \overline{\mathbb{B}^1_\tau} \subset [-\frac{\sigma}{2}, \frac{\sigma}{2}]$, we are able to formulate the identity

$$F^\wedge_{\mathbb{B}^1_\rho}(y) = \sum_{h\in\frac{1}{\sigma}\mathbb{Z}} F^\wedge_{\mathbb{B}^1_\rho}(h)\, K_{\mathbb{B}^1_\rho}(h-y). \tag{13.82}$$

Finally it should be noted that the standard form of Shannon sampling (cf. (13.74)) is provided by taking $\overline{\mathbb{B}^1_\rho} = \overline{\mathbb{B}^1_\tau} = [-\frac{\sigma}{2}, \frac{\sigma}{2}] = \overline{\mathcal{F}_{\sigma\mathbb{Z}}}$, so that

$$F^\wedge_{\mathcal{F}_{\sigma\mathbb{Z}}}(y) = \sum_{h\in\frac{1}{\sigma}\mathbb{Z}} F^\wedge_{\mathcal{F}_{\sigma\mathbb{Z}}}(h)\, K_{\mathcal{F}_{\sigma\mathbb{Z}}}(h-y). \tag{13.83}$$

Obviously, by choosing a small sampling density in (13.82) such that the interval $[-\frac{\sigma}{2}, \frac{\sigma}{2}]$ covers the compact support $\overline{\mathbb{B}^1_\rho}$ of the original signal $F^\wedge_{\mathbb{B}^1_\rho}$, the number of samples $F^\wedge_{\mathbb{B}^1_\rho}(h), h \in \frac{1}{\sigma}\mathbb{Z}$, for reconstruction is high, and vice versa. In practice, we are therefore required to find a compromise between the sampling density and the total number of samples. This compromise can be achieved by a choice of $\sigma\mathbb{Z}$ such that $[-\frac{\sigma}{2}, \frac{\sigma}{2}]$ covers $\overline{\mathbb{B}^1_\rho}$ tightly.

Perspectives. The 1D Poisson-type summation formula (13.51) opens different manifestations of Shannon-type sampling. Turning over to higher dimensions the weighted lattice point approach to the sampling theorem demands a more detailed attention: Unfortunately, our one-dimensional proof cannot be directly generalized in the same way. An essential reason is that the Fourier expansion of the lattice function does not converge absolutely and uniformly for lattices $\mathbb{Z}^q$ of dimension $q \geq 2$. Once more, we are confronted with the situation that the dimension $q = 2$ is "is not far away from the convergence", whereas the higher dimensions $q \geq 3$ become more and more critical. In other words, weighted Hardy-Landau identities in the plane turn out to be attackable (almost) similarly to the one-dimensional case, while this procedure becomes more and more problematic for increasing dimensions. In order to find a way out for higher dimensions we go over to lattice point identities in the sense of certain kinds of summability. More concretely, in our work, we shall use Gauss-Weierstrass means to verify lattice point identities leading to multivariate exponentially accelerated sampling. A fortunate accessory is decorrelation wavelet sampling to specify high frequency transitions in a signature to be investigated. Bessel-type summability originated by "lattice ball Euler summation formulas" instead of lattice point Euler summation formulas can be found in W. Freeden, M.Z. Nashed [**298**].

Our multivariate approach to lattice point sampling presented below shows a threefold interest, namely

- to extend the lattice point as well as sampling results from the unit to general lattices $\Lambda \subset \mathbb{R}^q$,
- to discuss general geometries for characterizing "base bands" in the form of regular regions $\mathcal{G}$ (finite bandwidth sampling) or the Euclidean space $\mathbb{R}^q$ itself (infinite bandwidth sampling),
- to decorrelate signature content by specifying high frequency transitions using some kind of wavelet sampling.

13.2. Multivariate lattice functions and summation formulas

In its customary formulation known in communication and electrical engineering, the *Shannon sampling theorem* is related to time-dependent signals given on a $1D$ lattice in Euclidean space $\mathbb{R}$, for which a condition between a bandwidth and sample rate has to be established (see C.E. Shannon [**764, 765**]).

Multivariate as iterated univariate concept. The 1D sampling formulation on the lattice $\Lambda = \sigma\mathbb{Z}$ has a canonical 1D iterated counterpart for the lattice $\Lambda = (\sigma\mathbb{Z})^q$, $\sigma > 0$, in Euclidean space $\mathbb{R}^q, q \geq 2$ (as discussed by E. Parzen [**681**]):

Any function $F^{\wedge}_{\mathcal{F}_\Lambda}$ bandlimited to the fundamental cell $\mathcal{F}_\Lambda = [-\frac{\sigma}{2}, \frac{\sigma}{2})^q$, given by

$$F^{\wedge}_{\mathcal{F}_\Lambda}(y) = \int_{\mathcal{F}_\Lambda} F(x) e^{-2\pi i x \cdot y}\, dx, \quad y \in \mathbb{R}^q, \tag{13.84}$$

may be reconstructed from the family of samples

$$\{F^{\wedge}_{\mathcal{F}_\Lambda}(h)\}_{h \in \Lambda^{-1}} \tag{13.85}$$

(note that $\Lambda^{-1} = (\frac{1}{\sigma}\mathbb{Z})^q$, $h = (\frac{1}{\sigma}n_1, \ldots, \frac{1}{\sigma}n_q)^T \in \Lambda^{-1}$, $n_i \in \mathbb{Z}$*) in the form*

$$F^{\wedge}_{\mathcal{F}_\Lambda}(y) = \sum_{h \in \Lambda^{-1}} F^{\wedge}_{\mathcal{F}_\Lambda}(h) \, \frac{1}{\sigma^q} \int_{\mathcal{F}_\Lambda} e^{2\pi i x \cdot (h-y)} \, dx, \quad y \in \mathbb{R}^q. \tag{13.86}$$

Explicitly written out (cf. E. Parzen [**681**]*), we obtain from* (13.86) *for the lattice* $\Lambda = \sigma\mathbb{Z}^q, q \geq 2$, *generated by the dilation with the factor* σ *from the canonical orthonormal vectors* $\varepsilon_1, \ldots, \varepsilon_q \in \mathbb{R}^q$

$$F^{\wedge}_{\mathcal{F}_\Lambda}(y) = \sum_{h \in \Lambda^{-1}} F^{\wedge}_{\mathcal{F}_\Lambda}(h) \, \frac{J_1(1; \pi|\varepsilon_1 \cdot (h-y)|)}{\pi|\varepsilon_1 \cdot (h-y)|} \cdot \ldots \cdot \frac{J_1(1; \pi|\varepsilon_q \cdot (h-y)|)}{\pi|\varepsilon_q \cdot (h-y)|}.$$

Our perspectives. An interesting multivariate sampling problem, however, is what happens if $F^{\wedge}_{\mathcal{F}_\Lambda}$ is replaced by a bandlimited function in $\mathbb{R}^q$ given by the Fourier transform

$$F^{\wedge}_{\mathcal{G}}(y) = \int_{\mathcal{G}} F(x) e^{-2\pi i x \cdot y} \, dx, \quad y \in \mathbb{R}^q, \tag{13.87}$$

where the "base band" $\mathcal{G}$ is a *regular region in* $\mathbb{R}^q$, i.e., different from the usually used lattice-determined $\mathcal{F}_\Lambda$ (as geoscientifically relevant 3D regular regions $\mathcal{G}$, we have in mind, for example, the (actual) Earth's body or parts of it, the interior of geoequipotential surfaces such as ball, ellipsoid, geoid, telluroid, but also tetrahedron, cube, other polyhedral bodies, etc.).

It is evident from our one-dimensional considerations that, for regions $\mathcal{G}, \mathcal{H} \subset \mathbb{R}^q$, the specific properties of a *cardinal series associated to* (13.87)

$$\sum_{h \in \Lambda^{-1}} \underbrace{\int_{\mathcal{G}} F(x) e^{-2\pi i h \cdot x} \, dx}_{=F^{\wedge}_{\mathcal{G}}(h)} \, \underbrace{\frac{1}{\|\mathcal{F}_\Lambda\|} \int_{\mathcal{H}} e^{2\pi i x \cdot (h-y)} \, dx}_{=K_{\mathcal{H}}(h-y)}, \quad y \in \mathbb{R}^q, \tag{13.88}$$

strongly depend on the geometric configuration of $\mathcal{G}, \mathcal{H}$ in relation to the fundamental cell $\mathcal{F}_\Lambda$ of a lattice $\Lambda \subset \mathbb{R}^q$, that actually determines the so-called *sampling density* $\frac{1}{\|\mathcal{F}_\Lambda\|}$. This is the reason why the connection between (13.87) and (13.88) is of particular significance. More concretely, we are interested in

- the (pointwise) convergence properties of the series (13.88) in $\mathbb{R}^q$,
- achievable cases of identifying (13.87) and (13.88),
- quantifying characterizations of under- and oversampling depending on the choice of $\mathcal{G}, \mathcal{H}$,
- detecting possible reproducing kernel Hilbert space structure of Paley-Wiener spaces $B_{\mathcal{H}}$ corresponding to regular regions $\mathcal{H} \subset \mathbb{R}^q$,
- explicitly representing the aliasing error.

Following W. Freeden [**268**], W. Freeden, M.Z. Nashed [**298**] the structural concept leading to (13.88) can be closely based on classical variants of the famous Hardy-Landau identities (see, e.g., [**410–413, 415, 436, 601–603**] for original formulations) in number theory. As a consequence, our methodological concept represents a natural continuation of the quite simple apparatus due to A. Papoulis [**678**] of realizing the Shannon sampling theorem, however, under the particular use of the rich equipment of recent number theoretical means and techniques involving lattice points manifestations for general geometric configurations.

Multivariate lattice functions and resulting summation formulas. Our goal is to explain the lattice point theoretic approach to Shannon-type sampling in a detailed way. To this end, we remember the univariate case, that already shows all specific features of the Shannon-type approach, but avoids any obstacle to guarantee the convergence of 1D cardinal series. So, in the multivariate case, we are first required to use summability techniques for the lattice point identities and afterwards to discuss the question of the necessity of summability in associated lattice point sampling identities and subsequently cardinal series. Altogether, the following sections of this chapter aim at demonstrating that lattice point identities, which are specifically imbedded in a certain metaharmonic framework, are eminently suitable for purposes of establishing lattice point sampling methodologies.

First, our purpose is to formulate the Euler summation formula in Euclidean space $\mathbb{R}^q$ for arbitrary lattices Λ and regular regions $\mathcal{G} \subset \mathbb{R}^q$ with respect to the second order operator $L = \Delta + 4\pi i a \cdot \nabla$, $a \in \mathbb{R}^q$, $q \geq 2$. For the operator L in $\mathbb{R}^q$ an elementary calculation shows that $(L+L^\wedge(h))e^{2\pi i h\cdot x} = 0, x \in \mathbb{R}^q$, holds true for $L^\wedge(h)$ given by $L^\wedge(h) = 4\pi^2((h+a)^2 - a^2)$. We shall say that λ is an eigenvalue of the lattice Λ with respect to L if there is a nontrivial solution U of the differential equation $(L+\lambda)U = 0, \lambda \in \mathbb{R}$, satisfying the "boundary condition of periodicity" $U(x+g) = U(x)$ for all $x \in \mathbb{R}^q$ and $g \in \Lambda$. The function U is called an eigenfunction of the lattice Λ with respect to the eigenvalue λ and the operator L. In analogy to the one-dimensional case we are able to verify that the spectrum is given by

$$\mathrm{Spect}_L(\Lambda) = \{L^\wedge(h) : L^\wedge(h) = 4\pi^2((h+a)^2 - a^2),\ h \in \Lambda^{-1}\}. \tag{13.89}$$

In accordance with Hilbert's approach (see D. Hilbert [**433**]), the Λ lattice function may be understood as the Green function for the operator $L+\lambda$, $\lambda \in \mathbb{R}$, in Euclidean space $\mathbb{R}^q$ corresponding to the "boundary condition of periodicity" with respect to the lattice Λ. Obviously, we have

$$\mathrm{Spect}_{L+\lambda}(\Lambda) = \{(L+\lambda)^\wedge(h) : (L+\lambda)^\wedge(h) = \lambda - L^\wedge(h), h \in \Lambda^{-1}\}. \tag{13.90}$$

Since the Fourier series (13.98) of $G(L+\lambda;\cdot)$ does not show pointwise convergence properties in $\mathbb{R}^q, q \geq 2$, as it was the case for its one-dimensional counterpart, we are required to introduce the lattice function $G(L+\lambda;\cdot)$ by its defining properties:

$G(L+\lambda;\cdot) : \mathbb{R}^q \setminus \Lambda \to \mathbb{C}$, $\lambda \in \mathbb{R}$ *fixed, is called the* Λ*-lattice function for the operator* $L+\lambda$ *if it satisfies the following properties:*

(i) *For all* $x \in \mathbb{R}^q \setminus \Lambda$ *and* $g \in \Lambda$

$$G(L+\lambda; x+g) = G(L+\lambda; x). \tag{13.91}$$

(ii) *For all* $x \in \mathbb{R}^q \setminus \Lambda$

$$\overline{G(L+\lambda; x)} = G(L+\lambda; -x). \tag{13.92}$$

(iii) $G(L+\lambda;\cdot)$ *is twice continuously differentiable for all* $x \in \mathbb{R}^q \setminus \Lambda$.
For $\lambda \notin \mathrm{Spect}_L(\Lambda)$,

$$(L+\lambda)G(L+\lambda; x) = 0. \tag{13.93}$$

For $\lambda \in \operatorname{Spect}_L(\Lambda)$,

$$(L+\lambda)G(L+\lambda;x) = -\frac{1}{\|\mathcal{F}_\Lambda\|} \sum_{\substack{(L+\lambda)^\wedge(h)=0 \\ h\in\Lambda^{-1}}} e^{2\pi i h\cdot x}. \tag{13.94}$$

(iv) *In a neighborhood of the origin*

$$G(L;x) + \frac{(2\pi)^{-\frac{q}{2}}(\sqrt{-\lambda-4\pi^2a^2})^{\frac{q-2}{2}} e^{-2\pi i a\cdot x}}{|x|^{\frac{q-2}{2}}} K_{\frac{q-2}{2}}(\sqrt{-\lambda-4\pi^2a^2}\,|x|) \tag{13.95}$$

is continuously differentiable (so that $G(L;x)$ *takes the singularity of the Kelvin function* $K_{\frac{q-2}{2}} = K_{\frac{q-2}{2}}(2;\cdot)$ *of order* $\frac{q-2}{2}$ *in the origin, hence, in all lattice points of* Λ *because of condition (ii)).*

(v) *For all* $h \in \Lambda^{-1}$ *with* $(L+\lambda)^\wedge(h) = 0$

$$\int_{\mathcal{F}_\Lambda} G(L+\lambda;x)e^{-2\pi i h\cdot x}\,dx = 0. \tag{13.96}$$

Green's theorem yields

$$\int_{\mathcal{F}_\Lambda} G(L+\lambda;x)e^{-2\pi i h\cdot x}\,dx = \begin{cases} 0 & , \quad (L+\lambda)^\wedge(h) = 0, \\ -\frac{1}{\|\mathcal{F}_\Lambda\|}\frac{1}{(L+\lambda)^\wedge(h)} & , \quad (L+\lambda)^\wedge(h) \neq 0, \end{cases} \tag{13.97}$$

such that $G(L+\lambda;x)$ possesses the required Fourier coefficients. Hence, for each number $\lambda \in \mathbb{R}$ there exists one and only one Λ-lattice function $G(L+\lambda;\cdot)$ satisfying the constituting conditions (i) – (v) listed above (the proof follows by the same arguments as for the operator $\Delta+\lambda$ by virtue of the Fredholm theory of weakly singular integral equations (see [**268**])).

In what follows, the major difficulty to be discussed in contrast to the 1D approach is the non-convergence of the bilinear expansion for the corresponding Λ-lattice (Euler-Green) function $G(L+\lambda;\cdot), \lambda \in \mathbb{R}$, in terms of the eigenfunction system of periodic polynomials in $\mathbb{R}^q, q \geq 2$, given by

$$G(L+\lambda;x-z) \sim \sum_{\substack{(L+\lambda)^\wedge(h)\neq 0 \\ h\in\Lambda^{-1}}} \frac{1}{-(L+\lambda)^\wedge(h)}\,\Phi_h(x)\,\overline{\Phi_h(z)}, \tag{13.98}$$

where the symbol of $L+\lambda$ is given by $(L+\lambda)^\wedge(h) = \lambda-(4\pi^2((h+a)^2-a^2)), h \in \Lambda^{-1}$.

Euler summation formula for the operator L. Let Λ be an arbitrary lattice in $\mathbb{R}^q$. Suppose that $\mathcal{G} \subset \mathbb{R}^q$ is a regular region. Let H be a function of class $C^{(2)}(\overline{\mathcal{G}}), \overline{\mathcal{G}} = \mathcal{G} \cup \partial\mathcal{G}$. Then, for every $\lambda \in \mathbb{R}$ and (sufficiently small) $\varepsilon > 0$, the second Green theorem in connection with the Gauss theorem in $\mathbb{R}^q$ give

$$\begin{aligned}
&\int_{\substack{x\in\overline{\mathcal{G}} \\ x\notin\mathbb{B}^q_\varepsilon+\Lambda}} \left\{H(x)\,(L+\lambda)G(L+\lambda;x) - G(L+\lambda;x)\left(\overline{L}+\lambda\right)H(x)\right\}\,dx \\
&= \int_{\substack{x\in\partial\mathcal{G} \\ x\notin\mathbb{B}^q_\varepsilon+\Lambda}} \left\{H(x)\left(\frac{\partial}{\partial\nu_x}+4\pi i a\cdot\nu(x)\right)G(L+\lambda;x) - G(L+\lambda;x)\frac{\partial}{\partial\nu_x}H(x)\right\}\,dS(x) \\
&\quad + \sum_{\substack{g\in\overline{\mathcal{G}} \\ g\in\Lambda}} \int_{\substack{|x-g|=\varepsilon \\ x\in\overline{\mathcal{G}}}} \left\{H(x)\left(\frac{\partial}{\partial\nu_x}+4\pi i a\cdot\nu(x)\right)G(L+\lambda;x) - G(L+\lambda;x)\frac{\partial}{\partial\nu_x}H(x)\right\}\,dS(x),
\end{aligned}$$

where $\frac{\partial}{\partial\nu}$ denotes the derivative in the direction of the outer normal ν on $\partial\mathcal{G}$. Using the differential equation (i.e., Condition (iii)) we obtain

$$\int_{\substack{x\in\overline{\mathcal{G}}\\ x\notin\mathbb{B}^q_\varepsilon+\Lambda}} H(x)\ (L+\lambda)G(L+\lambda;x)\,dx = -\frac{1}{\|\mathcal{F}_\Lambda\|}\int_{\substack{x\in\overline{\mathcal{G}}\\ x\notin\mathbb{B}^q_\varepsilon+\Lambda}} \sum_{\substack{(L+\lambda)^\wedge(h)=0\\ h\in\Lambda^{-1}}} H(x)\ e^{-2\pi i h\cdot x}\,dx,$$

where the sum on the right side is to be extended over all $h\in\Lambda^{-1}$ with $(L+\lambda)^\wedge(h)=\lambda-L^\wedge(h)=0$. Letting $\varepsilon\to0$ and observing the characteristic singularity of the Λ-lattice function (i.e., Condition (iv)) we are led to the desired *Euler-type summation formula* by virtue of standard potential theoretic arguments (see, e.g., O.D. Kellogg [**493**], S.G. Michlin [**584**]).

Let Λ be an arbitrary lattice in $\mathbb{R}^q$. Let $\mathcal{G}\subset\mathbb{R}^q$ be a regular region. Let H be twice continuously differentiable on $\overline{\mathcal{G}}$, $\overline{\mathcal{G}}=\mathcal{G}\cup\partial\mathcal{G}$. Then, for $\lambda\in\mathbb{R}$, we have

(13.99)

$$\begin{aligned}\sum_{\substack{g\in\overline{\mathcal{G}}\\ g\in\Lambda}}{}' H(g) &= \frac{1}{\|\mathcal{F}_\Lambda\|}\sum_{\substack{(L+\lambda)^\wedge(h)=0\\ h\in\Lambda^{-1}}}\int_{\mathcal{G}} H(x)e^{-2\pi i h\cdot x}\,dx\\ &\quad+\int_{\mathcal{G}} G(L+\lambda;x)\ (\overline{L}+\lambda)H(x)\ dx\\ &\quad+\int_{\partial\mathcal{G}}\left\{H(x)\left(\frac{\partial}{\partial\nu_x}+4\pi i a\cdot\nu(x)\right)G(L+\lambda;x)\right.\\ &\qquad\qquad\left.-G(L+\lambda;x)\frac{\partial}{\partial\nu_x}H(x)\right\}\ dS(x),\end{aligned}$$

From an algorithmic point of view it should be noted that the formula (13.99) expresses a sum in terms of integrals over $\mathcal{G}$ and its boundary $\partial\mathcal{G}$ involving derivatives of the function F up to the second order. For $\lambda\in\mathrm{Spect}_L(\Lambda)$ we have to expect some kind of resonance phenomena in the summation procedures. The classical case $a=0$, $\lambda=0$, i.e., the Euler summation formula corresponding to the Laplacian, turns out to be a "resonance case". Indeed, the formula for the Laplacian Δ is an immediate multi-dimensional generalization of the one-dimensional Euler summation formula, where $G(\Delta;\cdot)$ takes the multi-dimensional role of the Bernoulli polynomial of degree 2.

Obviously, the multivariate "lattice discrepancy"

$$\sum_{\substack{g\in\overline{\mathcal{G}}\\ g\in\Lambda}}{}'\,1\ -\ \|\mathcal{G}\| \tag{13.100}$$

in the resulting lattice point identity (with $a=0, H=1$ in (13.99))

$$\sum_{\substack{g\in\overline{\mathcal{G}}\\ g\in\Lambda}}{}'\,1\ =\|\mathcal{G}\|+\int_{\partial\mathcal{G}}\frac{\partial}{\partial\nu}G(\Delta;x)\ dS(x) \tag{13.101}$$

does not allow the pointwise representation as a multivariate alternating series (see [**268**] for more details).

By using elementary identities analogously to the univariate approach (13.39)–(13.42)

$$\Delta_x e^{2\pi i a\cdot x} H(x-y) = e^{2\pi i a\cdot x}\left(\Delta_x + 4\pi i a\cdot\nabla_x - 4\pi^2 a^2\right) H(x-y), \tag{13.102}$$

$$a\cdot\nabla_x e^{2\pi i a\cdot x} H(x-y) = e^{2\pi i a\cdot x}\left(a\cdot\nabla_x + 2\pi i a^2\right) H(x-y), \tag{13.103}$$

and

$$\left(\Delta_x - 4\pi i a\cdot\nabla_x\right) e^{2\pi i a\cdot x} H(x-y) \;=\; e^{2\pi i a\cdot x}\left(\Delta_x + 4\pi^2 a^2\right) H(x-y) \tag{13.104}$$

we are able to reformulate the Euler-type summation formula in such a way that it becomes useful for our later sampling purposes.

Let Λ be an arbitrary lattice in $\mathbb{R}^q$. Let $\mathcal{G}\subset\mathbb{R}^q$ be a regular region. Let H be twice continuously differentiable on $\overline{\mathcal{G}}$, $\overline{\mathcal{G}}=\mathcal{G}\cup\partial\mathcal{G}$. Then, for all $\lambda\in\mathbb{R}$ and $a,y\in\mathbb{R}^q$, we have

$$\begin{aligned}
&\sum_{\substack{g-y\in\overline{\mathcal{G}}\\ g\in\Lambda}}{}' \, e^{2\pi i a\cdot g} H(g-y)\\
&= \frac{1}{\|\mathcal{F}_\Lambda\|}\, e^{2\pi i a\cdot y} \sum_{\substack{4\pi^2\left((h+a)^2-a^2\right)=\lambda\\ h\in\Lambda^{-1}}} e^{2\pi i h\cdot y}\int_{\mathcal{G}} H(x) e^{2\pi i x\cdot(a+h)}\,dx\\
&\quad+ \int_{\substack{x-y\in\mathcal{G}\\ x\in\mathbb{R}^q}} G(L+\lambda;x) e^{2\pi i a\cdot x}\left(\Delta_x+\lambda+4\pi^2a^2\right) H(x-y)\,dx\\
&\quad- \int_{\substack{x-y\in\partial\mathcal{G}\\ x\in\mathbb{R}^q}} e^{2\pi i a\cdot x} G(L+\lambda;x)\frac{\partial}{\partial\nu_x} H(x-y)\,dS(x)\\
&\quad+ \int_{\substack{x-y\in\partial\mathcal{G}\\ x\in\mathbb{R}^q}} e^{2\pi i a\cdot x} H(x-y)\left(\frac{\partial}{\partial\nu_x} + 4\pi i a\cdot\nu(x)\right) G(L+\lambda;x)\,dS(x).
\end{aligned}$$

Euler and Poisson-type sampling attunements. Generally, for all regular regions $\mathcal{G}\subset\mathbb{R}^q, q\geq 2$, and twice continuously differentiable functions F on $\overline{\mathcal{G}}=\mathcal{G}\cup\partial\mathcal{G}$ and contrary to the 1D approach, the q-dimensional Euler summation formula with respect to the operator $L=\Delta+4\pi i a\cdot\nabla$ does not allow us both to guarantee the convergence of the *Hardy-type lattice point series* (even understood in spherical sense)

$$\begin{aligned}
&\frac{1}{\|\mathcal{F}_\Lambda\|}\sum_{h\in\Lambda^{-1}} e^{2\pi i a\cdot h}\int_{\mathcal{G}} F(x) e^{-2\pi i x\cdot(h-y)}\,dV(x)\\
&= \lim_{N\to\infty}\frac{1}{\|\mathcal{F}_\Lambda\|}\sum_{\substack{|h-y|\leq N\\ h\in\Lambda^{-1}}} e^{2\pi i a\cdot h}\int_{\mathcal{G}} F(x) e^{-2\pi i x\cdot(h-y)}\,dV(x), \qquad a,y\in\mathbb{R}^q,
\end{aligned} \tag{13.105}$$

and to identify (13.105) with the finite lattice point sum

$$\sum_{\substack{a+g\in\overline{\mathcal{G}}\\ g\in\Lambda}}{}' e^{2\pi i y\cdot(a+g)} F(a+g) = \sum_{\substack{a+g\in\overline{\mathcal{G}}\\ g\in\Lambda}} e^{2\pi i y\cdot(a+g)}\,\alpha_{\partial\mathcal{G}}(a+g)\,F(a+g),\ a,y\in\mathbb{R}^q, \tag{13.106}$$

where $\alpha_{\partial\mathcal{G}}(a+g)$ is the *solid angle subtended at* $a+g \in \overline{\mathcal{G}}$ *by the surface* $\partial\mathcal{G}$ (see, e.g., W. Freeden [**268**] for more potential theoretic details). Number theoretically important exceptions that we are able to prove (cf. W. Freeden [**248**]) the bivariate *Hardy-type lattice point identity*

$$\sum_{\substack{a+g\in\overline{\mathcal{G}}\\ g\in\Lambda}}{}' e^{2\pi i y\cdot(a+g)} F(a+g) = \lim_{N\to\infty} \frac{1}{\|\mathcal{F}_\Lambda\|} \sum_{\substack{|h-y|\le N\\ h\in\Lambda^{-1}}} e^{2\pi i a\cdot h} \int_{\mathcal{G}} F(x) e^{-2\pi i x\cdot(h-y)}\, dx, \quad a,y\in\mathbb{R}^2, \tag{13.107}$$

for *homogeneous weight functions on the boundary* $\partial\mathcal{G}$ or the special *choice of a convex region* $\mathcal{G}\subset\mathbb{R}^2$ containing the origin and possessing a boundary curve $\partial\mathcal{G}$ such that its normal field ν is continuously differentiable and its curvature is nonvanishing (so that asymptotic methods for the stationary phase become applicable).

Because of divergence (for dimensions $q \geq 3$) and/or slow convergence problems (even for the dimension $q = 2$ and the indicated convex regions $\mathcal{G}$), additional summability concepts for establishing (13.107) have to come into play. Our summability method under consideration here (see also [**268**]) is Gauss-Weierstrass summability. It turns out that, in the nomenclature of *Gauss-Weierstrass summability*, we are indeed able to deduce that

$$\sum_{\substack{a+g\in\overline{\mathcal{G}}\\ g\in\Lambda}}{}' e^{2\pi i y\cdot(a+g)} F(a+g) = \lim_{\substack{\tau\to 0\\ \tau>0}} \frac{1}{\|\mathcal{F}_\Lambda\|} \sum_{h\in\Lambda^{-1}} e^{-\tau\pi^2 h^2} e^{2\pi i h\cdot a} \int_{\mathcal{G}} F(x) e^{-2\pi i x\cdot(h-y)}\, dx, \quad a,y\in\mathbb{R}^q, \tag{13.108}$$

holds for all regular regions $\mathcal{G}\subset\mathbb{R}^q$ and continuous functions F on $\overline{\mathcal{G}}$.

In this respect it should be observed that, for every $y\in\mathbb{R}^q$, the series on the right side of (13.108) show Λ-periodicity as functions of the variable $a\in\mathbb{R}^q$. Hence, our Shannon-type sampling procedures obtained by formal integration over a measurable bounded region $\mathcal{H}\subset\mathbb{R}^q$ from the lattice point identity (13.108)

$$\int_{\mathcal{H}} \sum_{\substack{a+g\in\overline{\mathcal{G}}\\ g\in\Lambda}}{}' e^{-2\pi i y\cdot(a+g)} F(a+g)\, da = \lim_{\substack{\tau\to 0\\ \tau>0}} \sum_{h\in\Lambda^{-1}} e^{-\tau\pi^2 h^2} \underbrace{\int_{\mathcal{G}} F(x) e^{-2\pi i h\cdot x}\, dx}_{=F^\wedge_{\mathcal{G}}(h)} \underbrace{\frac{1}{\|\mathcal{F}_\Lambda\|} \int_{\mathcal{H}} e^{2\pi i a\cdot(h-y)}\, da}_{=K_{\mathcal{H}}(h-y)} \tag{13.109}$$

have many interesting properties: For example, the convergence of the cardinal-type series on the right side of (13.109) is exponentially accelerated, which also opens the way to multiscale formulations. All manifestations of over- and undersampling can be explicitly analyzed dependent on the geometric configurations of the chosen regions $\mathcal{G},\mathcal{H}$ (note that the integral conception allows us to make no notational difference, e.g., between $F^\wedge_{\mathcal{G}}$ and $F^\wedge_{\overline{\mathcal{G}}}$).

13.3. Bivariate lattice point identities and Shannon-type sampling

Next, as a preparation of more general results in higher dimensions, we show how to use two-dimensional lattice point identities to derive generalizations of the Shannon sampling theorem. We start with the pointwise *Fourier inversion formula for regular regions* $\mathcal{G} \subset \mathbb{R}^2$ (see W. Freeden [**251**] for the proof), which plays a particular role in our context.

Let $\mathcal{G}$ be a regular region in $\mathbb{R}^2$. Assume that F is a continuously differentiable function in $\overline{\mathcal{G}} = \mathcal{G} \cup \partial\mathcal{G}$. Then, for every $w \in \mathbb{R}^2$, the integral

$$\int_{\mathbb{R}^2} e^{2\pi i w\cdot x} \int_{\substack{y\in\mathcal{G}\\ y\in\mathbb{R}^2}} F(y)\, e^{-2\pi i x\cdot y}\, dy\, dx \tag{13.110}$$

exists in the sense $\int_{\mathbb{R}^2} \cdots = \lim_{N\to\infty} \int_{\substack{|x|\leq N\\ x\in\mathbb{R}^2}} \cdots$, *and we have*

$$\begin{aligned}\int_{\mathbb{R}^2} e^{2\pi i w\cdot x} \int_{\substack{y\in\mathcal{G}\\ y\in\mathbb{R}^2}} F(y)\, e^{-2\pi i x\cdot y}\, dy\, dx &= \lim_{N\to\infty} N \int_{\substack{y\in\mathcal{G}\\ y\in\mathbb{R}^2}} F(y) \frac{J_1(2; 2\pi|x-y|N)}{|w-y|}\, dy\\ &= \alpha_{\partial\mathcal{G}}(w) F(w),\end{aligned} \tag{13.111}$$

where $\alpha_{\partial\mathcal{G}}(w)$, $w \in \mathbb{R}^2$, is the solid angle subtended by $\partial\mathcal{G}$ at w.

As in one-dimensional theory, our aim is to establish two-dimensional lattice point identities, from which sampling results can be obtained in the already indicated way. Before we start with our investigations, some remarks should be made:

The problem of the proof of weighted two-dimensional (Hardy-Landau) lattice point sums over regular regions $\mathcal{G}$ is twofold, namely

- to guarantee the pointwise convergence of the series expansion

$$\frac{1}{\|\mathcal{F}_\Lambda\|} \sum_{h\in\Lambda^{-1}} e^{2\pi i a\cdot h} \underbrace{\int_{\substack{x\in\mathcal{G}\\ x\in\mathbb{R}^2}} F(x) e^{-2\pi i x\cdot(h-y)}\, dV(x)}_{=F^\wedge_{\mathcal{G}}(h-y)}, \quad a \in \mathbb{R}^2, \tag{13.112}$$

- to identify (13.112) with the function

$$\sum_{\substack{a+g\in\overline{\mathcal{G}}\\ g\in\Lambda}}{}' e^{2\pi i y\cdot(a+g)} F(a+g), \quad a \in \mathbb{R}^2. \tag{13.113}$$

The standard example in analytic theory of numbers (see, e.g., G.H. Hardy [**410**], G.H. Hardy, E. Landau [**415**], C. Müller [**601, 602**]) is a constant weight function (i.e., $F = 1$), the choice $a = y = 0$, and the restriction to circles $\mathcal{G} = \mathbb{B}^2_R, R > 0$), where the convergence of the series on the right side of (13.112) can be realized by estimates known from the theory of Bessel functions (cf. G.H. Hardy [**410**]). However, it seems that the convergence of the series (13.112) for "general regular regions" $\mathcal{G}$ and twice continuously differentiable functions F, i.e., "general weights" on $\overline{\mathcal{G}}$, can be guaranteed only under the restrictive assumption that asymptotic procedures of the metaharmonic theory as presented in the last section (see also C. Müller [**600**], G.N. Watson [**857**], W. Freeden [**268**]) for $|y| \to \infty$ are applicable to the crucial curve integral

$$-\frac{1}{4\pi^2 y^2} \int_{\partial\mathcal{G}} F(x) \left(\frac{\partial}{\partial \nu_x} e^{-2\pi i x\cdot y} \right) dS(x),\ F \in C^{(2)}\left(\overline{\mathcal{G}}\right). \tag{13.114}$$

In fact, as already pointed out earlier, this integral is the critical ingredient for the bandlimited function $F^{\wedge}_{\mathcal{G}}$, as we can see from the second Green theorem

$$\begin{aligned} F^{\wedge}_{\mathcal{G}}(y) = & -\frac{1}{4\pi^2 y^2} \int_{\substack{x\in\mathcal{G} \\ a\in\mathbb{R}^2}} e^{-2\pi i x\cdot y} \left(\Delta F(x)\right) dx \\ & -\frac{1}{4\pi^2 y^2} \int_{\substack{x\in\partial\mathcal{G} \\ x\in\mathbb{R}^2}} F(x) \left(\frac{\partial}{\partial \nu_x} e^{-2\pi i x\cdot y} \right) dS(x) \\ & +\frac{1}{4\pi^2 y^2} \int_{\substack{x\in\partial\mathcal{G} \\ x\in\mathbb{R}^2}} \left(\frac{\partial}{\partial \nu} F(x) \right) e^{-2\pi i x\cdot y}\, dS(x) \end{aligned} \tag{13.115}$$

for $|y| \to \infty$. Only extensions of the Hardy-Landau identity for constant weight (e.g., $F = 1$) to regular (convex) regions around the origin with smooth boundary curves $\partial\mathcal{G}$ possessing a nonvanishing curvature are known, yet (see, e.g., V.K. Ivanow [**457, 458**]).

Nonetheless, our intention is a two-fold generalization of the classical two-dimensional Hardy-Landau identity (cf. G.H. Hardy [**410**], G.H. Hardy, E. Landau [**415**]) in the following sense:

- the generalization from a "circle geometry" to a "general", i.e., regular geometry $\mathcal{G}$.
- the generalization from a constant to a general, i.e., twice continuously differentiable weight function F.

Unfortunately, in contrast to a lattice ball approach (cf. [**268**]), it must be confessed that our lattice point approach does not succeed in both generalizations (without summability techniques), and it may be questioned from the methodological point of view if this is at all possible. A compromise in lattice point summation was formulated in W. Freeden [**248, 249, 268**], where general weights for circular geometry in spherical summation of the two-dimensional series (13.112) had been realized under the particular choice $a = y = 0$. The concept in this work – having the connection of lattice point summation and two-dimensional sampling in mind – needs lattice point identities of the more general nature (13.112), (13.113) involving free parameters $\lambda \in \mathbb{R}, a, y \in \mathbb{R}^q$, and within this framework it amounts to the following progress even in weighted Hardy-Landau lattice point theory:

- the extension to "general", i.e., regular geometries, but under homogeneous "boundary weights",
- the extension to "general weights", however, restricted to convex regions allowing methods of the stationary point asymptotics (see Section 2.2) or metaharmonic theory (see Section 8.3) for the boundary curve.

Within this context, two-dimensional generalizations of the Shannon sampling theorem become immediately attackable by our lattice point identities. Indeed, reflecting the internal structure of (13.112), (13.113) W. Freeden [**248, 249**] was able to give a direct proof analogously to the one-dimensional theory.

Circles and general weights. We are going to discuss weighted lattice point sums in circles, where the weights are generated by an arbitrary twice continuously differentiable function $F : \overline{\mathbb{B}^2_R} \to \mathbb{R}$, $R > 0$. The essential tool is the metaharmonic theory. In order to verify a "weighted version of the Hardy-Landau identity on circles" $\mathbb{B}^2_R$, we base our considerations on the two-dimensional Euler-type summation formula for general lattices $\Lambda \subset \mathbb{R}^2$, but now on the bandlimited function $F^{\wedge}_{\mathbb{B}^2_R}$ given by

$$F^{\wedge}_{\mathbb{B}^2_R}(y) = \int_{\substack{|x| \le R \\ x \in \mathbb{R}^2}} F(x)\, e^{-2\pi i x \cdot y}\, dx, \quad y \in \mathbb{R}^2, \tag{13.116}$$

where F is assumed to be of class $C^{(2)}\left(\overline{\mathbb{B}^2_R}\right)$, i.e., $\overline{\mathcal{G}} = \mathbb{B}^2_R$ (see **[251]**, **[268]**). Obviously, $F^{\wedge}_{\mathbb{B}^2_R}$ is real analytic, hence, infinitely differentiable in $\mathbb{R}^2$.

More explicitly, for every $N > 0$, $\lambda \in \mathbb{R}$, and all $a, y \in \mathbb{R}^2$, we have

$$\begin{aligned}
&\sum_{\substack{|g-y| \le N \\ g \in \Lambda}} e^{2\pi i a \cdot g}\, F^{\wedge}_{\mathbb{B}^2_R}(x - y) \\
&= \frac{1}{\|\mathcal{F}_\Lambda\|}\, e^{2\pi i a \cdot y} \sum_{\substack{4\pi^2\left((h+a)^2 - a^2\right) = \lambda \\ h \in \Lambda^{-1}}} e^{2\pi i\, h \cdot y} \int_{\substack{|x| \le N \\ x \in \mathbb{R}^2}} F^{\wedge}_{\mathbb{B}^2_R}(x) e^{2\pi i\, x \cdot (a+h)} dx \\
&\quad + \int_{\substack{|x-y| \le N \\ x \in \mathbb{R}^q}} G(L + \lambda; x)\, e^{2\pi i a \cdot x} \left(\Delta_x + \lambda + 4\pi^2 a^2\right) F^{\wedge}_{\mathbb{B}^2_R}(x - y)\, dx \\
&\quad - \int_{\substack{|x-y| = N \\ x \in \mathbb{R}^q}} e^{2\pi i a \cdot x} G(L + \lambda; x) \frac{\partial}{\partial \nu_x} F^{\wedge}_{\mathbb{B}^2_R}(x - y)\, dS(x) \\
&\quad + \int_{\substack{|x-y| = N \\ x \in \mathbb{R}^q}} e^{2\pi i a \cdot x} F^{\wedge}_{\mathbb{B}^2_R}(x - y) \left(\frac{\partial}{\partial \nu_x} + 4\pi i a \cdot \nu(x)\right) G(L + \lambda; x)\, dS(x) \\
&\quad + \frac{1}{2} \sum_{\substack{|g-y| = N \\ g \in \Lambda}} e^{2\pi i a \cdot g}\, F^{\wedge}_{\mathbb{B}^2_R}(g - y).
\end{aligned} \tag{13.117}$$

From the second Green theorem (cf. (13.115)) we obtain for all $y \in \mathbb{R}^2$ with $y \neq 0$, by observation of the continuity of the first and second order derivative of F on $\overline{\mathbb{B}^2_R}$,

$$F^{\wedge}_{\mathbb{B}^2_R}(y) = \frac{i}{2\pi y^2} \int_{\substack{|x| = R \\ x \in \mathbb{R}^2}} F(x)\, (\nu(x) \cdot y)\, e^{-2\pi i y \cdot x}\, dS(x) + O\left(\frac{1}{|y|^2}\right). \tag{13.118}$$

The integral on the right side of (13.118) can be formulated as an integral over the unit circle $\mathbb{S}^1$ in $\mathbb{R}^2$ by use of polar coordinates $a = R\eta$, $y = |y|\xi$

$$\int_{\substack{|x| = R \\ x \in \mathbb{R}^2}} F(x)\, (\nu(x) \cdot y)\, e^{-2\pi i y \cdot x}\, dS(a) = R|y| \int_{\eta \in \mathbb{S}^1} F(R\eta)\, (\eta \cdot \xi)\, e^{-2\pi i R|y|(\xi \cdot \eta)}\, dS(\eta). \tag{13.119}$$

The last integral is an entire solution of the Helmholtz equation (in the variable $y \in \mathbb{R}^2$); hence, integral asymptotics in metaharmonic theory and/or the theory of stationary phase provide the asymptotic relation

$$\int_{\eta\in\mathbb{S}^1} F(R\eta)\ (\eta\cdot\xi)\ e^{-2\pi i R|y|(\xi\cdot\eta)}\ dS(\eta)\ =\ O\left(|y|^{-\frac{1}{2}}\right) \tag{13.120}$$

for $|y|\to\infty$. In connection with (13.118) and (13.119) we therefore obtain from (13.120) the asmptotic relation $F^{\wedge}_{\mathbb{B}^2_R}(y) = O\left(|y|^{-\frac{3}{2}}\right)$, $|y|\to\infty$. In addition, we are able to verify the following lemma:

For all multi-indices $\alpha = (\alpha_1,\alpha_2)^T$ with $\alpha_i\in\mathbb{N}_0$, $i=1,2$, we have

$$(\nabla_y)^\alpha F^{\wedge}_{\mathbb{B}^2_R}(y)\ =\ O\left(|y|^{-\frac{3}{2}}\right),\ |y|\to\infty. \tag{13.121}$$

In connection with the already mentioned result due to C.F. Gauss [**344**] it follows that

$$\frac{1}{2}\sum_{\substack{|g|=N\\ g\in\Lambda}} F^{\wedge}_{\mathbb{B}^2_R}(g) = O\left(N^{-\frac{3}{2}}\sum_{\substack{|g|=N\\ g\in\Lambda}} 1\right) = O\left(N^{-\frac{1}{2}}\right),\ N\to\infty. \tag{13.122}$$

In connection with (13.121) we are able to show that

$$\int_{\substack{|x-y|=N\\ x\in\mathbb{R}^2}} e^{2\pi i a\cdot x} G(L;x)\frac{\partial}{\partial\nu_x} F^{\wedge}_{\mathbb{B}^2_R}(x-y)\ dS(x) = O(N^{-\frac{1}{2}}), \tag{13.123}$$

and

$$\int_{\substack{|x-y|=N\\ x\in\mathbb{R}^2}} F^{\wedge}_{\mathbb{B}^2_R}(x-y)\left(\frac{\partial}{\partial\nu_x} + 4\pi i a\cdot\nu(x)\right) G(L;x)\ dS(x) = O(N^{-\frac{1}{2}}), \tag{13.124}$$

for $N\to\infty$. Altogether, for every $\lambda\in\mathbb{R}$, and all $a,y\in\mathbb{R}^2$ it follows from (13.117) in connection with (13.122), (13.123) and (13.124) that

(13.125)

$$\begin{aligned}
&\sum_{\substack{|g-y|\le N\\ g\in\Lambda}} e^{2\pi i a\cdot g}\ F^{\wedge}_{\mathbb{B}^2_R}(x-y)\\
&\quad = \frac{1}{\|\mathcal{F}_\Lambda\|}\ e^{2\pi i a\cdot y} \sum_{\substack{4\pi^2\left((h+a)^2-a^2\right)=\lambda\\ h\in\Lambda^{-1}}} e^{2\pi i h\cdot y}\int_{\substack{|x|\le N\\ x\in\mathbb{R}^2}} F^{\wedge}_{\mathbb{B}^2_R}(x)\ e^{2\pi i x\cdot(a+h)}dx\\
&\quad + \int_{\substack{|x-y|\le N\\ x\in\mathbb{R}^q}} G(L+\lambda;x)\ e^{2\pi i a\cdot x}\left(\Delta_x+\lambda+4\pi^2a^2\right) F^{\wedge}_{\mathbb{B}^2_R}(x-y)\ dx\\
&\quad + O(N^{-\frac{1}{2}}),\quad N\to\infty.
\end{aligned}$$

Choosing especially $\lambda = 4\pi^2\left(R^2-a^2\right)$ we are able to conclude by virtue of the second Green theorem that $\left(\Delta_x+4\pi^2R^2\right)\ F^{\wedge}_{\mathbb{B}^2_R}(x-y)\ =\ O\left(|x-y|^{-\frac{5}{2}}\right)$. Therefore, the absolute convergence of the integral

$$\int_{\mathbb{R}^q} G(L+4\pi^2\left(R^2-a^2\right);x)\ e^{2\pi i a\cdot x}\left(\Delta_x+4\pi^2R^2\right) F^{\wedge}_{\mathbb{B}^2_R}(x-y)\ dx \tag{13.126}$$

can be guaranteed, and we have

$$\begin{aligned}\int_{\substack{|x-y|\le N\\ x\in\mathbb{R}^q}} & G\left(L+4\pi^2\left(R^2-a^2\right);x\right)\ e^{2\pi i a\cdot x}\left(\Delta_x+4\pi^2R^2\right)F^{\wedge}_{\mathbb{B}^2_R}(x-y)\,dx\\ &=\frac{1}{\|\mathcal{F}_\Lambda\|}\sum_{\substack{|h+a|\neq R\\ h\in\Lambda^{-1}}}\frac{1}{(4\pi^2(R^2-(h+a)^2))^2}\\ &\qquad\times\int_{\substack{|x-y|\le N\\ y\in\mathbb{R}^2}} e^{2\pi i x\cdot(a+h)}(\Delta_x+4\pi^2R^2)^2F^{\wedge}_{\mathbb{B}^2_R}(x-y)\,dy.\end{aligned}$$

It can be readily seen that, for all $h\in\Lambda^{-1}$ with $|h+a|\neq R$

(13.127)

$$\begin{aligned}\int_{\substack{|x-y|\le N\\ y\in\mathbb{R}^2}} & e^{2\pi i x\cdot(a+h)}(\Delta_x+4\pi^2R^2)^2F^{\wedge}_{\mathbb{B}^2_R}(x-y)\,dy\\ &=\int_{\substack{|x-y|\le N\\ y\in\mathbb{R}^2}} e^{2\pi i x\cdot(a+h)}\int_{\substack{|x|\le R\\ \xi\in\mathbb{R}^2}}(4\pi^2R^2-4\pi^2z^2)^2F(z)e^{-2\pi i z\cdot(x-y)}\,dz\,dx\\ &=e^{2\pi i y\cdot(a+h)}\int_{\substack{|x|\le N\\ x\in\mathbb{R}^2}} e^{2\pi i x\cdot(a+h)}\int_{\substack{|x|\le R\\ \xi\in\mathbb{R}^2}}(4\pi^2R^2-4\pi^2z^2)^2F(z)e^{-2\pi i z\cdot x}dz\,dx.\end{aligned}$$

Again we introduce an auxiliary function, namely $\tilde{F}:\overline{\mathbb{B}^2_R}\to\mathbb{R}$ given by

$$\tilde{F}(z)=(4\pi R^2-4\pi^2z^2)^2F(z),\quad z\in\overline{\mathbb{B}^2_R}. \tag{13.128}$$

The Fourier inversion formula tells us that, for every $\varepsilon>0$, there exists an integer $M(=M(\varepsilon))$ such that, for all $N\ge M$,

$$\left|\int_{\substack{|x|\le N\\ x\in\mathbb{R}^2}} e^{2\pi i\ x\cdot(a+h)}\left(\int_{\substack{|z|\le R\\ z\in\mathbb{R}^2}}\tilde{F}(y)\ e^{-2\pi i x\cdot z}\ dz\right)\ dx\right|\le\varepsilon \tag{13.129}$$

holds uniformly for all $h\in\Lambda^{-1}$ with $|a+h|\ge R_0>R$. Therefore, for all $N\ge M$, we have

$$\begin{aligned}&\left|\sum_{\substack{|h+a|\ge R_0\\ h\in\Lambda^{-1}}}\frac{1}{(4\pi^2(R^2-(h+a)^2))^2}e^{2\pi i y\cdot(a+h)}\right.\\ &\qquad\left.\times\int_{\substack{|x-y|\le N\\ y\in\mathbb{R}^2}} e^{2\pi i\ x\cdot(a+h)}\left(\int_{\substack{|z|\le R\\ z\in\mathbb{R}^2}}\tilde{F}(y)\ e^{-2\pi i x\cdot z}\ dz\right)\ dx\right|\\ &\qquad\le\varepsilon\,C\sum_{\substack{|h+a|\ge R_0\\ h\in\Lambda^{-1}}}\frac{1}{4\pi^2(a^2-(a+h)^2)}.\end{aligned} \tag{13.130}$$

In other words, for $N \to \infty$, we are able to show that

$$\sum_{\substack{|g-y|\leq N\\ g\in\Lambda}} e^{2\pi i a\cdot g}\ F^{\wedge}_{\mathbb{B}^2_R}(x-y) \tag{13.131}$$
$$= \frac{1}{\|\mathcal{F}_\Lambda\|} \sum_{\substack{|h+a|=R\\ h\in\Lambda^{-1}}} e^{2\pi i(a+h)\cdot y} \int_{\substack{|x|\leq N\\ x\in\mathbb{R}^2}} F^{\wedge}_{\mathbb{B}^2_R}(x) e^{2\pi i x\cdot(a+h)} dx$$
$$+ \frac{1}{\|\mathcal{F}_\Lambda\|} \sum_{\substack{|h+a|\neq R\\ |h+a|\leq R_0, h\in\Lambda^{-1}}} \frac{e^{2\pi i y\cdot(a+h)}}{(4\pi^2(R^2-(a+h)^2))^2}$$
$$\times \int_{\substack{|x|\leq N\\ x\in\mathbb{R}^2}} e^{2\pi i x\cdot(a+h)} \int_{\mathcal{G}} (4\pi^2R^2 - 4\pi^2 z^2)^2 F(z) e^{-2\pi i z\cdot x}\ dz\ dx$$
$$+ o(1), \quad N\to\infty.$$

Using the Fourier inversion formula and following already known arguments we are therefore led to the following theorem:

Let Λ be an arbitrary lattice in $\mathbb{R}^2$. Then, for every function $F \in C^{(2)}\left(\overline{\mathbb{B}^2_R}\right)$, $R > 0$, and all $a, y \in \mathbb{R}^2$, the series

$$\sum_{g\in\Lambda} e^{2\pi i a\cdot g} \underbrace{\int_{\substack{|x|\leq R\\ x\in\mathbb{R}^2}} F(x) e^{-2\pi i x\cdot(g-y)}\ dx}_{=F^{\wedge}_{\mathbb{B}^2_R}(g-y)} \tag{13.132}$$

converges in the spherical sense $\sum_{g\in\Lambda} \ldots = \lim_{N\to\infty} \sum_{\substack{|g-y|\leq N\\ g\in\Lambda}} \ldots$ *, and we have*

$$\frac{1}{\|\mathcal{F}_\Lambda\|} \sum_{\substack{|a+h|\leq R\\ h\in\Lambda^{-1}}} e^{2\pi i y\cdot(a+h)} F(a+h) = \sum_{g\in\Lambda} e^{2\pi i a\cdot g} \int_{\substack{|x|\leq R\\ x\in\mathbb{R}^2}} F(x) e^{-2\pi i x\cdot(g-y)}\ dx. \tag{13.133}$$

Replacing again $\Lambda \subset \mathbb{R}^2$ by $\Lambda^{-1} \subset \mathbb{R}^2$ we finally obtain the following extension of the Hardy-Landau identity (cf. W. Freeden [**248**]).

Let Λ be an arbitrary lattice in $\mathbb{R}^2$. Suppose that F is twice continuously differentiable in $\overline{\mathbb{B}^2_R}$, $R > 0$. Then, for all $a, y \in \mathbb{R}^2$,

$$\sum_{\substack{|a+g|\leq R\\ g\in\Lambda}}{}' e^{2\pi i y\cdot(a+g)} F(a+g) = \lim_{N\to\infty} \frac{1}{\|\mathcal{F}_\Lambda\|} \sum_{\substack{|h-y|\leq N\\ h\in\Lambda^{-1}}} e^{2\pi i a\cdot h} \int_{\substack{|x|\leq R\\ x\in\mathbb{R}^2}} F(x)\ e^{-2\pi i x\cdot(h-y)}\ dx. \tag{13.134}$$

Choosing $F = 1$ we obtain a generalization of the famous *Hardy-Landau identity on circular disks* (see [**410**], [**517**], [**602**]) involving the jinc-function on the right side

$$e^{2\pi i a\cdot y} \sum_{\substack{|g+a|\leq R\\ g\in\Lambda}}{}' e^{2\pi i g\cdot y} = \lim_{N\to\infty} \frac{\pi R^2}{\|\mathcal{F}_\Lambda\|} \sum_{\substack{|h-y|\leq N\\ h\in\Lambda^{-1}}} e^{2\pi i a\cdot h} \frac{J_1(2; 2\pi|h-y|R)}{\pi|h-y|R}. \tag{13.135}$$

For $a = y = 0$ we get the classical Hardy-Landau identity [**415**]

$$\sum_{\substack{|g|\leq R \\ g\in\Lambda}}{}' 1 = \lim_{N\to\infty} \frac{\pi R^2}{\|\mathcal{F}_\Lambda\|} \sum_{\substack{|h|\leq N \\ h\in\Lambda^{-1}}} \frac{J_1(2;2\pi|h|R)}{\pi|h|R}, \tag{13.136}$$

that expresses the number of lattice points inside a circle around the origin with radius R in terms of a two-dimensional alternating series involving the Bessel function $J_1 = J_1(2;\cdot)$. Note that $J_1(2;r) = \frac{r}{2} + \ldots$ so that (13.136) can be written in the standard form

$$\sum_{\substack{|g|\leq R \\ g\in\Lambda}}{}' 1 = \frac{\pi}{\|\mathcal{F}_\Lambda\|} R^2 + \lim_{N\to\infty} \frac{R}{\|\mathcal{F}_\Lambda\|} \sum_{\substack{0<|h|\leq N \\ h\in\Lambda^{-1}}} \frac{J_1(2;2\pi|h|R)}{|h|}. \tag{13.137}$$

in use for the circle problem. More explicitly, in connection with (8.3), Hardy's (unproven) conjecture (see (8.17)) suggests that the alternating (spherically understood) series on the right side of (13.136) satisfies the asymptotic law

$$\frac{R}{\|\mathcal{F}_\Lambda\|} \sum_{\substack{|h>0 \\ h=\Lambda^{-1}}} \frac{J_1(2;2\pi|h|R)}{|h|} = O(R^{\frac{1}{2}+\varepsilon}) \tag{13.138}$$

for *every* $\varepsilon > 0$.

General geometry and homogeneous boundary weight. Next we use the 2D Euler-type summation formula to derive a "weighted Hardy-Landau-type identity" on (general) $\mathcal{G} \in \mathbb{R}^2$, however, under the restrictive assumption of a homogeneous boundary weight, i.e., $F|\partial\mathcal{G} = 0$. Indeed, under the restriction that $F \in C^{(2)}(\overline{\mathcal{G}})$ with $F|\partial\mathcal{G} = 0$, the " critical curve integral" vanishes such that the continuity of F and its derivatives $\nabla F, \Delta F$ on $\overline{\mathcal{G}}$ via (13.115) imply the asymptotic result

$$F^{\wedge}_{\mathcal{G}}(y) = O\left(\frac{1}{|y|^2}\right), \quad |y| \to \infty. \tag{13.139}$$

In accordance with the arguments known from the preceding case of circles (cf. [**268**]) we are led to the following result:

Let Λ be a lattice in $\mathbb{R}^2$. Suppose that $F \in C^{(2)}\left(\overline{\mathcal{G}}\right)$ satisfies the homogeneous boundary condition $F|\partial\mathcal{G} = 0$. Then we have

$$\begin{aligned} \frac{1}{\|\mathcal{F}_\Lambda\|} \sum_{\substack{a+h\in\overline{\mathcal{G}} \\ h\in\Lambda^{-1}}}{}' e^{2\pi i y\cdot(a+h)} F(a+h) \\ = \sum_{g\in\Lambda} e^{2\pi i a\cdot g} \int_{\mathcal{G}} F(x) e^{-2\pi i x\cdot(g-y)}\, dx. \end{aligned} \tag{13.140}$$

Replacing $\Lambda \subset \mathbb{R}^2$ by its inverse lattice $\Lambda^{-1} \subset \mathbb{R}^2$ we obtain the following extended version of the two-dimensional Poisson summation formula on a regular region $\mathcal{G}$:

Let Λ be a lattice in $\mathbb{R}^2$. Let $\mathcal{G} \subset \mathbb{R}^2$ be an arbitrary regular region. Suppose that F is twice continuously differentiable on $\overline{\mathcal{G}}$ with $F|\partial\mathcal{G} = 0$. Then, for $a, y \in \mathbb{R}^2$, we have

$$\begin{aligned}(13.141) \qquad & \sum_{\substack{a+g\in\overline{\mathcal{G}} \\ g\in\Lambda}} {}' e^{2\pi i y\cdot(a+g)} F(a+g) \\ &= \lim_{N\to\infty} \frac{1}{\|\mathcal{F}_\Lambda\|} \sum_{\substack{|h-y|\le N \\ h\in\Lambda^{-1}}} e^{2\pi i a\cdot h} \int_{\mathcal{G}} F(x) e^{-2\pi i x\cdot(h-y)}\, dy.\end{aligned}$$

Convex "smooth" regions and general weights. We deal with weighted lattice point sums in convex regions $\mathcal{G}$ including the origin and possessing a smooth boundary curve $\partial\mathcal{G}$ with nonvanishing curvature (see V.K. Ivanow [**457**] for the case of constant weight). The essential tool is the theory of stationary phase (see Section 2.2) applied to the critical curve integral occurring in the 2D Euler-type summation formula (for details see [**268**]).

Let Λ be an arbitrary lattice in $\mathbb{R}^2$. Let $\mathcal{G} \subset \mathbb{R}^2$ be a convex region containing the origin and possessing a boundary curve ∂G such that its normal field ν is continuously differentiable and its curvature is nonvanishing. Suppose that F is of class $C^{(2)}(\overline{\mathcal{G}})$. Then, for all $a, y \in \mathbb{R}^2$, the series

$$(13.142) \qquad \sum_{g\in\Lambda} e^{2\pi i a\cdot g} \underbrace{\int_{\mathcal{G}} F(x)\, e^{-2\pi i x\cdot(g-y)}\, dx}_{=F^\wedge_{\mathcal{G}}(g-y)}$$

converges in the spherical sense, and we have

$$(13.143) \qquad \frac{1}{\|\mathcal{F}_\Lambda\|} \sum_{\substack{a+h\in\overline{\mathcal{G}} \\ h\in\Lambda^{-1}}} {}' e^{2\pi i y\cdot(a+h)} F(a+h) = \lim_{N\to\infty} \sum_{\substack{|g-y|\le N \\ g\in\Lambda}} e^{2\pi i a\cdot g} \int_{\mathcal{G}} F(x) e^{-2\pi i x\cdot(g-y)} dx.$$

Replacing the lattice Λ by its inverse lattice Λ^{-1} we are led to the following corollary:

Under the aforementioned assumptions we have

$$(13.144) \qquad \sum_{\substack{a+g\in\overline{\mathcal{G}} \\ g\in\Lambda}} {}' e^{-2\pi i y\cdot(a+g)} F(a+g) = \frac{1}{\|\mathcal{F}_\Lambda\|} \lim_{N\to\infty} \sum_{\substack{|h|\le N \\ h\in\Lambda^{-1}}} F^\wedge_{\mathcal{G}}(h)\, e^{2\pi i a\cdot(h-y)}.$$

Two-dimensional Shannon sampling. Let $\mathcal{G}$ be an arbitrary regular region in $\mathbb{R}^2$. Suppose that Λ is a lattice in $\mathbb{R}^2$ generated by the basis vectors $g_1, g_2 \in \mathbb{R}^2$. Then it follows that

$$(13.145) \qquad \int_{\mathcal{F}_\Lambda} \sum_{\substack{a+g\in\overline{\mathcal{G}} \\ g\in\Lambda}} {}' e^{-2\pi i y\cdot(a+g)} F(a+g)\, da = \int_{\mathcal{G}} F(x)\, e^{-2\pi i x\cdot y} dx = F^\wedge_{\mathcal{G}}(y).$$

This leads to the following result:

Under the above assumptions,

$$F_{\mathcal{G}}^{\wedge}(y) = \sum_{h\in\Lambda^{-1}} F_{\mathcal{G}}^{\wedge}(h)\, \frac{1}{\|\mathcal{F}_\Lambda\|} \int_{\mathcal{F}_\Lambda} e^{2\pi i a\cdot(h-y)}\, da, \tag{13.146}$$

holds for all $y \in \mathbb{R}^2$, *where*

(13.147)

$$\int_{\mathcal{F}_\Lambda} e^{2\pi i a\cdot(h-y)}\, da \;=\|\mathcal{F}_\Lambda\| \;\frac{J_1(1;\pi|g_1\cdot(y-h)|)}{\pi|g_1\cdot(y-h)|}\;\frac{J_1(1;\pi|g_2\cdot(y-h)|)}{\pi|g_2\cdot(y-h)|},\; y\in\mathbb{R}^2.$$

For a given measurable bounded region $\mathcal{H}\subset\mathbb{R}^2$ and a regular region $\mathcal{G}\subset\mathbb{R}^2$ with $\overline{\mathcal{G}},\overline{\mathcal{H}}\subset\mathcal{F}_\Lambda$ we obtain

$$\begin{aligned}
&\int_{\mathcal{H}} \sum_{\substack{a+g\in\overline{\mathcal{G}}\\ g\in\Lambda^{-1}}}{}' e^{-2\pi i y\cdot(a+g)} F(a+g)\, da \\
&\quad= \sum_{h\in\Lambda^{-1}} \underbrace{\int_{\mathcal{G}} F(x)\, e^{-2\pi i x\cdot h}\, dV(x)}_{=F_{\mathcal{G}}^{\wedge}(h)}\; \underbrace{\frac{1}{\|\mathcal{F}_\Lambda\|}\int_{\mathcal{H}} e^{2\pi i a\cdot(h-y)}\, da}_{=K_{\mathcal{H}}(h-y)},
\end{aligned} \tag{13.148}$$

where the series on the right side is absolutely and uniformly convergent on $\mathbb{R}^2$. The integral on the left side of (13.148) can be calculated in more detail

$$\int_{\mathcal{H}} \sum_{\substack{a+g\in\overline{\mathcal{G}}\\ g\in\Lambda^{-1}}}{}' e^{-2\pi i y\cdot(a+g)} F(a+g)\, da = \int_{\mathcal{G}\cap\mathcal{H}} F(a) e^{-2\pi i y\cdot a} da = F_{\mathcal{G}\cap\mathcal{H}}^{\wedge}(y) \tag{13.149}$$

provided that $\mathcal{G},\mathcal{H}$ are, in addition, subsets of $\mathcal{F}_\Lambda$ (for more details the reader is referred to W. Freeden, M.Z. Nashed [**288**]).

By virtue of the formula (13.149) we are led to distinguish two particular situations:

Oversampling. $\mathcal{G}$ is a subset of $\mathcal{H}\subset\mathcal{F}_\Lambda$. Then we are able to conclude from (13.149) that

$$F_{\mathcal{G}}^{\wedge}(y) = \sum_{h\in\Lambda^{-1}} F_{\mathcal{G}}^{\wedge}(h) K_{\mathcal{H}}(h-y),$$

i.e., $F_{\mathcal{G}}^{\wedge}$ can be reconstructed from its lattice point values in Λ^{-1}. However, if $\mathcal{H}$ is not equal to $\mathcal{G}$, the bandwidth for sampling is larger than required. Moreover, for a lattice point $h'\in\Lambda^{-1}$, we are linked to interpolation.

Aliasing. $\mathcal{G}\cap\mathcal{H}$ is a strict subset of $\mathcal{G}$. Then we are led to the identity

$$F_{\mathcal{G}\cap\mathcal{H}}^{\wedge}(y) = \sum_{h\in\Lambda^{-1}} F_{\mathcal{G}}^{\wedge}(h) K_{\mathcal{H}}(h-y),$$

i.e., the values $F_{\mathcal{G}}^{\wedge}(h)$, $h\in\Lambda^{-1}$, sum up to $F_{\mathcal{G}\cap\mathcal{H}}^{\wedge}$, so that $F_{\mathcal{G}}^{\wedge}$ cannot be reconstructed under this assumption. Therefore, the sampling of $F_{\mathcal{G}}^{\wedge}$ by the right side of (13.150) is erroneous. This phenomenon is known as *aliasing.*

As a special case we mention $\mathcal{G} \cap \mathbb{B}^2_R \subset \mathbb{B}^2_R = \mathcal{H} = \mathbb{B}^2_R \subset \mathcal{F}_\Lambda$. Then it follows that

$$F^\wedge_{\mathcal{G}\cap\mathbb{B}^2_R}(y) = \frac{R^2}{\|\mathcal{F}_\Lambda\|} \sum_{h\in\Lambda^{-1}} F^\wedge_{\mathcal{G}}(h) \, \frac{J_1(2; 2\pi|h-y|R)}{|h-y|R}, \quad y \in \mathbb{R}^2. \tag{13.150}$$

Obviously, over- and undersampling, respectively, occur dependent on the geometric configuration as described above. It should be pointed out that one-dimensional oversampling is known to accelerate the convergence of the sampling series (see, e.g., R.J. Marks II [**561**], C. Blatter [**99**]). This phenomenon can be also detected in our two-dimensional framework (as shown in W. Freeden, M.Z. Nashed [**288**]).

13.4. Multivariate Gauss-Weierstrass integral transform

The Fourier theory of integral transforms in Euclidean spaces is well understood in the usual $L^1(\mathbb{R}^q)$-nomenclature. In our pointwise oriented approach based on the asymptotic relations of Section 2.2, however, we are led to discuss some weighted variants of the (inverse) Fourier transform, namely the Gauss and Weierstrass integral transform as well as the Poisson and Abel integral transform in their particular interrelation to the Fourier transform. In doing so we are essentially inspired by C. Müller [**607**] to transfer the theory of Fourier transforms (defined as spherical principal values) to so-called spherically continuous functions. As results we are led to structures and settings that are adequate in inversion procedures of lattice point as well as sampling theory.

The following integrals involving Bessel functions $J_n(q;\cdot)$ of degree n and dimension q are well known in the literature (cf. [**607**]: For $t \geq 0$, $a \in \mathbb{R}$, we have

$$\|\mathbb{S}^{q-1}\| \int_0^\infty e^{-2\pi tr} J_0(q; 2\pi ar)\, r^{q-1}\, dr = \frac{2}{\|\mathbb{S}^q\|} \frac{t}{(a^2+t^2)^{\frac{q+1}{2}}}. \tag{13.151}$$

For $\rho > 0, n \in \mathbb{N}_0$, and $\sigma \in \mathbb{C}$ with $\Re(\sigma) > 0$ we have

$$\int_0^\infty e^{-\pi\sigma r^2} J_n(q; 2\pi r\rho)\, r^{n+q-1}\, dr = \frac{1}{2}\Gamma\left(\frac{q}{2}\right) \frac{\rho^n}{\pi^{\frac{q}{2}}\sigma^{n+\frac{q}{2}}}\, e^{-\frac{\pi\rho^2}{\sigma}}. \tag{13.152}$$

In particular with $x, y \in \mathbb{R}^q$ and $\rho = |x-y|$ we obtain the identity

$$\|\mathbb{S}^{q-1}\| \int_0^\infty e^{-\pi\sigma r^2} J_0(q; 2\pi|x-y|r) r^{q-1}\, dr = \left(\frac{1}{t}\right)^{\frac{q}{2}} e^{-\frac{\pi}{\sigma}(x-y)^2}. \tag{13.153}$$

There are many more interesting integrals involving Bessel functions that can be evaluated in a similar way (see, e.g., W. Freeden [**268**]). Nevertheless, the important cases to be needed for our Fourier inversion techniques here are covered by the following examples.

Gauss-Weierstrass transform over Euclidean spaces. We start from the *"Weierstrass function"* $W : \mathbb{R}^q \to \mathbb{C}$ given by

$$W(x) = e^{2\pi i x\cdot z} e^{-\pi\tau x^2}, \quad z \in \mathbb{R}^q, \ \tau > 0. \tag{13.154}$$

An elementary calculation yields

$$W^{\wedge}_{\mathbb{R}^q}(x) = \int_{\mathbb{R}^q} W(y) e^{-2\pi i x\cdot y}\, dy = \int_{\mathbb{R}^q} e^{-\pi\tau y^2} e^{2\pi i (z-x)\cdot y}\, dy. \tag{13.155}$$

With $y = \rho\eta$ we find

$$W^{\wedge}_{\mathbb{R}^q}(x) = \int_0^{\infty} \rho^{q-1} e^{-\pi\tau\rho^2} \left(\int_{\mathbb{S}^{q-1}} e^{2\pi i\rho (z-x)\cdot\eta}\, dS_{(q-1)}(\eta) \right) d\rho. \tag{13.156}$$

Remembering the definition of the Bessel function of order 0, we find

$$\begin{aligned} &\int_{\mathbb{S}^{q-1}} e^{2\pi i\rho((z-x)\cdot\eta)}\, dS_{(q-1)}(\eta) \\ &\quad = \|\mathbb{S}^{q-1}\| \int_{-1}^{1} e^{2\pi i\rho|z-x|s} (1-s^2)^{\frac{q-3}{2}}\, ds \\ &\quad = \|\mathbb{S}^{q-1}\|\, J_0(q; 2\pi|z-x|\rho). \end{aligned} \tag{13.157}$$

For $\tau > 0$, $a \in \mathbb{R}$, we know that

$$\|\mathbb{S}^{q-1}\| \int_0^{\infty} e^{-\pi\tau r^2} J_0(q; 2\pi a r) r^{q-1}\, dr = \tau^{-\frac{q}{2}} e^{-\frac{\pi}{\tau} a^2}. \tag{13.158}$$

In other words, with $x, z \in \mathbb{R}^q$ and $a = |x - z|$, we have

$$\|\mathbb{S}^{q-1}\| \int_0^{\infty} e^{-\pi\tau r^2} J_0(q; 2\pi|x-z|r)\, r^{q-1}\, dr = \tau^{-\frac{q}{2}} e^{-\frac{\pi}{\tau}(x-z)^2}. \tag{13.159}$$

Altogether this shows us that

$$W^{\wedge}_{\mathbb{R}^q}(x) = \left(\frac{1}{\tau}\right)^{\frac{q}{2}} e^{-\frac{\pi}{\tau}(x-z)^2}. \tag{13.160}$$

Next we consider the *"Gauss function"* $G : \mathbb{R}^q \to \mathbb{C}$ given by

$$G(x) = \left(\frac{1}{\tau}\right)^{\frac{q}{2}} e^{-\frac{\pi}{\tau}(z-x)^2}, \quad z \in \mathbb{R}^q,\ \tau > 0. \tag{13.161}$$

Now we have

$$\begin{aligned} G^{\wedge}_{\mathbb{R}^q}(-x) &= \left(\frac{1}{\tau}\right)^{\frac{q}{2}} \int_{\mathbb{R}^q} e^{-\frac{\pi}{\tau}(z-y)^2} e^{2\pi i x\cdot y}\, dy \\ &= \left(\frac{1}{\tau}\right)^{\frac{q}{2}} e^{2\pi i x\cdot z} \int_{\mathbb{R}^q} e^{-\frac{\pi}{\tau}(y-z)^2} e^{2\pi i x\cdot(y-z)}\, dy. \end{aligned} \tag{13.162}$$

By use of polar coordinates $y = z + \rho\eta$ the integral can be transformed to

$$\begin{aligned} &\int_0^{\infty} \rho^{q-1} e^{-\frac{\pi}{\tau}\rho^2} \left(\int_{\mathbb{S}^{q-1}} e^{2\pi i\rho(x\cdot\eta)}\, dS_{(q-1)}(\eta) \right) d\rho \\ &\quad = \|\mathbb{S}^{q-1}\| \int_0^{\infty} \rho^{q-1} e^{-\frac{\pi}{\tau}\rho^2} J_0(q; 2\pi|x|\rho)\, d\rho. \end{aligned} \tag{13.163}$$

The last integral is known. For $x \in \mathbb{R}^q$ we have

$$G^{\wedge}_{\mathbb{R}}(-x) = e^{2\pi i x\cdot z} e^{-\pi\tau x^2} = W(x). \tag{13.164}$$

In particular, we have

$$G^{\wedge}_{\mathbb{R}}(0) = \ 1 \ = W(0). \tag{13.165}$$

Summarizing our results we therefore obtain the following relations:

For all $x \in \mathbb{R}^q$ we have

$$G^{\wedge}_{\mathbb{R}^q}(-x) = W(x) = \int_{\mathbb{R}^q} \underbrace{W^{\wedge}_{\mathbb{R}^q}(y)}_{=G(y)} e^{2\pi i x\cdot y} \, dy, \tag{13.166}$$

$$W^{\wedge}_{\mathbb{R}^q}(x) = G(x) = \int_{\mathbb{R}^q} \underbrace{G^{\wedge}_{\mathbb{R}^q}(y)}_{=W(-y)} e^{2\pi i x\cdot y} \, dy. \tag{13.167}$$

We are now in the position to introduce the Gauss transform and Weierstrass transform as integrals involving the kernels G and W, respectively:

For $\tau > 0$, the Gauss transform $\mathcal{G}_\tau(F)$ of $F : \mathbb{R}^q \to \mathbb{C}$ at $x \in \mathbb{R}^q$ is defined by

$$\mathcal{G}_\tau(F)(x) = \left(\frac{1}{\tau}\right)^{q/2} \int_{\mathbb{R}^q} e^{-\frac{\pi}{\tau}(x-y)^2} F(y) \, dy, \tag{13.168}$$

while the Weierstraß transform $\mathcal{W}_\tau(F)$ of $F : \mathbb{R}^q \to \mathbb{C}$ at $x \in \mathbb{R}^q$ is defined by

$$\mathcal{W}_\tau(F)(x) = \int_{\mathbb{R}^q} e^{-\pi\tau y^2} e^{2\pi i x\cdot y} F(y) \, dy. \tag{13.169}$$

The relation between the Gauss transform and the Weierstrass transform is characterized by the following identities (see also [**268**]).

For $F \in L^1(\mathbb{R}^q)$, $\tau > 0$, and $z \in \mathbb{R}^q$

$$\mathcal{W}_\tau(F^{\wedge}_{\mathbb{R}^q})(z) = (\mathcal{W}_\tau(F))^{\wedge}_{\mathbb{R}^q}(z) = \mathcal{G}_\tau(F)(z), \tag{13.170}$$

$$\mathcal{G}_\tau(F^{\wedge}_{\mathbb{R}^q})(z) = (\mathcal{G}_\tau(F))^{\wedge}_{\mathbb{R}^q}(z) = \mathcal{W}_\tau(F)(-z). \tag{13.171}$$

Next our aim is to show that the "integral means" $\mathcal{W}_\tau(F)(x)$ and $\mathcal{G}_\tau(F)(x)$ tend to $F(x)$ as $\tau \to 0$, if suitable conditions are imposed on the position x as well as the function F under consideration.

Suppose that F is continuous and uniformly bounded in $\mathbb{R}^q$ such that, for $z \in \mathbb{R}^q$, the "spherical principal value" of $F^{\wedge}_{\mathbb{R}^q}(z)$, i.e., the limit

$$\lim_{N\to\infty} \int_{\substack{|x|\le N\\ x\in\mathbb{R}^q}} F(x) \, e^{-2\pi i z\cdot x} \, dx \tag{13.172}$$

exists. Then

$$\lim_{\substack{\tau\to 0\\ \tau>0}} \mathcal{W}_\tau(F)(z) = \lim_{N\to\infty} \int_{\substack{|x|\le N\\ x\in\mathbb{R}^q}} F(x) \, e^{-2\pi i x\cdot z} \, dx. \tag{13.173}$$

Before we discuss the Gauss-Weierstrass transform, we introduce the notion of "spherical continuity" at a point $z \in \mathbb{R}^q$.

A function $F : \mathbb{R}^q \to \mathbb{C}$ is called *spherically continuous* at $z \in \mathbb{R}^q$ if the spherical mean

$$\Phi(r; z) = \frac{1}{\|\mathbb{S}^{q-1}\|} \int_{\mathbb{S}^{q-1}} F(z + r\xi)\, dS_{(q-1)}(\xi) \tag{13.174}$$

exists for all values $r > 0$ and is continuous for $r \to 0$ with $\Phi(0; z) = F(z)$.

Suppose that $F \in L^1(\mathbb{R}^q)$ is spherically continuous at $z \in \mathbb{R}^q$. Then

$$\lim_{\substack{\tau \to 0\\ \tau > 0}} \mathcal{G}_\tau(F)(z) = \lim_{\substack{\tau \to 0\\ \tau > 0}} \mathcal{W}_\tau(F^\wedge_{\mathbb{R}^q})(z) = F(z). \tag{13.175}$$

It should be pointed out that the limit relation (13.175) allows the interpretation as *pointwise Gaussian inversion formula.*

Suppose that $F \in L^1(\mathbb{R}^q)$ is spherically continuous at $z \in \mathbb{R}^q$. Furthermore, assume that the "spherical principal value"

$$\lim_{N \to \infty} \int_{\substack{|y| \le N\\ y \in \mathbb{R}^q}} e^{2\pi i z \cdot y} F^\wedge_{\mathbb{R}^q}(y)\, dy \tag{13.176}$$

exists. Then we have

$$F(z) = \lim_{N \to \infty} \int_{\substack{|y| \le N\\ y \in \mathbb{R}^q}} e^{2\pi i z \cdot y} F^\wedge_{\mathbb{R}^q}(y)\, dy. \tag{13.177}$$

If the spherical principal value does not exist, the limit relation

$$F(z) = \lim_{\substack{\tau \to 0\\ \tau > 0}} \int_{\mathbb{R}^q} e^{-\pi\tau y^2} e^{2\pi i z \cdot y} F^\wedge_{\mathbb{R}^q}(y)\, dy \tag{13.178}$$

remains valid as a "mean representation" of F at the point z of spherical continuity.

Gauss-Weierstrass transform over regular regions. In what follows, we are led to structures and settings that are adequate for our pointwise efforts in sampling theory involved with discontinuous functions relative to boundaries of regular regions. We introduce the *bandlimited Gauss-Weierstrass transform as integrals over a regular region $\mathcal{D}$* involving the kernels G and W, respectively.

Let $\mathcal{D}$ be a regular region. For $\tau > 0$, the *Gauss transform* $\mathcal{G}^{\mathcal{D}}_\tau(F)$ of $F \in C^{(0)}(\overline{\mathcal{D}})$ at $x \in \mathbb{R}^q$ is defined by

$$\mathcal{G}^{\mathcal{D}}_\tau(F)(x) = \left(\frac{1}{\tau}\right)^{q/2} \int_{\mathcal{D}} e^{-\frac{\pi}{\tau}(x-y)^2} F(y)\, dy, \tag{13.179}$$

while the *Weierstrass transform* $\mathcal{W}^{\mathcal{D}}_\tau(F)$ of $F \in C^{(0)}(\overline{\mathcal{D}})$ at $x \in \mathbb{R}^q$ is defined by

$$\mathcal{W}^{\mathcal{D}}_\tau(F)(x) = \int_{\mathcal{D}} e^{-\pi\tau y^2}\, e^{2\pi i x \cdot y}\, F(y)\, dy. \tag{13.180}$$

The relation between the Gauss and the Weierstrass transform is documented by the following identities: *For $\tau > 0$, and $z \in \mathbb{R}^q$,*

$$\mathcal{W}_\tau(F^\wedge_{\mathcal{D}})(z) = \mathcal{G}^{\mathcal{D}}_\tau(F)(z), \tag{13.181}$$

$$\mathcal{G}_\tau(F^\wedge_{\mathcal{D}})(z) = \mathcal{W}^{\mathcal{D}}_\tau(F)(-z). \tag{13.182}$$

The "integral mean" $\mathcal{W}_\tau(F^\wedge_\mathcal{D})(w)$ tends to $\alpha_{\partial\mathcal{D}}(w)\ F(w)$ as $\tau \to 0$, where $\alpha_{\partial\mathcal{D}}(w)$ is the solid angle at $w \in \mathbb{R}^q$ subtended by the surface $\partial\mathcal{D}$ (see [**298**]).

Let $\mathcal{D}$ be a regular region in $\mathbb{R}^q$. Suppose that F is of class $F \in C^{(0)}(\overline{\mathcal{D}})$. Then

$$\lim_{\substack{\tau\to 0\\ \tau>0}} \mathcal{G}^{\mathcal{D}}_\tau(F)(w) = \lim_{\substack{\tau\to 0\\ \tau>0}} \mathcal{W}_\tau(F^\wedge_\mathcal{D})(w) = \alpha_{\partial\mathcal{D}}(w)\ F(w), \quad w \in \overline{\mathcal{D}}. \tag{13.183}$$

In order to understand (13.183) we start from

$$\begin{aligned} \mathcal{G}^{\mathcal{D}}_\tau(F)(w) &= \left(\frac{1}{\tau}\right)^{\frac{q}{2}} \int_\mathcal{D} e^{-\frac{\pi}{\tau}(w-x)^2} F(x)\ dx \\ &= \left(\frac{1}{\tau}\right)^{\frac{q}{2}} \int_{\mathbb{R}^q} e^{-\frac{\pi}{\tau}(w-x)^2} \tilde{F}(x)\ dx, \end{aligned} \tag{13.184}$$

where $\tilde{F}$ is the function given by

$$\tilde{F}(x) = \begin{cases} F(x) & , \quad x \in \overline{\mathcal{D}} \\ 0 & , \quad x \notin \overline{\mathcal{D}}. \end{cases} \tag{13.185}$$

By introducing polar coordinates $x = w + r\xi$, $r > 0$, $\xi \in \mathbb{S}^{q-1}$, we obtain

$$\mathcal{G}^{\mathcal{D}}_\tau(F)(w) = \left(\frac{1}{\tau}\right)^{\frac{q}{2}} \int_0^\infty e^{-\frac{\pi}{\tau}r^2} \Phi(r;w)\ r^{q-1}\ dr, \tag{13.186}$$

where

$$\Phi(r;w) = \int_{\mathbb{S}^{q-1}} \tilde{F}(w + r\xi)\ dS_{(q-1)}(\xi). \tag{13.187}$$

Observing that $\tilde{F} \in L^1(\mathbb{R}^q)$ we find that

$$\Phi(0;z) = \|\mathbb{S}^{q-1}\|\ \alpha_{\partial\mathcal{D}}(w)\ F(w). \tag{13.188}$$

Remembering the convergence behavior of spherical principle integral means we therefore obtain in connection with a known 1D limit relation (see (2.105))

$$\lim_{\substack{\tau\to 0\\ \tau>0}} \left(\frac{1}{\tau}\right)^{\frac{q}{2}} \int_0^\infty e^{-\frac{\pi}{\tau}r^2} \Phi(r;w)\ r^{q-1}\ dr = \frac{\Phi(0;w)}{\|\mathbb{S}^{q-1}\|} = \alpha_{\partial\mathcal{D}}(w)\ F(w). \tag{13.189}$$

Abel-Poisson transform over Euclidean spaces. For $\tau > 0$, $z \in \mathbb{R}^q$, we consider the *"Abel function"* A given by

$$A(x) = e^{-2\pi\tau|x|} e^{2\pi i z\cdot x}, \quad x \in \mathbb{R}^q. \tag{13.190}$$

For $\tau > 0$, $a \in \mathbb{R}$, it is well known (see, e.g., [**268**]) that

$$\|\mathbb{S}^{q-1}\| \int_0^\infty e^{-2\pi\tau r} J_0(q; 2\pi a r) r^{q-1}\ dr = \frac{2}{\|\mathbb{S}^q\|} \frac{\tau}{(a^2+\tau^2)^{\frac{q-1}{2}}}. \tag{13.191}$$

For $y \in \mathbb{R}^q$, this gives

$$\begin{aligned} A^\wedge_{\mathbb{R}^q}(y) &= \int_{\mathbb{R}^q} e^{-2\pi\tau|x|} e^{2\pi i x\cdot z} e^{-2\pi i x\cdot y}\ dx \\ &= \frac{2}{\|\mathbb{S}^q\|} \frac{\tau}{(\tau^2 + (y-z)^2)^{\frac{q+1}{2}}} = P(y), \end{aligned} \tag{13.192}$$

where P stands for the *"Poisson function"*. Furthermore, we know that
(13.193)

$$P^{\wedge}_{\mathbb{R}^q}(y) = \frac{2}{\|\mathbb{S}^q\|} \int_{\mathbb{R}^q} \frac{\tau}{(\tau^2 + (x-z)^2)^{\frac{q+1}{2}}} e^{-2\pi i x \cdot y}\, dy = e^{-2\pi\tau|y|} e^{-2\pi i y \cdot z} = A(-y).$$

These identities motivate the following definitions:

For $\tau > 0$, the *Abel transform* $\mathcal{A}_\tau(F)$ of $F : \mathbb{R}^q \to \mathbb{C}$ at $x \in \mathbb{R}^q$ is defined by

$$\mathcal{A}_\tau(F)(x) = \int_{\mathbb{R}^q} e^{-2\pi\tau|y|} e^{2\pi i y \cdot x} F(y)\, dy, \tag{13.194}$$

while the *Poisson transform* $\mathcal{P}_\tau(F)$ of $F : \mathbb{R}^q \to \mathbb{C}$ at $x \in \mathbb{R}^q$ is defined by

$$\mathcal{P}_\tau(F)(x) = \frac{2}{\|\mathbb{S}^q\|} \int_{\mathbb{R}^q} \frac{\tau}{(\tau^2 + (x-y)^2)^{\frac{q+1}{2}}} F(y)\, dy. \tag{13.195}$$

By analogous arguments as those used for Gauss-Weierstrass means we find the following relations:

For $F \in L^1(\mathbb{R}^q)$, $\tau > 0$, *and* $z \in \mathbb{R}^q$

$$\mathcal{A}_\tau(F^{\wedge}_{\mathbb{R}^q})(z) = (\mathcal{A}_\tau(F))^{\wedge}_{\mathbb{R}^q}(z) = \mathcal{P}_\tau(F)(z), \tag{13.196}$$

$$\mathcal{P}_\tau(F^{\wedge}_{\mathbb{R}^q})(z) = (\mathcal{P}_\tau(F))^{\wedge}_{\mathbb{R}^q}(z) = \mathcal{A}_\tau(F)(-z). \tag{13.197}$$

For the Poisson transform we are able to verify the following result:

Suppose that $F \in L^1(\mathbb{R}^q)$ *is spherically continuous at* $z \in \mathbb{R}^q$. *Then*

$$\lim_{\substack{\tau\to 0\\ \tau>0}} \mathcal{P}_\tau(F)(z) = \lim_{\substack{\tau\to 0\\ \tau>0}} \mathcal{A}_\tau(F^{\wedge}_{\mathbb{R}^q})(z) = F(z). \tag{13.198}$$

In fact, we have

$$\mathcal{P}_\tau(z) = \frac{2}{\|\mathbb{S}^q\|} \int_{\mathbb{R}^q} \frac{\tau}{(\tau^2 + (x-z)^2)^{\frac{q+1}{2}}} F(x)\, dx. \tag{13.199}$$

Polar coordinates $x = z + r\xi$, $r > 0$, $\xi \in \mathbb{S}^{q-1}$, yield

$$\mathcal{P}_\tau(F)(z) = 2\frac{\|\mathbb{S}^{q-1}\|}{\|\mathbb{S}^q\|} \int_0^\infty \frac{\tau}{(\tau^2 + r^2)^{\frac{q+1}{2}}} \Phi(r; z) r^{q-1}\, dr, \tag{13.200}$$

where

$$\Phi(r; z) = \frac{1}{\|\mathbb{S}^{q-1}\|} \int_{\mathbb{S}^{q-1}} F(z + r\xi)\, dS_{(q-1)}(\xi) \tag{13.201}$$

and

$$\Phi(0; z) = F(z). \tag{13.202}$$

Hence,

$$\lim_{\substack{\tau\to 0\\ \tau>0}} \frac{2\|\mathbb{S}^{q-1}\|}{\|\mathbb{S}^q\|} \int_0^\infty \frac{\tau}{(\tau^2 + r^2)^{\frac{q+1}{2}}} \Phi(r; z) r^{q-1}\, dr = \Phi(0; z) = F(z). \tag{13.203}$$

It should be mentioned that the formulation of the pointwise (Fourier) inversion formula remains unchanged by this alternative limit relation (more precisely, by use of $\mathcal{A}_\tau$ instead of $\mathcal{W}_\tau$).

Abel-Poisson transform over regular regions. Let $\mathcal{D}$ be a regular region. For $\tau > 0$, the *Abel and Poisson transform*, respectively, is defined by

$$\mathcal{A}_\tau^{\mathcal{D}}(F)(x) = \int_{\mathcal{D}} e^{-2\pi\tau|y|} e^{2\pi iy\cdot x} F(y)\, dy, \tag{13.204}$$

and

$$\mathcal{P}_\tau^{\mathcal{D}}(F)(x) = \frac{2}{\|\mathbb{S}^q\|} \int_{\mathcal{D}} \frac{\tau}{(\tau^2 + (x-y)^2)^{\frac{q+1}{2}}} F(y)\, dy. \tag{13.205}$$

For $F \in C^0(\overline{\mathcal{D}})$, $\tau > 0$, $z \in \mathbb{R}^q$,

$$\mathcal{A}_\tau(F_{\mathcal{D}}^\wedge))(z) = \mathcal{P}_\tau^{\mathcal{D}}(F)(z), \tag{13.206}$$

$$\mathcal{P}_\tau(F_{\mathcal{D}}^\wedge))(z) = \mathcal{A}_\tau^{\mathcal{D}}(F)(-z), \tag{13.207}$$

and we have

$$\lim_{\substack{\tau\to 0\\ \tau>0}} \mathcal{P}_\tau^{\mathcal{D}}(F)(w) = \lim_{\substack{\tau\to 0\\ \tau>0}} \mathcal{A}_\tau(F_{\mathcal{D}}^\wedge))(w) = \alpha_{\partial\mathcal{D}}(w)\, F(w). \tag{13.208}$$

13.5. Multivariate Gauss-Weierstrass summability Shannon sampling

After these preparations our interest now is in justifying the Poisson summation formula over regular regions $\mathcal{G}$ in $\mathbb{R}^q$, however, under Gauss-Weierstrass summability conditions (note that Abel-Poisson summability can be used in similar way).

Our purpose is to deal with the interesting question that arises from our one- and two-dimensional work, in what respect the infinite series

$$\lim_{N\to\infty} \frac{1}{\|\mathcal{F}_\Lambda\|} \sum_{\substack{|h-y|\le N\\ h\in\Lambda^{-1}}} e^{2\pi a\cdot h} \int_{\mathcal{G}} F(x) e^{-2\pi ix\cdot(h-y)}\, dx \tag{13.209}$$

can be "summed" to the finite sum

$$\sum_{\substack{a+g\in\overline{\mathcal{G}}\\ g\in\Lambda}}{}' e^{2\pi iy\cdot(a+g)} F(a+g) = \sum_{\substack{a+g\in\overline{\mathcal{G}}\\ g\in\Lambda}} \alpha_{\partial\mathcal{G}}(a+g) e^{2\pi iy\cdot(a+g)} F(a+g), \tag{13.210}$$

where $\alpha_{\partial\mathcal{G}}(g+a)$ denotes the solid angle subtended by the boundary $\partial\mathcal{G}$ at $g+a$. We already know that the answer cannot be given directly because of serious convergence problems of the series (13.209) and the discontinuous behavior of (13.210) as a function of a. Thus, our basic idea is to identify (13.209) and (13.210) in the sense of the Gauss-Weierstrass means. Essential tools are the counterparts of the Fourier inversion formula and the functional equation of the Theta function in $\mathbb{R}^q$.

For arbitrary points $a, x \in \mathbb{R}^q$ and lattices $\Lambda \subset \mathbb{R}^q$ we call $\vartheta^{(q)}(\cdot\,; a, x; \Lambda)$ given by

$$\vartheta^{(q)}(\sigma; a, x; \Lambda) = \sum_{g\in\Lambda} e^{-\pi\sigma|g-a|^2} e^{2\pi ig\cdot x}, \quad \sigma \in \mathbb{C},\ \Re(\sigma) > 0, \tag{13.211}$$

the *Theta function of dimension* q *on* Λ.

From the Poisson summation formula under its ordinary convergence criteria (cf. (8.42)) we obtain

$$\begin{aligned}\vartheta^{(q)}(\sigma; a, x; \Lambda) &= \sum_{g\in\Lambda} e^{-\pi\sigma|g-a|^2}\, e^{2\pi i g\cdot x} \\ &= \frac{\|\mathbb{S}^{q-1}\|\,\Gamma\left(\frac{q}{2}\right)}{2\|\mathcal{F}_\Lambda\|\,\pi^{\frac{q}{2}}\sigma^{\frac{q}{2}}} e^{2\pi i a\cdot x} \sum_{h\in\Lambda^{-1}} e^{-\frac{\pi}{\sigma}|h+x|^2} e^{2\pi i h\cdot a}.\end{aligned} \tag{13.212}$$

With the explicit value $\|\mathbb{S}^{q-1}\| = 2\pi^{\frac{q}{2}}\left(\Gamma\left(\frac{q}{2}\right)\right)^{-1}$ for the area of the unit sphere $\mathbb{S}^{q-1}$ in $\mathbb{R}^q$ this leads to the *functional equation of the Theta function of dimension* q with respect to Λ.

For all values $\sigma \in \mathbb{C}$ with $\Re(\sigma) > 0$, the Theta function $\vartheta^{(q)}(\cdot; a, x; \Lambda)$ is holomorphic, and we have

$$\vartheta^{(q)}(\sigma; a, x; \Lambda) = \frac{1}{\|\mathcal{F}_\Lambda\|} e^{2\pi i a\cdot x} \sigma^{-\frac{q}{2}}\, \vartheta^{(q)}\left(\frac{1}{\sigma}; -x, a; \Lambda^{-1}\right). \tag{13.213}$$

For all values $\tau \in \mathbb{R}$ with $\tau > 0$ and $x \in \mathbb{R}^q$ we readily obtain from the functional equation of the qD Theta function that

$$\begin{aligned}&\frac{1}{\|\mathcal{F}_\Lambda\|} \int_{\mathcal{G}} \vartheta^{(q)}\left(\tau; 0, x-a; \Lambda^{-1}\right)\, e^{2\pi i y\cdot x} F(x)\, dx \\ &\qquad = \tau^{\frac{q}{2}} \int_{\mathcal{G}} \vartheta^{(q)}\left(\frac{1}{\tau}; a-x; 0; \Lambda\right)\, e^{2\pi i y\cdot x}\, F(x)\, dx\end{aligned} \tag{13.214}$$

holds for every function $F \in C^{(0)}(\mathcal{G})$. Explicitly written out in series notation we therefore have

$$\begin{aligned}&\frac{1}{\|\mathcal{F}_\Lambda\|} \int_{\mathcal{G}} \sum_{h\in\Lambda^{-1}} e^{-\tau\pi h^2} e^{2\pi i y\cdot x} F(x) e^{2\pi i h\cdot(a-x)}\, dx \\ &\qquad = \tau^{-\frac{q}{2}} \int_{\mathcal{G}} \sum_{g\in\Lambda} e^{-\frac{\pi}{\tau}|x-(g+a)|^2} e^{2\pi i y\cdot x} F(x)\, dx\end{aligned} \tag{13.215}$$

for all $\tau > 0$ and all $x \in \mathbb{R}^q$, for arbitrary $a \in \mathbb{R}^q$, and all regular regions $\mathcal{G}$ as well as for all arbitrary (weight) functions F of class $C^{(0)}(\overline{\mathcal{G}})$. Interchanging sums and integrals we obtain

$$\begin{aligned}&\tau^{-\frac{q}{2}} \sum_{g\in\Lambda} \int_{\mathcal{G}} e^{-\frac{\pi}{\tau}|x-(g+a)|^2} e^{2\pi i y\cdot x} F(x)\, dx \\ &\qquad = \frac{1}{\|\mathcal{F}_\Lambda\|} \sum_{h\in\Lambda^{-1}} e^{-\tau\pi h^2} e^{2\pi i h\cdot a} \int_{\mathcal{G}} F(x) e^{-2\pi i x\cdot(h-y)}\, dx.\end{aligned} \tag{13.216}$$

In fact, by observing the limit relation

$$\lim_{\substack{\tau\to 0\\ \tau>0}} \tau^{-\frac{q}{2}} \int_{\mathcal{G}} e^{-\frac{\pi}{\tau}|x-(g+a)|^2} e^{2\pi i y\cdot x} F(x)\, dx = \alpha_{\partial\mathcal{G}}(a+g) e^{2\pi i y\cdot(a+g)} F(a+g) \tag{13.217}$$

we are led to the main result, i.e., the Poisson summation formula in Gauss-Weierstrass summability (the proof can be found in W. Freeden [**268**]).

Let $\mathcal{G} \subset \mathbb{R}^q$ be a regular region. Let Λ be a lattice in $\mathbb{R}^q$. Suppose that F is a member of class $C^{(0)}(\overline{\mathcal{G}})$. Then, for all $\tau > 0$ and all $a, y \in \mathbb{R}^q$, the series

$$\sum_{h\in\Lambda^{-1}} e^{-\tau\pi h^2} e^{2\pi i h\cdot a} \int_{\mathcal{G}} F(x) e^{-2\pi i x\cdot(h-y)}\, dx \tag{13.218}$$

is absolutely convergent. Moreover, for all $a \in \mathbb{R}^q$, we have the lattice point identity

$$\begin{aligned} \sum_{\substack{a+g\in\overline{\mathcal{G}}\\ g\in\Lambda}}{}' e^{2\pi i y\cdot(a+g)} F(a+g) \\ = \lim_{\substack{\tau\to 0\\ \tau>0}} \frac{1}{\|\mathcal{F}_\Lambda\|} \sum_{h\in\Lambda^{-1}} e^{-\tau\pi h^2} e^{2\pi i h\cdot a} \int_{\mathcal{G}} F(x) e^{-2\pi i x\cdot(h-y)}\, dx, \end{aligned} \tag{13.219}$$

where

$$\sum_{\substack{a+g\in\overline{\mathcal{G}}\\ g\in\Lambda}}{}' e^{2\pi i y\cdot(a+g)} F(a+g) = \sum_{\substack{a+g\in\overline{\mathcal{G}}\\ g\in\Lambda}} \alpha_{\partial\mathcal{G}}(a+g) e^{2\pi i y\cdot(a+g)} F(a+g) \tag{13.220}$$

and $\alpha_{\partial\mathcal{G}}(g+a)$ denotes the solid angle subtended by the boundary $\partial\mathcal{G}$ at $g+a$ (note that a similar result is available in the Abel-Poisson summability framework).

Theoretically, it is interesting to investigate under which assumptions Gauss-Weierstrass convergence allows the transition to ordinary convergence, i.e., the limit $\lim_{\tau\to 0}$ can be interchanged with the sum $\sum_{h\in\Lambda^{-1}}$. We know some conditions (from W. Freeden [**268**]) under which Gauss-Weierstrass summability can be avoided, i.e., under which conditions the following formula holds true. We only mention the following example:

Let Λ be an arbitrary lattice in $\mathbb{R}^q$. Let $\mathcal{G} \subset \mathbb{R}^q$ be a regular region. Suppose that F is of class $C^{(0)}(\overline{\mathcal{G}})$. Furthermore, assume that

$$\lim_{N\to\infty} \frac{1}{\|\mathcal{F}_\Lambda\|} \sum_{\substack{|h-y|\le N\\ h\in\Lambda^{-1}}} e^{2\pi i a\cdot h} \int_{\mathcal{G}} F(x) e^{-2\pi i x\cdot(h-y)} dx \tag{13.221}$$

is convergent. Then we have

$$\sum_{\substack{a+g\in\overline{\mathcal{G}}\\ g\in\Lambda}}{}' e^{2\pi i y\cdot(a+g)} F(a+g) = \lim_{N\to\infty} \frac{1}{\|\mathcal{F}_\Lambda\|} \sum_{\substack{|h-y|\le N\\ h\in\Lambda^{-1}}} e^{2\pi i a\cdot h} \int_{\mathcal{G}} F(x) e^{-2\pi i x\cdot(h-y)} dx.$$

Furthermore, summabilty can be avoided for (13.221) on regular regions $\mathcal{G} \subset \mathbb{R}^q$ by applying certain homogeneous conditions on the boundary $\partial\mathcal{G}$ (for more details see W. Freeden [**268**]).

Our goal now is to document that (13.220) offers the entry point for purposes of Shannon-type sampling, namely by providing a cardinal series resulting from the right side of (13.220) and summing up functional values of the Fourier transform of F over $\mathcal{G}$ in y-shifted lattice points (cf. [**288**].

Fourier transformed values based Shannon-type sampling (in Gauss-Weierstrass summability). Integrating (13.220) over a measurable bounded region $\mathcal{H} \subset \mathbb{R}^q$, $q \geq 2$, we obtain the following lattice point formula, that serves as the point of departure for our Shannon sampling approach. In other words,

$$\int_{\mathcal{H}} \sum_{\substack{a+g \in \overline{G} \\ g \in \Lambda}}{}' e^{-2\pi i y\cdot(a+g)} F(a+g)\, da \qquad (13.222)$$

$$= \lim_{\substack{\tau\to 0 \\ \tau>0}} \sum_{h\in\Lambda^{-1}} e^{-\tau\pi h^2} \underbrace{\int_{\mathcal{G}} F(x) e^{-2\pi i h\cdot x}\, dx}_{=F_{\mathcal{G}}^{\wedge}(h)} \underbrace{\frac{1}{\|\mathcal{F}_\Lambda\|} \int_{\mathcal{H}} e^{2\pi i a\cdot(h-y)}\, da}_{=K_{\mathcal{H}}(h-y)}$$

is valid for all points $y \in \mathbb{R}^q$, where the lattice constant $\frac{1}{\|\mathcal{F}_\Lambda\|}$ can be regarded as an indicator of the sampling density. We denote by $\Lambda(\mathcal{H})$ the finite set of lattice points $g' \in \Lambda \subset \mathbb{R}^q$ such that $\mathcal{F}_\Lambda + \{g'\}$ contains points of $\mathcal{H}$, i.e., in mathematical formalism

$$\Lambda(\mathcal{H}) = \{g' \in \Lambda : (\mathcal{F}_\Lambda + \{g'\}) \cap \mathcal{H} \neq \emptyset\}. \qquad (13.223)$$

Keeping this abbreviation in mind we are able to see that

$$\int_{\mathcal{H}} \sum_{\substack{a+g \in \overline{\mathcal{G}} \\ g \in \Lambda}}{}' e^{-2\pi i y\cdot(a+g)} F(a+g)\, da$$

$$= \sum_{g'\in\Lambda(\mathcal{H})} \int_{\mathcal{H}_{g'}} \sum_{\substack{a+g \in \overline{\mathcal{G}} \\ g \in \Lambda}}{}' e^{-2\pi i y\cdot(a+g)} F(a+g)\, da, \qquad (13.224)$$

where the set $\mathcal{H}_{g'}, g' \in \Lambda(\mathcal{H})$, is given by

$$\mathcal{H}_{g'} = (\mathcal{F}_\Lambda + \{g'\}) \cap \mathcal{H}, \quad g' \in \Lambda(\mathcal{H}). \qquad (13.225)$$

Since the function

$$a \mapsto \sum_{\substack{a+g \in \overline{\mathcal{G}} \\ g \in \Lambda}}{}' e^{-2\pi i y\cdot(a+g)} F(a+g), \quad a \in \mathbb{R}^q, \qquad (13.226)$$

is Λ–periodic, the right hand side of (13.224) can be rewritten in the form

$$\sum_{g'\in\Lambda(\mathcal{H})} \int_{\mathcal{H}_{g'}} \sum_{\substack{g+a \in \overline{\mathcal{G}} \\ g \in \Lambda}}{}' e^{-2\pi i y\cdot(a+g)} F(a+g)\, da \qquad (13.227)$$

$$= \sum_{g'\in\Lambda(\mathcal{H})} \int_{\mathcal{H}^0_{g'}} \sum_{\substack{a+g \in \overline{G} \\ g \in \Lambda}}{}' e^{-2\pi i y\cdot(a+g)} F(a+g)\, da$$

$$= \sum_{g'\in\Lambda(\mathcal{H})} \int_{\mathcal{G}} \chi_{\mathcal{H}^0_{g'}+\Lambda}(x)\, F(x)\, e^{-2\pi i y\cdot x}\, dx,$$

where $\mathcal{H}^0_{g'}, g' \in \Lambda(\mathcal{H})$, is the shift of $\mathcal{H}_{g'}$ to the fundamental cell, i.e.,

$$\mathcal{H}^0_{g'} = \mathcal{H}_{g'} - \{g'\},\ g' \in \Lambda(\mathcal{H}), \qquad (13.228)$$

is a subset of $\mathcal{F}_\Lambda$, and the characteristic function $x \mapsto \chi_{\mathcal{H}^0_{g'}+\Lambda}(x)$, $x \in \mathbb{R}^q$, is given by

$$\chi_{\mathcal{H}^0_{g'}+\Lambda}(x) = \begin{cases} 1 & , \quad x \in \bigcup_{g\in\Lambda} \left(H^0_{g'} + \{g\}\right) \\ 0 & , \quad \text{otherwise.} \end{cases} \tag{13.229}$$

Consequently we obtain the identity

$$\begin{aligned} &\int_{\mathcal{H}} \sum_{\substack{a+g\in\overline{\mathcal{G}} \\ g\in\Lambda}}{}' e^{-2\pi i y\cdot(a+g)} F(a+g)\, da \\ &= \sum_{g'\in\Lambda(\mathcal{H})} \int_{\mathcal{G}} \chi_{\mathcal{H}^0_{g'}+\Lambda}(x) F(x) e^{-2\pi i y\cdot x}\, dx, \\ &= \sum_{\substack{(\mathcal{F}_\Lambda+\{g'\})\cap\mathcal{H}\neq\emptyset \\ g'\in\Lambda}} \int_{\mathcal{G}\cap \bigcup_{g\in\Lambda}(((\mathcal{H}\cap(\mathcal{F}_\Lambda+\{g'\}))-\{g'\})+\{g\})} F(x)\, e^{-2\pi i y\cdot x}\, dx. \end{aligned} \tag{13.230}$$

Collecting our results we obtain the following Shannon-type sampling theorem in Gauss–Weierstrass summability:

Let $\mathcal{G}$ be a regular region in $\mathbb{R}^q$. Suppose that F is a member of the class $C^{(0)}(\overline{\mathcal{G}})$. Then, for every measurable bounded $\mathcal{H} \subset \mathbb{R}^q$, we have

(13.231)

$$\begin{aligned} &\sum_{\substack{(\mathcal{F}_\Lambda+\{g'\})\cap\mathcal{H}\neq\emptyset \\ g'\in\Lambda}} F^\wedge_{\mathcal{G}\cap \bigcup_{g\in\Lambda}(((\mathcal{H}\cap(\mathcal{F}_\Lambda+\{g'\}))-\{g'\})+\{g\})}(y) \\ &\quad - \sum_{\substack{(\mathcal{F}_\Lambda+\{g'\})\cap\mathcal{H}\neq\emptyset \\ g'\in\Lambda}} \int_{\mathcal{G}\cap \bigcup_{g\in\Lambda}(((\mathcal{H}\cap(\mathcal{F}_\Lambda+\{g'\}))-\{g'\})+\{g\})} F(x)\, e^{-2\pi i y\cdot x}\, dx \\ &\quad = \lim_{\substack{\tau\to 0 \\ \tau>0}} \sum_{h\in\Lambda^{-1}} e^{-\tau\pi h^2} \underbrace{\int_{\mathcal{G}} F(x) e^{-2\pi i h\cdot x}\, dV(x)}_{=F^\wedge_{\mathcal{G}}(h)} \underbrace{\frac{1}{\|\mathcal{F}_\Lambda\|} \int_{\mathcal{H}} e^{2\pi i a\cdot(h-y)}\, da}_{=K_{\mathcal{H}}(h-y)}. \end{aligned}$$

This formula includes all manifestations of over- and undersampling involving the "base bands" $\mathcal{G}, \mathcal{H} \subset \mathbb{R}^q$. Some geometric configurations of regular regions $\mathcal{G}, \mathcal{H}$ should be studied: We begin with $\overline{\mathcal{H}} \subset \overline{\mathcal{F}_\Lambda}$ and $\mathcal{G}$ arbitrary. In this case $\Lambda(\overline{\mathcal{H}}) = \{0\}$ and $\mathcal{H}^0_0$ is the interior of $\mathcal{F}_\Lambda$ such that

$$\begin{aligned} &\sum_{\substack{(\mathcal{F}_\Lambda+\{g'\})\cap\mathcal{H}\neq\emptyset \\ g'\in\Lambda}} F^\wedge_{\mathcal{G}\cap \bigcup_{g\in\Lambda}(((\mathcal{H}\cap(\mathcal{F}_\Lambda+\{g'\}))-\{g'\})+\{g\})}(y) \\ &= F^\wedge_{\mathcal{G}\cap\bigcup_{g\in\Lambda}(\mathcal{H}+\{g\})}(y) \\ &= \lim_{\substack{\tau\to 0 \\ \tau>0}} \sum_{h\in\Lambda^{-1}} e^{-\tau\pi h^2} \underbrace{\int_{\mathcal{G}} F(x) e^{-2\pi i h\cdot x}\, dx}_{=F^\wedge_{\mathcal{G}}(h)} \underbrace{\frac{1}{\|\mathcal{F}_\Lambda\|} \int_{\mathcal{H}} e^{2\pi i a\cdot(h-y)}\, da}_{=K_{\mathcal{H}}(h-y)}. \end{aligned} \tag{13.232}$$

For $\overline{\mathcal{H}} = \overline{\mathcal{F}_\Lambda}$ we therefore obtain

$$\begin{aligned}(13.233)\qquad &\sum_{\substack{(\mathcal{F}_\Lambda+\{g'\})\cap\mathcal{H}\neq\emptyset\\ g'\in\Lambda}} F^\wedge_{\mathcal{G}\cap\bigcup_{g\in\Lambda}(((\mathcal{H}\cap(\mathcal{F}_\Lambda+\{g'\}))-\{g'\})+\{g\})}(y)\\ &= F^\wedge_{\mathcal{G}}(y)\\ &= \lim_{\substack{\tau\to 0\\ \tau>0}} \sum_{h\in\Lambda^{-1}} e^{-\tau\pi h^2} \underbrace{\int_{\mathcal{G}} F(x)e^{-2\pi i h\cdot x}\,dx}_{=F^\wedge_{\mathcal{G}}(h)} \underbrace{\frac{1}{\|\mathcal{F}_\Lambda\|}\int_{\mathcal{F}_\Lambda} e^{2\pi i a\cdot(h-y)}\,da}_{=K_{\mathcal{F}_\Lambda}(h-y)}.\end{aligned}$$

We continue with $\overline{\mathcal{G}} \subset \overline{\mathcal{F}_\Lambda}$ and $\mathcal{H}$ arbitrary. This yields the identity

$$\begin{aligned}(13.234)\qquad &\sum_{\substack{(\mathcal{F}_\Lambda+\{g'\})\cap\mathcal{H}\neq\emptyset\\ g'\in\Lambda}} F^\wedge_{\mathcal{G}\cap\bigcup_{g\in\Lambda}(((\mathcal{H}\cap(\mathcal{F}_\Lambda+\{g'\}))-\{g'\})+\{g\})}(y)\\ &= \sum_{\substack{(\mathcal{F}_\Lambda+\{g'\}\cap\mathcal{H}\neq\emptyset)\\ g'\in\Lambda}} F^\wedge_{\mathcal{G}\cap((\mathcal{H}\cap(\mathcal{F}_\Lambda+\{g'\}))-\{g'\})}(y)\\ &= \lim_{\substack{\tau\to 0\\ \tau>0}} \sum_{h\in\Lambda^{-1}} e^{-\tau\pi h^2} \underbrace{\int_{\mathcal{G}} F(x)e^{-2\pi i h\cdot x}\,dx)}_{=F^\wedge_{\mathcal{G}}(h)} \underbrace{\frac{1}{\|\mathcal{F}_\Lambda\|}\int_{\mathcal{H}} e^{2\pi i a\cdot(h-y)}\,da}_{=K_{\mathcal{H}}(h-y)}.\end{aligned}$$

For $\overline{\mathcal{G}}, \overline{\mathcal{H}} \subset \overline{\mathcal{F}_\Lambda}$ we have

$$\begin{aligned}(13.235)\qquad &\sum_{\substack{(\mathcal{F}_\Lambda+\{g'\})\cap\mathcal{H}\neq\emptyset\\ g'\in\Lambda}} F^\wedge_{\mathcal{G}\cap\bigcup_{g\in\Lambda}(((\mathcal{H}\cap(\mathcal{F}_\Lambda+\{g'\}))-\{g'\})+\{g\})}(y)\\ &= F^\wedge_{\mathcal{G}\cap\mathcal{H}}(y)\\ &= \lim_{\substack{\tau\to 0\\ \tau>0}} \sum_{h\in\Lambda^{-1}} e^{-\tau\pi h^2} F^\wedge_{\mathcal{G}}(h)\;\; K_{\mathcal{H}}(h-y).\end{aligned}$$

For $\overline{\mathcal{G}} \subset \overline{\mathcal{H}} \subset \overline{\mathcal{F}_\Lambda}$ we have

$$(13.236)\qquad F^\wedge_{\mathcal{G}}(y) = \lim_{\substack{\tau\to 0\\ \tau>0}} \sum_{h\in\Lambda^{-1}} e^{-\tau\pi h^2} F^\wedge_{\mathcal{G}}(h)\;\; K_{\mathcal{H}}(h-y),$$

whereas, for $\overline{\mathcal{H}} \subset \overline{\mathcal{G}} \subset \overline{\mathcal{F}}_\Lambda$,

$$(13.237)\qquad F_{\mathcal{H}}(y) = \lim_{\substack{\tau\to 0\\ \tau>0}} \sum_{h\in\Lambda^{-1}} e^{-\tau\pi h^2} F^\wedge_{\mathcal{G}}(h)\;\; K_{\mathcal{H}}(h-y).$$

In particular, for $\overline{\mathcal{G}} = \overline{\mathcal{H}} \subset \overline{\mathcal{F}_\Lambda}$, we are able to formulate the following identity

$$(13.238)\qquad F^\wedge_{\mathcal{G}}(y) = \frac{1}{\|\mathcal{F}_\Lambda\|} \lim_{\substack{\tau\to 0\\ \tau>0}} \sum_{h\in\Lambda^{-1}} e^{-\tau\pi h^2} F^\wedge_{\mathcal{G}}(h)\; K_{\mathcal{G}}(h-y).$$

Obviously, by choosing a small sampling density such that $\overline{\mathcal{F}_\Lambda}$ covers the compact support $\overline{\mathcal{G}}$ of the original signal $F^\wedge_{\mathcal{G}}$, the number of samples $F^\wedge_{\mathcal{G}}(h), h \in \Lambda^{-1}$, for reconstruction is high, and vice versa. In practice, we are therefore required to find a compromise between sampling density and total number of samples. This can be achieved by a choice of Λ such that $\overline{\mathcal{F}_\Lambda}$ covers $\overline{\mathcal{G}}$ tightly.

For $\overline{\mathcal{G}} \subset \overline{\mathcal{F}_\Lambda} = \overline{\mathcal{H}}$ and the lattice Λ generated by the vectors $g_1, \ldots, g_q \in \mathbb{R}^q$ we have in explicitly written form, i.e., an exponentially accelerated cardinal series of E. Parzen's result (cf. [**681**])

(13.239)
$$F_{\mathcal{G}}^{\wedge}(y) = \lim_{\substack{\tau\to 0\\ \tau>0}} \sum_{h\in\Lambda^{-1}} e^{-\tau\pi h^2} F_{\mathcal{G}}^{\wedge}(h) \frac{J_1(1;\pi|g_1\cdot(h-y)|)}{\pi|g_1\cdot(h-y)|} \cdot\ldots\cdot \frac{J_1(1;\pi|g_q\cdot(h-y)|)}{\pi|g_q\cdot(h-y)|}.$$

In other words, for sufficiently small $\tau > 0$, $F_{\mathcal{G}}^{\wedge}$ can be expressed by the series on the right side of (13.239) in exponential convergence, i.e.,

(13.240)
$$(F^{(\tau)})_{\mathcal{G}}^{\wedge}(y) \simeq \sum_{h\in\Lambda^{-1}} e^{-\tau\pi h^2} F_{\mathcal{G}}^{\wedge}(h) \frac{J_1(1;\pi|g_1\cdot(h-y)|)}{\pi|g_1\cdot(h-y)|} \cdot\ldots\cdot \frac{J_1(1;\pi|g_q\cdot(h-y)|}{\pi|g_q\cdot(h-y)|}.$$

Replacing the lattice Λ by its dilated lattice $\sigma\Lambda$, $\sigma > 1$, we find that

$$\begin{aligned} F_{\mathcal{G}}^{\wedge}(y) = \lim_{\substack{\tau\to 0\\ \tau>0}} \sigma^q \sum_{h\in\Lambda^{-1}} e^{-\tau\pi\left(\frac{h}{\sigma}\right)^2} F_{\mathcal{G}}^{\wedge}\left(\frac{h}{\sigma}\right) \frac{J_1(1;\pi|\sigma g_1\cdot(\frac{h}{\sigma}-y)|)}{\pi|\sigma g_1\cdot(\frac{h}{\sigma}-y)|} \\ \cdot\ldots\cdot \frac{J_1(1;\pi|\sigma g_q\cdot(\frac{h}{\sigma}-y)|)}{\pi|\sigma g_q\cdot(\frac{h}{\sigma}-y)|}. \end{aligned} \tag{13.241}$$

The standard form of sampling in Gauss-Weierstrass summability is provided by taking $\overline{\mathcal{G}} = \overline{\mathcal{H}} = \overline{\mathcal{F}_\Lambda}$, i.e.,

$$F_{\mathcal{F}_\Lambda}^{\wedge}(y) = \lim_{\substack{\tau\to 0\\ \tau>0}} \sum_{h\in\Lambda^{-1}} e^{-\tau\pi h^2} F_{\mathcal{F}_\Lambda}^{\wedge}(h)\ K_{\mathcal{F}_\Lambda}(h-y),\ \ y\in\mathbb{R}^q. \tag{13.242}$$

Sampling using finite unions of fundamental cells. The algebraic structure of a lattice closely relates Shannon sampling to lattice point theory. New types of lattice point identities are detectable which, in turn, give an explicit insight into over- and undersampling processes, even for general geometries.. The focus of our sampling activities was the Shannon-type sampling theorem (13.231). Even more, this theorem allows us to exhibit a special sampling theorem for regular regions covered by finite unions of fundamental cells: Let $\mathcal{G}$ be an arbitrary regular region. Suppose that $\mathcal{H}$ is a finite union of periodic cells such that $\mathcal{G} \subset \mathcal{H}$. Under these assumptions the Shannon-type sampling theorem can be rewritten as follows:

$$\left(\sum_{\substack{g\in\mathcal{H}\\ g\in\Lambda}} 1\right) F_{\mathcal{G}}^{\wedge}(y) = \lim_{\substack{\tau\to 0\\ \tau>0}} \sum_{h\in\Lambda^{-1}} e^{-\tau\pi h^2} F_{\mathcal{G}}^{\wedge}(h)\ K_{\mathcal{H}}(h-y), \tag{13.243}$$

where

$$K_{\mathcal{H}}(h-y) = \sum_{\substack{g'\in\mathcal{H}\\ g'\in\Lambda}} K_{\mathcal{F}_\Lambda+\{g'\}}(h-y) = \left(\sum_{\substack{g'\in\mathcal{H}\\ g'\in\Lambda}} e^{2\pi i g'\cdot y}\right) K_{\mathcal{F}_\Lambda}(h-y). \tag{13.244}$$

Thus, for all $y \in \mathbb{R}^q$, we obtain

$$F_{\mathcal{G}}^{\wedge}(y) = \frac{\left(\sum_{\substack{g' \in \mathcal{H} \\ g' \in \Lambda}} e^{2\pi i g' \cdot y}\right)}{\left(\sum_{\substack{g \in \mathcal{H} \\ g \in \Lambda}} 1\right)} \sum_{h \in \Lambda^{-1}} F_{\mathcal{G}}^{\wedge}(h)\, K_{\mathcal{F}_\Lambda}(h-y). \tag{13.245}$$

As an interesting example we discuss the spherical case, i.e., $\mathcal{H}$ is a union of periodic cells such that $\mathbb{B}_R^q \subset \mathcal{H}$, $R > 0$. Then, for all $y \in \mathbb{R}^q$, we obtain

$$F_{\mathbb{B}_R^q}^{\wedge}(y) = \frac{\left(\sum_{\substack{g' \in \mathcal{H} \\ g' \in \Lambda}} e^{2\pi i g' \cdot y}\right)}{\left(\sum_{\substack{g \in \mathcal{H} \\ g \in \Lambda}} 1\right)} \sum_{h \in \Lambda^{-1}} F_{\mathbb{B}_R^q}^{\wedge}(h)\, K_{\mathcal{F}_\Lambda}(h-y), \tag{13.246}$$

whereas the choice of a ball $\mathbb{B}_R^q$ in (13.231) provides the identity

$$\begin{aligned} &\sum_{\substack{\mathcal{F}_\Lambda + \{g'\} \cap \overline{\mathbb{B}_R^q} \neq \emptyset \\ g' \in \Lambda}} F^{\wedge}_{\mathcal{G} \cap \bigcup_{g \in \Lambda} (((\overline{\mathbb{B}_R^q} \cap \mathcal{F}_\Lambda + \{g'\}) - \{g'\}) + \{g\})}(y) \\ &\quad = \frac{\|\mathbb{S}^{q-1}\| R^q}{\|\mathcal{F}_\Lambda\|} \lim_{\substack{\tau \to 0 \\ \tau > 0}} \sum_{h \in \Lambda^{-1}} e^{-\tau \pi h^2} F_{\mathcal{G}}^{\wedge}(h) \frac{J_1(q; 2\pi |h-y| R)}{2\pi |h-y| R}, \end{aligned} \tag{13.247}$$

for an arbitrary regular region $\mathcal{G} \subset \mathbb{R}^q$. If $\overline{\mathbb{B}_R^q}$ is a subset of $\overline{\mathcal{F}_\Lambda}$, we especially have for an arbitrary regular $\mathcal{G} \subset \mathbb{R}^q$

$$F^{\wedge}_{\mathcal{G} \cap \bigcup_{g \in \Lambda} (\mathbb{B}_R^q + \{g\})}(y) = \frac{\|\mathbb{S}^{q-1}\| R^q}{\|\mathcal{F}_\Lambda\|} \lim_{\substack{\tau \to 0 \\ \tau > 0}} \sum_{h \in \Lambda^{-1}} e^{-\tau \pi h^2} F_{\mathcal{G}}^{\wedge}(h) \frac{J_1(q; 2\pi |h-y| R)}{2\pi |h-y| R}, \tag{13.248}$$

such that the additional condition $\overline{\mathcal{G}} \subset \overline{\mathcal{F}_\Lambda}$ leads back to the case (already discussed in our two-dimensional context, but now for general $\mathcal{G} \subset \mathbb{R}^q$ and in Gauss-Weierstrass nomenclature)

$$F^{\wedge}_{\mathcal{G} \cap \overline{\mathbb{B}_R^q}}(y) = \frac{\|\mathbb{S}^{q-1}\| R^q}{\|\mathcal{F}_\Lambda\|} \lim_{\substack{\tau \to 0 \\ \tau > 0}} \sum_{h \in \Lambda^{-1}} e^{-\tau \pi h^2} F_{\mathcal{G}}^{\wedge}(h) \frac{J_1(q; 2\pi |h-y| R)}{2\pi |h-y| R}. \tag{13.249}$$

Finally, letting $\mathcal{G} = \mathbb{B}_R^q$ we have

$$F^{\wedge}_{\mathbb{B}_R^q}(y) = \frac{\|\mathbb{S}^{q-1}\| R^q}{\|\mathcal{F}_\Lambda\|} \lim_{\substack{\tau \to 0 \\ \tau > 0}} \sum_{h \in \Lambda^{-1}} e^{-\tau \pi h^2} F_{\mathbb{B}_R^q}^{\wedge}(h) \frac{J_1(q; 2\pi |h-y| R)}{2\pi |h-y| R}. \tag{13.250}$$

so that, in connection with (13.246), we arrive at the identity

$$
\begin{aligned}
(13.251) \qquad & F^{\wedge}_{\mathbb{B}^q_R}(y) \\
&= \frac{\left(\sum\limits_{\substack{g'\in\mathcal{H}\\ g'\in\Lambda}} e^{2\pi i g'\cdot y}\right)}{\left(\sum\limits_{\substack{g\in\mathcal{H}\\ g\in\Lambda}} 1\right)} \sum_{h\in\Lambda^{-1}} F^{\wedge}_{\mathbb{B}^q_R}(h)\, K_{\mathcal{F}_\Lambda}(h-y) \\
&= \frac{\|\mathbb{S}^{q-1}\| R^q}{\|\mathcal{F}_\Lambda\|} \lim_{\substack{\tau\to 0\\ \tau>0}} \sum_{h\in\Lambda^{-1}} e^{-\tau\pi h^2} F^{\wedge}_{\mathbb{B}^q_R}(h)\, \frac{J_1(q; 2\pi|h-y|R)}{2\pi|h-y|R}.
\end{aligned}
$$

Down- and upsampling. Given an arbitrary lattice $\Lambda \subset \mathbb{R}^q$ and an arbitrary regular region $\mathcal{H} \subset \mathbb{R}^q$, we are able to find a constant $\sigma \in \mathbb{R}$ such that $\overline{\mathcal{H}} \subset \overline{\mathcal{F}_{\sigma\Lambda}}$ as tightly as possible. Under these circumstances we obtain

$$
\begin{aligned}
(13.252) \qquad & F^{\wedge}_{\mathcal{G}\cap \bigcup\limits_{g\in\sigma\Lambda}(\mathcal{H}+\{g\})}(y) \\
&= \lim_{\substack{\tau\to 0\\ \tau>0}} \sum_{h\in(\sigma\Lambda)^{-1}} e^{-\tau\pi h^2} F^{\wedge}_{\mathcal{G}}(h) K_{\mathcal{H}}(h-y) \\
&= \lim_{\substack{\tau\to 0\\ \tau>0}} \frac{1}{\sigma^q} \sum_{h\in\Lambda^{-1}} e^{-\tau\pi(\frac{h}{\sigma})^2} F^{\wedge}_{\mathcal{G}}\left(\frac{h}{\sigma}\right) K_{\mathcal{H}}\left(\frac{h}{\sigma}-y\right).
\end{aligned}
$$

In comparison to sampling with respect to the lattice Λ the identity (13.252) related to $\sigma\Lambda$ provides *upsampling* for $\sigma > 1$ or *downsampling* for $\sigma < 1$, respectively. Of particular significance are choices of σ such that Λ is a sublattice of $\sigma\Lambda, \sigma > 1$, or $\sigma\Lambda, \sigma < 1$, is a sublattice of Λ.

Parseval identity. Suppose that $\mathcal{G}, \mathcal{H}$ are regular regions. Assume that $F = 1$ on $\overline{\mathcal{G}}$. Then, for all $y \in \mathbb{R}^q$, we have

$$
\begin{aligned}
(13.253) \qquad & \int_{\mathcal{H}} \sum_{\substack{a+g\in\overline{\mathcal{G}}\\ g\in\Lambda}}{}' e^{-2\pi i y\cdot(a+g)}\, da \\
&= \sum_{\substack{(\mathcal{F}_\Lambda+\{g'\})\cap\mathcal{H}\neq\emptyset\\ g\in\Lambda}} \int_{\mathcal{G}\cap\bigcup\limits_{g\in\Lambda}(((\mathcal{H}\cap(\mathcal{F}_\Lambda+\{\Gamma'\})-\{g'\})+\{g\}))} e^{-2\pi i y\cdot x}\, dx \\
&= \frac{1}{\|\mathcal{F}_\Lambda\|} \lim_{\substack{\tau\to 0\\ \tau>0}} \sum_{h\in\Lambda^{-1}} e^{-\tau\pi h^2} \int_{\mathcal{G}} e^{-2\pi i h\cdot x}\, dx \int_{\mathcal{H}} e^{2\pi i x\cdot(h-y)}\, dx,
\end{aligned}
$$

such that, especially for $\mathcal{H} = \mathcal{G}$,

$$\begin{aligned}(13.254)\qquad &\int_{\mathcal{G}} \sum_{\substack{a+g\in\overline{\mathcal{G}}\\ g\in\Lambda}}{}' e^{-2\pi i y\cdot(a+g)}\, da \\ &= \sum_{\substack{(\mathcal{F}_\Lambda+\{g'\})\cap\mathcal{G}\neq\emptyset\\ g\in\Lambda}} \int_{\mathcal{G}\cap \bigcup\limits_{g\in\Lambda}(((\mathcal{G}\cap(\mathcal{F}_\Lambda+\{\Gamma'\})-\{g'\})+\{g\}))} e^{-2\pi i y\cdot x}\, dx \\ &= \frac{1}{\|\mathcal{F}_\Lambda\|} \lim_{\substack{\tau\to 0\\ \tau>0}} \sum_{h\in\Lambda^{-1}} e^{-\tau\pi h^2} \int_{\mathcal{G}} e^{-2\pi i h\cdot x}\, dx \int_{\mathcal{G}} e^{2\pi i x\cdot(h-y)}\, dx.\end{aligned}$$

In particular, for $y = 0$, we obtain

$$\begin{aligned}(13.255)\qquad \int_{\mathcal{H}} \sum_{\substack{a+g\in\overline{\mathcal{G}}\\ g\in\Lambda}}{}' 1\, da &= \sum_{\substack{(\mathcal{F}_\Lambda+\{g'\})\cap\mathcal{H}\neq\emptyset\\ g\in\Lambda}} \int_{\mathcal{G}\cap \bigcup\limits_{g\in\Lambda}(((\mathcal{H}\cap(\mathcal{F}_\Lambda+\{g'\})-\{g'\})+\{g\}))} 1\, dx \\ &= \frac{1}{\|\mathcal{F}_\Lambda\|} \lim_{\substack{\tau\to 0\\ \tau>0}} \sum_{h\in\Lambda^{-1}} e^{-\tau\pi h^2} \int_{\mathcal{G}} e^{-2\pi i h\cdot x}\, dx \int_{\mathcal{H}} e^{2\pi i x\cdot h}\, dx\end{aligned}$$

and, especially for $\mathcal{H} = \mathcal{G}$,

$$\begin{aligned}(13.256)\qquad &\int_{\mathcal{G}} \sum_{\substack{a+g\in\overline{\mathcal{G}}\\ g\in\Lambda}}{}' 1\, da \\ &= \sum_{\substack{(\mathcal{F}_\Lambda+\{g'\})\cap\mathcal{G}\neq\emptyset\\ g'\in\Lambda}} \int_{\mathcal{G}\cap \bigcup\limits_{g\in\Lambda}(((\mathcal{G}\cap(\mathcal{F}_\Lambda+\{g'\})-\{g'\})+\{g\}))} 1\, dx \\ &= \lim_{\substack{\tau\to 0\\ \tau>0}} \frac{1}{\|\mathcal{F}_\Lambda\|} \sum_{h\in\Lambda^{-1}} e^{-\tau\pi h^2} \left| \int_{\mathcal{G}} e^{-2\pi i h\cdot x}\, dx \right|^2.\end{aligned}$$

The identities (13.255) and (13.256) are the canonical preparations to turn over to the Parseval identity involving regions $\mathcal{G}, \mathcal{H} \subset \mathbb{R}^q$ within the Gauss-Weierstrass framework. The formula (13.256) was already used in the two-dimensional Euclidean space $\mathbb{R}^2$. In fact, C. Müller [**604**] noticed that the Parseval-type identity

$$(13.257)\qquad \int_{\mathcal{G}} \sum_{\substack{a+g\in\overline{\mathcal{G}}\\ g\in\mathbb{Z}^2}}{}' 1\ da = \sum_{h\in\mathbb{Z}^2} \left| \int_{\mathcal{G}} e^{-2\pi i h\cdot x}\, dx \right|^2$$

holds (without summability conditions) for all symmetric (with respect to the origin) and convex regions $\mathcal{G} \subset \mathbb{R}^2$. An easy consequence of (13.257) is the inequality

$$(13.258)\qquad \int_{\mathcal{G}} \sum_{\substack{a+g\in\overline{\mathcal{G}}\\ g\in\mathbb{Z}^2}}{}' 1\ da \geq \|\mathcal{G}\|^2 = \left(\int_{\mathcal{G}} dx \right)^2,$$

such that under the assumption $\|\mathcal{G}\| > 4$, an essential step towards *Minkowski's Theorem* [**587**] can be made by considering simultaneously the lattice $2\mathbb{Z}^2$ (see, e.g., E. Hlawka [**436, 437**], E. Hlawka, J. Schoißengeier, R. Taschner [**439**]). Indeed, the inequality

$$\int_{\mathcal{G}} \sum_{\substack{a+g\in\overline{\mathcal{G}} \\ g\in 2\mathbb{Z}^2}}{}' \, 1 \; da \;\geq\; \frac{\|\mathcal{G}\|}{4}\,\|\mathcal{G}\| \;>\; \|\mathcal{G}\| \tag{13.259}$$

is the key to guarantee that a symmetric (with respect to the origin) and convex region $\mathcal{G} \subset \mathbb{R}^2$ with $\|\mathcal{G}\| > 4$ contains lattice points of $\mathbb{Z}^2$ different from the origin. It should be noted that C. Müller's two-dimensional approach (cf. [**601**–**603**]) will be generalized here in various ways (see also [**268**]). For example, the Parseval identity of type (13.257) can be extended to arbitrary lattices $\Lambda \subset \mathbb{R}^q$ and regular regions $\mathcal{G} \subset \mathbb{R}^q$. Moreover, arbitrary continuous weight functions can be included instead of constant weights. Furthermore, Gauss-Weierstrass summability can be avoided.

For this purpose we formulate the extended Parseval identity in Gauss-Weierstrass summability:

Let Λ be a lattice in $\mathbb{R}^q$. Let $\mathcal{G}, \mathcal{H} \subset \mathbb{R}^q$ be arbitrary regular regions. Suppose that F is of class $C^{(0)}(\overline{\mathcal{G}})$ and G is of class $C^{(0)}(\overline{\mathcal{H}})$, respectively. Then we have

$$\begin{aligned} &\lim_{\substack{\tau\to 0\\ \tau>0}} \frac{1}{\|\mathcal{F}_\Lambda\|} \sum_{h\in\Lambda^{-1}} e^{-\tau\pi h^2} \int_{\mathcal{G}} F(x) e^{-2\pi i x\cdot h} dx \overline{\int_{\mathcal{H}} G(x) e^{-2\pi i x\cdot h}\, dx} \\ &= \lim_{\substack{\tau\to 0\\ \tau>0}} \frac{1}{\|\mathcal{F}_\Lambda\|} \sum_{h\in\Lambda^{-1}} e^{-\tau\pi h^2} F_{\mathcal{G}}^{\wedge}(h)\, \overline{G_{\mathcal{H}}^{\wedge}(h)} \\ &= \int_{\mathcal{H}} \sum_{\substack{a+g\in\overline{\mathcal{G}}\\ g\in\Lambda}}{}' F(a+g)\, \overline{G(a)}\, da. \end{aligned} \tag{13.260}$$

Now, if $\overline{\mathcal{G}}, \overline{\mathcal{H}}$ are subsets of $\overline{\mathcal{F}_\Lambda}$, then we obtain from the extended Parseval identity (13.260) the formula

$$\lim_{\substack{\tau\to 0\\ \tau>0}} \frac{1}{\|\mathcal{F}_\Lambda\|} \sum_{g\in\Lambda^{-1}} e^{\tau\pi h^2} F_{\mathcal{G}}^{\wedge}(h)\, \overline{C_{\mathcal{H}}^{\wedge}(h)} = \int_{\mathcal{G}\cap\mathcal{H}} F(a)\, \overline{G(a)}\, da. \tag{13.261}$$

If $\overline{\mathcal{G}} \subset \overline{\mathcal{H}} \subset \overline{\mathcal{F}_\Lambda}$, then

$$\lim_{\substack{\tau\to 0\\ \tau>0}} \frac{1}{\|\mathcal{F}_\Lambda\|} \sum_{g\in\Lambda^{-1}} e^{-\tau\pi h^2} F_{\mathcal{G}}^{\wedge}(h)\, \overline{G_{\mathcal{H}}^{\wedge}(h)} = \int_{\mathcal{G}} F(a)\, \overline{G(a)}\, da. \tag{13.262}$$

Thus, the special configuration $\overline{\mathcal{G}} = \overline{\mathcal{H}} = \overline{\mathcal{F}_\Lambda}$ leads back to the conventional Parseval identity of classical Fourier theory (however, in Gauss-Weierstrass summability).

With $\mathcal{H} = \mathcal{G}$ we arrive at the following corollary in Gauss-Weierstrass summability.

Let Λ be a lattice in $\mathbb{R}^q$. Let $\mathcal{G} \subset \mathbb{R}^q$ be a regular region. Suppose that F, G are of class $C^{(0)}(\overline{\mathcal{G}})$. Then the following variant of the Parseval identity holds true:

$$\lim_{\substack{\tau\to 0\\ \tau>0}} \frac{1}{\|\mathcal{F}_\Lambda\|} \sum_{h\in\Lambda^{-1}} e^{-\tau\pi h^2} F_{\mathcal{G}}^{\wedge}(h)\, \overline{G_{\mathcal{G}}^{\wedge}(h)} = \int_{\mathcal{G}} \sum_{\substack{a+g\in\overline{\mathcal{G}}\\ g\in\Lambda}}{}' F(a+g)\, \overline{G(a)}\, da. \tag{13.263}$$

In particular, we have

$$\lim_{\substack{\tau\to 0\\ \tau>0}} \frac{1}{\|\mathcal{F}_\Lambda\|} \sum_{h\in\Lambda^{-1}} e^{-\tau\pi h^2} \left|F_{\mathcal{G}}^{\wedge}(h)\right|^2 = \int_{\mathcal{G}} \sum_{\substack{a+g\in\overline{\mathcal{G}}\\ g\in\Lambda}}{}' F(a+g)\overline{F(a)}\, da. \tag{13.264}$$

13.6. Multivariate Shannon-type nonsummability sampling

Following arguments due to W. Freeden, M.Z. Nashed [**298**], the Parseval identity allows us to conclude that limit and sum on the right of (13.260) may be interchanged. In other words, we obtain the Parseval-type identity for regular regions in standard way, i.e., in a version of nonsummability.

Under the above assumptions we have

$$\frac{1}{\|\mathcal{F}_\Lambda\|} \sum_{h\in\Lambda^{-1}} F_{\mathcal{G}}^{\wedge}(h)\, \overline{G_{\mathcal{H}}^{\wedge}(h)} = \int_{\mathcal{H}} \sum_{\substack{a+g\in\overline{\mathcal{G}}\\ g\in\Lambda}}{}' F(a+g)\, \overline{G(a)}\, da. \tag{13.265}$$

In particular, if $\overline{\mathcal{G}} = \overline{\mathcal{H}}$ and $F = G$, then

$$\frac{1}{\|\mathcal{F}_\Lambda\|} \sum_{h\in\Lambda^{-1}} \left|F_{\mathcal{G}}^{\wedge}(h)\right|^2 = \int_{\mathcal{G}} \sum_{\substack{a+g\in\overline{\mathcal{G}}\\ g\in\Lambda}}{}' F(a+g)\, \overline{F(a)}\, da). \tag{13.266}$$

Under the choice $\overline{\mathcal{G}} = \overline{\mathcal{H}} \subset \overline{\mathcal{F}_\Lambda}$ and $F = G$ we find that

$$\frac{1}{\|\mathcal{F}_\Lambda\|} \sum_{h\in\Lambda^{-1}} \left|F_{\mathcal{G}}^{\wedge}(h)\right|^2 = \int_{\mathcal{G}} |F(a)|^2\, da. \tag{13.267}$$

Fourier transformed values based sampling (in nonsummability context). It is remarkable that the Parseval-type identity also enables us to formulate the Shannon-type sampling theorem in ordinary sense avoiding Gauss-Weierstrass nomenclature (by letting $G(x) = 1$ for $x \in \mathcal{H}$ and replacing $F(x)$ by $F(x)e^{-2\pi i y\cdot x}$ for $x \in \mathcal{G}$).

Let $\mathcal{G}$ be a regular region in $\mathbb{R}^q$. Suppose that F is a member of the class $C^{(0)}(\overline{\mathcal{G}})$. Then, for every measurable bounded subset $\mathcal{H}$ of $\mathbb{R}^q$, we have

$$\begin{aligned} &\int_{\mathcal{H}} \sum_{\substack{a+g\in\overline{G}\\ g\in\Lambda}}{}' e^{-2\pi i y\cdot(a+g)} F(a+g)\, da \\ &= \sum_{\substack{(\mathcal{F}_\Lambda+\{g'\})\cap\mathcal{H}\neq\emptyset\\ g'\in\Lambda}} F^{\wedge}_{\mathcal{G}\cap\bigcup_{g\in\Lambda}(((\mathcal{H}\cap(\mathcal{F}_\Lambda+\{g'\}))-\{g'\})+\{g\})}(y) \\ &= \sum_{h\in\Lambda^{-1}} F_{\mathcal{G}}^{\wedge}(h)\, K_{\mathcal{H}}(h-y) \end{aligned}$$

for all $y \in \mathbb{R}^q$. Especially, for $\overline{\mathcal{G}} \subset \overline{\mathcal{H}} \subset \overline{\mathcal{F}_\Lambda}$, we have

$$F_{\mathcal{G}}^{\wedge}(y) = \sum_{h\in\Lambda^{-1}} F_{\mathcal{G}}^{\wedge}(h)\;\; K_{\mathcal{H}}(h-y),\;\; y\in\mathbb{R}^q. \tag{13.268}$$

Summarizing our results on metaharmonically reflected sampling we come to the following conclusion: The famous Shannon sampling theorem gives an answer to the question of how a one-dimensional time-dependent bandlimited signal can be reconstructed from discrete values in lattice points. Multivariate Hardy-type lattice point identities over regular regions can be formulated from which space-dependent Shannon-type sampling theorems are obtainable by straightforward integration over certain regular regions. An answer is given to the problem of how a signal bandlimited to a regular region in q-dimensional Euclidean space allows a reconstruction from discrete values in the lattice points of a (general) q-dimensional lattice. Weighted Hardy-type lattice point formulas are derived to allow explicit characterizations of over- and undersampling, thereby specifying not only the occurrence, but also the type of aliasing in a thorough mathematical description. An essential tool for the proof of Hardy-type identities in lattice point theory is the extension of the Euler summation formula to second order Helmholtz-type operators involving associated Green functions with respect to the "boundary condition" of periodicity. In order to circumvent convergence difficulties and/or slow convergence in multivariate Hardy-type lattice point summation, some summability methods are necessary, such that Gauß-Weierstraß averaging come into play. As a consequence, multivariate Shannon-type lattice sampling becomes available in a proposed summability context to accelerate the summation of the associated cardinal-type series. Finally (see Section 15.2), some aspects of constructive approximation in a resulting Paley-Wiener framework can be indicated, such as the recovery of a finite set of lost samples, the reproducing Hilbert space context of spline interpolation, and Shannon-type spline sampling.

13.7. Gauss-Weierstrass wavelet sampling

In what follows we are concerned with multiscale Gauss-Weierstrass sampling methods, which aim at two methodologies in sampling processes, namely accelerating the convergence of the cardinal series and decorrelating characteristic signature phenomena.

Gauss-Weierstrass summability wavelet sampling. We choose a positive, monotonically decreasing sequence $\{\tau_j\}_{j\in\mathbb{N}_0}$, so that $\lim_{j\to\infty}\tau_j = 0$. As a particularly important example of a sequence we may take the *dyadic sequence* given by $\tau_j = C\,2^{-j}$, $j\in\mathbb{N}_0$, $C = \text{const}$. In this case, it is easy to see that $2\tau_{j+1} = \tau_j$, $j\in\mathbb{N}_0$, which describes the relation between two consecutive elements of the dyadic sequence. We consider the *sampling scaling functions* $\{S^{\mathcal{H}}_{\tau_j}(F^{\wedge}_{\mathcal{G}})\}_{j\in\mathbb{N}_0}$ given by

$$S^{\mathcal{H}}_{\tau_j}(F^{\wedge}_{\mathcal{G}})(y) = \sum_{h\in\Lambda^{-1}} e^{-\tau_j\pi h^2} F^{\wedge}_{\mathcal{G}}(h)\;\; K_{\mathcal{H}}(h-y), \tag{13.269}$$

Clearly, we have

$$\lim_{j\to\infty} S^{\mathcal{H}}_{\tau_j}(F^{\wedge}_{\mathcal{G}})(y) \; = S^{\mathcal{H}}(F^{\wedge}_{\mathcal{G}})(y), \tag{13.270}$$

where

$$S^{\mathcal{H}}(F_{\mathcal{G}}^{\wedge})(y) = \sum_{\substack{(\mathcal{F}_\Lambda+\{g'\})\cap\mathcal{H}\neq\emptyset \\ g'\in\Lambda}} \int_{\mathcal{G}\cap\bigcup_{g\in\Lambda}(((\mathcal{H}\cap(\mathcal{F}_\Lambda+\{g'\}))-\{g'\})+\{g\})} F(x)\, e^{-2\pi i y\cdot x}\, dx$$

$$= \lim_{j\to\infty} \sum_{h\in\Lambda^{-1}} e^{-\tau_j \pi h^2} \underbrace{\int_{\mathcal{G}} F(x) e^{-2\pi i h\cdot x}\, dV(x)}_{=F_{\mathcal{G}}^{\wedge}(h)} \underbrace{\frac{1}{\|\mathcal{F}_\Lambda\|}\int_{\mathcal{H}} e^{2\pi i a\cdot(h-y)}\, da}_{=K_{\mathcal{H}}(h-y)}.$$

The *Gauss-Weierstrass summability sampling wavelets* $\{W_{\tau_j}^{\mathcal{H}}(F_{\mathcal{G}}^{\wedge})\}_{j\in\mathbb{N}_0}$ are to be understood as the difference

$$W_{\tau_j}^{\mathcal{H}}(F_{\mathcal{G}}^{\wedge}) = S_{\tau_{j+1}}^{\mathcal{H}}(F_{\mathcal{G}}^{\wedge}) - S_{\tau_j}^{\mathcal{H}}(F_{\mathcal{G}}^{\wedge}). \tag{13.271}$$

From (13.271) it follows that multiscale sampling is designed by forming difference wavelets in such a way that a lowpass filtering $S_{\tau_j}^{\mathcal{H}}(F_{\mathcal{G}}^{\wedge})$ to any scale j plus the bandpass filtering $W_{\tau_j}^{\mathcal{H}}(F_{\mathcal{G}}^{\wedge})$ to the same scale j is equal to the lowpass filtering $S_{\tau_{j+1}}^{\mathcal{H}}(F_{\mathcal{G}}^{\wedge})$ to the scale $j+1$. Conversely, the difference between the lowpass filtering to the scale $j+1$ and the lowpass filtering to the scale j is the bandpass filtering to the scale j.

Now, let $J > J_0$, $J, J_0 \in \mathbb{N}_0$, be any scale. Then it is easily seen that

$$S_{\tau_J}^{\mathcal{H}}(F_{\mathcal{G}}^{\wedge})(y) = S_{\tau_{J_0}}^{\mathcal{H}}(F_{\mathcal{G}}^{\wedge})(y) + \sum_{j=0}^{J-1} W_{\tau_j}^{\mathcal{H}}(F_{\mathcal{G}}^{\wedge})(y), \quad y\in\mathbb{R}^q. \tag{13.272}$$

The formula (13.272) explains the interaction of scaling and wavelet functions in a multiscale sampling reconstruction of $S^{\mathcal{H}}(F_{\mathcal{G}}^{\wedge})(y), y \in \mathbb{R}^q$. Obviously, the difference of two filterings ("smoothings") results in a decorrelated band signature information, from which it can be expected that better and better signature transitions will be provided with increasing scales. So, the aim of the above multiscale decorrelation technique is to obtain ever more precise band information about the signature under consideration by using as sampling input always the values of $F_{\mathcal{G}}^{\wedge}$ at lattice points of the inverse lattice Λ^{-1}. In order to obtain more locally detailed bandpass signature information we are additionally able to use lattices showing, in parallel, an increase in width (e.g., by choosing $C\ \sigma_j\ \Lambda^{-1}$ for an appropriate sequence $\{\sigma_j\}_{j\in\mathbb{N}_0}$). This, however, requires that corresponding input lattice point data sets are available.

Gauss-Weierstrass mollifier wavelet sampling. We again start from the *Poisson-type summation formula in summability context*: Let Λ be a lattice in $\mathbb{R}^q$. Let $\mathcal{G} \in \mathbb{R}^q$ be a regular region. Suppose that F is a member of class $C^{(0)}(\overline{\mathcal{G}})$. Then, for all $\tau > 0$ and all $a, y \in \mathbb{R}^q$, we have the limit formula

$$\sum_{\substack{a+g\in\overline{\mathcal{G}} \\ g\in\Lambda}}{}' e^{-2\pi i y\cdot(a+g)} F(a+g) = \lim_{\substack{\tau\to 0 \\ \tau>0}} \frac{1}{\|\mathcal{F}_\Lambda\|} \sum_{h\in\Lambda^{-1}} e^{-\tau\pi h^2}\, F_{\mathcal{G}}^{\wedge}(h)\, e^{2\pi i(h-y)\cdot a}. \tag{13.273}$$

Gauss mollifier sampling. For all $\sigma > 0$ and all $a, z \in \mathbb{R}^q$, integration of the Weierstrass function over $\mathbb{R}^q$ leads to a lattice point identity of the form

(13.274)
$$\sum_{\substack{a+g\in\overline{\mathcal{G}}\\ g\in\Lambda}}{}' \int_{\mathbb{R}^q} e^{2\pi i y\cdot z}\ e^{-\pi\sigma y^2}\ e^{-2\pi i y\cdot(a+g)}\ dy\ F(a+g)$$
$$= \lim_{\substack{\tau\to 0\\ \tau>0}} \frac{1}{\|\mathcal{F}_\Lambda\|} \sum_{h\in\Lambda^{-1}} e^{-\tau\pi h^2}\ F^\wedge_{\mathcal{G}}(h) \int_{\mathbb{R}^q} e^{2\pi i y\cdot z}\ e^{-\pi\sigma y^2}\ e^{2\pi i(h-y)\cdot a}\ dy,$$

so that

$$(13.275)\quad \left(\frac{1}{\sigma}\right)^{q/2} \int_{\mathcal{H}} \sum_{\substack{a+g\in\overline{\mathcal{G}}\\ g\in\Lambda}}{}' e^{-\frac{\pi}{\sigma}|(z-(a+g))|^2}\ F(a+g)\ da$$
$$= \lim_{\substack{\tau\to 0\\ \tau>0}} \frac{1}{\|.\mathcal{F}_\Lambda\|} \sum_{h\in\Lambda^{-1}} e^{-\tau\pi h^2} \left(\frac{1}{\sigma}\right)^{q/2} F^\wedge_{\mathcal{G}}(h) \int_{\mathcal{H}} e^{2\pi i h\cdot a} e^{-\frac{\pi}{\sigma}|z-a|^2}\ da.$$

This shows that, for all $\sigma > 0$ and all $z \in \mathbb{R}^q$, we have in the Gauss-Weierstrass nomenclature

(13.276)
$$\sum_{\substack{(\mathcal{F}_\Lambda+\{g'\})\cap\mathcal{H}\neq\emptyset\\ g'\in\Lambda}} \mathcal{G}_\sigma\left(F^\wedge_{\mathcal{G}\cap\bigcup_{g\in\Lambda}(((\mathcal{H}\cap(\mathcal{F}_\Lambda+\{g'\}))-\{g'\})+\{g\})}\right)(z)$$
$$= \sum_{\substack{(\mathcal{F}_\Lambda+\{g'\})\cap\mathcal{H}\neq\emptyset\\ g'\in\Lambda}} \mathcal{W}_\sigma^{\mathcal{G}\cap\bigcup_{g\in\Lambda}(((\mathcal{H}\cap(\mathcal{F}_\Lambda+\{g'\}))-\{g'\})+\{g\})}(F)(-z)$$
$$= \lim_{\substack{\tau\to 0\\ \tau>0}} \frac{1}{\|\mathcal{F}_\Lambda\|} \sum_{h\in\Lambda^{-1}} e^{-\tau\pi h^2}\ F^\wedge_{\mathcal{G}}(h) \left(\frac{1}{\sigma}\right)^{q/2} \int_{\mathcal{H}} e^{2\pi i h\cdot a} e^{-\frac{\pi}{\sigma}|z-a|^2}\ da,$$

and the series on the right side is absolutely and uniformly convergent on $\mathbb{R}^q$.

As a special case, for given regular regions $\mathcal{G}, \mathcal{H} \subset \mathbb{R}^q$ with $\overline{\mathcal{G}}, \overline{\mathcal{H}} \subset \overline{\mathcal{F}_\Lambda}$, we therefore obtain

(13.277)
$$\mathcal{G}_\sigma\left(F^\wedge_{\mathcal{G}\cap\mathcal{H}}\right)(z)$$
$$= \lim_{\substack{\tau\to 0\\ \tau>0}} \frac{1}{\|\mathcal{F}_\Lambda\|} \sum_{h\in\Lambda^{-1}} e^{-\tau\pi h^2}\ F^\wedge_{\mathcal{G}}(h) \left(\frac{1}{\sigma}\right)^{q/2} \int_{\mathcal{H}} e^{2\pi i h\cdot a} e^{-\frac{\pi}{\sigma}|z-a|^2}\ da.$$

Moreover, for given regular regions $\mathcal{G}, \mathcal{H} \subset \mathbb{R}^q$ with , $\overline{\mathcal{H}} = \overline{\mathcal{G}} \subset \overline{\mathcal{F}_\Lambda}$ we are led to the following σ-parameter dependent Shannon-type sampling identity

(13.278)
$$\mathcal{G}_\sigma\left(F^\wedge_{\mathcal{G}}\right)(z) = \lim_{\substack{\tau\to 0\\ \tau>0}} \frac{1}{\|\mathcal{F}_\Lambda\|} \sum_{h\in\Lambda^{-1}} e^{-\tau\pi h^2}\ F^\wedge_{\mathcal{G}}(h) \underbrace{\left(\frac{1}{\sigma}\right)^{q/2} \int_{\mathcal{G}} e^{2\pi i h\cdot a} e^{-\frac{\pi}{\sigma}|z-a|^2}\ da.}_{=K_{\sigma;\mathcal{G}}(h,z)}$$

For sufficiently small $\sigma > 0$, the identity (13.278) approximately coincides with

$$(13.279)\qquad F_{\mathcal{G}}^{\wedge}(y) = \lim_{\substack{\tau\to 0\\ \tau>0}} \sum_{h\in\Lambda^{-1}} e^{-\tau\pi h^2} F_{\mathcal{G}}^{\wedge}(h) \frac{1}{\|\mathcal{F}_\Lambda\|} \int_{\mathcal{F}_\Lambda} e^{2\pi i a\cdot(h-y)}\, da,\ y\in\mathbb{R}^q.$$

By a "mollification", the sharp features in a "rough signature" are smoothed, but they remain close to the original non-smooth signature, which is of tremendous importance. Usually, a sampling mollifier is not only one kernel function but a sequence, or usually a one-parameter (i.e., scale dependent) family of kernels so that the kernels of a mollifier have to tend to the Shannon kernels. Mollifier generated filtering methods refer to procedures, where highly oscillatory behavior (in our case, of the Shannon kernel) are eliminated. In fact, lowpass filters generated by convolution against Gauss-Weierstrass (or Abel-Poisson) kernels can, of course, be used to control high oscillations. As a consequence, filtering (convolution) with mollifier wavelets constituting the difference of two subsequent mollifier kernels provides bandpass information which, with increasing scales, deliver better and better details up to a certain scale.

Although mollifiers are of particular significance in modern mathematics, they have had no deeper significance in sampling so far. Nevertheless, the mollifier multiscale procedures based on (13.276) are very helpful in lattice sampling.

In our approach, the Shannon kernel $K_{\mathcal{H}}(x,z)$ is replaced by a "Gaussian mollifier Shannon kernel function" of the form

$$(13.280)\qquad (x,z)\mapsto K_{\mathcal{H}}^{\mathcal{G}_\sigma}(x,z) = \left(\frac{1}{\sigma}\right)^{q/2} \int_{\mathcal{H}} e^{-\frac{\pi}{\sigma}|z-a|^2}\, e^{2\pi i x\cdot a}\, da, \qquad x,z\in\mathbb{R}^q,$$

and the σ-dependent kernel (13.280) is used in the cardinal series to produce a filtered function

$$(13.281)\qquad \sum_{\substack{(\mathcal{F}_\Lambda+\{g'\})\cap\mathcal{H}\neq\emptyset\\ g'\in\Lambda}} \mathcal{G}_\sigma\left(F^{\wedge}_{\mathcal{G}\cap\bigcup\limits_{g\in\Lambda}(((\mathcal{H}\cap(\mathcal{F}_\Lambda+\{g'\}))-\{g'\})+\{g\})}\right)(z)$$

which is "close" to

$$(13.282)\qquad \sum_{\substack{(\mathcal{F}_\Lambda+\{g'\})\cap\mathcal{H}\neq\emptyset\\ g'\in\Lambda}} F^{\wedge}_{\mathcal{G}\cap\bigcup\limits_{g\in\Lambda}(((\mathcal{H}\cap(\mathcal{F}_\Lambda+\{g'\}))-\{g'\})+\{g\})}(z)$$

but "smoother". We choose a positive, monotonically decreasing sequence $\{\sigma_j\}_{j\in\mathbb{N}_0}$ tending to 0, i.e., $\lim_{j\to\infty}\sigma_j = 0$. Corresponding to the sequence $\{\sigma_j\}_{j\in\mathbb{N}_0}$ there is a scale-discrete sequence $\{K_{\mathcal{H}}^{\mathcal{G}_{\sigma_j}}(\cdot,\cdot)\}_{j\in\mathbb{N}_0}$ of sampling mollifier kernels $K_{\mathcal{H}}^{\mathcal{G}_{\sigma_j}}(\cdot,\cdot), j\in\mathbb{N}_0$. The *sampling mollifier wavelets* $\{(WK)_{\mathcal{H}}^{\mathcal{G}_{\sigma_j}}\}_{j\in\mathbb{N}_0}$ are to be understood as the difference between two consecutive sampling mollifier kernels

$$(13.283)\qquad (WK)_{\mathcal{H}}^{\mathcal{G}_{\sigma_j}}(h,z) = K_{\mathcal{H}}^{\mathcal{G}_{\sigma_{j+1}}}(h,z)\ -\ K_{\mathcal{H}}^{\mathcal{G}_{\sigma_j}}(h,z).$$

We consider the *sampling mollifier scaling functions*

(13.284)

$$\sum_{\substack{(\mathcal{F}_\Lambda+\{g'\})\cap\mathcal{H}\neq\emptyset \\ g'\in\Lambda}} \mathcal{G}_{\sigma_j}\left(F^{\wedge}_{\mathcal{G}\cap\bigcup_{g\in\Lambda}(((\mathcal{H}\cap(\mathcal{F}_\Lambda+\{g'\}))-\{g'\})+\{g\})}\right)(z)$$
$$= \lim_{\substack{\tau\to 0\\ \tau>0}} \frac{1}{\|\mathcal{F}_\Lambda\|}\sum_{h\in\Lambda^{-1}} e^{-\tau\pi h^2}\, F^{\wedge}_{\mathcal{G}}(h)\left(\frac{1}{\sigma_j}\right)^{q/2}\int_{\mathcal{H}} e^{2\pi i h\cdot a} e^{-\frac{\pi}{\sigma_j}|z-a|^2}\, da,$$

and the corresponding *sampling mollifier wavelets*

$$\sum_{\substack{(\mathcal{F}_\Lambda+\{g'\})\cap\mathcal{H}\neq\emptyset \\ g'\in\Lambda}} (W\mathcal{G})_{\sigma_j}\left(F^{\wedge}_{\mathcal{G}\cap\bigcup_{g\in\Lambda}(((\mathcal{H}\cap(\mathcal{F}_\Lambda+\{g'\}))-\{g'\})+\{g\})}\right) \tag{13.285}$$
$$= \sum_{\substack{(\mathcal{F}_\Lambda+\{g'\})\cap\mathcal{H}\neq\emptyset \\ g'\in\Lambda}} \mathcal{G}_{\sigma_{j+1}}\left(F^{\wedge}_{\mathcal{G}\cap\bigcup_{g\in\Lambda}(((\mathcal{H}\cap(\mathcal{F}_\Lambda+\{g'\}))-\{g'\})+\{g\})}\right)$$
$$- \sum_{\substack{(\mathcal{F}_\Lambda+\{g'\})\cap\mathcal{H}\neq\emptyset \\ g'\in\Lambda}} \mathcal{G}_{\sigma_j}\left(F^{\wedge}_{\mathcal{G}\cap\bigcup_{g\in\Lambda}(((\mathcal{H}\cap(\mathcal{F}_\Lambda+\{g'\}))-\{g'\})+\{g\})}\right).$$

So, multiscale mollifier sampling is designed by difference wavelets in such a way that a lowpass filtering to any scale j plus the bandpass filtering to the same scale j is equal to the lowpass filtering to the scale $j+1$. Conversely, the difference between the lowpass filtering to the scale $j+1$ and the lowpass filtering to the scale j is the bandpass filtering to the scale j. Now, let $J_0, J \in \mathbb{N}$ with $J_0 < J$ be any scales. By an elementary calculation we obtain the formula

(13.286)

$$\sum_{\substack{(\mathcal{F}_\Lambda+\{g'\})\cap\mathcal{H}\neq\emptyset \\ g'\in\Lambda}} \mathcal{G}_{\sigma_J}\left(F^{\wedge}_{\mathcal{G}\cap\bigcup_{g\in\Lambda}(((\mathcal{H}\cap(\mathcal{F}_\Lambda+\{g'\}))-\{g'\})+\{g\})}\right)$$
$$= \sum_{\substack{(\mathcal{F}_\Lambda+\{g'\})\cap\mathcal{H}\neq\emptyset \\ g'\in\Lambda}} \mathcal{G}_{\sigma_{J_0}}\left(F^{\wedge}_{\mathcal{G}\cap\bigcup_{g\in\Lambda}(((\mathcal{H}\cap(\mathcal{F}_\Lambda+\{g'\}))-\{g'\})+\{g\})}\right)$$
$$+ \sum_{j=J_0}^{J-1}\sum_{\substack{(\mathcal{F}_\Lambda+\{g'\})\cap\mathcal{H}\neq\emptyset \\ g'\in\Lambda}} (W\mathcal{G})_{\sigma_j}\left(F^{\wedge}_{\mathcal{G}\cap\bigcup_{g\in\Lambda}(((\mathcal{H}\cap(\mathcal{F}_\Lambda+\{g'\}))-\{g'\})+\{g\})}\right).$$

The formula (13.286) shows the interaction of the occurring scaling and wavelet functions in a multiscale sampling reconstruction. Obviously, the difference of two filterings ("smoothings") results in a decorrelated band signature information, from which it can be expected that better and better signature details will be provided with increasing scales.

Weierstrass mollifier sampling. From the *Poisson-type summation formula in Gauss-Weierstrass summability* we obtain, for all $\sigma > 0$ and all $a, z \in \mathbb{R}^q$,

$$\int_{\mathcal{H}} \sum_{\substack{a+g\in\overline{\mathcal{G}}\\ g\in\Lambda}}{}' e^{-\pi\sigma|a+g|^2}\, e^{-2\pi i z\cdot(a+g)}\, F(a+g)\, da$$
$$= \lim_{\substack{\tau\to 0\\ \tau>0}} \frac{1}{\|\mathcal{F}_\Lambda\|} \sum_{h\in\Lambda^{-1}} e^{-\tau\pi h^2}\, F_{\mathcal{G}}^{\wedge}(h) \int_{\mathcal{H}} e^{-\pi\sigma|a|^2}\, e^{-2\pi i(z-h)\cdot a}\, da. \tag{13.287}$$

This shows that, for all $\sigma > 0$ and all $z \in \mathbb{R}^q$, we have

$$\sum_{\substack{(\mathcal{F}_\Lambda+\{g'\})\cap\mathcal{H}\neq\emptyset\\ g'\in\Lambda}} \int_{\mathcal{G}\cap\bigcup_{g\in\Lambda}(((\mathcal{H}\cap(\mathcal{F}_\Lambda+\{g'\}))-\{g'\})+\{g\})} e^{-\pi\sigma|x|^2}\, e^{-2\pi i z\cdot x}\, F(x)\, dx$$
$$= \lim_{\substack{\tau\to 0\\ \tau>0}} \frac{1}{\|\mathcal{F}_\Lambda\|} \sum_{h\in\Lambda^{-1}} e^{-\tau\pi h^2}\, F_{\mathcal{G}}^{\wedge}(h) \int_{\mathcal{H}} e^{-\pi\sigma|a|^2}\, e^{-2\pi i(z-h)\cdot a}\, da. \tag{13.288}$$

Rewritten in Weierstrass nomenclature we therefore have

$$\sum_{\substack{(\mathcal{F}_\Lambda+\{g'\})\cap\mathcal{H}\neq\emptyset\\ g'\in\Lambda}} \mathcal{W}_\sigma\left(F^{\wedge}_{\mathcal{G}\cap\bigcup_{g\in\Lambda}(((\mathcal{H}\cap(\mathcal{F}_\Lambda+\{g'\}))-\{g'\})+\{g\})}\right)(z)$$
$$= \sum_{\substack{(\mathcal{F}_\Lambda+\{g'\})\cap\mathcal{H}\neq\emptyset\\ g'\in\Lambda}} \mathcal{G}_\sigma^{\mathcal{G}\cap\bigcup_{g\in\Lambda}(((\mathcal{H}\cap(\mathcal{F}_\Lambda+\{g'\}))-\{g'\})+\{g\})}(F)(z)$$
$$= \lim_{\substack{\tau\to 0\\ \tau>0}} \frac{1}{\|\mathcal{F}_\Lambda\|} \sum_{h\in\Lambda^{-1}} e^{-\tau\pi h^2}\, F_{\mathcal{G}}^{\wedge}(h) \int_{\mathcal{H}} e^{-\pi\sigma|a|^2}\, e^{-2\pi i(z-h)\cdot a}\, da. \tag{13.289}$$

and the series on the right side is absolutely and uniformly convergent on $\mathbb{R}^q$. In other words, the Shannon kernel $K_{\mathcal{H}}(x, z)$ is replaced by a "Weierstrass mollifier kernel function" of the form

$$(x, z) \mapsto K_{\mathcal{H}}^{\mathcal{W}_\sigma}(x, z) = \int_{\mathcal{H}} e^{-\pi\sigma|a|^2}\, e^{-2\pi i(z-h)\cdot a}\, da, \quad x, z \in \mathbb{R}^q. \tag{13.290}$$

As a consequence, we are confronted with a one-parameter (i.e., scale dependent) family of mollifier kernels, which may be used to act as constituting structures in a mollifier multiscale methodology in the already described way.

CHAPTER 14

Polymetaharmonic infinite bandwidth sampling

Next, asymptotic properties for functions at infinity are studied in the framework of Euler summation formulas leading to infinite bandwidth lattice sampling methodologies and explicit representations of the aliasing error.

14.1. Univariate lattice point theory based sampling

For purposes of a better understanding, we start with applying the 1D Euler-type summation formula with respect to the operator $L+\lambda$, $L=\Delta+4\pi i a\cdot\nabla$, $a\in\mathbb{R}$, especially to sums of spherical type, i.e., to 1D intervals $\mathbb{B}^1_N(y)$ with arbitrarily chosen $y\in\mathbb{R}$ and radii $N>0$. We are interested in investigating the asymptotic behavior with increasing radii N of the summation formula

$$
\begin{aligned}
(14.1)\qquad & \sum_{\substack{|g-y|\le N\\ g\in\Lambda}}{}' e^{2\pi i a\cdot g} F^{\wedge}_{\mathbb{B}^1_R}(g-y)\\
&= \sum_{\substack{4\pi^2((h+a)^2-a^2)=\lambda\\ h\in\Lambda^{-1}}} e^{2\pi i y\cdot(h+a)} \int_{\substack{|x|\le N\\ x\in\mathbb{R}}} F^{\wedge}_{\mathbb{B}^1_R}(x)\, e^{2\pi i x\cdot(h+a)}\, dx\\
&\quad + \int_{\substack{|x-y|\le N\\ x\in\mathbb{R}}} G(L+\lambda;x)\, e^{2\pi i a\cdot x}\left(\Delta_x+\lambda+4\pi^2a^2\right) F^{\wedge}_{\mathbb{B}^1_R}(x-y)\, dx\\
&\quad + \left\{e^{2\pi i a\cdot x} G(L+\lambda;x)\nabla_x F^{\wedge}_{\mathbb{B}^1_R}(x-y)\right\}\Big|_{-N+y}^{N+y}\\
&\quad - \left\{F^{\wedge}_{\mathbb{B}^1_R}(x-y)\, e^{2\pi i a\cdot x}(\nabla_x+4\pi i a)G(L+\lambda;x)\right\}\Big|_{-N+y}^{N-+y}.
\end{aligned}
$$

to derive sufficient criteria for the validity of the Poisson-type summation formula over the whole Euclidean space $\mathbb{R}$. Questions of the convergence of (14.1) as $N\to\infty$ require asymptotic estimates of the constituting elements of the "remainder term" $R(N)$ given by

$$
\begin{aligned}
(14.2)\qquad R(N) &= \frac{1}{2}\sum_{\substack{|g-y|=N\\ g\in\Lambda}} e^{2\pi i a\cdot g} F(g-y)\\
&\quad + \left\{e^{2\pi i a\cdot x} G(L+\lambda;x)\nabla_x F^{\wedge}_{\mathbb{B}^1_R}(x-y)\right\}\Big|_{-N+y}^{N+y}\\
&\quad - \left\{F^{\wedge}_{\mathbb{B}^1_R}(x-y) e^{2\pi i a\cdot x}(\nabla_x+4\pi i a)G(L+\lambda;x)\right\}\Big|_{-N+y}^{N-+y}.
\end{aligned}
$$

The estimates,

$$|G(L+\lambda;x)| \,|_{-N+y}^{N+y} = O(1), \quad N\to\infty \tag{14.3}$$

$$|\nabla_x G(L+\lambda;x)| \,|_{-N+y}^{N+y} = O(1), \quad N\to\infty, \tag{14.4}$$

which essentially follow from the boundedness of the lattice function $G(L+\lambda;\cdot)$ and its derivative, play an important role in the development of convergence criteria: First, the properties (14.3) and (14.4) are used to formulate 1D convergence criteria for spherically reflected series:

Function space $X_\lambda^{(2)}(\mathbb{R})$. A function F is called of class $X^{(2)}(\mathbb{R})$, if F is of class $C^{(2)}(\mathbb{R})$ satisfying

$$|F(x)| \;=\; o\,(1)\,, \quad |x|\to\infty \tag{14.5}$$

and

$$|\nabla_x F(x)| \;=\; o\,(1)\,, \quad |x|\to\infty. \tag{14.6}$$

Observing the definition of $X^{(2)}(\mathbb{R})$ in (14.2) we are immediately led to the following result:

For $N\to\infty$ we have $R(N)=o(1)$, so that

$$\begin{aligned}\sum_{\substack{|g-y|\le N\\ g-y\in\Lambda}} & e^{2\pi i a\cdot g} F(g-y) \\ &= \frac{1}{\|\mathcal{F}_\Lambda\|}\; e^{2\pi i a\cdot y} \sum_{\substack{4\pi^2\left((h+a)^2-a^2\right)=\lambda\\ h\in\Lambda^{-1}}} e^{2\pi i h\cdot y}\int_{|x|\le N} F(x)e^{2\pi i x\cdot(a+h)}dx \\ &\quad + \int_{\substack{|x-y|\le N\\ x\in\mathbb{R}^q}} G(L+\lambda;x)\; e^{2\pi i a\cdot x}\left(\Delta_x+\lambda+4\pi^2a^2\right)F(x-y)\;dx \\ &\quad + o(1).\end{aligned} \tag{14.7}$$

Next we discuss the second term of the right side of (14.7):

Function space $Y_\lambda^{(2)}(\mathbb{R})$. For an appropriately specified $\lambda\in\mathbb{R}$, F is called of class $Y_\lambda^{(2)}(\mathbb{R})$, if F is of class $C^{(2)}(\mathbb{R})$ satisfying

$$(\Delta_x+\lambda+4\pi^2a^2)F(x) = O\left(|x|^{-(1+\varepsilon)}\right), \quad \varepsilon>0,\ |x|\to\infty. \tag{14.8}$$

In connection with the definition of $Y_\lambda^{(2)}(\mathbb{R})$ we immediately obtain the following result:

The integral

$$\int_{\mathbb{R}} G(L+\lambda;x)\; e^{2\pi i a\cdot x}\left(\Delta_x+\lambda+4\pi^2a^2\right)F(x-y)\;dx \tag{14.9}$$

is absolutely convergent.

Combining the last two results we are able to formulate the following limit relation:

Suppose that, for a preassigned value $\lambda \in \mathbb{R}$, F *is of class* $X^{(2)}(\mathbb{R}) \cap Y_\lambda^{(2)}(\mathbb{R})$. *Then*

$$\lim_{N\to\infty} \left(\sum_{\substack{|g-y|\le N \\ g\in\Lambda}} e^{2\pi i a\cdot g} F(g-y) \right. \tag{14.10}$$

$$\left. - \frac{1}{\|\mathcal{F}_\Lambda\|} e^{2\pi i a\cdot y} \sum_{\substack{4\pi^2\left((h+a)^2-a^2\right)=\lambda \\ h\in\Lambda^{-1}}} e^{2\pi i h\cdot y} \int_{\substack{|x|\le N \\ x\in\mathbb{R}}} F(x)\, e^{2\pi i x\cdot(a+h)} dx \right)$$

$$= \int_{\mathbb{R}} G(L+\lambda;x)\, e^{2\pi i a\cdot x} \left(\Delta_x + \lambda + 4\pi^2 a^2\right) F(x-y)\, dx.$$

The formula (14.10) demonstrates that the convergence of 1D sums (understood in spherical summation) is closely related to the spectrum. In fact, for the class of noneigenvalues, i.e., for values $\lambda \notin \mathrm{Spect}_\Delta(\Lambda)$, (14.10) immediately guarantees the convergence of the infinite series on the left side. In order to ensure the convergence in the case of eigenvalues, however, we have to require an additional convergence condition for the remaining volume integrals:

Function space $Z_\lambda^{(2)}(\mathbb{R})$. *For a given* $\lambda \in \mathrm{Spect}_\Delta(\Lambda)$, $F \in C^{(2)}(\mathbb{R})$ *is said to be of class* $Z_\lambda^{(2)}(\mathbb{R})$, *if the integrals* $\int_{\mathbb{R}} F(x)\, e^{2\pi i x\cdot(a+h)} dx$ *exist for all* $h \in \Lambda^{-1}$ *with* $4\pi^2\left((h+a)^2 - a^2\right) = \lambda$ *in the (spherical) sense*

$$\int_{\mathbb{R}} F(x)\, e^{2\pi i x\cdot(a+h)} dx = \lim_{N\to\infty} \int_{\substack{|x|\le N \\ x\in\mathbb{R}}} F(x)\, e^{2\pi i x\cdot(a+h)} dx. \tag{14.11}$$

Observing the characteristics of $Z_\lambda^{(2)}(\mathbb{R})$ we are able to formulate the following convergence theorem:

For a preassigned value $\lambda \in \mathrm{Spect}_\Delta(\Lambda)$, *let* F *be a member of class* $X^{(2)}(\mathbb{R}) \cap Y_\lambda^{(2)}(\mathbb{R}) \cap Z_\lambda^{(2)}(\mathbb{R})$. *Then the series*

$$\lim_{N\to\infty} \sum_{\substack{|g-y|\le N \\ g\in\Lambda}} e^{2\pi i a\cdot g} F(g-y) \tag{14.12}$$

is convergent (in the indicated spherical sense). In more detail, for given $\lambda \notin \mathrm{Spect}_\Delta(\Lambda)$ *and* $F \in X^{(2)}(\mathbb{R}) \cap Y_\lambda^{(2)}(\mathbb{R})$, *we have*

$$\lim_{N\to\infty} \sum_{\substack{|g-y|\le N \\ g\in\Lambda}} e^{2\pi i a\cdot g} F(g-y)$$

$$= \int_{\mathbb{R}} G(L+\lambda;x)\, e^{2\pi i a\cdot x} \left(\Delta_x + \lambda + 4\pi^2 a^2\right) F(x-y)\, dx,$$

whereas, for given $\lambda \in \operatorname{Spect}_\Delta(\Lambda)$ *and* $F \in X^{(2)}(\mathbb{R}) \cap Y_\lambda^{(2)}(\mathbb{R}) \cap Z_\lambda^{(2)}(\mathbb{R})$, *we have*

$$\lim_{N\to\infty} \sum_{\substack{|g-y|\le N \\ g\in\Lambda}} e^{2\pi i a\cdot g} F(g-y)$$

$$= \lim_{N\to\infty} \frac{1}{\|\mathcal{F}_\Lambda\|} \; e^{2\pi i a\cdot y} \sum_{\substack{4\pi^2\left((h+a)^2-a^2\right)=\lambda \\ h\in\Lambda^{-1}}} e^{2\pi i h\cdot y} \int_{\substack{|x|\le N \\ x\in\mathbb{R}}} F(x)\, e^{2\pi i x\cdot(a+h)} dx$$

$$+ \int_{\mathbb{R}} G(L+\lambda;x)\, e^{2\pi i a\cdot x} \left(\Delta_x + \lambda + 4\pi^2 a^2\right) F(x-y)\; dx.$$

All in all, our approach canonically implies to the following useful convention.

Function space $C_\lambda^{(2)}(\mathbb{R})$. *For a given* $\lambda \in \mathbb{R}$, *a function* $F \in C^{(2)}(\mathbb{R})$ *is said to belong to the class* $C_\lambda^{(2)}(\mathbb{R})$, *if one of the alternative conditions is satisfied:*

(i) F *is of class* $X^{(2)}(\mathbb{R}) \cap Y_\lambda^{(2)}(\mathbb{R})$, *provided that* $\lambda \notin \operatorname{Spect}_\Delta(\Lambda)$,

(ii) F *is of class* $X^{(2)}(\mathbb{R}) \cap Y_\lambda^{(2)}(\mathbb{R}) \cap Z_\lambda^{(2)}(\mathbb{R})$, *provided that* $\lambda \in \operatorname{Spect}_\Delta(\Lambda)$.

The convergence criteria enable us to derive an extension of the classical Poisson summation formula. As the Λ-lattice function $G((L+\lambda);\cdot)$ permits an absolutely and uniformly convergent Fourier series in $\mathbb{R}$, Lebesgue's theorem of dominated convergence allows us to interchange summation and integration such that

$$\text{(14.13)} \quad \int_{\mathbb{R}} G((L+\lambda);x)\, e^{2\pi i a\cdot x} \left(\Delta_x + \lambda + 4\pi^2 a^2\right) F(x-y)\; dx$$

$$= \frac{1}{\|\mathcal{F}_\Lambda\|} \sum_{\substack{4\pi^2\left((h+a)^2-a^2\right)\neq\lambda \\ h\in\Lambda^{-1}}} \frac{1}{\lambda - 4\pi^2\left((h+a)^2-a^2\right)}$$

$$\times \int_{\mathbb{R}} e^{2\pi i(a+h)\cdot x} \left(\Delta_x + \lambda + 4\pi^2 a^2\right) F(x-y)\; dx$$

holds for a function $F \in C_\lambda^{(2)}(\mathbb{R})$. Moreover, integration by parts, i.e., Green's theorem yields the identity

$$\text{(14.14)} \quad \int_{\mathbb{R}} e^{2\pi i(a+h)\cdot x} \left(\Delta_x + \lambda + 4\pi^2 a^2\right) F(x-y)\; dx$$

$$= \int_{\mathbb{R}} \left(\Delta_x + \lambda + 4\pi^2 a^2\right) e^{2\pi i(a+h)\cdot x}\; F(x-y)\; dx$$

$$= \left(\lambda - 4\pi^2\left((h+a)^2-a^2\right)\right) \int_{\mathbb{R}} e^{2\pi i(a+h)\cdot x}\; F(x-y)\; dx.$$

Inserting (14.14) into (14.13) we therefore find that

$$\text{(14.15)} \quad \int_{\mathbb{R}} G((L+\lambda);x)\, e^{2\pi i a\cdot x} \left(\Delta_x + \lambda + 4\pi^2 a^2\right) F(x-y)\; dx$$

$$= \frac{1}{\|\mathcal{F}_\Lambda\|} \; e^{2\pi i a\cdot y} \sum_{\substack{4\pi^2\left((h+a)^2-a^2\right)\neq\lambda \\ h\in\Lambda^{-1}}} e^{2\pi i h\cdot y} \int_{\mathbb{R}} F(x)\, e^{2\pi i x\cdot(a+h)} dx.$$

By combination of all results we finally obtain the 1D *Poisson-type summation formula*:

For a preassigned $\lambda \in \mathbb{R}$, suppose that the function F is of class $C_\lambda^{(2)}(\mathbb{R})$. Then the limit

$$\lim_{N\to\infty} \sum_{\substack{|g-y|\leq N \\ g\in\Lambda}} e^{2\pi i a\cdot g} F(g-y) \tag{14.16}$$

is convergent, and we have

$$\lim_{N\to\infty} \sum_{\substack{|g-y|\leq N \\ g\in\Lambda}} e^{2\pi i a\cdot g} F(g-y) = \sum_{h\in\Lambda^{-1}} \int_{\mathbb{R}} F(x)\ e^{2\pi i x\cdot(a+h)} dx\ \frac{1}{\|\mathcal{F}_\Lambda\|} e^{2\pi i(a+h)\cdot y}. \tag{14.17}$$

Analogously to the context implied by the Poisson-type summation formula for finite intervals, the identity (14.17) offers the entry point for Shannon-type sampling characterized by a resulting cardinal series on the left side of (14.17) extended over functional values of $F \in C_\lambda^{(2)}(\mathbb{R})$ in y-shifted lattice points of Λ.

Functional values based Shannon-type sampling (infinite bandwidth). From the identity (14.17) we obtain by integration over $\mathbb{B}_\tau^1$ the following Shannon-type sampling theorem (the proof is analogous to that presented in the multivariate part of this book).

Assume that Λ is an arbitrary lattice in $\mathbb{R}$. Suppose that F is a member of the class $C_\lambda^{(2)}(\mathbb{R}^q)$. Then, for all $y \in \mathbb{R}$, we have

$$\begin{aligned} &\sum_{\substack{(\mathcal{F}_\Lambda+\{g'\})\cap\mathbb{B}_\tau^1\neq\emptyset \\ g'\in\Lambda}} \int_{\bigcup_{g\in\Lambda}(((\mathbb{B}_\tau^1\cap(\mathcal{F}_\Lambda+\{g'\}))-\{g'\})+\{g\})} F_{\mathbb{R}}^\wedge(w)\ e^{2\pi i y\cdot w}\ dw \\ &\quad = \lim_{N\to\infty} \sum_{\substack{|h|\leq N \\ h\in\Lambda^{-1}}} F(h)\ \ K_{\mathbb{B}_\tau^1}(h+y), \end{aligned} \tag{14.18}$$

We choose $\Lambda = \sigma\mathbb{Z}$ and consider as special case $\overline{\mathbb{B}_\tau^1} \subset \overline{\mathcal{F}_{\sigma\mathbb{Z}}}$: It follows that

$$\int_{\bigcup_{g\in\sigma\mathbb{Z}}(\mathbb{B}_\tau^1+\{g\})} F_{\mathbb{R}}^\wedge(x)\ e^{2\pi i y\cdot x}\ dx = \lim_{N\to\infty} \sum_{\substack{|h|\leq N \\ h\in\frac{1}{\sigma}\mathbb{Z}}} F(h)\ \ K_{\mathbb{B}_\tau^1}(h+y). \tag{14.19}$$

Thus we are confronted with the following observation: The closer $\overline{\mathbb{B}_\tau^1}$ is to the fundamental cell of $\sigma\mathbb{Z}$, the smaller is the aliasing error by this variant of Shannon-type sampling.

For $\overline{\mathbb{B}_\tau^1} = \overline{\mathcal{F}_{\sigma\mathbb{Z}}}$, we obtain by application of the Fourier inversion formula

$$F(y) = \lim_{N\to\infty} \sum_{\substack{|h|\leq N \\ h\in\frac{1}{\sigma}\mathbb{Z}}} F(h)\ \ K_{\mathcal{F}_{\sigma\mathbb{Z}}}(h+y). \tag{14.20}$$

In other words, the not necessarily bandlimited functions F of the class $C_\lambda^{(2}(\mathbb{R})$ can be sampled in the standard way by choosing the fundamental cell $\mathcal{F}_{\sigma\mathbb{Z}}$ to act as the kernel generator $K_{\mathcal{F}_{\sigma\mathbb{Z}}}$ of the cardinal series in Shannon sampling

$$F(y) = \lim_{N\to\infty} \sum_{\substack{|h|\leq N \\ h\in\frac{1}{\sigma}\mathbb{Z}}} F(h)\ K_{\mathcal{F}_{\sigma\mathbb{Z}}}(h+y), \quad y\in\mathbb{R}. \tag{14.21}$$

Altogether, in the case of a lattice cell (or finite unions of them) aliasing can be avoided. In all other cases, a way is found in not-necessarily bandlimited sampling to detect the explicit aliasing error.

14.2. Multivariate summation and lattice point identities

Analogously to the 1D case, we apply the multi-dimensional Euler-type summation formulas with respect to the operator $L+\lambda$ especially to sums over balls $\overline{\mathbb{B}^q_N(y)}$ with $y\in\mathbb{R}^q$ arbitrary, but fixed, and N tending to infinity. In other words, our interest is in investigating the asymptotic behavior with increasing radii N to derive *sufficient criteria for the validity of a Poisson-type summation formulas over the whole Euclidean space* $\mathbb{R}^q$, which do not necessarily imply the absolute convergence of the infinite multivariate sums on both sides (as it is the case in the approach, e.g., proposed by [**790**]). The resulting criteria, therefore, are of particular importance in the context of multivariate alternating sum convergence. As essential tools we need some integral mean asmptotics for iterated Λ-lattice (Euler-Green) functions, which are formulated in similarity to the bivariate deductions (see W. Freeden [**248, 250, 251**]). The resulting variants of the Poisson-type summation formulas over $\mathbb{R}^q$ act as transitional identities to provide Shannon-type sampling realizations.

Integral means for iterated Λ-lattice functions. The following integral estimates (for the proof see W. Freeden [**248, 268**]) play an important part in the discussion of series over alternating functions in $\mathbb{R}^q, q\geq 2$:

For all $m\in\mathbb{N}$, $q\geq 2$, all lattices $\Lambda\subset\mathbb{R}^q$, and for all $\lambda\in\mathbb{R}$ and $a,y\in\mathbb{R}^q$, the asymptotic expansions

$$\int_{\substack{|x-y|=N \\ x\in\mathbb{R}^q}} |G((L+\lambda)^m;x)|\ dS_{(q-1)}(x) = O(N^{q-1}) \tag{14.22}$$

and

$$\int_{\substack{|x-y|=N \\ x\in\mathbb{R}^q}} \left|\frac{\partial G((L+\lambda)^m;x)}{\partial\nu}\right|\ dS_{(q-1)}(x) = O(N^{q-1}) \tag{14.23}$$

hold true for $N\to\infty$.

Euler-type summation formula over increasing balls involving iterated Euler-Green functions. We assume that the function F under consideration admits a certain differentiability order, so that iterated Λ-lattice functions may be used in an Euler-type summation context. Essential tools for the asymptotic handling of extended Euler-type summation formulas are the integral estimates (14.22) and (14.23). Our approach naturally implies that the convergence criteria justifying Poisson-type summation formulas over Euclidean spaces $\mathbb{R}^q$ have to be understood in spherical sense.

To be more concrete, for arbitrary lattices $\Lambda \subset \mathbb{R}^q$, arbitrary vectors $a, y \in \mathbb{R}^q$, and $2m$-times continuously differentiable functions F on closed balls $\overline{\mathbb{B}_N^q(y)}$, the Euler-type summation formula leads to the identity

(14.24)

$$\begin{aligned}
&\sum_{\substack{|g-y|\leq N\\ g\in\Lambda}} e^{2\pi i a\cdot g} F(g-y)\\
&\quad = \frac{1}{\|\mathcal{F}_\Lambda\|}\, e^{2\pi i a\cdot y} \sum_{\substack{4\pi^2\left((h+a)^2-a^2\right)=\lambda\\ h\in\Lambda^{-1}}} e^{2\pi i h\cdot y} \int_{\substack{|x|\leq N\\ x\in\mathbb{R}^q}} F(x) e^{2\pi i x\cdot(a+h)}\, dx\\
&\qquad + \int_{\substack{|x-y|\leq N\\ x\in\mathbb{R}^q}} G((L+\lambda)^m; x)\, e^{2\pi i a\cdot x} \left(\Delta_x + \lambda + 4\pi^2 a^2\right)^m F(x-y)\, dx\\
&\qquad + R_m(N),
\end{aligned}$$

where the *"remainder term"* is given by

(14.25)

$$\begin{aligned}
&R_m(N)\\
&\quad = \frac{1}{2} \sum_{\substack{|g-y|=N\\ g\in\Lambda}} e^{2\pi i a\cdot g} F(y-y)\\
&\qquad - \sum_{k-0}^{m-1} \int_{\substack{|x-y|=N\\ x\in\mathbb{R}^q}} e^{2\pi i a\cdot x} G((L+\lambda)^{k+1}; x) \frac{\partial}{\partial\nu_x} \left(\Delta_x+\lambda+4\pi^2 a^2\right)^k F(x-y)\, dS(x)\\
&\qquad + \sum_{k=0}^{m-1} \int_{\substack{|x-y|=N\\ x\in\mathbb{R}^q}} e^{2\pi i a\cdot x} \left(\Lambda_x + \lambda + 4\pi^2 a^2\right)^k F(x-y)\\
&\qquad \times \left(\frac{\partial}{\partial\nu_x} + 4\pi i a\cdot\nu(x)\right) G((L+\lambda)^{k+1}; x)\, dS(x).
\end{aligned}$$

Spherically reflected convergence criteria for multivariate alternating series. Questions of the convergence as $N \to \infty$ require estimates of the remainder term (14.25), where the integral estimates for Λ-lattice functions to iterated $(L+\lambda)$-operators (14.22) and (14.23) play an important part.

For a given $\lambda \in \mathbb{R}$, the function spaces $X_\lambda^{(2m)}(\mathbb{R}^q)$, $Y_\lambda^{(2m)}(\mathbb{R}^q)$, respectively, characterizing "admissible candidates" are introduced as follows:

Function space $X_\lambda^{(2m)}(\mathbb{R}^q)$: A function F is said to be of class $X_\lambda^{(2m)}(\mathbb{R}^q)$, if F is of class $C^{(2m)}(\mathbb{R}^q)$ and satisfies the asymptotic relations

(14.26) $$\left(\Delta_x + \lambda + 4\pi^2 a^2\right)^k F(x) = o\left(|x|^{1-q}\right), \quad |x| \to \infty.$$

and

$$\left|\nabla_x \left(\Delta_x + \lambda + 4\pi^2 a^2\right)^k F(x)\right| = o\left(|x|^{1-q}\right), \quad |x| \to \infty, \tag{14.27}$$

for $k = 0, \ldots, m-1$ *(note that the case* $m = 1$ *is independent of* $\lambda \in \mathbb{R}$*, so that* $X_\lambda^{(2)}(\mathbb{R}^q)$ *simply coincides with* $X^{(2)}(\mathbb{R}^q)$*).*

Function space $Y_\lambda^{(2m)}(\mathbb{R}^q)$*: A function* F *is said to be of class* $Y_\lambda^{(2m)}(\mathbb{R}^q)$*, if* F *is of class* $C^{(2m)}(\mathbb{R}^q)$ *and satisfies the asymptotic rule*

$$\left(\Delta_x + \lambda + 4\pi^2 a^2\right)^m F(x) = O\left(|x|^{-(q+\varepsilon)}\right), \quad \varepsilon > 0, \quad |x| \to \infty \tag{14.28}$$

Function space $Z_\lambda^{(2m)}(\mathbb{R}^q)$*: A function* $F \in C^{(2m)}(\mathbb{R}^q)$ *is said to be of class* $Z_\lambda^{(2m)}(\mathbb{R}^q)$*, if the integrals* $\int_{\mathbb{R}^q} F(x)\, \overline{\Phi_h(x)}\, dx$ *exist for all lattice points* $h \in \Lambda^{-1}$ *with* $L^\wedge(h) = 4\pi^2\left((h+a)^2 - a^2\right) = \lambda$ *in the (spherical) sense*

$$\int_{\mathbb{R}^q} F(x)\overline{\Phi_h(x)}\, dx = \lim_{N\to\infty} \int_{\substack{|x|\le N\\ x\in\mathbb{R}^q}} F(x)\, \overline{\Phi_h(x)}\, dx. \tag{14.29}$$

Based on these propositions W. Freeden [**268**] derived asymptotic convergence criteria for multi-dimensional alternating series from a straightforward discussion of (14.24) and its remainder term (14.25): First, in connection with the already known Gaussian estimate (8.1), we see that, for a function F of class $X_\lambda^{2m}(\mathbb{R}^q)$, we have

$$\frac{1}{2} \sum_{\substack{|g-y|=N\\ g\in\Lambda}} e^{2\pi i a\cdot g} F(g-y) = o\left(N^{1-q} \sum_{\substack{|g-y|=N\\ g\in\Lambda}} 1\right) = o(1), \quad N \to \infty. \tag{14.30}$$

In addition, we have

$$\begin{aligned}
&-\sum_{k=0}^{m-1} \int_{\substack{|x-y|=N\\ x\in\mathbb{R}^q}} e^{2\pi i a\cdot x} G((L+\lambda)^{k+1};x) \frac{\partial}{\partial\nu_x}\left(\Delta_x + \lambda + 4\pi^2 a^2\right)^k F(x-y)\, dS(x)\\
&\quad + \sum_{k=0}^{m-1} \int_{\substack{|x-y|=N\\ x\in\mathbb{R}^q}} e^{2\pi i a\cdot x}\left(\Delta_x + \lambda + 4\pi^2 a^2\right)^k F(x-y)\\
&\qquad\qquad \times \left(\frac{\partial}{\partial\nu_x} + 4\pi i a\cdot\nu(x)\right) G((L+\lambda)^{k+1};x)\, dS(x)\\
&= o\left(N^{1-q}\left(\sum_{k=0}^{m-1} \int_{\substack{|x|=N\\ x\in\mathbb{R}^q}} \left|G((L+\lambda)^{k+1};x)\right| \right.\right.\\
&\qquad\qquad \left.\left. + \left|\frac{\partial}{\partial\nu} G((L+\lambda)^{k+1};x)\right| dS(x)\right)\right)
\end{aligned} \tag{14.31}$$

for $N \to \infty$. By virtue of the integral estimates (14.22) and (14.23) we therefore obtain

$$
\begin{aligned}
&-\sum_{k=0}^{m-1} \int_{\substack{|x-y|=N \\ x\in\mathbb{R}^q}} e^{2\pi i a\cdot x} G((L+\lambda)^{k+1};x) \frac{\partial}{\partial \nu_x} \left(\Delta_x + \lambda + 4\pi^2 a^2\right)^k F(x-y)\, dS(x) \\
&\quad + \sum_{k=0}^{m-1} \int_{\substack{|x-y|=N \\ x\in\mathbb{R}^q}} e^{2\pi i a\cdot x} \left(\Delta_x + \lambda + 4\pi^2 a^2\right)^k H(x-y) \\
&\qquad\qquad \times \left(\frac{\partial}{\partial \nu_x} + 4\pi i a \cdot \nu(x)\right) G((L+\lambda)^{k+1};x)\, dS(x) \\
&= o(1)
\end{aligned} \tag{14.32}
$$

for $N \to \infty$. Collecting all details we arrive at the following desired result:

Assume that $F \in C^{(2m)}(\mathbb{R}^q), m \in \mathbb{N}$, is of class $X_\lambda^{(2m)}(\mathbb{R}^q)$. Then,

$$
R_m(N) = o(1), \quad N \to \infty. \tag{14.33}
$$

Next we come to the discussion of the volume integral in (14.24) containing the product of $G((L+\lambda)^m;x)\ e^{2\pi i a\cdot x}$ and $\left(\Delta_x + \lambda + 4\pi^2 a^2\right)^m F(x-y)$ as integrand. Here, we are able to guarantee that, for a function F of class $Y_\lambda^{(2m)}(\mathbb{R}^1)$,

$$
\begin{aligned}
&\left| \int_{\substack{M\le|x-y|\le N \\ x\in\mathbb{R}^q}} G((L+\lambda)^m;x) \left(\Delta_x + \lambda + 4\pi^2 a^2\right)^m F(x-y)\, dx \right| \\
&\quad = O\left(\int_M^N \frac{1}{(1+r)^{q+\varepsilon}} \left(\int_{\substack{|x-y|=r \\ x\in\mathbb{R}^q}} |G((L+\lambda)^m;x)|\ dS(x) \right) dr \right) \\
&\quad = O\left(\int_M^N \frac{r^{q-1}}{(1+r)^{q+\varepsilon}}\, dr \right).
\end{aligned} \tag{14.34}
$$

holds true for sufficiently large positive constants M, N. Consequently, the absolute convergence of the integral written down in (14.35) is obvious:

Assume that, for a preassigned $\lambda \in \mathbb{R}$, the function F is of class $Y_\lambda^{(2m)}(\mathbb{R}^q)$. Then the volume integral

$$
\int_{\mathbb{R}^q} G((L+\lambda)^m;x)\ e^{2\pi i a\cdot x} \left(\Delta_x + \lambda + 4\pi^2 a^2\right)^m F(x-y)\, dx \tag{14.35}
$$

is absolutely convergent.

Combining (14.33) and (14.35) we obtain the following limit formula as an immediate consequence:

Assume that, for a preassigned value $\lambda \in \mathbb{R}$, F is of class $X_\lambda^{(2m)}(\mathbb{R}^q) \cap Y_\lambda^{(2m)}(\mathbb{R}^q)$. Then

$$\lim_{N\to\infty} \left(\sum_{\substack{|g-y|\le N \\ g\in\Lambda}} e^{2\pi i a\cdot g} F(g-y) - \frac{1}{\|\mathcal{F}_\Lambda\|} e^{2\pi i a\cdot y} \sum_{\substack{4\pi^2\left((h+a)^2-a^2\right)=\lambda \\ h\in\Lambda^{-1}}} e^{2\pi i h\cdot y} \int_{\substack{|x|\le N \\ x\in\mathbb{R}^q}} F(x)\, e^{2\pi i x\cdot(a+h)}\, dx \right) \tag{14.36}$$
$$= \int_{\mathbb{R}^q} G((L+\lambda)^m; x)\, e^{2\pi i a\cdot x} \left(\Delta_x + \lambda + 4\pi^2 a^2\right)^m F(x-y)\, dx.$$

It should be remarked that, in the case that $\lambda \notin \mathrm{Spect}_\Delta(\Lambda)$, i.e., that there is no lattice point $h \in \Lambda^{-1}$ satisfying $L^\wedge(h) = 4\pi^2\left((h+a)^2 - a^2\right) = \lambda$, the finite sum

$$\frac{1}{\|\mathcal{F}_\Lambda\|} e^{2\pi i a\cdot y} \sum_{\substack{4\pi^2\left((h+a)^2-a^2\right)=\lambda \\ h\in\Lambda^{-1}}} e^{2\pi i h\cdot y} \int_{\substack{|x|\le N \\ x\in\mathbb{R}^q}} F(x)\, e^{2\pi i x\cdot(a+h)}\, dx \tag{14.37}$$

is zero, so that

$$\lim_{N\to\infty} \sum_{\substack{|g-y|\le N \\ g\in\Lambda}} e^{2\pi i a\cdot g} F(g-y) \tag{14.38}$$
$$= \int_{\mathbb{R}^q} G((L+\lambda)^m; x)\, e^{2\pi i a\cdot x} \left(\Delta_x + \lambda + 4\pi^2 a^2\right)^m F(x-y)\, dx.$$

In other words, the convergence of multi-dimensional alternating sums (understood in spherical summation)

$$\sum_{g\in\Lambda} e^{2\pi i a\cdot g} F(g-y) = \lim_{N\to\infty} \sum_{\substack{|g-y|\le N \\ g\in\Lambda}} e^{2\pi i a\cdot g} F(g-y) \tag{14.39}$$

is closely related to the spectrum $\mathrm{Spect}_\Delta(\Lambda)$. For the class of noneigenvalues, i.e., $\lambda \notin \mathrm{Spect}_\Delta(\Lambda)$, (14.38) already guarantees the convergence of the series (14.39) in spherical sense. In order to guarantee the convergence of the series (14.39) for an eigenvalue $\lambda \in \mathrm{Spect}_\Delta(\Lambda)$, however, the space $Z_\lambda^{(2m)}(\mathbb{R}^q)$ has to come into play. In fact, for $\lambda \in \mathrm{Spect}_\Delta(\Lambda)$, we are led to formulate the following convergence theorem:

Let F be a of class $X_\lambda^{(2m)}(\mathbb{R}^q) \cap Y_\lambda^{(2m)}(\mathbb{R}^q) \cap Z_\lambda^{(2m)}(\mathbb{R}^q)$. Then the series

$$\sum_{g\in\Lambda} e^{2\pi i a\cdot g} F(g-y) = \lim_{N\to\infty} \sum_{\substack{|g-y|\le N \\ g\in\Lambda}} e^{2\pi i a\cdot g} F(g-y) \tag{14.40}$$

is convergent (in the indicated spherical sense).

More concretely, for $\lambda \notin \mathrm{Spect}_\Delta(\Lambda)$ and $F \in X_\lambda^{(2m)}(\mathbb{R}^q) \cap Y_\lambda^{(2m)}(\mathbb{R}^q)$, we have

$$\begin{aligned} &\lim_{N\to\infty} \sum_{\substack{|g-y|\le N\\ g\in\Lambda}} e^{2\pi i a\cdot g} F(g-y) \\ &\qquad = \int_{\mathbb{R}^q} G((L+\lambda)^m; x)\, e^{2\pi i a\cdot x} \left(\Delta_x + \lambda + 4\pi^2 a^2\right)^m F(x-y)\, dx. \end{aligned}$$

whereas, for $\lambda \in \mathrm{Spect}_\Delta(\Lambda)$ and $F \in X^{(2)}(\mathbb{R}^q) \cap Y_\lambda^{(2)}(\mathbb{R}^q) \cap Z_\lambda^{(2)}(\mathbb{R}^q)$,

$$\begin{aligned} &\lim_{N\to\infty} \sum_{\substack{|g-y|\le N\\ g\in\Lambda}} e^{2\pi i a\cdot g} F(g-y) \\ &= \lim_{N\to\infty} \frac{1}{\|\mathcal{F}_\Lambda\|}\, e^{2\pi i a\cdot y} \sum_{\substack{4\pi^2\left((h+a)^2-a^2\right)=\lambda\\ h\in\Lambda^{-1}}} e^{2\pi i h\cdot y} \int_{\substack{|x|\le N\\ x\in\mathbb{R}^q}} F(x) e^{2\pi i x\cdot(a+h)}\, dx \\ &= \int_{\mathbb{R}^q} G((L+\lambda)^m; x)\, e^{2\pi i a\cdot x} \left(\Delta_x + \lambda + 4\pi^2 a^2\right)^m F(x-y)\, dx. \end{aligned}$$

Altogether we are led to the following convention:

Function space $C_\lambda^{(2)}(\mathbb{R}^q)$. For a given $\lambda \in \mathbb{R}$, a function $F \in C^{(2m)}(\mathbb{R}^q), m \in \mathbb{N}$, is said to belong to the class $C_\lambda^{(2m)}(\mathbb{R}^q)$, if one of the following conditions is satisfied:

(i) F is of class $X_\lambda^{(2m)}(\mathbb{R}^q) \cap Y_\lambda^{(2m)}(\mathbb{R}^q)$, provided that $\lambda \notin \mathrm{Spect}_\Delta(\Lambda)$,

(ii) F is of class $X_\lambda^{(2m)}(\mathbb{R}^q) \cap Y_\lambda^{(2m)}(\mathbb{R}^q) \cap Z_\lambda^{(2)}(\mathbb{R}^q)$, provided that $\lambda \in \mathrm{Spect}_\Delta(\Lambda)$.

Poisson-type summation formulas over Euclidean spaces. The convergence criteria, which led to (14.40), enable us to derive a multi-dimensional analogue of the Poisson summation formula. In the context of continuous $(L+\lambda)^m$-lattice functions satisfying the condition $m > \frac{q}{2}$, we are able to do this with mathematical rigor. The price, however, is the additional smoothness imposed on the function F under consideration. In more detail, for the integers $m \in \mathbb{N}$ with $m > \frac{q}{2}$, the iterated Λ-lattice function $G((L+\lambda)^m; \cdot)$ permits an absolutely and uniformly convergent Fourier series in $\mathbb{R}^q$. Thus, Lebesgue's theorem allows us to interchange the order of summation and integration. It follows that

$$\begin{aligned} (14.41) \quad &\int_{\mathbb{R}^q} G((L+\lambda)^m; x)\, e^{2\pi i a\cdot x} \left(\Delta_x + \lambda + 4\pi^2 a^2\right)^m F(x-y)\, dx \\ &= \frac{1}{\|\mathcal{F}_\Lambda\|} \sum_{\substack{4\pi^2\left((h+a)^2-a^2\right)\neq\lambda\\ h\in\Lambda}} \frac{1}{(\lambda - 4\pi^2\left((h+a)^2-a^2\right))^m} \\ &\quad \times \int_{\mathbb{R}^q} e^{2\pi i (a+h)\cdot x} \left(\Delta_x + \lambda + 4\pi^2 a^2\right)^m F(x-y)\, dx \end{aligned}$$

holds true for a function $F \in C_\lambda^{(2m)}(\mathbb{R}^q)$. Moreover, in connection with (14.26) and (14.27), repeated application of Green's theorem yields the identity

$$\begin{aligned}(14.42)\qquad &\int_{\mathbb{R}^q} e^{2\pi i(a+h)\cdot x}\, \left(\Delta_x + \lambda + 4\pi^2 a^2\right)^m F(x-y)\, dx \\ &= \int_{\mathbb{R}^q} \left(\Delta_x + \lambda + 4\pi^2 a^2\right)^m e^{2\pi i(a+h)\cdot x}\, F(x-y)\, dx \\ &= (\lambda - 4\pi^2 \left((h+a)^2 - a^2\right))^m \int_{\mathbb{R}^q} e^{2\pi i(a+h)\cdot x}\, F(x-y)\, dx.\end{aligned}$$

Inserting (14.42) into (14.41) we therefore find that

$$\begin{aligned}(14.43)\qquad &\int_{\mathbb{R}^q} G((L+\lambda)^m; x)\, e^{2\pi i a\cdot x} \left(\Delta_x + \lambda + 4\pi^2 a^2\right)^m F(x-y)\, dx \\ &= \frac{1}{\|\mathcal{F}_\Lambda\|}\, e^{2\pi i a\cdot y} \sum_{\substack{4\pi^2\left((h+a)^2 - a^2\right) \neq \lambda \\ h\in\Lambda^{-1}}} e^{2\pi i h\cdot y} \int_{\mathbb{R}^q} F(x)\, e^{2\pi i x\cdot(a+h)} dx.\end{aligned}$$

In combination with (14.40) this finally yields to the desired *multi-dimensional Poisson-type summation formula.*

Let Λ be an arbitrary lattice in the Euclidean space $\mathbb{R}^q$. For a preassigned $\lambda \in \mathbb{R}$, suppose that the function F is of class $C_\lambda^{(2m)}(\mathbb{R}^q)$, $m > \frac{q}{2}$. Then the limit

$$(14.44)\qquad \sum_{g\in\Lambda} e^{2\pi i a\cdot g} F(g-y) = \lim_{N\to\infty} \sum_{\substack{|g-y|\le N \\ g\in\Lambda}} e^{2\pi i a\cdot g} F(g-y)$$

exists, and we have

$$\begin{aligned}(14.45)\qquad &\sum_{g\in\Lambda} e^{2\pi i a\cdot g} F(g-y) \\ &= \frac{1}{\|\mathcal{F}_\Lambda\|} \sum_{h\in\Lambda^{-1}} e^{2\pi i y\cdot(a+h)} \int_{\mathbb{R}^q} F(x)\, e^{2\pi i x\cdot(a+h)}\, dx.\end{aligned}$$

It should be alluded that the sum on the left side of the identity (14.45)) is not necessarily absolutely convergent in $\mathbb{R}^q$, so the spherical summation must be taken into account.

Once more, it should be pointed out that the "wave number" $\lambda \in \mathbb{R}$ reflects the specific character of a function of the class $C_\lambda^{(2m)}(\mathbb{R}^q)$, $m > \frac{q}{2}$, i.e.,

- $C_\lambda^{(2m)}(\mathbb{R}^q) = X_\lambda^{(2m)}(\mathbb{R}^q) \cup Y_\lambda^{(2m)}(\mathbb{R}^q)$, if $\lambda \notin Spect_\Delta(\Lambda)$,
- $C_\lambda^{(2m)}(\mathbb{R}^q) = X_\lambda^{(2m)}(\mathbb{R}^q) \cup Y_\lambda^{(2m)}(\mathbb{R}^q) \cap Z_\lambda^{(2m)}(\mathbb{R}^q)$, if $\lambda \in Spect_\Delta(\Lambda)$.

Hence, our approach shows that it is a helpful prerequisite to adapt the parameter $\lambda \in \mathbb{R}$ specifically to the oscillating properties of the function under consideration. In doing so, our sufficient conditions to establish the validity of the Poisson-type summation formula are particularly suited for the discussion of multi-dimensional alternating series and integrals. Convincing case scenarios within the framework of sampling identities can be found below.

14.3. Multivariate lattice sampling

We base our infinite bandwidth sampling obligations on the lattice point identity (14.45) derived in the last section. It offers the entry point for Shannon-type sampling so that the summation of the resulting cardinal series on the left side of (14.45) is extended over functional values of $F \in C^{(2m)}_{\lambda}(\mathbb{R}^q)$, $m > \frac{q}{2}$, in y-shifted lattice points of Λ.

Functional values based Shannon-type sampling. For $y \in \mathbb{R}^q$ and $F \in C^{(2m)}_{\lambda}(\mathbb{R}^q)$, $m > \frac{q}{2}$, we obtain from the identity (14.45) by integration over a measurable bounded $\mathcal{H} \subset \mathbb{R}^q$

$$\begin{aligned}(14.46)\qquad &\int_{\mathcal{H}} \sum_{g\in\Lambda} e^{2\pi i y\cdot(g-a)} \int_{\mathbb{R}^q} F(x)\, e^{-2\pi i x\cdot(g-a)}\, dx\, da \\ &= \lim_{N\to\infty} \sum_{\substack{|h|\le N\\ h\in\Lambda^{-1}}} F(h)\, \frac{1}{\|\mathcal{F}_\Lambda\|} \int_{\mathcal{H}} e^{2\pi i a\cdot(h+y)}\, da.\end{aligned}$$

We are interested in a modification of the left side of (14.46) such that the right side of (14.46) serves as a cardinal series in a Shannon-type sampling process. We again denote by $\Lambda(\mathcal{H})$ the set of lattice points $g' \in \Lambda \subset \mathbb{R}^q$ such that $\mathcal{F}_\Lambda + \{g'\}$ contains points of $\mathcal{H}$, i.e.,

$$(14.47)\qquad \Lambda(\mathcal{H}) = \{g' \in \Lambda : (\mathcal{F}_\Lambda + \{g'\}) \cap \mathcal{H} \neq \emptyset\}.$$

Using this abbreviation we are able to write

$$\begin{aligned}&\sum_{g'\in\Lambda(\mathcal{H})} \int_{\mathcal{H}_{g'}} \sum_{g\in\Lambda} e^{2\pi i y\cdot(g-a)} \int_{\mathbb{R}^q} F(x)\, e^{-2\pi i x\cdot(g-a)}\, dx\, da \\ (14.48)\qquad &= \lim_{N\to\infty} \sum_{\substack{|h|\le N\\ h\in\Lambda^{-1}}} F(h)\quad K_{\mathcal{H}}(h+y),\end{aligned}$$

where the set $\mathcal{H}_{g'}, g' \in \Lambda(\mathcal{H})$, is given by

$$(14.49)\qquad \mathcal{H}_{g'} = (\mathcal{F}_\Lambda + \{y'\}) \cap \mathcal{H},\ \ g' \in \Lambda(\mathcal{H}).$$

Since the function

$$(14.50)\qquad a \mapsto \sum_{g\in\Lambda} e^{2\pi i y\cdot(g-a)} \int_{\mathbb{R}^q} F(x)\, e^{-2\pi i x\cdot(g-a)} dx,\quad a \in \mathbb{R}^q,$$

is Λ-periodic, we are able to reformulate (14.48) as follows:

$$\begin{aligned}(14.51)\qquad &\sum_{g'\in\Lambda(\mathcal{H})} \int_{\mathcal{H}_{g'}} \sum_{g\in\Lambda} e^{2\pi i y\cdot(g-a)} \int_{\mathbb{R}^q} F(x)\, e^{-2\pi i x\cdot(g-a)}\, dx\, da \\ &= \sum_{g'\in\Lambda(\mathcal{H})} \int_{\mathcal{H}^0_{g'}} \sum_{g\in\Lambda} e^{2\pi i y\cdot(g-a)} \int_{\mathbb{R}^q} F(x)\, e^{-2\pi i x\cdot(g-a)}\, dx\, da \\ &= \lim_{N\to\infty} \sum_{\substack{|h|\le N\\ h\in\Lambda^{-1}}} F(h)\quad K_{\mathcal{H}}(h+y),\end{aligned}$$

where $\mathcal{H}^0_{g'}, g' \in \Lambda(\mathcal{H})$, is the shift of $\mathcal{H}_{g'}$ to the fundamental cell, i.e.,

$$(14.52)\qquad \mathcal{H}^0_{g'} = \mathcal{H}_{g'} - \{g'\},\ g' \in \Lambda(\mathcal{H}),$$

forms a subset of $\mathcal{F}_\Lambda$. In doing so we arrive at the identity

$$\sum_{g'\in\Lambda(\mathcal{H})} \int_{\mathcal{H}^0_{g'}} \sum_{g\in\Lambda} e^{2\pi i y\cdot(g-a)} \int_{\mathbb{R}^q} F(x)\, e^{-2\pi i x\cdot(g-a)}\, dx\, da$$
$$= \sum_{g'\in\Lambda(\mathcal{H})} \int_{\mathbb{R}^q} e^{2\pi i y\cdot w}\ \chi_{\mathcal{H}^0_{g'}+\Lambda}(w) \int_{\mathbb{R}^q} F(x)\ e^{-2\pi i x\cdot w}\, dx\, dw$$
$$= \lim_{N\to\infty} \sum_{\substack{|h|\le N\\ h\in\Lambda^{-1}}} F(h)\quad K_{\mathcal{H}}(h+y), \tag{14.53}$$

where the characteristic function $x\mapsto \chi_{\mathcal{H}^0_{g'}+\Lambda}(x)$, $x\in\mathbb{R}^q$, is, as usual, given by

$$\chi_{\mathcal{H}^0_{g'}+\Lambda}(x) = \begin{cases} 1, & x\in \bigcup\limits_{g\in\Lambda} \left(\mathcal{H}^0_{g'}+\{g\}\right)\\ 0, & \text{otherwise.}\end{cases} \tag{14.54}$$

Summarizing our results we therefore obtain the following *lattice point Shannon-type sampling formula.*

Let $\mathcal{H}$ be a measurable bounded region in $\mathbb{R}^q$. Assume that Λ is an arbitrary lattice in $\mathbb{R}^q$. Suppose that F is a member of the class $C^{(2m)}_\lambda(\mathbb{R}^q)$, $m>\frac{q}{2}$. Then, for all $y\in\mathbb{R}^q$, we have

$$\sum_{\substack{(\mathcal{F}_\Lambda+\{g'\})\cap\mathcal{H}\neq\emptyset\\ g'\in\Lambda}} \int_{\mathbb{R}^q} e^{2\pi i y\cdot w}\ \chi_{\mathcal{H}^0_{g'}+\Lambda}(w) \underbrace{\int_{\mathbb{R}^q} F(x)\ e^{-2\pi i x\cdot w}\, dx}_{=F^\wedge_{\mathbb{R}^q}(w)}\, dw$$
$$= \sum_{\substack{(\mathcal{F}_\Lambda+\{g'\})\cap\mathcal{H}\neq\emptyset\\ g'\in\Lambda}} \int_{\bigcup\limits_{g\in\Lambda}(((\mathcal{H}\cap(\mathcal{F}_\Lambda+\{g'\}))-\{g'\})+\{g\})} F^\wedge_{\mathbb{R}^q}(w)\ e^{2\pi i y\cdot w}\, dw$$
$$= \lim_{N\to\infty} \sum_{\substack{|h|\le N\\ h\in\Lambda^{-1}}} F(h)\quad K_{\mathcal{H}}(h+y).$$

We distinguish the three cases $\overline{\mathcal{H}}\subsetneq\overline{\mathcal{F}_\Lambda}$, $\overline{\mathcal{H}}=\overline{\mathcal{F}_\Lambda}$, and $\overline{\mathcal{H}}\supsetneq\overline{\mathcal{F}_\Lambda}$:

For $\overline{\mathcal{H}}\subsetneq\overline{\mathcal{F}_\Lambda}$ we find that

$$\int_{\bigcup\limits_{g\in\Lambda}(\mathcal{H}+\{g\})} F^\wedge_{\mathbb{R}^q}(x)\ e^{2\pi i y\cdot x}\, dx = \lim_{N\to\infty} \sum_{\substack{|h|\le N\\ h\in\Lambda^{-1}}} F(h)\ \frac{1}{\|\mathcal{F}_\Lambda\|}\quad K_{\mathcal{H}}(h+y). \tag{14.55}$$

Especially, by choosing $\overline{\mathcal{H}}=\overline{\mathbb{B}^q_\tau}\subset\overline{\mathcal{F}_\Lambda}$, we obtain

$$\int_{\bigcup\limits_{g\in\Lambda}(\mathbb{B}^q_\tau+\{g\})} F^\wedge_{\mathbb{R}^q}(x)\ e^{2\pi i y\cdot x}\, dx \tag{14.56}$$
$$= \lim_{N\to\infty} \sum_{\substack{|h|\le N\\ h\in\Lambda^{-1}}} F(h)\ \tau^q\ \frac{\|\mathbb{S}^{q-1}\|}{\|\mathcal{F}_\Lambda\|}\quad \frac{J_1(q;2\pi|h+y|\tau)}{2\pi|h+y|\tau}.$$

Thus we are confronted with the following observation: The closer $\overline{\mathcal{H}}$ is to a finite (closed) union of lattice cells of Λ, the smaller is the aliasing error by this variant of Shannon-type sampling. For $\overline{\mathcal{H}} = \overline{\mathcal{F}_\Lambda}$, we obtain by application of the Fourier inversion formula

$$F(y) = \lim_{N\to\infty} \sum_{\substack{|h|\leq N \\ h\in\Lambda^{-1}}} F(h) \quad K_{\mathcal{F}_\Lambda}(h+y).$$

As a consequence, not-necessarily bandlimited functions F of the class $C_\Lambda^{(2m)}(\mathbb{R}^q)$, $m > \frac{q}{2}$, may be sampled in the standard way by choosing the fundamental cell $\mathcal{F}_\Lambda$ of the lattice Λ to determine the kernel $K_{\mathcal{F}_\Lambda}$

$$F(y) = \lim_{N\to\infty} \sum_{\substack{|h|\leq N \\ h\in\Lambda^{-1}}} F(h) \quad K_{\mathcal{F}_\Lambda}(h+y), \quad y \in \mathbb{R}^q. \tag{14.57}$$

For $\overline{\mathcal{H}} \supsetneq \overline{\mathcal{F}_\Lambda}$

$$\begin{aligned} \sum_{\substack{(\mathcal{J}_\Lambda + \{g'\})\cap\mathcal{H}\neq\emptyset \\ g'\in\Lambda}} \int_{\bigcup_{g\in\Lambda}(((\mathcal{H}\cap(\mathcal{F}_\Lambda+\{g'\}))-\{g'\})+\{g\})} F^\wedge_{\mathbb{R}^q}(x)\ e^{2\pi i y\cdot x}\ dx \\ = \lim_{N\to\infty} \sum_{\substack{|h|\leq N \\ h\in\Lambda^{-1}}} F(h) \ K_{\mathcal{H}}(h+y). \end{aligned} \tag{14.58}$$

All in all, we are led to the conclusion that, in the case where $\mathcal{H}$ consists of a finite union of lattice cells, aliasing can be avoided. So, the functional based Shannon-type sampling as discussed in this section does not provide a cornucopia of results comparable to the bandlimited case involving finite regions. Nevertheless, a way is found in not-necessarily bandlimited sampling to detect the explicit aliasing error and to formulate Shannon-type approximate identities.

Part 7

Sampling examples

CHAPTER 15

Sampling methodologies in technology

In this chapter we discuss two important sampling applications in modern technology, namely the *inversion of the 3D gravimetry problem* yielding the density distribution from uniform data by use of certain expressions allowing to reproduce exponentials (Section 15.1) and the *inversion of the qD antenna problem* by specific Paley-Wiener and Slepian spline structures, providing the aperture distribution from a given far-field radiation pattern (Section 15.2). As a consequence, the recovery philosophy in its constituents, i.e., regularization and sampling, finds a convincing manifestation in both of the above examples.

15.1. Exponential reproduction inversion in gravimetry

Our concept is essentially motivated by the work proposed by J. Murray-Bruce, P.L. Dragotti [**608**]. While these authors essentially deal with sensor network strategies for solving inverse source problems for diffusion and wave fields, our special interest is in the Earth's gravitational potential field (in continuation of the context already presented in Chapter 7). We demonstrate that the unknown signature of the density distribution inside an area $\mathcal{B}$ of the Earth can be recovered from a sequence of so-called generalized measurements by using frequency estimation techniques. The weights of the sums correspond to certain expressions, which reproduce multidimensional exponentials (similarly to the Gaussian approach (7.378)). Explicit formulas are derived for the sequence of weights, that map gravimetric samples to the sequence of generalized measurements, thereby observing the fact in case of uniform data that the fundamental solution of the Laplace differential operator satisfies the Strang-Fix condition.

Inverse gravimetry sampling decorrelation problem. In exploration geophysics, the inverse source problem of gravimetry is playing an increasingly important role, as it is suitable in future to replace the much more expensive seismic methods in theory and measurement. Inverse gravimetry (IG) has already been discussed as an inverse problem in Chapter 7. The point of departure in the case of IG is Poisson's differential equation, which is using the Laplace operator Δ, so that $\rho = -\Delta V$ in $\mathcal{B}$, where according to the Newtonian theory, V is the gravitational geopotential to the (sufficiently smooth) density distribution ρ. The general solution of the Poisson equation can be calculated with the help of the fundamental solution (Newtonian monopole kernel) of the Laplace operator Δ in Euclidean space $\mathbb{R}^3$

$$F_3^{(1)}(|x-y|) = \frac{1}{4\pi}\frac{1}{|x-y|}, \quad x, y \in \mathbb{R}^3,\ x \neq y. \tag{15.1}$$

by the improper (volumetric) Newtonian integral representation

$$V(x) = \int_{\mathcal{B}} F_3^{(1)}(|x-y|)\, \rho(y)\, dy, \quad x \in \mathbb{R}^3. \tag{15.2}$$

Moreover, in the mollifier nomenclature presented in Chapter 7, V can be substituted by $V_{K_{\tau_J}}$ given by the proper integral

$$V_{K_{\tau_J}}(x) = \int_{\mathcal{B}} \left(F_3^{(1)}\right)_{K_{\tau_J}} (|x-y|)\, \rho(y)\, dy \tag{15.3}$$

with J sufficiently large and $\left(F_3^{(1)}\right)_{K_{\tau_J}}$ given by (7.313). So, inverse gravimetry (IG) amounts to the determination of the density distribution

$$\rho_{K_{\tau_J}}(x) = \int_{\mathcal{B}} K_{\tau_J}(|x-y|)\, \rho(y)\, dy, \quad x \in \mathcal{B}, \tag{15.4}$$

using the integral equation (15.3) corresponding to the gravitational potential $V_{K_{\tau_J}}$.

In the generalized lattice sampling paradigm (see, e.g., [**21**], [**836**] and the references therein) the primary goal is to reconstruct some functions of a continuous variable from a discrete set of measurements collected on a lattice (i.e., uniform grid). Our sampling approach below is based on canonical assumptions on today's gravimetry: Suppose that a gravimeter collects a uniform sample of values of the (mollifier approximation of the) Earth's gravitational potential $V_{K_{\tau_J}}$. Furthermore, we assume that a basic version of the Earth's gravitational potential is known, e.g., from an available Earth's gravitational model EGM (see, e.g., [**682**]). Mathematically, we interpret the a priori known information from the EGM-model in mollifier nomenclature as low-pass filter $V_{K_{\tau_{J_0}}}$ for some scale $J_0 < J$.

The inverse gravimetry sampling problem to be solved is as follows: Given the potential difference

$$(WV)_{J_0}^{J} = V_{K_{\tau_J}} - V_{K_{\tau_{J_0}}}, \quad J_0 < J, \tag{15.5}$$

for a (pointset of a) lattice $\Lambda \subset \mathbb{R}^3$, we intend to estimate the unknown density signature $\rho_{K_{\tau_J}}$ in $\mathcal{B}$.

The reason for using mollifier bandpass structured samples $(WV)_{J_0}^{J}$ is that the associated Newtonian mollifier bandpass kernel

$$\left(WF_3^{(1)}\right)_{J_0}^{J} = \left(F_3^{(1)}\right)_{K_{\tau_J}} - \left(F_3^{(1)}\right)_{K_{\tau_{J_0}}}, \quad J_0 < J, \tag{15.6}$$

shows a local support (for further details the reader is referred to Chapter 7).

Context of exponential reproduction from uniform translates. The main focus of our considerations will be on the following mollifier parametrization (see Chapter 7) for the required density distribution

$$\rho(x) = \sum_{l=1}^{M} c_l\, \rho_{K_{\tau_J}}(|x-x_l|), \quad x \in \overline{\mathcal{B}}, \tag{15.7}$$

where $c_l \in \mathbb{R}$ is the intensity at the location $x_l \in \mathcal{B}$. For the sufficiently large scale J the kernel K_{τ_J} takes the role of the Dirac delta function. As a result of the parametrization (15.7), IG becomes one of estimating the unknown source parameters $\{c_l, x_l\}_{l=1,\dots,M}$, in the spirit of [**608**], but now in our mollifier nomenclature of Chapter 7, by the *framework of generalized measurements*

$$(15.8)\qquad (\rho_{K_{\tau_J}})^{\wedge}_{\mathcal{B}}(x) = \int_{\mathcal{B}} \rho_{K_{\tau_J}}(y)\; e^{-2\pi i\; x\cdot y}\; dy = \sum_{l=1}^{M} c_l\; e^{-2\pi i\; x\cdot x_l},\; x \in \mathbb{R}^3.$$

Obviously, we propose the use of exponentials with purely imaginary exponents as the proper choice for our sensing functions. The reasons are twofold: The first is that the left side of (15.8) becomes a system of superimposed sinusoids, which can be efficiently solved by using multidimensional extensions of Prony's frequency estimation methods. The second reason is because the use of imaginary exponentials improves the stability of the estimation problem, since in this case the magnitude of the terms in the generalized sequence remain bounded.

In order to compute the desired sequence of generalized measurements we again follow [**608**], i.e., we start from a weighted linear combination of the form

$$(15.9)\qquad \sum_{g\in\Lambda} \sigma_k(g)\; \left(WF_3^{(1)}\right)^{J}_{J_0} (|g-x|) = e^{-2\pi i\; k\cdot x},\quad x\in\mathbb{R}^3,\; k\in\mathbb{Z}^q,$$

(note that $\mathbb{N}^q$ is used in [**608**] instead of $\mathbb{Z}^q$).

Then, in connection with the Newtonian mollifier theory (cf. (15.3)), it follows that

$$\begin{aligned}(15.10)\qquad (\rho_{K_{\tau_J}})^{\wedge}_{\mathcal{B}}(k) &= \int_{\mathcal{B}} \rho_{K_{\tau_J}}(y)\; e^{-2\pi i\; k\cdot y}\; dy \\ &= \int_{\mathcal{B}} \rho_{K_{\tau_J}}(y) \sum_{g\in\Lambda} \sigma_k(g)\; \left(WF_3^{(1)}\right)^{J}_{J_0} (|g-y|)\; dy \\ &= \sum_{g} \sigma_k(g\in\Lambda) \int_{\mathcal{B}} \rho_{K_{\tau_J}}(y)\; \left(WF_3^{(1)}\right)^{J}_{J_0} (|g-y|)\; dy \\ &= \sum_{g\in\Lambda} \sigma_k(g)\; (WV)^{J}_{J_0}(g),\qquad k\in\mathbb{Z}^q.\end{aligned}$$

Thus, computing the set of generalized measurements $(\rho_{K_{\tau_J}})^{\wedge}_{\mathcal{B}}(k)$ by taking $\sigma_k(g)$-weighted linear combinations of the gravimetric data $(WV)^{J}_{J_0}(g)$ is equivalent to reproducing exponentials $e^{-2\pi i\; k\cdot x}$ from space translates of the Newton mollifier wavelet kernel (3D mollifier wavelet monopole). Altogether, we are confronted with an *exponential reproduction problem* (15.9), where the only missing piece is how to obtain the coefficients $\sigma_k(g)$. Surprisingly, these coefficients are known from the literature (see [**799**]), because, the Newton mollifier wavelet kernel (15.6) is locally supported. Therefore, the equations (15.9) satisfy the Strang-Fix conditions (cf. [**799**]), so that we are immediately able to deduce that

$$(15.11)\qquad \sigma_k(g) = \frac{e^{-2\pi i k\cdot g}}{\left(\left(WF_3^{(1)}\right)^{J}_{J_0}\right)^{\wedge}_{\mathbb{R}^3}(k)},\qquad k\in\mathbb{Z}^q,\; g\in\Lambda.$$

Estimating the parameters of superimposed exponentials. With the access to the desired coefficients, $\sigma_k(g)$, that are capable of reproducing exponentials from the translates of the Newtonian wavelet monopole kernel, all we have to do at the end is the evaluation of $\{(\rho_{K_{\tau_J}})^{\wedge}_{\mathcal{B}}(k)\}_{k\in\mathbb{Z}^q}$, to extract the unknown source parameters $\{c_l, x_l\}_{l=1,\ldots,M}$. In fact, the sequence (15.8) forms a superposition of damped complex sinusoids. Thus, we are confronted with a Prony-like system. This system is prevalent, for example, in spectral estimation and array processing applications. Many approaches have been put forward for recovering the unknown frequencies and amplitudes, particularly for the two-dimensional case (see, e.g., [**449**], [**557**], [**728**]). Solving the Prony-like system allows to recover the unknown source parameters by utilizing, e.g., an ESPRIT algorithm by S. Sahnoun et al. [**740**], which represents an update of the two-dimensional ESPRIT algorithm of [**728**]. Specifically, in the framework as presented here, an adapted ESPRIT algorithm can be found in [**608**].

Non-uniform gravimeter data placement. A simple, yet effective scheme for handling a non-uniform gravimeter data placement is to interpolate the field and resample it on a uniform grid. In so doing, we can obtain an approach that can still exploit the closed form expression as above even when the original data points are irregularly positioned. Besides avoiding the inversion of poorly-conditioned matrices occurring in linear systems for non-uniform gravimeter data, the interpolation approach is less intensive computationally compared to matrix inversion, particularly when the coefficient matrix of the linear system is large. In experimental results, 1D splines may be used as the interpolatory family.

Finally, it should be mentioned that sampling of inverse magnetometry (IM) can be handled analogously to the methodology for inverse gravimetry (IG).

15.2. Slepian spline inversion in antenna technology

In what follows, we continue in canonical way the 1D sampling theory presented in the beginning of this work (Chapter 1). However, instead of the fundamental cell of the unit lattice we choose a regular region inside the fundamental cell of an arbitrary multi-dimensional lattice Λ.

Paley-Wiener reproducing kernel Hilbert space. Let $\mathcal{G} \subset \mathbb{R}^q$ be a regular region with $\overline{\mathcal{G}} \subset \overline{\mathcal{F}_\Lambda}$, so that the continuous signal

$$F^{\wedge}_{\mathcal{G}}(y) = \int_{\mathcal{G}} F(w) e^{-2\pi i w\cdot y}\, dw,\ y \in \mathbb{R}^q, \tag{15.12}$$

is recovered from the sampled signal over lattice points of the inverse lattice $F^{\wedge}_{\mathcal{G}}(h)$, $h \in \Lambda^{-1}$, i.e.,

$$F^{\wedge}_{\mathcal{G}}(y) = \sum_{h\in\Lambda^{-1}} F^{\wedge}_{\mathcal{G}}(h)\ K_{\mathcal{G}}(h-y),\ y \in \mathbb{R}^q, \tag{15.13}$$

with

$$K_{\mathcal{G}}(x-y) = \frac{1}{\|\mathcal{F}_\Lambda\|} \int_{\mathcal{G}} e^{-2\pi i w\cdot(x-y)}\, dw,\ x, y \in \mathbb{R}^q. \tag{15.14}$$

We are able to deduce some interesting results in the area of *approximate integration* (see also [**278**]): As a matter of fact, the bandlimited function $F^{\wedge}_{\mathcal{G}}$ allows us to express its integral over the Euclidean space $\mathbb{R}^q$ by the product of the lattice density and the sum over all samples in points of the inverse lattice

$$\lim_{N\to\infty} \int_{\substack{|x|\leq N\\ x\in\mathbb{R}^q}} F_{\mathcal{G}}^{\wedge}(x)\, dx = \frac{1}{\|\mathcal{F}_\Lambda\|} \sum_{h\in\Lambda^{-1}} F_{\mathcal{G}}^{\wedge}(h). \tag{15.15}$$

Furthermore, the Parseval identity yields

$$\frac{1}{\|\mathcal{F}_\Lambda\|} \sum_{h\in\Lambda^{-1}} |F_{\mathcal{G}}^{\wedge}(h)|^2 = \int_{\mathcal{G}} |F(x)|^2\, dx. \tag{15.16}$$

From the Fourier theory it follows that

$$\int_{\mathcal{G}} |F(x)|^2\, dx = \int_{\mathbb{R}^q} \left|F_{\mathcal{G}}^{\wedge}(y)\right|^2\, dy. \tag{15.17}$$

In other words, if $F_{\mathcal{G}}^{\wedge}$, $\mathcal{G} \subset \overline{\mathcal{F}_\Lambda}$, belongs to the inner product space

$$B_{\mathcal{G}}^{(0)} = \left\{ \underbrace{\int_{\mathcal{G}} F(x)\, e^{-2\pi i x\cdot}\, dx}_{=F_{\mathcal{G}}^{\wedge}} : F \in C^{(0)}(\overline{\mathcal{G}}) \right\}, \tag{15.18}$$

then

$$\int_{\mathbb{R}^q} |F_{\mathcal{G}}^{\wedge}(y)|^2\, dy = \frac{1}{\|\mathcal{F}_\Lambda\|} \sum_{h\subset\Lambda^{-1}} |F_{\mathcal{G}}^{\wedge}(h)|^2. \tag{15.19}$$

Replacing Λ by its inverse lattice Λ^{-1} we find that

$$\int_{\mathbb{R}^q} |F_{\mathcal{G}}^{\wedge}(y)|^2\, dy = \|\mathcal{F}_\Lambda\| \sum_{g\in\Lambda} |F_{\mathcal{G}}^{\wedge}(g)|^2. \tag{15.20}$$

Looking at our above approach critically we notice that the sampling result is formulated on the reference set $B_{\mathcal{G}}^{(0)}$, that is a strict subset of the *Paley-Wiener space*

$$B_{\mathcal{G}} = \left\{ y \mapsto \int_{\mathcal{G}} F(x)\, e^{-2\pi i x\cdot y}\, dx,\ y \in \mathbb{R}^q : F \in L^2(\mathcal{G}) \right\}. \tag{15.21}$$

This observation, however, does not bother us very much, since every $F \in L^2(\mathcal{G})$ can be approximated (in $L^2(\mathcal{G})$–sense) by a function $F_\varepsilon \in C^{(0)}(\overline{\mathcal{G}})$ in ε-accuracy such that

$$\begin{aligned}
\sup_{y\in\mathbb{R}^q} \left|F_{\mathcal{G}}^{\wedge}(y) - (F_\varepsilon)_{\mathcal{G}}^{\wedge}(y)\right| &\leq \sup_{y\in\mathbb{R}^q} \left| \int_{\mathcal{G}} (F(x) - F_\varepsilon(x))\, e^{-2\pi i x\cdot y}\, dx \right| \\
&\leq \left(\int_{\mathcal{G}} |F(x) - F_\varepsilon(x)|^2\, dx \right)^{1/2} \left(\int_{\mathcal{G}} |e^{-2\pi i x\cdot y}|^2\, dx \right)^{1/2} \\
&\leq \left(\int_{\mathbb{R}^q} \left|F_{\mathcal{G}}^{\wedge}(x) - (F_\varepsilon)_{\mathcal{G}}^{\wedge}(x)\right|^2\, dx \right)^{1/2} \left(\int_{\mathcal{G}} |e^{-2\pi i x\cdot y}|^2\, dx \right)^{1/2} \\
&= \sqrt{\|\mathcal{G}\|}\ \varepsilon.
\end{aligned}$$

All in all, if $\mathcal{G} \subset \mathbb{R}^q$ is a regular region with $\overline{\mathcal{G}} \subset \overline{\mathcal{F}_\Lambda}$, then the *Paley–Wiener space* $B_{\overline{\mathcal{G}}}$ is the completion of the space $B^{(0)}_{\mathcal{G}}$ under the $L^2(\mathbb{R}^q)$-topology, i.e.,

$$B_{\mathcal{G}} = \overline{B^{(0)}_{\mathcal{G}}}^{\|\cdot\|_{L^2(\mathbb{R}^q)}}. \tag{15.22}$$

Under the aforementioned assumption that $\mathcal{G}$ is a regular region with $\overline{\mathcal{G}} \subset \overline{\mathcal{F}_\Lambda}$, the set $B_{\mathcal{G}}$ forms a *reproducing kernel space* with kernel (15.14). In fact, by virtue of (15.17), we see that for all $y \in \mathbb{R}^q$

$$|F^{\wedge}_{\mathcal{G}}(y)| \leq \sqrt{\|\mathcal{G}\|}\sqrt{\int_{\mathcal{G}} |F(x)|^2\, dx} = \sqrt{\|\mathcal{G}\|}\sqrt{\int_{\mathbb{R}^q} |F^{\wedge}_{\mathcal{G}}(w)|^2\, dw}. \tag{15.23}$$

Moreover, standard Fourier inversion yields

$$F^{\wedge}_{\mathcal{G}}(y) = \int_{\mathbb{R}^q} F^{\wedge}_{\mathcal{G}}(x) \left(\frac{1}{\|\mathcal{F}_\Lambda\|} \int_{\mathcal{G}} e^{2\pi i w\cdot(x-y)}\, dw \right) dx = \int_{\mathbb{R}^q} F^{\wedge}_{\mathcal{G}}(x)\, K_{\mathcal{G}}(x-y)\, dx \tag{15.24}$$

for all $y \in \mathbb{R}^q$, where $\int_{\mathbb{R}^q} \cdots = \lim_{N\to\infty} \int_{|x|\leq N, \, x\in\mathbb{R}^q} \cdots$. Hence, $B_{\mathcal{G}}$ is a reproducing kernel Hilbert space with kernel (15.14). Even more, going over to the Paley-Wiener space $B_{\mathcal{F}_\Lambda}$ we are able to guarantee, in addition, that the reproducing kernels form an orthonormal system with discrete othogonality property (see, e.g., M.Z. Nashed [**629**], M.Z. Nashed, Q. Sun [**637**] for the univariate case).

Paley-Wiener spline inversion. There are diverse important spline concepts in the context of lattice point sampling, for example, [**21**], [**103**], [**127**], [**152**], [**809**], mostly based on (iterated) univariate approaches. Here, in the line of the spline concepts developed in [**256**], [**257**],[**261**], [**263**], we propose a specific spline procedure resulting from the reproducing kernel structure of the Paley-Wiener space $B_{\overline{\mathcal{G}}}$. More concretely, let $\mathcal{G}$ be a regular region in $\mathbb{R}^q$ with $\overline{\mathcal{G}} \subset \overline{\mathcal{F}_\Lambda}$. Suppose that $\mathcal{Q} \subset \mathbb{R}^q$ is a finite region such that the total number $\sharp_{\Lambda^{-1}}(\mathcal{Q})$ of lattice points $h \in \Lambda^{-1}$ contained in $\mathcal{Q}$ is positive. Assume that the set of far field radiation patterns $\{F^{\wedge}_{\mathcal{G}}(h)\}_{h\in\mathcal{Q}}$ corresponding to the finite set of lattice points $h \in \mathcal{Q}$ are known.

Our goal is to deal with *Paley-Wiener spline interpolation*, i.e., minimum norm interpolation in the Paley-Wiener Hilbert space $B_{\mathcal{G}}$. We determine the spline interpolant of $F^{\wedge}_{\mathcal{G}}$ from the known values $F^{\wedge}_{\mathcal{G}}(h'), h' \in \mathcal{Q}$, in the following way: Suppose that the matrix $K_{\mathcal{Q}} = [K_{\mathcal{G}}(h'-h)]_{h',h\in\mathcal{Q}}$ is nonsingular (note that $K_{\mathcal{Q}}$ is a Gram matrix, so it suffices to assume that the functions $\{K_{\mathcal{G}}(\cdot - h)\}_{h\in\mathcal{Q}}$ are linearly independent). Then, any function $S_{\mathcal{Q}}$ of the form

$$S_{\mathcal{Q}}(y) = \sum_{h'\in\mathcal{Q}} a_{h'}\, K_{\mathcal{G}}(h'-y), \quad y \in \mathbb{R}^q, \tag{15.25}$$

is called a *$B_{\mathcal{G}}$-spline function relative to $\mathcal{Q}$*. The class of all $B_{\mathcal{G}}$ -splines relative to $\mathcal{Q}$ is denoted by $\mathrm{Spline}_{B_{\mathcal{G}}}(\mathcal{Q})$. Obviously, $\mathrm{Spline}_{B_{\mathcal{G}}}(\mathcal{Q})$ is a finite subset of $B_{\mathcal{G}}$ (for simplicity, we only use lattice points as nodal system for the spline. The construction of Paley-Wiener splines also using non-lattice data points is straightforward).

By virtue of the reproducing kernel structure it is easy to verify that the so-called *spline interpolation formula*

$$\sum_{h'\in\mathcal{Q}} a_{h'}P(h') = \sum_{h'\in\mathcal{Q}} a_{h'} \int_{\mathbb{R}^q} K_{\mathcal{G}}(h'-y)\,\overline{P(y)}\,dy = \int_{\mathbb{R}^q} S_{\mathcal{Q}}(y)\,\overline{P(y)}\,dy \tag{15.26}$$

holds true for all $P \in B_{\mathcal{G}}$. There exists one and only one spline function $\hat{S}^F_{\mathcal{Q}} \in \mathrm{Spline}_{B_{\mathcal{G}}}(\mathcal{Q})$ of the form

$$\hat{S}^F_{\mathcal{Q}}(y) = \sum_{h'\in\mathcal{Q}} \hat{a}^F_{h'} K_{\mathcal{G}}(h'-y), \quad y \in \mathbb{R}^q, \tag{15.27}$$

satisfying the linear equations $\hat{S}^F_{\mathcal{Q}}(h) = F^{\wedge}_{\mathcal{G}}(h), \quad h \in \mathcal{Q}$, i.e.,

$$\sum_{h'\subset\mathcal{Q}} \hat{a}^F_{h'}\; K_{\mathcal{G}}(h'-h) \;=\; F^{\wedge}_{\mathcal{G}}(h), \;\; h \in \mathcal{Q}. \tag{15.28}$$

From the linear equations it follows that

$$\int_{\mathbb{R}^q} \hat{S}^F_{\mathcal{Q}}(y)\,\overline{(\hat{S}^F_{\mathcal{Q}}(y) - P(y))}\,dy = \sum_{h'\in\mathcal{Q}} \hat{a}^F_{h'}\hat{S}^F_{\mathcal{Q}}(h) - \sum_{h'\in\mathcal{Q}} \hat{a}^F_{h'}P(h') \;=\; 0 \tag{15.29}$$

holds for all $P \in B_{\overline{\mathcal{G}}}$ with $P(h) = F^{\wedge}_{\mathcal{G}}(h)$, $h \in \mathcal{Q}$. This result implies that

$$\begin{aligned} \int_{\mathbb{R}^q} P(y)\,\overline{P(y)}\,dy &= \int_{\mathbb{R}^q} \hat{S}^F_{\mathcal{Q}}(y)\,\overline{\hat{S}^F_{\mathcal{Q}}(y)}\,dy \\ &\quad + \int_{\mathbb{R}^q} \Big(\hat{S}^F_{\mathcal{Q}}(y) - P(y)\Big)\,\overline{(\hat{S}^F_{\mathcal{Q}}(y) - P(y))}\,dy. \end{aligned} \tag{15.30}$$

Summarizing our results we obtain the *variational formulation of Paley-Wiener spline interpolation*:

Suppose that a finite discrete dataset $\{F^{\wedge}_{\mathcal{G}}(h), h \in \mathcal{Q}\}$ *is known. Then, the spline function given by*

$$\hat{S}^F_{\mathcal{Q}}(y) = \sum_{h'\in\mathcal{Q}} \hat{a}^F_{h'} \frac{1}{\|\mathcal{F}_\Lambda\|} \int_{\mathcal{G}} e^{2\pi i x\cdot(h'-y)}\,dx,\; y \in \mathbb{R}^q, \tag{15.31}$$

with coefficients $\hat{a}^F_{h'}$ *determined by the linear equations* (15.28) *fulfills the minimum norm property*

$$\left(\int_{\mathcal{G}} \left|\hat{S}^F_{\mathcal{Q}}(y)\right|^2 dy\right)^{1/2} = \min_{\substack{P\in B_{\overline{\mathcal{G}}}\\ P(h)=F^{\wedge}_{\mathcal{G}}(h)\\ h\in\Lambda^{-1}\cap\mathcal{Q}}} \left(\int_{\mathcal{G}} |P(y)|^2\,dy\right)^{1/2}. \tag{15.32}$$

In addition, the *Paley-Wiener spline interpolatory convergence* is guaranteed. More explicitly (see [**296**]), let $F^{\wedge}_{\mathcal{G}}$ be of class $B_{\overline{\mathcal{G}}}$. Suppose that $\{\Lambda_N^{-1}\}$ is a sequence of lattices Λ_N^{-1} such that $\Theta_{\mathcal{Q}}^{\Lambda_N^{-1}} = \sup_{y\in\mathbb{R}^q}\left(\min_{\substack{h\in\mathcal{Q}\\ h\in\Lambda^{-1}}} |y-h|\right) \to 0$ as $N \to \infty$.

Then the sequence $\{(\hat{S}_{\mathcal{Q}}^{F})^{\Lambda_N^{-1}}\}$ of the uniquely determined solutions $(\hat{S}_{\mathcal{Q}}^{F})^{\Lambda_N^{-1}}$ of the variational problems

$$\left(\int_{\mathcal{G}} \left|(\hat{S}_{\mathcal{Q}}^{F})^{\Lambda_N^{-1}}(y)\right|^2 dy\right)^{1/2} = \min_{\substack{P\in B_{\mathcal{G}}\\ P(h)=F_{\mathcal{G}}^{\wedge}(h)\\ h\in\Lambda_N^{-1}\cap\mathcal{Q}}} \left(\int_{\mathcal{G}} |P(y)|^2\, dy\right)^{1/2} \tag{15.33}$$

satisfies the limit relation $\lim_{N\to\infty}\ \sup_{y\in\overline{\mathcal{G}}} \mid (\hat{S}_{\mathcal{Q}}^{F})^{\Lambda_N^{-1}}(y) - F_{\mathcal{G}}^{\wedge}(y) \mid = 0$. In other words, we are able to approximate the far-field pattern $F_{\mathcal{G}}^{\wedge} \in B_{\mathcal{G}}$ in a constructive way using Paley-Wiener spline interpolation provided that the "interpolatory width" satisfies the condition $\Theta_{\mathcal{Q}}^{\Lambda_N^{-1}} \to 0$ as $N \to \infty$. Further spline properties (such as Sard's Theorem, Schönberg's Theorem, spline smoothing of noisy data, best approximation of linear functionals, etc.) can be realized in a canonical way (see, e.g., [**256**, **257**, **263**] for the methodological background) by use of the reproducing property. The details are omitted here. Moreover, combined spline interpolation/smoothing (as proposed in [**331**]) can be used instead of spline interpolation if the data are (partially) noisy.

Finally, knowing the far-field pattern $\hat{S}_{\mathcal{Q}}^{F}$, the function $F_{\mathcal{Q}}$ given by

$$F_{\mathcal{Q}}(x) = \sum_{h'\in\mathcal{Q}} \hat{a}_{h'}^{F} \frac{1}{\|\mathcal{F}_\Lambda\|}\, e^{2\pi i x\cdot h'},\ x \in \mathbb{R}^q, \tag{15.34}$$

can be considered as an *approximation of the desired aparture distribution* F, since

$$\int_{\mathcal{G}} F_{\mathcal{Q}}(x) e^{-2\pi i x\cdot y}\, dx = \hat{S}_{\mathcal{Q}}^{F}(y) = \sum_{h'\in\mathcal{Q}} \hat{a}_{h'}^{F} \frac{1}{\|\mathcal{F}_\Lambda\|} \int_{\mathcal{G}} e^{2\pi i x\cdot(h'-y)} dx, \quad y \in \mathbb{R}^q.$$

Slepian spline inversion. The uncertainty principle tells us that functions cannot be space- and bandlimited at the same time (see, e.g., [**155**], [**297**]). Finding signals that are optimally concentrated in both is a fundamental problem in information theory, for which fundamental results were given in the early 1960s by D. Slepian, H.J. Landau, and H.O. Pollak [**523**], [**778**], [**779**], [**782**]. The concentration problem has strong connections to the role of the uncertainty principle in the concept of multiscale approximation (see, e.g., I. Daubechies [**173**], W. Freeden, T. Gervens, M. Schreiner [**276**], W. Freeden [**266**], W. Freeden, M. Schreiner [**321**, **326**], W. Freeden, M.Z. Nashed, M. Schreiner [**297**]).

Our interest next is in what is usually referred to as Slepian functions (see, e.g., F.J. Simons [**776**], F.J. Simons, A. Plattner [**777**], W. Freeden et al. [**297**] and the references therein). In fact, the ideas and concepts of Slepian functions can be applied to the system $\{K_{\mathcal{G}}(h' - \cdot)\}_{h'\in\mathcal{Q}}$ (see also W. Freeden [**271**]). For this purpose we consider spline functions relative to $\mathcal{Q}$ of the form

$$S(x) = \sum_{h'\in\mathcal{Q}} b_{h'}\ K_{\mathcal{G}}(h' - x), \quad x \in \mathbb{R}^q, \tag{15.35}$$

with the aim that they maximize the *spatial concentration*

$$0 < \gamma_{\mathcal{G}}(S) = \frac{\displaystyle\int_{\mathcal{G}} |S(x)|^2\, dx}{\displaystyle\int_{\mathbb{R}^q} |S(x)|^2\, dx} < 1. \tag{15.36}$$

In antenna theory, the value $\gamma_{\mathcal{G}}(S)$ may be also interpreted as a *superdirective ratio*, i.e., a measure of the difficulty in realizing the aparture distribution from discrete far field knowledge, since the aparture distribution may have rapid oscillation (which imply that a portion of the input power would be radiated). With the realization condition in mind, the requirement (15.36) means a restriction of the aparture distribution (especially by the prescribed value 1).

The condition (15.36) leads to a generalized eigenvalue problem

$$\sum_{h'\in\mathcal{Q}} \langle K_{\mathcal{G}}(h-\cdot), K_{\mathcal{G}}(h'-\cdot)\rangle_{L^2(\mathcal{G})}\ b_{h'} \tag{15.37}$$

$$= \lambda \sum_{h'\in\mathcal{Q}} \langle K_{\mathcal{G}}(h-\cdot), K_{\mathcal{G}}(h'-\cdot)\rangle_{L^2(\mathbb{R}^3)}\ b_{h'},\ h\in\mathcal{Q}. \tag{15.38}$$

We assume that the eigenvalues are ordered such that $1 \geq \lambda_{\mathcal{G}}^{(1)} \geq \ldots \geq \lambda_{\mathcal{G}}^{\sharp_{\Lambda^{-1}}\mathcal{Q}} \geq 0$. The corresponding eigenvectors $b_{h'}^{(j)}$, $h' \in \mathcal{Q}$, $j = 1, \ldots, \sharp_{\Lambda^{-1}}\mathcal{Q}$, generate the *spline-type Slepian functions*

$$K^{(j)}(x) = \sum_{h'\in\mathcal{Q}} b_{h'}^{(j)}\ K_{\mathcal{G}}(h'-x),\quad j = 1,\ldots,\sharp_{\Lambda^{-1}}\mathcal{Q}. \tag{15.39}$$

They form an orthogonal system of functions represented by a linear combination of $\{K_{\mathcal{G}}(h'-\cdot)\}_{h'\in\mathcal{Q}}$. The smaller the index j is, the more concentrated is the function $K^{(j)}$ in the region $\mathcal{G}$.

Constructive approximation using Slepian functions (including noise) has enjoyed increasing popularity over the last 20 years (see, e.g., the geomathematical strategies explained by F.J. Simons [**776**] and also the references therein). This is the reason why we do not go into detail to specify coefficients c_j, $j = 1, \ldots, \sharp_{\Lambda^{-1}}\mathcal{Q}$, such that a linear combination

$$\sum_{j=1}^{\sharp_{\Lambda^{-1}}\mathcal{Q}} c_j\ K^{(j)} \tag{15.40}$$

is "close to" the far field $F_{\mathcal{G}}^{\wedge}$, so that

$$\begin{aligned} F_{\mathcal{G}}^{\wedge}(x) \approx \sum_{j=1}^{\sharp_{\Lambda^{-1}}\mathcal{Q}} c_j\ K^{(j)}(x) &= \sum_{j=1}^{\sharp_{\Lambda^{-1}}\mathcal{Q}} c_j \sum_{h'\in\mathcal{Q}} b_{h'}^{(j)}\ K_{\mathcal{G}}(h'-x), \\ &= \sum_{h'\in\mathcal{Q}} \underbrace{\sum_{j=1}^{\sharp_{\Lambda^{-1}}\mathcal{Q}} c_j\ b_{h'}^{(j)}}_{=d_{h'}}\ K_{\mathcal{G}}(h'-x). \end{aligned} \tag{15.41}$$

As a consequence of (15.41), the aperture distribution F admits an approximation of the form

$$F(x) \approx \sum_{h'\in\mathcal{Q}} d_{h'}\ e^{2\pi i x\cdot h'},\quad d_{h'} = \sum_{j=1}^{\sharp_{\Lambda^{-1}}\mathcal{Q}} c_j\ b_{h'}^{(j)}. \tag{15.42}$$

All in all, the antenna problem induced extensions of the Shannon-type sampling theorem as presented here allow many applications in engineering and physics.

Machine learning induced antenna sampling. Machine learning is thought of as a very powerful tool that can improve our ability to carry out scientific research. However, specific issues and suitable data amount need to be addressed before this impression becomes a reality. Our specific consideration below (influenced by work of the Princeton University (see, e.g., [**201**]) and the Massachusetts Institute of Technology (see, e.g. [**6**])) is especially devoted to the question: How can we integrate machine learning with antenna-based modeling to develop reliable physical results? As a matter of fact, we only provide an intuitive explanation for the reasons behind the success of machine learning in handling the multivariate antenna problem. Before getting into the integration of machine learning with antenna-based modeling, let us briefly discuss an issue that is in many people's mind: What is the magic behind machine learning? While this question is still at the center of very intensive studies in theoretical machine learning, some insight can already be gained from the following problem:

Let $\mathcal{G}$ be a regular region in $\mathbb{R}^q$. Suppose that $X_N = \{x_1, \ldots, x_N\}$ is a pointset in $\mathbb{R}^q$. Assume that the set of samples $\{F_{\mathcal{G}}^{\wedge}(x_k)\}_{k=1,\ldots,N}$ corresponding to the finite set X_N is known, learn $F_{\mathcal{G}}^{\wedge}$ (for simplicity, we neglect measurement noises).

Now, under suitable data availability, machine learning divides the data set $(x_k, F_{\mathcal{G}}^{\wedge}(x_k))_{k=1,\ldots,N}$ into two subsets, a training set used to train the model and a testing set used to test the model. Clearly, the approximation of functions is a classical topic in constructive approximation theory, where we typically use polynomials, splines wavelets, and other linear combinations of fixed basis functions to approximate a discretely given function. However, these kinds of approximations suffer from the "curse of dimensionality" (CoD)

$$F_{\mathcal{G}}^{\wedge} - (F_{\mathcal{G}}^{\wedge})_M^{\text{approx}} \sim M^{-\frac{\alpha}{q}} R(F_{\mathcal{G}}^{\wedge}), \tag{15.43}$$

where M is the number of free parameters in the approximation scheme, α is some fixed constant depending on the approximation scheme, $R(F_{\mathcal{G}}^{\wedge})$ is an error constant depending on the target function $F_{\mathcal{G}}^{\wedge}$ (take, for example $\alpha = 1$, if we are interested in reducing the error by a factor of 10, the number of parameters needed goes up by 10 q. This is obviously impractical when the dimensionality q is large). The one area, in which we have lots of experiences and successes in handling high-dimensional problems, is statistical physics, providing the computation of high-dimensional expectations by use of *Monte Carlo (MC) methods* (see, e.g., [**176**] for more details). Indeed, computing integrals of functions with millions of variables has become a routine practice in statistical physics:

Let H and μ be a function and a probability distribution in the Euclidean space $\mathbb{R}^q$, respectively, and let there be given

$$I(H) = E_{x\sim\mu} H(x), \quad I_M(H) = \frac{1}{M}\sum_{j=1}^{M} H(x_j), \tag{15.44}$$

where $\{x_j\}$ constitutes a system of i.i.d. samples of the probability distribution μ. Then we have

$$E\left(I(H) - I_M(H)\right)^2 = \frac{\mathrm{Var}(H)}{M}. \tag{15.45}$$

Note that the error rate $N^{-\frac{1}{2}}$ is independent of the dimensionality of the problem. Had we used grid-based algorithms such as the trapezoidal rule, we would have instead a situation characterized by

$$I(H) - I_M(H) \sim M^{-\frac{\alpha}{q}} R(H) \tag{15.46}$$

resulting in CoD (note that the variance can be very large in high dimensions, such that variance reduction is a central theme in today's MC).

The MC story is standard for dealing with high-dimensional problems. A natural question is as follows: Can we turn the function approximation problem, the key issue in supervised learning, into a methodologically related problem of MC-type integration? To gain some insight, let us consider the details involving the function $F_{\mathcal{G}}^{\wedge}$. Instead of discretized Fourier transform by cubature, functions of type

$$F(y) = \int_{\mathbb{R}^q} F_{\mathcal{G}}^{\wedge}(x)\; e^{2\pi i x\cdot y}\; \pi(dx) = E_{y\sim\pi} F_{\mathcal{G}}^{\wedge}(y)\; e^{-2\pi i x\cdot y} \tag{15.47}$$

are considered, where π is a probability distribution on Euclidean space $\mathbb{R}^q$. The canonical approximation is given by

$$F(x) \sim \frac{1}{M} \sum_{j=1}^{M} a_j \;\; e^{2\pi i x\cdot x_j}, \tag{15.48}$$

where $\{x_j\}$ is a sequence of i.i.d. samples of the probability distribution π . For reasons discussed earlier, this approximation does not suffer from CoD. In fact, the right hand side of (15.48) is an example of neural network functions with one hidden layer and activation function defined by $e^{2\pi i x\cdot}$ (here, $a_j = F_{\mathcal{G}}^{\wedge}(x_j)$). Following [**201**], if one wants a one-sentence description of the magic behind neural network models, the reasoning above should be a good candidate.

However, there is one key difference between the supervised learning problem and MC integration: In MC, the probability distribution μ is given, usually in fairly explicit form. In supervised learning, the probability distribution π is unknown. Instead we are only given information about the target function $F_{\mathcal{G}}^{\wedge}$ on a finite dataset. Therefore, it is infeasible to perform the approximation in (15.48). Instead one has to replace it by some optimization algorithms based on fitting the dataset.

Part 8

Concluding remarks

CHAPTER 16

Recovery as interconnecting whole

During the last decades a substantial amount of machinery from special function theory, functional analysis, approximation theory and numerical analysis has been brought to bear on the understanding of an inverse and/or sampling problem as a recovery problem (RP), and the interdisciplinary character of many recovery problems in their strong interrelations has emerged very clearly. As a result, the scientific situation has changed, so that the facets of the individual disciplines such as inverse and sampling problems could be formed into an interconnecting whole.

The common thread today among the scientific fields is a canonical problem, namely

recovering an object (function, signal, signature, picture)

from partial or indirect information about the object.

16.1. Regularization as constituting methodology

The vivid research activity has led to a vast literature on inverse and ill-posed problems. Throughout the book, we have referred to various papers and monographs including further details on several aspects. We list only a selection of contributions as suitable points of entry:

- *Textbooks*

  G. Anger [**44**], J. Baumeister [**70**], H.W. Engl [**218**], H. W. Engl, M. Hanke, A. Neubauer [**215**], C. W. Groetsch [**378, 379**], B. Hofmann [**442**], J. Kaipio, E. Somersalo [**484**], A. Kirsch [**497**], M.M. Lavrentiev [**527**], A.K. Louis [**547**], V. A. Morozov [**597**], [**598**], M.Z. Nashed [**617, 618**], L.E. Payne [**683**], A. G. Ramm [**707**], A. Rieder [**723**], T. Schuster [**758**], T. Schuster, B. Kaltenbacher, B. Hofmann, K.S. Kazimierski [**759**], A.N. Tikonov et al. [**821–823**], C.R. Vogel [**840**].
- *Conference reports, handbooks*

  Y. I. Alber [**15**], T.S. Angell, M.Z. Nashed [**43**], M. Bertero, C. DeMol, G.A. Viano [**81**], M. Burger, B. Kaltenbacher [**120**], H.W. Engl, A.K. Louis, W. Rundell [**217**], W. Freeden, F. Schneider, M. Schreiner [**316**], C.W. Groetsch [**377**], M. Ismail, M.Z. Nashed, A. Zayed, A. Ghaleb [**456**], M.Z. Nashed et al. [**613, 617, 622, 624, 627, 627, 629, 631, 632, 635, 637**], W.T. Reid [**716**].

In addition to the books we mention the journals:

- *Inverse Problems,*

- *Inverse Problems in Science and Engineering,*
- *Journal on Inverse and Ill-Posed Problems,*
- *Inverse Problems and Imaging,*
- *Mathematical Inverse Problems.*

The *GEM-International Journal on Geomathematics* is a forum in which geoscientifically relevant ill-posed problems gain appropriate recognition.

Many authors contributed results in different areas of ill-posed and inverse problems (note that he following list is rather incomplete, for more details the reader is referred to the references in the aforementioned textbooks or the below listed journal and handbook publications):

- *Generalized inverse and least squares problems*
  [10], **[89]**, **[91]**, **[182]**, **[198]**, **[204]**, **[205]**, **[210]**, **[211]**, **[218]**, **[234]**, **[293]**, **[332]**, **[328]**, **[364]**, **[367]**, **[401]**, **[443]**, **[555]**, **[566]**, **[559]**, **[586]**, **[574]**, **[588]**, **[589]**, **[613]**, **[615]**, **[618]**, **[622]**, **[624, 625]**, **[625]**, **[627]**, **[628]**, **[640]**, **[641]**, **[642]**, **[685]** **[685]**, **[709]**, **[706]**, **[724]**.
- *Truncated singular value, Tikhonov regularization, and discrepancy principles*
  [78], **[79]**, **[76]**, **[86]**, **[88]**, **[177]**, **[206]**, **[207]**, **[212]**, **[213]**, **[244]**, **[245]**, **[360]**, **[378]**, **[377]**, **[422]**, **[444]**, **[542]**, **[545]**, **[565]**, **[568]**, **[569]**, **[596]**, **[616]**, **[618]**, **[620]**, **[621]**, **[625]**, **[648]**, **[652]**, **[686]**, **[693]**, **[696]**, **[719]**, **[723]**, **[802]**, **[770]**, **[819]**, **[821]**, **[822]**, **[844]**.
- *Inversion in RKHS's*
  [243], **[434]**, **[873]**, **[625]**, **[640]**, **[641]**, **[642]**, **[741]**, **[771]**, **[859]**.
- *Iterative methods, finite element methods, other computational methods*
  [11], **[14]**, **[10]**, **[62]**, **[117]**, **[121]**, **[177]**, **[186]**, **[208]**, **[332]**, **[360]**, **[372]**, **[380]**, **[397]**, **[400]**, **[402]**, **[406]**, **[405]**, **[487]**, **[488]**, **[509]**, **[612]**, **[625]**, **[647]**, **[666]**, **[691]**, **[822]**, **[826]**, **[798]**, **[833]**, **[842]**, **[868]**, **[883]**, **[885]**, **[886]**.
- *Projection methods, moment problems*
  [14], **[15]**, **[134]**, **[135]**, **[206]**, **[207]**, **[497]**, **[212]**, **[625]**, **[619]**, **[646]**, **[788]**, **[867]**.
- *Mollifier methods*
  [100], **[215]**, **[296]**, **[291]**, **[312]**, **[270]**, **[326]**, **[547]**, **[548]**.
- *Multiscale methods*
  [100], **[187]**, **[192]**, **[193]**, **[281]**, **[284]**, **[302]**, **[313]**, **[286]**, **[289]**, **[270]**, **[359]**, **[525]**, **[549]**, **[565]**, **[567]**, **[577]**, **[578]**, **[579]**.
- *Variational methods, implicit function theorems, ill-posed problems in differential eqations*
  [4], **[13]**, **[17]**, **[18]**, **[44]**, **[188]**, **[138]**, **[191]**, **[202]**, **[204]**, **[209]**, **[212]**, **[213]**, **[541]**, **[632]**, **[633]**, **[634]** **[683]**, **[692]**, **[718]**, **[731]**, **[832]**.
- *Stochastic methods*
  [84], **[85]**, **[86]**, **[92]**, **[93]**, **[191]**, **[205]**,**[215]**, **[225]**, **[308]**, **[371]**, **[440]**, **[442]**, **[450]**, **[484]**, **[501]**, **[547]**, **[569]**, **[631]**, **[655]**, **[664]**, **[665]**, **[710]**, **[749]**, **[789]**, **[814]**.

- *Nonlinear methods*
  [**67**], [**98**], [**145**], [**151**], [**180**], [**184**], [**185**], [**216**], [**215**], [**399**], [**428**], [**441**], [**442**], [**486**], [**492**], [**535**], [**630**] , [**651**], [**667**], [**700**], [**701**], [**702**], [**723**], [**744**], [**745**], [**746**], [**760**], [**761**], [**815**], [**882**].

16.2. Sampling as constituting methodology

Application of sampling theory permeate many branches such as signal analysis, seismic imaging, radar as well as laser technology, and many others. Sampling theory in its classical manifestation naturally makes an offspring of interpolation. In modern concepts, which are not immediate advancements of the Shannon context, sampling is strongly involved in reproducing kernel structure of Hilbert and/or Banach space topology. So, W. Freeden, M.Z. Nashed [**288**], [**296**], [**271**] were canonically led to the following conclusion to characterize essential constituents of sampling theory:

> "Almost all routes to sampling expansions lead to reproducing kernels. Seminal ideas are often triggered by amazing insights. Many of such ideas are drastically generalized in new directions and applied in diverse fields, far beyond what anyone could have envisioned. The developments in sampling theory and the theory of ill-posed problems represent such examples."

In fact, sampling theory (in the sense of the representation of an analog signal in terms of its samples) has attracted considerable interest in the past three decades. Major advances in the theory and applications of sampling expansions in different function spaces have been made by mathematicians and engineers. Nowadays, sampling has its own "professional society". We quote from Wikipedia:

> "SampTA (Sampling Theory and Applications) is a biennial interdisciplinary conference for mathematicians, engineers, and applied scientists. The main purpose of SampTA is to exchange recent advances in sampling theory and to explore new trends and directions in related areas of applications. The SampTA conference series began as a small workshop in 1995 in Riga, Latvia, but the meetings grew into full-fledged conferences attracting an even mix of mathematicians and engineers as the interest in sampling theory blossomed."

The sampling field also sports its own journals:

- *Sampling Theory in Signal and Image Processing,*
- *Sampling Theory, Signal Processing, and Data Analysis.*

There are several other journals on signal processing and computational harmonic analysis that publish papers on sampling theory and its applications:

- *Journal of Mathematical Imaging and Vision,*
- *Journal of Fourier Analysis and Applications,*
- *SIAM Journal on Imaging Sciences,*
- *IEEE Tansactions on Image Processing,*
- *International Journal of Computer Vision,*

- *Inverse Problems and Imaging.*

All in all, sampling is a very active field of research in applied mathematics and engineering, with a fastly growing bibliography (of which we are not able to give a more detailed description). The following list only give points of entry:

- *WKS sampling and early sampling expansions*
  The most common form is sampling of a bandlimited signal. In Shannon sampling, perfect reconstruction of the signal from lattice samples is possible when the samples are taken at a certain rate related to the bandwidth (cf. E.T. Whittaker [**862–865**], V.A. Kotel'nikov [**508**], C.E. Shannon [**764**]). The Shannon sampling theorem has been generalized in many directions (for early generalizations see, e.g., P.L. Butzer and his group [**123**] - [**131**], J.R. Higgens [**430, 431**], J.A. Jerri [**463–465**], M. Unser [**827**]). Some of the aforementioned properties have served as key ingredients in the extensions and generalizations of the Shannon sampling theorem. In particular, it has been shown that many extensions have a formulation in appropriate reproducing kernel Hilbert spaces (RKHSs) (see, e.g., L.L. Campbell [**136**], R.F. Hoskins, P. de Sousa [**446**], T. Koornwinder, G.G. Walter [**504**], H.P. Kramer [**512**], A. Papoulis [**679**], M.D. Rawn [**711**], K. Yao [**872**], A. Zayed [**879**], and related references cited therein). They usually involve transforms other than the Fourier transform and include Sturm-Liouville transforms [**136**], [**512**], Jacobi transforms [**504**], Laguerre transforms [**464**], and Hankel transformation [**463**]. These as well as finite-order sampling theorems all have a setting in reproducing kernel Hilbert spaces. This was partially recognized by K. Yao [**872**] who showed that there is a relation between the sampling theorem and the reproducing kernel in some special cases. For various perspectives and surveys on sampling expansions the reader is referred to P.L. Butzer [**123, 126, 128**], J. R. Higgins [**430–432**], J.A. Jerri [**463–465**], M. Unser [**827**], M.Z. Nashed, G.G. Walter [**643, 644**], and many others. M.Z. Nashed, G.G. Walter [**643, 644**] have shown that there is a strong affinity between RKHS and sampling expansions. They have derived univariate sampling theorems for functions in RKHS which are subspaces of $L^2(\mathbb{R})$ closed in the Sobolev space $H^{-1}(\mathbb{R})$.
- *Various sampling methods*
  Methods of functional analysis and harmonic analysis have played a pivotal role to provide major advances in sampling theory and its foundational aspects. In particular, new directions have been proposed involving various function spaces that admit sampling expansions such as Sobolev spaces, reproducing Hilbert spaces, Wiener amalgam space, shift invariant spaces, translation-invariant spaces, convolution sampling, nonlinear sampling and other fundamental issues in sampling theory. For various perspectives the reader is referred to, e.g, [**21**], [**24**], [**26**], [**71**], [**73**], [**74**], [**84**], [**123**], [**126**], [**125**], [**128**], [**396**], [**430**], [**431**], [**432**], [**456**], [**465**], [**525**], [**561**], [**562**], [**564**], [**572**], [**629**], [**635**], [**636, 637**], [**643**], [**783**], [**769**], [**791**], [**792**], [**803**], [**805**], [**807**], [**808**], [**855**], [**858**], [**872**], [**879**], [**878**], [**880**], and the references therein. The papers A. Papoulis [**678,679**], A.V. Oppenheim, R.W. Schäfer [**664**], R.A. Haddad, T.W. Parsons [**392**] (see

also the references therein) deal with sampling aspects of engineering realization. Many of the basic ideas of sampling are drastically generalized in new directions and applied in diverse fields of sampling, far beyond what anyone could have envisioned in the early days of the sampling theory. So, there have been many advances in the sampling theory and its applications to signal and image processing. Many authors developed sampling theorems based on the theory of regular and singular boundary value problems and also by use of transforms other than the Fourier transform, including such techniques as the Sturm-Liouville, Legendre, Laguerre, Jacobi transform (see, e.g., M.Z. Nashed [**629**], M. Z. Nashed, Q. Sun [**636**], A. I. Zayed [**878**]) originated by H.P. Kramer's approach (see [**512**]).

- *Irregular and nonuniform sampling*

  Irregular sampling data sets are discussed by, e.g., A. Aldroubi, H.G. Feichtinger [**22**], H.G. Feichtinger [**229, 230**], H.G. Feichtinger, K. Gröchenig [**231**], K. Gröchenig [**382**], E. Acosta-Reyes et al. [**2**], and many others. Another field of research has been in nonuniform sampling (see, e.g., A. Aldroubi et al. [**21–23, 25, 26**], P. Jorgensen et al. [**471, 472, 477–479**], Q. Sun [**803**], and the references therein.
- *Sampling with finite rate of innovation*

  Signals with finite rate of innovation are those signals which allow a determination by finitely many samples per unit of time. The concept of signals with finite rate of innovation was studied by M. Vetterli and his school (see, e.g., [**837**], [**836**] and the literature therein). Contributions are also due to Q. Sun [**803, 805**].
- *Sampling and machine learning*

  What is the magic behind machine learning? While this question is still at the center of very intensive studies in theoretical machine learning, some insight can already be gained from specific work of the Princeton University (see, e.g., [**201**]) and the Massachusetts Institute of Technology (see, e.g. [**6**])). This work is especially devoted to the question: How can we integrate machine learning with sampling to develop reliable physical results?
- *RKHS sampling*

  Almost all of the sampling methodologies are in an intimate relation to a certain reproducing kernel structure. M. Z. Nashed et al. [**396, 572, 636, 643, 644**] showed how to construct a reproducing kernel Hilbert space from a function space that admits sampling expansions. In view of the affinity between reproducing kernel spaces and sampling expansions, it is not surprising that in the past 30 years, reproducing kernel Hilbert spaces have played major roles in signal analysis and applications in inverse problems and imaging (see, e.g. M.Z. Nashed, Q. Sun [**637**] and the literature therein). Mostly the literature so far has focused on the theory of kernel functions defined on continuous domains, either domains in Euclidean spaces or on complex domains. For these cases the Dirac distributions do not have a finite RKHS-norm, but for RHKSs over discrete point distributions the Dirac distributions indeed show finite norm (see, e.g., P. Jorgensen et al. [**475, 478**] and the references therein). The concept of reproducing kernel Hilbert spaces has a natural generalization to Banach spaces. Like the Hilbert space case, the sampling theory for a Banach space involves a Banach space frame for reconstruction und p-frames for determination (see, e.g., [**636**], [**478**]). Of interest in the literature (cf. O. Christensen, D. Stoeva [**150**], M.Z. Nashed, Q. Sun [**637**]) is the question, if determination by a p-frame in a Banach space implies reconstruction, and vice versa?

- *Spherical sampling*

    Another form of recovery, that is of particular importance in geosciences, is spherical sampling (which is not discussed in the present work). Indeed, spherical sampling reflects both the approximate shape of the Earth's surface and the typical satellite geometry of an low Earth orbiter (LEO), so that we are interested in reconstruction formulas corresponding to different types of point sets on spheres and various observables usually characterized by (rotation-invariant) pseudo-differential operators (see W. Freeden, M.Z. Nashed, M. Schreiner [**300**]). Large structured point sets, e.g., longitude-latitude distributions occur in observational methods by satellite, scattered point systems usually are the result of terrestrial measurements reflecting, e.g., topographic features. It turns out that the close relationship between spherical sampling and and reproducing kernel Hilbert spcace (RKHS) theory has its manifestations in various polynomial (i.e., spherical harmonic), spline, and wavelet settings. Spherical sampling has its own manifestation in the early monograph W. Freeden, T. Gervens, M. Schreiner [**276**] and in the recently published book by W. Freeden, M.Z. Nashed, M. Schreiner [**297**].

- *Multi-dimensional sampling*

    Multi-dimensional sampling by simply using iterated 1D Shannon kernels can be found in, e.g., E. Parzen [**681**]. A. Aldroubi et al.[**21**, **26**], H. Behmard, A. Faridani [**71**], P.L. Butzer [**123**], [**159**], D. Costarelli, G. Vinti [**154**], T. Kalker [**485**], R. J. Marks II [**561**], H.G. Feichtinger, K. Gröchenig [**231**] form many other contributions. Usually, the sampling in these approaches is restricted to the fundamental cell of $\mathbb{Z}^q, q \geq 2$, and based on iterated one-dimensional framework.

- *Polyharmonic lattice point sampling*

    Earliest published work about the idea of interpolating by use of fundamental solutions to powers of the Laplacian seems to date back to R.L. Harder, R.N. Demarais [**409**]. However, it is known that many mathematicians were aware of the idea and many consequences, either from the viewpoint of reproducing kernel Hilbert space theory or from the extension of the variational aspect of univariate spline theory to the multivariate case (see, e.g., J. Duchon [**195**], J. Meinguet [**573**]). Finally, it should be noted that "thin plate spline sampling can be realized in various way [**200**], [**199**], [**576**] (for more details see also the monograph M.D. Buhmann [**119**] for Euclidean and W. Freeden et al. [**276**] for spherical concepts). I.J. Schoenberg's theory of univariate cardinal splines of odd order (cf. [**755**]) may be regarded as a development of certain properties of those functions, which satisfy the polyharmonic differential equation on the complement of the integer lattice $\mathbb{Z}$ and show appropriate smoothness conditions on all of the real line $\mathbb{R}$. A canonical extension of this concept to the multivariate case is due to W. Madych, S. Nelson [**553**]. The note [**553**] is concerned with the basic properties of splines together with the problems of existence and uniqueness of cardinal interpolation.The variational framework of cardinal splines based on an appropriate integral formulas over $\mathbb{R}^q$ is presented in this work.

- *Polymetaharmonic lattice point sampling*
  The close relationship of multi-variate Shannon sampling to lattice point identities of geometric number theory seems to be a new field of interest. W. Freeden, M.Z. Nashed [**288**], [**296**], [**298**], W. Freeden [**271**] provide the point of departure for a palette of so-called *Shannon-type sampling formulas* and explicit over- and under-sampling characterizations for a diversity of geometries different from the fundamental cell of the chosen lattice. The practical usefulness, applicability, economy, and efficiency of multivariate lattice point sampling are evident (for example, in geo-exploration discussing Fredholm integral equations of the first kind from gravimetry [**270**] and magnetometry [**100**]).
- *Sampling by exponential sum reduction*
  In physics-driven inverse problems unknown field sources are recovered by a sampling framework using weighted sums in terms of exponentials. Different approaches have been proposed in the literature mostly leading to algorithms based on Prony's method. W. Hackbusch et al. [**389**], [**388**], G. Beylkin, L., Monzón [**82**], [**83**] propose different concepts of discretizing a potential field integral expression involving the fundamental solution of a potential operator such as the Laplacian to reduce the number of terms of Gaussian sums. The use of weighted sums of exponentials with purely imaginary exponents for solving inverse source problems of diffusion and wave fields (see, e.g., J. Murray-Bruce, P.L. Dragotti [**608**] and the references therein) is for two main reasons: The first is to generate a system of superimposed sinusoids (see, e.g., M. Vetterli et al. [**836**]), which can be efficiently solved by using multidimensional extensions of Prony's frequency estimation methods (cf. [**449**], [**557**]). The second reason is because the use of imaginary exponentials improves the stability of the estimation problem, since in this case the magnitude of the terms in the generalized sequence remain bounded. Explicit formulas are available for the sequence of weights, that map sensor samples to an error-free sequence of (generalized) measurements, thereby observing the fact in case of uniform data that the fundamental solution of the underlying differential operator satisfies the generalized Strang-Fix condition (see [**799**] and the extensions obtained in [**829**]). Some approaches have been put forward for recovering the unknown frequencies and amplitudes via Prony's method, particularly for the two-dimensional case (see, e.g., [**449**], [**557**], [**728**], [**740**], [**728**]) utilizing an ESPRIT algorithm. Concerning algorithms based on Prony's method we further mention the references [**28**], [**689**], and [**717**].

16.3. Recovery as common thread

The indispensable requirement for a work about recovery problems in its constituting building blocks of inverse and sampling problems is methodologically characterized by a specific circumstance, namely that the rationale in methods for solvability is the construction of a "solution"

- that is acceptable physically as a meaningful approximation,
- that is sufficiently stable from the computational standpoint.

As an immediate consequence, a common emphasis must be put on

- the distinction between "solution" and "resolution".

An essential aim of the book therefore was to show that the philosophy of resolution leads to different manifestations, viz. the use of algebraic methods versus function space methods, statistical versus deterministic approaches, strong versus weak noise, etc. Resulting mathematical schemes turn into algorithms once a resolution strategy can be effectively implemented. Unfortunately, this requires a determination of a suitable value, i.e., a certain parameter associated with the scheme (e.g., truncation parameter, mesh size, dimension of useful subspaces, classification of noise, etc.). As already mentioned, this is not a trivial problem since it involves a trade-off between accuracy and numerical stability, a situation that does not usually arise in problems not concerned with recovering an object from partial or indirect information about the object. So, the main dilemma of handling recovery problems is that a resolution must be numerically terminated at a suitable resolution stage, before essential inaccuracies creep into the resolution process.

As a matter of fact, it was our relief to be able to finalize this work without having to define what is meant by the term recovery problem. Until now, in spite of the strong interconnections, the understanding and handling of recovery problems is mostly separated in the scientific fields as inverse problems and sampling. Our overall purpose, however, was to provide a new generalizing outlook how resolution methodologies in recovery problems and resulting concepts and results can be better motivated, understood, and further developed from an interconnecting standpoint between the different disciplines. Within this general context, more transparent criteria and structural insights can be given relative to which the scope and limitations of the various methods can be assessed. This strategic outcome is important both in theory and practice since there is no cure-all method for resolution problems; therefore it is imperative to be able to clarify why a certain method works in some context as well as when not to use that method or to combine appropriately different procedures. For the convenience of the reader, our bibliography refers to various papers and monographs, particularly in the theories of inverse problems and sampling, without claiming to be complete. Our listing is only an attempt to get a gateway to various recovery intentions and obligations, and from there to access more scientific materials.

Recovery problems in their different manifestations arise in many branches of science and mathematics, including computer vision, natural language processing, machine learning, statistics, medical imaging (such as tomography and EEG/ERP), remote sensing, nondestructive testing, astronomy, geodesy, geophysics, geosystems mathematics (especially involving, e.g., gravimetric, geomagnetic, seismic exploration, and "downward continuation" in satellite observational technology), and many other fields. Altogether, this book develops a general framework for recovery methods. Our objective is to provide an unifying outlook within which relevant results can be better motivated, interpreted, and ordered in their relationships. The development is fairly general in scope and theory, and it is applicable to a wide range of recovery problems.

It is hoped that the viewpoints and approaches developed in this book would be found useful also in connection with other research areas in mathematics and engineering. Once more, our purpose was to provide a unifying outlook within which technical results can be better understood and interrelated. Within this framework, criteria can be given relative to which the scope and limitations of the various methods can be assessed. The work discusses at length the intuitive principles that underlie the various methods and establishes interconnecting results within this framework, thereby omitting technicalities of the proofs.

List of symbols

$\mathbb{N}_0$	set of non–negative integers
$\mathbb{N}$	set of positive integers
$\mathbb{Z}$	set of integers
$\mathbb{R}$	set of real numbers
$\mathbb{C}$	set of complex numbers
$\Re(s)$	real part of $s \in \mathbb{C}$
$\Im(s)$	imaginary part of $s \in \mathbb{C}$
$\mathbb{R}^q$	q–dimensional Euclidean space
x, y, z	elements of $\mathbb{R}^q$
$x \cdot y$	scalar product of vectors
$\|x\|$	Euclidean norm of $x \in \mathbb{R}^q$
$\varepsilon^i,\ i = 1, \ldots, q$	canonical orthonormal basis in $\mathbb{R}^q$
$\delta_{i,j}$	Kronecker symbol
$C^{(k)}, L^p$	classes of functions
F, G	scalar–valued functions
$\mathfrak{f}, \mathfrak{g}$	vector–valued functions
$\mathbf{i}$	identity tensor
$\mathbf{t}$	orthogonal matrix
$\mathbf{t}^T$	transpose of the matrix $\mathbf{t}$
$\det\ \mathbf{t}$	determinant of the matrix $\mathbf{t}$
$\mathcal{G}$	point set in $\mathbb{R}^q$
$\partial\mathcal{G}$	boundary of Γ
$\overline{\mathcal{G}}$	closure of Γ
$F\|M$	restriction of F to M
∇	gradient
$\nabla \cdot$	divergence
Δ	Laplace operator
$\Delta + \lambda$	Helmholtz operator
$\{\ldots\}$	set of the elements
$\emptyset$	empty set
$\mathbb{B}^q$	unit ball in $\mathbb{R}^q$ around 0
$\mathbb{B}^q_N(y)$	ball in $\mathbb{R}^q$ with radius N around y
$\mathbb{S}^{q-1}$	unit sphere in $\mathbb{R}^q$ around 0
$\mathbb{S}^{q-1}_N(y)$	sphere in $\mathbb{R}^q$ with radius N around y
t, θ, φ	polar coordinates in $\mathbb{R}^2$
ξ, η, ζ	elements of $\mathbb{S}^{q-1}$
$dx, dx_{(q)}$	volume element in $\mathbb{R}^q$
$dS, dS_{(q-1)}$	surface element in $\mathbb{R}^q$

$d\sigma$ .... curve element in $\mathbb{R}^2$
∇^* .... surface gradient in $\mathbb{R}^q$
$\nabla^* \cdot$ .... surface divergence in $\mathbb{R}^q$
Δ^* .... Laplace-Beltrami operator in $\mathbb{R}^q$
$\Delta^* + \lambda$ .... Helmholtz-Beltrami operator in $\mathbb{R}^q$
$(\Delta^*)^\wedge(n)$ .... $\mathbb{S}^{q-1}$–symbol of the Laplace-Beltrami operator in $\mathbb{R}^q$
$(\Delta^* + \lambda)^\wedge(n)$ .... $\mathbb{S}^{q-1}$–symbol of the Helmholtz-Beltrami operator in $\mathbb{R}^q$
$\mathbb{Z}^q$ .... unit lattice in $\mathbb{R}^q$
$\tau\mathbb{Z}^q$ .... dilated unit lattice in $\mathbb{R}^q$
$\mathbb{Z}^q + \{x\}$ .... translated unit lattice based at x in $\mathbb{R}^q$
Λ .... (general) lattice in $\mathbb{R}^q$
Λ^{-1} .... inverse lattice of Λ in $\mathbb{R}^q$
$\mathcal{F}_\Lambda$ .... fundamental cell of Λ in $\mathbb{R}^q$
$\mathcal{F}_\Lambda^{-1}$ .... fundamental cell of Λ^{-1} in $\mathbb{R}^q$
$\|\mathcal{F}_\Lambda\|$ .... volume of the fundamental cell Λ
g .... lattice points of Λ
h .... lattice points of Λ^{-1}
$\mathrm{Spect}_\Delta(\Lambda)$ .... eigenspectrum of the Laplace operator in $\mathbb{R}^q$
$\Phi_h,\ h \in \Lambda^{-1}$ .... (orthonormal) Λ–periodical polynomials in $\mathbb{R}^q$
$C_\Lambda^{(k)}, \mathrm{L}_\Lambda^p$ .... classes of Λ–periodical functions in $\mathbb{R}^q$
$F_\Lambda^\wedge(h)$ .... Fourier transform of F at $h \in \Lambda^{-1}$ with respect to Λ
$F_{\mathbb{R}^q}^\wedge(x)$ .... Fourier transform of F at $x \in \mathbb{R}^q$ with respect to $\mathbb{R}^q$
$\mathcal{X}_\mathcal{M}$ .... characteristic function for a set $\mathcal{M}$
$\tau\Lambda$ .... dilated point lattice in $\mathbb{R}^q$
$\Lambda - \{x\}$ .... shifted point lattice in $\mathbb{R}^q$
$(\Delta)^\wedge(h)$ .... Λ–symbol of the Laplace operator in $\mathbb{R}^q$
$(L+\lambda)^\wedge(h)$ .... Λ–symbol of the Helmholtz operator in $\mathbb{R}^q$
$G(\Delta\,;\cdot)$ .... Λ–lattice function for Δ
$G(L+\lambda\,;\cdot)$ .... Λ–lattice function for $L+\lambda$
$\mathcal{M} + \Lambda$ .... figure lattice associated to $\mathcal{M}$ by Λ
$\mathbb{B}_\tau^q + \Lambda$ .... ball lattice associated to $\mathbb{B}_\tau^q$ by Λ
$\mathcal{G}$ .... regular region in $\mathbb{R}^q$
$\overline{\mathcal{G}} = \mathcal{G} \cup \partial\mathcal{G}$ .... closure of a regular region in $\mathbb{R}^q$
$\partial\mathcal{G}$ .... boundary of the regular region $\mathcal{G}$ in $\mathbb{R}^q$
$\|\mathcal{G}\|$ .... volume of the regular region $\mathcal{G}$ in $\mathbb{R}^q$
$\alpha(x)(=\alpha_{\partial\mathcal{G}}(x))$ .... solid angle at x subtended by $\partial\mathcal{G}$
o, O .... Landau symbols
$\lfloor\delta\rfloor$ .... the largest integer $\leq \delta$ (floor-function)
$\lceil\delta\rceil$ .... the smallest integer $\geq \delta$ (ceil-function)
Γ .... Gamma function
ϑ .... Theta function
$\sharp_\Lambda(\overline{\mathcal{G}}) = \sum_{\substack{g\in\overline{\mathcal{G}}\\ g\in\Lambda}} 1$ .... total number of lattice points of Λ inside $\overline{\mathcal{G}}$

Bibliography

[1] J. Abhau, Z. Belhachmi, and O. Scherzer, *On a Decomposition Model for Optical Flow*, Energy Minimization Methods in Computer Vision and Pattern Recognition, Lectures Notes in Computer Science, Springer Verlag, Berlin, Heidelberg, **5681** (2009), 126–139

[2] E. Acosta-Reyes, A. Aldroubi, and I. Krishtal, *On stability of sampling-reconstruction models*, Adv. Comput. Math. **31** (2009), no. 1-3, 5–34, DOI 10.1007/s10444-008-9083-6. MR2511572

[3] M. Abramowitz, and I. A. Stegun, *Handbook of Mathematical Functions*, Dover Publications, Inc., New York, 1972

[4] R. Acar and C. R. Vogel, *Analysis of bounded variation penalty methods for ill-posed problems*, Inverse Problems **10** (1994), no. 6, 1217–1229. MR1306801

[5] R. A. Adams, *Sobolev Spaces*, New York, Academic Press, 1975

[6] A. Adler, M. Araya-Polo, T. Poggio, *Deep learning for seismic inverse problems: toward the acceleration of geophysical analysis workflows*, IEEE Signal Processing Magazine, **38** (2021), 1–28

[7] H. P. S. Ahluwalia, W. M. Boerner, *Application of electromagnetic inverse boundary conditions to profile characteristic inversion of condition spherical shapes*, IEEE Trans. Antenna Propagat., **22** (1974), 673–682

[8] D. Alpay, *The Schur algorithm, reproducing kernel spaces and system theory*, SMF/AMS Texts and Monographs, vol. 5, American Mathematical Society, Providence, RI; Société Mathématique de France, Paris, 2001. Translated from the 1998 French original by Stephen S. Wilson. MR1839648

[9] D. Alpay and P. Jorgensen, *Reproducing kernel Hilbert spaces generated by the binomial coefficients*, Illinois J. Math. **58** (2014), no. 2, 471–495. MR3367659

[10] A. Albert, *Regressions and the Moore-Penrose Pseudoinverse*, Academic Press, 1972

[11] Y. Alber, *The solution of nonlinear equations with monotone operators in Banach spaces*, Siberian Math. J., **16** (1975), 1–8

[12] Y. Alber, *Iterative regularization in Banach spaces*, Soviet Math. (Iz. VUZ), **30** (1986), 1–8

[13] Ya. I. Al′ber, *The regularization method for variational inequalities with nonsmooth unbounded operators in Banach space*, Appl. Math. Lett. **6** (1993), no. 4, 63–68, DOI 10.1016/0893-9659(93)90125-7. MR1348234

[14] Y. Alber, *Generalized projection operators in Banach spaces: properties and applications*, Functional Differential Equations, Proceedings of the Israel Seminar, **1** (1994), 1–21

[15] Y. I. Alber, *Metric and generalized projection operators in Banach spaces: properties and applications*, Theory and applications of nonlinear operators of accretive and monotone type, Lecture Notes in Pure and Appl. Math., vol. 178, Dekker, New York, 1996, pp. 15–50. MR1386667

[16] Ya. I. Al′ber and A. I. Notik, *Perturbed unstable variational inequalities with unbounded operators on approximately given sets*, Set-Valued Anal. **1** (1993), no. 4, 393–402, DOI 10.1007/BF01027828. MR1267205

[17] Ya. I. Al′ber and S. Reich, *An iterative method for solving a class of nonlinear operator equations in Banach spaces*, Panamer. Math. J. **4** (1994), no. 2, 39–54. MR1274188

[18] Y. Alber, I. Rjazanceva, *Variational inequalities with discontinuous monotone mappings*, Soviet Math. Dokl., **25** (1982), 206–210

[19] A. Aldroubi, *Non-uniform weighted average sampling and reconstruction in shift-invariant and wavelet spaces*, Appl. Comput. Harmon. Anal. **13** (2002), no. 2, 151–161, DOI 10.1016/S1063-5203(02)00503-1. MR1942749

[20] A. Aldroubi and K. Gröchenig, *Beurling-Landau-type theorems for non-uniform sampling in shift invariant spline spaces*, J. Fourier Anal. Appl. **6** (2000), no. 1, 93–103, DOI 10.1007/BF02510120. MR1756138

[21] A. Aldroubi and K. Gröchenig, *Nonuniform sampling and reconstruction in shift-invariant spaces*, SIAM Rev. **43** (2001), no. 4, 585–620, DOI 10.1137/S0036144501386986. MR1882684

[22] A. Aldroubi and H. Feichtinger, *Exact iterative reconstruction algorithm for multivariate irregularly sampled functions in spline-like spaces: the L^p-theory*, Proc. Amer. Math. Soc. **126** (1998), no. 9, 2677–2686, DOI 10.1090/S0002-9939-98-04319-6. MR1451788

[23] A. Aldroubi and C. Leonetti, *Non-uniform sampling and reconstruction from sampling sets with unknown jitter*, Sampl. Theory Signal Image Process. **7** (2008), no. 2, 187–195. MR2451766

[24] A. Aldroubi, Q. Sun, and W.-S. Tang, *p-frames and shift invariant subspaces of L^p*, J. Fourier Anal. Appl. **7** (2001), no. 1, 1–21, DOI 10.1007/s00041-001-0001-2. MR1812992

[25] A. Aldroubi, Q. Sun, and W.-S. Tang, *Nonuniform average sampling and reconstruction in multiply generated shift-invariant spaces*, Constr. Approx. **20** (2004), no. 2, 173–189, DOI 10.1007/s00365-003-0539-0. MR2036639

[26] A. Aldroubi, Q. Sun, and W.-S. Tang, *Convolution, average sampling, and a Calderon resolution of the identity for shift-invariant spaces*, J. Fourier Anal. Appl. **11** (2005), no. 2, 215–244, DOI 10.1007/s00041-005-4003-3. MR2131637

[27] A. Aldroubi, J. Davis, and I. Krishtal, *Dynamical sampling: time-space trade-off*, Appl. Comput. Harmon. Anal. **34** (2013), no. 3, 495–503, DOI 10.1016/j.acha.2012.09.002. MR3027915

[28] A. Aldroubi and I. Krishtal, *Krylov subspace methods in dynamical sampling*, Sampl. Theory Signal Image Process. **15** (2016), 9–20. MR3670914

[29] A. Aldroubi, C. Cabrelli, U. Molter, and S. Tang, *Dynamical sampling*, Appl. Comput. Harmon. Anal. **42** (2017), no. 3, 378–401, DOI 10.1016/j.acha.2015.08.014. MR3613395

[30] A. Aldroubi, L. Huang, I. Krishtal, A. Ledeczi, R. R. Lederman, and P. Volgyesi, *Dynamical sampling with additive random noise*, Sampl. Theory Signal Image Process. **17** (2018), no. 2, 153–182. MR3877746

[31] A. Aldroubi, K. Gröchenig, L. Huang, P. Jaming, I. Krishtal, and J. L. Romero, *Sampling the flow of a bandlimited function*, J. Geom. Anal. **31** (2021), no. 9, 9241–9275, DOI 10.1007/s12220-021-00617-0. MR4302220

[32] G. Alessandrini, *An identification problem for an elliptic equation in two variables*, Annali di matematica pura et applicata, **145** (2008), 265–295

[33] A. Aldroubi and M. Unser, *Sampling procedures in function spaces and asymptotic equivalence with Shannon's sampling theory*, Numer. Funct. Anal. Optim. **15** (1994), no. 1-2, 1–21, DOI 10.1080/01630569408816545. MR1261594

[34] D. Alpay and H. Dym, *On reproducing kernel spaces, the Schur algorithm, and interpolation in a general class of domains*, Operator theory and complex analysis (Sapporo, 1991), Oper. Theory Adv. Appl., vol. 59, Birkhäuser, Basel, 1992, pp. 30–77. MR1246809

[35] D. Alpay and H. Dym, *On a new class of structured reproducing kernel spaces*, J. Funct. Anal. **111** (1993), no. 1, 1–28, DOI 10.1006/jfan.1993.1001. MR1200633

[36] D. Alpay, V. Bolotnikov, A. Dijksma, and H. de Snoo, *On some operator colligations and associated reproducing kernel Hilbert spaces*, Operator extensions, interpolation of functions and related topics (Timişoara, 1992), Oper. Theory Adv. Appl., vol. 61, Birkhäuser, Basel, 1993, pp. 1–27. MR1246577

[37] D. Alpay, P. Jorgensen, R. Seager, and D. Volok, *On discrete analytic functions: products, rational functions and reproducing kernels*, J. Appl. Math. Comput. **41** (2013), no. 1-2, 393–426, DOI 10.1007/s12190-012-0608-2. MR3017129

[38] D. Alpay, P. Jorgensen, and D. Volok, *Relative reproducing kernel Hilbert spaces*, Proc. Amer. Math. Soc. **142** (2014), no. 11, 3889–3895, DOI 10.1090/S0002-9939-2014-12121-6. MR3251728

[39] H. W. Alt, *Lineare Funktionalanalysis*, Springer, Berlin, Heidelberg, 2012

[40] W. O. Amrein and A. M. Berthier, *On support properties of L^p-functions and their Fourier transforms*, J. Functional Analysis **24** (1977), no. 3, 258–267, DOI 10.1016/0022-1236(77)90056-8. MR0461025

[41] H. Ammari, E. Bonnetier, Y. Capdeboscq, M. Tanter, and M. Fink, *Electrical impedance tomography by elastic deformation*, SIAM J. Appl. Math. **68** (2008), no. 6, 1557–1573, DOI 10.1137/070686408. MR2424952

[42] L. C. Andrews, *Special Functions for Engineers and Applied Mathematicians*, Maxmillian, New York, 1985

[43] T. S. Angell, M. Z. Nashed, *Operator-theoretic and computational aspects of ill-posed problems in antenna theory*, Proc. Int. Symp. Math. Theory of Networks and Syst., Delft Univ. of Technology, The Netherlands, (1979), 499–511

[44] G. Anger, *Inverse Problems in Differential Equations*, Akademie-Verlag, Berlin, 1990

[45] N. Aoki, G. T. Schuster, *Fast least squares migration with a deblurring filter*, Proc. SEG Houston 2009 Annual Meeting, (2009), 2829–2833

[46] T. M. Apostol, *An elementary view of Euler's summation formula*, Amer. Math. Monthly **106** (1999), no. 5, 409–418, DOI 10.2307/2589145. MR1699259

[47] N. Aronszajn, *Theory of reproducing kernels*, Trans. Amer. Math. Soc. **68** (1950), 337–404, DOI 10.2307/1990404. MR51437

[48] N. Aronszajn, T. M. Creese, L. J. Lipkin, *Polyharmonic Functions*, Clarendon Press, Oxford, 1983

[49] G. Aubert and P. Kornprobst, *Mathematical problems in image processing*, 2nd ed., Applied Mathematical Sciences, vol. 147, Springer, New York, 2006. Partial differential equations and the calculus of variations; With a foreword by Olivier Faugeras. MR2244145

[50] M. Augustin, S. Eberle, and M. Grothaus, *An overview on tools from functional analysis*, Handbook of mathematical geodesy, Geosyst. Math., Birkhäuser/Springer, Cham, 2018, pp. 165–199. MR3792541

[51] M. Augustin, W. Freeden, and H. Nutz, *About the importance of the Runge-Walsh concept for gravitational field determination*, Handbook of mathematical geodesy, Geosyst. Math., Birkhäuser/Springer, Cham, 2018, pp. 517–560. MR3792546

[52] P. Aviles and Y. Giga, *Variational integrals on mappings of bounded variation and their lower semicontinuity*, Arch. Rational Mech. Anal. **115** (1991), no. 3, 201–255, DOI 10.1007/BF00380769. MR1106293

[53] M. L. Baart, *Methods for Solution of Fredholm Integral Equations of the First Kind*, CSIR, Pretoria, Techn. Rep., **104** (1979)

[54] G. E. Backus, F. Gilbert, *Numerical applications of a formalism for geophysical inverse problems*, Geophys. J.R. Astron. Soc., **13** (1967), 247–276

[55] G. E. Backus, F. Gilbert, *The resolving power of gross Earth data*, Geophys. J. R. Astron. Soc., **16** (1968), 169–205

[56] G. Backus and F. Gilbert, *Uniqueness in the inversion of inaccurate gross Earth data*, Philos. Trans. Roy. Soc. London Ser. A **266** (1970), no. 1173, 123–192, DOI 10.1098/rsta.1970.0005. MR449460

[57] R. Baer, *Linear Algebra and Projective Geometry*, Academic Press, New York, 1952

[58] G. E. Backus, *Non-uniqueness of the external geomagnetic field determined by surface intensity measurements*, J. Geophys. Res., **75** (1970), 6339–6341

[59] G. E. Backus, R. Parker, C. Constable, *Foundations of Geomagnetism*, Cambridge University Press, Cambridge, 1996

[60] S. Baker, D. Scharstein, J. P. Lewis, S. Roth, M. J. Black, R. A. Szeliski, *Database and evaluation methodology for optical flow*, Int. J. Comput. Vision, **92** (2011), 1–31

[61] A. B. Bakushinskii, *The problem of the convergence of the iteratively regularized Gauss-Newton method*, Comput. Math. Phys., **32** (1992), 1353–1359

[62] A. B. Bakuschinskii, *On the principle of iterative regularization*, USSR Comput. Math. and Math. Phys., **19** (1979), 256–260

[63] R. Balan, *A study of Weyl-Heisenberg and wavelet frames*, Ph.-D. Thesis, Princeton University, 1998

[64] R. Balian, *Un principe d'incertitude fort en théorie du signal ou en mécanique quantique* (French, with English summary), C. R. Acad. Sci. Paris Sér. II Méc. Phys. Chim. Sci. Univers Sci. Terre **292** (1981), no. 20, 1357–1362. MR644367

[65] J. M. Ball, *Convexity conditions and existence theorems in nonlinear elasticity*, Arch. Rat. Mech. Anal., **63** (1977), 337–403.

[66] C. T. H. Bakero, *The Numerical Treatment of Integral Equations*, Clarendon Press, Oxford, 1977

[67] A. B. Bakushinskii, *The problem of the convergence of the iteratively regularized Gauss-Newton method*, Comput. Math. Phys., **32** (1992) 1353–1359, 1992

[68] H. P. Baltes (Ed.), *Inverse Scattering in Optics (Topics in Current Physics)*, Springer, Berlin, Heidelberg, New York, 1979

[69] R. E. Barnhill, B. R. Piper, and K. L. Rescorla, *Interpolation to arbitrary data on a surface*, Geometric modeling, SIAM, Philadelphia, PA, 1987, pp. 281–289. MR936460

[70] J. Baumeister, *Stable Solution of Inverse Problems*, Vieweg, Braunschweig, 1987

[71] H. Behmard and A. Faridani, *Sampling of bandlimited functions on unions of shifted lattices*, J. Fourier Anal. Appl. **8** (2002), no. 1, 43–58, DOI 10.1007/s00041-002-0003-8. MR1882365

[72] J. J. Benedetto, M. E. Frazier (Eds.), *Wavelets: Mathematics and Applications*, CRC Press, Boca Raton, FL, 1993

[73] J. J. Benedetto and P. J. S. G. Ferreira (eds.), *Modern sampling theory*, Applied and Numerical Harmonic Analysis, Birkhäuser Boston, Inc., Boston, MA, 2001. Mathematics and applications, DOI 10.1007/978-1-4612-0143-4. MR1865678

[74] J. J. Benedetto and A. I. Zayed (eds.), *Sampling, wavelets, and tomography*, Applied and Numerical Harmonic Analysis, Birkhäuser Boston, Inc., Boston, MA, 2004, DOI 10.1007/978-0-8176-8212-5. MR2037639

[75] M. Benedicks, *On Fourier transforms of functions supported on sets of finite Lebesgue measure*, J. Math. Anal. Appl. **106** (1985), no. 1, 180–183, DOI 10.1016/0022-247X(85)90140-4. MR780328

[76] A. Ben-Israel, *A Newton-Raphson method for the solution of systems of equations*, J. Math. Anal. Appl. **15** (1966), 243–252, DOI 10.1016/0022-247X(66)90115-6. MR205445

[77] A. Ben-Israel and T. N. E. Greville, *Generalized inverses: theory and applications*, Pure and Applied Mathematics, Wiley-Interscience [John Wiley & Sons], New York-London-Sydney, 1974. MR0396607

[78] M. Bertero, C. de Mol, G. A. Viano, *Linear inverse problems with discrete data*, Inverse Problems, **4** (1985), 573-594

[79] M. Bertero, P. Brianzi, E. R. Pike, and L. Rebolia, *Linear regularizing algorithms for positive solutions of linear inverse problems*, Proc. Roy. Soc. London Ser. A **415** (1988), no. 1849, 257–275. MR932923

[80] M. Bertero, C. De Mol, and G. A. Viano, *On the problems of object restoration and image extrapolation in optics*, J. Math. Phys. **20** (1979), no. 3, 509–521, DOI 10.1063/1.524103. MR525114

[81] M. Bertero, C. De Mol, and G. A. Viano, *The stability of inverse problems.*, Inverse scattering problems in optics, Topics Current Phys., 20, , 1980, pp. 161–214,. MR0612324

[82] G. Beylkin and L. Monzón, *On approximation of functions by exponential sums*, Appl. Comput. Harmon. Anal. **19** (2005), no. 1, 17–48, DOI 10.1016/j.acha.2005.01.003. MR2147060

[83] G. Beylkin and L. Monzón, *Approximation by exponential sums revisited*, Appl. Comput. Harmon. Anal. **28** (2010), no. 2, 131–149, DOI 10.1016/j.acha.2009.08.011. MR2595881

[84] N. Bi, M. Z. Nashed, and Q. Sun, *Reconstructing signals with finite rate of innovation from noisy samples*, Acta Appl. Math. **107** (2009), no. 1-3, 339–372, DOI 10.1007/s10440-009-9474-9. MR2520024

[85] P. J. Bickel and E. Levina, *Regularized estimation of large covariance matrices*, Ann. Statist. **36** (2008), no. 1, 199–227, DOI 10.1214/009053607000000758. MR2387969

[86] A. Binder, H. W. Engl, C. W. Groetsch, A. Neubauer, and O. Scherzer, *Weakly closed nonlinear operators and parameter identification in parabolic equations by Tikhonov regularization*, Appl. Anal. **55** (1994), no. 3-4, 215–234, DOI 10.1080/00036819408840301. MR1381290

[87] B. L. Biondi, *Three-Dimensional Seismic Imaging*, Society of Exploration Geophysicists, Tulsa, 2006

[88] N. Bissantz, T. Hohage, A. Munk, and F. Ruymgaart, *Convergence rates of general regularization methods for statistical inverse problems and applications*, SIAM J. Numer. Anal. **45** (2007), no. 6, 2610–2636, DOI 10.1137/060651884. MR2361904

[89] A. Bjerhammar, *Rectangular reciprocal matrices with special reference to geodetic calculations*, Bull. Geod., (1951) 188–220

[90] A. Bjerhammar, *Gravity Reduction to an Internal Sphere*, Division of Geodesy, Stockholm, Sweden, 1962

[91] A. Bjerhammar, *A review of discrete methods in physical geodesy*, H. Moritz, H. Sünkel (Eds.), Approximation Methods in Physical Geodesy, Wichmann, Karlsruhe, (1978), 211–233

[92] A. Björck, *Numerical Methods for Least Squares Problems*, SIAM, Philadelphia, 1996.

[93] A. Björck, L. Elden, *Methods in Numerical Algebra for Ill-Posed Problems*, Ill-Posed Problems: Theory and Practice, Dordrecht and Boston, Reidel, 1981.

[94] A. Bhandari, F. Krahmer, and T. Poskitt, *Unlimited sampling from theory to practice: Fourier-Prony recovery and prototype ADC*, IEEE Trans. Signal Process. **70** (2022), 1131–1141, DOI 10.1109/TSP.2021.3113497. MR4396313

[95] R. J. Blakely, *Potential Theory in Gravity and Magnetic Application*, Cambridge University Press, Cambridge, 1996

[96] M. J. Black, P. Anandan, *Robust dynamic motion estimation over time*, Proc. IEEE Comp. Soc. Conf. on Computer Vision and Pattern Recognition. (CVPR '91, Maui, June 3-6, 1991), IEEE Computer Society Press: Los Alamitos, (1991), 292-302

[97] M. J. Black, P. Anandan, *The robust estimation of multiple motions: parametric and piecewise-smooth flow fields*, Computer Vision and Image Understanding, **63** (1996), 75-104

[98] B. Blaschke, A. Neubauer, and O. Scherzer, *On convergence rates for the iteratively regularized Gauss-Newton method*, IMA J. Numer. Anal. **17** (1997), no. 3, 421–436, DOI 10.1093/imanum/17.3.421. MR1459331

[99] C. Blatter, *Wavelets—Eine Einführung*, Advanced Lectures in Mathematics, Vieweg+Teubner, Wiesbaden, 2003

[100] C. Blick, W. Freeden, M. Z. Nashed, H. Nutz, M. Schreiner, *Inverse Magnetometry: Mollifier Magnetization Distribution from Geomagnetic Field Data*, Lecture Notes in Geosystems Mathematics and Computing, Birkhäuser, Springer International Publishing, Switzerland, Cham, 2021

[101] T. Blu, K. Dragotti, M. Veetterli, P. Marziliano, L. Coulot, *Sparse sampling of signal innovations*, IEEE Sign. Process. Mag., **25** (2008), 31-40

[102] T. Blu and M. Unser, *Approximation error for quasi-interpolators and (multi-) wavelet expansions*, Appl. Comput. Harmon. Anal. **6** (1999), no. 2, 219–251, DOI 10.1006/acha.1998.0249. MR1676842

[103] C. de Boor, K. Höllig, and S. Riemenschneider, *Bivariate cardinal interpolation by splines on a three-direction mesh*, Illinois J. Math. **29** (1985), no. 4, 533–566. MR806466

[104] C. de Boor, R. A. DeVore, and A. Ron, *Approximation from shift-invariant subspaces of* $L_2(\mathbf{R}^d)$, Trans. Amer. Math. Soc. **341** (1994), no. 2, 787–806, DOI 10.2307/2154583. MR1195508

[105] B. Borchers, Uram, T. J. Hendrix, *Tikhonov regularization of electrical conductivity depth profiles in field soils*, Soil Science Society of Amer. J., **61** (1997), 1004-1009

[106] W. M. Boerner, *Polarization utilization in electromagnetic inverse scattering: microwave imaging, radar target mapping, radon shape reconstruction and remote sensing*, Univ. Illinois at Chicago Circle, Chicago, Communications Lab., Rep. 78-3, 1978.

[107] W. M. Boerner, *Polarization Utilization in Electromagnetic Inverse Scattering*, Topics in Current Physics, H.B. Baltes, (Ed.), Springer, Berlin, Heidelberg, New York, 1979.

[108] T. L. Boullion and P. L. Odell, *Generalized inverse matrices*, Wiley-Interscience [A Division of John Wiley & Sons, Inc.], New York-London-Sydney, 1971. MR0338012

[109] A. Bonami, B. Demange, and P. Jaming, *Hermite functions and uncertainty principles for the Fourier and the windowed Fourier transforms*, Rev. Mat. Iberoamericana **19** (2003), no. 1, 23–55, DOI 10.4171/RMI/337. MR1993414

[110] K. Borre, P. Meissl, *Strength Analysis of Levelling-Type Networks*, Geodetic Institute, Copenhagen, Report No. 50, 1974

[111] M. Bauer, M. Grasmair, and C. Kirisits, *Optical flow on moving manifolds*, SIAM J. Imaging Sci. **8** (2015), no. 1, 484–512, DOI 10.1137/140965235. MR3317380

[112] C. J. Bouwkamp and N. G. de Bruijn, *The problem of optimum antenna current distribution*, Philips Research Rep. **1** (1946), 135–158. MR16268

[113] L. de Branges, *Hilbert spaces of entire functions*, Prentice-Hall, Inc., Englewood Cliffs, N.J., 1968. MR0229011

[114] O. Bratteli, P. Jorgensen, *Wavelets Through a Looking Glass*, Appl. Numeric. Harm. Analysis, Birkhäuser Boston, Inc., Boston, 2002

[115] J. L. Brown, S. D. Cabrera, *On well-posedness of the Papoulis generalized sampling expansion*, IEEE Trans. Circ. Syst., **38** (1991), no. 5, 554-556

[116] J. L. Brown Jr. and K. Sa-Ngsari, *Sampling reconstruction of N-dimensional band-limited images after multilinear filtering*, IEEE Trans. Circuits and Systems **36** (1989), no. 7, 1035–1038, DOI 10.1109/31.31344. MR998874

[117] R. E. Bruck Jr., *A strongly convergent iterative solution of $0 \in U(x)$ for a maximal monotone operator U in Hilbert space*, J. Math. Anal. Appl. **48** (1974), 114–126, DOI 10.1016/0022-247X(74)90219-4. MR361941

[118] E. C. Bullard, *The magnetic field within the Earth*, Proc. Cambridge Phil. Soc, **A197** (1949), 438-453

[119] M. D. Buhmann, *Radial Basis Functions: Theory and Implementation*, Cambridge University Press, Cambridge, 2003.

[120] M. Burger, B. Kaltenbacher, *Regularizing Newton-Kaczmarz Methods for Nonlinear Ill-Posed Problems*, SFB-Report 04-17, University of Linz, 2004

[121] M. Burger, E. Resmerita, and L. He, *Error estimation for Bregman iterations and inverse scale space methods in image restoration*, Computing **81** (2007), no. 2-3, 109–135, DOI 10.1007/s00607-007-0245-z. MR2354192

[122] H. C. Burschäpers, *Local Modeling of Gravitational Data*, Master Thesis, University of Kaiserslautern, Mathematics Department, Geomathematics Group, 2013

[123] P. L. Butzer, *A survey of the Whittaker-Shannon sampling theorem and some of its extensions*, J. Math. Res. Exposition **3** (1983), no. 1, 185–212. MR724869

[124] P. L. Butzer, R. Nessel, *Fourier Analysis and Approximation Theory*, Birkhäuser, Basel, 1971.

[125] P. L. Butzer and R. L. Stens, *The Euler-MacLaurin summation formula, the sampling theorem, and approximate integration over the real axis*, Linear Algebra Appl. **52/53** (1983), 141–155, DOI 10.1016/0024-3795(83)80011-1. MR709348

[126] P. L. Butzer, W. Splettstösser, and R. L. Stens, *The sampling theorem and linear prediction in signal analysis*, Jahresber. Deutsch. Math.-Verein. **90** (1988), no. 1, 70. MR928745

[127] P. L. Butzer, W. Engels, S. Ries, and R. L. Stens, *The Shannon sampling series and the reconstruction of signals in terms of linear, quadratic and cubic splines*, SIAM J. Appl. Math. **46** (1986), no. 2, 299–323, DOI 10.1137/0146020. MR833479

[128] P. L. Butzer and R. L. Stens, *Sampling theory for not necessarily band-limited functions: a historical overview*, SIAM Rev. **34** (1992), no. 1, 40–53, DOI 10.1137/1034002. MR1156288

[129] P. L. Butzer, A. Fischer, and R. L. Stens, *Generalized sampling approximation of multivariate signals; general theory*, Atti Sem. Mat. Fis. Univ. Modena **41** (1993), no. 1, 17–37. MR1225669

[130] P. L. Butzer, J. R. Higgins, and R. L. Stens, *Classical and approximate sampling theorems: studies in the $L^P(\mathbb{R})$ and the uniform norm*, J. Approx. Theory **137** (2005), no. 2, 250–263, DOI 10.1016/j.jat.2005.07.011. MR2186950

[131] P. L. Butzer, G. Schmeisser, and R. L. Stens, *The classical and approximate sampling theorems and their equivalence for entire functions of exponential type*, J. Approx. Theory **179** (2014), 94–111, DOI 10.1016/j.jat.2013.11.010. MR3148887

[132] H. S. Cabayan, H. S. Murphy, Pavlasek, T. J. S., *Numerical stabiliy and near field reconstruction*, IEEE Trans. Antenna Propagat., **21** (1973), 346–351

[133] J. A. Cadzow, *Signal enhancement - a composite property mapping algorithm*, IEEE Transactions on Acoustics, Speech and Signal Processing, **36** (1988), 49–62

[134] L. Cavalier and Yu. Golubev, *Risk hull method and regularization by projections of ill-posed inverse problems*, Ann. Statist. **34** (2006), no. 4, 1653–1677, DOI 10.1214/009053606000000542. MR2283712

[135] L. Cavalier, G. K. Golubev, D. Picard, and A. B. Tsybakov, *Oracle inequalities for inverse problems*, Ann. Statist. **30** (2002), no. 3, 843–874, DOI 10.1214/aos/1028674843. Dedicated to the memory of Lucien Le Cam. MR1922543

[136] L. L. Campbell, *A comparison of the sampling theorems of Kramer and Whittaker*, J. Soc. Indust. Appl. Math. **12** (1964), 117–130. MR164173

[137] E. J. Candès, J. K. Romberg, and T. Tao, *Stable signal recovery from incomplete and inaccurate measurements*, Comm. Pure Appl. Math. **59** (2006), no. 8, 1207–1223, DOI 10.1002/cpa.20124. MR2230846

[138] B. D. Craven and M. Z. Nashed, *Generalized implicit function theorems when the derivative has no bounded inverse*, Nonlinear Anal. **6** (1982), no. 4, 375–387, DOI 10.1016/0362-546X(82)90023-2. MR654813

[139] P. G. Casazza, D. Han, and D. R. Larson, *Frames for Banach spaces*, The functional and harmonic analysis of wavelets and frames (San Antonio, TX, 1999), Contemp. Math., vol. 247, Amer. Math. Soc., Providence, RI, 1999, pp. 149–182, DOI 10.1090/conm/247/03801. MR1738089

[140] P. G. Casazza, M. Fickus, D. G. Mixon, J. Peterson, and I. Smalyanau, *Every Hilbert space frame has a Naimark complement*, J. Math. Anal. Appl. **406** (2013), no. 1, 111–119, DOI 10.1016/j.jmaa.2013.04.047. MR3062405

[141] J. W. S. Cassels, *An Introduction to the Geometry of Numbers*, Springer, Berlin, Heidelberg, New York, 1968

[142] F. Chamizo and H. Iwaniec, *On the sphere problem*, Rev. Mat. Iberoamericana **11** (1995), no. 2, 417–429, DOI 10.4171/rmi/178. MR1344899

[143] K. Chandrasekharan and R. Narasimhan, *Hecke's functional equation and the average order of arithmetical functions*, Acta Arith. **6** (1960/61), 487–503, DOI 10.4064/aa-6-4-487-503. MR126423

[144] J.-r. Chen, *The lattice-points in a circle*, Sci. Sinica **12** (1963), 633–649. MR154855

[145] X. Chen, Z. Nashed, and L. Qi, *Convergence of Newton's method for singular smooth and nonsmooth equations using adaptive outer inverses*, SIAM J. Optim. **7** (1997), no. 2, 445–462, DOI 10.1137/S1052623493246288. MR1443628

[146] W. Cheney, W. Light, *A Course in Approximation Theory*, Brooks/Cole, 2000

[147] H. Cheng, L. Greengard, and V. Rokhlin, *A fast adaptive multipole algorithm in three dimensions*, J. Comput. Phys. **155** (1999), no. 2, 468–498, DOI 10.1006/jcph.1999.6355. MR1723309

[148] J. G. Christensen, *Sampling in reproducing kernel Banach spaces on Lie groups*, J. Approx. Theory, **64** (2012), 79-203

[149] O. Christensen, *Moment problems and stability results for frames with applications to irregular sampling and Gabor frames*, Appl. Comput. Harmon. Anal. **3** (1996), no. 1, 82–86, DOI 10.1006/acha.1996.0007. MR1374398

[150] O. Christensen and D. T. Stoeva, *p-frames in separable Banach spaces*, Adv. Comput. Math. **18** (2003), no. 2-4, 117–126, DOI 10.1023/A:1021364413257. Frames. MR1968115

[151] M. T. Chu, *On a numerical treatment for the curve-tracing of the homotopy method*, Numer. Math. **42** (1983), no. 3, 323–329, DOI 10.1007/BF01389577. MR723629

[152] C. K. Chui, K. Jetter, and J. D. Ward, *Cardinal interpolation by multivariate splines*, Math. Comp. **48** (1987), no. 178, 711–724, DOI 10.2307/2007838. MR878701

[153] J. Claerbout, *Basic Earth Imaging*, Stanford University, Standford, 2009

[154] F. Cluny, D. Costarelli, A Minotti, G. Vinti, *Enhancement of thermographic images as tool for structural analysis in earthquake engineering*, NDT & E International, **70** (2015), 60–72

[155] L. Cohen, *Time-frequency distribution - a review*, Proc. IEEE, **77** (1989), 941–981

[156] R. Courant, D. Hilbert, *Methoden der Mathematischen Physik I,II*, Springer, Berlin, 2nd ed., 1968

[157] I. Cohen, *Nonlinear variational method for optical flow computation*, Proc. Eighth Scandi navian Conf. on Image Analysis (SCIA '93, Tromsö, May 25-28, 1993), **1** (1993), 523–530

[158] L. Cohen, *Time-Frequency Analysis*, Prentice Hall, Englewood Cliffs, New Jersey, 1995

[159] D. Costarelli and G. Vinti, *Approximation by nonlinear multivariate sampling Kantorovich type operators and applications to image processing*, Numer. Funct. Anal. Optim. **34** (2013), no. 8, 819–844, DOI 10.1080/01630563.2013.767833. MR3175598

[160] R. E. Collins, F. J. Zucker, *Antenna Theory*, McGraw-Hill, New York, 1969

[161] D. Colton, R. Kress, *Integral Equation Methods in Scattering Theory*, Wiley, New York, 1983

[162] D. Colton, R. Kress, *Inverse Acoustic and Electromagnetic Scattering Theory*, 2nd ed., Springer, Heidelberg, 1998

[163] D. D. Cox, *Asymptotics for M-type smoothing splines*, Ann. Statist. **11** (1983), no. 2, 530–551. MR696065

[164] C. K. Chui, *An Introduction to Wavelets*, Academic Press, San Diego, London, 1992

[165] C. K. Chui and L. Zhong, *Polynomial interpolation and Marcinkiewicz-Zygmund inequalities on the unit circle*, J. Math. Anal. Appl. **233** (1999), no. 1, 387–405, DOI 10.1006/jmaa.1999.6337. MR1684394

[166] F. Cucker and S. Smale, *On the mathematical foundations of learning*, Bull. Amer. Math. Soc. (N.S.) **39** (2002), no. 1, 1–49, DOI 10.1090/S0273-0979-01-00923-5. MR1864085

[167] J. Cui and W. Freeden, *Equidistribution on the sphere*, SIAM J. Sci. Comput. **18** (1997), no. 2, 595–609, DOI 10.1137/S1064827595281344. MR1433797

[168] P. Craven and G. Wahba, *Smoothing noisy data with spline functions. Estimating the correct degree of smoothing by the method of generalized cross-validation*, Numer. Math. **31** (1978/79), no. 4, 377–403, DOI 10.1007/BF01404567. MR516581

[169] B. Dacorogna, *Weak Continuity and Weak Lower Semi-Continuity of Nonlinear Functionals*, Springer, Berlin, 1982

[170] B. Dacorogna, *Direct Methods in the Calculus of Variations*, Springer, Berlin, 1989

[171] S. B. Damelin and V. Maymeskul, *On point energies, separation radius and mesh norm for s-extremal configurations on compact sets in* $\mathbb{R}^n$, J. Complexity **21** (2005), no. 6, 845–863, DOI 10.1016/j.jco.2005.04.005. MR2182448

[172] J. W. Daniel, *The Approximate Minimization of Functions*, Prentice-Hall, Englewood Cliffs, N.J., 1971

[173] I. Daubechies, *Time-frequency localization operators: a geometric phase space approach*, IEEE Transactions on Informat. Theory, **34** (1988), 961–1005

[174] I. Daubechies, *Ten Lectures on Wavelets*, SIAM, Philadelphia, 1992

[175] P. J. Davis, *Interpolation and Approximation*, Blaisdell, New York, 1963

[176] P. J. Davis and P. Rabinowitz, *Numerical integration*, Blaisdell Publishing Co. [Ginn and Co.], Waltham, Mass.-Toronto, Ont.-London, 1967. MR0211604

[177] M. Defrise and C. De Mol, *A note on stopping rules for iterative regularization methods and filtered SVD*, Inverse problems: an interdisciplinary study (Montpellier, 1986), Adv. Electron. Electron Phys., Suppl. 19, Academic Press, London, 1987, pp. 261–268. MR1005575

[178] G. Deidda, E. Bonomi, D. Manzi, *Inversion of electrical conductivity data with Tikhonov regularization approach: some considerations*, Annals for Geophysics, **46** (2003), 549–558

[179] B. Demange, *Principes D'incertitude Associés á des Formes Quadratiques non Dégénérées*, Thése de l'université d'Orléans, 2004

[180] J.-P. Dedieu and M.-H. Kim, *Newton's method for analytic systems of equations with constant rank derivatives*, J. Complexity **18** (2002), no. 1, 187–209, DOI 10.1006/jcom.2001.0612. MR1895083

[181] J. A. DeSanto, A. W. Saenz, Zachary, W. W. (Eds.), *Mathematical Methods and Applications of Scattering Theory*, Springer, New York, 1980

[182] L. Desbat and D. Girard, *The "minimum reconstruction error" choice of regularization parameters: some more efficient methods and their application to deconvolution problems*, SIAM J. Sci. Comput. **16** (1995), no. 6, 1387–1403, DOI 10.1137/0916080. MR1355499

[183] G. A. Deschamps, H. S. Cabayan, *Antenna synthesis and solution of inverse problems by regularization methods*, IEEE Trans. Antennas Propagat., **20** (1972), 268–274

[184] P. Deuflhard, H. W. Engl, and O. Scherzer, *A convergence analysis of iterative methods for the solution of nonlinear ill-posed problems under affinely invariant conditions*, Inverse Problems **14** (1998), no. 5, 1081–1106, DOI 10.1088/0266-5611/14/5/002. MR1654603

[185] P. Deuflhard, A. Hohmann, *Numerische Mathematik I: Eine algorithmisch orientierte Einführung*, Walter de Gruyter, Berlin, 1993

[186] J. B. Diaz and F. T. Metcalf, *On iteration procedures for equations of the first kind,* $Ax = y$, *and Picard's criterion for the existence of a solution*, Math. Comp. **24** (1970), 923–935, DOI 10.2307/2004626. MR281376

[187] V. Dicken and P. Maass, *Wavelet-Galerkin methods for ill-posed problems*, J. Inverse Ill-Posed Probl. **4** (1996), no. 3, 203–221, DOI 10.1515/jiip.1996.4.3.203. MR1401487

[188] V. I. Dmitriev, A. S. Il′inskiĭ, and A. G. Svešnikov, *Development of mathematical methods for the study of direct and inverse problems of electrodynamics* (Russian), Uspehi Mat. Nauk **31** (1976), no. 6(192), 123–141. MR0502977

[189] J. Dienstel, *The Geometry of Banach Spaces*, Lecture Notes Math., Springer, Heidelberg, **485**, 1975

[190] M. Dobrowolski, *Angewandte Funktionalanalysis*, Springer, Berlin, Heidelberg, 2010

[191] D. Dobson and O. Scherzer, *Analysis of regularized total variation penalty methods for denoising*, Inverse Problems **12** (1996), no. 5, 601–617, DOI 10.1088/0266-5611/12/5/005. MR1413421

[192] D. L. Donoho, *Nonlinear solution of linear inverse problems by wavelet-vaguelette decomposition*, Appl. Comput. Harmon. Anal. **2** (1995), no. 2, 101–126, DOI 10.1006/acha.1995.1008. MR1325535

[193] D. L. Donoho and I. M. Johnstone, *Minimax estimation via wavelet shrinkage*, Ann. Statist. **26** (1998), no. 3, 879–921, DOI 10.1214/aos/1024691081. MR1635414

[194] H. Duan, B. P. Ng, C. M. S. See, and J. Fang, *Broadband interference suppression performance of minimum redundancy arrays*, IEEE Trans. Signal Process. **56** (2008), no. 6, 2406–2416, DOI 10.1109/TSP.2007.916130. MR2516643

[195] J. Duchon, *Splines minimizing rotation-invariant semi-norms in Sobolev spaces.*, Constructive theory of functions of several variables (Proc. Conf., Math. Res. Inst., Oberwolfach, 1976), Lecture Notes in Math., Vol. 571, , 1977, pp. 85–100. MR0493110

[196] R. J. Duffin and A. C. Schaeffer, *A class of nonharmonic Fourier series*, Trans. Amer. Math. Soc. **72** (1952), 341–366, DOI 10.2307/1990760. MR47179

[197] N. Dunford, J. Schwarz, *Linear Operators*, Vol. II, Wiley-Interscience, New York, 1963

[198] C. S. Duris, *Optimal Quadrature Formulas Using Generalized Inverses. Part I. General Theory and Minimum Variance Formulas*, Math. Comp., **25** (1971), 495-504

[199] N. Dyn, D. Levin, and S. Rippa, *Numerical procedures for surface fitting of scattered data by radial functions*, SIAM J. Sci. Statist. Comput. **7** (1986), no. 2, 639–659, DOI 10.1137/0907043. MR833927

[200] N. Dyn, *Interpolation and approximation by radial and related functions*, Approximation theory VI, Vol. I (College Station, TX, 1989), Academic Press, Boston, MA, 1989, pp. 211–234. MR1090994

[201] E. Weinan, J. Han, L. Zhang, *Integrating Machine Learning with Physics-Based Modeling*, `arXiv:2006.02619v1`, 2020

[202] P. P. B. Eggermont, *Maximum entropy regularization for Fredholm integral equations of the first kind*, SIAM J. Math. Anal. **24** (1993), no. 6, 1557–1576, DOI 10.1137/0524088. MR1241158

[203] P. P. B. Eggermont, V. N. LaRicca, *Maximum Penalized Likelihood Estimation*, Vol. II, Regression, Springer, New York, 2009

[204] P. P. B. Eggermont, V. N. LaRiccia, and M. Z. Nashed, *On weakly bounded noise in ill-posed problems*, Inverse Problems **25** (2009), no. 11, 115018, 14, DOI 10.1088/0266-5611/25/11/115018. MR2558678

[205] P. P. B. Eggermont, V. N. LaRiccla, M. Z. Nashed, *Noise Models for Ill-Posed Problems*, Handbook of Geomathematics, Vol. 2, W. Freeden, M. Z. Nashed, T. Sonar (Eds.), Springer, Heidelberg, (2010), 741–762

[206] P. P. B. Eggermont, V. N. LaRiccia, and M. Z. Nashed, *Moment discretization for ill-posed problems with discrete weakly bounded noise*, GEM Int. J. Geomath. **3** (2012), no. 2, 155–178, DOI 10.1007/s13137-012-0037-2. MR3000076

[207] P. P. B. Eggermont, V. N. LaRiccia, M. Z. Nashed, *Noise models for ill-posed problems*, Handbook of Geomathematics, W. Freeden, M. Z. Nashed, T. Sonar (Eds.), 2nd ed., Springer, New York, 2015

[208] B. Eicke, *Iteration methods for convexly constrained ill-posed problems in Hilbert space*, Numer. Funct. Anal. Optim. **13** (1992), no. 5-6, 413–429, DOI 10.1080/01630569208816489 MR1187903

[209] I. Ekeland, R. Temam, *Convex analysis and variational problem*, S. North Holland, Amsterdam, 1976

[210] L. Eldén, *Algorithms for the regularization of ill-conditioned least squares problems*, Nordisk Tidskr. Informationsbehandling (BIT) **17** (1977), no. 2, 134–145, DOI 10.1007/bf01932285. MR474912

[211] L. Eldén, *An algorithm for the regularization of ill-conditioned, banded least squares problems*, SIAM J. Sci. Statist. Comput. **5** (1984), no. 1, 237–254, DOI 10.1137/0905017. MR731894

[212] H. W. Engl and M. Z. Nashed, *Stochastic projectional schemes for random linear operator equations of the first and second kinds*, Numer. Funct. Anal. Optim. **1** (1979), no. 5, 451–473, DOI 10.1080/01630567908816029. MR546127

[213] H. W. Engl and M. Z. Nashed, *New extremal characterizations of generalized inverses of linear operators*, J. Math. Anal. Appl. **82** (1981), no. 2, 566–586, DOI 10.1016/0022-247X(81)90217-1. MR629778

[214] H. W. Engl, *Discrepancy principles for Tikhonov regularization of ill-posed problems leading to optimal convergence rates*, J. Optim. Theory Appl. **52** (1987), no. 2, 209–215, DOI 10.1007/BF00941281. MR879198

[215] H. W. Engl, M. Hanke, A. Neubauer, *Regularization of inverse problems*, Kluwer, Dordrecht, 1996

[216] H. W. Engl, K. Kunisch, and A. Neubauer, *Convergence rates for Tikhonov regularisation of nonlinear ill-posed problems*, Inverse Problems **5** (1989), no. 4, 523–540. MR1009037

[217] H. Engl, A. K. Louis, Rundell, W. (Eds.), *Inverse Problems in Geophysical Applications*, SIAM, Philadelphia, 1997

[218] H. Engl, *Integralgleichungen*, Springer, Berlin, 1997

[219] P. Erdös, P. M. Gruber, and J. Hammer, *Lattice points*, Pitman Monographs and Surveys in Pure and Applied Mathematics, vol. 39, Longman Scientific & Technical, Harlow; copublished in the United States with John Wiley & Sons, Inc., New York, 1989. MR1003606

[220] G. I. Eskin, *Boundary value problems for elliptic pseudodifferential equations*, Translations of Mathematical Monographs, vol. 52, American Mathematical Society, Providence, R.I., 1981. Translated from the Russian by S. Smith. MR623608

[221] L. Euler, *Methodus universalis serierum convergentium summas quam proxime inveniendi*, Commentarii Academiae Scientiarum Petropolitanae, 8:3-9, Opera Omnia (XIV), 101–107, 1736

[222] L. Euler, *Methodus universalis series summandi ulterius promota*, Commentarii Academiae Scientarium Petropolitanae, 8:147-158, Opera Omnia (XIV), 124–137, 1736

[223] L. C. Evans, M. Gariepy, *Measure Theory and Fine Properties of Functions*, CRC Press, Boca Raton, 1992

[224] L. D. Evans, *Partial Differential Equation*, American Mathematical Society (AMS), Providence, R.I., 3rd ed., 2002

[225] T. D. Fadale, A. V. Nenarokomov, A. F. Emery, *Uncertainties in parameter estimation: the inverse problem*, Internat. J. Heat and Mass Transfer, **38** (1995), 511–518

[226] R. L. Fante, J. T. Mayhan, *Bounds on the electric field outside a radiating system*, IEEE Trans. Antennas Propagat, **16** (1968), 712–717

[227] R. L. Fante, J. T. Mayhan, *Bounds on the electric field outside a radiating system-II*, IEEE Trans. Antennas Propagat., **18** (1970), 64–68

[228] W. G. Faris, *Inequalities and uncertainty principles*, J. Mathematical Phys. **19** (1978), no. 2, 461–466, DOI 10.1063/1.523667. MR484134

[229] H. G. Feichtinger, *New results on regular and irregular sampling based on Wiener amalgams*, Function Spaces, K. Jarosz (Ed.), Marcel Dekker, New York, (1992a), 107–121

[230] H. G. Feichtinger, *Wiener amalgams over Euclidean spaces and some of their applications*, Function Spaces, K. Jarosz (Ed.), Marcel Dekker, New York, (1992b), 123–137

[231] H. G. Feichtinger and K. Gröchenig, *Iterative reconstruction of multivariate band-limited functions from irregular sampling values*, SIAM J. Math. Anal. **23** (1992), no. 1, 244–261, DOI 10.1137/0523013. MR1145171

[232] M. J. Fengler, W. Freeden, and M. Gutting, *The spherical Bernstein wavelet*, Int. J. Pure Appl. Math. **31** (2006), no. 2, 209–230. MR2268146

[233] H. R. Fernández-Morales, A. G. García, and G. Pérez-Villalón, *Generalized sampling in $L^2(\mathbb{R}^d)$ shift-invariant subspaces with multiple stable generators*, Multiscale signal analysis and modeling, Springer, New York, 2013, pp. 51–80, DOI 10.1007/978-1-4614-4145-8_3. MR3024464

[234] J. Flemming and B. Hofmann, *A new approach to source conditions in regularization with general residual term*, Numer. Funct. Anal. Optim. **31** (2010), no. 1-3, 254–284, DOI 10.1080/01630561003765721. MR2677255

[235] G. B. Folland, *Introduction to Partial Differential Equations*, Princeton University Press, Princeton, 1995

[236] G. B. Folland and A. Sitaram, *The uncertainty principle: a mathematical survey*, J. Fourier Anal. Appl. **3** (1997), no. 3, 207–238, DOI 10.1007/BF02649110. MR1448337

[237] O. M. Fomenko, *On the Problem of Gauss*, Acta. Arith., **6** (1961), 277–284

[238] T. S. Fong, *On the problem of optimum antenna aperture distribution*, J. Franklin Inst. **283** (1967), 235–249, DOI 10.1016/0016-0032(67)90026-9. MR210379

[239] I. Fonseca and G. A. Francfort, *Relaxation in BV versus quasiconvexification in $W^{1,p}$; a model for the interaction between fracture and damage*, Calc. Var. Partial Differential Equations **3** (1995), no. 4, 407–446, DOI 10.1007/BF01187895. MR1385294

[240] I. Fonseca and S. Müller, *Quasi-convex integrands and lower semicontinuity in L^1*, SIAM J. Math. Anal. **23** (1992), no. 5, 1081–1098, DOI 10.1137/0523060. MR1177778

[241] I. Fonseca and S. Müller, *Relaxation of quasiconvex functionals in* $\mathrm{BV}(\Omega, \mathbf{R}^p)$ *for integrands* $f(x, u, \nabla u)$, Arch. Rational Mech. Anal. **123** (1993), no. 1, 1–49, DOI 10.1007/BF00386367. MR1218685

[242] M. Fornasier and H. Rauhut, *Continuous frames, function spaces, and the discretization problem*, J. Fourier Anal. Appl. **11** (2005), no. 3, 245–287, DOI 10.1007/s00041-005-4053-6. MR2167169

[243] H. Frankenberger and M. Hanke, *Kernel polynomials for the solution of indefinite and ill-posed problems*, Numer. Algorithms **25** (2000), no. 1-4, 197–212, DOI 10.1023/A:1016641914524. Mathematical journey through analysis, matrix theory and scientific computation (Kent, OH, 1999). MR1827154

[244] J. N. Franklin, *Well-posed stochastic extensions of ill-posed linear problems*, J. Math. Anal. Appl. **31** (1970), 682–716, DOI 10.1016/0022-247X(70)90017-X. MR267654

[245] J. N. Franklin, *On Tikhonov's method for ill-posed problems*, Math. Comp. **28** (1974), 889–907, DOI 10.2307/2005354. MR375817

[246] I. Fredholm, *Sur une nouvelle méthode pour la résolution du problème de Dirichlet*, Översigt Kongl. Vetenskaps-Akademiens Förhandlingar **57** (1900), 39–46

[247] I. Fredholm, *Sur une classe d'équations fonctionnelles* (French), Acta Math. **27** (1903), no. 1, 365–390, DOI 10.1007/BF02421317. MR1554993

[248] W. Freeden, *Eine Verallgemeinerung der Hardyschen Identität*, Ph.-D. Thesis, RWTH Aachen, 1975

[249] W. Freeden, *An application of a summation formula to numerical computation of integrals over the sphere*, Bull. Géodésique **52** (1978), no. 3, 165–175, DOI 10.1007/BF02521770. MR509269

[250] W. Freeden, *Über gewichtete Gitterpunktsummen in kreisförmigen Bereichen*, Mitt. Math. Seminar Giessen, **132** (1978b), 1–22

[251] W. Freeden, *Über eine Verallgemeinerung der Hardy-Landauschen Identität*, Manuscr. Math., **24** (1978c), 205–216

[252] W. Freeden, *Über eine Klasse von Integralformeln der Mathematischen Geodäsie*, Veröff. Geod. Inst. RWTH Aachen, Report 27, 1979

[253] W. Freeden, H. Kersten, *The Geodetic Boundary-Value Problem Using the Known Surface of the Earth*, Veröff. Geod. Inst. RWTH Aachen, Report 29, 1980a

[254] W. Freeden, *On the approximation of external gravitational potential with closed systems of (trial) functions*, Bull. Geod., **54** (1980b), 1–20

[255] W. Freeden, *On integral formulas of the (unit) sphere and their application to numerical computation of integrals*, Computing, **25** (1980c), 131–146

[256] W. Freeden, *On spherical spline interpolation and approximation*, Math. Meth. Appl. Sci., **3** (1981a), 551–575

[257] W. Freeden, *On approximation by harmonic splines*, Manuscr. Geod., **6** (1981b), 193–244

[258] W. Freeden, *Interpolation and best approximation by harmonic spline functions - theoretical and computational aspects*, Boll. Geod. Sci. Affini, **41** (1982a), 106–120

[259] W. Freeden, *Multidimensional Euler summation formulas and numerical cubature*, ISNM, **57** (1982b), 77–88

[260] W. Freeden, *Least Squares Approximation by Linear Combinations of (Multi)-Poles*, Department of Geodetic Science and Surveying, The Ohio State University, Columbus, Report 344, 1983

[261] W. Freeden, *Spherical spline interpolation—basic theory and computational aspects*, J. Comput. Appl. Math. **11** (1984), no. 3, 367–375, DOI 10.1016/0377-0427(84)90011-6. MR777111

[262] W. Freeden, *A spline interpolation method for solving boundary value problems of potential theory from discretely given data*, Numer. Methods Partial Differential Equations **3** (1987), no. 4, 375–398, DOI 10.1002/num.1690030408. MR1108128

[263] W. Freeden, *Interpolation by multidimensional periodic splines*, J. Approx. Theory **55** (1988), no. 1, 104–117, DOI 10.1016/0021-9045(88)90114-1. MR963538

[264] W. Freeden, *Spherical spline approximation and its application in physical geodesy*, Geophysical Data Inversion Methods and Applications, A. Vogel, C.O. Ofeagbu, R. Gorenflo, B. Ursin (Eds.), Vieweg Publication, (1990), 79–104

[265] W. Freeden and H. Kersten, *A constructive approximation theorem for the oblique derivative problem in potential theory*, Math. Methods Appl. Sci. **3** (1981), no. 1, 104–114, DOI 10.1002/mma.1670030108. MR606851

[266] W. Freeden, *Multiscale Modelling of Spaceborne Geodata*, B. G. Teubner, Stuttgart, Leipzig, 1999

[267] W. Freeden, *Geomathematik, was ist das überhaupt?* (German, with English and German summaries), Jahresber. Deutsch. Math.-Verein. **111** (2009), no. 3, 125–152. MR2559914

[268] W. Freeden, *Metaharmonic Lattice Point Theory*, Chapman and Hall, CRC Press, Boca Raton, London, New York, 2011

[269] W. Freeden, *Geomathematics: Its Role, Its Aim, and Its Potential*, Handbook of Geomathematics, W. Freeden, M. Z. Nashed, T. Sonar, (Eds.), Springer, Heidelberg, Vol. 1, 2nd ed., 3-78, 2015

[270] W. Freeden, *Decorrelative Mollifier Gravimetry - Basics, Concepts, Examples, and Perspectives*, Geosystems Mathematics, Birkhäuser, Cham, 2021a

[271] W. Freeden, *Antenna problem induced regularization and sampling strategies*, Deterministic and Stochastic Optimal Control and Inverse Problems, B. Jadamba, A. A. Khan, S. Migorski, M. Sama (Eds.), CRC Press, Boca Raton, London, New York, 319-353, 2021b

[272] W. Freeden, C. Blick, *Signal decorrelation by means of multiscale methods*, World of Mining, **65** (2013), 1–15

[273] W. Freeden, T. Fehlinger, M. Klug, D. Mathar, K. Wolf, *Classical globally reflected gravity field determination in modern locally oriented multiscale framework*, J. Geod., **83** (2009), 1171–1191

[274] W. Freeden, C. Gerhards, *Geomathematically Oriented Potential Theory*, Chapman and Hall, CRC Press, Boca Raton, London, New York, 2013

[275] W. Freeden, T. Gervens, and J. C. Mason, *A minimum norm interpolation method for determining the displacement field of a homogeneous isotropic elastic body from discrete data*, IMA J. Appl. Math. **44** (1990), no. 1, 55–76, DOI 10.1093/imamat/44.1.55. MR1044255

[276] W. Freeden, T. Gervens, M. Schreiner, *Constructive Approximation on the Sphere (With Applications to Geomathematics)*, Oxford Science Publications, Clarendon, Oxford, 1998

[277] W. Freeden, M. Gutting, *Special Functions of Mathematical (Geo)Physics*, Birkhäuser, Basel, 2013

[278] W. Freeden, M. Gutting, *Integration and Cubature Methods*, Chapman and Hall, CRC Press, Boca Raton, London, New York, 2018

[279] W. Freeden and K. Hesse, *On the multiscale solution of satellite problems by use of locally supported kernel functions corresponding to equidistributed data on spherical orbits*, Studia Sci. Math. Hungar. **39** (2002), no. 1-2, 37–74, DOI 10.1556/SScMath.39.2002.1-2.3. MR1909147

[280] W. Freeden and T. Maier, *On multiscale denoising of spherical functions: basic theory and numerical aspects*, Electron. Trans. Numer. Anal. **14** (2002), 56–78. Orthogonal polynomials, approximation theory, and harmonic analysis (Inzel, 2000). MR1929056

[281] W. Freeden and T. Maier, *Spectral and multiscale signal-to-noise thresholding of spherical vector fields*, Comput. Geosci. **7** (2003), no. 3, 215–250, DOI 10.1023/A:1025581622035. MR2007579

[282] W. Freeden and C. Mayer, *Wavelets generated by layer potentials*, Appl. Comput. Harmon. Anal. **14** (2003), no. 3, 195–237, DOI 10.1016/S1063-5203(03)00003-4. MR1984547

[283] W. Freeden and C. Mayer, *Modeling tangential vector fields on regular surfaces by means of MIE potentials*, Int. J. Wavelets Multiresolut. Inf. Process. **5** (2007), no. 3, 417–449, DOI 10.1142/S0219691307001835. MR2342792

[284] W. Freeden, V. Michel, *Multiscale Potential Theory (With Applications to Geoscience)*, Birkhäuser, Boston, 2004

[285] W. Freeden, D. Michel, V. Michel, *Local multiscale approximation of geostrophic oceanic flow: theoretical background and aspects of scientific computing*, Marine Geodesy, **28** (2005), 313–329

[286] W. Freeden, V. Michel, and H. Nutz, *Satellite-to-satellite tracking and satellite gravity gradiometry (advanced techniques for high-resolution geopotential field determination)*, J. Engrg. Math. **43** (2002), no. 1, 19–56, DOI 10.1023/A:1016577524288. MR1921451

[287] W. Freeden, V. Michel, and F. J. Simons, *Spherical harmonics based special function systems and constructive approximation methods*, Handbook of mathematical geodesy, Geosyst. Math., Birkhäuser/Springer, Cham, 2018, pp. 753–819. MR3792551

[288] W. Freeden and M. Z. Nashed, *Multi-variate Hardy-type lattice point summation and Shannon-type sampling*, GEM Int. J. Geomath. **6** (2015), no. 2, 163–249, DOI 10.1007/s13137-015-0076-6. MR3412548

[289] W. Freeden and M. Z. Nashed, *Ill-posed problems: operator methodologies of resolution and regularization*, Handbook of mathematical geodesy, Geosyst. Math., Birkhäuser/Springer, Cham, 2018, pp. 201–314. MR3792542

[290] W. Freeden and M. Z. Nashed, *Inverse gravimetry as an ill-posed problem in mathematical geodesy*, Handbook of mathematical geodesy, Geosyst. Math., Birkhäuser/Springer, Cham, 2018, pp. 641–685. MR3820919

[291] W. Freeden and M. Z. Nashed, *Inverse gravimetry: background material and multiscale mollifier approaches*, GEM Int. J. Geomath. **9** (2018), no. 2, 199–264, DOI 10.1007/s13137-018-0103-5. MR3861426

[292] W. Freeden and M. Z. Nashed, *Operator-theoretic and regularization approaches to ill-posed problems*, GEM Int. J. Geomath. **9** (2018), no. 1, 1–115, DOI 10.1007/s13137-017-0100-0. MR3780190

[293] W. Freeden and M. Z. Nashed, *Ill-posed problems: operator methodologies of resolution and regularization*, Handbook of mathematical geodesy, Geosyst. Math., Birkhäuser/Springer, Cham, 2018, pp. 201–314. MR3792542

[294] W. Freeden and M. Z. Nashed, *Inverse gravimetry as an ill-posed problem in mathematical geodesy*, Handbook of mathematical geodesy, Geosyst. Math., Birkhäuser/Springer, Cham, 2018, pp. 641–685. MR3820919

[295] W. Freeden, M. Z. Nashed, (Eds.), *Handbook of Mathematical Geodesy*, Springer International Publishing, Birkhäuser, Basel, New York, Heidelberg, Geosystems Mathematics, 2018g

[296] W. Freeden, M. Z. Nashed, *From Gaussian circle problem to multivariate Shannon sampling*, Frontiers in Orthogonal Polynomials and q-Series. M.Z. Nashed, X. Li (Eds.), World Scientific, Singapore, (2018b), 213–238

[297] W. Freeden, M. Z. Nashed, and M. Schreiner, *Spherical sampling*, Geosystems Mathematics, Birkhäuser/Springer, Cham, 2018, DOI 10.1007/978-3-319-71458-5. MR3791815

[298] W. Freeden, M. Z. Nashed, *Lattice Point Identities and Shannon-Type Sampling*, CRC Press, Taylor & Francis, Boca Raton, 2020a

[299] W. Freeden, M. Z. Nashed, *Inverse Gravimetry: Density Signatures from Gravitational Potential Data*, Handbuch der Geodäsie, Mathematische Geodäsie/Mathematical Geodesy, W. Freeden (Band Hrsg.), W. Freeden, R. Rummel (Hrsg.), Springer Spektrum, Heidelberg, 969-1052, 2020b

[300] W. Freeden, M. Z. Nashed, M. Schreiner, *Multiscale Tree Sampling Regularization of Inverse Spherical Pseudodifferental Equations*, Sampling, Approximation, and Signal Analysis-Harmonic Analysis in the Spirit of J. R. Higgins, S.D. Casey, M. Dodson, P. J. S. G. Ferreira, A. Zayed (Eds.), Applied and Numerical Harmonic Analysis, Springer, 2023

[301] W. Freeden, Z. Nashed, Sonar, T. (Eds.), *Handbook of Geomathematics*, Springer, New-York, 2nd ed., Vol. 1,2, and 3, 2015

[302] W. Freeden and H. Nutz, *Satellite gravity gradiometry as tensorial inverse problem*, GEM Int. J. Geomath. **2** (2011), no. 2, 177–218, DOI 10.1007/s13137-011-0026-x. MR2854366

[303] W. Freeden, H. Nutz, *Mathematische Methoden*, Handbuch Tiefe Geothermie, M. Bauer, W. Freeden, H. Jacobi, T. Neu (Eds.), Springer, Heidelberg, 2014

[304] W. Freeden and H. Nutz, *Mathematik als Schlüsseltechnologie zum Verständnis des Systems "tiefe Geothermie"* (German, with German summary), Jahresber. Dtsch. Math.-Ver. **117** (2015), no. 1, 45–84, DOI 10.1365/s13291-014-0107-5. MR3311949

[305] W. Freeden and H. Nutz, *Geodetic observables and their mathematical treatment in multiscale framework*, Handbook of mathematical geodesy, Geosyst. Math., Birkhäuser/Springer, Cham, 2018, pp. 315–458. MR3792543

[306] W. Freeden, H. Nutz, *Inverse Probleme der Geodäsie: Ein Abriss mathematischer Lösungsstrategien*, Handbuch der Geodäsie, Mathematische Geodäsie/Mathematical Geodesy, W. Freeden (Bd. Hrsg.), W. Freeden, R. Rummel (Hrsg.), Springer Spektrum, Heidelberg, (2020), 65–90

[307] W. Freeden, H. Nutz, R. Rummel, M. Schreiner, *Satellite Gravity Gradiometry (SGG): Methodological Foundation and Geomathematical Advances*, Handbuch der Geodäsie, Mathematische Geodäsie/Mathematical Geodesy, W. Freeden (Bd. Hrsg.), W. Freeden, R. Rummel (Hrsg.), Springer Spektrum, Heidelberg, Vol. 2, 1185-1256, 2020

[308] W. Freeden, S. Pereverzev, *Spherical Tikhonov regularization wavelets in satellite gravity gradiometry with random noise*, J. Geodesy, **74** (2001), 730-736

[309] W. Freeden and R. Reuter, *A class of multidimensional periodic splines*, Manuscripta Math. **35** (1981), no. 3, 371–386, DOI 10.1007/BF01263270. MR636463

[310] W. Freeden and R. Reuter, *Exact computation of spherical harmonics* (English, with German summary), Computing **32** (1984), no. 4, 365–378, DOI 10.1007/BF02243779. MR748937

[311] W. Freeden, R. Reuter, *An efficient algorithm for the generation of homogeneous harmonic polynomials*, Scientific Software Systems, M.G. Cox, Mason, J.C. (Eds.), Chapman & Hall, London, (1990), 166–180

[312] W. Freeden, F. Sansó, *Interactions, acquisitions, and open problems*, IX Hotine-Marussi Symposium on Mathematical Geodesy, P. Novák, M. Crespi, N. Sneeuw, F. Sansó (Eds.), Springer, (2018) 219–250

[313] W. Freeden and F. Schneider, *Regularization wavelets and multiresolution*, Inverse Problems **14** (1998), no. 2, 225–243, DOI 10.1088/0266-5611/14/2/002. MR1619437

[314] W. Freeden, F. Schneider, *An integrated wavelet concept of physical geodesy*, J. Geodesy, **72** (1998b), 259–281

[315] W. Freeden and F. Schneider, *Wavelet approximations on closed surfaces and their application to boundary-value problems of potential theory*, Math. Methods Appl. Sci. **21** (1998), no. 2, 129–163, DOI 10.1002/(SICI)1099-1476(19980125)21:2¡129::AID-MMA942¿3.0.CO;2-7. MR1600014

[316] W. Freeden, F. Schneider, and M. Schreiner, *Gradiometry—an inverse problem in modern satellite geodesy*, Inverse problems in geophysical applications (Yosemite, CA, 1995), SIAM, Philadelphia, PA, 1997, pp. 179–239. MR1427993

[317] W. Freeden and M. Schreiner, *Non-orthogonal expansions on the sphere*, Math. Methods Appl. Sci. **18** (1995), no. 2, 83–120, DOI 10.1002/mma.1670180202. MR1311372

[318] W. Freeden and M. Schreiner, *Orthogonal and nonorthogonal multiresolution analysis, scale discrete and exact fully discrete wavelet transform on the sphere*, Constr. Approx. **14** (1998), no. 4, 493–515, DOI 10.1007/s003659900087. MR1646543

[319] W. Freeden, M. Schreiner, *Local multiscale modelling of geoid undulations from deflections of the vertical*, J. Geodesy, **79** (2006), 641–651

[320] W. Freeden and M. Schreiner, *Biorthogonal locally supported wavelets on the sphere based on zonal kernel functions*, J. Fourier Anal. Appl. **13** (2007), no. 6, 693–709, DOI 10.1007/s00041-006-6905-0. MR2350444

[321] W. Freeden, M. Schreiner, *Spherical Functions of Mathematical Geosciences - A Scalar, Vecterial, and Tensorial Setup*, 1st ed., Springer, Heidelberg, 2009

[322] W. Freeden, M. Schreiner, *Satellite gravity gradiometry (sgg): from scalar to tensorial solution*, Handbook of Geomathematics, Vol. 1, W. Freeden, M. Z. Nashed, T. Sonar (Eds.), Springer, Heidelberg, 269-302, 2010

[323] W. Freeden, M. Schreiner, *Satellite gravity gradiometry (sgg): from scalar to tensorial solution*, Handbook of Geomathematics, W. Freeden, M. Z. Nashed, T. Sonar (Eds.), Springer, 2nd ed., New-York, Heidelberg, 2015

[324] W. Freeden, M. Schreiner, *Mathematical geodesy: its role, its aim, and its potential*, Handbuch der Geodäsie, Mathematische Geodäsie/Mathematical Geodesy, W. Freeden (Bd. Hrsg.), W. Freeden, R. Rummel (Hrsg.), Springer Spektrum, Heidelberg, 3-64, 2020a

[325] W. Freeden, M. Schreiner, *Spherical harmonics, splines, and wavelets - definitoric constituents, strategic perspectives, specific applicability and applications*, Handbuch der Geodäsie, Mathematische Geodäsie/Mathematical Geodesy, W. Freeden (Bd. Hrsg.), W. Freeden, R. Rummel (Hrsg.), Springer Spektrum, Heidelberg, (2020b), 147–194

[326] W. Freeden and M. Schreiner, *Spherical functions of mathematical geosciences—a scalar, vectorial, and tensorial setup*, 2nd ed., Geosystems Mathematics, Birkhäuser/Springer, Berlin, 2022, DOI 10.1007/978-3-662-65692-1. MR4516554

[327] W. Freeden, M. Schreiner, and R. Franke, *A survey on spherical spline approximation*, Surveys Math. Indust. **7** (1997), no. 1, 29–85. MR1470453

[328] W. Freeden, T. Sonar, and B. Witte, *Gauss as scientific mediator between mathematics and geodesy from the past to the present*, Handbook of mathematical geodesy, Geosyst. Math., Birkhäuser/Springer, Cham, 2018, pp. 1–163. MR3792540

[329] W. Freeden and U. Windheuser, *Spherical wavelet transform and its discretization*, Adv. Comput. Math. **5** (1996), no. 1, 51–94, DOI 10.1007/BF02124735. MR1381282

[330] W. Freeden and U. Windheuser, *Combined spherical harmonic and wavelet expansion—a future concept in Earth's gravitational determination*, Appl. Comput. Harmon. Anal. **4** (1997), no. 1, 1–37, DOI 10.1006/acha.1996.0192. MR1429676

[331] W. Freeden and B. Witte, *A combined (spline-) interpolation and smoothing method for the determination of the external gravitational potential from heterogeneous data*, Bull. Géodésique **56** (1982), no. 1, 53–62, DOI 10.1007/BF02525607. MR658963

[332] W. Freeden, B. Witte, *From Gaussian least squares approximation to today's operator-theoretic regularization of ill-posed problems*, Handbuch der Geodäsie, Mathematische Geodäsie/Mathematical Geodesy, W. Freeden (Bd. Hrsg.), W. Freeden, R. Rummel (Hrsg.), Springer Spektrum, Heidelberg, Vol. 2, (2020), 827–930

[333] W. Freeden and K. Wolf, *Klassische Erdschwerefeldbestimmung aus der Sicht moderner Geomathematik* (German, with German summary), Math. Semesterber. **56** (2009), no. 1, 53–77, DOI 10.1007/s00591-008-0046-2. MR2480942

[334] F. Fricker, *Geschichte des Kreisproblems*, Mitt. Math. Sem. Giessen, **111** (1975), 1–34

[335] F. Fricker, *Einführung in die Gitterpunktlehre*, Birkhäuser, Basel, 1982

[336] K. Friedrich, *Allgemeine für die Rechenpraxis geeignete Lösung für die Aufgaben der kleinsten Absolutsumme und der günstigsten Gewichtsverteilung*, Z. Vermessungsw. (ZfV), (1937), 337–358

[337] K. O. Friedrichs, *The identity of weak and strong extensions of differential operators*, Trans. Amer. Math. Soc. **55** (1944), 132–151, DOI 10.2307/1990143. MR9701

[338] H. Funk, *Beiträge zur Theorie der Kugelfunktionen*, Math. Ann., **77** (1916), 136–152

[339] F. Frühauf, C. Pontow, and O. Scherzer, *Texture enhancing based on variational image decomposition*, Mathematical image processing, Springer Proc. Math., vol. 5, Springer, Heidelberg, 2011, pp. 127–140, DOI 10.1007/978-3-642-19604-1_7. MR2867523

[340] H. Gajewski, K. Gröger, and K. Zacharias, *Nichtlineare Operatorgleichungen und Operatordifferentialgleichungen* (German), Mathematische Lehrbücher und Monographien, II. Abteilung, Mathematische Monographien, Band 38, Akademie-Verlag, Berlin, 1974. MR0636412

[341] B. G. Galerkin, *Expansions in stability problems for elastic rods and plates* (in Russian), Vestnik Inzkenorov, **19** (1915), 897–908.

[342] A. Galle, *Über die geodätischen Arbeiten von Gauss, I. Abschnitt: Die Methode der kleinsten Quadrate*, Berlin, Gauss Werke, Bd. XI 2. Abteilung, 1924

[343] A. Garcia, *Orthogonal sampling formulas: a unified approach*, SIAM Rev., **4** (2000), 499–512

[344] C. F. Gauss, *Untersuchungen über höhere Arithmetik* (German), Chelsea Publishing Co., New York, 1965. Deutsch herausgegeben von H. Maser. MR0188045

[345] C. F. Gauss, *Theoria Motus Corporum Coelestium in Sectionibus Conicis Solem Ambientium*, Hamburg, 1809, Werke 7, Translated into English by C.H. Davis, 1963

[346] C. F. Gauss, *Disquisitio de Elementis Ellipticis Paladis ex Opositionibus Annorum 1803, 1804, 1805, 1807, 1808, 1809*, Göttingen, 1810

[347] C. F. Gauss, *Disquisitiones Generales Circa Seriem Infinitam* $1 + ((\alpha \cdot \beta)/\ (1-\gamma))x + \ldots$ *etc.*, Commentationes Societatis Regiae Scientarium Gottingensis Recentiores, 2. Werke Ill, (1812), 123–162

[348] C. F. Gauss, *Theoria attractionis corporum sphaeroidicorum ellipticorum homogeneorum methodo nova tractata*, Commentationes Societatis Regiae Scientiarium Gottingensis Recentiores, **2** (1813), 355–378

[349] C. F. Gauss, *Methodus nova integralium valores per approximationem inveniendi*, Commentationes Societatis Regiae Scientarium Gottingensis Recentiores. 3, Werke III, (1814), 163–196

[350] C. F. Gauss, *Theoria combinations observationum erroribus minimus obnoxiae*, Werke 4, Authorized French translation by J. Bertrand. English translation from the French by H.F. Trotter, Gauss'work (1803-1826) on the Theory of Least Squares. Technical Report Nop. 5, Statistical Techniques Research Group, Department of Math., Princeton University, Princeton, 1957

[351] C. F. Gauss, *Allgemeine Auflösung der Aufgabe: Die Teile einer gegebenen Fläche auf einer andern gegebenen Fläche so abzubilden, dass die Abbildung dem Abgebildeten in den kleinsten Theilen ähnlich wird*, Astronomische Abhandlungen, H. C. Schumacher (Hrsg.), Heft 3, Altona 1825, (1822), 5–30

[352] C. F. Gauss, *Theoria Combinationis Observationum Erroribus Minimis Obnoxiae*, 2. Teil, Göttingen, 1823

[353] C. F. Gauss, *Supplementum Theoriae Combinationis Observationum Erroribus Minimis Obnoxiae*, Göttingen, 1826

[354] C. F. Gauss, *Werke*, Vierter Band. Hrsg. Königl. Gesellschaft der Wissenschaft zu Göttingen, 109–117, B.G. Teubner, Leipzig, (1880).

[355] A. V. Gavrilov, *On best quadrature formulas in the reproducing kernel Hilbert space*, Siberian J. Numer. Math., **1** (1998), 313–320

[356] B. Gebauer and O. Scherzer, *Impedance-acoustic tomography*, SIAM J. Appl. Math. **69** (2008), no. 2, 565–576, DOI 10.1137/080715123. MR2465856

[357] I. M. Gel'fand and G. E. Shilov, *Generalized functions. Vol. I: Properties and operations*, Academic Press, New York-London, 1964. Translated by Eugene Saletan. MR0166596

[358] I. M. Gelfand, N. Ya. Vilenkin, *Generalized Functions*, Vol. 4, Academic Press, New York, 1964

[359] C. Gerhards, *Spherical Multiscale Methods in Terms of Locally Supported Wavelts: Theory and Application to Geomagnetic Modeling*, Ph.-D. Thesis, Geomathematics Group, TU Kaiserslautern, 2011

[360] H. Gfrerer, *An a posteriori parameter choice for ordinary and iterated Tikhonov regularization of ill-posed problems leading to optimal convergence rates*, Math. Comp., **49** (1987a), 523–542

[361] D. Gilbarg, N. S. Trudinger, *Elliptic Partial Differential Equations of Second Order*, Springer, New York, 2001

[362] E. Giusti, *Minimal Surfaces and Functions of Bounded Variations*, Birkhäuser, Boston, 1984

[363] H.-J. Glaeske and O. I. Maričev, *The Laguerre transform of some elementary functions* (English, with German and Russian summaries), Z. Anal. Anwendungen **3** (1984), no. 3, 237–244, DOI 10.4171/ZAA/104. MR753460

[364] A. Gholami, H. R. Siahkoohi, *Regularization of Linear and Nonlinear Geophysical Ill-Posed Problems with Joint Sparsity Constraints*, Geophys. J. Int., **180** (2010), 871–882

[365] I. Gohberg, S. Goldberg, and M. A. Kaashoek, *Classes of linear operators. Vol. II*, Operator Theory: Advances and Applications, vol. 63, Birkhäuser Verlag, Basel, 1993, DOI 10.1007/978-3-0348-8558-4_1. MR1246332

[366] M. Golomb and H. F. Weinberger, *Optimal approximation and error bounds*, On numerical approximation. Proceedings of a Symposium, Madison, April 21-23, 1958, Publication of the Mathematics Research Center, U.S. Army, the University of Wisconsin, no. 1, University of Wisconsin Press, Madison, Wis., 1959, pp. 117–190. Edited by R. E. Langer. MR0121970

[367] G. H. Golub, C. F. Van Loan, *Matrix Computations*, John Hopkins Univ. Press, 1983.

[368] G. H. Golub, M. Heath, and G. Wahba, *Generalized cross-validation as a method for choosing a good ridge parameter*, Technometrics **21** (1979), no. 2, 215–223, DOI 10.2307/1268518. MR533250

[369] T. N. T. Goodman, C. A. Micchelli, G. Rodriguez, and S. Seatzu, *On the limiting profile arising from orthonormalizing shifts of exponentially decaying functions*, IMA J. Numer. Anal. **18** (1998), no. 3, 331–354, DOI 10.1093/imanum/18.3.331. MR1631962

[370] T. N. T. Goodman, C. A. Micchelli, and Z. Shen, *Riesz bases in subspaces of* $L_2(\mathbf{R}_+)$, Constr. Approx. **17** (2001), no. 1, 39–46, DOI 10.1007/s003650010019. MR1794800

[371] E. W. Grafarend, *Six Lectures on Geodesy and Global Geodynamics*, Proceedings of the Third International Summer School in the Mountains, H. Moritz, H. Sünkel (Eds.), (1982), 531–685

[372] J. Graves and P. M. Prenter, *Numerical iterative filters applied to first kind Fredholm integral equations*, Numer. Math. **30** (1978), no. 3, 281–299, DOI 10.1007/BF01411844. MR502805

[373] L. Greengard and V. Rokhlin, *A new version of the fast multipole method for the Laplace equation in three dimensions*, Acta numerica, 1997, Acta Numer., vol. 6, Cambridge Univ. Press, Cambridge, 1997, pp. 229–269, DOI 10.1017/S0962492900002725. MR1489257

[374] U. Grenander, *Abstract Inference*, Wiley, New York, 1981

[375] T. N. E. Greville, *Introduction to spline functions*, Theory and Applications of Spline Functions (Proceedings of Seminar, Math. Research Center, Univ. of Wisconsin, Madison, Wis., 1968), Academic Press, New York, 1969, pp. 1–35. MR0240521

[376] C. W. Groetsch, *Generalized Inverses of Linear Operators*, Marcel Dekker. Inc., New York, 1977

[377] C. W. Groetsch, *Comments on Morozov's discrepancy principle*, Improperly posed problems and their numerical treatment (Oberwolfach, 1982), Internat. Schriftenreihe Numer. Math., vol. 63, Birkhäuser, Basel, 1983, pp. 97–104. MR726767

[378] C. W. Groetsch, *The Theory of Tikhonov Regularization for Fredholm Equations of the First Kind*, Pitman, London-Boston, 1984

[379] C. W. Groetsch, *Inverse Problems in the Mathematical Science*, Vieweg, Braunschweig, 1993

[380] C. W. Groetsch and O. Scherzer, *Iterative stabilization and edge detection*, Inverse problems, image analysis, and medical imaging (New Orleans, LA, 2001), Contemp. Math., vol. 313, Amer. Math. Soc., Providence, RI, 2002, pp. 129–141, DOI 10.1090/conm/313/05373. MR1940994

[381] K. Gröchenig, *Describing functions, frames versus atomic decompositions*, Monatshefte für Mathematik, **112** (1991), 1–41

[382] K. Gröchenig, *Wiener's Lemma: Theme and Variations, an Introduction to Spectral Invariance and its Applications*, Four Short Courses on Harmonic Analysis: Wavelets, Frames, Time-Frequency Methods, and Applicatory to Signal and Image Analysis. P. Massopust and B. Förster (Eds.), Birkhäuser, Boston 2010

[383] E. Grosswald, *Representations of integers as sums of squares*, Springer-Verlag, New York, 1985, DOI 10.1007/978-1-4613-8566-0. MR803155

[384] M. Gutting, *Fast Multipole Methods for Oblique Derivative Problems*, PhD thesis, Geomathematics Group, University of Kaiserslautern, 2007

[385] M. Gutting, *Fast multipole accelerated solution of the oblique derivative boundary value problem*, GEM Int. J. Geomath. **3** (2012), no. 2, 223–252, DOI 10.1007/s13137-012-0038-1. MR3000079

[386] W. Freeden, M. Z. Nashed, and T. Sonar (eds.), *Handbook of geomathematics. Vol. 1, 2, 3*, 2nd ed., Springer, Heidelberg, 2015, DOI 10.1007/978-3-642-54551-1. MR3523845

[387] A. Haar, *Zur Theorie der orthogonalen Funktionensysteme* (German), Math. Ann. **69** (1910), no. 3, 331–371, DOI 10.1007/BF01456326. MR1511592

[388] W. Hackbusch, *Entwicklungen nach Exponentialsummen*, Technical Report. Max-Planck-Institut für Mathematik in den Naturwissenschaften, Leipzig, 2010

[389] W. Hackbusch, B. N. Khoromoskij, A. Klaus, *Approximation of Functions by Exponential Sums Based on the Newton-Type Optimisation*, Technical Report, Max-Planck-Institut für Mathematik in den Naturwissenschaften, Leipzig, 2005

[390] J. Hadamard, *Sur les problémes aux dérivés partielles et leur signification physique*, Princeton Univ. Bull., **13** (1902), 49–52

[391] J. Hadamard, *Lectures on the Cauchy Problem in Linear Partial Differential Equations*, Yale University Press, New Haven, 1923

[392] R. A. Haddad, T. W. Parsons, *Digital Signal Processing: Theory, Applications and Hardware*, Computer Science Press, 1991

[393] S. Haeseler, M. Keller, D. Lenz, J. Masamune, M. Schmidt, *Global properties of Dirichlet forms in terms of Green's formula*, arXiv e-prints, 2014

[394] D. Han and D. R. Larson, *Frames, bases and group representations*, Mem. Amer. Math. Soc. **147** (2000), no. 697, x+94, DOI 10.1090/memo/0697. MR1686653

[395] D. Han and A. I. Zayed, *Sampling expansions for functions having values in a Banach space*, Proc. Amer. Math. Soc. **133** (2005), no. 12, 3597–3607, DOI 10.1090/S0002-9939-05-08163-3. MR2163595

[396] D. Han, M. Z. Nashed, and Q. Sun, *Sampling expansions in reproducing kernel Hilbert and Banach spaces*, Numer. Funct. Anal. Optim. **30** (2009), no. 9-10, 971–987, DOI 10.1080/01630560903408606. MR2589760

[397] M. Hanke, *Conjugate Gradient Type Methods for Ill-Posed Problems*, Pitman Research Notes in Mathematics, Longman House, Harlow, Essex, 1995

[398] M. Hanke, *Limitations of the L-curve method in ill-posed problems*, BIT **36** (1996), no. 2, 287–301, DOI 10.1007/BF01731984. MR1432249

[399] M. Hanke, A. Neubauer, and O. Scherzer, *A convergence analysis of the Landweber iteration for nonlinear ill-posed problems*, Numer. Math. **72** (1995), no. 1, 21–37, DOI 10.1007/s002110050158. MR1359706

[400] M. Hanke and P. C. Hansen, *Regularization methods for large-scale problems*, Surveys Math. Indust. **3** (1993), no. 4, 253–315. MR1253521

[401] M. Hanke and O. Scherzer, *Inverse problems light: numerical differentiation*, Amer. Math. Monthly **108** (2001), no. 6, 512–521, DOI 10.2307/2695705. MR1840657

[402] M. Hanke and C. R. Vogel, *Two-level preconditioners for regularized inverse problems. I. Theory*, Numer. Math. **83** (1999), no. 3, 385–402, DOI 10.1007/s002110050455. MR1715577

[403] P. C. Hansen, *The truncated SVD as a method for regularization*, BIT, **27** (1987), 543–553

[404] P. C. Hansen, *Rank-Deficient and Discrete Ill-Posed Problems*, SIAM, Philadelphia, 1998

[405] R. J. Hanson, *A numerical method for solving Fredholm integral equations of the first kind*, SIAM J. Numer. Anal., **8** (1971), 616–662

[406] P. C. Hansen, *Analysis of discrete ill-posed problems by means of the* L*-curve*, SIAM Rev. **34** (1992), no. 4, 561–580, DOI 10.1137/1034115. MR1193012

[407] P. C. Hansen, *Regularization Tools version 4.0 for Matlab 7.3*, Numer. Algorithms **46** (2007), no. 2, 189–194, DOI 10.1007/s11075-007-9136-9. MR2358249

[408] P. C. Hansen and D. P. O'Leary, *The use of the L-curve in the regularization of discrete ill-posed problems*, SIAM J. Sci. Comput. **14** (1993), no. 6, 1487–1503, DOI 10.1137/0914086. MR1241596

[409] R. L. Harder, R. N. Demarais, *Interpolation using surface splines*, J. Aircraft, **9** (1972), 189-191

[410] G. H. Hardy, *On the expression of a number as the sum of two squares*, Quart. J. Math. (Oxford), **46** (1915), 263–283

[411] G. H. Hardy, *The Average Order of the Arithmetical Functions P(x) and delta(x)*, Proc. London Math. Soc. (2) **15** (1916), 192–213, DOI 10.1112/plms/s2-15.1.192. MR1576556

[412] G. H. Hardy, *A Theorem Concerning Fourier Transforms*, J. London Math. Soc. **8** (1933), no. 3, 227–231, DOI 10.1112/jlms/s1-8.3.227. MR1574130

[413] G. H. Hardy, *Notes on special systems of orthogonal functions. IV. The orthogonal functions of Whittaker's cardinal series*, Proc. Cambridge Philos. Soc. **37** (1941), 331–348, DOI 10.1017/s0305004100017977. MR5145

[414] G. H. Hardy, *Divergent Series*, Clarendon Press, Oxford, 1949.

[415] G. H. Hardy, E. Landau, *The Lattice Points of a Circle*, Proc. Royal Society, A, **105** (1924), 244–258

[416] G. H. Hardy and E. M. Wright, *Einführung in die Zahlentheorie* (German), R. Oldenbourg, Münich, 1958. MR0094309

[417] R. F. Harrington, *Field Computation by Moment Methods*, Macmillan, New York, 1986

[418] V. P. Havin and B. Jöricke, *The uncertainty principle in harmonic analysis [MR1129019 (93e:42001)]*, Commutative harmonic analysis, III, Encyclopaedia Math. Sci., vol. 72, Springer, Berlin, 1995, pp. 177–259, 261–266. MR1447633

[419] D. R. Heath-Brown, *Lattice points in the sphere*, Number theory in progress, Vol. 2 (Zakopane-Kościelisko, 1997), de Gruyter, Berlin, 1999, pp. 883–892. MR1689550

[420] E. Hecke, *Über orthogonal-invariante Integralgleichungen*, Math. Ann., **78** (1918), 398-404

[421] H. Hedenmalm and P. J. Nieminen, *The Gaussian free field and Hadamard's variational formula*, Probab. Theory Related Fields **159** (2014), no. 1-2, 61–73, DOI 10.1007/s00440-013-0501-4. MR3201917

[422] M. Hegland, *Variable Hilbert scales and their interpolation inequalities with applications to Tikhonov regularization*, Appl. Anal. **59** (1995), no. 1-4, 207–223, DOI 10.1080/00036819508840400. MR1378036

[423] W. A. Heiskanen, H. Moritz, *Physical Geodesy*, Freeman, San Francisco, 1967

[424] L. L. Helms, *Introduction to Potential Theory*, Wiley-Interscience, New York, 1969

[425] J. M. H. Hendrickx, B. Borchers, D. L. Corwin, S. M. Lesch, S. M. Hilgendorf, J. Schlue, *Inversion of soil conductivity profiles from electromagnetic induction measurements: theory and experimental verification*, Soil Science Soc. Amer. J., **66** (2002), 673–685

[426] G. T. Herman, *Fundamentals of computerized tomography*, 2nd ed., Advances in Pattern Recognition, Springer, Dordrecht, 2009. Image reconstruction from projections, DOI 10.1007/978-1-84628-723-7. MR2848263

[427] M. R. Hestenes and E. Stiefel, *Methods of conjugate gradients for solving linear systems*, J. Research Nat. Bur. Standards **49** (1952), 409–436 (1953). MR0060307

[428] F. Hettlich and W. Rundell, *A second degree method for nonlinear inverse problems*, SIAM J. Numer. Anal. **37** (2000), no. 2, 587–620, DOI 10.1137/S0036142998341246. MR1740766

[429] H. Heuser, *Funktionalanalysis*, 4. Auflage, Teubner, Stuttgart, 1975

[430] J. R. Higgins, *Five short stories about the cardinal series*, Bull. Amer. Math. Soc. (N.S.) **12** (1985), no. 1, 45–89, DOI 10.1090/S0273-0979-1985-15293-0. MR766960

[431] J. R. Higgins, *Sampling Theory in Fourier and Signal Analysis. Vol. 1: Foundations*, Oxford University Press, Oxford, 1996

[432] J. R. Higgins, R. L. Stens, *Sampling Theory in Fourier and Signal Analysis. Vol. 2: Advanced Topics*, Oxford Science Publications, Oxford, 2000

[433] D. Hilbert, *Grundzüge einer allgemeinen Theorie der linearen Integralgleichungen*, Teubner, Leipzig, 1912

[434] E. Hille, *Introduction to general theory of reproducing kernels*, Rocky Mountain J. Math. **2** (1972), no. 3, 321–368, DOI 10.1216/RMJ-1972-2-3-321. MR315109

[435] F. Hirzebruch and W. Scharlau, *Einführung in die Funktionalanalysis* (German), B. I. Hochschultaschenbücher, No. 296*, Bibliographisches Institut, Mannheim-Vienna-Zurich, 1971. MR0463864

[436] E. Hlawka, *Zur Geometrie der Zahlen* (German), Math. Z. **49** (1943), 285–312, DOI 10.1007/BF01174201. MR9782

[437] E. Hlawka, *Grundbegriffe der Geometrie der Zahlen* (German), Jber. Deutsch. Math.-Verein. **57** (1954), no. Abt. 1, 37–55. MR63409

[438] E. Hlawka, *Theorie der Gleichverteilung*, Bibliographisches Institut, B.I., Mannheim, 1979

[439] E. Hlawka, J. Schoissengeier, and R. Taschner, *Geometric and analytic number theory*, Universitext, Springer-Verlag, Berlin, 1991. Translated from the 1986 German edition by Charles Thomas, DOI 10.1007/978-3-642-75306-0. MR1123023

[440] M. E. Hochstenbach, L. Reichel, and G. Rodriguez, *Regularization parameter determination for discrete ill-posed problems*, J. Comput. Appl. Math. **273** (2015), 132–149, DOI 10.1016/j.cam.2014.06.004. MR3239242

[441] B. Hofmann, *On Ill-Posed and Local Ill-Posedness of Operator Equation Hilbert Spaces*, Preprint 97-8, Fakultät für Mathematik, Technische Universität Chemnitz-Zwickau, Chemnitz, 1997

[442] B. Hofmann, *Inverse Probleme*, Teubner, Stuttgart, 1999

[443] B. Hofmann, P. Mathé, and H. von Weizsäcker, *Regularization in Hilbert space under unbounded operators and general source conditions*, Inverse Problems **25** (2009), no. 11, 115013, 15, DOI 10.1088/0266-5611/25/11/115013. MR2558673

[444] T. Hohage and M. Pricop, *Nonlinear Tikhonov regularization in Hilbert scales for inverse boundary value problems with random noise*, Inverse Probl. Imaging **2** (2008), no. 2, 271–290, DOI 10.3934/ipi.2008.2.271. MR2395144

[445] B. K. P. Horn, *Robot Vision*, MIT Electrical Engineering and Computer Sciences Series, MIT Press, 1986

[446] R. F. Hoskins and J. de Sousa Pinto, *Sampling expansions for functions band-limited in the distributional sense*, SIAM J. Appl. Math. **44** (1984), no. 3, 605–610, DOI 10.1137/0144042. MR745116

[447] G. C. Hsiao and W. L. Wendland, *A finite element method for some integral equations of the first kind*, J. Math. Anal. Appl. **58** (1977), no. 3, 449–481, DOI 10.1016/0022-247X(77)90186-X. MR461963

[448] L.-K. Hua, *Die Abschätzung von Expotentialsummen und ihre Anwendung in der Zahlentheorie*, Enzyklopädie der Mathematischen Wissenschaften, Bd. 12, Heft 13, Teil 1, (1959), 105–107

[449] Y. Hua and T. K. Sarkar, *Matrix pencil method for estimating parameters of exponentially damped/undamped sinusoids in noise*, IEEE Trans. Acoust. Speech Signal Process. **38** (1990), no. 5, 814–824, DOI 10.1109/29.56027. MR1051029

[450] G. Huang, S. Noschese, and L. Reichel, *Regularization matrices determined by matrix nearness problems*, Linear Algebra Appl. **502** (2016), 41–57, DOI 10.1016/j.laa.2015.12.008. MR3490784

[451] G. Hulot, C. C. Finlay, C. Constable, N. Olsen, M. Mandea, *The magnetic field of planet Earth*, Space Sci. Rev., **152** (2010), 159–222,

[452] W. A. Hurwitz, *On the pseudo-resolvent to the kernel of an integral equation*, Trans. Amer. Math. Soc. **13** (1912), no. 4, 405–418, DOI 10.2307/1988578. MR1500925

[453] M. N. Huxley, *Exponential sums and lattice points. III*, Proc. London Math. Soc. (3) **87** (2003), no. 3, 591–609, DOI 10.1112/S0024611503014485. MR2005876

[454] M. Ilyasov, *A Tree Algorithm for Helmholtz Potential Wavelets on Non-smooth Surfaces: Theoretical Background and Application to Seismic Data Processing*, Ph.-D. Thesis, Geomathematics Group, University of Kaiserslautern, 2011.

[455] A. Imiya, H. Sugaya, A. Torii, Y. Mochizuki, *Variational Analysis of Spherical Images*, Computer Analysis of Images and Patterns, A Gagalowicz, W. Philips, (Eds.), Vol 3691, Lecture Notes in Computer Science, Springer, Berlin Heidelberg, (2005), 104–111

[456] M. E. H. Ismail, M. Z. Nashed, A. I. Zayed, and A. F. Ghaleb (eds.), *Mathematical analysis, wavelets, and signal processing*, Contemporary Mathematics, vol. 190, American Mathematical Society, Providence, RI, 1995, DOI 10.1090/conm/190. MR1354842

[457] V. K. Ivanov, *A generalization of the Voronoĭ-Hardy identity* (Russian), Sibirsk. Mat. Ž. **3** (1962), 195–212. MR0146148

[458] V. K. Ivanow, *Higher-dimensional generalization of the Euler summation formula* (in Russian), Izv. Vyss. Ucebn. Zaved. Mathematika, **37** (1963), no. 6, 72–80

[459] V. V. Īvanov and V. Ju. Kudrinskiĭ, *Approximate solution of linear operator equations in Hilbert space by the method of least squares. I* (Russian), Ž. Vyčisl. Mat i Mat. Fiz. **6** (1966), 831–841. MR208816

[460] A. Ivić, E. Krätzel, M. Kühleitner, and W. G. Nowak, *Lattice points in large regions and related arithmetic functions: recent developments in a very classic topic*, Elementare und analytische Zahlentheorie, Schr. Wiss. Ges. Johann Wolfgang Goethe Univ. Frankfurt am Main, vol. 20, Franz Steiner Verlag Stuttgart, Stuttgart, 2006, pp. 89–128. MR2310176

[461] M. Jacobsen, P. C. Hansen, and M. A. Saunders, *Subspace preconditioned LSQR for discrete ill-posed problems*, BIT **43** (2003), no. suppl., 975–989, DOI 10.1023/B:BITN.0000014547.88978.05. MR2058879

[462] L. Jantscher, *Distributionen*, Walter de Gruyter, Berlin, New York, 1971

[463] A. J. Jerri, *On the application of some interpolating functions in physics*, J. Res. Nat. Bur. Standards Sect. B **73B** (1969), 241–245. MR256026

[464] A. J. Jerri, *Sampling expansion for a Laguerre-L_ν^α transform*, J. Res. Nat. Bur. Standards Sect. B **80B** (1976), no. 3, 415–418. MR425527

[465] J. A. Jerri, *The Shannon sampling theorem - its various extensions and applications: a tutorial review*, IEEE Proc., **65** (1977), 1565–1596

[466] R.-Q. Jia, *Shift-invariant spaces and linear operator equations*, Israel J. Math. **103** (1998), 259–288, DOI 10.1007/BF02762276. MR1613580

[467] R.-Q. Jia, *Stability of the shifts of a finite number of functions*, J. Approx. Theory **95** (1998), no. 2, 194–202, DOI 10.1006/jath.1998.3215. MR1652860

[468] R. Q. Jia and C. A. Micchelli, *On linear independence for integer translates of a finite number of functions*, Proc. Edinburgh Math. Soc. (2) **36** (1993), no. 1, 69–85, DOI 10.1017/S0013091500005903. MR1200188

[469] F. John, *Continuous dependence on data for solutions of partial differential equations with a presribed bound*, Comm. Pure Appl. Math. **13** (1960), 551–585, DOI 10.1002/cpa.3160130402. MR130456

[470] D. S. Jones, *Numerical methods for antenna problems*, IEEE Proc., **12** (1974), 573–582

[471] P. E. T. Jorgensen and E. P. J. Pearse, *A Hilbert space approach to effective resistance metric*, Complex Anal. Oper. Theory **4** (2010), no. 4, 975–1013, DOI 10.1007/s11785-009-0041-1. MR2735315

[472] P. E. T. Jorgensen and M.-S. Song, *Comparison of discrete and continuous wavelet transforms*, Computational complexity. Vols. 1–6, Springer, New York, 2012, pp. 513–526, DOI 10.1007/978-1-4614-1800-9_34. MR3074509

[473] P. E. T. Jorgensen and M.-S. Song, *Compactification of infinite graphs and sampling*, Sampl. Theory Signal Image Process. **12** (2013), no. 2-3, 139–158. MR3285408

[474] P. E. T. Jorgensen and E. P. J. Pearse, *A discrete Gauss-Green identity for unbounded Laplace operators, and the transience of random walks*, Israel J. Math. **196** (2013), no. 1, 113–160, DOI 10.1007/s11856-012-0165-2. MR3096586

[475] P. Jorgensen, F. Tian, *Graph Laplacians and Discrete Reproducing Kernel Hilbert Spaces from Restrictions*, arXiv e-prints, 2015a

[476] P. Jorgensen, F. Tian, *Generalized Gramians: Creating Frame Vectors in Maximal Subspaces*, arXiv e-prints, 2015b

[477] P. Jorgensen and F. Tian, *Infinite networks and variation of conductance functions in discrete Laplacians*, J. Math. Phys. **56** (2015), no. 4, 043506, 27, DOI 10.1063/1.4918646. MR3390972

[478] P. Jorgensen and F. Tian, *Nonuniform sampling, reproducing kernels, and the associated Hilbert spaces*, Sampl. Theory Signal Image Process. **15** (2016), 37–72. MR3670916

[479] P. Jorgensen, S. Pedersen, and F. Tian, *Restrictions and extensions of semibounded operators*, Complex Anal. Oper. Theory **8** (2014), no. 3, 591–663, DOI 10.1007/s11785-012-0241-y. MR3167762

[480] P. Jorgensen, M.-S. Song, J. Tian, *Operator Theory, Kernels, and Feedforward Neural Networks*, `arXiv:2301.01327v2` [cs.LG], 2023

[481] M. L. Joy, A. Nachman, K. F. Hasanov, R. S. Yoon, A. W. Ma, *A New Approach to Current Density Impedance Imaging (CDII)*, Proc. ISMRM, No. 2356, Kyoto, Japan, 2004

[482] S. Khan, R. M. Goodall, and R. Dixon, *Non-uniform sampling strategies for digital control*, Internat. J. Systems Sci. **44** (2013), no. 12, 2234–2254, DOI 10.1080/00207721.2012.687785. MR3172778

[483] S. Kahn, J. Lefevre, H. Ammari, S. Baillet, *Feature detection and tracking in optical flow on non-flat manifolds*, Pattern Recogn. Lett., **32** (2011), 2047–2052

[484] J. Kaipio, E. Somersalo, *Statistical and Computational Inverse Problems*, Springer, Berlin, 2005

[485] T. Kalker, *On Multidimensional Sampling*, Digital Signal Processing Handbook, V. K. Madisetti, Williams D.B. (Eds.), CRC Press, Boca Raton, 1999

[486] B. Kaltenbacher, *A posteriori parameter choice strategies for some Newton type methods for the regularization of nonlinear ill-posed problems*, Numer. Math. **79** (1998), no. 4, 501–528, DOI 10.1007/s002110050349. MR1631677

[487] W. J. Kammerer and M. Z. Nashed, *On the convergence of the conjugate gradient method for singular linear operator equations*, SIAM J. Numer. Anal. **9** (1972), 165–181, DOI 10.1137/0709016. MR319368

[488] W. J. Kammerer and M. Z. Nashed, *Iterative methods for best approximate solutions of linear integral equations of the first and second kinds*, J. Math. Anal. Appl. **40** (1972), 547–573, DOI 10.1016/0022-247X(72)90002-9. MR320677

[489] L. W. Kantorowitsch and G. P. Akilow, *Funktionalanalysis in normierten Räumen* (German), Mathematische Lehrbücher und Monographien, II. Abteilung, Mathematische Monographien, Band XVII, Akademie-Verlag, Berlin, 1964. In deutscher Sprache herausgegeben von P. Heinz Müller. MR0177273

[490] S. Karlin and Z. Ziegler, *Some inequalities of total positivity in pure and applied mathematics*, Total positivity and its applications (Jaca, 1994), Math. Appl., vol. 359, Kluwer Acad. Publ., Dordrecht, 1996, pp. 247–261, DOI 10.1007/978-94-015-8674-0_12. MR1421605

[491] T. Kato, *Perturbation theory for nullity, deficiency and other quantities of linear operators*, J. Analyse Math. **6** (1958), 261–322, DOI 10.1007/BF02790238. MR107819

[492] C. T. Kelley, *Iterative methods for linear and nonlinear equations*, Frontiers in Applied Mathematics, vol. 16, Society for Industrial and Applied Mathematics (SIAM), Philadelphia, PA, 1995. With separately available software, DOI 10.1137/1.9781611970944. MR1344684

[493] O. D. Kellogg, *Foundations of Potential Theory*, Frederick Ungar Publishing Company, New York, 1929

[494] S. Kim, O. Kwon, J. K. Seo, and J.-R. Yoon, *On a nonlinear partial differential equation arising in magnetic resonance electrical impedance tomography*, SIAM J. Math. Anal. **34** (2002), no. 3, 511–526, DOI 10.1137/S0036141001391354. MR1970881

[495] C. Kirisits, L. F. Lang, and O. Scherzer, *Decomposition of optical flow on the sphere*, GEM Int. J. Geomath. **5** (2014), no. 1, 117–141, DOI 10.1007/s13137-013-0055-8. MR3189272

[496] C. Kirisits, L. F. Lang, and O. Scherzer, *Optical flow on evolving surfaces with space and time regularisation*, J. Math. Imaging Vision **52** (2015), no. 1, 55–70, DOI 10.1007/s10851-014-0513-4. MR3337515

[497] A. Kirsch, *An Introduction to the Mathematical Theory of Inverse Problems*, Springer, Heidelberg, 1996

[498] A. Kirsch, R. Kress, P. Monk, and A. Zinn, *Two methods for solving the inverse acoustic scattering problem*, Inverse Problems **4** (1988), no. 3, 749–770. MR965646

[499] F. Klein, *Elementarmathematik III*, Die Grundlagen der Mathematischen Wissenschaften, Band 16, Springer, Berlin, 1928

[500] R. E. Kleinman, *Some Application of Functional Analysis in Classical Scattering*, Mathematical Methods and Applications of Scattering Theory, DeSanto et al. (Eds.), Springer, New York, 17–23, 1980.

[501] K.-R. Koch, *Introduction to Bayesian Statistics*, 2nd ed., Springer, Heidelberg, 2010

[502] T. Kohlberger, E. Memin, Schnörr, Ch., *Variational dense motion estimation using the Helmholtz decomposition*, Scale Space Methods in Computer Vision, L.D. Griffin, M. Lillholm (Eds.), Vol. 2695, Lecture Notes in Computer Science, Springer, Berlin, (2003), 432–448

[503] G. Kolesnik, *On the method of exponent pairs*, Acta Arith. **45** (1985), no. 2, 115–143, DOI 10.4064/aa-45-2-115-143. MR797257

[504] T. H. Koornwinder and G. G. Walter, *The finite continuous Jacobi transform and its inverse*, J. Approx. Theory **60** (1990), no. 1, 83–100, DOI 10.1016/0021-9045(90)90075-2. MR1028896

[505] O. Kovrijkine, *Some results related to the Logvinenko-Sereda theorem*, Proc. Amer. Math. Soc. **129** (2001), no. 10, 3037–3047, DOI 10.1090/S0002-9939-01-05926-3. MR1840110

[506] E. Kreyszig, *Introductory Functional Analysis with Applications*, John Wiley and Sons, New York, 1978

[507] A. Korganoff and M. Pavel-Parvu, *Méthodes de calcul numérique. Tome II: Éléments de théorie des matrices carrées et rectangles en analyse numérique* (French), Dunod, Paris, 1967. MR0207184

[508] V. A. Kotel'nikov, *On the Carrying Capacity of the Ether and Wire in Telecommunications* (in Russian), Material for the First All-Union Conference on Questions of Communication, Izd. Red. Upr. Svyazi RKKA, Moscow, 1933

[509] R. Kowar, O. Scherzer, *Convergence Analysis of a Landweber-Koczmarz Method for Solving Nonlinear Ill-Posed Problems*, Ill-Posed and Inverse Problems, S. Romanov, S. I. Kabanikhin, Y. E. Anikonov, A. L. Bukhgeim, (Eds.), VSP Publishers, Zeist, 2002

[510] E. Krätzel, *Lattice Points*, Kluwer Dordrecht, Boston, London, 1988

[511] E. Krätzel, *Analytische Funktionen in der Zahlentheorie*, Teubner, Stuttgart, Leipzig, Wiesbaden, 2000

[512] H. P. Kramer, *A generalized sampling theorem*, J. Math. and Phys. **38** (1959/60), 68–72, DOI 10.1002/sapm195938168. MR103786

[513] R. Kress, *Linear Integral Equations*, Springer, Berlin, 1989

[514] S. Kulkarni and G. Harman, *An elementary introduction to statistical learning theory*, Wiley Series in Probability and Statistics, John Wiley & Sons, Inc., Hoboken, NJ, 2011, DOI 10.1002/9781118023471. MR2908346

[515] O. Kwon, J.-Y. Lee, and J.-R. Yoon, *Equipotential line method for magnetic resonance electrical impedance tomography*, Inverse Problems **18** (2002), no. 4, 1089–1100, DOI 10.1088/0266-5611/18/4/310. MR1929284

[516] O. Kwon, E. J. Woo, J. R. Yoon, J. K. Seo, *Magnetic resonance electric impedance tomography (MREIT): simulation study of J-substitution algorithm*, IEEE Trans. Biomed. Eng., **49** (2002), 160–167

[517] E. Landau, *Über die Gitterpunkte in einem Kreis (Erste Mitteilung)*, Gött. Nachr., (1915), 148–160

[518] E. Landau, *Über die Gitterpunkte in einem Kreis IV*, Gött. Nachr., (1924), 58–65

[519] E. Landau, *Über die Gitterpunkte in mehrdimensionalen Ellipsoiden (Zweite Abhandlung)*, Math. Z., **24** (1925), 299–310

[520] P. G. L. Dirichlet, *Vorlesungen über Zahlentheorie* (German), Chelsea Publishing Co., New York, 1968. Herausgegeben und mit Zusätzen versehen von R. Dedekind; Vierte, umgearbeitete und vermehrte Auflage. MR0237283

[521] H. J. Landau, *Necessary density conditions for sampling and interpolation of certain entire functions*, Acta Math. **117** (1967), 37–52, DOI 10.1007/BF02395039. MR222554

[522] H. J. Landau and H. O. Pollak, *Prolate spheroidal wave functions, Fourier analysis and uncertainty. II*, Bell System Tech. J. **40** (1961), 65–84, DOI 10.1002/j.1538-7305.1961.tb03977.x. MR140733

[523] H. J. Landau and H. O. Pollak, *Prolate spheroidal wave functions, Fourier analysis and uncertainty. III. The dimension of the space of essentially time- and band-limited signals*, Bell System Tech. J. **41** (1962), 1295–1336, DOI 10.1002/j.1538-7305.1962.tb03279.x. MR147686

[524] P. S. de Laplace, *Theorie des attractions des sphéroides et de la figure des planètes*, Mèm. de l'Acad., Paris, 1785

[525] D. R. Larson, P. Massopust, Z. Nashed, M. C. Nguyen, M. Papadakis, and A. Zayed (eds.), *Frames and operator theory in analysis and signal processing*, Contemporary Mathematics, vol. 451, American Mathematical Society, Providence, RI, 2008. Papers from the AMS-SIAM Special Session held in San Antonio, TX, January 12–15, 2006, DOI 10.1090/conm/451. MR2422280

[526] S. Lata and V. Paulsen, *The Feichtinger conjecture and reproducing kernel Hilbert spaces*, Indiana Univ. Math. J. **60** (2011), no. 4, 1303–1317, DOI 10.1512/iumj.2011.60.4358. MR2975345

[527] M. M. Lavrentiev, *Some Improperly Posed Problems of Mathematical Physics*, Izdat. Sibirsk. Otdel, Akad. Nauk. SSSR, Novosibirsk, 1962, Englisch Transl., Springer Tracts in Natural Philosophy, Vol. 11, Springer-Verlag, Berlin, 1967

[528] J.-Y. Lee, *A reconstruction formula and uniqueness of conductivity in MREIT using two internal current distributions*, Inverse Problems **20** (2004), no. 3, 847–858, DOI 10.1088/0266-5611/20/3/012. MR2067504

[529] J. Lees, R. S. Crosson, *Bayesian ART Versus Conjugate Gradient Methods in Tomographic Seismic Imaging: An Application at Mount St. Helens, Washington*, Spatial Statistics and Imaging, A. Possolo (Ed.), Hayward, Calif. Institute of Math. Statistics, 186-208, 1991

[530] J. Lefevre, S. Baillet, *Optical Flow and Advection on 2-Riemannian Manifold. A Common Framwork*, IEEE Trans. Pattern Anal. Mach. Intell., **30** (2008), 1081–1092

[531] A. M. Legendre, *Recherches sur l'attraction des sphèroides homogènes*, Mèm. math. phys. près. à l'Acad. Aci. par. divers savantes, **10** (1785), 411–434

[532] C. G. Lekkerkerker, *Geometry of Numbers*, North Holland, Amsterdam, London, 1969

[533] O. V. Lepskii, *On a Problem of Adaptive Estimation in Gaussian White Noise*, Theory Prob. Appl., **35** (1990), 454–466

[534] B. Ya. Levin, *Lectures on entire functions*, Translations of Mathematical Monographs, vol. 150, American Mathematical Society, Providence, RI, 1996. In collaboration with and with a preface by Yu. Lyubarskii, M. Sodin and V. Tkachenko; Translated from the Russian manuscript by Tkachenko, DOI 10.1090/mmono/150. MR1400006

[535] Y. Levin and A. Ben-Israel, *A Newton method for systems of m equations in n variables*, Proceedings of the Third World Congress of Nonlinear Analysts, Part 3 (Catania, 2000), Nonlinear Anal. **47** (2001), no. 3, 1961–1971, DOI 10.1016/S0362-546X(01)00325-X. MR1977075

[536] X. Li, Y Xu, B He, *Imaging Electrical Impedance from Acoustic Measurements by Means of Magnetoacoustic Tomography with Magnetic Induction (MAT-MI)*, IEEE Transact. Biomed. Engineering, **54** (2007), 323–330

[537] L. A. Lusternik and V. J. Sobolev, *Elements of functional analysis*, Frederick Ungar Publishing Co., New York, 1961. Translated from the Russian by Anthony E. Labarre, Jr., Herbert Izbicki, H. Ward Crowley. MR0141966

[538] W. A. Light and E. W. Cheney, *Quasi-interpolation with translates of a function having noncompact support*, Constr. Approx. **8** (1992), no. 1, 35–48, DOI 10.1007/BF01208904. MR1142692

[539] J. Lindenstrauss and L. Tzafriri, *Classical Banach spaces. II.*, Springer-Verlag, Berlin-New York,, 1979. Function spaces. MR0540367

[540] J.-L. Lions and E. Magenes, *Non-homogeneous boundary value problems and applications. Vol. I*, Die Grundlehren der mathematischen Wissenschaften, Band 181, Springer-Verlag, New York-Heidelberg, 1972. Translated from the French by P. Kenneth. MR0350177

[541] O. A. Liskovets, *Regularization of variational inqualities with pseudo-monotone operators on approximately given sets*, Differ. Equations, **11** (1989), 1970–1977

[542] F. Liu and M. Z. Nashed, *Tikhonov regularization of nonlinear ill-posed problems with closed operators in Hilbert scales*, J. Inverse Ill-Posed Probl. **5** (1997), no. 4, 363–376, DOI 10.1515/jiip.1997.5.4.363. MR1473637

[543] Y. M. Liu and G. G. Walter, *Irregular sampling in wavelet subspaces*, J. Fourier Anal. Appl. **2** (1995), no. 2, 181–189, DOI 10.1007/s00041-001-4027-2. MR1365204

[544] Y. T. Lo, S. W. Lee, O. H. Lee, *Optimization of directivity and signal-to-noise ratio of an arbitrary antenna aray*, IEEE Proc., **54** (1966), 1033–1045

[545] J. Locker and P. M. Prenter, *Regularization with differential operators. I. General theory*, J. Math. Anal. Appl. **74** (1980), no. 2, 504–529, DOI 10.1016/0022-247X(80)90145-6. MR572669

[546] V. N. Logvinenko and Ju. F. Sereda, *Equivalent norms in spaces of entire functions of exponential type* (Russian), Teor. Funkciĭ Funkcional. Anal. i Priložen. **Vyp. 20** (1974), 102–111, 175. MR0477719

[547] A. K. Louis, *Inverse und schlecht gestellte Probleme*, Teubner, Stuttgart, 1989

[548] A. K. Louis and P. Maass, *A mollifier method for linear operator equations of the first kind*, Inverse Problems **6** (1990), no. 3, 427–440. MR1057035

[549] Louis, A. K., P. Maass, A. Rieder, *Wavelets: Theorie und Anwendungen*, Teubner, Suttgart, 1998.

[550] F. Low, *Complete sets of wave packets*, A Passion for Physics-Essays in Honor of Geoffrey Chew, C., DeTar et al. (Eds.), World Scientific, Singapore, (1985), 17–22

[551] D. S. Lubinsky, *Marcinkiewicz-Zygmund inequalities: methods and results*, Recent progress in inequalities (Niš, 1996), Math. Appl., vol. 430, Kluwer Acad. Publ., Dordrecht, 1998, pp. 213–240. MR1609947

[552] H. D. Lüke, *The origins of the sampling theorem*, IEEE Communications Magazine, **37** (1999), 106–108

[553] W. R. Madych and S. A. Nelson, *Polyharmonic cardinal splines*, J. Approx. Theory **60** (1990), no. 2, 141–156, DOI 10.1016/0021-9045(90)90079-6. MR1033167

[554] W. Magnus, F. Oberhettinger, and R. P. Soni, *Formulas and theorems for the special functions of mathematical physics*, Third enlarged edition, Die Grundlehren der mathematischen Wissenschaften, Band 52, Springer-Verlag New York, Inc., New York, 1966. MR0232968

[555] B. A. Mair and F. H. Ruymgaart, *Statistical inverse estimation in Hilbert scales*, SIAM J. Appl. Math. **56** (1996), no. 5, 1424–1444, DOI 10.1137/S0036139994264476. MR1409127

[556] S. Mallat, *Applied mathematics meets signal processing*, Proceedings of the International Congress of Mathematicians, Berlin, Vol. I, Documenta Mathematica, (1998), 319–338.

[557] I. Maravić and M. Vetterli, *Exact sampling results for some classes of parametric nonbandlimited 2-D signals*, IEEE Trans. Signal Process. **52** (2004), no. 1, 175–189, DOI 10.1109/TSP.2003.819984. MR2044733

[558] E. Martensen, *Potentialtheorie* (German), Leitfäden der Angewandten Mathematik und Mechanik, Band 12, B. G. Teubner, Stuttgart, 1968. MR0247116

[559] J. T. Marti, *An algorithm for computing minimum norm solutions of Fredholm integral equations of the first kind*, SIAM J. Numer. Anal. **15** (1978), no. 6, 1071–1076, DOI 10.1137/0715071. MR512683

[560] J. T. Marti, *On the convergence of an algorithm computing minimum-norm solutions of ill-posed problems*, Math. Comp. **34** (1980), no. 150, 521–527, DOI 10.2307/2006100. MR559200

[561] R. J. Marks II, *Introduction to Shannon Sampling and Interpolation Theory*, Springer, Berlin, 1991

[562] R. J. Marks II, *Advanced Topics in Shannon Sampling and Interpolation Theory*, Springer, Berlin, Heidelberg, 1993

[563] D. L. Marks, *A family of approximations spanning the Born and Rytov scattering series*, Optics Express, **14** (2013), 8837–8848

[564] F. Marvasti (ed.), *Nonuniform sampling*, Information Technology: Transmission, Processing and Storage, Kluwer Academic/Plenum Publishers, New York, 2001. Theory and practice; With 1 CD-ROM, DOI 10.1007/978-1-4615-1229-5. MR1875674

[565] P. Mathé, *The Lepskiĭ principle revisited*, Inverse Problems **22** (2006), no. 3, L11–L15, DOI 10.1088/0266-5611/22/3/L02. MR2235633

[566] P. Mathé and B. Hofmann, *How general are general source conditions?*, Inverse Problems **24** (2008), no. 1, 015009, 5, DOI 10.1088/0266-5611/24/1/015009. MR2384768

[567] P. Mathé and S. V. Pereverzev, *Geometry of linear ill-posed problems in variable Hilbert scales*, Inverse Problems **19** (2003), no. 3, 789–803, DOI 10.1088/0266-5611/19/3/319. MR1984890

[568] P. Mathé and S. V. Pereverzev, *The discretized discrepancy principle under general source conditions*, J. Complexity **22** (2006), no. 3, 371–381, DOI 10.1016/j.jco.2005.06.008. MR2229899

[569] P. Mathé and S. V. Pereverzev, *Regularization of some linear ill-posed problems with discretized random noisy data*, Math. Comp. **75** (2006), no. 256, 1913–1929, DOI 10.1090/S0025-5718-06-01873-4. MR2240642

[570] J. R. Mautz and R. F. Harrington, *Computational methods for antenna pattern synthesis*, IEEE Trans. Antennas and Propagation **AP-23** (1975), 507–512, DOI 10.1109/TAP.1975.1141126. MR373472

[571] V. G. Maz'ja, *Sobolev spaces*, Springer Series in Soviet Mathematics, Springer-Verlag, Berlin, 1985. Translated from the Russian by T. O. Shaposhnikova, DOI 10.1007/978-3-662-09922-3. MR817985

[572] C. V. M. van der Mee, M. Z. Nashed, and S. Seatzu, *Sampling expansions and interpolation in unitarily translation invariant reproducing kernel Hilbert spaces*, Adv. Comput. Math. **19** (2003), no. 4, 355–372, DOI 10.1023/A:1024233232215. MR1989621

[573] J. Meinguet, *Multivariate interpolation at arbitrary points made simple* (English, with French summary), Z. Angew. Math. Phys. **30** (1979), no. 2, 292–304, DOI 10.1007/BF01601941. MR535987

[574] P. A. Meissl, *Hilbert Spaces and Their Applications to Geodetic Least Squares Problems*, Boll. Geod. Sci. Aff., 1:181-210, 1976

[575] Y. Meyer, *Oscillating patterns in image processing and nonlinear evolution equations*, University Lecture Series, vol. 22, American Mathematical Society, Providence, RI, 2001. The fifteenth Dean Jacqueline B. Lewis memorial lectures, DOI 10.1090/ulect/022. MR1852741

[576] C. A. Micchelli, *Interpolation of scattered data: distance matrices and conditionally positive definite functions*, Constr. Approx. **2** (1986), no. 1, 11–22, DOI 10.1007/BF01893414. MR891767

[577] V. Michel, *A Multiscale Method for the Gravimetry Problem: Theoretical and Numerical Aspects of Harmonic and Anharmonic Modelling*, Ph.D. Thesis, Geomathematics Group, University of Kaiserslautern, Shaker, Aachen, 1999

[578] V. Michel, *Scale continuous, scale discretized and scale discrete harmonic wavelets for the outer and the inner space of a sphere and their application to an inverse problem in geomathematics*, Appl. Comp. Harm. Anal. (ACHA), **12** (2002a), 77–99

[579] V. Michel, *A Multiscale Approximation for Operator Equations in Separable Hilbert Spaces – Case Study: Reconstruction and Description of the Earth's Interior*, Habilitation Thesis, Geomathematics Group, University of Kaiserslautern, Shaker, Aachen, 2002b

[580] V. Michel, *Regularized wavelet-based multiresolution recovery of the harmonic mass density distribution from data of the Earth's gravitational field at satellite height*, Inverse Problems **21** (2005), no. 3, 997–1025, DOI 10.1088/0266-5611/21/3/013. MR2146818

[581] V. Michel, *Lectures on Constructive Approximation: Fourier, Spline, and Wavelet Methods on the Real Line, the Sphere, and the Ball.*, Birkhäuser, Basel, Boston, 2013

[582] V. Michel and A. S. Fokas, *A unified approach to various techniques for the non-uniqueness of the inverse gravimetric problem and wavelet-based methods*, Inverse Problems **24** (2008), no. 4, 045019, 25, DOI 10.1088/0266-5611/24/4/045019. MR2425886

[583] S. G. Michlin, *The Problem of the Minimum of a Quadratic Functional*, Holden-Day, Amsterdam, San Francisco, 1970.

[584] S. G. Michlin, *Mathematical Physics, an Advanced Course*, North Holland, Amsterdam, London, 1970

[585] S. G. Michlin, *Lehrgang der Mathematischen Physik*, 2nd ed., Akademie Verlag, Berlin, 1975

[586] K. Miller, *Least squares methods for ill-posed problems with a prescribed bound*, SIAM J. Math. Anal. **1** (1970), 52–74, DOI 10.1137/0501006. MR260196

[587] H. Minkowski, *Geometrie der Zahlen*, Teubner, Leipzig, 1896

[588] E. H. Moore, *On the reciprocal of the general algebraic matrix*, Bull. Amer. Mathe. Soc., **26** (1920), 394-395

[589] E. H. Moore, *General analysis*, Memoirs Amer. Math. Soc., **1** (1935), 197-209

[590] L. J. Mordell, *Poisson's summation formula in several variables and some applications to the theory of numbers*, Cambr. Phil. Soc., **25** (1928a), 412–420

[591] L. J. Mordell, *Some applications of Fourier series in the analytic theory of numbers*, Cambr. Phil. Soc., **24** (1928b), 585–595

[592] L. J. Mordell, *Poisson's summation formula and the Riemann zeta function*, J. London Math. Soc., **4** (1929), 285-296

[593] L. J. Mordell, *On some arithmetical results in the geometry of numbers*, Compositio Math. **1** (1935), 248–253. MR1556892

[594] L. J. Mordell, *Diophantine Equations*, Academic Press, 1969.

[595] V. A. Morozov, *The choice of parameter when solving functional equations by regularization*, Doklady Akademii Nauk SSSR, **175** (1962), 1225-1228

[596] V. A. Morozov, *On the solution of functional equations by the method of regularization*, Soviet Math. Dokl. **7** (1966), 414–417. MR0208819

[597] V. A. Morozov, *Regularization Methods for the Solution of Ill-Posed Problem*, Moscow: Moscow Univ., (English Transl., Springer, New York), 1975

[598] V.A. Morozov, *Methods for Solving Incorrectly Posed Problems*, Springer, New York, 1984.

[599] C. B. Morrey Jr., *Quasi-convexity and the lower semicontinuity of multiple integrals*, Pacific J. Math. **2** (1952), 25–53. MR54865

[600] C. Müller, *Über die ganzen Lösungen der Wellengleichung* (German), Math. Ann. **124** (1952), 235–264, DOI 10.1007/BF01343563. MR56814

[601] C. Müller, *Eine Verallgemeinerung der Eulerschen Summenformel und ihre Anwendung auf Fragen der analytischen Zahlentheorie*, Abh. Math. Sem. Univ. Hamburg, **19** (1954a), 41–61

[602] C. Müller, *Eine Formel der analytischen Zahlentheorie*, Abh. Math. Sem. Univ. Hamburg, **19** (1954b), 62–65

[603] C. Müller, *Eine Erweiterung der Hardyschen Identität*, Abh. Math. Sem. Univ. Hamburg, **19** (1954c), 66–76

[604] C. Müller, *Die Grundprobleme der Geometrie der Zahlen* (German), Math.-Phys. Semesterber. **5** (1956), 63–70. MR84529

[605] C. Müller, *Spherical harmonics*, Lecture Notes in Mathematics, vol. 17, Springer-Verlag, Berlin-New York, 1966. MR0199449

[606] C. Müller, *Foundations of the Mathematical Theory of Electromagnetic Waves*, Springer, Berlin, 1969

[607] C. Müller, *Analysis of spherical symmetries in Euclidean spaces*, Springer, New York, Berlin, Heidelberg, 1998

[608] J. Murray-Bruce and P. L. Dragotti, *A sampling framework for solving physics-driven inverse source problems*, IEEE Trans. Signal Process. **65** (2017), no. 24, 6365–6380, DOI 10.1109/TSP.2017.2742983. MR3743557

[609] A. Nachman, A. Tamasan, and A. Timonov, *Conductivity imaging with a single measurement of boundary and interior data*, Inverse Problems **23** (2007), no. 6, 2551–2563, DOI 10.1088/0266-5611/23/6/017. MR2441019

[610] A. Nachman, A. Tamasan, and A. Timonov, *Recovering the conductivity from a single measurement of interior data*, Inverse Problems **25** (2009), no. 3, 035014, 16, DOI 10.1088/0266-5611/25/3/035014. MR2480184

[611] H.-H. Nagel, *On the estimation of optical flow: relations between new approaches and some new results*, Artif. Intell., **33** (1987), 299–324

[612] M. Z. Nashed, *Steepest descent for singular linear operator equations*, SIAM J. Numer. Anal. **7** (1970), 358–362, DOI 10.1137/0707027. MR269093

[613] M. Z. Nashed, *Generalized inverses, normal solvability and iteration for singular operator equations*, Nonlinear Functional Analysis and Applications, L.B. Rall, (Ed.), Academic Press, New York, (1971a), 311–359

[614] M. Z. Nashed, *Differentiability and related properties of nonlinear operators: some aspects of the role of differentials in nonlinear functional analyis*, Nonlinear Functional Analysis and Applications, L.B. Rall (Ed.), Academic Press, New York, (1971b), 103–309

[615] M. Z. Nashed, *Some aspects of regularization and approximations of solutions of ill-posed operator equations*, Proc. 1972 Army Numerical Analysis Conf., Edgewood, MD, (1972), 163–181

[616] M. Z. Nashed, *Approximate Regularized Solutions to Improperly Posed Linear Integral and Operator Equations*, Constructive and Computational Methods for Differential and Integral Equations, D.L. Colton and R.P. Gilbert (Eds.), Springer, New York, LNM 430, (1974), 289–322

[617] M. Z. Nashed (Ed.), *Generalized Inverses and Applications*, Academic Press, New York, 1976a

[618] M. Z. Nashed, *Aspects of Generalized Inverses in Analysis and Regularization*, Generalized Inverses and Applications, Academic Press, New York, (1976b), 193–244

[619] M. Z. Nashed, *On moment-discretization and least-squares solutions of linear integration equations of the first kind*, J. Math. Anal. Appl., **53** (1976c), 359–366

[620] M. Z. Nashed, *Perturbations and approximation for generalized inverses and linear operators*, Generalized Inverses and Applications, Academic Press, New York, (1976b), 325–396

[621] M. Z. Nashed, *Regularization and approximation of ill-posed problems in system theory*, Proc. 1979 Conf. on Information Sciences and Systems, G.G.L. Meyer, C.R. Westgate (Eds.), The Johns Hopkins University, New York, (1979), 568–575

[622] M. Z. Nashed, *New Applications of Generalized Inverses in System and Control Theory*, Proc. 1980 Conf. on Information Sciences and Systems, J.B. Thomas (Ed.), Princeton, NJ, Princeton, (1980), 353–358

[623] M. Z. Nashed, *Analysis and Implementation of Projection-Regularization Methods for Ill-Posed Problems*, Ill-Posed Problems: Theory and Practice, Reidel, Dordrecht and Boston, 1981a

[624] M. Z. Nashed, *Continuous and Semicontinuous Analogous of Iterative Method of Cimmino and Kaczmarz with Applications to the Inverse Radon Transform*, Mathematical Aspects of Computerized Tomography, G.T. Herman, F. Natterer (Eds.), Springer, New York, (1981b), 160–178

[625] M. Z. Nashed (Ed.), *Ill-Posed Problems*, Theory and Practice, Reidel, Dordrecht and Boston, 1981c

[626] M. Z. Nashed, *Operator-theoretic and computational approaches to ill-posed problems with applications to antenna theory*, IEEE Trans. Antennas Propagation, **29** (1981d), 220-231

[627] H. W. Engl and C. W. Groetsch (eds.), *Inverse and ill-posed problems*, Notes and Reports in Mathematics in Science and Engineering, vol. 4, Academic Press, Inc., Boston, MA, 1987. Papers from the Alpine-U.S. Seminar held in Sankt Wolfgang, June 1986. MR1020304

[628] M. Z. Nashed, *Inner, outer, and generalized inverses in Banach and Hilbert spaces*, Numer. Funct. Anal. Optim., **9** (1987b), 261–326

[629] A. H. Siddiqi, R. C. Singh, and P. Manchanda (eds.), *Mathematics in science and technology*, World Scientific Publishing Co. Pte. Ltd., Hackensack, NJ, 2011. Mathematical methods, models and algorithms in science and technology, DOI 10.1142/8063. MR2893670

[630] M. Z. Nashed and X. Chen, *Convergence of Newton-like methods for singular operator equations using outer inverses*, Numer. Math. **66** (1993), no. 2, 235–257, DOI 10.1007/BF01385696. MR1245013

[631] M. Z. Nashed and H. W. Engl, *Random generalized inverses and approximate solutions of random operator equations*, Approximate solution of random equations, North-Holland, New York-Amsterdam, 1979, pp. 149–210. MR597714

[632] M. Z. Nashed and F. Liu, *On nonlinear ill-posed problems. II. Monotone operator equations and monotone variational inequalities*, Theory and applications of nonlinear operators of accretive and monotone type, Lecture Notes in Pure and Appl. Math., vol. 178, Dekker, New York, 1996, pp. 223–240. MR1386679

[633] M. Z. Nashed and O. Scherzer, *Stable approximation of nondifferentiable optimization problems with variational inequalities*, Recent developments in optimization theory and nonlinear analysis (Jerusalem, 1995), Contemp. Math., vol. 204, Amer. Math. Soc., Providence, RI, 1997, pp. 155–170, DOI 10.1090/conm/204/02671. MR1443001

[634] M. Z. Nashed and O. Scherzer, *Stable approximations of a minimal surface problem with variational inequalities*, Abstr. Appl. Anal. **2** (1997), no. 1-2, 137–161, DOI 10.1155/S1085337597000316. MR1604173

[635] M. Z. Nashed and O. Scherzer (eds.), *Inverse problems, image analysis, and medical imaging*, Contemporary Mathematics, vol. 313, American Mathematical Society, Providence, RI, 2002, DOI 10.1090/conm/313. MR1940985

[636] M. Z. Nashed and Q. Sun, *Sampling and reconstruction of signals in a reproducing kernel subspace of $L^p(\mathbb{R}^d)$*, J. Funct. Anal. **258** (2010), no. 7, 2422–2452, DOI 10.1016/j.jfa.2009.12.012. MR2584749

[637] M. Z. Nashed and Q. Sun, *Function spaces for sampling expansions*, Multiscale signal analysis and modeling, Springer, New York, 2013, pp. 81–104, DOI 10.1007/978-1-4614-4145-8_4. MR3024465

[638] M. Z. Nashed, Q. Sun, and W.-S. Tang, *Average sampling in L^2* (English, with English and French summaries), C. R. Math. Acad. Sci. Paris **347** (2009), no. 17-18, 1007–1010, DOI 10.1016/j.crma.2009.07.011. MR2554566

[639] M. Z. Nashed and G. F. Votruba, *A unified operator theory of generalized inverses*, Generalized inverses and applications (Proc. Sem., Math. Res. Center, Univ. Wisconsin, Madison, Wis., 1973), Publ. Math. Res. Center Univ. Wisconsin, No. 32, Academic Press, New York, 1976, pp. 1–109. MR0493448

[640] M. Z. Nashed and G. Wahba, *Regularization and approximation of linear operator equations in reproducing kernel spaces*, Bull. Amer. Math. Soc. **80** (1974), 1213–1218, DOI 10.1090/S0002-9904-1974-13684-0. MR365192

[641] M. Z. Nashed and G. Wahba, *Convergence rates of approximate least squares solutions of linear integral and operator equations of the first kind*, Math. Comp. **28** (1974), 69–80, DOI 10.2307/2005817. MR461895

[642] M. Z. Nashed, G. Wahba, *Rates of Convergence of Approximate Least Squares Solutions of Linear Integral and Operator Equations*, Math. Comp., **28** (1974c), 69–80

[643] M. Z. Nashed and G. G. Walter, *General sampling theorems for functions in reproducing kernel Hilbert spaces*, Math. Control Signals Systems **4** (1991), no. 4, 363–390, DOI 10.1007/BF02570568. MR1128261

[644] M. Z. Nashed and G. G. Walter, *Reproducing kernel Hilbert spaces from sampling expansions*, Mathematical analysis, wavelets, and signal processing (Cairo, 1994), Contemp. Math., vol. 190, Amer. Math. Soc., Providence, RI, 1995, pp. 221–226, DOI 10.1090/conm/190/02304. MR1354856

[645] I. P. Natanson, *Constructive Function Theory*, Frederick Ungar Publ. Co., New York, 1965

[646] M. N. Le Roux, *Méthode d'éléments finis pour la résolution numérique de problèmes extérieurs en dimension* 2 (French, with English summary), RAIRO Anal. Numér. **11** (1977), no. 1, 27–60, 112, DOI 10.1051/m2an/1977110100271. MR448954

[647] F. Natterer, *Regularisierung schlecht gestellter Probleme durch Projektionsverfahren* (German, with English summary), Numer. Math. **28** (1977), no. 3, 329–341, DOI 10.1007/BF01389972. MR488721

[648] F. Natterer, *Error bounds for Tikhonov regularization in Hilbert scales*, Applicable Anal. **18** (1984), no. 1-2, 29–37, DOI 10.1080/00036818408839508. MR762862

[649] F. L. Nazarov, *Local estimates for exponential polynomials and their applications to inequalities of the uncertainty principle type* (Russian, with Russian summary), Algebra i Analiz **5** (1993), no. 4, 3–66; English transl., St. Petersburg Math. J. **5** (1994), no. 4, 663–717. MR1246419

[650] E. Nelson, *Kernel functions and eigenfunction expansions*, Duke Math. J., **25** (1957), 15-27

[651] A. Neubauer, *On Landweber iteration for nonlinear ill-posed problems in Hilbert scales*, Numer. Math. **85** (2000), no. 2, 309–328, DOI 10.1007/s002110050487. MR1754723

[652] A. Neubauer, *On converse and saturation results for Tikhonov regularization of linear ill-posed problems*, SIAM J. Numer. Anal. **34** (1997), no. 2, 517–527, DOI 10.1137/S0036142993253928. MR1442926

[653] F. Neumann, *Vorlesungen über die Theorie des Potentials und der Kugelfunktionen*, B.G. Teubner, Leipzig, 1887

[654] A. Noble, *A Bibliography on: Methods for Solving Integral Equations – Subject Listing*, MRC Technical Summary Report Nr. 1177, Mathematics Research Center, University of Wisconsin, 1971

[655] S. Noschese and L. Reichel, *Inverse problems for regularization matrices*, Numer. Algorithms **60** (2012), no. 4, 531–544, DOI 10.1007/s11075-012-9576-8. MR2946480

[656] P. Novikoff, *Sur le problème inverse du potentiel*, Comptes Rendus de l'Académie des Sciences de l'URSS, **18** (1938), 165–168

[657] H. Nyquist, *Certain factors affecting telegraph speed*, Bell Syst. Tech. J., **3** (1924), 324–346

[658] H. Nyquist, *Certain topics in telegraph transmission theory*, Trans. AIEE., **47** (1928), 617–644

[659] H. Nutz, *A Unified Setup of Gravitational Field Observables*, Ph.-D. Thesis, University of Kaiserslautern, Geomathematics Group, Shaker, Aachen, 2002

[660] K. Ogura, *On a certain transcendental integral function in the theory of interpolation*, Tôhoku. Math. J., **17** (1920), 64–72

[661] N. Olsen, G. Hulot, T. J. Sabaka, *Sources of the Geomagnetic Field and the Modern Data that Enable Their Investigation*, Handbook of Geomathematics, 1st. ed., W. Freeden, Z. Nashed, T. Sonar (Eds.), Springer, Reference, Heidelberg, New York, Dordrecht, London, Vol. 1, (2010), 106–124

[662] K. A. Okoudjou, R. S. Strichartz, and E. K. Tuley, *Orthogonal polynomials on the Sierpinski gasket*, Constr. Approx. **37** (2013), no. 3, 311–340, DOI 10.1007/s00365-013-9187-1. MR3054607

[663] A. Olevskii and A. Ulanovskii, *Almost integer translates. Do nice generators exist?*, J. Fourier Anal. Appl. **10** (2004), no. 1, 93–104, DOI 10.1007/s00041-004-8006-2. MR2045527

[664] A. V. Oppenheim, R. W. Schäfer, *Digital Signal Processing*, Prentice-Hall, New Jersey, 1989

[665] A. V. Oppenheim, A. S. Willsky, *Signals and Systems*, 2nd ed., Prentice-Hall, New Jersey, 1997

[666] J. M. Ortega and W. C. Rheinboldt, *On discretization and differentiation of operators with application to Newton's method*, SIAM J. Numer. Anal. **3** (1966), no. 1, 143–156, DOI 10.1137/0703010. MR205450

[667] J. M. Ortega, W. C. Rheinboldt, *Iterative solution of nonlinear equations in several variables*, (First Published by Academic Press, New York, London, 1977), SIAM Publ., Philadelphia, 2000

[668] J. Ortega-Cerdà and B. Pridhnani, *Beurling-Landau's density on compact manifolds*, J. Funct. Anal. **263** (2012), no. 7, 2102–2140, DOI 10.1016/j.jfa.2012.07.004. MR2956935

[669] J. Ortega-Cerdà and B. Pridhnani, *Carleson measures and Logvinenko-Sereda sets on compact manifolds*, Forum Math. **25** (2013), no. 1, 151–172, DOI 10.1515/form.2011.110. MR3010851

[670] J. Ortega-Cerdà and J. Saludes, *Marcinkiewicz-Zygmund inequalities*, J. Approx. Theory **145** (2007), no. 2, 237–252, DOI 10.1016/j.jat.2006.09.001. MR2312468

[671] S. Osher, A. Solé, and L. Vese, *Image decomposition and restoration using total variation minimization and the H^{-1} norm*, Multiscale Model. Simul. **1** (2003), no. 3, 349–370, DOI 10.1137/S1540345902416247. MR2030155

[672] A. Oya, J. Navarro-Moreno, and J. C. Ruiz-Molina, *Numerical evaluation of reproducing kernel Hilbert space inner products*, IEEE Trans. Signal Process. **57** (2009), no. 3, 1227–1233, DOI 10.1109/TSP.2008.2010424. MR3027802

[673] S. W. Paddock, *Confocal laser scanning microscopy*, BioTechniques, **27** (1999), 992–1007

[674] C. C. Paige and M. A. Saunders, *Towards a generalized singular value decomposition*, SIAM J. Numer. Anal. **18** (1981), no. 3, 398–405, DOI 10.1137/0718026. MR615522

[675] R. E. A. C. Paley, N. Wiener, *Fourier Transforms in the Complex Domain*, Colloquium Publicatios, Vol. 29, Amer. Math. Soc. (AMS), Providence, R.I., 1934

[676] B. Paneah, *Support dependent weighted norm estimates for Fourier transforms*, J. Math. Anal. Appl. **189** (1995), no. 2, 552–574, DOI 10.1006/jmaa.1995.1035. MR1312061

[677] B. Paneah, *Support-dependent weighted norm estimates for Fourier transforms*, II Duke Math. J., **92** (1998), 350–353

[678] A. Papoulis, *Signal Analysis*, McGraw-Hill, New York, 1971

[679] A. Papoulis, *Generalized sampling expansion*, IEEE Trans. Circuits and Systems **CAS-2** (1977), no. 11, 652–654, DOI 10.1109/TCS.1977.1084284. MR465493

[680] E. Parzen, *Statistical inference on time series by RKHS methods*, Proc. 12th Biennial Canadian Mathematical Seminar, R. Pyke (Ed.), Amer. Math. Soc. (AMS), Providence, RI, (1971), 1–37

[681] E. Parzen, *A Simple Proof and Some Extensions of the Sampling Theorem*, Technical Report No. 7, Department of Statistics, Stanford University, (1956), 1–10

[682] N. K. Pavlis, S. A. Holmes, S. C. Kenyon, K. John, J. K. Factor, *The Development and Evaluation of the Earth Gravitational Model 2008 (EGM2008)*, J. Geophys. Research: Solid Earth (1978-2012), 117, Issue B4, 2012

[683] L. E. Payne, *Improperly Posed Problems in Partial Differential Equations*, SIAM Publications, Philadelphia, 1975

[684] R. Penrose, *A generalized inverse for matrices*, Proc. Cambridge Philos. Soc. **51** (1955), 406–413. MR69793

[685] R. Penrose, *On best approximate solutions of linear matrix equations*, Proceedings of the Cambridge Philosphical Society, **25** (1956), 17–19

[686] S. Pereverzev and E. Schock, *On the adaptive selection of the parameter in regularization of ill-posed problems*, SIAM J. Numer. Anal. **43** (2005), no. 5, 2060–2076, DOI 10.1137/S0036142903433819. MR2192331

[687] W. L. Perry, *On the Bojarski-Lewis Inverse Scattering Method*, IEEE Trans. Antennas Propagat., **6** (1974), 826–829

[688] W. L. Perry, *Approximate solution of inverse problems with piecewise continuous solution*, Radio Science, **12** (1977), 634–642

[689] T. Peter and G. Plonka, *A generalized Prony method for reconstruction of sparse sums of eigenfunctions of linear operators*, Inverse Problems **29** (2013), no. 2, 025001, 21, DOI 10.1088/0266-5611/29/2/025001. MR3011964

[690] D. P. Petersen and D. Middleton, *Sampling and reconstruction of wave-number-limited functions in N-dimensional Euclidean spaces*, Information and Control **5** (1962), 279–323. MR151331

[691] W. V. Petryshyn, *On generalized inverses and on the uniform convergence of $(I - \beta K)^n$ with application to iterative methods*, J. Math. Anal. Appl. **18** (1967), 417–439, DOI 10.1016/0022-247X(67)90036-4. MR208381

[692] G. I. Petrov, *Appliation of Galerkin's Method to a Problem of the Stability of the Flow of a Viscous Fluid (in Russian)*, Priklad. Mate. Mekh., **4** (1940), 3–12

[693] D. L. Phillips, *A technique for the numerical solution of certain integral equations of the first kind*, J. Assoc. Comput. Mach. **9** (1962), 84–97, DOI 10.1145/321105.321114. MR134481

[694] M. A. Pinsky, *Introduction to Fourier Analysis and Wavelets*, Brooks/Cole, Thomson Learning, Pacific Grove, CA., 2002

[695] R. L. Plackett, *A historical note on the method of least squares*, Biometrika **36** (1949), 458–460, DOI 10.1093/biomet/36.3-4.458. MR33476

[696] R. Plato, *Optimal algorithms for linear ill-posed problems yield regularization methods*, Numer. Funct. Anal. Optim. **11** (1990), no. 1-2, 111–118, DOI 10.1080/01630569008816364. MR1058780

[697] D. Potts, G. Steidl, and M. Tasche, *Fast Fourier transforms for nonequispaced data: a tutorial*, Modern sampling theory, Appl. Numer. Harmon. Anal., Birkhäuser Boston, Boston, MA, 2001, pp. 247–270. MR1865690

[698] J. F. Price, *Inequalities and local uncertainty principles*, J. Math. Phys. **24** (1983), no. 7, 1711–1714, DOI 10.1063/1.525916. MR709504

[699] J. F. Price and A. Sitaram, *Local uncertainty inequalities for locally compact groups*, Trans. Amer. Math. Soc. **308** (1988), no. 1, 105–114, DOI 10.2307/2000952. MR946433

[700] Q.-N. Jin, *The analysis of a discrete scheme of the iteratively regularized Gauss-Newton method*, Inverse Problems **16** (2000), no. 5, 1457–1476, DOI 10.1088/0266-5611/16/5/319. MR1800605

[701] Q.-n. Jin, *Error estimates of some Newton-type methods for solving nonlinear inverse problems in Hilbert scales*, Inverse Problems **16** (2000), no. 1, 187–197, DOI 10.1088/0266-5611/16/1/315. MR1741236

[702] Q.-N. Jin, *On the iteratively regularized Gauss-Newton method for solving nonlinear ill-posed problems*, Math. Comp. **69** (2000), no. 232, 1603–1623, DOI 10.1090/S0025-5718-00-01199-6. MR1680887

[703] M. Ha Quang, S. H. Kang, and T. M. Le, *Image and video colorization using vector-valued reproducing kernel Hilbert spaces*, J. Math. Imaging Vision **37** (2010), no. 1, 49–65, DOI 10.1007/s10851-010-0192-8. MR2607639

[704] A. Campilho and M. Kamel (eds.), *Image analysis and recognition. Part I*, Lecture Notes in Computer Science, vol. 8814, Springer, Cham, 2014, DOI 10.1007/978-3-319-11758-4. MR3278634

[705] H. Raabe, *Untersuchungen an der wechselseitigen Mehrfachübertragung (Multiplexübertragung)*, Elektrische Nachrichtentechnik, **6** (1939), 213–228

[706] R. Rado, *Note on Generalized Inverses of Matrices*, Proc. Cambridge Phil. Soc., **52** (1956), 600–601

[707] A. G. Ramm, *Inverse Problems: Mathematical and Analytical Techniques With Applications to Engineering*, Springer, Heidelberg, 2005

[708] H. Ramsin and P.-Å. Wedin, *A comparison of some algorithms for the nonlinear least squares problem*, Nordisk Tidskr. Informationsbehandling (BIT) **17** (1977), no. 1, 72–90, DOI 10.1007/bf01932400. MR443343

[709] C. R. Rao and S. K. Mitra, *Generalized inverse of matrices and its applications*, John Wiley & Sons, Inc., New York-London-Sydney, 1971. MR0338013

[710] C. E. Rasmussen and C. K. I. Williams, *Gaussian processes for machine learning*, Adaptive Computation and Machine Learning, MIT Press, Cambridge, MA, 2006. MR2514435

[711] M. D. Rawn, *Generalized sampling theorems for Bessel-type transforms of bandlimited functions and distributions*, SIAM J. Appl. Math. **49** (1989), no. 2, 638–649, DOI 10.1137/0149036. MR988623

[712] L. Rayleigh, *The Theory of Sound*, London, 1896

[713] M. Reed, B. Simon, *Functional Analysis*, I. Academic Press, New York, 1972.

[714] S. Reich, *Constructive Techniques for Accretive and Monotone Operators*, Appl. Nonlinear Anal., Academic Press, New York, (1979), 335–335

[715] L. Reichel and G. Rodriguez, *Old and new parameter choice rules for discrete ill-posed problems*, Numer. Algorithms **63** (2013), no. 1, 65–87, DOI 10.1007/s11075-012-9612-8. MR3040403

[716] W. T. Reid, *Generalized Inverses of Differential and Integral Operators*, Theory and Applications of Generalized Inverses of Matrices. T.L. Boullion and P.L. Odell (Eds.), Symposium Proceedings, Texas Tech. University Math. Series, Lubbock; Texas, Vol. 4., (1990), 1–25

[717] G. Reise, G. Matz, and K. Gröchenig, *Distributed field reconstruction in wireless sensor networks based on hybrid shift-invariant spaces*, IEEE Trans. Signal Process. **60** (2012), no. 10, 5426–5439, DOI 10.1109/TSP.2012.2205918. MR2979004

[718] G. Ribière, *Régularisation d'opérateurs* (French), Rev. Française Informat. Recherche Opérationnelle **1** (1967), no. 5, 57–79. MR0224267

[719] G. R. Richter, *Numerical solution of integral equations of the first kind with nonsmooth kernels*, SIAM J. Numer. Anal. **15** (1978), no. 3, 511–522, DOI 10.1137/0715033. MR474923

[720] F. Riesz and B. Sz.-Nagy, *Functional analysis*, Frederick Ungar Publishing Co., New York, 1955. Translated by Leo F. Boron. MR0071727

[721] A. Rieder, *A wavelet multilevel method for ill-posed problems stabilized by Tikhonov regularization*, Numer. Math. **75** (1997), no. 4, 501–522, DOI 10.1007/s002110050250. MR1431213

[722] W. Ritz, *Über lineare Funktionalgleichung*, Acta Math., **41** (1918), 71–98

[723] A. Rieder, *Keine Probleme mit Inversen Problemen*, Vieweg, 2003

[724] D. W. Robinson, *Gauss and generalized inverses* (English, with German summary), Historia Math. **7** (1980), no. 2, 118–125, DOI 10.1016/0315-0860(80)90032-4. MR572273

[725] R. T. Rockafellar, *On the maximal monotonicity of subdifferential mappings*, Pacific J. Math. **33** (1970), 209–216. MR262827

[726] G. Rodriguez and S. Seatzu, *Numerical solution of the finite moment problem in a reproducing kernel Hilbert space*, J. Comput. Appl. Math. **33** (1990), no. 3, 233–244, DOI 10.1016/0377-0427(90)90049-6. MR1092115

[727] G. Rodriguez and D. Theis, *An algorithm for estimating the optimal regularization parameter by the L-curve*, Rend. Mat. Appl. (7) **25** (2005), no. 1, 69–84. MR2142125

[728] S. Rouquette, M. Najim, *Estimation of frequencies and damping factors by two-dimensional ESPRIT type methods*, IEEE Transactions on Signal Processing, **49** (2001), 237–245

[729] M. Rudelson and R. Vershynin, *On sparse reconstruction from Fourier and Gaussian measurements*, Comm. Pure Appl. Math. **61** (2008), no. 8, 1025–1045, DOI 10.1002/cpa.20227. MR2417886

[730] L. I. Rudin, *Functional Analysis*, Mc Graw-Hill, New York, 1973

[731] L. I. Rudin, S. Osher, and E. Fatemi, *Nonlinear total variation based noise removal algorithms*, Phys. D **60** (1992), no. 1-4, 259–268, DOI 10.1016/0167-2789(92)90242-F. Experimental mathematics: computational issues in nonlinear science (Los Alamos, NM, 1991). MR3363401

[732] R. Rummel, *Satellite Gradiometry*, Lecture Notes in Earth Sciences - Mathematical and Numerical Techniques in Physical Geodesy, H. Sünkel (Ed.), Springer, Berlin, (1986), 318–363

[733] R. Rummel, M. van Gelderen, *Spectral analysis of the full gravity tensor*, Geophys. J. Int, **111** (1992), 159–169

[734] R. Rummel, *Spherical spectral properties of the Earth's gravitational potential and its first and second derivatives*, Lecture Notes in Earth Science, Springer, Berlin, R. Rummel, F. Sanso (Eds.), **65** (1997), 359–404

[735] R. Rummel, *GOCE: Gravitational Gradiometry in a Satellite*, Handbook of Geomathematics, W. Freeden, M.Z. Nashed, T. Sonar (Eds.), Springer, Heidelberg, Vol 1, (2010), 98–105

[736] R. Rummel, *GOCE: Gravitational Gradiometry in a Satellite*, Handbook of Geomathematics. W. Freeden, M.Z. Nashed, T. Sonar (Eds.), Springer, New York, 2nd ed., Vol. 1, (2015), 211–226

[737] R. Rummel, F. Sansò, (Eds.), *Geodetic Boundary Value Problems in View of the One Centimeter Geoid*, Lecture Notes in Earth Sciences, Vol. 65, Springer, Berlin, Heidelberg, 1997

[738] R. Rummel, M. van Gelderen, R. Koop, E. Schrama, F. Sansò, M. Brovelli, F. Miggliaccio, F. Sacerdote, *Spherical Harmonic Analysis of Satellite Gradiometry*, Publications on Geodesy, Delft, 1993

[739] C. Runge, *Zur Theorie der eindeutigen analytischen Funktionen*, Acta Math., **6** (1885), 229–234

[740] S. Sahnoun, K. Usevich, P. Comon, *Multi-Dimensional ESPRIT: Algorithm, Computations, and Perturbation*, Analysis, `https://hal.archives-ouvertes.fr/hal-01360438`, 2016

[741] S. Saitho, *Theory of Reproducing Kernels and its Applications*, Longman, New York, 1988

[742] R. W. Saltus, R. J. Blakely, *Unique geologic insights from "non-unique" gravity and magnetic interpretation*, GSA Today, **21** (2011), 4–11

[743] A. Sard, *The measure of the critical values of differentiable maps*, Bull. Amer. Math. Soc. **48** (1942), 883–890, DOI 10.1090/S0002-9904-1942-07811-6. MR7523

[744] O. Scherzer, *The use of Morozov's discrepancy principle for Tikhonov regularization for solving nonlinear ill-posed problems* (English, with English and German summaries), Computing **51** (1993), no. 1, 45–60, DOI 10.1007/BF02243828. MR1242658

[745] O. Scherzer, *An iterative multi-level algorithm for solving nonlinear ill-posed problems*, Numer. Math. **80** (1998), no. 4, 579–600, DOI 10.1007/s002110050379. MR1650043

[746] O. Scherzer, H. W. Engl, and K. Kunisch, *Optimal a posteriori parameter choice for Tikhonov regularization for solving nonlinear ill-posed problems*, SIAM J. Numer. Anal. **30** (1993), no. 6, 1796–1838, DOI 10.1137/0730091. MR1249043

[747] B. Schlkopf, A. J. Smola, *Learning with Kernels: Support Vector Machines, Regularization, Optimization, and Beyond (Adaptive Computation and Machine Learning)*, MIT Press, 1st ed., 12, 2001

[748] G. Schmeisser and F. Stenger, *Sinc approximation with a Gaussian multiplier*, Sampl. Theory Signal Image Process. **6** (2007), no. 2, 199–221. MR2343406

[749] P. Schneider, K. Bunte, H. Stiekema, B. Hammer, T. Villmann, M. Biehl, *Regularization in Matrix Relevance Learning*, IEEE Transactions on Neural Networks, **21** (2010), 831- 840

[750] Ch. Schnörr, *Determining optical flow for irregular deomains by minimizing quadratic functionals of a certain class*, Int. J. Comput. Vision, **6** (1991), 25–38

[751] Ch. Schnörr, *Segmentation of Visual Motion by Minimizing Convex Non- Quadratic Functionals*, Proc. 12th Int. Conf. on Pattern Recognition, Jerusalem: Israel, Vol. A, (1994), 661–663

[752] I. J. Schoenberg, *Metric spaces and completely monotone functions*, Ann. of Math. (2) **39** (1938), no. 4, 811–841, DOI 10.2307/1968466. MR1503439

[753] I. J. Schoenberg, *On trigonometric spline interpolation*, J. Math. Mech. **13** (1964), 795–825. MR0165300

[754] I. J. Schoenberg, *Spline functions and the problem of graduation*, Proc. Nat. Acad. Sci. U.S.A. **52** (1964), 947–950, DOI 10.1073/pnas.52.4.947. MR167768

[755] I. J. Schoenberg, *Cardinal spline interpolation*, Conference Board of the Mathematical Sciences Regional Conference Series in Applied Mathematics, No. 12, Society for Industrial and Applied Mathematics, Philadelphia, Pa., 1973. MR0420078

[756] O. Schramm and S. Sheffield, *A contour line of the continuum Gaussian free field*, Probab. Theory Related Fields **157** (2013), no. 1-2, 47–80, DOI 10.1007/s00440-012-0449-9. MR3101840

[757] T. Schuster and J. Weickert, *On the application of projection methods for computing optical flow fields*, Inverse Probl. Imaging **1** (2007), no. 4, 673–690, DOI 10.3934/ipi.2007.1.673. MR2350221

[758] T. Schuster, *The Method of Approximate Inverse: Theory and Applications*, Lecture Notes in Mathematics, Springer, Berlin, 2007

[759] T. Schuster, B. Kaltenbacher, B. Hofmann, and K. S. Kazimierski, *Regularization methods in Banach spaces*, Radon Series on Computational and Applied Mathematics, vol. 10, Walter de Gruyter GmbH & Co. KG, Berlin, 2012, DOI 10.1515/9783110255720. MR2963507

[760] T. I. Seidman, *Nonconvergence results for the application of least-squares estimation to ill-posed problems*, J. Optim. Theory Appl. **30** (1980), no. 4, 535–547, DOI 10.1007/BF01686719. MR572154

[761] T. I. Seidman and C. R. Vogel, *Well-posedness and convergence of some regularisation methods for nonlinear ill posed problems*, Inverse Problems **5** (1989), no. 2, 227–238. MR991919

[762] K. Seip, *Interpolation and sampling in spaces of analytic functions*, University Lecture Series, vol. 33, American Mathematical Society, Providence, RI, 2004, DOI 10.1090/ulect/033. MR2040080

[763] I. Serb, *Estimates for the modulus of smoothness and convexity of a Banach space*, Mathematica, **34** (1992), 61–70

[764] C. E. Shannon, *Communication in the presence of noise*, Proc. Institute of Radio Engineers, **37** (1949a), 10–21

[765] C. E. Shannon, *The Mathematical Theory of Communication*, University of Illinois Press, Urbana, 1949b

[766] H. S. Shapiro, *Topics in Approximation Theory*, Lecture Notes in Mathematics, Vol. 187, Springer, Berlin, Heidelberg, New York, 1981

[767] H. S. Shapiro, *Uncertainty principles for bases in $L^2(\mathbb{R})$*, Proceedings of the Conference on Harmonic Analysis and Number Theory. CIRM, Marseille-Luminy, L. Habsieger, A. Plagne, B. Saffari (Eds.), 2005

[768] J. Shawe-Taylor, N. Cristianini, *Kernel Methods for Pattern Analysis*, Cambridge University Press, 2004

[769] X. Shen, A. I. Zayed, (Eds.), *Multiscale Signal Analysis and Modeling*, Springer, New York, 2013.

[770] N. Shinozaki, M. Sibuya, and K. Tanabe, *Numerical algorithms for the Moore-Penrose inverse of a matrix: direct methods*, Ann. Inst. Statist. Math. **24** (1972), 193–203, DOI 10.1007/BF02479751. MR324888

[771] D. W. Showalter and A. Ben-Israel, *Representation and computation of the generalized inverse of a bounded linear operator between Hilbert spaces* (English, with Italian summary), Atti Accad. Naz. Lincei Rend. Cl. Sci. Fis. Mat. Nat. (8) **48** (1970), 184–194. MR273432

[772] C. E. Shin and Q. Sun, *Stability of localized operators*, J. Funct. Anal. **256** (2009), no. 8, 2417–2439, DOI 10.1016/j.jfa.2008.09.011. MR2502521

[773] L. Shure, R. L. Parker, G. E. Backus, *Harmonic Splines for Geomagnetic Modeling*, Phys. Earth Planet. Inter., **28** (1982), 215–229

[774] C. L. Siegel, *Über die analytische Theorie der quadratischen Formen, III*, Annals of Mathematics, **38** (1937), 212–291

[775] M. M. J.-A. Simeoni, *Functional Inverse Problems on Spheres: Theory, Algorithms, and Applications*, Ph.D.-thesis, No. 7174, EPFL, Lausanne, 2020

[776] F. J. Simons, *Slepian functions and their use in signal estimation and spectral analysis*, Handbook of Geomathematics, W. Freeden, M. Z. Nashed, T. Sonar (Eds.), 1st ed., Springer, Heidelberg, Vol.1, (2010), 891–923

[777] F. J. Simons, A. Plattner, *Scalar and vector slepian functions, spherical signal estimation and spectral analysis*, Handbook of Geomathematics, W. Freeden, M. Z. Nashed, T. Sonar (Eds.), 2nd ed., Vol. 3, (2015), 2563–2608

[778] D. Slepian and H. O. Pollak, *Prolate spheroidal wave functions, Fourier analysis and uncertainty. I*, Bell System Tech. J. **40** (1961), 43–63, DOI 10.1002/j.1538-7305.1961.tb03976.x. MR140732

[779] D. Slepian, *Prolate spheroidal wave functions, Fourier analysis and uncertainty. IV. Extensions to many dimensions; generalized prolate spheroidal functions*, Bell System Tech. J. **43** (1964), 3009–3057, DOI 10.1002/j.1538-7305.1964.tb01037.x. MR181766

[780] D. Slepian and H. O. Pollak, *Prolate spheroidal wave functions, Fourier analysis and uncertainty. I*, Bell System Tech. J. **40** (1961), 43–63, DOI 10.1002/j.1538-7305.1961.tb03976.x. MR140732

[781] D. Slepian, *Prolate spheroidal wave functions, Fourier analysis and uncertainty. IV. Extensions to many dimensions; generalized prolate spheroidal functions*, Bell System Tech. J. **43** (1964), 3009–3057, DOI 10.1002/j.1538-7305.1964.tb01037.x. MR181766

[782] D. Slepian, *Some comments on Fourier analysis, uncertainty and modeling*, SIAM Rev. **25** (1983), no. 3, 379–393, DOI 10.1137/1025078. MR710468

[783] S. Smale and D.-X. Zhou, *Shannon sampling and function reconstruction from point values*, Bull. Amer. Math. Soc. (N.S.) **41** (2004), no. 3, 279–305, DOI 10.1090/S0273-0979-04-01025-0. MR2058288

[784] S. Smale and D.-X. Zhou, *Online learning with Markov sampling*, Anal. Appl. (Singap.) **7** (2009), no. 1, 87–113, DOI 10.1142/S0219530509001293. MR2488871

[785] F. Smithies, *Integral Equations*, Cambridge Tracts in Math. and Math. Phys., 49, Cambridge Univ. Press, London, 1985

[786] R. Snieder, *The perturbation method in elastic wave scattering and inverse scattering in pure and applied science*, General Theory of Elastic Waves, Academic Press, San Diego, (2002), 528–542

[787] A. Sommerfeld, *Vorlesungen der Theoretischen Physik II*, 6th ed., Verlag Harri Deutsch, Frankfurt, 1978

[788] M. Song, *Regularization-Projection Methods and Finite Element Approximations for Ill-Posed Linear Operator Equations*, Ph.-D. Thesis, University of Michigan, 1978

[789] H. W. Sorensen, *Parameter Estimation: Principles and Problems*, Marcel Dekker, Inc. New York, 1980

[790] E. M. Stein and G. Weiss, *Introduction to Fourier analysis on Euclidean spaces*, Princeton Mathematical Series, No. 32, Princeton University Press, Princeton, N.J., 1971. MR0304972

[791] F. Stenger, *Approximations via Whittaker's cardinal function*, J. Approximation Theory **17** (1976), no. 3, 222–240, DOI 10.1016/0021-9045(76)90086-1. MR481786

[792] F. Stenger, *Numerical methods based on Whittaker cardinal, or sinc functions*, SIAM Rev. **23** (1981), no. 2, 165–224, DOI 10.1137/1023037. MR618638

[793] F. Stenger, *Numerical Methods Based on Sinc and Analytic Functions*, Springer-Verlag, New York, 1993.

[794] P. Sternberg and W. P. Ziemer, *Generalized motion by curvature with a Dirichlet condition*, J. Differential Equations **114** (1994), no. 2, 580–600, DOI 10.1006/jdeq.1994.1162. MR1303041

[795] J. Stewart, *Positive definite functions and generalizations, an historical survey*, Rocky Mountain J. Math. **6** (1976), no. 3, 409–434, DOI 10.1216/RMJ-1976-6-3-409. MR430674

[796] M. H. Stone, *On unbounded operators in Hilbert space*, J. Indian Math. Soc. (N.S.) **15** (1951), 155–192 (1952). MR52042

[797] V. N. Strakhov, *Solution of Linear Inverse Problems of Gravimetry and Magnetometry*, Dokl. Akad. Nauk SSSR, **310** (1990) 1358–1362

[798] O. N. Strand, *Theory and methods related to the singular-function expansion and Landweber's iteration for integral equations of the first kind*, SIAM J. Numer. Anal. **11** (1974), 798–825, DOI 10.1137/0711066. MR362966

[799] G. Strang, G. Fix, *A Fourier analysis of the finite element variational method*, Constructive Aspects of Functional Analysis, G. Geymonat, Ed., C.I.M.E., (1973), 793–840

[800] R. S. Strichartz, *Transformation of spectra of graph Laplacians*, Rocky Mountain J. Math., **6** (2010), 2037-2062

[801] R. S. Strichartz and A. Teplyaev, *Spectral analysis on infinite Sierpiński fractafolds*, J. Anal. Math. **116** (2012), 255–297, DOI 10.1007/s11854-012-0007-5. MR2892621

[802] V. N. Sudakov and L. A. Halfin, *A statistical approach to the correctness of the problems of mathematical physics* (Russian), Dokl. Akad. Nauk SSSR **157** (1964), 1058–1060. MR0165578

[803] Q. Sun, *Nonuniform average sampling and reconstruction of signals with finite rate of innovation*, SIAM J. Math. Anal. **38** (2006/07), no. 5, 1389–1422, DOI 10.1137/05063444X. MR2286012

[804] Q. Sun, *Wiener's lemma for infinite matrices*, Trans. Amer. Math. Soc. **359** (2007), no. 7, 3099–3123, DOI 10.1090/S0002-9947-07-04303-6. MR2299448

[805] Q. Sun, *Frames in spaces with finite rate of innovation*, Adv. Comput. Math. **28** (2008), no. 4, 301–329, DOI 10.1007/s10444-006-9021-4. MR2390281

[806] Q. Sun, *Wiener's lemma for localized integral operators*, Appl. Comput. Harmon. Anal. **25** (2008), no. 2, 148–167, DOI 10.1016/j.acha.2007.10.006. MR2436767

[807] Q. Sun, *Local reconstruction for sampling in shift-invariant spaces*, Adv. Comput. Math. **32** (2010), no. 3, 335–352, DOI 10.1007/s10444-008-9109-0. MR2595176

[808] Q. Sun, *Localized nonlinear functional equations and two sampling problems in signal processing*, Adv. Comput. Math. **40** (2014), no. 2, 415–458, DOI 10.1007/s10444-013-9314-3. MR3194713

[809] W. Sun and X. Zhou, *Average sampling in spline subspaces*, Appl. Math. Lett. **15** (2002), no. 2, 233–237, DOI 10.1016/S0893-9659(01)00123-9. MR1880763

[810] W. Sun and X. Zhou, *Reconstruction of functions in spline subspaces from local averages*, Proc. Amer. Math. Soc. **131** (2003), no. 8, 2561–2571, DOI 10.1090/S0002-9939-03-07082-5. MR1974656

[811] S. L. Svensson, *Pseudodifferential operators - a new approach to the boundary value problems of physical geodesy*, Manuscr. Geod., **8** (1983), 1–40

[812] G. Szegö, *Orthogonal polynomials*, American Mathematical Society Colloquium Publications, Vol. 23, American Mathematical Society, Providence, R.I., 1959. Revised ed. MR0106295

[813] E. Tadmor, S. Nezzar, and L. Vese, *A multiscale image representation using hierarchical (BV, L^2) decompositions*, Multiscale Model. Simul. **2** (2004), no. 4, 554–579, DOI 10.1137/030600448. MR2113170

[814] A. Tarantola, *Inverse problem theory and methods for model parameter estimation*, Society for Industrial and Applied Mathematics (SIAM), Philadelphia, PA, 2005, DOI 10.1137/1.9780898717921. MR2130010

[815] U. Tautenhahn, *On a general regularization scheme for nonlinear ill-posed problems*, Inverse Problems **13** (1997), no. 5, 1427–1437, DOI 10.1088/0266-5611/13/5/020. MR1474375

[816] A. E. Taylor, D. Lay, *Functional Analysis*, Wiley, New York, 1979

[817] A. N. Tikhonov, *On the stability of inverse problems*, Dokl. Akad. Nauk. SSSR, **39** (1943), 195–198

[818] A. N. Tikhonov, *On the stability of inverse problems*, Dokl. Akad. Nauk SSSR, **39** (1943), 195–198

[819] A. N. Tihonov, *On the solution of ill-posed problems and the method of regularization* (Russian), Dokl. Akad. Nauk SSSR **151** (1963), 501–504. MR0162377

[820] A. N. Tikhonov, *On methods of solving incorrect problems*, Amer. Math. Soc. Transl., **70** (1968), 222–224

[821] A. N. Tikhonov and V. Y. Arsenin, *Solutions of ill-posed problems*, Scripta Series in Mathematics, V. H. Winston & Sons, Washington, D.C.; John Wiley & Sons, New York-Toronto-London, 1977. Translated from the Russian; Preface by translation editor Fritz John. MR0455365

[822] A. N. Tikhonov, A. V. Goncharsky, V. V. Stepanov, A. G. Yagola, *Numerical Methods for the Solution of Ill-Posed Problems*, Kluwer, Dordrecht. 1995

[823] A. N. Tikhonov, A. S. Leonov, and A. G. Yagola, *Nonlinear ill-posed problems. Vol. 1, 2*, Applied Mathematics and Mathematical Computation, vol. 14, Chapman & Hall, London, 1998. Translated from the Russian, DOI 10.1007/978-94-017-5167-4. MR1630660

[824] W. Torge and J. Müller, *Geodesy*, 4th ed., De Gruyter Textbook, Walter de Gruyter & Co., Berlin, 2012, DOI 10.1515/9783110250008. MR2962515

[825] V. K. Tuan and M. Z. Nashed, *Stable recovery of analytic functions using basic hypergeometric series*, J. Comput. Anal. Appl. **3** (2001), no. 1, 33–51, DOI 10.1023/A:1011587816931. MR1824499

[826] S. Twomey, *On the numerical solution of Fredholm integral equations of the first kind by the inversion of the linear system produced by quadrature*, J. Assoc. Comput. Mach. **10** (1963), 97–101, DOI 10.1145/321150.321157. MR148249

[827] M. Unser, *Sampling - 50 years after Shannon*, Proc. IEEE, **88** (2000), 569–587

[828] M. Unser, A. Aldroubi, *A general sampling theory for nonideal acqisition devices*, IEEE Transact. Sign. Processing, **42** (1994), 2915–2925

[829] J. A. Uriguen, T. Blu, P. L. Dragotti, *FRI sampling with arbitrary kernels*, IEEE Transact. Sign. Processing, **61** (2013), 5310–5323

[830] J. G. van der Corput, *Zum Teilerproblem* (German), Math. Ann. **98** (1928), no. 1, 697–716, DOI 10.1007/BF01451619. MR1512430

[831] C. F. Van Loan, *Generalizing the singular value decomposition*, SIAM J. Numer. Anal. **13** (1976), no. 1, 76–83, DOI 10.1137/0713009. MR411152

[832] M. M. Vainberg, *Variational Method and Method of Monotone Operators*, Wiley, New York, 1973

[833] J. M. Varah, *On the numerical solution of ill-conditioned linear systems with applications to ill-posed problems*, SIAM J. Numer. Anal. **10** (1973), 257–267, DOI 10.1137/0710025. MR334486

[834] L. A. Vese and S. J. Osher, *Modeling textures with total variation minimization and oscillating patterns in image processing*, J. Sci. Comput. **19** (2003), no. 1-3, 553–572, DOI

10.1023/A:1025384832106. Special issue in honor of the sixtieth birthday of Stanley Osher. MR2028858

[835] I. N. Vekua, *Über die Vollständigkeit des Systems harmonischer Polynome im Raum*, (in Russisch), Dokl. Akad. Nauk, **90** (1953), 495–498

[836] M. Vetterli, J. Kovacevic, V. K. Goyal, *Foundations of Signal Processing*, Cambridge University Press, 2014

[837] M. Vetterli, P. Marziliano, and T. Blu, *Sampling signals with finite rate of innovation*, IEEE Trans. Signal Process. **50** (2002), no. 6, 1417–1428, DOI 10.1109/TSP.2002.1003065. MR1930786

[838] M. Vuletić, *The Gaussian free field and strict plane partitions* (English, with English and French summaries), 25th International Conference on Formal Power Series and Algebraic Combinatorics (FPSAC 2013), Discrete Math. Theor. Comput. Sci. Proc., AS, Assoc. Discrete Math. Theor. Comput. Sci., Nancy, 2013, pp. 1041–1052. MR3091062

[839] I. M. Vinogradov, *Verbesserungen der asymptotischen Formeln für die Anzahl der ganzzahligen Punkte in dreidimensionalen Bereichen (in Russisch)*, Izvestija Akad. Nauk SSSR, Ser. Mat., **19** (1955), 3–9

[840] C. R. Vogel, *Computational Methods for Inverse Problems*, SIAM, Philadelphia, 2002

[841] C. R. Vogel, *Non-Convergence of the L-Curve Regularization Parameter Selction Method*, Inverse Problems, 12, 1996

[842] C. R. Vogel and M. E. Oman, *Iterative methods for total variation denoising*, SIAM J. Sci. Comput. **17** (1996), no. 1, 227–238, DOI 10.1137/0917016. Special issue on iterative methods in numerical linear algebra (Breckenridge, CO, 1994). MR1375276

[843] G. Voronoi, *Sur un problème du calcul des fonctions asymptotiques* (French), J. Reine Angew. Math. **126** (1903), 241–282, DOI 10.1515/crll.1903.126.241. MR1580627

[844] G. Wahba, *Convergence rates of certain approximate solutions to Fredholm integral equations of the first kind*, J. Approximation Theory **7** (1973), 167–185, DOI 10.1016/0021-9045(73)90064-6. MR346453

[845] G. Wahba, *Convergence Properties of the Method of Regularization for Noisy Linear Operator Equations*, MRC Technical Summary Report, 1132, Mathematics Research Center, The University of Wisconsin, 1973

[846] G. Wahba, *A class of approximate solutions to linear operator equations*, J. Approximation Theory **9** (1973), 61–77, DOI 10.1016/0021-9045(73)90112-3. MR386284

[847] G. Wahba, *Practical approximate solutions to linear operator equations when the data are noisy*, SIAM J. Numer. Anal. **14** (1977), no. 4, 651–667, DOI 10.1137/0714044. MR471299

[848] G. Wahba, *Spline Models for Observational Data*, SIAM Philadelphia, 1990

[849] A. Walfisz, *Teilerprobleme*, Math. Z., **26** (1927), 66–88

[850] A. Walfisz, *Gitterpunkte in mehrdimensionalen Kugeln*, Panstwowe Wydawnictwo Naukowe, Warschau, 1957

[851] A. Walfisz, *Über Gitterpunkte in vierdimensionalen Ellipsoiden*, Math. Z., **72** (1960), 259–278

[852] J. L. Walsh, *The approximation of harmonic functions by harmonic polynomials and by harmonic rational functions*, Bull. Amer. Math. Soc. **35** (1929), no. 4, 499–544, DOI 10.1090/S0002-9904-1929-04753-0. MR1561772

[853] W. Walter, *Einführung in die Potentialtheorie*, BI Hochschulskripten, 765/765a, 1971

[854] W. Walter, *Einführung in die Theorie der Distributionen*, 3rd ed., BI-Wissenschaftsverlag, Mannheim, 1994

[855] G. G. Walter, *A sampling theorem for wavelet subspaces*, IEEE Trans. Inform. Theory **38** (1992), no. 2, 881–884, DOI 10.1109/18.119745. MR1162226

[856] G. G. Walter and X. Shen, *Wavelets and other orthogonal systems*, 2nd ed., Studies in Advanced Mathematics, Chapman & Hall/CRC, Boca Raton, FL, 2001. MR1887929

[857] G. N. Watson, *A Treatise on the Theory of Bessel Functions*, 2nd ed., Cambridge University Press, Cambridge, 1944

[858] E. Weber, *The Geometry of Sampling on Unions of Lattices*, Proc. Amer. Math. Soc., 2002

[859] H. W. Weiner, (Eds.), *Reproducing Kernel Hilbert Spaces. Applications in Statistical Signal Processing*, Hutchinson Ross, Stroudsburg, PA, 1982

[860] D. Weiskopf, G. Erlebacher, *Overview of Flow Visualization*, The Visualization Handbook, C. D. Hansen, C. R. Johnson (Eds.), Elsevier, Amsterdam, (2005), 261–278

[861] D. Werner, *Funktionalanalysis*, 4. Auflage; Springer, Berlin, 2002

[862] E. T. Whittaker, *On the functions which are represented by the expansions of the interpolation theory*, Proc. Royal Soc. Edinburgh, Sec. A, **35** (1915), 181–194

[863] E. T. Whittaker, *On a new method of graduation*, Proc. Edinb. Math. Soc., **41** (1923), 63–75

[864] E. T. Whittaker, G. N. Watson, *A Course of Modern Analysis*, 4th ed., Cambridge University Press, 1927

[865] E. T. Whittaker, *The Fourier theory of cardinal functions*, Proc. Math. Cox. Edinburgh, **1** (1929), 169–176

[866] J. Wloka, *Partial Differential Equations*, Cambridge University Press., Cambridge, 1992

[867] X. G. Xia and M. Z. Nashed, *The Backus-Gilbert method for signals in reproducing kernel Hilbert spaces and wavelet subspaces*, Inverse Problems **10** (1994), no. 3, 785–804. MR1278587

[868] X.-G. Xia and M. Z. Nashed, *A method with error estimates for band-limited signal extrapolation from inaccurate data*, Inverse Problems **13** (1997), no. 6, 1641–1661, DOI 10.1088/0266-5611/13/6/015. MR1484008

[869] J. Xian, *Error estimates from noise samples for iterative algorithm in shift-invariant signal spaces*, Abstr. Appl. Anal., posted on 2010, 1–9, DOI 10.1155/2010/214213. MR2737926

[870] J. Xian and S. Li, *Sampling set conditions in weighted multiply generated shift-invariant spaces and their applications*, Appl. Comput. Harmon. Anal. **23** (2007), no. 2, 171–180, DOI 10.1016/j.acha.2006.10.004. MR2344609

[871] J. Xian and W. Lin, *Sampling and reconstruction in time-warped spaces and their applications*, Appl. Math. Comput. **157** (2004), no. 1, 153–173, DOI 10.1016/j.amc.2003.08.031. MR2085530

[872] K. Yao, *Applications of reproducing kernel Hilbert spaces-bandlimited signal models*, Inform. and Control, **11** (1967), 429–444

[873] Y. Lin and L. D. Brown, *Statistical properties of the method of regularization with periodic Gaussian reproducing kernel*, Ann. Statist. **32** (2004), no. 4, 1723–1743, DOI 10.1214/009053604000000454. MR2089140

[874] R. M. Young, *An Introduction to Nonharmonic Fourier Series*, Academic Press, New York, 1980

[875] K. Yoshida, *Functional Analysis*, Springer, Berlin, 1965

[876] H. Zhang, Y. Xu, and Q. Zhang, *Refinement of operator-valued reproducing kernels*, J. Mach. Learn. Res. **13** (2012), 91–136. MR2913695

[877] M. Zakai, *Band-limited functions and the sampling theorem*, Information and Control **8** (1965), 143–158. MR174403

[878] A. Zayed, *Advances in Shannon's Sampling Theory*, CRC Press, Boca Raton, London, New York, 1993

[879] A. I. Zayed, G. Hinsen, and P. L. Butzer, *On Lagrange interpolation and Kramer-type sampling theorems associated with Sturm-Liouville problems*, SIAM J. Appl. Math. **50** (1990), no. 3, 893–909, DOI 10.1137/0150053. MR1050920

[880] A. I. Zayed, G. Schmeisser, *New Perspectives on Approximation and Sampling*, Appl. Harm. Anal., Birkhäuser, Basel 2014

[881] E. Zeidler, *Nonlinear Functional Analysis and its Applications III: Variational Mathods and Optimization*, Springer, New York, 1984

[882] Z. Zeng, *A Newton's Iteration Converges Quadratically to Nonisolated Solutions Too*, Preprint, 2021

[883] L. Zhou, X. Li, and F. Pan, *Gradient-based iterative identification for Wiener nonlinear systems with non-uniform sampling*, Nonlinear Dynam. **76** (2014), no. 1, 627–634, DOI 10.1007/s11071-013-1156-5. MR3189198

[884] A. Zygmund, *Trigonometric Series*, Cambridge University Press, New York, 1957

[885] M. Zwaan, *Approximation of the solution to the moment problem in a Hilbert space*, Numer. Funct. Anal. Optim., **11** (1990), 601–612

[886] M. Zwaan, *Dynamic MRI reconstruction as a moment problem*, Math. Methods Appl. Sci. **15** (1992), no. 9, 661–675, DOI 10.1002/mma.1670150906. MR1193160

Index

Selected Published Titles in This Series

274 **Willi Freeden and M. Zuhair Nashed,** Recovery Methodologies: Regularization and Sampling, 2023
273 **Gennadiy Feldman,** Characterization of Probability Distributions on Locally Compact Abelian Groups, 2023
272 **Tadashi Ochiai,** Iwasawa Theory and Its Perspective, Volume 1, 2023
271 **Hideto Asashiba,** Categories and Representation Theory, 2022
270 **Leslie Hogben, Jephian C.-H. Lin, and Bryan L. Shader,** Inverse Problems and Zero Forcing for Graphs, 2022
269 **Ralph S. Freese, Ralph N. McKenzie, George F. McNulty, and Walter F. Taylor,** Algebras, Lattices, Varieties, 2022
268 **Ralph S. Freese, Ralph N. McKenzie, George F. McNulty, and Walter F. Taylor,** Algebras, Lattices, Varieties, 2022
267 **Dragana S. Cvetković Ilić,** Completion Problems on Operator Matrices, 2022
266 **Kate Juschenko,** Amenability of Discrete Groups by Examples, 2022
265 **Nabil H. Mustafa,** Sampling in Combinatorial and Geometric Set Systems, 2022
264 **J. Scott Carter and Seiichi Kamada,** Diagrammatic Algebra, 2021
263 **Vugar E. Ismailov,** Ridge Functions and Applications in Neural Networks, 2021
262 **Ragnar-Olaf Buchweitz,** Maximal Cohen–Macaulay Modules and Tate Cohomology, 2021
261 **Shiri Artstein-Avidan, Apostolos Giannopoulos, and Vitali D. Milman,** Asymptotic Geometric Analysis, Part II, 2021
260 **Lindsay N. Childs, Cornelius Greither, Kevin P. Keating, Alan Koch, Timothy Kohl, Paul J. Truman, and Robert G. Underwood,** Hopf Algebras and Galois Module Theory, 2021
259 **William Heinzer, Christel Rotthaus, and Sylvia Wiegand,** Integral Domains Inside Noetherian Power Series Rings, 2021
258 **Pramod N. Achar,** Perverse Sheaves and Applications to Representation Theory, 2021
257 **Juha Kinnunen, Juha Lehrbäck, and Antti Vähäkangas,** Maximal Function Methods for Sobolev Spaces, 2021
256 **Michio Jimbo, Tetsuji Miwa, and Fedor Smirnov,** Local Operators in Integrable Models I, 2021
255 **Alexandre Boritchev and Sergei Kuksin,** One-Dimensional Turbulence and the Stochastic Burgers Equation, 2021
254 **Karim Belabas and Henri Cohen,** Numerical Algorithms for Number Theory, 2021
253 **Robert R. Bruner and John Rognes,** The Adams Spectral Sequence for Topological Modular Forms, 2021
252 **Julie Déserti,** The Cremona Group and Its Subgroups, 2021
251 **David Hoff,** Linear and Quasilinear Parabolic Systems, 2020
250 **Bachir Bekka and Pierre de la Harpe,** Unitary Representations of Groups, Duals, and Characters, 2020
249 **Nikolai M. Adrianov, Fedor Pakovich, and Alexander K. Zvonkin,** Davenport–Zannier Polynomials and Dessins d'Enfants, 2020
248 **Paul B. Larson and Jindrich Zapletal,** Geometric Set Theory, 2020
247 **István Heckenberger and Hans-Jürgen Schneider,** Hopf Algebras and Root Systems, 2020
246 **Matheus C. Bortolan, Alexandre N. Carvalho, and José A. Langa,** Attractors Under Autonomous and Non-autonomous Perturbations, 2020

For a complete list of titles in this series, visit the
AMS Bookstore at **www.ams.org/bookstore/survseries/**.